SECOND EDITION (REVISED)

Wetland Plants *of* Minnesota

A COMPLETE GUIDE TO THE WETLAND AND AQUATIC PLANTS OF THE NORTH STAR STATE

STEVE W. CHADDE

WETLAND PLANTS OF MINNESOTA

A complete guide to the wetland and aquatic plants of the North Star state

SECOND EDITION (REVISED)

Steve W. Chadde

Printed in the United States of America.

ISBN-13: 978-1477645178
ISBN-10: 1477645179

The Second Edition incorporates the 2012 wetland indicator status ratings developed by the National Wetland Plant List; see: *https://wetland_plants.usace.army.mil.*

Second Edition (Revised, October 2013): Updated distribution maps added based on data from the Biota of North America Program (BONAP). The author gratefully acknowledges the Program for permission to use their data to generate the maps.

NOTE: A Third Edition of this book is planned for early 2014, incorporating taxonomic changes and other updates.

The author can be reached via email: steve@chadde.net

VERSION 2.5

Showy lady's-slipper
(Cypripedium reginae)
STATE FLOWER OF MINNESOTA

WETLAND PLANTS OF MINNESOTA is a comprehensive manual on the identification, habitats, and distribution of approximately 900 vascular plants found in the state's aquatic and wetland environments. The intended user of this guide is the botanist, ecologist, natural resource manager, consultant, or other scientist engaged in wetland studies. However, this book was also written for the person simply interested in plants, or in the ecology, conservation, or aesthetic beauty provided by wetlands. To this end, the use of technical terminology has been minimized when possible, while still allowing the reader to accurately identify unknown plant specimens. For definitions of the more technical terms, refer to the Glossary, page 623.

Nomenclature of plant families, genera and species generally conforms to that of the published volumes of The Flora of North America series, the USDA PLANTS database (online at http://plants.usda.gov), and a recent (October 2010) checklist of the state's flora prepared by Dr. Anita F. Cholewa of the University of Minnesota. In some cases, for example within the genus Scirpus, traditional names have been retained, followed by the newer names. Synonyms are also included in the species descriptions.

Plants were first chosen for inclusion in this work according to their **wetland indicator status**, a classification system developed by the U.S. Fish and Wildlife Service (Reed 1988). As of June 2012, a revised and updated **National Wetland Plant List** has been prepared and is incorporated into this second edition (Lichvar and Kartesz 2012).

Species occurring in Minnesota with a wetland status indicator of **Obligate** (OBL) or **Facultative Wetland** (FACW) were included. In addition, a number of species more typical of drier habitats (**Facultative** species [FAC] and **Facultative Upland** species [FACU]) were included, based on published reports of their occurrence in wetlands, or on the author's own encounters with them in the state's wetlands (see page 41 for more information on the wetland classification system).

The result is a floristic guide for Minnesota that is applicable to the aquatic plants of lakes, ponds, rivers, and streams, the plants of the adjoining wetlands, and to the plants found in the transition area, or ecotone, between wetlands and the surrounding uplands. For convenience, the terms wetland or wetland plant are sometimes used to also include aquatic situations and aquatic plants.

Wetlands are home to a large number of uncommon plant species. The website of the Minnesota County Biological Survey (Minnesota Dept. of Natural Resources; *www.dnr.state.mn.us/eco/mcbs*) maintains a list of species listed as endangered, threatened, or of special concern; report sightings of these uncommon taxa to them.

Conservation status

- **Endangered** A species is considered endangered if the species is threatened with extinction throughout all or a significant portion of its range within Minnesota.
- **Threatened** A species is considered threatened if the species is likely to become endangered within the foreseeable future throughout all or a significant portion of its range within Minnesota.
- **Special concern** Species not endangered or threatened, but which are uncommon in Minnesota, or have unique or highly specific habitat requirements. Species on the periphery of their range that are not listed as threatened may be included in this category along with species that were once threatened or endangered but now have increasing or protected stable populations.

Wetlands defined

In general, wetlands are lands where water saturation is the primary factor determining the nature of soil that will develop and the types of plant and animals that will occur there (Cowardin and others 1979). Wetlands occur as transitional areas between upland and aquatic ecosystems, and are covered by shallow water, or have a water table that is usually at or near the ground surface.

Wetlands will have at least one, and often all of the following three attributes regarding their vegetation, soils and hydrology:

- The land supports (at least periodically) plants that are predominantly "**hydrophytes.**" Hydrophytes are plants growing in water, or on soil that is at least periodically deficient in oxygen as a result of high water content.
- Soils are largely undrained, "**hydric**" soils. Hydric soils are soils that are flooded or saturated long enough during the growing season so that anaerobic conditions develop in the upper portion of the soil.
- Water will either permanently or periodically cover the area during some or all of the growing season each year.

In Minnesota, wetlands are concentrated in the northeastern portion of the state. This is evident on the map at right; wetlands are indicated by the darker shading. Notable is the extensive **Red Lake peatland** located in north-central Minnesota (see page 488). Southward, large-scale conversion of the original forest and prairie vegetation to agriculture and urban development has occurred, including the draining and filling of many wetlands.

Aquatic plants

Plants of open water areas—**aquatic plants**—include those species found in lakes, ponds, rivers, streams, and springs. They occur as either free-floating plants on the water surface, such as the duckweeds (*Lemna*) and water-meal (*Wolffia*), or float below the water surface, as in coontail (*Ceratophyllum*) and common bladderwort (*Utricularia vulgaris*). Aquatic species also occur as submergent plants anchored to the bottom substrate by roots or rhizomes. Submergent species may be entirely underwater as in water-milfoil (*Myriophyllum*), or have leaves floating on the water surface (water-lilies, *Nuphar*). Some aquatic species have differently shaped underwater and floating leaves, as in a number of pondweeds (*Potamogeton*).

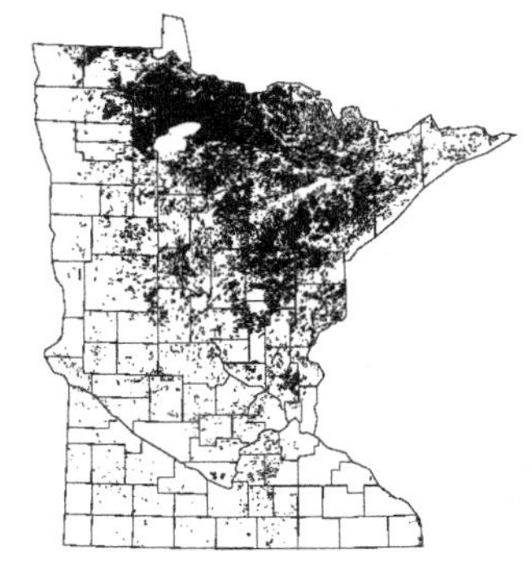

Current distribution of Minnesota wetlands
(GREAT LAKES INFORMATION NETWORK)

Wetland plants

Wetland species are intermediate between truly aquatic species and plants adapted to the moist or dry conditions of upland environments. Wetland species occur in habitats generally referred to as bogs, fens, marshes, swamps, thickets, and wet meadows, among others. Soils of these habitats may be organic (composed of decaying plant remains, or "peat"), of mineral origin, or a combination of both. Wetlands may be covered with shallow water for all or part of the year, or simply wet for a portion of the growing season.

On sites covered by shallow water, a group of wetland plant species termed **emergent** occur. These plants are rooted in soil but their stems and leaves are partly or entirely above the water surface. This large group includes many of the sedges (*Carex*), bulrushes (*Scirpus*), bur-reeds (*Sparganium*), rushes (*Juncus*), grasses (Poaceae) and the cat-tails (*Typha*).

A second large group of wetland species are those occurring on soils which are moist or even saturated throughout the year but are rarely covered by standing water. These sites include wetlands such as moist or wet meadows and low prairie. Many of the herbaceous dicots ("wildflowers") included here occur in these types of wetlands, especially on the drier margins. Species typically found in drier habitats are occasionally found in these wetlands as well, but are often restricted to hummocks slightly elevated above the surrounding lower, wetter areas.

Using this Book

■ MAJOR GROUPS

Presented first are the group of families comprising the **Ferns and Fern Allies** (true ferns, horsetails, clubmosses), followed by the **Gymnosperms** (conifers), and the **Angiosperms** (which make up the majority of the species described). The Angiosperms are subdivided into two groups—the **Dicots** and the **Monocots**. Dicots include many familiar trees, shrubs, and "wildflowers", such as those of the Aster Family; within the Monocots are the grass, sedge, rush and orchid families.

■ SPECIES DESCRIPTIONS

A standard format is used to describe each species: the lifeform (annual, perennial, herb, grass, etc.), stem and leaf characteristics, features of the flower and fruit, and a range of months when the species generally flowers are described for each species. The species' nativity to Minnesota is also noted (native to the state or introduced from elsewhere). **Synonyms** (other formerly accepted scientific names) are listed in italics. **Habitats** where the plant generally occurs are noted, as is the species' **conservation status** (endangered, threatened, or special concern in Minnesota).

A **map**, generated from data from the Biota of North America Program, shows county distribution patterns within the state. Note that the maps do not include every population of a species within Minnesota, only documented collections or sightings.

■ WETLAND INDICATOR STATUS

Updated (June 2012) wetland indicator status ratings are listed for each species. Five categories are defined: **OBL**, **FACW**, **FAC**, **FACU** and **UPL** (see page 41 for more information on the classification system). Unlike the earlier ratings, more than one rating may be applicable to a species, depending on the ecological region of the occurrence. Minnesota contains portions of three regions (see maps, page 42):

- **NC** Northcentral and northeast region
- **MW** Midwest
- **GP** Great Plains

■ ADDITIONAL RESOURCES

Key to all Families included in the guide, page 14; key to just the **Woody Plants** (trees, shrubs, vines), page 35. See page 623 for a **Glossary** of technical terms used in the Flora. **Abbreviations** used throughout the text are listed on page 634. The **Index** (page 639) includes the scientific and common names for all the described plant families, genera and species; synonyms are listed in *italics*.

Non-vascular plants (mosses, liverworts) are not included in this guide, although their importance as indicators of water chemistry and hydrologic conditions should be considered when conducting wetland studies. The standard reference for the region is *Mosses of Eastern North America* (Crum and Anderson 1981). A macro-algae of hardwater lakes, *Chara*, is illustrated on page 511.

Also excluded here are species which occur only rarely in wetlands. For example, a number of upland plants are found on sites which may be wet for only a short time in the spring following snowmelt or runoff. Under the **National Wetland Plant List** classification system, these would generally be classified as either Facultative Upland (FACU) or Obligate Upland (UPL) species.

Plant names

Each species of plant has a unique name made up of three parts. The first part is the **genus** to which a species belong; the second part is termed the **specific epithet**. Part three is the name of the person or persons who first named a species (the **authority**), and are often abbreviated. For example, "L." refers to Carolus Linnaeus, the 18th-century Swedish botanist considered the father of modern plant taxonomy. Together, these conventions form the scientific, or botanical, name of the species.

Over time, scientific names of plants sometimes change to better reflect relationships between various groups and species. For example, alkali-bulrush, long known as *Scirpus maritimus*, was renamed *Schoenoplectus maritimus*, then treated as *Bolboschoenus maritimus* (Flora of North America, Vol. 23). However, name changes are not always universally accepted and different names continue to be used for the same plant. To hopefully minimize confusion, these older, traditionally used botanical names, or **synonyms**, are included in the species descriptions, or, in some cases, are retained as the primary name (this simply reflects the author's preference).

Identifying unknown plants

The **keys** are the first step in identifying an unknown plant specimen. The keys are termed "dichotomous," meaning that the reader must choose between one of two leads which form a "couplet" at each step in the key. The lead is chosen which best describes the plant being identified and the user proceeds to the number following the lead in the key. The process is continued until the family, genus, and ultimately species are identified.

For a completely unknown plant (one in which even the family is unknown), start with the **Family Keys** (page 14). For families having only one genus or species in the guide, these will be identified here as well. Once the family is identified, turn to the **genus key**; once the genus has been determined, use the **species key** to identify the specimen. In addition to the keys, use the description of the plant and its habitat and range information to verify your identification. With practice, the characters of many families and genera will be recognizable on sight, greatly reducing the time spent keying.

Equipment needs are minimal for successful identification—a 10–20 power **hand lens** and a **ruler** with mm scale (a ruler is also printed on the last page of the index). For some species, as in the Sedge Family (Cyperaceae), a more powerful **dissecting microscope** is useful for examining the smaller plant parts often needed for positive identification. When a microscope is available, plants may be collected, pressed and dried, and identified at a later date.

Whenever possible, plants should be examined by leaving them rooted in place rather than uprooting or tearing off pieces to examine at eye level. This is especially important when working in wetlands, which are home to a disproportionate number of rare and uncommon plant species.

Vegetation of Minnesota

Minnesota may be divided into four broad vegetation zones:

- **Northern coniferous forest**
- **Eastern deciduous forest**
- **Aspen parkland**
- **Prairie**

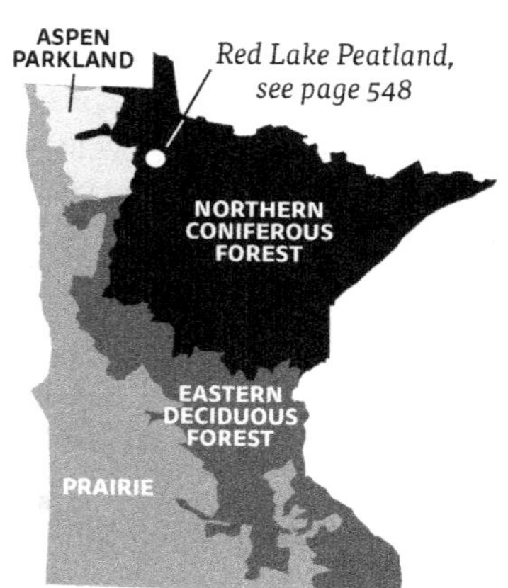

■ **Northern coniferous forest**
Coniferous forest covers about one-third of the state, primarily in the northeast. Dominant trees include balsam fir, white spruce, black spruce, paper birch and quaking aspen. Drier sites had a larger component of pines (white pine, red pine) and on the driest or most fire-prone sites, jack pine is common.
Wetlands occupy large portions of the coniferous forest zone. Notable are the extensive Red Lake peatlands of north-central Minnesota, supporting wet forests of black spruce and tamarack, smaller areas of northern white cedar, balsam-poplar, paper birch, and quaking aspen, and open bogs, patterned peatlands, and other types of shrub and herbaceous wetlands.

■ **Eastern deciduous forest** The original deciduous forest of Minnesota has been largely altered by agriculture and urban development. This forest was the westernmost extension of the deciduous forest formation of eastern North America. Typical components of the forest are sugar maple, basswood, oak and elm. Northward, yellow birch becomes important. Along major rivers, floodplain forests of silver maple, green ash, black willow, American elm, and cottonwood occur; poison ivy and stinging nettle are common. In the south, swamp white oak and buttonbush reach their northern limit.

■ **Aspen parkland** Located in northwestern Minnesota, aspen parkland forms an ecotone between the coniferous forest to the east and north and prairie to the west and south. The parkland occupies a flat, sometimes poorly drained landscape left by Glacial Lake Agassiz. Low-lying areas feature sedge meadow and low prairie wetlands; uplands may be dominated by extensive groves of quaking aspen mixed with other trees such as bur oak and northern pin oak, and various shrubs.

■ **Prairie** Southern and western Minnesota once supported tallgrass prairie which covered about one-third of the state. Dominant grasses included big bluestem (*Andropogon gerardii*) and Indian grass (*Sorghastrum nutans*). Due to conversion to agriculture, less than one percent of undisturbed prairie remains. Also lost have been many of the small, scattered low-lying depressions called "prairie pothole wetlands", with less than 10 percent of these wetlands remaining. Various sedges and grasses such as bluejoint (*Calamagrostis canadensis*) and prairie cordgrass (*Spartina pectinata*) were common.

Wetland Types of Minnesota

Wetlands occur across Minnesota in a wide variety of physical settings and under a large range of climatic conditions. As a result, wetlands vary markedly in their size, appearance, and composition. Wetlands of the northern coniferous forest region of the state differ from those of the deciduous forest and prairie vegetation vegetation to the south and west. Northerly wetlands, for example, feature extensive conifer swamps of black spruce (*Picea mariana*), tamarack (*Larix laricina*), and northern white cedar (*Thuja occidentalis*), and by various types of shrub-dominated and herbaceous wetlands, including alder thickets, sedge meadows, and open bogs.

Southward, floodplain forests along major rivers, and marshes and sedge meadows are more typical. In southern and western Minnesota, calcareous fens associated with upwelling groundwater are an unusual wetland type and home to a large number of uncommon plants. In north-central Minnesota, the Red Lake area supports a vast peatland. Continuing west onto the eastern edge of the Great Plains, rolling grasslands are interspersed with sedge meadows, low prairie, and prairie pothole ponds.

INTRODUCTION

The following key is a basic classification of the wetland types found in Minnesota, with applicability to Wisconsin and Michigan as well. It is based in part on Vegetation of Wisconsin (Curtis 1971) and *Wetland Plants and Plant Communities of Minnesota and Wisconsin* (Eggers and Reed 1997), and *Minnesota's Native Vegetation: A Key to Natural Communities* (Minnesota DNR 1993). More detailed classifications have been prepared by the Minnesota Dept. of Natural Resources; see *www.dnr.state.mn.us/npc*.

WETLAND TYPES OF MINNESOTA & THE GREAT LAKES REGION

1 Open water areas of lakes, ponds, rivers and streams; vegetation is absent or aquatic plants with underwater or floating leaves are present .... OPEN WATER COMMUNITIES

1 Wetland types of shallow water areas or on saturated or moist soil; trees, shrubs or herbaceous plants are dominant 2

2 Wetlands of shallow depressions, flats, or lakeshores; standing water may be present for several weeks each year, then drying; herbaceous plants dominant .... SEASONALLY WET BASIN OR SHORE

2 Standing water present, or soils saturated for all or most of growing season 3

3 Woody plants (trees and shrubs) dominant (greater than about 50 percent canopy cover) 4

3 Herbaceous plants dominant, woody plants sparse; trees, if present, are stunted and mostly small in diameter (less than about 15 cm wide) and less than 5–6 m tall 11

4 Trees dominant; overstory canopy ± closed by mature trees 5

4 Shrubs dominant; mature trees absent or sparse and not forming a closed canopy 8

5 Hardwood trees dominant; soils are alluvial deposits, poorly drained mineral soils, or sometimes organic .... (WET HARDWOOD FOREST) 6

5 Conifers such as tamarack (*Larix laricina*), black spruce (*Picea mariana*), or northern white cedar (*Thuja occidentalis*) dominant; soils usually highly organic 7

6 Forests adjacent to rivers on alluvial soils, mostly central and southern portions of the region. Silver maple (*Acer saccharinum*), American elm (*Ulmus americana*), river birch, (*Betula nigra*) green ash (*Fraxinus pennsylvanica*), black willow (*Salix nigra*) or eastern cottonwood (*Populus deltoides*) are typical dominants .... FLOODPLAIN FOREST

6 Forests of low-lying basins, soils poorly drained, mineral or organic; mostly central and northern areas. Black ash (*Fraxinus nigra*), red maple (*Acer rubrum*) or silver maple (*Acer saccharinum*) are typical dominants; northern white cedar (*Thuja occidentalis*) may be common ... HARDWOOD SWAMP

7 Tamarack and/or black spruce dominant; sphagnum moss forms a nearly continuous carpet; soils are organic and acid . . . . CONIFER BOG

7 Northern white cedar or tamarack dominant; sphagnum moss may be present but not a forming a continuous ground cover; soils with a high organic content, usually neutral to alkaline . . . . CONIFER SWAMP

8 Tall shrubs (mostly more than 1 m high) dominant; sphagnum moss may be present but not forming a continuous mat . . . . . . (SHRUB WETLAND) 9

8 Low shrubs (mostly less than 1 m high) dominant; sphagnum moss present, or sparse or absent 10

9 Speckled or tag alder (*Alnus incana*) dominant. . . . . ALDER THICKET

9 Dogwoods (*Cornus sericea, Cornus amomum*), willows (*Salix*), or other tall shrubs dominant . . . . SHRUB-CARR

10 Evergreen shrubs of the Heath Family (Ericaceae) such as leatherleaf (*Chamaedaphne calyculata*), bog-rosemary (*Andromeda glaucophylla*), Labrador-tea (*Ledum groenlandicum*), or small cranberry (*Vaccinium oxycoccos*) dominant or common; sphagnum moss forming a ± continuous carpet . . . . OPEN BOG (POOR FEN)

10 Deciduous shrubs, such as shrubby cinquefoil (*Potentilla fruticosa*), common; sites often sloping and maintained by a spring-fed supply of calcium-rich water; other calcium-indicating plants present; sphagnum moss absent or sparse . . . . CALCAREOUS FEN

11 Dominant plants sedges (*Carex*) and other members of the Cyperaceae, cat-tails (*Typha*), and large emergent species such as bur-reed (*Sparganium*), bulrush (*Scirpus*), and water plantain (*Alisma*) 12

11 Dominant plants are grasses or plants of calcium-rich sites 14

12 Cat-tails, bulrushes (*Scirpus*), water plantain (*Alisma*), arrowheads (*Sagittaria*), pickerelweed (*Pontederia*), giant bur-reed, or emergent lake sedges (for example, *Carex lacustris, C. utriculata*) dominant; sites have standing water for all or most of growing season (sometimes drying to surface in summer or fall) . . . . MARSH

12 Sedges (primarily *Carex*) dominant; water level often falling below ground surface during growing season 13

13 Sedges dominant; sphagnum moss may be present but not a continuous carpet; soils saturated most of growing season, acid to alkaline . . . . SEDGE MEADOW

13 Sphagnum moss forming a ± continuous mat; soils are acid peats; pitcher plants (*Sarracenia purpurea*), sundews (*Drosera*), sedges (such as *Carex oligosperma*), cottongrass (*Eriophorum*), and shrubs of the Heath Family (Ericaceae) are typically present . . . . OPEN BOG (POOR FEN)

14 Wetlands often on sloping sites, with a spring-fed supply of calcium-rich water; species indicative of calcium-rich sites are typical: fen star sedge (*Carex sterilis*), marsh-muhly (*Muhlenbergia glomerata*), needle beak-rush (*Rhynchospora capillacea*), beaked spike-rush (*Eleocharis rostellata*), tufted bulrush (*Scirpus cespitosus*), American grass of-parnassus (*Parnassia glauca*) and brook lobelia (*Lobelia kalmii*) .... CALCAREOUS FEN

14 Wetlands with water mainly from rainfall, springs, or surface drainage; calcium-indicating species not dominant; sites have standing water or are saturated during all or part of growing season 15

15 Wetlands with saturated soils (rarely with standing water; grasses such as bluejoint (*Calamagrostis canadensis*), redtop (*Agrostis stolonifera*), and reed canarygrass (*Phalaris arundinacea*) dominant .... WET MEADOW

15 Wetlands with standing water or saturated soil during the growing season; prairie grasses such as big bluestem (*Andropogon gerardii*), prairie cordgrass (*Spartina pectinata*), and prairie lowland species present.... LOW PRAIRIE

OPEN WATER COMMUNITIES

These plant communities occur in shallow to deep water of lakes, ponds, rivers and streams (photo, page 40). Plants are free-floating, or submergent (anchored to the bottom and with underwater and/or floating leaves). A controlling factor of these communities is that water levels remain deep enough so that emergent vegetation typical of marshes is unable to establish.

MARSHES

Marshes are dominated by emergent plants growing in permanent or nearly permanent shallow to deep water (photo, page 231). Shallow water marshes are covered with up to about 15 cm of water for all or most of the growing season, but water levels may sometimes drop to the soil surface. Deep marshes have standing water between 15 and 100 cm or more deep during most of the growing season. A mix of emergent species and aquatic species of the adjoining open water are present.

SEDGE MEADOWS

Sedge meadows are dominated by members of the Sedge Family (Cyperaceae), with grasses (Poaceae) and rushes (*Juncus*) often present. Other herbaceous species are usually present only as scattered individuals. Because of differences in their species composition, sedge meadows can be subdivided into those occurring in northern portions of the region and those occurring southward. Soils are typically organic deposits (peat or muck), and saturated throughout the growing season. Common tussock sedge (*Carex stricta*) is a major component of sedge meadows, forming large hummocks composed of a mass of roots and rhizomes (photo, page 376). Periodic fires may help maintain the dominance of the sedges by killing invading shrubs and trees.

WET MEADOWS

Wet meadows occur on saturated soils, and are dominated by grasses (especially *Calamagrostis canadensis*) and other types of perennial herbaceous plants. Soils may be inundated for brief periods (1–2 weeks) in the spring following snowmelt or during floods. Shrubs occur only as scattered plants.

LOW PRAIRIE

Low prairies are dominated by grasses and grasslike plants. These communities are similar to wet meadows but are dominated by native grasses and other herbaceous species characteristic of the prairie. Low prairie communities primarily occur in southern and western Minnesota, but are occasional in the north on sandy plains or in wet swales.

CALCAREOUS FENS

Calcareous fens are an uncommon wetland type, both in Minnesota and worldwide. They typically develop on sites that are sloping and with a steady flow of groundwater rich in calcium and magnesium bicarbonates (Curtis 1971). The calcium and magnesium bicarbonates precipitate out at the ground surface, leading to a highly alkaline soil. Such conditions are tolerated by a fairly small group of plants termed "calciphiles." Sphagnum mosses are absent or sparse. Due to their uniqueness, calcareous fens support a large number of rare plant species such as these members of the Sedge Family (Cyperaceae): *Carex sterilis, Eleocharis rostellata, Rhynchospora capillacea,* and *Scleria verticillata*.

OPEN BOGS

Open bogs (photo, page 268) are found most commonly in northern portions of Minnesota, but also occur as small relict features to the south. Technically, true bogs are extremely nutrient-poor, receiving nutrients only from precipitation. Many of the region's bogs may better be termed "poor fens," but given the widespread use of the term among scientists and general public alike, "bog" is used throughout this book. Bogs have a characteristic, nearly continuous carpet of sphagnum moss, and are dominated by shrubs of the Heath Family (Ericaceae), especially leatherleaf (*Chamaedaphne calyculata*), members of the Sedge Family (Cyperaceae), and scattered herbs such as pitcher plant (*Sarracenia purpurea*) and sundew (*Drosera*). Stunted trees of black spruce (*Picea mariana*) and tamarack (*Larix laricina*) may be present; this type of bog is sometimes termed "muskeg." Soils are composed entirely of saturated organic peat.

CONIFER BOGS

Conifer bogs are similar to open bogs except that trees of black spruce and/or tamarack are predominant. Sphagnum moss carpets the ground surface. Shrubs of the Heath Family (Ericaceae), sedges, and a small number of other species occur. Soils are organic, saturated peats.

SHRUB-CARRS

Shrub-carrs are dominated by tall shrubs. Soils are organic peat or muck, or alluvial soils of floodplains, and are often saturated throughout the growing season, and sometimes inundated during floods. Willows (*Salix*) and dogwoods (*Cornus*) are especially characteristic.

ALDER THICKETS

Alder thickets (photo, page 39) are a tall, deciduous shrub community similar to shrub-carrs except that speckled or tag alder (*Alnus incana*) is dominant, and often forming dense colonies. Alder thickets are common in northern Minnesota along rivers and streams, or along the wet margins of marshes, sedge meadows and bogs.

HARDWOOD SWAMPS

Hardwood swamps (photo, page 231) are dominated by deciduous trees such as black ash (*Fraxinus nigra*), red maple (*Acer rubrum*), and yellow birch (*Betula alleghaniensis*). Southward, silver maple (*Acer saccharinum*) increases in importance; northward, northern white cedar (*Thuja occidentalis*) may be present. American elm (*Ulmus americana*) may be present, but in greatly reduced numbers than formerly due to losses from Dutch elm disease. Soils are saturated during much of the growing season, and may be covered with shallow water in spring.

CONIFER SWAMPS

Conifer swamps are forested wetlands dominated primarily by tamarack (*Larix laricina*) and black spruce (*Picea mariana*). In areas where the water is not as stagnant and in areas underlain by limestone, northern white cedar (*Thuja occidentalis*) may form dense stands. Conifer swamps occur almost entirely in northern Minnesota. Soils are usually peat or muck, and vary in reaction from acidic and nutrient-poor, to neutral or alkaline and relatively fertile. Tamarack is most abundant on acid soils, while northern white cedar is most common on neutral or alkaline soils. Soils are typically saturated much of the growing season, and sometimes covered with shallow water. Sphagnum moss may be present, but does not usually form a continuous mat.

FLOODPLAIN FORESTS

Floodplain forests are dominated by deciduous hardwood trees growing on alluvial soils along rivers. These forests often have standing water during the spring, with water levels dropping to the surface (or below) in summer and fall. Typical trees include silver maple (*Acer saccharinum*), green ash (*Fraxinus pennsylvanica*), black willow (*Salix nigra*), eastern cottonwood (*Populus deltoides*), and river birch (*Betula nigra*). The diversity of the undergrowth is high, as flood waters remove and deposit alluvium, creating microhabitats suitable for many species.

SEASONALLY FLOODED BASINS

Seasonally flooded basins are poorly drained, shallow depressions in glacial deposits (kettles), low spots in outwash plains, or depressions in floodplains. The basins may have standing water for a few weeks in the spring or for short periods following heavy rains. Soils are typically dry, however, for much of the growing season. In western and southern Minnesota, attempts are made to cultivate the basins, and in combination with the fluctuating water level, these areas are often dominated by a wide array of annual or weedy plant species.

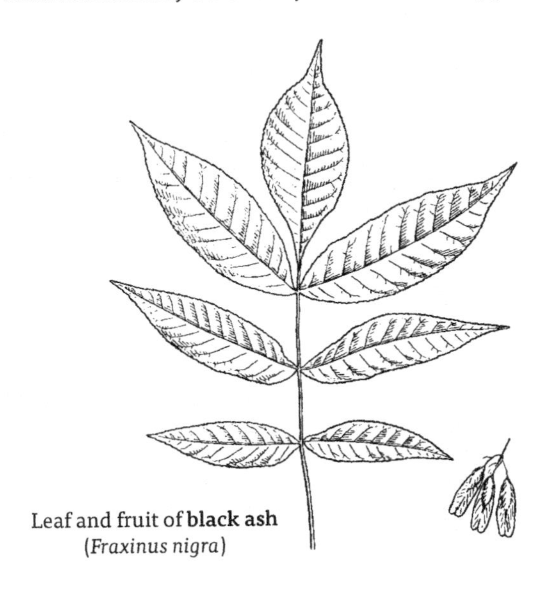

Leaf and fruit of **black ash**
(*Fraxinus nigra*)

FAMILY KEYS

THESE KEYS will lead to each plant family included in *Wetland Plants of Minnesota*. In cases where the key is specific to a particular genus or species, these are noted in italics. Begin identifying an unknown plant by first using the **Key to Groups and Families** below. After identifying the family, turn to the family page and continue keying the plant using the genus and species keys.

KEY TO GROUPS AND FAMILIES

1 Plants herbaceous (not woody), reproducing by spores .... **Key 1 - Ferns and Fern Allies**
1 Trees, shrubs or herbs; reproducing by seeds...(conifers and flowering plants) 2

2 Trees or shrubs with needlelike or scalelike, usually evergreen leaves; seeds in a dry cone, not inside an ovary .... **Key 2 - Conifers**
2 Trees, shrubs, or herbs; leaves usually deciduous; seeds inside an ovary which matures into the fruit.......... (ANGIOSPERMS) 3

3 Plants with unusual or specialized features; the Lemnaceae (Duckweed Family), Asteraceae (Aster Family), Orchidaceae (Orchid Family), and Asclepiadaceae (Milkweed Family) will key here 4
3 "Typical" trees, shrubs, or herbs 7

4 Plants very small, floating in still water or stranded on wet shores; stem and leaf not differentiated; roots absent or 1–several from leaf underside.......... LEMNACEAE (Duckweed Family) p. 512
4 Plants larger; stems and leaves present.......... 5

5 Flowers clustered into a head on a plate-like receptacle, the head resembling a single flower .... ASTERACEAE (Aster Family) p. 101
5 Flowers not as above; stamens, style and stigma are highly modified and joined into a special structure at center of flower.......... 6

6 Flowers irregular, the lower petal different than the other 2 petals (or rarely the upper petal different); ovary inferior .... ORCHIDACEAE .... (Orchid Family) p. 524
6 Flowers regular, the 5 petals alike; ovary superior; fruit a podlike follicle containg many seeds, the seeds tufted with hairs ..... ASCLEPIADACEAE .... (Milkweed Family) p. 100

7 Herbaceous plants with undivided, usually narrow leaves, main veins parallel; petals and sepals in 3s or multiples of 3 .... **Key 3 - Monocots**
7 Plants herbaceous or woody; leaves simple or divided, veins usually in a net-like pattern; petals and sepals usually equal to or in multiples of 2, 4 or 5 .... **Key 4 - Dicots**

KEY 1 - FERNS AND FERN ALLIES

Divisions Lycopodiophyta, Equisetophyta, and Pteridophyta (Clubmosses, Horsetails and Ferns)

1 Plants 1–2 cm long, floating on water surface; leaves very small (*Azolla*) . . . . AZOLLACEAE (Azolla Family) p. 44
1 Plants rooted in soil or muck; leaves large or small 2

2 Leaves small and narrow, 1-2 mm wide, simple, often scalelike 3
2 Leaves broader, 3 cm or more long, often large and dissected 6

3 Stems hollow, grooved lengthwise and jointed, usually easily pulled apart at the nodes; leaves reduced to a whorl of papery scales in at each node; plants with simple or branched stems; sporangia in a terminal cone (*Equisetum*) . . . . EQUISETACEAE (Horsetail Family) p. 55
3 Stems and leaves not as above; spores at plant base, in leaf axils, or in a terminal cone 4

4 Leaves grasslike, 5–15 cm long, joined in a cormlike base; spores borne in a pocket at base of each leaf; plants rooted in mud and often underwater (*Isoetes*) . . . . ISOETACEAE (Quillwort Family) p. 63
4 Leaves small, less than 2 cm long; plants spreading by rhizomes or trailing stems and resembling large mosses; spores in upper leaf axils or in terminal cones 5

5 Spores in 4-sided cones; of 2 types, either of many very small male spores, or of a few larger female spores; leaves with a ligule (small projection near base of leaf) . . . . SELAGINELLACEAE (Selaginella Family) p. 76
5 Spores in round cones (in cross-section) or in upper leaf axils; only 1 type of spore present; leaves without a ligule. . . . . LYCOPODIACEAE . . . . (Clubmoss Family) p. 65

6 Leaves on long stalks; divided like a 4-leaf clover, sometimes floating on water surface; plants rooted in mud (*Marsilea*). . . . . MARSILACEAE . . . . (Water-clover Family) p. 68
6 Leaves not clover-like 7

7 Plants with a small, erect stem and fleshy roots; spores borne on a specialized fertile spike or panicle that rises from the upper surface of the leaf . . . . OPHIOGLOSSACEAE (Adder's tongue Family) p. 69
7 Plants larger, with fibrous roots and spreading by short or long rhizomes; spores borne in clusters (sori) on leaf undersides or on completely fertile leaves 8

8 Spores borne in clusters on modified leaves, not in clusters on leaf undersides (*Osmunda*) . . . . OSMUNDACEAE (Royal Fern Family) p. 73

8 Spores borne in clusters (sori) on leaf undersides 9

9 Leaf midribs, and veins with sparse to dense covering of transparent hairs; fronds arising singly from widely creeping rhizomes .... THELYPTERIDACEAE (Marsh Fern Family) p. 77
9 Leaf midribs and veins lacking transparent hairs; in most species (all but *Onoclea sensibilis*), fronds clumped from short rootstocks.. DRYOPTERIDACEAE (Wood Fern Family) p. 45

KEY 2 - CONIFERS

Division Pinophyta | Gymnosperms

1 Leaves scalelike, pressed flat to the stem (*Thuja occidentalis*) .... CUPRESSACEAE (Cypress Family) p. 80
1 Leaves needlelike, often borne in clusters .. PINACEAE (Pine Family) p. 81

KEY 3 - MONOCOTS

Division Magnoliophyta | Flowering Plants
(Class Liliopsida | Monocots)

1 Sepals and petals (perianth) absent or reduced to scales or bristles, never petal-like in size or color 2
1 Sepals and petals present 17

2 Flowers in axils of chaffy scales larger than the flower; sepals and petals absent or reduced to bristles or small scales; flowers in regular heads or spikes 3
2 Flowers not in chaffy bracts, or if bracts present, the flower equal or larger in size and not hidden by the bracts 5

3 Plants with a basal cluster of narrow linear leaves and an upright stalk tipped by a single, button-like head (*Eriocaulon aquaticum*) .... ERIOCAULACEAE (Pipewort Family) p. 489
3 Plants with leaves on stem (these sometimes reduced to scales), or plants with several to many heads or spikes 4

4 Leaves usually 2-ranked around stem; stems usually hollow, round, or flat, and never triangular in cross-section; leaf sheath usually open on side opposite blade .... POACEAE (Grass Family) p. 549
4 Leaves usually 3-ranked (sometimes reduced to scales); stems usually solid and pithy and triangular in cross-section; leaf sheath usually closed .... CYPERACEAE (Sedge Family) p. 390

5 Plants aquatic, leaves submerged or floating and becoming limp when withdrawn from water; flowers underwater, floating, or sometimes raised slightly above water surface 6
5 Plants of land or shallow water, leaves and flowers normally above water surface 11

6 Flowers small and inconspicuous, single to several from leaf axils 7
6 Flowers in heads or spikes 10

7 Leaves alternate, or sometimes uppermost leaves opposite 8
7 Leaves all opposite (or whorled) 9

8 Plants of mostly fresh water; flowers with 4 sepals and 4 stamens . . . . POTAMOGETONACEAE (Pondweed Family) p. 591
8 Uncommon plants of brackish water; western and southern Minnesota (*Ruppia cirrhosa*). . . . . RUPPIACEAE (Ditch-grass Family) p. 610

9 Leaves 1–4 cm long, tapering from base to a long point, margins with minute, spiny teeth; flowers with 1 ovary (*Najas*) . . . . NAJADACEAE . . . . (Water-nymph Family) p. 521
9 Leaves 3–10 cm long, threadlike, margins not spiny; flowers usually with 4 ovaries. . . . . ZANNICHELLIACEAE (Horned Pondweed Family) p. 622

10 Flowers in globe-shaped heads; upper heads male (and deciduous), lower heads female (and persistent) (*Sparganium*) . . . . SPARGANIACEAE . . . . (Bur-reed Family) p. 612
10 Flower heads or spikes all alike; flowers perfect (with both male and female parts) . . . . POTAMOGETONACEAE (Pondweed Family) p. 591

11 Flowers tiny, in spikes or heads surrounded by a large white or co-lored bract (spathe); leaves broad . . . . ARACEAE (Arum Family) p. 385
11 Flowers not surrounded by a large spathe; leaves ± linear. 12

12 Flowers many, in long spikes. 13
12 Flowers in globe-shaped heads, racemes, or loose, open clusters 15

13 Spike appearing to be from side of stem (*Acorus calamus*) . . . ACORACEAE . . . . (Sweet Flag Family) p. 378
13 Spike erect at end of stem 14

14 Top of spike slender, made up of male flowers; lower spike broader and of female flowers (*Typha*). . . . . TYPHACEAE (Cat-tail Family) p. 618
14 Spike uniform; flowers both male and female (*Triglochin*) . . . . JUNCAGINACEAE (Arrow-grass Family) p. 510

15 Flowers either male or female, with male in upper clusters, female in lower clusters (*Sparganium*) . . . . SPARGANIACEAE (Bur-reed Family) p. 612
15 Flowers with both male and female parts **16**

16 Flowers with 1 ovary; fruit a 3-parted capsule (*Juncus*) . . . . JUNCACEAE . . . . (Rush Family) p. 497
16 Flowers with 3 or 6 ovaries; fruit separating into segments when mature (*Triglochin*) . . . . JUNCAGINACEAE (Arrow-grass Family) p. 510

17 Flowers either male or female, on same or separate plants. **18**
17 Flowers with both male and female parts **20**

18 Flowers with 3 green sepals and 3 white or pinkish petals . . . . ALISMATACEAE (Water-Plantain Family) p. 379
18 Sepals and petals ± same color **19**

19 Aquatic plants, mostly underwater; stamens 3–12. . HYDROCHARITACEAE . . . . (Frog's-bit Family) p. 490
19 Land plants; stamens 3–6 . . . . LILIACEAE (Lily Family) p. 517

20 Ovary inferior (located below sepals and petals) **21**
20 Ovary or ovaries superior (attached above sepals and petals); plants aquatic and on land **23**

21 Aquatic plants; leaves underwater or floating . . . . . HYDROCHARITACEAE . . . . (Frog's-bit Family) p. 490
21 Plants emergent in shallow water or on wet soils **22**

22 Stamens 3 . . . . IRIDACEAE (Iris Family) p. 493
22 Stamens 6 . . . . LILIACEAE (Lily Family) p. 517

23 Flowers with 1 ovary **24**
23 Flowers with 2 or more ovaries **28**

24 Flowers radially symmetric (regular) **25**
24 Flowers irregular, not radially symmetric . . . . PONTEDERIACEAE . . . . (Water-hyacinth Family) p. 588

25 Sepals green; petals variously colored **26**
25 Sepals and petals with same appearance **27**

26 Stamens 3 (or sometimes 2) (*Xyris*). . . . . XYRIDACEAE . . . . (Yellow-eyed Grass Family) p. 620
26 Stamens 6 . . . . LILIACEAE (Lily Family) p. 517

27 Petals 6, united into a tube; stamens 3; plants underwater or on muddy shores . . . . PONTEDERIACEAE (Water-hyacinth Family) p. 538

27 Stamens 6; plants of drier sites . . . . LILIACEAE (Lily Family) p. 517588

28 Leaves all from base of plant and alternate on stem; flowers with 3 pistils (*Scheuchzeria palustris*). . . . . SCHEUCHZERIACEAE . . . . (Scheuchzeria Family) p. 611

28 Leaves all from base of plant; pistils 3 to many **29**

29 Flowers with 3 green sepals and 3 white or pinkish petals; flowers in panicles or umbels . . . . . . ALISMATACEAE (Water-Plantain Family) p. 379

29 Flowers with 6 pink petals, the outer 3 smaller and darker pink than the inner; flowers in an umbel (*Butomus umbellatus*) . . . . BUTOMACEAE . . . . (Flowering Rush Family) p. 389

KEY 4 - DICOTS

Division Magnoliophyta | Flowering Plants
(Class Magnoliopsida | Dicots)

1 Trees, shrubs, or woody vines . . . . **Group 1** p. 21

1 Plants herbaceous **2**

2 Flowers either male or female (imperfect) . . . . **Group 2** p. 25

2 Flowers with both male and female parts (perfect) **3**

3 Petals and sepals absent . . . . **Group 3** p. 26

3 Petals and sepals present **4**

4 Flowers with either sepals or petals but not both **5**

4 Flowers with both sepals and petals **6**

5 Ovary inferior (below sepals and petals) . . . . **Group 4** p. 26

5 Ovary superior (above sepals and petals). . . . . **Group 5** p. 27

6 Flowers with 2 or more ovaries . . . . **Group 6** p. 29

6 Flowers with 1 ovary; styles or stigmas 2 or more **7**

7 Ovary inferior . . . . **Group 7** p. 29

7 Ovary superior **8**

8 Stamens many and more than petals **9**

8 Stamens few; same or fewer than number of petals **10**

9 Flowers regular . . . . **Group 8** p. 31

9 Flowers irregular . . . . **Group 9** p. 32

10 Flower petals separate, not joined .... **Group 10** p. 32
10 Petals united 11

11 Flowers regular and stamens same number as petals or petal lobes .... **Group 11** p. 33
11 Flowers either irregular, or number fewer than corolla lobes ... **Group 12** .... p. 34

DICOT GROUP 1

Trees, shrubs, and woody vines.

1 Leaves opposite or whorled 2
1 Leaves alternate 15

2 Flowers fully developed before leaves expand 3
2 Flowers develop with or after leaves expand 6

3 Flowers with both petals and sepals 4
3 Only petals or sepals present (not both), or absent 5

4 Trees, petals separate, not joined; flowers either male or female; stamens 8; ovary superior (*Acer*) .... ACERACEAE (Maple Family) p. 86
4 Shrubs; flowers perfect, with both male and female parts; stamens 5; ovary inferior (*Lonicera*) .... CAPRIFOLIACEAE (Honeysuckle Family) p. 166

5 Stamens usually 2; ovary not lobed (*Fraxinus*) .... OLEACEAE .... (Olive Family) p. 259
5 Stamens usually 8; ovary distinctly 2-lobed (*Acer*) .... ACERACEAE .... (Maple Family) p. 86

6 Leaves compound 7
6 Leaves simple 10

7 Plants vinelike and trailing (*Clematis*) .... RANUNCULACEAE .... (Buttercup Family) p. 295
7 Plants not vinelike 8

8 Petals present (*Sambucus*) .... CAPRIFOLIACEAE .... (Honeysuckle Family) p. 166
8 Petals absent 9

9 Stamens 2; ovary not lobed (*Fraxinus*) .... OLEACEAE (Olive Family) p. 259
9 Stamens 8; ovary 2-lobed (*Acer negundo*).................. ACERACEAE .... (Maple Family) p. 86

10 More stamens than petals or petal lobes 11
10 Stamens equal or less than number of petals or petal lobes 13

11 Petals joined .... ERICACEAE (Heath Family) p. 195
11 Petals separate 12

12 Stamens 8–10 (*Decodon verticillatus*) .... LYTHRACEAE .... (Loosestrife Family) p. 247
12 Stamens numerous (*Hypericum*) .... HYPERICACEAE .... (St. John's-wort Family) p. 227

13 Petals separate (*Cornus*) .... CORNACEAE (Dogwood Family) p. 183
13 Petals joined 14

14 Flowers in many-flowered, globe-shaped heads; leaves entire (*Cephalanthus occidentalis*) .... RUBIACEAE (Madder Family) p. 325
14 Flowers not in dense heads; leaves entire, toothed, or lobed .... CAPRIFOLIACEAE (Honeysuckle Family) p. 166

15 Plants with either male or female flowers but not both on same plant (dioecious) 16
15 Plants with male and female flowers on same plant; flowers either perfect (with both male and female parts), or imperfect (flowers single sex only) 22

16 Woody climbing vines (*Vitis vulpina*) .... VITACEAE (Grape Family) p. 375
16 Trees or shrubs 17

17 Flowers in catkins; individual flowers small 18
17 Flowers not in catkins; flowers often large and showy 19

18 Trees or shrubs; twigs not covered with resinous dots ...... SALICACEAE .... (Willow Family) p. 330
18 Shrubs; twigs densely covered with resinous dots (*Myrica gale*) .... MYRICACEAE (Bayberry Family) p. 254

19 Small shrub; leaves cylindric, short, 3–8 mm long; rare in ne Minnesota (*Empetrum nigrum*) .... EMPETRACEAE (Crowberry Family) p. 194
19 Mostly larger shrubs; leaves much larger 20

20 Flowers in a large terminal panicle (*Toxicodendron vernix*) .... ANACARDIACEAE (Cashew Family) p. 89
20 Flowers 1 to several from leaf axils 21

21 Style very short .... AQUIFOLIACEAE (Holly Family) p. 98
21 Style relatively long .... RHAMNACEAE (Buckthorn Family) p. 309

22 Flowers small, either male or female, never perfect; male flowers in catkins or round heads **23**
22 Flowers mostly perfect, sometimes large and showy, not grouped into catkins or globe-shaped heads **25**

23 Female flowers 1, or several in clusters (*Quercus*) . . . . FAGACEAE . . . . (Beech Family) p. 208
23 Female flowers more numerous; cone-like, or arranged into catkins or heads **24**

24 Female flowers 2 or 3 behind each catkin scale . . . . BETULACEAE . . . . (Birch Family) p. 142
24 Female flowers 1 behind each catkin scale (*Myrica gale*) . . . . MYRICACEAE . . . . (Bayberry Family) p. 254

25 Petals and sepals absent or in a single series, the petals and sepals not distinct from one another **26**
25 Petals and sepals present and clearly distinct from each other **30**

26 Styles 2–3. . . . . ULMACEAE (Elm Family) p. 362
26 Style 1, sometimes branched **27**

27 Vines (*Vitis vulpina*) . . . . VITACEAE (Grape Family) p. 375
27 Shrubs or small trees **28**

28 Flowers in clusters at ends of branches (*Cornus*) . . . . CORNACEAE . . . . (Dogwood Family) p. 183
28 Flowers from leaf axils or on short lateral branches **29**

29 Style not branched, stigma 1 . . . . AQUIFOLIACEAE (Holly Family) p. 98
29 Style 2–4-parted, stigmas 2–4 . . RHAMNACEAE (Buckthorn Family) p. 309

30 Ovaries 3 or more; stamens more than 10 . ROSACEAE (Rose Family) p. 311
30 Ovary 1 . . . . **31**

31 Petals 1 (*Amorpha fruticosa*). . . . . . . FABACEAE (Pea or Bean Family) p. 204
31 Petals more than 1; flowers ± regular **32**

32 Petals joined **33**
32 Petals free. **35**

33 Stamens more in number than corolla lobes . . . . ERICACEAE . . . . (Heath Family) p. 195
33 Stamens equal to number of corolla lobes **34**

34 Style well-developed . . . . ERICACEAE (Heath Family) p. 181
34 Style short, the stigma ± stalkless. . . AQUIFOLIACEAE (Holly Family) p. 92

35 Ovary inferior 36
35 Ovary superior 39

36 Stamens 2x or more than number of petals 37
36 Stamens equal number of petals 38

37 Style 1 (*Vaccinium*) . . . . ERICACEAE (Heath Family) p. 181
37 Styles 2–5. . . . . ROSACEAE (Rose Family) p. 288

38 Petals 4 . . . . CORNACEAE (Dogwood Family) p. 170
38 Petals 5 (*Ribes*). . . . . GROSSULARIACEAE (Gooseberry Family) p. 200

39 Leaves short and cylindric, less than 1 cm long (*Empetrum nigrum*). . . . . . .
. . . . EMPETRACEAE (Crowberry Family) p. 180
39 Leaves flat and broader 40

40 Leaves compound (*Toxicodendron vernix*). . . . . ANACARDIACEAE
. . . . (Cashew Family) p. 83
40 Leaves simple 41

41 Stamens number more than petals . . . . ERICACEAE (Heath Family) p. 181
41 Stamens number as many as petals 42

42 Flowers in clusters at ends of stems; leaf margins rolled under, leaf underside densely hairy (*Ledum groenlandicum*). . . . . ERICACEAE
. . . . (Heath Family) p. 181
42 Flowers from leaf axils; leaf margins not rolled under, leaves not densely hairy below 43

43 Style 3-parted; stamens opposite the petals (*Rhamnus*). . . . RHAMNACEAE
. . . . (Buckthorn Family) p. 286

43 Style very short; stamens alternate with the petals . . . . . AQUIFOLIACEAE
. . . . (Holly Family) p. 92

DICOT GROUP 2

Herbaceous plants; flowers either male or female (imperfect).

1 Leaves absent or reduced to small scales (*Salicornia rubra*) .... CHENOPODIACEAE (Goosefoot Family) p. 179
1 Leaves not reduced to scales 2

2 Aquatic plants; leaves dissected into threadlike segments 3
2 Leaves not dissected into threadlike segments 4

3 Leaves pinnately dissected .... HALORAGACEAE .... (Water-milfoil Family) p. 220
3 Leaves palmately dissected (*Ceratophyllum*)........ CERATOPHYLLACEAE .... (Hornwort Family) p. 177

4 Leaves simple, not compound 5
4 Leaves compound 17

5 Leaves all from base of plant.. 6
5 Leaves all or mostly along stem 7

6 Flowers in short or long spikes .... PLANTAGINACEAE .... (Plantain Family) p. 269
6 Flowers in small panicles (*Rumex*).................. POLYGONACEAE .... (Smartweed Family) p. 274

7 Leaves whorled or opposite... 8
7 Leaves alternate.. 13

8 Flowers single 9
8 Flowers in clusters from leaf axils or at ends of stems 10

9 Leaves whorled (*Hippuris vulgaris*) .... HIPPURIDACEAE .... (Mare's-tail Family) p. 226
9 Leaves opposite....... CALLITRICHACEAE (Water-starwort Family) p. 159

10 Flower clusters from leaf axils 11
10 Flowers in clusters at ends of stems.................................. 12

11 Leaf margins entire .... PLANTAGINACEAE (Plantain Family) p. 269
11 Leaf margins toothed .... URTICACEAE (Nettle Family) p. 363

12 Stamens 3, style 1............. VALERIANACEAE (Valerian Family) p. 367
12 Stamens 10, styles mostly 5 (*Silene*) .... CARYOPHYLLACEAE .... (Pink Family) p. 172

13 Sepals and petals present 14
13 Petals absent 16

14 Plants vining and with tendrils . CURCURBITACEAE (Gourd Family) p. 188
14 Plants not vining; tendrils absent 15

15 Flowers in a panicle at end of stem (*Napaea dioica*) . . . . MALVACEAE . . . . (Mallow Family) p. 252
15 Flowers single (*Rubus*) . . . . ROSACEAE (Rose Family) p. 311

16 Sepals dry and chaffy. . . . . . . AMARANTHACEAE (Amaranth Family) p. 88
16 Sepals often absent, herbaceous (*Atriplex, Chenopodium*) . . . . CHENOPODIACEAE (Goosefoot Family) p. 179

17 Leaves divided into 3 leaflets (*Clematis*). . . . . RANUNCULACEAE . . . . (Buttercup Family) p. 295
17 Leaves divided into more than 3 leaflets 18

18 Stem leaves alternate (*Thalictrum*) . . . . RANUNCULACEAE . . . . (Buttercup Family) p. 295
18 Stem leaves opposite. . . . . VALERIANACEAE (Valerian Family) p. 367

DICOT GROUP 3

Herbaceous plants; flowers with both male and female parts (perfect); flowers without petals or sepals.

1 Plants underwater or sometimes stranded on shores . . . HALORAGACEAE . . . . (Water-milfoil Family) p. 220
1 Land plants or of shallow water only 2

2 Leaves deeply lobed or compound. . . . . RANUNCULACEAE . . . . (Buttercup Family) p. 295
2 Leaves simple, linear, margins entire (*Hippuris vulgaris*) . . . . HIPPURIDACEAE (Mare's-tail Family) p. 226

DICOT GROUP 4

Herbaceous plants; flowers with both male and female parts (perfect); petals and sepals not different (in 1 series); ovary inferior.

1 Stamens number more than sepals and petals 2
1 Stamens less than or equal to number of sepals and petals. 3

2 Leaves broadly ovate (*Chrysosplenium americanum*). . . . . SAXIFRAGACEAE . . . . (Saxifrage Family) p. 347

2 Leaves linear or divided into linear segments .... HALORAGACEAE .... (Water-milfoil Family) p. 205

3 Leaves from base of plant or alternate.................................. 4
3 Leaves opposite or whorled.. 7

4 Stamens 4 or less; perianth lobes 3 or 4................................ 5
4 Stamens 5; perianth lobes 5... 8

5 Stamens 3; perianth lobes 3......................... HALORAGACEAE .... (Water-milfoil Family) p. 220
5 Stamens 4; perianth lobes 4 6

6 Leaves broadly ovate, margins with rounded teeth (*Chrysosplenium americanum*) .... SAXIFRAGACEAE (Saxifrage Family) p. 347
6 Leaves linear or lance-shaped, margins entire (*Ludwigia*) .. ONAGRACEAE .... (Evening-primrose Family) p. 261

7 Flowers single in leaf axils; leaf margins entire (*Geocaulon lividum*) .... SANTALACEAE (Sandalwood Family) p. 345
7 Flowers in heads or umbels; leaves usually dissected .... APIACEAE .... (Carrot Family) p. 91

8 Flowers densely clustered into heads at ends of stems (*Cornus*) .... CORNACEAE (Dogwood Family) p. 183
8 Flowers not in dense heads 9

9 Leaves whorled (*Galium*) .... RUBIACEAE (Madder Family) p. 325
9 Leaves opposite... 10

10 Flowers in clusters at ends of stems; stamens 3 .... VALERIANACEAE .... (Valerian Family) p. 367
10 Flowers single or few from leaf axils; stamens 4 11

11 Style 1 (*Ludwigia*) .... ONAGRACEAE (Evening-primrose Family) p. 261
11 Styles 2 (*Chrysosplenium americanum*) .... SAXIFRAGACEAE .... (Saxifrage Family) p. 347

DICOT GROUP 5

Herbaceous plants; flowers with both male and female parts (perfect); petals and sepals not different (in 1 series); ovary superior.

1 Each flower with more than 1 ovary; these sometimes joined up to their middle, free above... 2
1 Flowers with 1 ovary 3

2 Ovaries of flowers joined up to their middle (*Penthorum sedoides*) . . . . SAXIFRAGACEAE (Saxifrage Family) p. 347
2 Ovaries distinct, not joined . . RANUNCULACEAE (Buttercup Family) p. 295

3 Stamens more than 2x number of perianth lobes 4
3 Stamens 2x or fewer than perianth lobes 5

4 Leaves all at base of plant, modified into water-holding pitchers (*Sarracenia purpurea*) . . . . SARRACENIACEAE (Pitcher-plant Family) p. 346
4 Leaves not pitcher-like, large and heart-shaped, usually floating on water surface . . . . NYMPHAEACEAE (Water-Lily Family) p. 256

5 Styles 2 or more 6
5 Styles absent or 1 10

6 Leaves small and scalelike; plants of salty habitats (*Salicornia rubra*) . . . . CHENOPODIACEAE (Goosefoot Family) p. 179
6 Leaves not scalelike 7

7 Leaves opposite or whorled 8
7 Leaves alternate 9

8 Leaf margins with shallow, rounded teeth (*Chrysosplenium americanum*) . . . . SAXIFRAGACEAE (Saxifrage Family) p. 347
8 Leaf margins entire . . . . CARYOPHYLLACEAE (Pink Family) p. 172

9 Stipules present at base of each leave and sheathing stem . . . . POLYGONACEAE (Smartweed Family) p. 274
9 Stipules absent . . . . CHENOPODIACEAE (Goosefoot Family) p. 179

10 Stamens more numerous than perianth lobes 11
10 Stamens as many or fewer than number of perianth lobes 12

11 Leaves opposite (*Ammania robusta*) . . . . LYTHRACEAE . . . . (Loosestrife Family) p. 247
11 Leaves alternate (*Cardamine*) . . . . BRASSICACEAE (Mustard Family) p. 149

12 Flowers in clusters at ends of stems . . . . CARYOPHYLLACEAE . . . . (Pink Family) p. 172
12 Flowers in clusters from leaf axils 13

13 Perianth with 4 lobes . . . . LYTHRACEAE (Loosestrife Family) p. 247
13 Perianth with 5 lobes (*Glaux maritima*) . . . . PRIMULACEAE . . . . (Primrose Family) p. 287

DICOT GROUP 6

Herbaceous plants; flowers with both male and female parts (perfect); petals and sepals present; each flower with 2 or more ovaries.

1 One style for each flower 2
1 Styles as many as ovaries, or styles absent 4

2 Petals ± separate, not joined; stamens many; ovaries 5 or more . . . . MALVACEAE (Mallow Family) p. 252
2 Petals joined; stamens 2–5; ovaries 4 3

3 Leaves opposite; stamens 2 or 4 . . . . LAMIACEAE (Mint Family) p. 232
3 Leaves alternate; stamens 5 . . . . . . BORAGINACEAE (Borage Family) p. 146

4 Sepals 3; petals 3 5
4 Sepals and petals each more than 3 6

5 Aquatic plants; leaves entire and floating, or leaves underwater and dissected . . . . CABOMBACEAE (Water-Shield Family) p. 158
5 Plants of muddy shores; leaves opposite (*Crassula aquatica*) . . . . CRASSULACEAE (Stonecrop Family) p. 187

6 Leaves round, attached to petiole at center; flowers single, large, 1–2 dm wide (*Nelumbo lutea*) . . . . NELUMBONACEAE (Lotus-Lily Family) p. 255
6 Leaves not attached to petiole at center 7

7 Sepals not joined to form a cup . . . . RANUNCULACEAE (Buttercup Family) p. 295
7 Sepals joined and cuplike 8

8 Pistils as many or more than number of petals . . . . ROSACEAE (Rose Family) p. 311
8 Pistils fewer than petals . . . . . . SAXIFRAGACEAE (Saxifrage Family) p. 347

DICOT GROUP 7

Herbaceous plants; flowers with both male and female parts (perfect); petals and sepals present; ovary inferior, 1 in each flower.

1 Stamens more than number of petals 2
1 Stamens less than or equal to number of petals 6

2 Style 1 3
2 Styles 2 5

3 Plants of moist soils ..... ONAGRACEAE (Evening-primrose Family) p. 261
3 Aquatic plants or plants of wet, muddy shores 4

4 Flowers small, not showy, white or green (*Myriophyllum*) ....
.... HALORAGACEAE (Water-milfoil Family) p. 220
4 Flowers large and showy, yellow (*Ludwigia*) .... ONAGRACEAE
.... (Evening-primrose Family) p. 261

5 Styles 2..................... SAXIFRAGACEAE (Saxifrage Family) p. 347
5 Styles 3 (*Montia chamissoi*) .... PORTULACACEAE (Purslane Family) p. 286

6 Petals separate.. 7
6 Petals joined for most of their length 13

7 Petals 2; stamens 2 (*Circaea alpina*) .... ONAGRACEAE
.... (Evening-primrose Family) p. 261
7 Petals 4 or 5; stamens 4 or 5....................................... 8

8 Petals 4 9
8 Petals 5... 11

9 Plants with underwater leaves, these dissected into narrow segments (*Myriophyllum*)........... HALORAGACEAE (Water-milfoil Family) p. 220
9 Underwater leaves absent, or if present, the leaves entire............. 10

10 Leaves with normal blade (*Ludwigia*)................... ONAGRACEAE
.... (Evening-primrose Family) p. 261
10 Leaves reduced to small scales (*Myriophyllum*) .... HALORAGACEAE
.... (Water-milfoil Family) p. 220

11 Leaves dissected or compound .... APIACEAE (Carrot Family) p. 91
11 Leaves entire 12

12 Flowers in panicles .... SAXIFRAGACEAE (Saxifrage Family) p. 347
12 Flowers in heads or umbels .... APIACEAE (Carrot Family) p. 91

13 Stem leaves alternate 14
13 Stem leaves alternate or whorled, or leaves all from base of plant...... 15

14 Flowers irregular (*Lobelia*) .. CAMPANULACEAE (Bellflower Family) p. 161
14 Flowers regular (*Campanula aparinoides*)............. CAMPANULACEAE
.... (Bellflower Family) p. 161

15 Leaves in whorls of 3–8 (*Galium*) ..... RUBIACEAE (Madder Family) p. 325
15 Leaves opposite along stem or all from base of plant................ 16

16 Stamens 3 . . . . VALERIANACEAE (Valerian Family) p. 367
16 Stamens 4 or 5 (*Linnaea borealis*). . . . . CAPRIFOLIACEAE
. . . . (Honeysuckle Family) p. 166

DICOT GROUP 8

Herbaceous plants; flowers with both male and female parts (perfect); petals and sepals present; flowers regular, with a single superior ovary; stamens number more than petals or corolla lobes.

1 Stamens more than 2x number of petals 2
1 Stamens 2x number of petals or less 4

2 Aquatic plants with large leaves from base of plant. . . . . NYMPHAEACEAE
. . . . (Water-Lily Family) p. 256
2 Land plants with alternate or opposite leaves along stem 3

3 Leaves alternate . . . . MALVACEAE (Mallow Family) p. 252
3 Leaves opposite . . . . HYPERICACEAE (St. John's-wort Family) p. 227

4 Stamens less than 2x number of petals 5
4 Stamens exactly 2x number of petals 8

5 Leaves alternate; style 1 . . . . BRASSICACEAE (Mustard Family) p. 149
5 Leaves opposite or whorled; styles 2–5 6

6 Flowers yellow . . . . HYPERICACEAE (St. John's-wort Family) p. 227
6 Flowers not yellow 7

7 Stamens grouped into 3 clusters of 3 each (*Triadenum*) . . . HYPERICACEAE
. . . . (St. John's-wort Family) p. 227
7 Stamens not grouped together . . CARYOPHYLLACEAE (Pink Family) p. 172

8 Petals 3 (*Crassula aquatica*) . . . . CRASSULACEAE (Stonecrop Family) p. 187
8 Petals 4 or more 9

9 Leaves compound or divided to base (*Oxalis montana*). . . . . OXALIDACEAE
. . . . (Wood Sorrel Family) p. 267
9 Leaves simple, margins entire or shallowly toothed 10

10 Style 1 . . . . PYROLACEAE (Shinleaf Family) p. 293
10 Styles 2 or more 11

11 Ovary lobed, each lobe tipped by a style . . . . SAXIFRAGACEAE
. . . . (Saxifrage Family) p. 347
11 Ovary not lobed, the styles all from tip of ovary 12

12 Flowers yellow (*Triadenum*) . . . . HYPERICACEAE . . . . (St. John's-wort Family) p. 227
12 Flowers white or pink-tinged . . . . CARYOPHYLLACEAE (Pink Family) p. 172

DICOT GROUP 9

Herbaceous plants; flowers with both male and female parts (perfect); petals and sepals present; flowers irregular, with a single superior ovary; stamens number more than petals or corolla lobes.

1 Sepals petal-like, or long and spurlike 2
1 Sepals not petal-like, often green 3

2 One sepal a spur or sac; leaf margins shallowly toothed BALSAMINACEAE . . . . (Touch-me-not Family) p. 140
2 Sepals not spurlike; leaves entire (*Polygala cruciata*) . . . . . POLYGALACEAE . . . . (Milkwort Family) p. 293

3 Lower 2 petals joined for their length along lower margins, enclosing the stamens . . . . FABACEAE (Pea or Bean Family) p. 204
3 Lower 2 petals not joined, or petals 1 4

4 Styles 2 (*Saxifraga pensylvanica*). . . . . SAXIFRAGACEAE . . . . (Saxifrage Family) p. 347
4 Style 1 (*Lythrum*) . . . . LYTHRACEAE (Loosestrife Family) p. 247

DICOT GROUP 10

Herbaceous plants; flowers with both male and female parts (perfect), with a single superior ovary; petals and sepals present; petals separate, not joined; stamens fewer or equal to number of petals.

1 Leaves opposite 2
1 Leaves alternate or all from base of plant 7

2 Sepals 2 or 3; petals 2 or 3 (*Elatine*) . . . . ELATINACEAE . . . . (Waterwort Family) p. 193
2 Sepals and petals each 4–6 or more 3

3 Style 1 4
3 Styles 2–5 6

4 Flowers with well-developed hypanthium (cuplike structure around ovary) . . . . LYTHRACEAE (Loosestrife Family) p. 247
4 Hypanthium absent 5

5 Stamens opposite the petal lobes. PRIMULACEAE (Primrose Family) p. 287
5 Stamens alternate with petal lobes .... GENTIANACEAE
.... (Gentian Family) p. 209

6 Flowers yellow (*Hypericum*) .... HYPERICACEAE
.... (St. John's-wort Family) p. 227
6 Flowers white to pink .... CARYOPHYLLACEAE (Pink Family) p. 172

7 Styles 2 or more .... DROSERACEAE (Sundew Family) p. 190
7 Styles 1 or absent 8

8 Hypanthium tube-shaped (*Lythrum*) .... LYTHRACEAE
.... (Loosestrife Family) p. 247
8 Hypanthium absent 9

9 Flowers irregular, sometimes spurred .. VIOLACEAE (Violet Family) p. 370
9 Flowers regular, spur absent 10

10 Flowers single at ends of stems (*Parnassia*) .... SAXIFRAGACEAE
.... (Saxifrage Family) p. 347
10 Flowers in a cluster at ends of stems (*Lysimachia*) .... PRIMULACEAE
.... (Primrose Family) p. 287

DICOT GROUP 11

Herbaceous plants; flowers with both male and female parts (perfect), with a single superior ovary; petals and sepals present; petals joined; stamens equal to number of petal lobes.

1 Leaves all from base of plant; flowers at end of naked stalk 2
1 Leaves mostly along stem 3

2 Flowers 4-parted, chaffy, in spikes or heads .... PLANTAGINACEAE
.... (Plantain Family) p. 269
2 Flowers 5-parted, not chaffy, in umbel at end of stem (*Primula mistassinica*)
.... PRIMULACEAE (Primrose Family) p. 287

3 Ovary deeply parted into 2 or 4 sections 4
3 Ovary not parted 5

4 Leaves opposite (*Mentha, Pycnanthemum*) .... LAMIACEAE
.... (Mint Family) p. 232
4 Leaves alternate .... BORAGINACEAE (Borage Family) p. 146

5 Leaves opposite or whorled 6
5 Leaves alternate 11

6 Flowers in crowded heads or spikes; corolla 4-lobed 7
6 Flowers mostly not in dense heads (sometimes in short racemes); corolla 4–12-lobed 8

7 Flowers chaffy; leaves linear. . PLANTAGINACEAE (Plantain Family) p. 269
7 Flowers petal-like; leaves not linear (*Phyla lanceolata*) . . . . VERBENACEAE . . . . (Vervain Family) p. 368

8 Stamens attached opposite the corolla lobes. . . . . PRIMULACEAE . . . . (Primrose Family) p. 287
8 Stamens alternate with corolla lobes 9

9 Corolla lobes 4 or 6–12. . . . . GENTIANACEAE (Gentian Family) p. 209
9 Corolla lobes 5 10

10 Stigmas 1. . . . . GENTIANACEAE (Gentian Family) p. 209
10 Stigmas 3. . . . . POLEMONIACEAE (Phlox Family) p. 251

11 Leaves small and scalelike; corolla 4-lobed (*Bartonia virginica*) . . . . GENTIANACEAE (Gentian Family) p. 209
11 Leaves compound or dissected into leaflets. 12

12 Leaves divided into 3 leaflets (*Menyanthes trifoliata*) . . MENYANTHACEAE . . . . (Buckbean Family) p. 253
12 Leaves pinnately compound (*Polemonium occidentale*). . POLEMONIACEAE . . . . (Phlox Family) p. 271

DICOT GROUP 12

Herbaceous plants; flowers with both male and female parts (perfect), with a single superior ovary; petals and sepals present; petals joined. Flowers irregular, or stamens less than number of petal lobes.

1 Corolla base a spur or sac. 2
1 Corolla not spurred or saclike. 4

2 Sepals 2-lobed (*Utricularia*). . . . . LENTIBULARIACEAE . . . . (Bladderwort Family) p. 241
2 Sepals joined, deeply 5-lobed 3

3 Leaves all from base of plant; flowers single atop a naked stalk (*Pinguicula vulgaris*) . . . . LENTIBULARIACEAE (Bladderwort Family) p. 241
3 Leaves mostly along stem. . . SCROPHULARIACEAE (Figwort Family) p. 351

4 Leaves alternate or all from base of plant 5
4 Leaves opposite or whorled. 6

5 Stamens 2 (*Veronica*) .... SCROPHULARIACEAE (Figwort Family) p. 351
5 Stamens 4 .... SCROPHULARIACEAE (Figwort Family) p. 351

6 Plants usually aromatic when rubbed; stems often 4-sided; ovary deeply 4-parted .... LAMIACEAE (Mint Family) p. 232
6 Plants not strongly scented; stems rarely 4-angled; ovary not deeply 4-parted. **7**

7 Stamens 2 .... SCROPHULARIACEAE (Figwort Family) p. 351
7 Stamens 4. .... **8**

8 Corolla ±regular (*Verbena hastata*). .... VERBENACEAE .... (Vervain Family) p. 368
8 Corolla lipped or irregular. **9**

9 Flower in clusters from leaf axils (*Phyla lanceolata*) .... VERBENACEAE .... (Vervain Family) p. 368
9 Flowers in clusters at ends of stems, or single from leaf axils **10**

10 Upper corolla lip absent (*Teucrium canadense*)... .... LAMIACEAE .... (Mint Family) p. 232
10 Upper corolla lip well-developed .... SCROPHULARIACEAE .... (Figwort Family) p. 351

Flowers of **leatherleaf** (*Chamaedaphne calyculata*), a member of the Heath Family (Ericaceae) and an important component of bogs in Minnesota and the northern Great Lakes region.

THE FOLLOWING KEY includes the trees, shrubs and vines found in Minnesota's wetlands. In some cases, only the genus is identified.

TREES

1 Leaves needlelike or scalelike . . . . **(Conifers)** 2
1 Leaves broad and flat . . . . **(Hardwoods)** 5

2 Leaves in bundles of 10 or more and shed in the fall . . . . TAMARACK (*Larix laricina*) p. 82
2 Leaves single and persistent 3

3 Leaves overlapping and scalelike . . . . NORTHERN WHITE CEDAR (*Thuja occidentalis*) p. 90
3 Leaves needlelike or strap-shaped 4

4 Leaves stiff, 4-sided in cross-section . . . . SPRUCE (*Picea*) p. 83
4 Leaves soft, flat in cross-section . . . . . . BALSAM FIR (*Abies balsamea*) p. 81

5 Leaves compound (with 3 or more leaflets) 6
5 Leaves simple 7

6 Leaflets 3-5; fruit a paired samara . . . . BOXELDER (*Acer negundo*) p. 86
6 Leaflets 7-11; samaras single . . . . ASH (*Fraxinus*) p. 259

7 Leaves opposite . . . . MAPLE (*Acer*) p. 86
7 Leaves alternate 8

8 Leaves lobed . . . . OAK (*Quercus*) p. 208
8 Leaves toothed but not lobed 9

9 Leaves asymmetrical at base (one lobe lower than other) . . . . AMERICAN ELM (*Ulmus americana*) p. 362
9 Leaves symmetrical at base (lobes equal) 10

10 Leaves at least 4x as long as wide . . . . WILLOW (*Salix*) p. 332
10 Leaves less than 4x as long as wide 11

11 Leaves as wide as long or wider . . . . COTTONWOOD (*Populus*) p. 330
11 Leaves longer than wide 12

12 Leaf margins doubly toothed (the teeth themselves toothed) . . . . . . BIRCH (*Betula*) p. 144
12 Leaf margins singly toothed . . . . COTTONWOOD (*Populus*) p. 330

WOODY PLANT KEY

SHRUBS AND VINES

1 Leaves evergreen and persistent on plant. 2
1 Leaves deciduous, shed in the fall 8

2 Leaves narrow, less than 5 mm wide 3
2 Leaves broader, 5 mm or more wide 4

3 Plants small and trailing; leaves elliptic, pointed or blunt-tipped; fruit a red cranberry; plants of sphagnum peatlands . . . . CRANBERRY . . . . . . (*Vaccinium macrocarpon, V. oxycoccos, V. vitis-idaea*) p. 201, 202, 203
3 Plants forming mats; leaves narrow with inrolled margins; fruit a black berry; uncommon . . . . BLACK CROWBERRY (*Empetrum nigrum*) p. 194

4 Leaves opposite or in whorl of 3. . . . . BOG-LAUREL (*Kalmia polifolia*) p. 198
4 Leaves alternate. 5

5 Leaf underside with a dense covering of white or brown hairs 6
5 Leaf underside with only scattered hairs or with small scales 7

6 Leaf underside with brown hairs; flowers cream-colored, in upright clusters . . . . LABRADOR-TEA (*Ledum groenlandicum*) p. 199
6 Leaf underside with short white hairs; flowers white to pink, urn-shaped and drooping . . . . BOG-ROSEMARY (*Andromeda glaucophylla*) p. 195

7 Upright shrubs; leaf underside with scales; flowers 5-parted . . . . LEATHERLEAF (*Chamaedaphne calyculata*) p. 196
7 Small, trailing shrubs, leaf underside with scattered brown bristly hairs; flowers 4-parted. . . . . CREEPING SNOWBERRY (*Gaultheria hispidula*) p. 197

8 Leaves opposite or whorled. 9
8 Leaves alternate on stem 14

9 Leaves compound, divided into leaflets 10
9 Leaves simple 11

10 Vines. . . . . CLEMATIS (*Clematis*) p. 297
10 Shrubs. . . . . ELDER (*Sambucus*) p. 168

11 Leaf margins distinctly lobed . . SQUASH-BERRY, HIGH-BUSH CRANBERRY . . . . (*Viburnum*) p. 170
11 Leaf margins not lobed 12

12 Leaves opposite or in whorls of 3 (or occasionally 4), flowers numerous, in a ball-shaped head atop a long stalk; fruits brown and nutlike. . . . . BUTTONBUSH (*Cephalanthus occidentalis*) p. 325

12 Leaves strictly opposite; fruit white or colored and berrylike **13**

13 Leaf lateral veins noticeably curved toward tip; Flowers white, 4-parted, stalkless with 4 white bracts, or in stalked clusters at ends of branches ... . DOGWOOD (*Cornus*) p. 183
13 Leaf lateral veins not curved toward tip; flowers light yellow, 5-parted, borne in pairs from leaf axils .... HONEYSUCKLE (*Lonicera*) p. 167

14 Leaves compound, divided into leaflets **15**
14 Leaves simple **19**

15 Leaflets 3 or 5, palmate or pinnate **16**
15 Leaflets 6 or more, pinnate **18**

16 Leaflets 3; margins entire or with a few coarse teeth; flowers many in a branched inflorescence; fruit a whitish drupe; prickles absent .... COMMON POISON-IVY (*Toxicodendron radicans*) p. 89
16 Leaflets 3 or 5; margins entire or coarselty toothed; flowers white, pink, or yellow; prickles sometimes present on stems **17**

17 Leaf margins coarsely toothed; fruit a berry .. RASPBERRY, BLACKBERRY, ... DEWBERRY (*Rubus*) p. 319
17 Leaflets narrow with entire margins, leaflets mostly 5, with upper 3 joined at base .... SHRUBBY CINQUEFOIL (*Potentilla fruticosa*) p. 317

18 Low, much-branched shrub less than 1 m high; leaflets 1–2 cm long, narrow; flowers yellow and showy, 1–2.5 cm wide; fruit a capsule .... SHRUBBY CINQUEFOIL (*Potentilla fruticosa*) p. 317
18 Taller shrubs; leaves 5 or more cm long; flowers small and green-yellow; fruit white and berrylike..... POISON-SUMAC (*Toxicodendron vernix*) p. 89

19 Leaves deeply or shallowly lobed **20**
19 Leaves not lobed **23**

20 Woody vines .... RIVERBANK GRAPE (*Vitis vulpina*) p. 375
20 Shrubs **21**

21 Stems thorny or prickly..................... GOOSEBERRY (*Ribes*) p. 215
21 Stems without thorns or prickles **22**

22 Stems with bark peeling into papery strips; flowers white, many in terminal clusters; fruit a dry brown pod.................... EASTERN NINEBARK .... (*Physocarpus opuliformis*) p. 315
22 Bark not peeling into papery strips; flowers cream white, yellow or green-purple, in small clusters from leaf axils; fruit a red to black berry .... CURRANT (*Ribes*) p. 215

23 Leaf margin entire **24**
23 Leaf margin toothed or wavy **28**

24 Leaves with resinous dots on both sides (especially underside) ....
.... BLACK HUCKLEBERRY (*Gaylussacia baccata*) p. 197
24 Leaves without resinous dots **25**

25 Large shrubs, often over 1 m tall; buds covered by a single scale; flowers in catkins; male and female flowers on separate plants; fruit a capsule .....
.... WILLOW (*Salix*) p. 332
25 Bud scales absent or buds covered by 2 or more scales; flowers single or in several to many flowered clusters **26**

26 Smaller shrubs, typically much less than 1 m; flowers bell-shaped, waxy white; fruit a blue or blue-black berry with many small seeds ....
.... BLUEBERRY (*Vaccinium*) p. 200
26 Taller shrubs, usually 2 m or more; flowers small; fruit a red-purple or crimson with several large seeds **27**

27 Leaves tipped with a small, sharp point; flowers and fruit single on very thin, long stalks .... MOUNTAIN-HOLLY (*Ilex mucronata*) p. 98
27 Leaves pointed but without a small, sharp tip; flowers single or several in leaf axils; fruit on short stalks.................... GLOSSY BUCKTHORN
.... (*Rhamnus frangula*) p. 310

28 Buds covered by a single scale; flowers in catkins; fruit a capsule........
.... WILLOW (*Salix*) p. 332
28 Buds covered by 2 or more scales; flowers various; if catkins present, fruit is hard and nutlike **29**

29 Leaves aromatic when rubbed; leaves with rounded, toothed tip, lower leaf margin entire, dotted on both sides with yellow glands ..... SWEET GALE
.... (*Myrica gale*) p. 254
29 Leaves not aromatic **30**

30 Young twigs usually with glands; leaf margins coarsely toothed; fruit in conelike, deciduous catkins .... SWAMP OR BOG BIRCH
.... (*Betula pumila*) p. 145
30 Glands absent; leaves various.................................. **31**

31 Leaf margins coarsely double-toothed or wavy .... ALDER (*Alnus*) p. 1422
31 Leaf margins not double-toothed or wavy **32**

32 Leaf midrib on upper surface of leaf with small dark glands ....
.... CHOKEBERRY (*Aronia*) p. 312
32 Glands absent from leaf blades or petioles **33**

33 Leaves with very short petioles, usually less than 5 mm long. **34**
33 Leaves with longer petioles, 5-30 mm long **35**

34 Flowers single or several, bell-shaped; fruit a blue to blue-black berry . . . . BLUEBERRY (*Vaccinium*) p. 200
34 Flowers small and numerous in upright, terminal clusters; fruit a persistent capsule. . . . . SPIRAEA (*Spiraea*) p. 323

35 Flowers larger, pink-white, 1–4 at ends of short branches, or single and terminal with single flowers from leaf axils; stalks 1–2 cm long . . . . JUNEBERRY (*Amelanchier bartramiana*) p. 312
35 Flowers small and yellow-green, single or in several-flowered clusters along branches, flower stalks short or absent **36**

36 Leaf margin with incurved, forward-pointing teeth; stipules persistent, dark-colored; fruit bright red, nearly stalkless . . . . WINTERBERRY . . . . (*Ilex verticillata*) p. 99
36 Leaf margin with rounded, forward-pointing teeth; leaves with pronounced raised veins on underside; stipules present, narrow, but falling before fruits mature; fruit purple-black and berrylike, stalked. . . . . BUCKTHORN . . . . (*Rhamnus*) p. 309

Alder thicket, summer. Speckled or Tag alder (*Alnus incana*) is a common, colony-forming shrub adjacent to streams, lakes, bogs, and marshes, especially within the coniferous forest region of Minnesota. This wetland type is best developed on well-aerated soils where water is not stagnant. Grasses such as bluejoint (*Calamagrostis canadensis*) and reed canary-grass (*Phalaris arundinacea*), and large sedges such as beaked sedge (*Carex utriculata*) and fringed sedge (*Carex crinita*) are common.

Open water, marsh, and sedge meadow communities, summer. Many aquatic and wetland plant species are present in this productive pond, including white water-lily (*Nymphaea odorata*), water-shield (*Brasenia schreberi*), buckbean (*Menyanthes trifoliata*), various species of pondweed (*Potamogeton*), softstem bulrush (*Scirpus validus*), three-way sedge (*Dulichium arundinaceum*), and common cat-tail (*Typha latifolia*). Adjacent to the pond is a sedge meadow dominated by common tussock sedge (*Carex stricta*), common lakeshore sedge (*Carex lacustris*), and bluejoint (*Calamagrostis canadensis*); sweet gale (*Myrica gale*) is a common shrub at the water's edge.

WETLAND INDICATOR STATUS

FOR EACH SPECIES DESCRIBED IN THE GUIDE, the wetland indicator status is provided (codes and definitions below). The indicator status ratings are based on the latest ***National Wetland Plant List*** (Lichvar and Kartesz 2012) and are widely used in wetland delineation studies. Several species with an indicator status in brackets were not ranked for Minnesota. The 2012 classification no longer uses positive or negative signs.

- **OBL (Obligate Wetland)**—Plants that almost always occur in wetlands (i.e. almost always in standing water or seasonally saturated soils.
- **FACW (Facultative Wetland)**—Plants that usually occur in wetlands, but may occur in non-wetlands.
- **FAC (Facultative)**—Plants that occur in wetlands and non-wetland habitats.
- **FACU (Facultative Upland)**—Plants that usually occur in non-wetlands but may occur in wetlands.
- **UPL (Obligate Upland)**— Plants that almost never occur in wetlands (or in standing water or saturated soils).

Regional delineations within Minnesota are based on those of the US Army Corps of Engineers (see maps, page 42). Three regions are present in Minnesota:

- **NC** (Northcentral and Northeast)
- **MW** (Midwest)
- **GP** (Great Plains)

A species' status rating may vary based on the region of its occurrence, as in the example below:

region code *status indicator*

Thelypteris palustris Schott
MARSH-FERN

NC **FACW** | MW **OBL** | GP **OBL**

For more information, see the website of the National Wetland Plant list: *https://wetland_plants.usace.army.mil.*

Full citation: Robert W. Lichvar and John T. Kartesz. 2009. North American Digital Flora: National Wetland Plant List, version 2.4.0 (*https://wetland_plants.usace.army.mil*). U.S. Army Corps of Engineers, Engineer Research and Development Center, Cold Regions Research and Engineering Laboratory, Hanover, NH, and BONAP, Chapel Hill, NC.

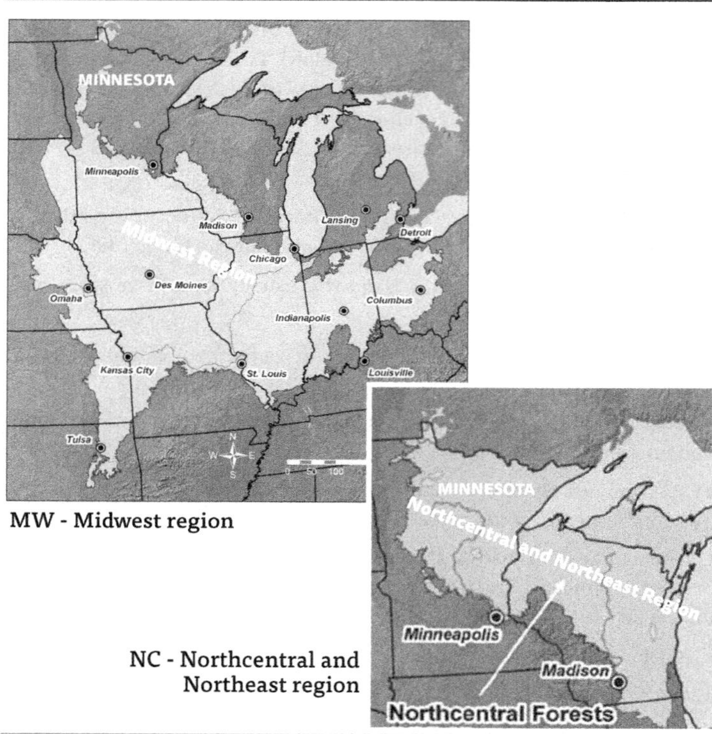

MW - Midwest region

NC - Northcentral and Northeast region

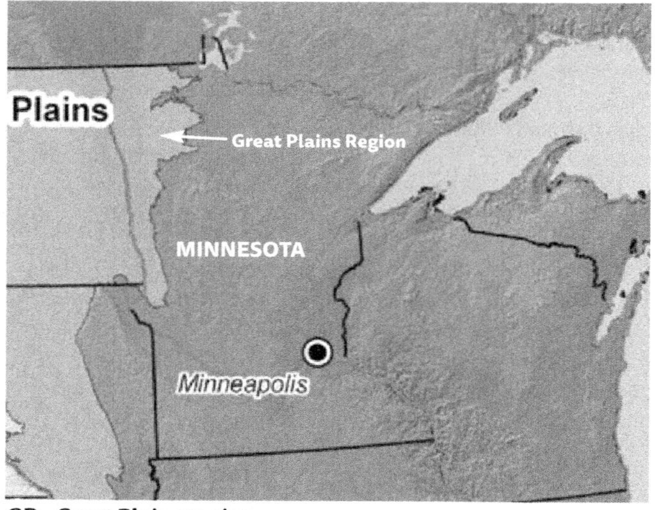

GP - Great Plains region

Ferns and Fern Allies

THE FERNS AND THEIR ALLIES in Minnesota can be classified into three major divisions (or phyla): **Pteridophyta** (true ferns), **Lycopodiophyta** (clubmosses), and **Equisetophyta** (horsetails).

The Pteridophyta are represented by six families in Minnesota's wetlands: **Azollaceae, Dryopteridaceae, Marsilaceae, Ophioglossaceae, Osmundaceae** and **Thelypteridaceae**. Three families are included in the Lycopodiophyta: **Lycopodiaceae, Selaginellaceae, Isoetaceae**. The Equisetophyta are represented by a single family (**Equisetaceae**).

All ferns and fern-allies reproduce by spores rather than seeds. The spore germinates and produces a tiny flattened thallus termed the gametophyte. The larger, leafy, spore-bearing plant (sporophyte) results from the fertilized egg produced by the female part of the gametophyte. The uncurling of the young leaf or frond is characteristic. The Lycopodiophyta (clubmosses) bear their spores on specialized leaves (sporophylls). The Equisetophyta (horsetails) bear spores in terminal cone-like structures, either on typical green stems, or on specialized fertile stems which lack chlorophyll (as in ***Equisetum arvense***, right).

See page 16 for key to ferns and fern-allies.

Pteridophyta
(*Phegopteris connectilis*)

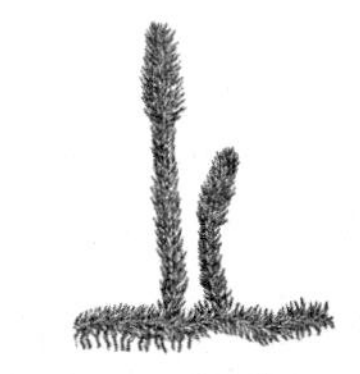

Lycopodiophyta
(*Lycopodiella inundata*)

Equisetophyta
(*Equisetum arvense*)

AZOLLA | *Mosquito-fern*

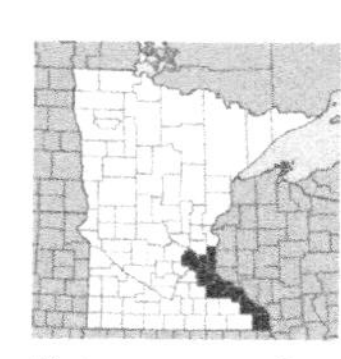

Azolla mexicana Presl.

MEXICAN MOSQUITO-FERN
NC **OBL** | MW **OBL** | GP **OBL**

Small annual aquatic fern, native; plants free-floating or forming floating mats several cm thick, sometimes stranded on mud; roots few and unbranched. **Stems** lying flat, 1–1.5 cm long, green or red, covered with small, alternate, overlapping leaves in 2 rows; **leaves** 2-lobed, the upper lobe to 1 mm long, emergent; lower lobe underwater and larger than the upper. **Sporangia** of 2 kinds; larger female megaspores (to 0.6 mm long) and tiny male spores (microsporangia), and borne in separate sporocarps. **Sporocarps** usually in pairs on underwater lobes of some leaves. **Habitat** Quiet water of marshes, ponds, streams and ditches. Near Miss River in se Minn, sw Wisc, Iowa and Ill.

Azolla caroliniana Willd. *reported for Minnesota, but distribution unclear. Sporocarps are needed to distinguish between Azolla mexicana and Azolla caroliniana, but these are rarely collected and require magnification to see differences.*

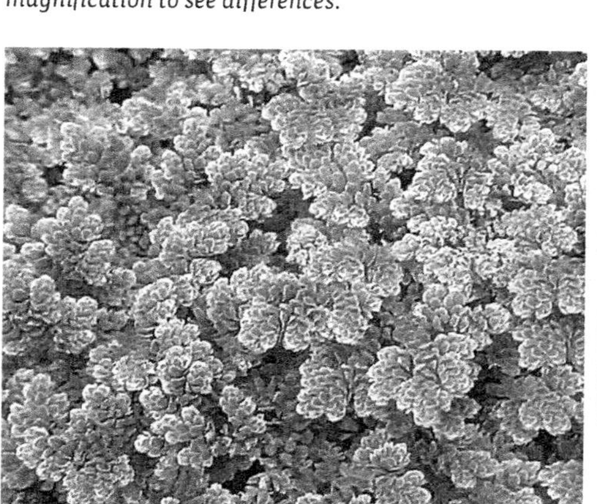

Azolla mexicana (MEXICAN MOSQUITO-FERN)

MEDIUM TO LARGE FERNS with scaly rhizomes. Fertile and sterile **leaves** mostly similar, usually divided or compound; **sori** (clusters of spore containers) present on lower leaf surface; **indusia** (covering over sori) present or absent.

DRYOPTERIDACEAE | WOOD FERN FAMILY

1 Leaves of two very different types (fertile and sterile) 2
1 Leaves all similar. 3

2 Leaves in large, vaselike clumps; fertile leaves 1-pinnate, surrounded by tall sterile leaves; veins of sterile leaves not netlike. . . . . *Matteuccia struthiopteris* | OSTRICH FERN
2 Leaves single or scattered along creeping rhizomes; fertile leaves 2-pinnate; veins of sterile leaves netlike . . . . *Onoclea sensibilis* . . . . SENSITIVE FERN

3 Sori elongate, either straight or hook-shaped, indusia present 4
3 Sori ± round, indusia present or absent. 5

4 Sori straight, blade 1-pinnate. . . . . *Deparia acrostichioides* . . . . SILVERY GLADE FERN
4 Sori curved or hook-shaped, blade 2-pinnate . . . . *Athyrium filix-femina* . . . . LADY FERN

5 Blade elongate, often with 'pealike' bulblets; indusium a hoodlike cover arching over spore clusters, attached at its base on side toward pinna midrib. . . . . *Cystopteris bulbifera* | BULBLET FERN; BLADDER FERN
5 Blade various, bulblets absent; indusium round or kidney-shaped, attached at its center, or absent . . . . *Dryopteris* | WOOD FERN

ATHYRIUM | *Lady fern*

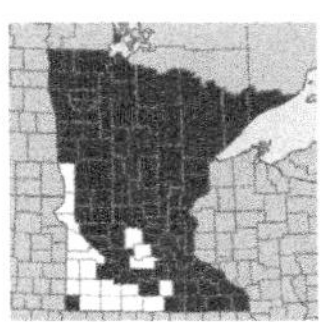

Athyrium filix-femina L.

LADY FERN
NC **FAC** | MW **FAC** | GP **FAC**

Clumped native fern, rhizomes short and ascending. **Leaves** deciduous, sterile and fertile leaves similar; petioles with brown, linear scales; blades elliptic, 2-pinnate, broadest at middle or slightly below middle; pinnae short-stalked or stalkless. **Sori** generally somewhat curved to hook-shaped, less often straight. **Synonyms** *Athyrium angustum*. **Habitat** Common; moist deciduous woods, shrub thickets, streambanks, hummocks in swamps, wetland margins, and shaded rock outcrops. Wide-ranging circumboreal species.

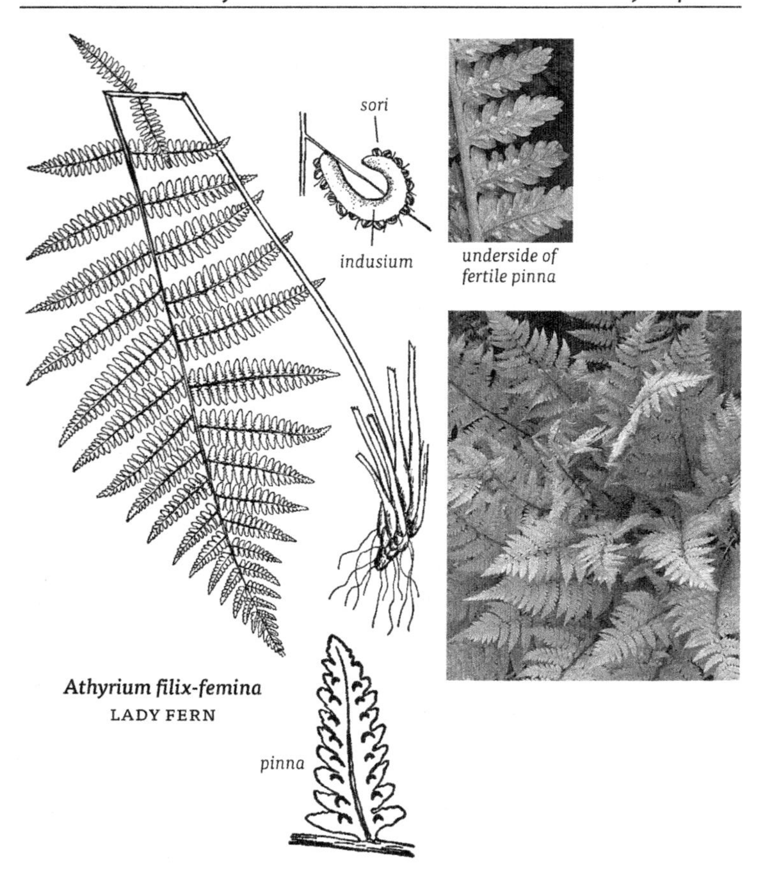

Athyrium filix-femina
LADY FERN

CYSTOPTERIS | *Bladder fern*

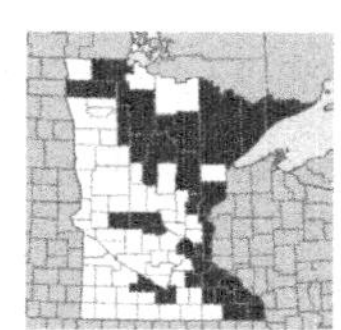

Cystopteris bulbifera (L.) Bernh.
BULBLET FERN; BLADDER FERN
NC **FACW** | MW **FACW** | GP **FACW**

Clumped native fern, rhizomes short and thick. **Leaves** deciduous, 30–100 cm long, sterile and fertile leaves similar but sterile blades usually shorter than fertile; petioles much shorter than blades; **blades** lance-shaped, 6–15 cm wide at base, long tapered to tip, with 20–30 pairs of pinnae; the veins ending in a notch (sinus). **Sori** round, on a small vein; **indusia** hoodlike and attached at its base, covered with scattered, short-stalked glands. **Green bulblets**, 4–5 mm wide, produced on lower side of rachis (main stem of leaf) toward upper end of blade, these falling and forming new plants. **Habitat** Rocky streambanks, ravines, and moist, shaded, often calcium-rich rocks and cliffs. In Minnesota, most common in se portion of state.

Distinguished from **fragile fern** *(Cystopteris fragilis), a common fern of moist woods, by the blade broadest at base, most veins ending in a notch, and the small bulblets on underside of rachis. In fragile fern, the blade is broadest above its base, most veins end in a tooth, and bulblets are absent.*

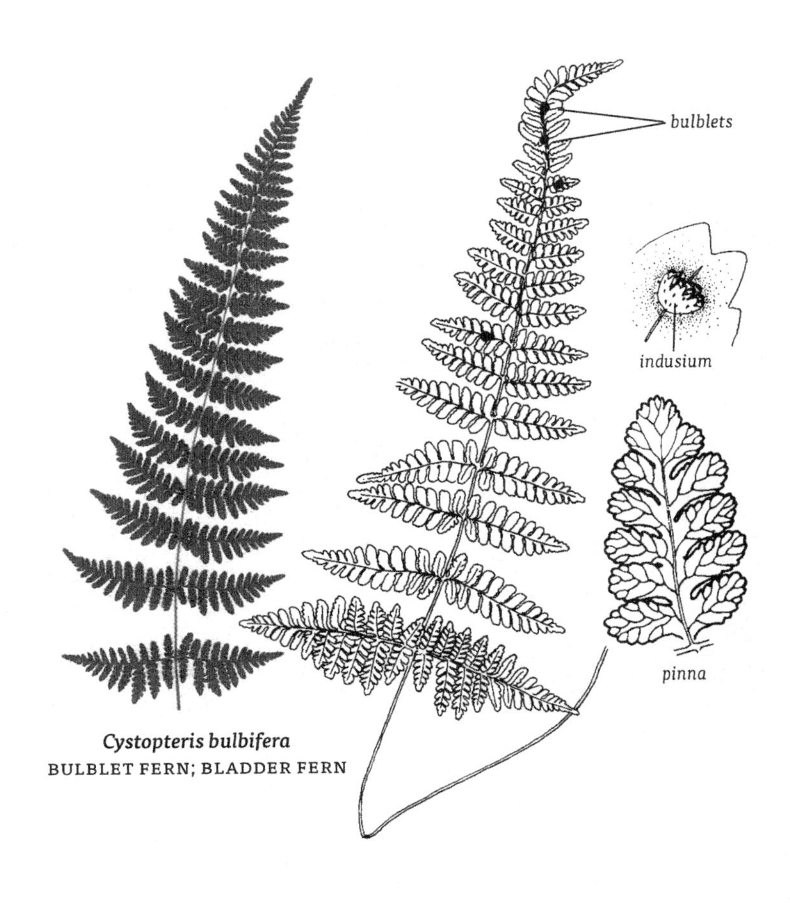

Cystopteris bulbifera
BULBLET FERN; BLADDER FERN

DEPARIA | *Glade fern*

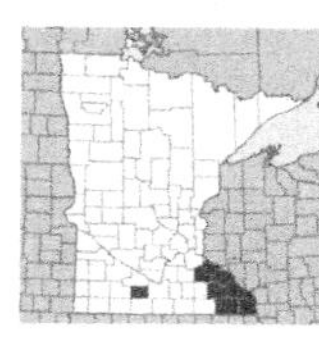

Deparia acrostichioides (Swartz) M. Kato
SILVERY GLADE FERN; SILVERY SPLEENWORT
NC **FAC** | MW **FAC**

Large native fern from creeping rhizomes. **Leaves** deciduous, 50–100 cm long, sterile and fertile leaves alike; petioles straw-colored (but dark red-brown at base), with brown lance-shaped scales; blades lance-shaped to oblong in outline, tapered at tip and distinctly narrowed toward base; deeply lobed, the segments blunt to somewhat tapered at their tip, margins entire to slightly lobed. **Sori** crowded, elongate, straight or sometimes curved, the **indusia** silvery and shiny when young. **Synonyms** *Athyrium thelypterioides*. **Habitat** Moist, rich deciduous woods, especially in wetter swales and depressions; streambanks; damp shady slopes.

Deparia acrostichioides
SILVERY GLADE FERN;
SILVERY SPLEENWORT

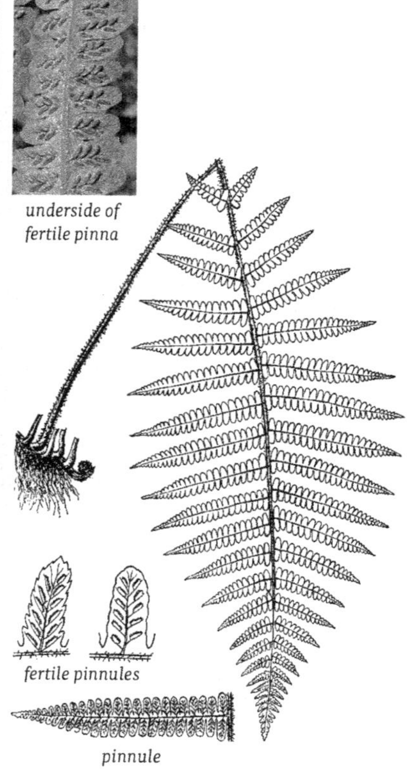

DRYOPTERIS | *Wood-fern*

Medium to large ferns; rhizomes short, stout and scaly, often covered with old petiole bases. **Leaves** dark green, sometimes evergreen; petioles shorter than blades, straw-colored or green, with chaffy scales near base. Sterile and fertile leaves alike or slightly different; sterile leaves sometimes persisting over winter; blades 1–3 pinnate, the smallest segments commonly toothed or lobed, veins simple to 1- or 2-branched. **Sori** round, on underside veins of pinnae; **indusia** round to kidney-shaped.

DRYOPTERIS | WOOD-FERN

1 Blades narrowly oblong, the lowest pair of pinna triangular in outline; sterile and fertile leaves somewhat different, sterile leaves smaller than fertile leaves .... *Dryopteris cristata* | CRESTED WOOD-FERN
1 Blades broader; sterile and fertile leaves alike 2

2 Blade less dissected, 1–2-pinnate .... *Dryopteris goldiana* .... GOLDIE'S WOOD-FERN
2 Blade more dissected, 2–3-pinnate................................ 3

3 Leaves deciduous late in season; lowermost inner pinnule longer than next outer one and longer and wider than opposite upper pinnule .... *Dryopteris carthusiana* | SPINULOSE WOOD-FERN
3 Leaves evergreen; lowermost inner pinnule shorter or equal to adjacent lower pinnule, not distinctly longer and wider than opposite upper pinnule .... *Dryopteris intermedia* | FANCY WOOD-FERN

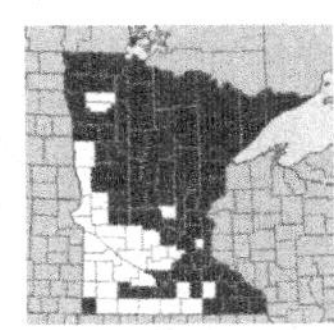

Dryopteris carthusiana (Villars) H. P. Fuchs

SPINULOSE WOOD-FERN
NC **FACW** | MW **FACW** | GP **FACW**

Clumped native fern, rhizomes short-creeping. **Leaves** all alike, deciduous, smooth except for chaffy, pale brown scales near base of petioles; **blades** 2- to nearly 3-pinnate, 2–6 dm long and 1–4 dm wide, tapered to tip, slightly narrowed at base; pinnae usually 10–15 pairs, alternate to nearly opposite, narrowly lance-shaped; pinnules toothed to deeply lobed, mostly 5–40 mm long and 3–10 mm wide, the teeth tipped with a small spine; innermost lower pinnule longer than next outer one and 2–3x longer than opposite upper pinnule. **Sori** halfway between midvein and margin; **indusia** 1 mm wide, without stalked glands. **Synonyms** *Dryopteris spinulosa*. **Habitat** Moist to wet woods, hummocks in swamps, thickets; also drier sand dunes and ridges.

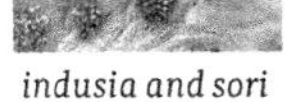

indusia and sori

Dryopteris cristata (L.) A. Gray
CRESTED WOOD-FERN
NC **OBL** | MW **OBL** | GP **OBL**

Clumped native fern, rhizomes short-creeping with ascending tips. Sterile and fertile **leaves** somewhat different, the outer sterile leaves waxy, persistent and smaller than inner fertile leaves; fertile leaves deciduous, 3–8 dm long. **Blades** 1-pinnate to nearly 2-pinnate, narrowly lance-shaped, 2–6 dm long and 7–15 cm wide, tapered to tip, narrowed at base; pinnae 5–9 cm long and to 4 cm wide, typically twisted to a nearly horizontal position, giving a "venetian blind" appearance to blades; pinnae segments to 20 mm long and 8 mm wide, with small spine-tipped teeth; **petioles** with sparse, pale brown, long-tapered scales. **Sori** round, midway between midvein and margin; **indusia** smooth, 1 mm wide.
Habitat Swamps, thickets, open bogs, fens and seeps.

indusia and sori

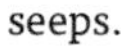

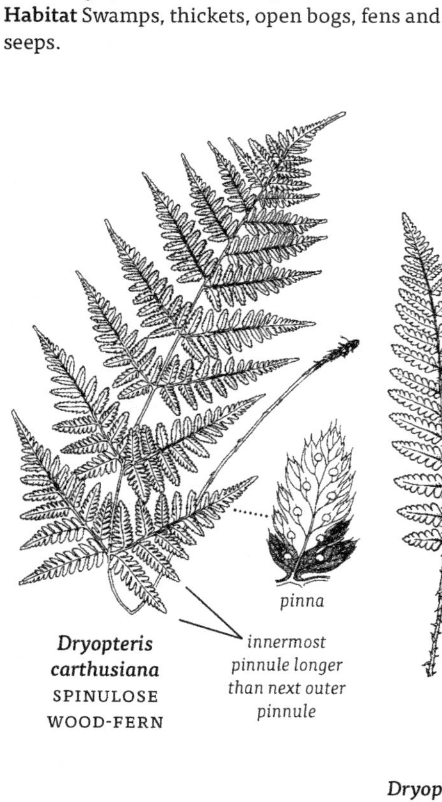

Dryopteris carthusiana
SPINULOSE WOOD-FERN

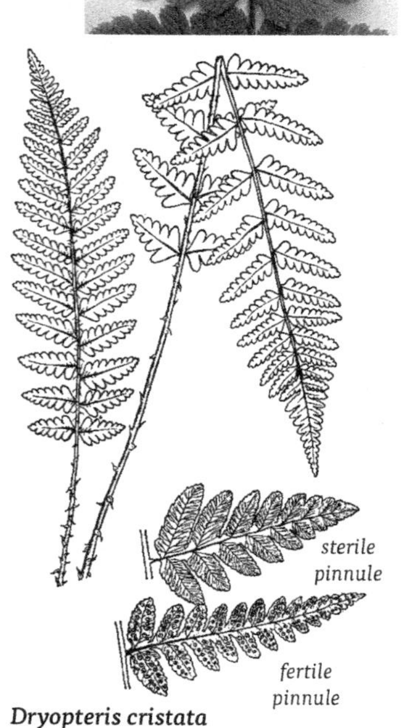

Dryopteris cristata
Crested wood-fern

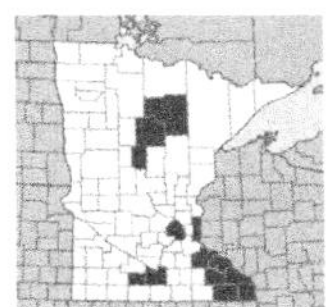

Dryopteris goldiana (Hook.) A. Gray
GOLDIE'S WOOD-FERN
NC **FAC** | MW **FAC** | *special concern*

Clumped native fern; rhizomes short-creeping, to 1 cm thick, densely scaly. **Leaves** to 1 m long; **blades** 30–60 cm long and 20–40 cm wide, deciduous late in season, the upper part abruptly narrowed to a small, tapered tip, the tip often mottled with white; pinnae with small, often rounded teeth; **petioles** brown, slightly shorter than blades, with narrow, pale brown scales 1–2 cm long, lower scales with a dark midstripe. **Sori** close to midveins, with a smooth **indusia** 1–2 mm across. **Habitat** Moist hardwood forests, shaded streambanks, talus slopes; soils rich in humus and usually neutral. **Status** Special concern.

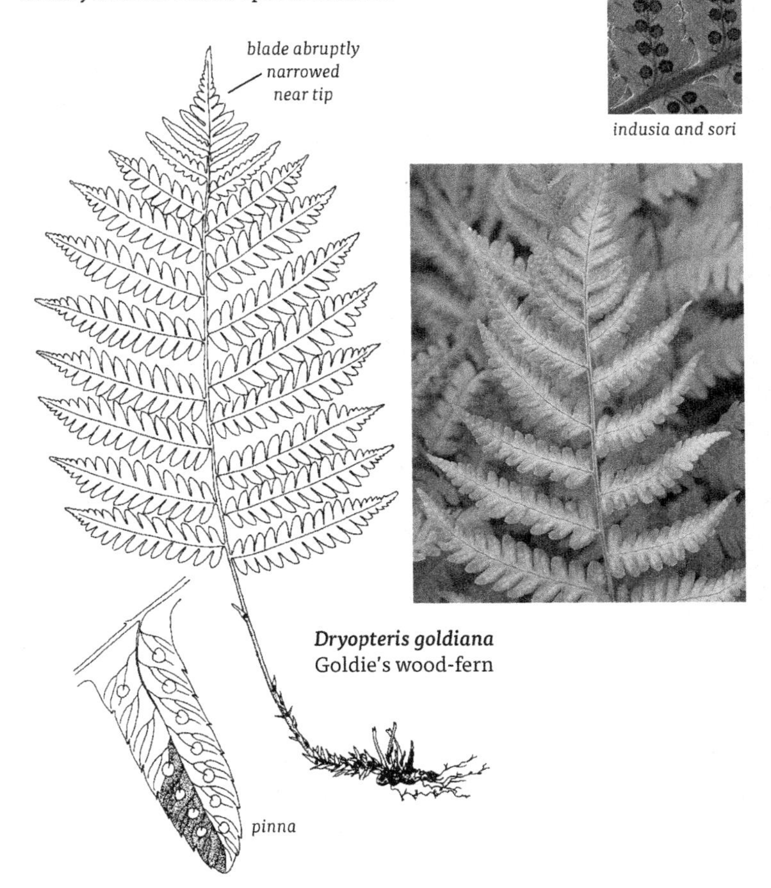

indusia and sori

Dryopteris goldiana
Goldie's wood-fern

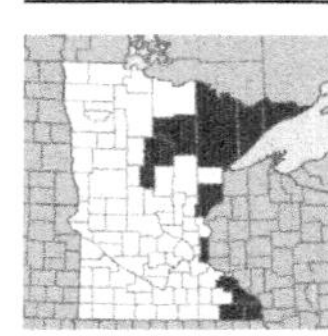

Dryopteris intermedia (Muhl.) A. Gray
FANCY WOOD-FERN
NC **FAC** | MW **FAC** | GP **FACW**

Clumped native fern, rhizomes ascending. **Leaves** in an open vaselike cluster of evergreen leaves; **blades** broadest just above base and abruptly tapered near tip, 2–5 dm long and 1–2 dm wide, 2-pinnate; pinnae at right angles to stem, lowermost inner pinnule usually shorter than next outer pinnule, pinnules toothed and tipped with small spines; **petioles** 1/3 as long as blade, with pale brown scales with a darker center, petioles and stems with small, gland-tipped hairs. **Sori** midway between midvein and margin, the **indusia** 1 mm wide, covered with stalked glands. **Habitat** Moist hardwood and mixed hardwood-conifer forests, hummocks in swamps; soils rich in humus and slightly acid to neutral.

indusia and sori

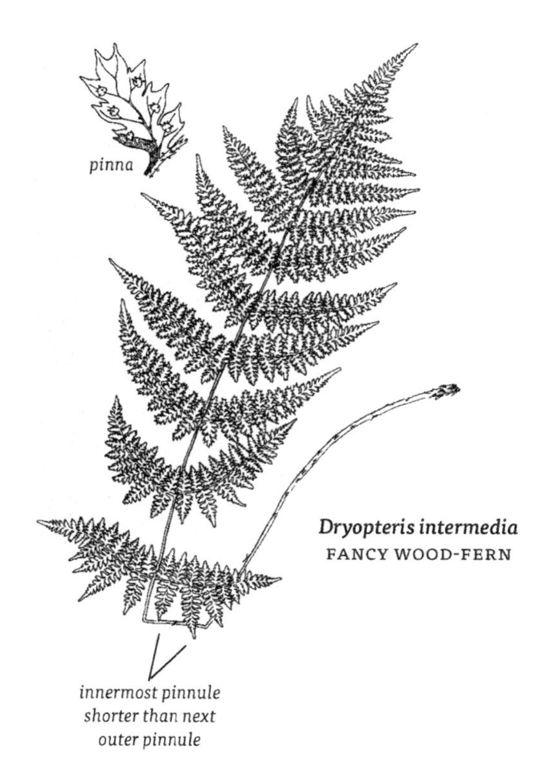

Dryopteris intermedia
FANCY WOOD-FERN

MATTEUCCIA | *Ostrich fern*

Matteuccia struthiopteris (L.) Todaro

OSTRICH FERN

NC **FAC** | MW **FACW** | GP **FACW**

Large, colony-forming native fern; rhizomes deep and long-creeping, black, scaly, producing erect leafy crowns. **Sterile leaves** upright, 1-pinnate, to 2 m tall and 15–50 cm wide; blades much longer than petioles, abruptly narrowed to tip, gradually tapered to base, stems ± hairy; each pinnae deeply divided into 20 or more pairs of pinnules, these 3–6 mm wide at base and rounded at tip; veins not netlike. **Fertile leaves** stiff and erect within a circle of sterile leaves, green at first, turning brown or black, much shorter than sterile leaves (to 6 dm tall), produced in mid to late summer and often persisting into following year; fertile blades 1-pinnate, pinnae upright or appressed, 2–6 cm long and 2–4 mm wide, the margins inrolled and covering the **sori**; **indusia** with a jagged margin. **Synonyms** *Pteretis pensylvanica*. **Habitat** Wet and swampy woods, streambanks, seeps, and ditches.

young shoots - "fiddleheads"

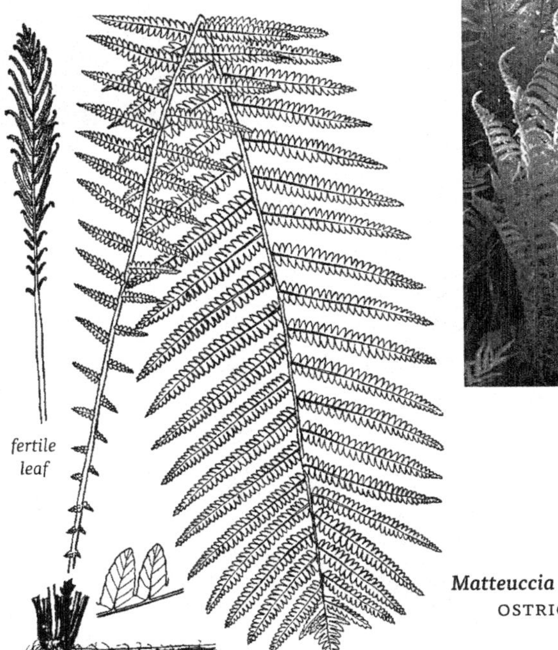

Matteuccia struthiopteris
OSTRICH FERN

ONOCLEA | *Sensitive fern*

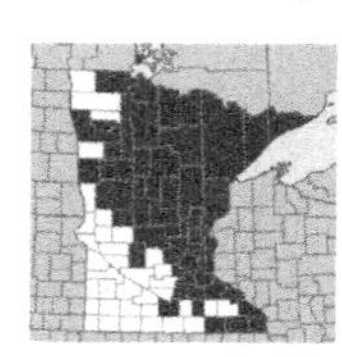

Onoclea sensibilis L.

SENSITIVE FERN

NC **FACW** | MW **FACW** | GP **FACW**

Medium-sized native fern, in clumps of several leaves, spreading by branching rhizomes and forming large patches. **Leaves** upright, with petioles about as long as blades. **Sterile leaves** deciduous, 1-pinnate at base, deeply cleft upward; the stem broader- winged toward the tip; blades 15–40 cm long and 15–35 cm wide, with 8–12 pairs of opposite pinnae, these deeply wavy-margined or coarsely toothed, 1–5 cm wide, with scattered white hairs on underside veins, the veins joined and netlike. **Fertile leaves** produced in late summer and persisting over winter, shorter than sterile leaves; fertile blades 1-pinnate, pinnae upright, divided into beadlike pinnules with inrolled margins covering the sori; veins not joined. **Sori** round and covered by a hoodlike **indusia**, becoming dry and hard. **Habitat** Swampy woods and low places in forests, wet meadows, calcareous fens, roadside ditches, wet or moist wheel ruts; sometimes weedy.

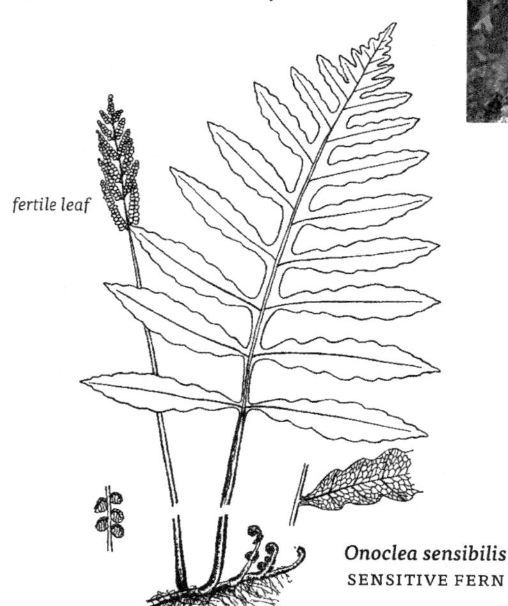

Onoclea sensibilis
SENSITIVE FERN

EQUISETUM | *Horsetail, Scouring-rush*

Rushlike herbs with dark rhizomes. **Stems** annual or perennial, grooved, usually with large central cavity and smaller outer cavities, unbranched or with whorls of branches at nodes. **Leaves** reduced to scales, united into a sheath at each node; top of sheath divided into dark-colored teeth. **Spores** in cones at tips of green or brown fertile stems.

***Stem cross-sections** are provided for each species; these are useful for identifying Equisetum in the field.*

EQUISETUM | HORSETAIL, SCOURING-RUSH

1 Stems evergreen; unbranched or with a few scattered branches, branches not in regular whorls (**scouring rushes**) 2
1 Stems annual; usually with regular whorls of branches, sometimes unbranched (**horsetails**) 5

2 Stems solid (central cavity absent); stems small, slender and sprawling. . . . . *Equisetum scirpoides* | DWARF SCOURING-RUSH
2 Stems hollow (central cavity present); stems larger, usually upright. 3

3 Stems 1–3 dm tall, with 5–12 ridges, central cavity to 1/3 diameter of stem . . . . . *Equisetum variegatum* | VARIEGATED SCOURING RUSH
3 Stems usually taller, with 16–50 ridges, central cavity more than half diameter of stem 4

4 Cones with a distinct, small sharp tip; stem sheaths with a black band at tip and base . . . . . *Equisetum hyemale* | COMMON SCOURING-RUSH
4 Cones blunt-tipped, sheaths with black band at tip only . . . . . *Equisetum laevigatum* | SMOOTH SCOURING-RUSH

5 Stems unbranched 6
5 Stems with regular whorls of branches. 10

6 Stems green 7
6 Stems brown or flesh-colored. 8

7 Stems with 9–25 shallow ridges; central cavity more than half diameter of stem; sheath teeth entirely black or with narrow white margins . . . . . *Equisetum fluviatile* | WATER-HORSETAIL
7 Stems with 5–10 strongly angled ridges; central cavity less than 1/3 diameter of stem; sheath teeth with white margins and dark centers. . . . . *Equisetum palustre* | MARSH-HORSETAIL

8 Sheath teeth papery and red-brown, teeth joined and forming several broad lobes . . . . . *Equisetum sylvaticum* | WOODLAND-HORSETAIL

8 Sheath teeth black or brown, not papery, separate or joined in more than 4 small groups **9**

9 Stems withering after spores mature, remaining unbranched . . . . *Equisetum arvense* | COMMON OR FIELD HORSETAIL
9 Stems persistent, becoming branched and green . . . . . *Equisetum pratense* . . . . MEADOW-HORSETAIL

10 First internode of each branch shorter than the subtending sheath of the main stem **11**
10 First internode of each branch equal or longer than the subtending sheath of the main stem **12**

11 Stems with 9–25 shallow ridges; central cavity more than half diameter of stem; sheath teeth more than 12, entirely black or with narrow white margins . . . . *Equisetum fluviatile* | WATER-HORSETAIL
11 Stems with 5–10 strongly angled ridges; central cavity about same size as outer cavities; sheath teeth 5–6, with white margins and dark centers . . . . *Equisetum palustre* | MARSH-HORSETAIL

12 Stem branches themselves branched; sheath teeth papery and red-brown, teeth joined and forming several broad lobes . . . . *Equisetum sylvaticum* | WOODLAND-HORSETAIL
12 Stem branches unbranched; sheath teeth black or brown, not papery, separate or joined in more than 4 small groups **13**

13 Stem branches ascending; teeth of branch sheaths gradually tapering to a slender tip . . . . *Equisetum arvense* | COMMON OR FIELD HORSETAIL
13 Stem branches spreading; teeth of branch sheaths broadly triangular . . . . *Equisetum pratense* | MEADOW-HORSETAIL

Equisetum arvense L.

COMMON OR FIELD HORSETAIL
NC **FAC** | MW **FAC** | GP **FAC**

Native; **stems** annual, upright from creeping, branched, tuber-bearing rhizomes covered with dark hairs. Sterile and fertile stems unalike; **sterile stems** appearing in spring as fertile wither, green, regularly branched, 1–6 dm tall and 2–5 mm wide, with 10–14 shallow ridges, the ridges usually rough-to-touch; central cavity 1/3–2/3 stem diameter; **sheaths** with 6–14 persistent, black-brown teeth 1–2 mm long; **branches** numerous in dense whorls, usually without branchlets, upright or spreading, 3–5-angled, solid. **Fertile stems** flesh-colored, shorter than sterile stems and with larger sheaths, maturing in early spring and soon withering, unbranched, to 3 dm tall and 8 mm wide; sheaths with 8–12 dark brown teeth. **Cones** blunt-tipped, long-stalked at

stem section

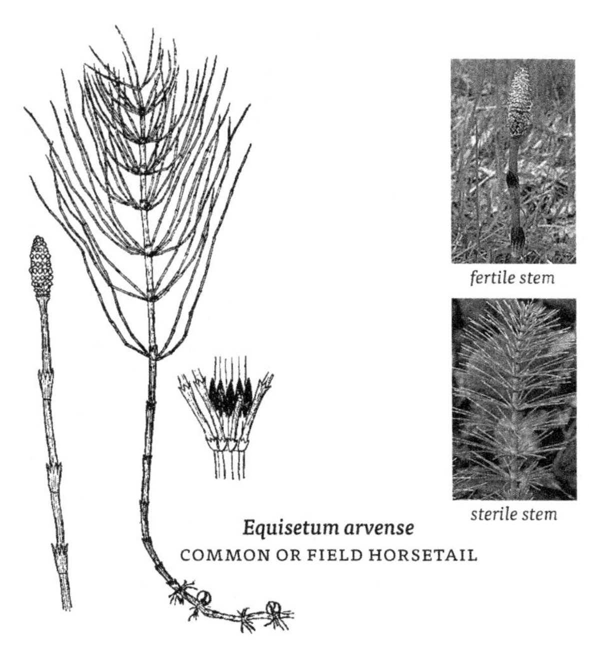

fertile stem

sterile stem

Equisetum arvense
COMMON OR FIELD HORSETAIL

end of stem, 0.5–3 cm long. **Habitat** Very common; streambanks, meadows, moist woods, ditches, roadsides and along railroads; calcareous fens.

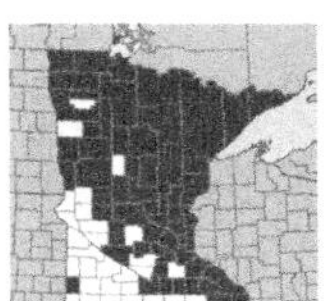

Equisetum fluviatile L.

WATER-HORSETAIL
NC **OBL** | MW **OBL** | GP **OBL**

Native; **stems** annual, fertile and sterile stems alike, to 1 m or more tall, from smooth, shiny, light brown, creeping rhizomes. Stems with 9–25 shallow, smooth ridges; central cavity large, about 4/5 stem diameter; stem **sheaths** green, 6–10 mm long; teeth 12–24, persistent, 2–3 mm long, dark brown to black, sometimes with narrow white margins; **branches** none or few, to many and regularly whorled from middle nodes, spreading, without branchlets, 4–6-angled, hollow. **Cones** 1–2 cm long at tips of stems, long-stalked, blunt-tipped, deciduous, maturing in summer. **Habitat** In standing water of marshes, ponds, peatlands, ditches and swales.

stem section

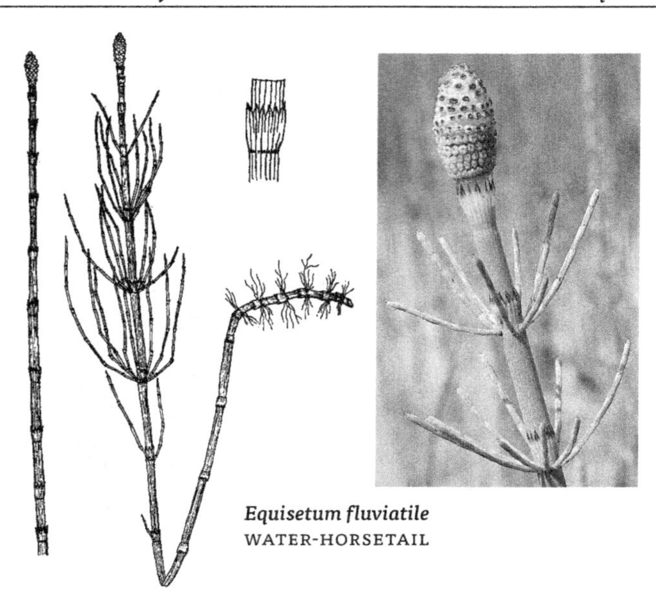

Equisetum fluviatile
WATER-HORSETAIL

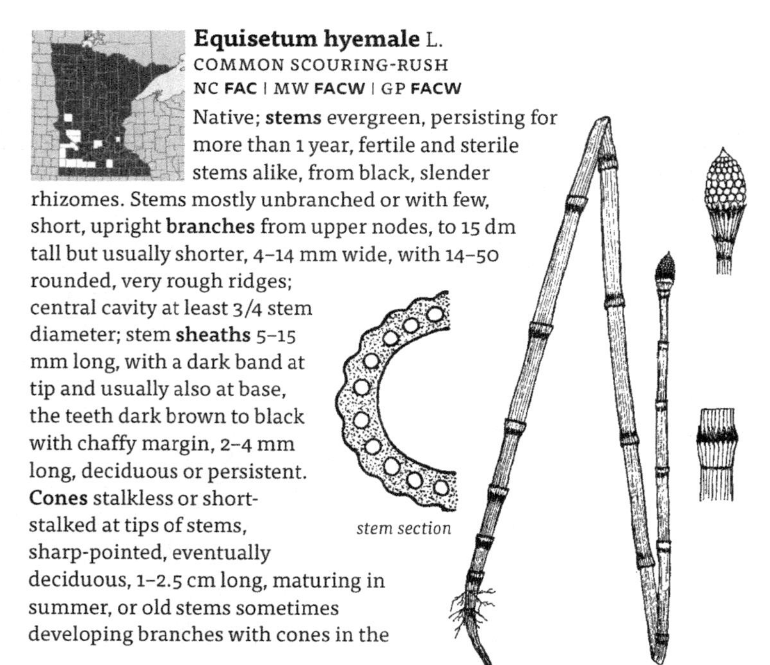

Equisetum hyemale L.
COMMON SCOURING-RUSH
NC **FAC** | MW **FACW** | GP **FACW**

Native; **stems** evergreen, persisting for more than 1 year, fertile and sterile stems alike, from black, slender rhizomes. Stems mostly unbranched or with few, short, upright **branches** from upper nodes, to 15 dm tall but usually shorter, 4–14 mm wide, with 14–50 rounded, very rough ridges; central cavity at least 3/4 stem diameter; stem **sheaths** 5–15 mm long, with a dark band at tip and usually also at base, the teeth dark brown to black with chaffy margin, 2–4 mm long, deciduous or persistent. **Cones** stalkless or short-stalked at tips of stems, sharp-pointed, eventually deciduous, 1–2.5 cm long, maturing in summer, or old stems sometimes developing branches with cones in the following spring. Ours are subspecies *affine*. **Synonyms** *Equisetum affine*. **Habitat** Often forming dense colonies in seeps, wet to moist meadows, shores and streambanks, ditches, roadsides and along railroads; usually where sandy or gravelly.

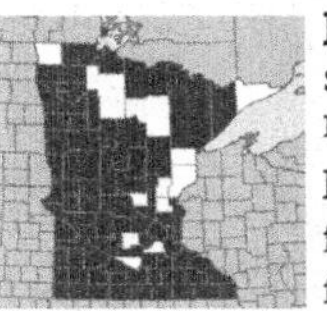

Equisetum laevigatum A. Braun
SMOOTH SCOURING-RUSH
NC **FACW** | MW **FACW** | GP **FAC**

Native; **stems** mostly annual, fertile and sterile stems alike, from brown or black rhizomes. Stems mostly unbranched or with a few upright branches, 3–10 dm tall and 3–8 mm wide, smooth and rather soft, with 10–32 ridges; central cavity 2/3–3/4 stem diameter; stem **sheaths** with a single dark band at tip, or rarely lowest sheaths with a dark band at base or entirely black; teeth dark brown or black with chaffy margins, free or partly joined in pairs, 1–4 mm long, soon deciduous. **Cones** short-stalked at tips of stems, rounded with a small sharp point, maturing in early summer and eventually deciduous. **Synonyms** *Equisetum kansanum*. **Habitat** Wet meadows, low prairie, streambanks, floodplains, seeps, and ditches, often where sandy or gravelly.

stem section

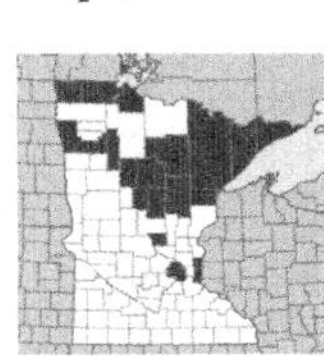

Equisetum palustre L.
MARSH-HORSETAIL
NC **FACW** | MW **FACW** | GP **FACW**

Native; **stems** annual, erect, fertile and sterile stems alike, from creeping, branched, shiny black rhizomes. Stems 2–8 dm tall, with 5–10 pronounced ridges, the ridges mostly smooth; central cavity small, 1/6–1/3 stem diameter; **sheaths** green, loose and flared upward; teeth 5–6, free or partly joined, persistent, 3–7 mm long, brown to black, with pale, translucent margins; **branches** few and irregular, to many and whorled at upper nodes, upright, without branchlets, 5–6-angled, hollow. **Cones** long-stalked at tips of stems, 1–3 cm long, blunt-tipped, maturing in summer, deciduous. **Habitat** In shallow water, along wetland margins, streambanks and fens.

stem section

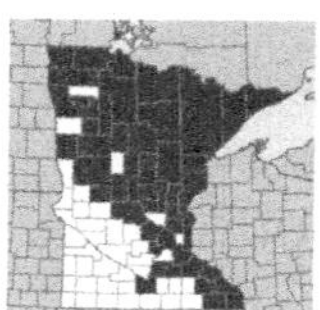

Equisetum pratense Ehrh.

MEADOW-HORSETAIL
NC **FACW** | MW **FACW** | GP **FACW**

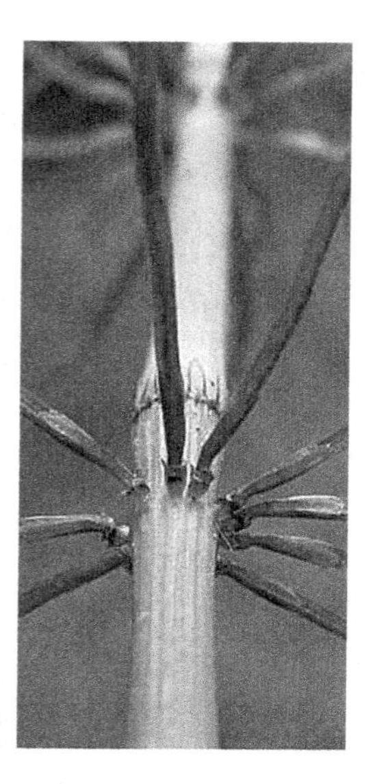

Native; **stems** annual and erect, sterile and fertile stems unalike, from creeping, dull black rhizomes. Sterile stems regularly branched, 2–5 dm tall and 1–3 mm wide; 8–18-ridged, the ridges roughened by silica on middle and upper stem; central cavity 1/3–1/2 stem diameter; main stem **sheaths** 2–6 mm long, the teeth persistent, 1–2 mm long, free or partly joined in pairs, brown with white margins and a dark midstripe; **branches** slender, many in regular whorls from middle and upper nodes, without branchlets, horizontal or drooping, mostly 3-angled, solid. **Fertile stems** uncommon, appearing in early spring before sterile stems and persisting, at first unbranched, fleshy and brown (without chlorophyll), later becoming green at nodes and producing many small green branches, mostly 1–3 dm tall; sheaths and teeth about twice as long as

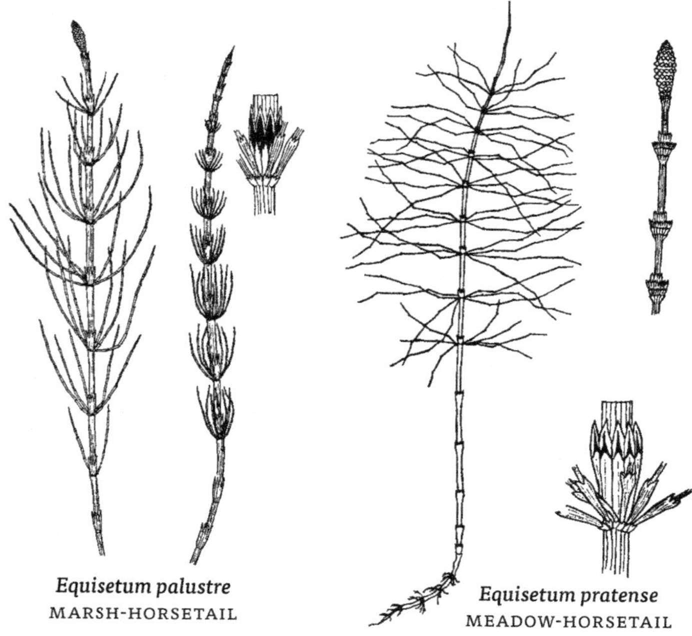

Equisetum palustre
MARSH-HORSETAIL

Equisetum pratense
MEADOW-HORSETAIL

on sterile stems. **Cones** long-stalked at tips of stems, to 2.5 cm long, blunt-tipped, deciduous. **Habitat** Moist woods, streambanks and meadows.

Equisetum scirpoides Michx.
DWARF SCOURING-RUSH
NC **FAC** | MW **FAC** | GP **FAC**

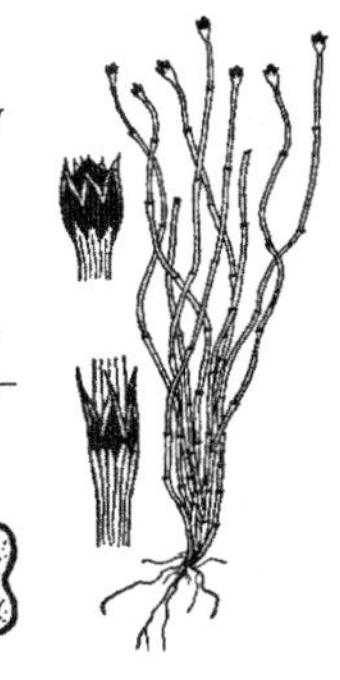

Native; **stems** evergreen, very slender, fertile and sterile stems alike, from widely branching rhizomes. Stems 5–30 cm long and only 0.5–1 mm wide, in dense clusters, usually unbranched and zigzagged, upright or trailing; central cavity absent, 3 small outer cavities present; **sheaths** green with broad black band at tip, loose and flared above, with 3–4 teeth; teeth with white, chaffy margin, ± persistent, but tips usually soon deciduous. **Cones** black, small, 3–5 mm long, sharp-tipped. **Habitat** Mossy places and moist, shaded woods, the stems often partly buried in humus.

stem section

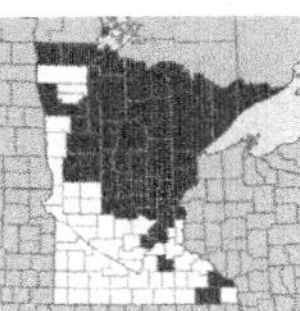

Equisetum sylvaticum L.
WOODLAND-HORSETAIL
NC **FACW** | MW **FACW** | GP **FACW**

Stems annual, erect, sterile and fertile stems unalike, from creeping, shiny light brown rhizomes, tubers occasionally present. Sterile stems green, 3–7 dm tall and 1.5–3 mm wide, with 10–18 ridges, rough-to-touch with sharp, hooked silica spines; central cavity 1/2–2/3 stem diameter; **sheaths** green at base, red-brown and flaring at tip; teeth brown, 3–5 mm long, joined in 3–5 broad lobes. Stems densely branched in regular whorls from the nodes, the **branches** themselves branched, often curving downward, 4–5-angled, solid. **Fertile stems** at first pink-brown (without chlorophyll), fleshy, unbranched, becoming green and branched as in sterile stems; sheaths and teeth larger than in sterile stems. **Cones** 1.5–3 cm long, stalked, blunt-tipped, deciduous. **Habitat** Wet or swampy woods, thickets, usually in partial shade.

stem section

Equisetum sylvaticum
WOODLAND-HORSETAIL

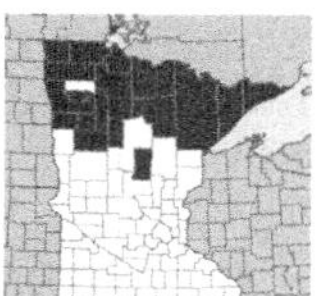

Equisetum variegatum Schleicher
VARIEGATED SCOURING RUSH
NC **FACW** | MW **FACW** | GP **FACW**

Native; **stems** evergreen, fertile and sterile stems alike, from creeping, much-branched, smooth rhizomes. Stems 1–3 dm tall and 1–2.5 mm wide, with 5–12 shallow, rough ridges, branched near base and otherwise usually unbranched; central cavity 1/4–1/3 stem diameter, smaller outer cavities present; **sheaths** green at base with a broad black band above; teeth persistent, with a dark brown or black midstripe and wide white margins, abruptly narrowed to a hairlike, deciduous tip 0.5–1 mm long. **Cones** to 1 cm long, strongly sharp-tipped, maturing in summer or persisting unopened until following spring. **Habitat** Lakeshores, streambanks, wet woods, moist meadows, fens, and ditches, moist sandy soil.

stem section

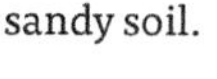

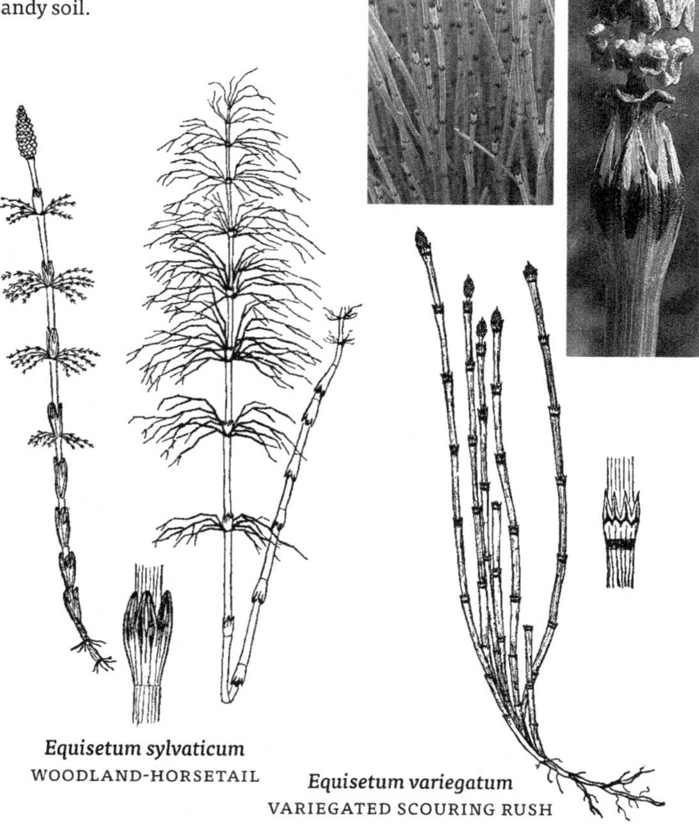

Equisetum sylvaticum
WOODLAND-HORSETAIL

Equisetum variegatum
VARIEGATED SCOURING RUSH

ISOETES | *Quillwort*

Perennial aquatic or emergent herbs. **Leaves** simple, entire, linear, from a 2–3 lobed rhizome (corm). Outermost and innermost leaves typically sterile. Outer fertile leaves have a pocketlike structure (sporangia) bearing whitish spores (**megaspores**; illustrated; magnification needed to see features); inner fertile leaves have numerous small microspores.

ISOETES | QUILLWORT

1 Megaspores conspicuously covered with small spines . . . . *Isoetes echinospora* | SPINY-SPORED QUILLWORT
1 Megaspores not spiny 2

2 Plants normally underwater; leaves without outer fibrous strands . . . . *Isoetes lacustris* | LAKE QUILLWORT
2 Plants emergent and on drying shores; leaves with 4 or more fibrous strands . . . . *Isoetes melanopoda* | BLACK-FOOT QUILLWORT

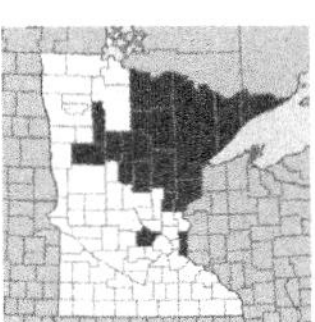

Isoetes echinospora Durieu

SPINY-SPORED QUILLWORT
NC **OBL** | MW **OBL** | GP **OBL**

Native; **leaves** linear, 7–25 or more, 5–15 cm long and 0.5–1.5 mm wide, usually erect, soft, bright green to yellow-green, tapered from base to a very long, slender tip, without peripheral strands from base; corm 2-lobed. **Sporangium** 4–8 mm long, usually brown-spotted when mature, half or more covered by a membranous flap (velum). **Megaspores** round, white, 0.3–0.6 mm wide, covered with short, sharp to blunt spines. **Synonyms** *Isoetes braunii, Isoetes muricata, Isoetes tenella*. **Habitat** Shallow water (to 1 m deep) of lakes, ponds and slow-moving rivers; plants rooted in mud, sand, or gravel.

megaspore

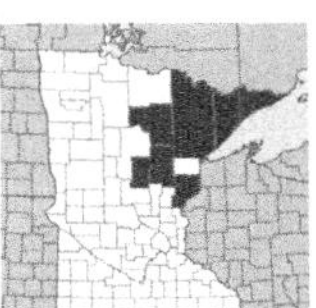

Isoetes lacustris L.

LAKE QUILLWORT
NC **OBL** | MW **OBL**

Native; **leaves** several to many, 5–20 cm long and 1–2 mm wide, stiff and erect or with leaf tips curved downward, dark green, fleshy and twisted, peripheral strands from base usually absent; corm 2-lobed. **Sporangium** to 5 mm long, usually not spotted; membranous flap (velum) covering up to half of sporangium. **Megaspores** round, white, 0.6–0.8 mm wide, with ridges forming an irregular netlike pattern. **Synonyms** *Isoetes macrospora*. **Habitat** Underwater in shallow to deep water of cold lakes, ponds and streams.

megaspore

Isoetes melanopoda Gay & Durieu
BLACK-FOOT QUILLWORT
NC **OBL** | MW **OBL** | GP **OBL** | *endangered*

Native; **leaves** 10–50 cm long and 0.5–2 mm wide, black at base with a pale line down middle of inner side, 4 peripheral strands from base usually present. **Sporangia** 5–20 mm long, brown-spotted when mature, up to 2/3 covered by membranous flap (velum). **Megaspores** 0.3–0.5 mm wide, covered with short, low ridges. **Habitat** Underwater or emergent in temporary ponds, wet streambanks, ditches and swales. **Status** Endangered; rare in sw Minnesota.

megaspore

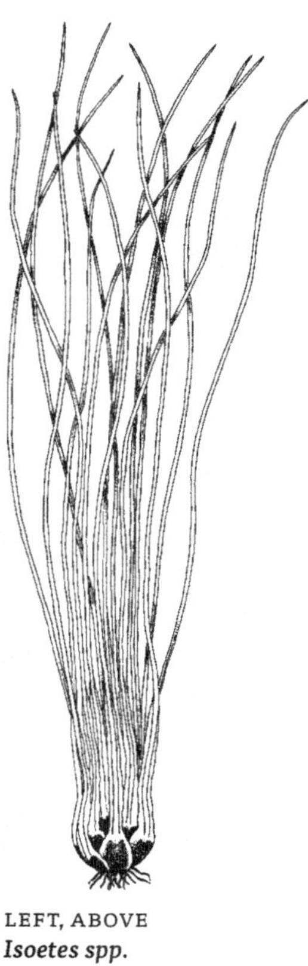

LEFT, ABOVE
Isoetes spp.
(QUILLWORT)

LOW, TRAILING, EVERGREEN HERBS resembling large mosses. **Leaves** needlelike or scalelike, alternate or opposite on stem. **Spore-bearing leaves** (sporophylls) similar to vegetative leaves or in conelike clusters at tips of upright stems.

LYCOPODIACEAE | CLUBMOSS FAMILY

1 Leafy horizontal stems absent; upright stems in clusters .... *Huperzia* .... FIR-MOSS
1 Leafy horizontal stems present and creeping on ground surface; upright stems borne singly along horizontal stems .... *Lycopodiella* .... BOG CLUBMOSS

HUPERZIA | *Fir-moss*

Low evergreen perennials with erect shoots; **leaves** spreading or appressed and upright. **Spores** borne at base of upper leaves.

HUPERZIA | FIR-MOSS

1 Leaves narrowly obovate with 1-8 irregular teeth; common plants of moist to wet conifer woods .... *Huperzia lucidula* | SHINING FIR-MOSS
1 Leaves lance-shaped to oblong lance-shaped, margins entire or with 1-3 small teeth; uncommon species of wetland margins ..... *Huperzia selago* .... NORTHERN FIR-MOSS

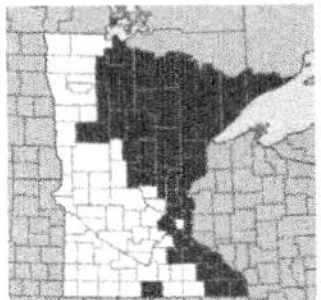

Huperzia lucidula (Michaux) Trev.

SHINING FIR-MOSS
NC **FAC** | MW **FACW** | GP **FACW**

Native; **stems** light green, creeping and rooting, upcurving stems forked several times, to 25 cm high , crowded with shiny dark green leaves which persist for more than one season. **Leaves** in mostly 6 rows, spreading or curved downward, in alternating groups of longer sterile and shorter fertile leaves, giving shoots a ragged look. **Sterile leaves** 6–12 mm long, toothed and broadest above middle; **sporophylls** barely widened and with small teeth or entire at tip. Small two-lobed buds (gemmae) produced in some upper leaf-axils; these may sprout into new plants after falling onto moist humus. **Synonyms** *Lycopodium lucidulum*. **Habitat** Moist to wet conifer and hardwood forests.

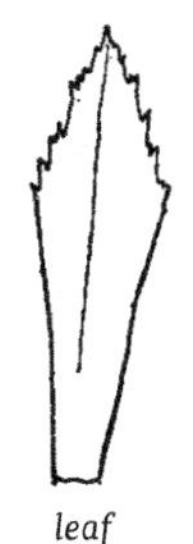
leaf

Huperzia selago L.
NORTHERN FIR-MOSS
NC **FACU** | MW **FACU**

Native; horizontal **stems** short; upright stems forked from base, 6–20 cm long and 2–3 mm wide (stem only). **Leaves** persistent, yellow-green, in 8–10 rows, 3–6 mm long and to 1 mm wide, swollen and concave at base, gradually tapered to tip, mostly without teeth, uniform in length; leaves appressed to stem, giving stems a smooth, cylindric outline. **Sporophylls** similar to vegetative leaves; sporangia produced early in season in leaf axils, followed later by sterile leaves. Upper axils produce small, 2-lobed reproductive buds (gemmae). **Synonyms** *Lycopodium selago*. **Habitat** An arctic tundra species of thickets, streambanks, cold woods and bog margins.

A hybrid between Huperzia selago and Huperzia lucidula known as ***Huperzia buttersii*** *sometimes occurs.*

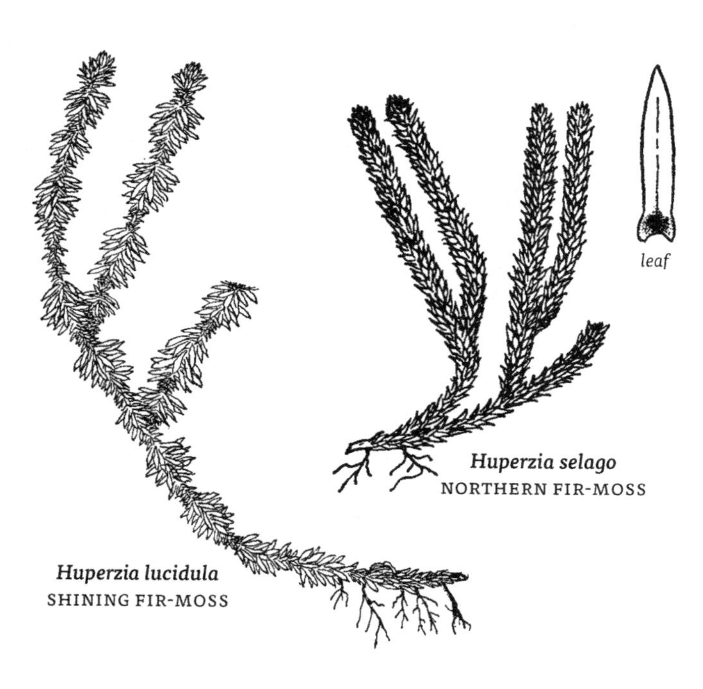

LYCOPODIELLA | *Bog clubmoss*

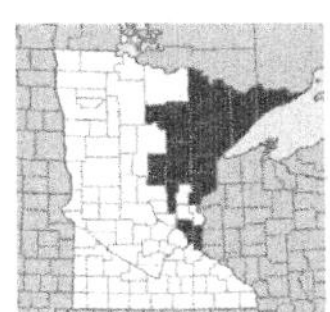

Lycopodiella inundata L. Holub

NORTHERN BOG CLUBMOSS
NC **OBL** | MW **OBL**

Low, creeping native perennial of wet habitats. **Stems** elongate and trailing, deciduous but with evergreen buds at tips, rooting throughout. Upright **shoots** unbranched, leafy, scattered along horizontal stems. **Leaves** in 8–10 rows, those on underside of trailing stems twisted upward, narrowly lance-shaped , margins ± entire. **Fertile branches** few, erect, to 1 dm high, with spreading leaves. **Spores** borne in terminal, leafy **cones**; the cones 1.5–5 cm long and 6–12 mm wide; sporophylls green, base widened and with a pair of teeth. **Synonyms** *Lycopodium inundata*. **Habitat** Acid, open sphagnum bogs, wet sandy shores and streambanks; disturbed wetlands.

Lycopodiella inundata
NORTHERN BOG CLUBMOSS

MARSILEA | *Water-clover*

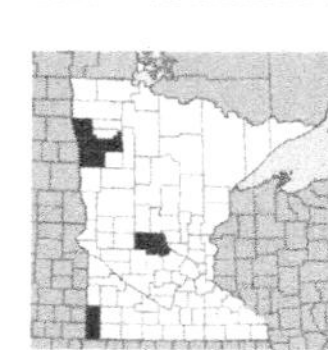

Marsilea vestita Hook. & Grev.

HAIRY WATER-CLOVER
MW **OBL** | GP **OBL** | *endangered*

Aquatic, creeping, perennial herb, native. **Stem** a slender rhizome, rooting in mud and only at nodes; **petioles** 2–20 cm long, sparsely hairy; **blades** floating or emergent, divided into 4 pinnae resembling a 4-leaf clover, 5–20 mm long and 5–15 mm wide, usually hairy on both sides. **Sporangia** in a single oval sporocarp borne on a short stalk at or near base of petiole, the sporocarp brown, 4–7 mm long, covered with stiff, flat hairs; sori in 2 rows inside the sporocarp. **Synonyms** *Marsilea mucronata*. **Habitat** Shallow water or mud of temporary ponds, floodplains and ditches. **Status** Endangered; rare in western Minnesota.

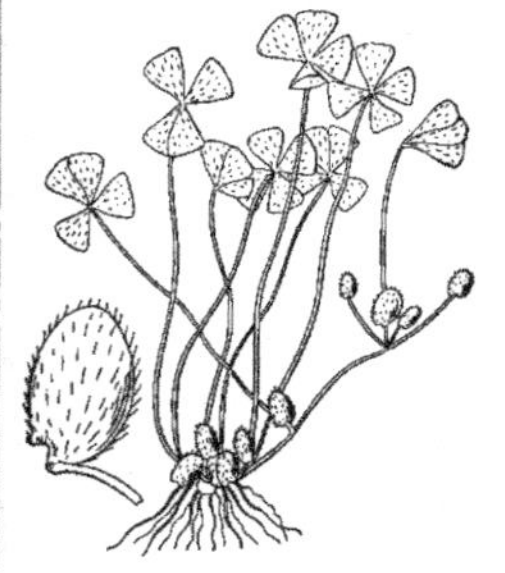

Marsilea vestita (HAIRY WATER-CLOVER)

PERENNIAL HERBS from short, erect rhizomes having several fleshy roots. Plants produce 1 leaf each year on a single stalk (stipe), with bud for next year's leaf at base of stipe. **Leaves** divided into a fertile segment (sporophyll) and a sterile expanded blade. Sterile blades entire (*Ophioglossum*), or lobed or 1–3x pinnately divided (*Botrychium*). **Spores** in numerous round sporangia borne on simple or branched fertile blades.

OPHIOGLOSSACEAE | ADDER'S-TONGUE FAMILY

1 Leaves dissected, veins forked but not joined; sporangia normally borne in a panicle . . . . *Botrychium* | GRAPE-FERN; MOONWORT
1 Leaves simple and entire, veins joined, forming a network; sporangia borne in a spike. . . . . *Ophioglossum pusillum* | NORTHERN ADDER'S-TONGUE

BOTRYCHIUM | *Grape-fern, Moonwort*

Mostly small plants with one leaf, the **blade** divided into sterile and fertile segments. **Sterile portion** of blade pinnately divided or lobed, **fertile portion** branched to form a panicle bearing the sporangia.

BOTRYCHIUM | GRAPE-FERN, MOONWORT

1 Plants 50 cm or more tall; blade large and broadly triangular in outline . . . . *Botrychium virginianum* | RATTLESNAKE FERN
1 Plants smaller, less than 25 cm tall 2

2 Blade broadly triangular in outline, dissected into narrow toothed segments (pinnae). . . . . *Botrychium lanceolatum* . . . . LANCE-LEAVED GRAPE-FERN
2 Blade divided into 3–6 fan-shaped pairs. . . . . *Botrychium lunaria* . . . . MOONWORT

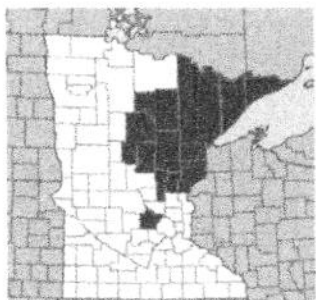

Botrychium lanceolatum (S. G. Gmelin) Angström

LANCE-LEAVED GRAPE-FERN
NC **FACW** | MW **FACW** | GP **FACW** | *threatened*

Native plants 6–30 cm tall, dark green, smooth, appearing in early summer and persisting to fall. **Stems** about 5x longer than blades; **blades** triangular in outline, 1–8 cm long and 1–5 cm wide, stalkless or on a short stalk to 6 mm long, divided into 2–5 pairs of sharp-pointed, toothed pinnae, lowermost pair the largest. Fertile segment 2–9 cm long, mostly twice pinnate, on a stalk 1–3 cm long. Ours are var. *angustisegmentum*. **Habitat** Moist humus-rich woods, hummocks in swamps, streambanks. **Status** Threatened.

***Var. angustisegmentum** occurs from Nfld and Ont to Minn, becoming increasingly rare south to New Jersey and Ohio.*

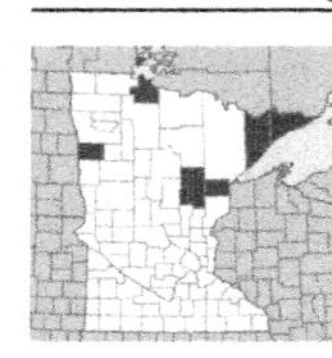

Botrychium lunaria (L.) Swartz

MOONWORT

NC **FACW** | MW **FACW** | GP **FAC** | *threatened*

Native plants 3–20 cm tall, rubbery-textured, appearing in late spring and withering in summer. **Leaf blades** 1.5–7 cm long and 1–3 cm wide, stalkless or on a short stalk to 5 mm long; pinnately divided into 3–6 pairs of stalkless pinnae, the pinnae fan-shaped, wider than long and without a midrib; petioles 1.5–3 cm long. Fertile segments 0.5–7 cm long, on stalks about as long as the segments. **Habitat** Grassy meadows, sandy or gravelly lakeshores and streambanks, rock ledges and mossy talus; most common on ± neutral soils. **Status** Threatened; uncommon in inland northern Minnesota and along Lake Superior.

Several similar species of Botrychium have been described by W. H. Wagner including ***Botrychium campestre, B. minganense,*** *and* ***B. mormo****. For a complete discussion, see Flora of North America, Vol. 2 (1993).*

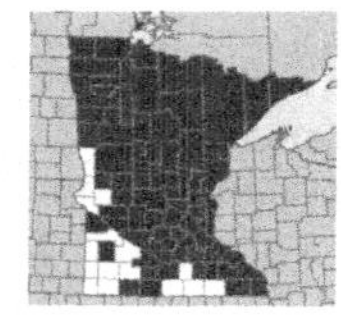

Botrychium virginianum (L.) Swartz

RATTLESNAKE FERN

NC **FACW** | MW **FACW** | GP **FACW**

Native plants 40–75 cm tall, appearing in spring, withering in autumn, not overwintering. **Blade** (trophophore) triangular, sessile, to 25 cm long and to 1.5x as wide, 3–4x pinnate, thin and herbaceous. Pinnae to 12 pairs, usually somewhat overlapping and slightly ascending; pinnules lance-shaped and deeply lobed, the lobes linear, sharply toothed and pointed at tip Spore-bearing portion (sporophore) 2-pinnate, 0.5–1.5x length of trophophore. **Habitat** Occasional in swamps of cedar and black spruce; more common in moist deciduous woods.

Rattlesnake fern *is the most widespread species of Botrychium in North America, occurring across Canada and the USA.*

Botrychium virginianum
RATTLESNAKE FERN

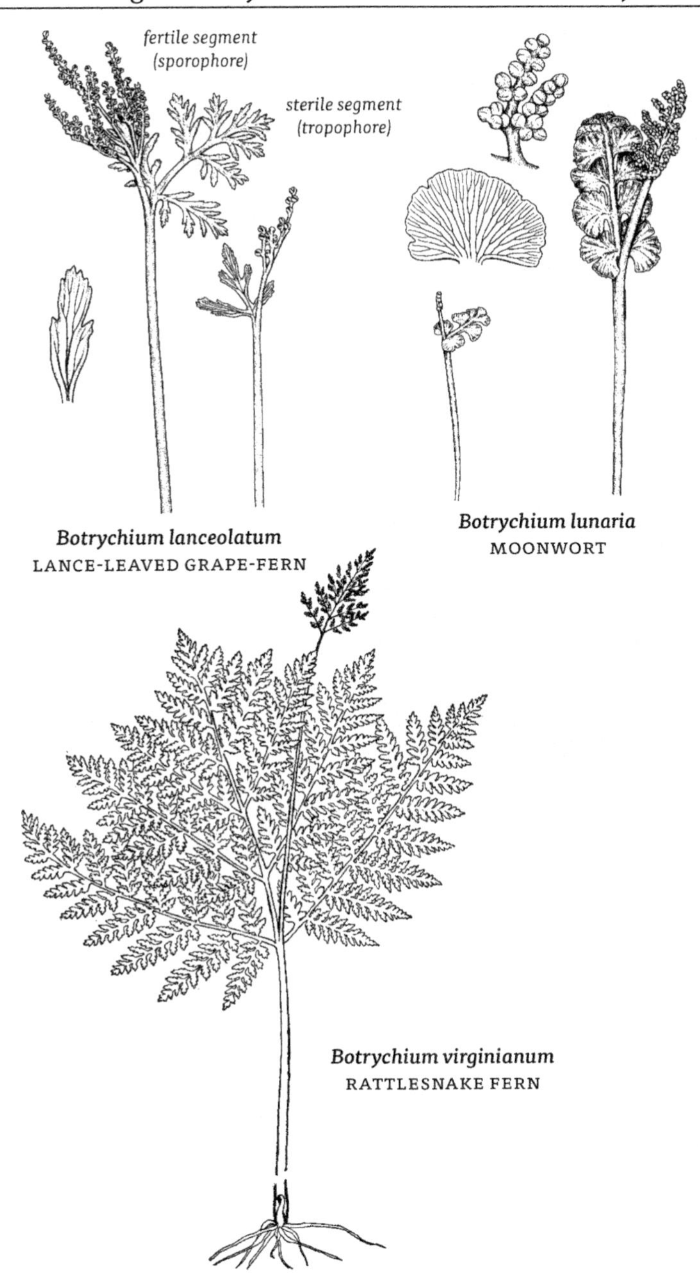

Botrychium lanceolatum
LANCE-LEAVED GRAPE-FERN

Botrychium lunaria
MOONWORT

Botrychium virginianum
RATTLESNAKE FERN

OPHIOGLOSSUM | *Adder's-tongue*

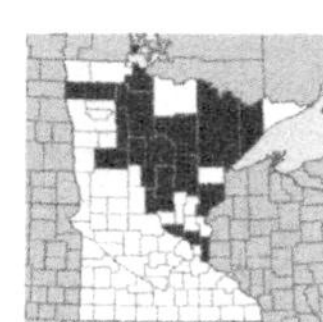

Ophioglossum pusillum Raf.
NORTHERN ADDER'S-TONGUE
NC **FACW** | MW **FACW** | GP **FACW**

Plants erect, 7–30 cm tall, from slender rhizomes, native. **Leaves** 1, entire, on a stalk 3–15 cm long; **blades** upright, oval to ovate, rounded to acute at tip, 3–8 cm long and 1–4 cm wide, conspicuously net-veined. **Sporangia** in 2 rows in a terminal, unbranched fertile segment, 1–5 cm long and 2–4 mm wide, on a stalk 6–15 cm long. **Synonyms** *Ophioglossum pycnostichium, Ophioglossum vulgatum*. **Habitat** Wet sandy meadows and prairies, moist depressions, fens and wetland margins.

Ophioglossum pusillum
NORTHERN ADDER'S-TONGUE

OSMUNDA | *Royal fern*

Perennial ferns with large rootstocks and exposed crowns covered with old roots and stalks, sending up tufts of coarse leaves. **Leaves** 1–2-pinnate, differentiated into sterile and fertile segments. **Sporangia** in round clusters, spores green.

OSMUNDA | ROYAL FERN

1 Leaves 2-pinnate, pinnae ± entire; sporangia on upper half of fertile leaves . . . . *Osmunda regalis* | ROYAL FERN

1 Leaves 1-pinnate, sterile pinnae deeply cleft; sporangia only near middle of fertile leaves, or fertile and sterile leaves separate 2

2 Fertile and sterile leaves separate, fertile leaves cinnamon-colored, sterile leaves with a tuft of wool in axil of pinnae . . . . *Osmunda cinnamomea* . . . . CINNAMON-FERN

2 Fertile pinnae near middle of vegetative leaves, with sterile pinnae above and below fertile portion, fertile portion green-black, pinnae mostly without tuft of wool in axil . . . *Osmunda claytoniana* | INTERRUPTED FERN

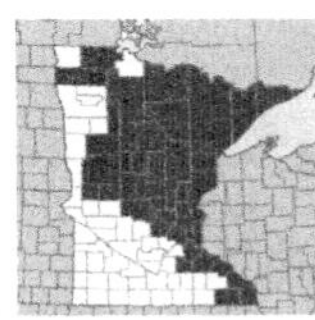

Osmunda cinnamomea L.

CINNAMON-FERN

NC **FACW** | MW **FACW** | GP **FACW**

Large clumped native fern, to 1 m or more tall. **Blades** of sterile leaves to 30 cm wide, gradually tapered to tip, 1-pinnate, with conspicuous tuft of white or brown woolly hairs at base of each pinna, pinnae stalkless and deeply cleft into segments, with fringe of short hairs on margins; **petioles** densely hairy when young. **Fertile leaves** at center of crown, surrounded by taller sterile leaves, without leafy tissue, arising in spring or early summer and turning cinnamon brown, withering and inconspicuous by midsummer. **Synonyms** *Osmundastrum cinnamomeum*. **Habitat** Swamps, bog-margins, wooded streambanks, and low wet places; soils acid.

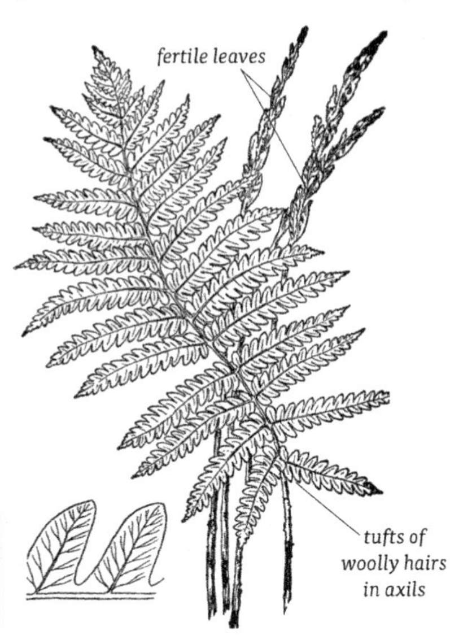

Osmunda claytoniana L.
INTERRUPTED FERN
NC **FAC** | MW **FAC** | GP **FAC**

Clumped native fern to 1 m or more tall; often forming large colonies. Outer **leaves** usually sterile, inner leaves larger and with 2–5 pairs of fertile pinnae in middle of blade; fertile segments to 6 cm long and 2 cm wide and much smaller than vegetative segments above and below them; **sporangia** clusters at first green-black, turning dark brown and withering. **Blades** 4–10 dm long and 15–30 cm wide; pinnae stalkless and deeply cut into segments, with smooth or slightly hairy margins. **Petioles** covered with tufts of woolly hairs when young, becoming smooth or sparsely hairy with age, the hars not forming tufts at pinna-bases (as in *Osmunda cinnamomea*). **Habitat** Moist or seasonally wet depressions in forests, hummocks in swamps, low prairie, wet roadsides; often in drier places than cinnamon fern or royal fern.

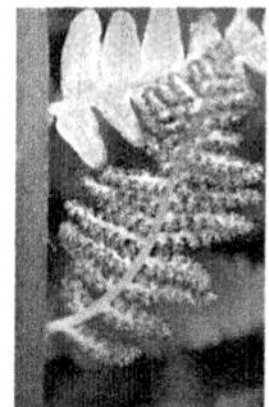
fertile pinna

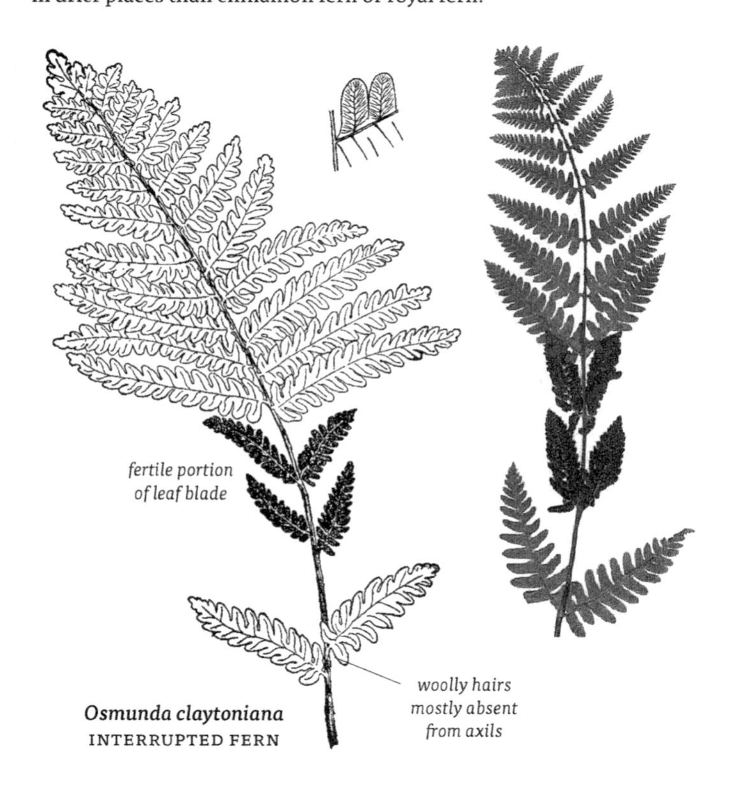

Osmunda claytoniana
INTERRUPTED FERN

Osmunda regalis L.

ROYAL FERN

NC **OBL** | MW **OBL** | GP **OBL**

Large native fern to 1 m or more tall. **Blades** broadly ovate in outline, 4–8 dm long and to 3–5 dm wide, 2-pinnate into ± opposite divisions (pinnules), these well-spaced, oblong, rounded at tips, with entire or finely toothed margins. **Fertile leaves** with uppermost several pinnae replaced by sporangia clusters. **Petioles** smooth, green or red-green, to 3/4 length of blade. **Habitat** Bogs, swamps, alder thickets and shallow pools; soils usually acidic.

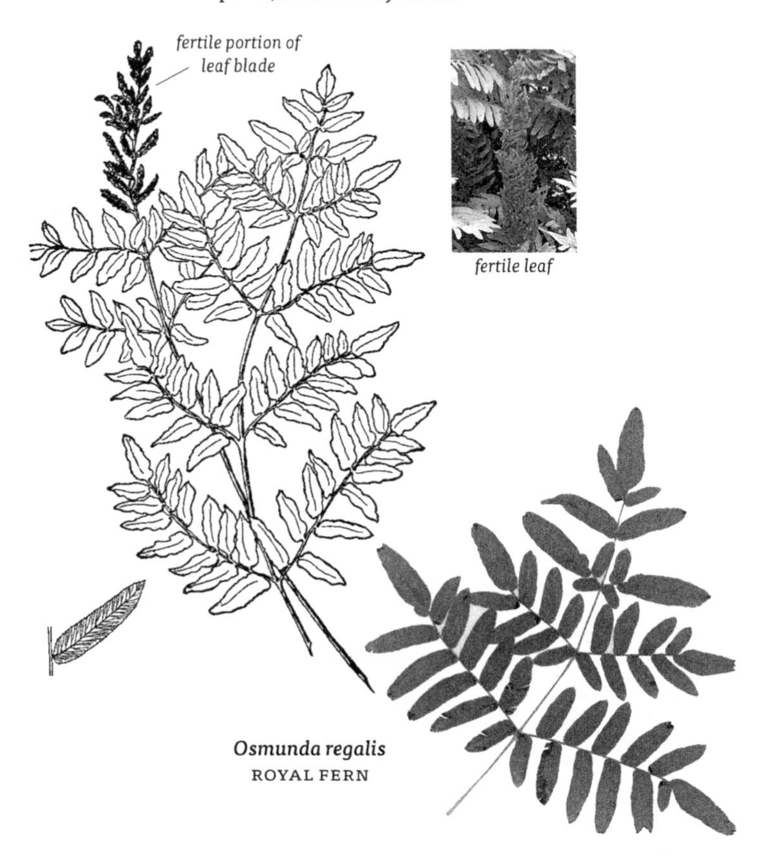

Osmunda regalis
ROYAL FERN

SELAGINELLA | *Spikemoss*

Selaginella selaginoides (L.) Link

NORTHERN SPIKEMOSS

NC **FACW** | *endangered*

Trailing, evergreen herbs forming small mats, native. **Stems** branched, leafy, rooting at branching points. Sterile **stems** prostrate, 2–5 cm long; fertile stems upright, deciduous, 5–10 cm high and 0.5 mm wide (stem only), changing upward into broader sporophylls. **Leaves** overlapping in multiple spiral rows, all alike, 2–4 mm long and 1 mm wide, with sharp tips and sparsely hairy margins; spore-bearing leaves similar to vegetative leaves and clustered in cones at ends of branches; **cones** ± cylindric but with 4 rounded angles, 1.5–3 cm long and to 5 mm wide. **Megaspores** 4 in each sporangium, yellow-white, with low rounded projections on the 3 flat surfaces. **Habitat** In Minnesota, known only from crevices in bedrock outcrops along Lake Superior shore; often associated with *Scirpus cespitosus*. **Status** Endangered.

Selaginella selaginoides
NORTHERN SPIKEMOSS

FERNS WITH CREEPING STEMS and not forming vase-like clumps; **blades** 1-pinnate. **Sori** round to oblong; **indusia** absent in *Phegopteris*; present and tan-colored in *Thelypteris*.

THELYPTERIDACEAE | MARSH FERN FAMILY

1 Leaf blades broadly triangular in outline, broadest at base, lowermost pinnae directed downward; indusia absent. . . . . *Phegopteris connectilis* . . . . NORTHERN BEECH-FERN

1 Blades lance-shaped in outline, broadest above base; indusia present . . . . *Thelypteris palustris* | MARSH-FERN

PHEGOPTERIS | *Beech-fern*

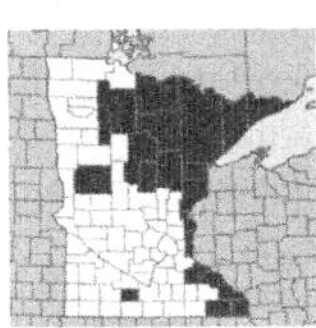

Phegopteris connectilis (Michaux) Watt
NORTHERN BEECH-FERN
NC **FACU** | MW **FACU** | GP **FACU**

Native fern; rhizomes long, slender, scaly and densely hairy. **Leaves** triangular, 15–25 cm long and 6–15 cm wide; **blades** 1-pinnate, the pinnalike divisions joined by a wing along rachis, except for lowermost pair which are free and angled downward; pinnules oblong, rounded at tip, and usually hairy; **petioles** longer than blades, hairy, with narrow, brown scales. **Synonyms** *Dryopteris phegopteris, Thelypteris phegopteris*. **Habitat** Cool moist woods, thickets, streambanks, sphagnum moss hummocks, shaded rock crevices.

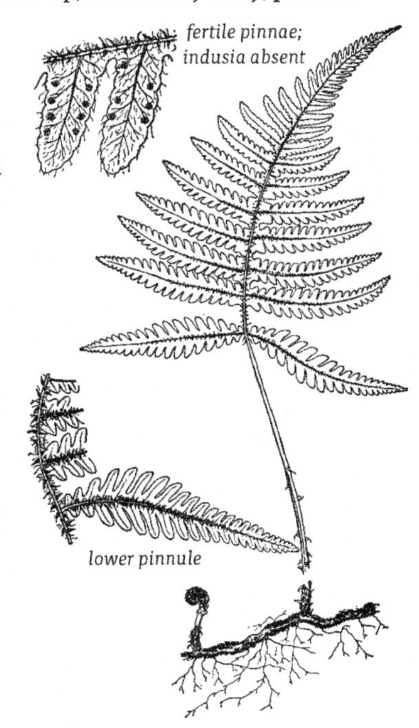

Phegopteris connectilis
NORTHERN BEECH FERN

THELYPTERIS | *Marsh-fern*

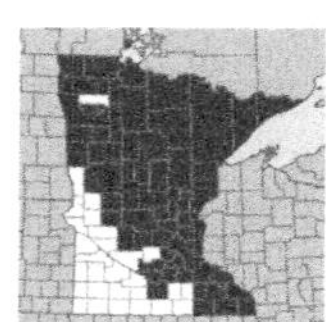

Thelypteris palustris Schott
MARSH-FERN
NC **FACW** | MW **OBL** | GP **OBL**

Small to medium native fern from slender rhizomes; rhizomes slender, spreading and branching. **Leaves** deciduous, ± hairy, erect, 20–60 cm long and to 15 cm wide; **blades** broadly lance-shaped, short-hairy on rachis and midveins, tapered to tip and only slightly narrowed at base; 1-pinnate, pinnae in 10–25 pairs, mostly alternate, narrowly lance-shaped, to 2 cm wide. Sterile and fertile leaves only slightly different; sterile leaves thin and delicate, pinnules blunt-tipped, 3–5 mm wide, veins once-forked. Fertile leaves longer than sterile leaves; pinnules oblong, 2-4 mm wide, the margins rolled under, veins mostly 1-forked; **petioles** longer than blades, black at base, hairless and without scales. **Sori** round, located halfway between midvein and margin, sometimes partly covered by the rolled under margin; **indusia** irregular in shape, usually with a fringe of hairs. **Synonyms** *Dryopteris thelypteris*. **Habitat** Swamps, low areas in forests, sedge meadows, forest depressions, open bogs, calcareous fens, marshes.

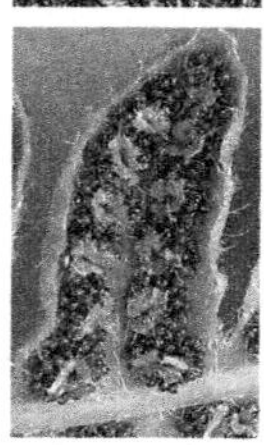

indusia and sori

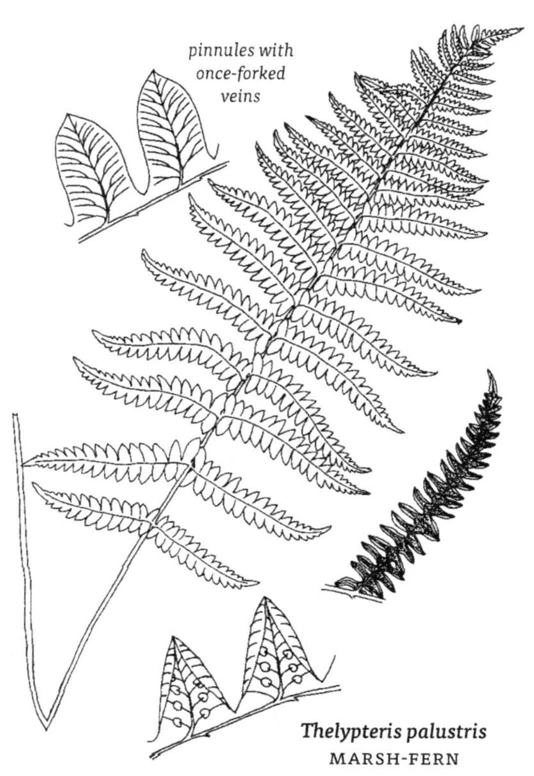

Thelypteris palustris
MARSH-FERN

Conifers

CONIFERS belong to the Gymnosperms (Latin gymn- "naked," Greek sperma, "seed"), within the division Pinophyta (also termed the Gymnospermae). In contrast to the Angiosperms, the seeds are not enclosed in an ovary (which develops into a fruit), but are exposed and attached to a cone or other structure.

Two families are represented in Minnesota wetlands: **Cupressaceae** (*Thuja occidentalis*), and **Pinaceae** (*Abies, Larix, Pinus, Picea*). In *Thuja*, the leaves are small, scalelike and appressed to the branches. The female cones are small, becoming brown with age. In Pinaceae, the leaves are needlelike, and either separate or grouped into bundles of two or more. The male and female flowers are borne separately on the same tree. The male flowers are borne in a herbaceous cone; the female flowers are in a woody cone with woody, overlapping scales.

Black spruce
(Picea mariana)

THUJA | *Arbor-vitae*

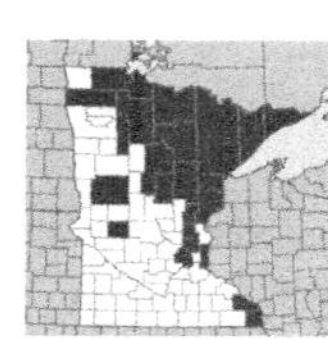

Thuja occidentalis L.

NORTHERN WHITE CEDAR; ARBOR-VITAE

NC **FACW** | MW **FACW** | GP **FACW**

Shade-tolerant native tree to 20 m tall, cone-shaped with widely spreading branches, sometimes layered at base, trunk to 1 m wide or more, bark reddish or gray-brown, in long shreddy strips; **twigs** flattened, in fanlike sprays. **Leaves** scalelike and overlapping, 3–6 mm long and 1–2 mm wide, yellow-green, aromatic, persisting for 1–2 years. **Seed cones** small, brown, 1 cm long, maturing in fall and persisting over winter. **Habitat** Cold, poorly drained swamps where *Thuja* may form dense stands; soils neutral or basic, usually highly organic, water not stagnant. Also along streams, on gravelly and sandy shores of Great Lakes, and dry soils over limestone.

Thuja occidentalis
NORTHERN WHITE CEDAR;
ARBOR-VITAE

cones and leaves

RESINOUS TREES with evergreen or deciduous, needlelike **leaves**. Male and female cones separate but borne on same tree. **Male cones** small and soft, falling after pollen is shed. **Female cones** larger, with woody scales arranged in a spiral. **Seeds** on upper surface of scales.

Although usually on well-drained soils in Minnesota, ***jack pine*** *(Pinus banksiana, right) and* ***eastern white pine*** *(Pinus strobus) are occasionally found in boggy habitats. When present they occur as stunted trees atop mossy hummocks. North of Minnesota, jack pine is more typically a wetland species.* ***Eastern hemlock*** *(Tsuga canadensis) sometimes occurs on swamp margins and in seasonally wet forest depressions.*

PINACEAE | PINE FAMILY

1 Leaves in clusters of 10–20; deciduous .... *Larix laricina*
.... TAMARACK; EASTERN LARCH
1 Leaves single and alternate on branches; persistent 2

2 Leaves flattened in cross-section, soft; cones upright ..... *Abies balsamea*
.... BALSAM FIR
2 Leaves 4-sided in cross-section, stiff; cones drooping...... *Picea* | SPRUCE

ABIES | *Fir*

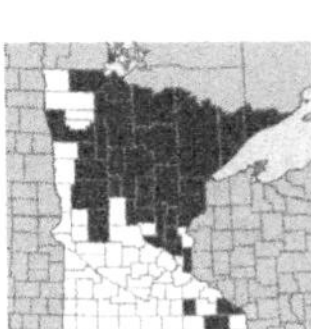

Abies balsamea (L.) Miller

BALSAM FIR

NC **FAC** | MW **FACW** | GP **FAC**

Shade-tolerant native tree to 25 m tall, crown spirelike, trunk to 6 dm wide; **bark** thin, smooth and gray, becoming brown and scaly with age; lower **branches** often drooping; **twigs** sparsely short-hairy. **Leaves** evergreen, linear, 12–25 mm long and 1–2 mm wide, blunt or with a small notch at tip, flat in cross-section, twisted at base and arranged in 1 plane (especially on lower branches), or spiraled on twigs. **Seed cones** 5–10 cm long and 1.5–3 cm wide, with broadly rounded scales. **Habitat** Cold boreal forests, swamps, and moist forests in n Minnesota; in se Minnesota, mostly restricted to fens.

Abies balsamea
BALSAM FIR

LARIX | *Larch*

Larix laricina (Duroi) K. Koch
TAMARACK; EASTERN LARCH
NC **FACW** | MW **FACW** | GP **FACW**

Shade-intolerant native tree to 20 m tall, crown narrow, trunk to 6 dm wide; **bark** smooth and gray when young, becoming scaly and red-brown; **twigs** yellow-brown, ± horizontal or with upright tips. **Leaves** deciduous, in clusters of 10–20, linear, 1–2.5 cm long and less than 1 mm wide, soft, blunt-tipped, bright green, turning yellow in fall. **Seed cones** 1–2 cm long and 0.5–1 cm wide, ripening in fall and persisting on trees for 1 year. **Habitat** Cold, poorly drained swamps, bogs and wet lakeshores; s in Minnesota, confined to wet depressions.

Larix laricina
TAMARACK;
EASTERN LARCH

Abies balsamea
BALSAM FIR

PICEA | *Spruce*

Evergreen trees; **bark** thin and scaly, resin blisters common in white spruce (*Picea glauca*). **Leaves** linear, square in cross-section, stiff, spreading in all directions around twig. **Cones** borne on last year's branches, drooping. **Seeds** wing-margined.

PICEA | SPRUCE

1 Leaves mostly 6–15 mm long; cones 1.5–3 cm long; twigs with fine hairs . . . . *Picea mariana* | BLACK SPRUCE

1 Leaves mostly 15–20 cm long; cones 2.5–6 cm long; twigs mostly without hairs . . . . *Picea glauca* | WHITE SPRUCE

Picea glauca (Moench) Voss

WHITE SPRUCE

NC **FACU** | MW **FACU** | GP **FACU**

Moderately shade-tolerant native tree to 30 m tall (often smaller), crown conelike, trunk to 60 cm or more wide; **bark** thin, gray-brown; **branches** slightly drooping, hairless. **Leaves** evergreen, linear, 1.5–2 cm long, 4-angled in cross-section, stiff, waxy blue-green, sharp-tipped. **Cones** 2.5–6 cm long, scales fan-shaped, rounded at tip, the tip entire. **Habitat** Moist to sometimes wet forests; absent from wetlands where water is stagnant.

Picea mariana (Miller) BSP.

BLACK SPRUCE

NC **FACW** | MW **FACW** | GP **FACW**

Moderately shade-tolerant native tree to 25 m tall (often smaller), crown narrow, often clublike at top, trunk to 25 cm wide; **bark** thin, scaly, gray-brown; **branches** short and drooping, often layered at base. **Leaves** evergreen, linear, 6–18 mm long, 4-angled in cross-section, stiff, waxy blue-green, mostly blunt-tipped. **Seed cones** 1.5–3 cm long, scales irregularly toothed, persisting for many years. **Habitat** Cold, acid, sphagnum bogs, swamps, and lakeshores; often where water is slow-moving and low in oxygen; less common in calcium-rich, well-aerated swamps dominated by northern white cedar (*Thuja occidentalis*).

Black spruce *can be distinguished from* **white spruce** *by its shorter needles, the branches with fine, white to red-brown hairs, the smaller, rounded seed cones with toothed scale margins, and its occurrence in generally wetter (and sometimes stagnant) habitats.*

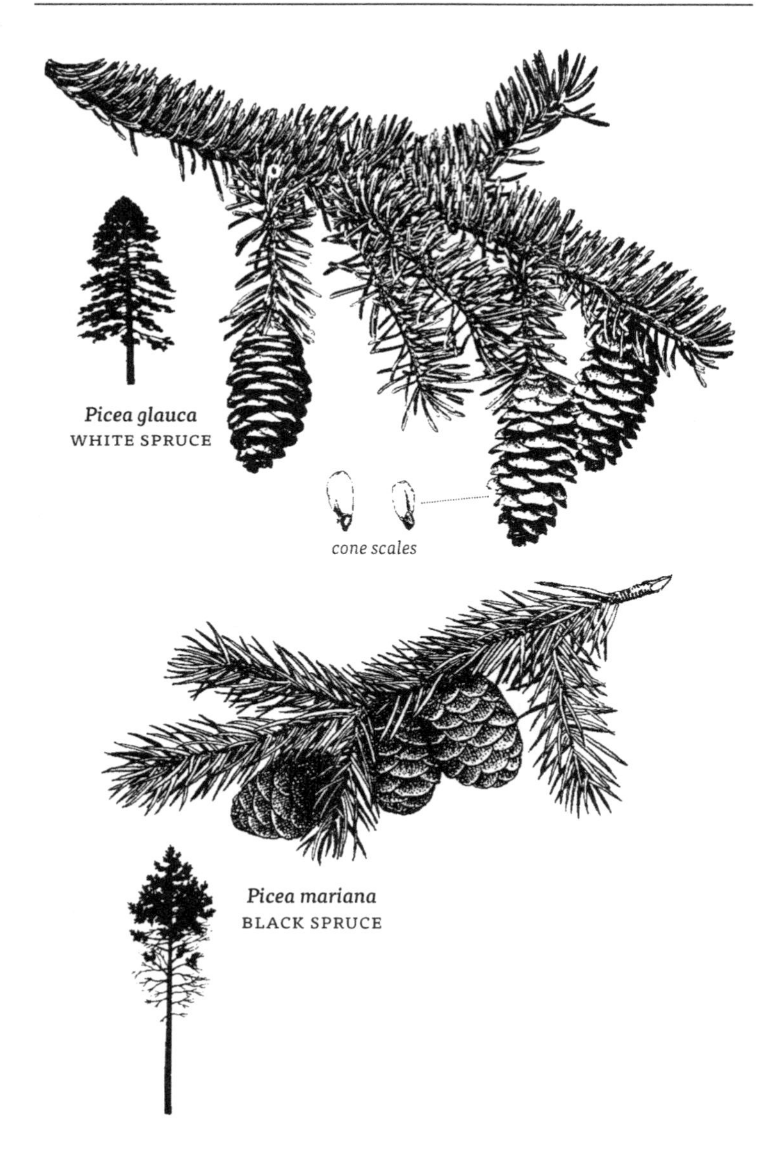

Picea glauca
WHITE SPRUCE

Picea mariana
BLACK SPRUCE

Dicots

ANGIOSPERMS (angion "vessel," sperm "seed") form the world's largest group of vascular plants. Rather than **cones** as in the gymnosperms, in angiosperms, the reproductive structures are **flowers**, and seeds are enclosed in fruits that typically develop from the ovary. Angiosperms are divided into two classes, the **Dicotyledoneae** (sometimes termed the Magnoliopsida and often shortened to "Dicots") and the **Monocotyledoneae** (or Liliopsida, the Monocots, page 377). The names are derived from the presence of either one or two "seed leaves" or cotyledons. The Dicots are a large and diverse group of plants, and include a variety of trees, shrubs, and herbaceous species (these often simply called "wildflowers"). Monocots include familiar families such as the orchids (Orchidaceae), grasses (Poaceae), and sedges (Cyperaceae). Although not always clear-cut, Dicots and Monocots may be distinguished from one another by a combination of the following characters:

Dicots

- *Embryo with two cotyledons*
- *Flower parts in multiples of 4 or 5*
- *Major leaf veins netlike*
- *Stem vascular bundles in a ring*
- *Roots develop from radicle*
- *Secondary growth often present*

Monocots

- *Embryo with single cotyledon*
- *Flower parts in multiples of 3*
- *Major leaf veins parallel*
- *Stem vacular bundles scattered*
- *Roots adventitious*
- *Secondary growth (wood) absent*

ACER | *Maple*

TREES OR SHRUBS. **Leaves** opposite, simple or compound. Male and female flowers borne on same or separate plants. **Flowers** with 5 sepals and 5 petals (sometimes absent), clustered into a raceme or umbel. **Fruit** a samara with 2 winged achenes joined at base.

ACER | MAPLE

1 Leaves compound. . . . . . . *Acer negundo* | BOXELDER; ASH-LEAVED MAPLE
1 Leaves simple. 2

2 Leaves shallowly lobed, the terminal lobe broadest at its base; flowers with petals . . . . *Acer rubrum* | RED MAPLE
2 Leaves deeply lobed to middle of blade or below, the terminal lobe narrowed at its base; flowers without petals . . . . *Acer saccharinum* . . . . SILVER-MAPLE; SOFT MAPLE

Acer negundo L.

BOXELDER; ASH-LEAVED MAPLE
NC **FAC** | MW **FAC** | GP **FAC**

Native tree to 20 m tall, the trunk soon dividing into widely spreading branches; **bark** brown, ridged when young, becoming deeply furrowed; **twigs** smooth, green and often waxy-coated. **Leaves** opposite, compound, leaflets 3–7, oval to ovate, coarsely toothed or shallowly lobed, upper surface light green and smooth, underside pale green and smooth or hairy. **Flowers** either male or female and on separate trees, appearing with leaves in spring; petals absent; male flowers in drooping, umbel-like clusters, female flowers in drooping racemes. **Fruit** a paired samara 3–4.5 cm long. **Habitat** Common; floodplain forests, streambanks, shores; also fencerows, drier woods and disturbed areas.

Acer negundo
BOXELDER; ASH-LEAVED MAPLE

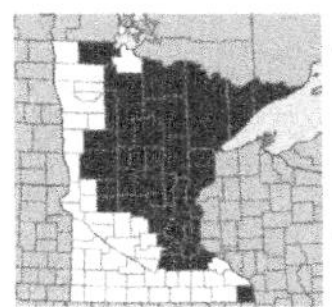

Acer rubrum L.
RED MAPLE
NC **FAC** | MW **FAC** | GP **FAC**

Native tree to 25 m tall; **bark** gray and smooth when young, becoming darker and scaly; **twigs** smooth, reddish with pale lenticels. **Leaves** opposite, 3–5-lobed (but not lobed to middle of blade), coarsely doubly toothed or with a few small lobes, upper surface green and smooth, underside pale green to white, smooth or hairy. **Flowers** either male or female, usually on different trees but sometimes on same tree, in dense clusters, opening before leaves in spring; sepals oblong, 1 mm long, petals narrower and slightly longer. **Fruit** a paired samara, 1–2.5 cm long. **Habitat** Floodplain forests, swamps; also common in drier forests.

Acer rubrum
RED MAPLE

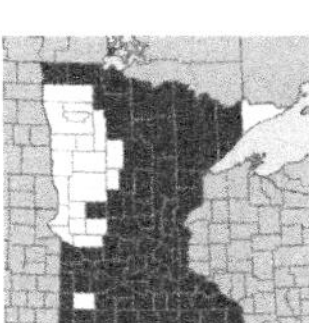

Acer saccharinum L.
SILVER-MAPLE; SOFT MAPLE
NC **FACW** | MW **FACW** | GP **FAC**

Native tree to 30 m tall; **bark** gray or silvery when young, becoming scaly; **twigs** red-brown, smooth. **Leaves** opposite, deeply 5-lobed to below middle of blade, sharply toothed, upper surface pale green and smooth, underside silvery white; petioles usually red-tinged. **Flowers** either male or female, usually on different trees but sometimes on same tree, in dense clusters, opening before leaves in spring. **Fruit** a paired samara, each fruit 3–5 cm long, falling in early to mid-summer. **Habitat** Floodplain forests, swamps, streambanks, shores, low areas in moist forests.

Acer saccharinum
SILVER-MAPLE; SOFT MAPLE

AMARANTHUS | *Amaranth*

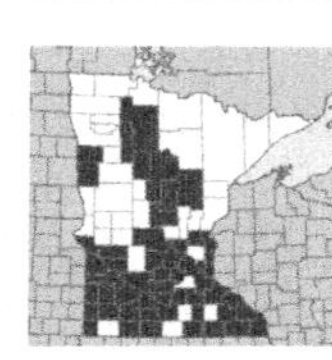

Amaranthus tuberculatus (Moq.) Sauer
WATER HEMP
NC **OBL** | MW **OBL** | GP **OBL**

Native annual herb. **Stems** erect to spreading, usually much-branched, 2–15 dm tall, usually hairless. **Leaves** alternate, ovate to lance-shaped, variable in size, larger leaves 4–10 cm long, smaller leaves 1–4 cm long. **Flowers** either male or female flowers and on different plants, in spikes from leaf axils and at ends of stems; male flowers with 5 sepals, 2–3 mm long and 5 stamens; female flowers without sepals or petals (rarely with 1–2 small sepals). **Fruit** a utricle 1–2 mm long; seeds red-brown, 1 mm wide. July–Sept. **Synonyms** *Acnida altissima*. **Habitat** Exposed muddy shores, streambanks, swamps, wet meadows, ditches.

Amaranthus tuberculatus
WATER-HEMP

TOXICODENDRON | *Poison-sumac*

Shrubs or vines. **Leaves** alternate, divided into 3 or more leaflets. **Flowers** in branched inflorescence; petals 5, green-white to yellowish. **Fruit** a white to yellowish drupe.

TOXICODENDRON | POISON-SUMAC

1 Leaflets 3; margins entire or toothed . . . . *Toxicodendron radicans* . . . . COMMON POISON-IVY

1 Leaflets 7–13; margins entire . . . . . . *Toxicodendron vernix* | POISON-SUMAC

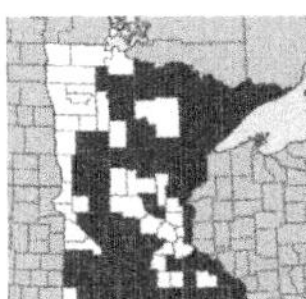

Toxicodendron radicans (L.) Kuntze

COMMON POISON-IVY

NC **FAC** | MW **FAC** | GP **FACU** | *Caution - skin irritant*

Vine or low shrub, native. **Leaves** alternate, divided into 3 dull to shiny ovate to elliptic leaflets, 5–15 cm long; margins entire or with a few coarse teeth. **Flowers** green-white or yellowish, 25 or more in a branched inflorescence from leaf axil; petals 5. **Fruit** a white drupe, 3–5 mm wide, these often drooping. May–July. **Synonyms** *Rhus radicans*. **Habitat** In n Minnesota, occasional in swamps and moist rocky forests, more common along roadsides, clearings, and on sand dunes. Southward, in floodplain forests, swamps and drier upland woods.

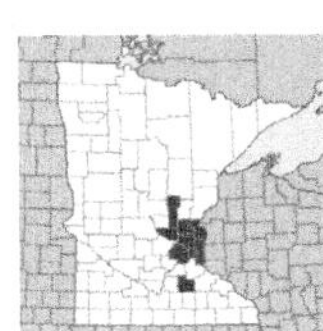

Toxicodendron vernix (L.) Kuntze

POISON-SUMAC

NC **OBL** | MW **OBL** | *Caution - skin irritant*

Native shrub or small tree to 5 m tall, often branched from base. **Leaves** alternate, divided into 7–13 leaflets, the leaflets oblong to oval, 4–6 cm long, tapered to a pointed tip; margins entire, smooth. **Flowers** small, white or green, in panicles to 2 dm long; sepals 5, joined at base; petals 5, not joined; stamens 5. **Fruit** a round, gray-white drupe, 4–5 mm wide. June–July. **Synonyms** *Rhus vernix*. **Habitat** Tamarack swamps, thickets, floating bog mats and bog margins, often in partial shade.

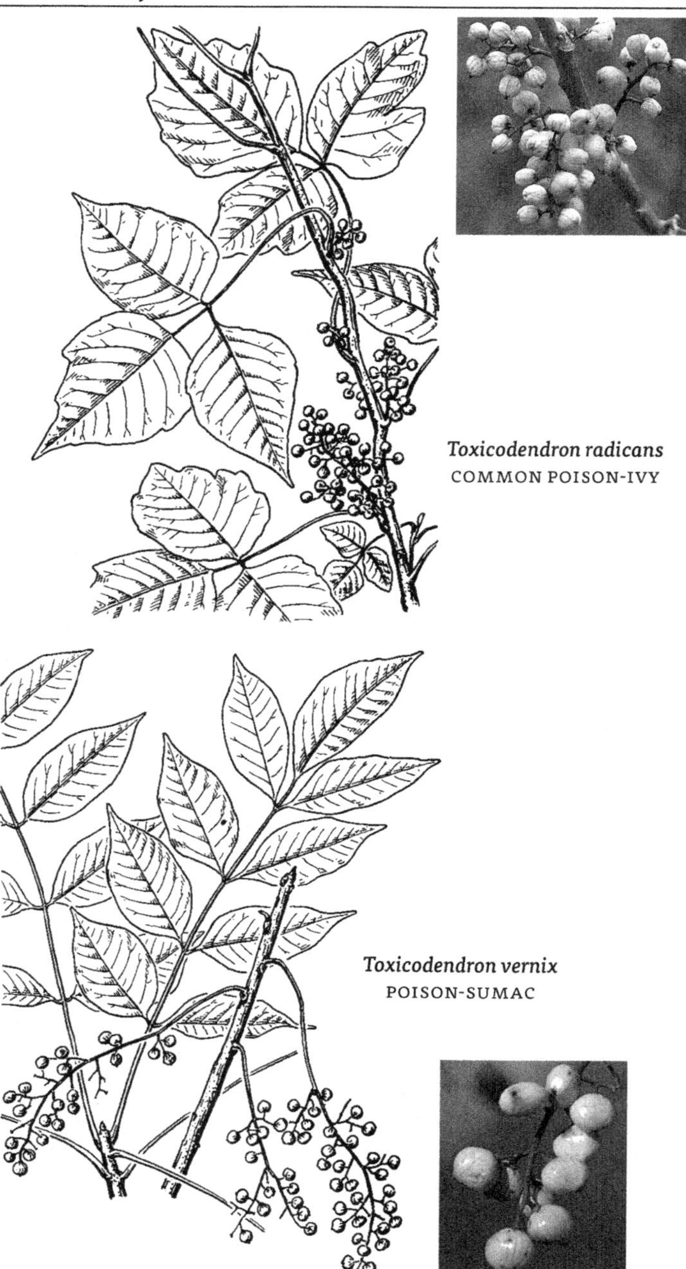

Toxicodendron radicans
COMMON POISON-IVY

Toxicodendron vernix
POISON-SUMAC

APIACEAE
Carrot Family

BIENNIAL OR PERENNIAL, aromatic herbs with hollow stems. **Leaves** alternate and sometimes also from base of plant, mostly compound; petioles sheathing stems. **Flowers** small, perfect (with both male and female parts), regular, in flat-topped or rounded umbrella-like clusters (umbels); sepals 5 or absent; petals 5, white or greenish (yellow in wild parsnip, *Pastinaca sativa*). **Fruit** 2-chambered, separating into 2, 1-seeded fruits when mature.

APIACEAE | CARROT FAMILY

1 Flowers yellow .... *Pastinaca sativa* | WILD PARSNIP
1 Flowers white or green-white 2

2 Leaves simple .... *Hydrocotyle americana* | MARSH PENNYWORT
2 Leaves divided or compound 3

3 Stems covered with woolly hairs; fruits and ovaries hairy. ....
.... *Heracleum lanatum* | COW-PARSNIP
3 Stems not covered with woolly hairs; fruits and ovaries hairless........ 4

4 Lateral veins of the leaflets end in the notches (lobes or sinuses) between the teeth; or bulblets in upper leaf axils .... *Cicuta* | WATER-HEMLOCK
4 Lateral veins of the leaflets end in the teeth; bulblets absent 5

5 Leaves finely divided, the smallest divisions deeply pinnately lobed; fruits winged, longer than wide; may occur in Minnesota (but not confirmed)..
.... *Conioselinum chinense* | HEMLOCK-PARSLEY
5 Smallest divisions of leaves toothed; fruits various 6

6 Divisions of stem leaves linear to narrowly lance-shaped, more than 4x longer than wide 7
6 Divisions of stem leaves broader, less than 4x longer than wide 8

7 Leaflets entire or with several coarse, irregularly spaced teeth ....
.... *Oxypolis rigidior* | COMMON WATER-DROPWORT
7 Leaflets finely toothed .... *Sium suave* | WATER-PARSNIP

8 Petiole sheaths large and inflated.. .... *Angelica atropurpurea*
.... PURPLESTEM-ANGELICA
8 Petiole sheaths not inflated 9

9 Large plants more than 1 m tall; stems with many purple spots; fruit with prominent ribs .... *Conium maculatum* | POISON HEMLOCK
9 Smaller plants, less than 1 m tall, often partially underwater; stems not purple spotted; fruit with indistinct ribs.................. *Berula erecta*
.... CUT-LEAF WATER PARSNIP

ANGELICA | *Angelica*

Angelica atropurpurea L.
PURPLESTEM-ANGELICA
NC **OBL** | MW **OBL** | *Caution - skin irritant*

Native perennial herb. **Stems** stout, 2–3 m tall, ± smooth, often streaked with purple and green. **Leaves** alternate, lower leaves 3-parted, 1–3 dm long, on long petioles; upper leaves smaller, less compound, on shorter petioles, or reduced to bladeless sheaths; leaflets ovate to lance-shaped, smooth, 4–10 cm long; margins sharp toothed. **Flowers** in rounded small clusters (umbelets), these grouped into large rounded umbels 1–2 dm wide; petals white to green-white. **Fruit** oval, 4–6 mm long, winged. May–July. **Habitat** Springs, seeps, calcareous fens, streambanks, shores, marshes, sedge meadows, wet depressions in forests; often where calcium-rich.

BERULA | *Water parsnip*

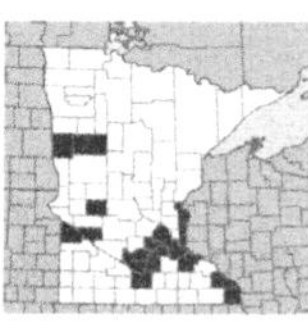

Berula erecta (Huds.) Cov.
CUT-LEAF WATER PARSNIP
NC **OBL** | MW **OBL** | GP **OBL**

Native perennial herb. **Stems** erect to trailing, sparsely branched, 4–8 dm long, often rooting along trailing portion. **Leaves** alternate, 1-pinnate, basal leaves larger and less dissected than stem leaves, oblong, 5–20 cm long and 2–10 cm wide; leaflets lance-shaped to ovate; margins toothed or lobed. **Flowers** grouped into 5–15 small clusters (umbelets) 1 cm wide, these grouped into umbels 3–6 cm across; flowers white, 1–2 mm wide; sepals small or absent. **Fruit** oval or round, slightly flattened, 1–2 mm long, but seldom maturing. July–Sept. **Synonyms** *Berula pusilla*. **Habitat** Shallow water, springs, spring-fed streams, marshes, swamps, often where calcium-rich.

CICUTA | *Water-hemlock*

Biennial or perennial toxic herbs. **Leaves** alternate, 2–3-pinnate; leaflets narrow or lance-shaped, entire or toothed; leaf veins ending in the lobes (sinuses) and not at teeth as in other members of family. **Flowers** white or green, in few to many umbels. **Fruit** oval or round, ribbed.

The tuberous roots, chambered stem base and young shoots of **common water-hemlock** *(Cicuta maculata) are especially poisonous if eaten.*

CICUTA | WATER-HEMLOCK

1 Upper leaflet axils usually with bulblets; leaflets narrow, to 5 mm wide . . . . *Cicuta bulbifera* | BULBLET-BEARING WATER-HEMLOCK

1 Bulblets absent; leaflets usually much more than 5 mm wide . . . . *Cicuta maculata* | COMMON WATER-HEMLOCK

Cicuta bulbifera L.

BULBLET-BEARING WATER-HEMLOCK
NC **OBL** | MW **OBL** | GP **OBL** | *Caution - toxic*

Biennial or perennial native herb; fibrous-rooted or with a few thickened, tuberlike roots. **Stems** slender, upright, 3–10 dm tall, not thickened at base. **Leaves** alternate along stem, to 15 cm long and 10 cm wide, pinnately divided; **leaflets** mostly linear, 1–5 mm wide, margins sparsely toothed to entire; upper leaves reduced in size, undivided or with few segments, with 1 to several bulblets 1–3 mm long, in axils. **Flowers** white, in umbels 2–4 cm wide. **Fruit** round, 1–2 mm wide, but rarely maturing. Aug–Sept. **Habitat** Streambanks, lake and pond shores, marshes, swamps, open bogs, thickets, springs and ditches.

Cicuta bulbifera
BULBLET-BEARING WATER-HEMLOCK

Cicuta maculata L.
COMMON WATER-HEMLOCK
NC **OBL** | MW **OBL** | GP **OBL** | *Caution - toxic*

Biennial or perennial native herb. **Stems** single or several together, often branched, 1–2 m long, distinctly hollow above the chambered and tuberous-thickened base. **Leaves** from base of plant and alternate on stem, mostly 10–30 cm long and 5–20 cm wide; basal leaves larger and longer stalked than stem leaves; leaflets linear to lance-shaped, 3–10 cm long and 5–35 mm wide; margins toothed. **Flowers** white, in several to many umbels, these 6–12 cm wide in fruit, on stout stalks 5–15 cm long. **Fruit** round to ovate, 2–4 mm long, with prominent ribs. June–Sept. **Synonyms** *Cicuta douglasii*. **Habitat** Wet meadows, marshes, swamps, moist to wet forests, thickets, shores, streambanks, springs.

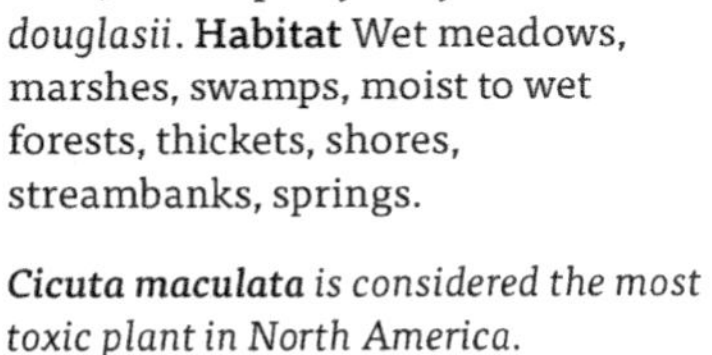

Cicuta maculata *is considered the most toxic plant in North America.*

CONIOSELINUM | *Hemlock-parsley*

Conioselinum chinense (L.) BSP.
HEMLOCK-PARSLEY

Native perennial herb. **Stems** erect, slender to stout, 5–15 dm tall, smooth or often with short, rough hairs in flower head. **Leaves** alternate, triangular in outline, 1–3x pinnate, on short, winged stalks; leaflets lance-shaped, deeply lobed, 2–4 cm long. **Flowers** white, in long-stalked umbels 3–12 cm wide. **Fruit** oval or oblong, 2–5 mm long. Aug–Sept. **Habitat** Tamarack swamps, forest seeps and discharge areas, streambanks, fens.

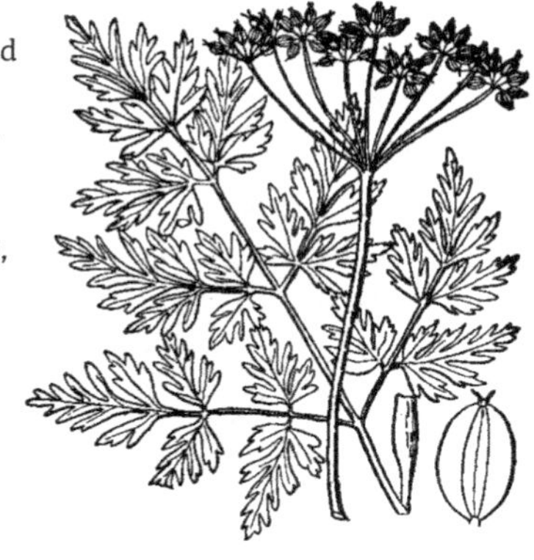

Hemlock-parsley *known from Wisconsin and Iowa; may be present in Minnesota but not verified.*

CONIUM | *Poison hemlock*

Conium maculatum L.

Caution - toxic

POISON HEMLOCK

NC **FACW** | MW **FACW** | GP **FACW** | *Caution - toxic*

Introduced biennial herb. **Stems** stout, branched, purple-spotted, 1–2 m long. **Leaves** alternate, 2–4 dm long, 3–4x pinnately divided, the leaflets toothed or sharply lobed. **Flowers** white, in many umbelets, these grouped in umbels to 6 cm wide. **Fruit** ovate, ribbed, 3 mm long. June–July. **Habitat** Weed of shores, streambanks, waste ground and roadsides, especially on moist, fertile soil. Introduced; now throughout most of s Canada and USA; likely more widespread in Minnesota than indicated on map.

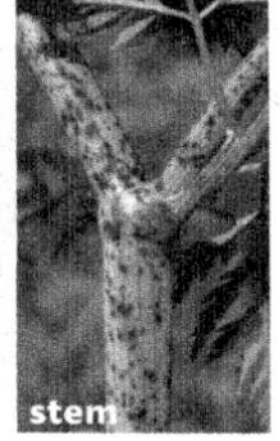

Plants of **poison hemlock** *have a strong, unpleasant odor. Caution - toxic if eaten, also a skin irritant.*

HERACLEUM | *Cow-parsnip*

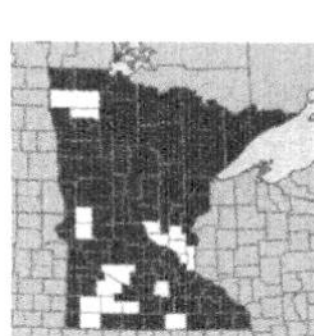

Heracleum lanatum Michx.

COW-PARSNIP

NC **FACW** | MW **FACW** | GP **FAC** | *Caution - skin irritant*

Large native perennial herb. **Stems** stout, hairy, 1–2 m long. **Leaves** alternate, nearly round in outline, divided into 3 leaflets; **leaflets** 1–4 dm long and as wide, margins coarsely toothed. **Flowers** white, in large umbels, the terminal umbel 1–2 dm wide. **Fruit** obovate, 8–12 mm long and nearly as wide, often hairy. May–July. **Synonyms** *Heracleum maximum, Heracleum sphondylium* subsp. *montanum*. **Habitat** Streambanks, thickets, wet meadows, moist forest openings and disturbed areas.

HYDROCOTYLE | *Pennywort*

Hydrocotyle americana L.

MARSH PENNYWORT | *special concern*

NC **OBL** | MW **OBL**

Small native perennial herb. **Stems** slender and creeping, 10–20 cm long, often rooting at nodes. **Leaves** round to kidney-shaped, 1–5 cm wide; margins with 7–12 shallow lobes; petioles long. **Flowers** small, white, in ± stalkless umbels from nodes; umbels 2–7-flowered. **Fruit** of 2 compressed carpels, ± round in outline, 1–2 mm wide, ribbed. June–Sept. **Habitat** In Minn, known only from wet edges of small, cold streams where the streams emerge from ravines. **Status** Special concern.

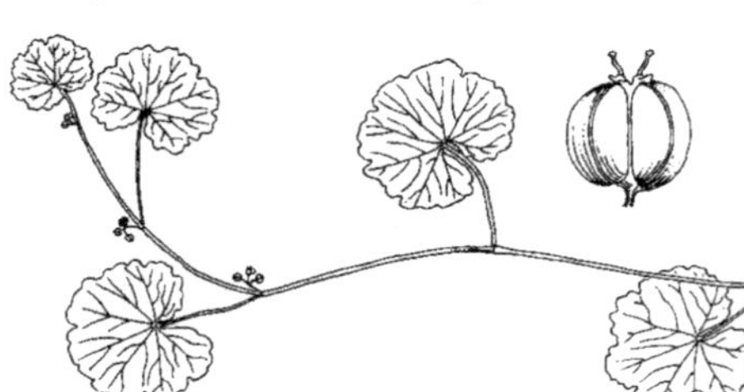

Hydrocotyle americana
MARSH PENNYWORT

OXYPOLIS | *Water-dropwort*

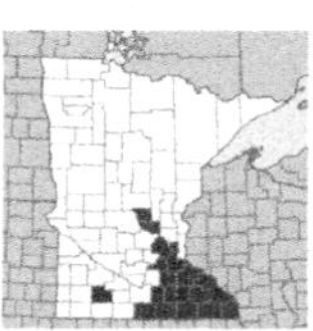

Oxypolis rigidior (L.) Raf.

COMMON WATER-DROPWORT

NC **OBL** | MW **OBL** | GP **OBL**

Native perennial herb. **Stems** stout or slender, to 1.5 m long, with few branches and leaves. **Leaves** 1-pinnate; leaflets 5–9, linear to oblong, 5–15 cm long and 5–40 mm wide; margins entire or with scattered coarse teeth. **Flowers** white, on stalks 5–20 mm long, in loose umbels to 15 cm wide. **Fruit** rounded at ends, 4–6 mm long and 3–4 mm wide. July–Sept. **Synonyms** *Oxypolis turgida*. **Habitat** Swamps, thickets, marshes, moist or wet prairie, calcareous fens.

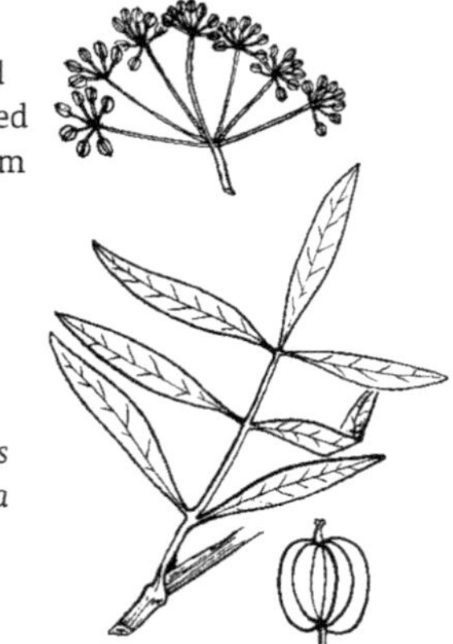

Similar to **water-parsnip** *(Sium suave) but differs in having entire to irregularly toothed leaves and a slightly grooved stem, while water-parsnip has finely toothed leaf margins and a more deeply grooved stem.*

PASTINACA | *Wild parsnip*

Pastinaca sativa L.
WILD PARSNIP
[NC **FAC** | MW **FAC** | GP **FAC**]| *Caution - skin irritant*

Biennial taprooted herb, introduced and invasive. **Stems** erect, 0.5–2 m long, branched, ± smooth to finely hairy, conspicuously angled and grooved. **Leaves** oblong to ovate in outline, 1-pinnate, the leaflets 5–11, coarsely toothed and lobed or divided. **Flowers** yellow, in flat-topped compound umbels. **Fruit** 4–6 mm wide, oblong to round. Jul–Aug. **Habitat** Roadsides, disturbed areas. Native of Eurasia,sporadic across nearly all of North America.

Wild parsnip *(Pastinaca sativa) is an invasive weed found in a variety of sunny, dry to moist habitats. In wetlands the species may occur in wet to moist prairie and calcareous fens. Contact with wild parsnip may cause mild to moderate cases of skin irritation. Gloves should be worn when handling plants.*

SIUM | *Water-parsnip*

Sium suave Walter
WATER-PARSNIP
NC **OBL** | MW **OBL** | GP **OBL**

Perennial emergent herb, native. **Stems** single, smooth, 5–20 dm long, strongly ribbed upward; stem base thickened and hollow with cross-partitions. **Leaves** 1-pinnate, on long, hollow stalks (shorter stalked above); leaflets 7–17 per leaf, linear to lance-shaped, 5–10 cm long and 3–15 mm wide; margins with fine, sharp, forward-pointing teeth; finely dissected underwater leaves often present from spring to midsummer. **Flowers** white or green-white, 1–2 mm wide, in stalked umbels 4–12 cm wide at ends of stems and from side branches. **Fruit** oval, 2–3 mm long, with prominent ribs. July–Sept. **Habitat** Wet forest depressions, marshes, swamps, streambanks, pond and lake margins, ditches; usually in shallow water.

Distinguished from the toxic **water-hemlock** *(Cicuta maculata) by its leaves only once compound; water-hemlock has leaves 2-3 times pinnately compound.*

ILEX | *Holly*

Shrubs with alternate, simple **leaves**. **Flowers** from leaf axils, 4–8-parted, usually either male or female, sometimes perfect, on same (*Ilex verticillata*) or different (*Ilex mucronata*) plants. **Fruit** a berrylike drupe.

ILEX | HOLLY

1 Leaves tipped with a short, sharp point; margins mostly entire or with a few scattered teeth; petals linear . . . . *Ilex mucronata* . . . . MOUNTAIN-HOLLY; CATBERRY

1 Leaves not tipped with a short, sharp point; margins toothed; petals oblong . . . . *Ilex verticillata* | WINTERBERRY

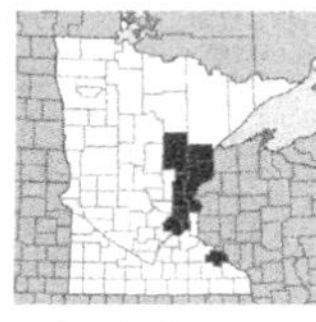

Ilex mucronata (L.) M. Powell, Savol. & S. Andrews
MOUNTAIN-HOLLY; CATBERRY
NC **OBL** | MW **OBL**

Much-branched native shrub to 3 m tall; young twigs purple-tinged. **Leaves** deciduous, alternate, oval or ovate, 3–6 cm long and 2–3 cm wide, bright green above, dull and paler below, tip of leaf with a small, sharp point; margins entire or with small scattered teeth, on purple-red stalks 1 cm long. **Flowers** very small, yellow-white, on threadlike stalks from leaf axils; male flowers usually in small groups, female flowers single. **Fruit** a purple-red berrylike drupe, 5–6 mm wide. May–June. **Synonyms** *Nemopanthus mucronatus*. **Habitat** Swamps, thickets, wet depressions in forests, lakeshores.

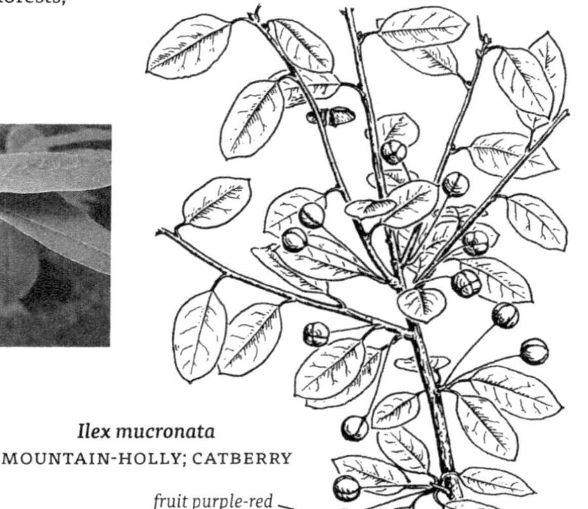

Ilex mucronata
MOUNTAIN-HOLLY; CATBERRY

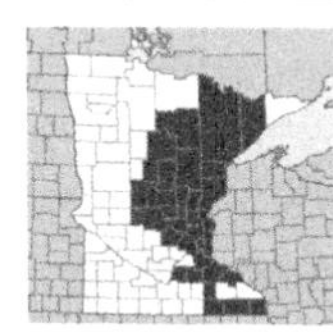

Ilex verticillata (L.) A. Gray
WINTERBERRY
NC **FACW** | MW **FACW**

Native shrub to 5 m tall; twigs smooth, finely ridged. **Leaves** deciduous, alternate, obovate to oval, tapered to a tip, dull green above, paler below; margins with incurved teeth. **Flowers** small, green-white, on short stalks from leaf axils, opening before leaves fully expanded in spring; male flowers in crowded clusters, female flowers 1 or several in a group. **Fruit** a berrylike drupe, orange or red, 5–6 mm wide and persisting into winter. June. **Habitat** Swamps, open bogs, thickets, shores and streambanks.

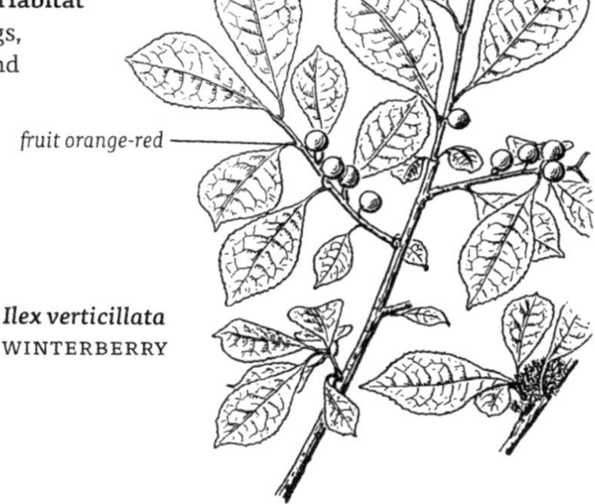

Ilex verticillata
WINTERBERRY

ASCLEPIAS | *Milkweed*

Asclepias incarnata L.

SWAMP-MILKWEED

NC **OBL** | MW **OBL** | GP **FACW**

Native perennial herb, from thick rhizomes; plants with milky juice. **Stems** stout, to 1.5 m long, branched above, smooth except for short, appressed hairs on upper stem. **Leaves** opposite, simple, mostly lance-shaped, 6–15 cm long and 1–5 cm wide, tapered to a sharp tip, margins entire, petioles short. **Flowers** pink to purple-red, numerous in umbels at ends of stems and from upper leaf axils, perfect, regular; sepals 5, spreading; petals 5, 4–6 mm long and curved downward; stamens 5; flowers with 5 petal-like "hoods," each with an awl-shaped "horn" projecting from the opening. **Fruit** a follicle (1-chambered and opening on 1 side only) with many seeds, the seeds having tufts of white hairs. June–Aug. **Habitat** Openings in conifer swamps, marshes, beaver ponds, streambanks, ditches, open bogs and fens; plants often in shallow water.

Asclepias incarnata
SWAMP-MILKWEED

ASTERACEAE
Aster Family

ANNUAL, BIENNIAL, OR PERENNIAL HERBS. **Leaves** simple or compound, opposite, alternate, or whorled. **Flowers** perfect (with both male and female parts) or single-sexed (sometimes sterile) and of 2 types: ray (or ligulate) and disk (or tubular). Ray flowers joined at base and with a long, flat, segment above (the ray); disk flowers tube-shaped with 5 lobes or teeth at tip.

Flowers clustered in one of 3 types of heads which resemble a single flower and are attached to a common surface (the receptacle): **ray flowers only** (as in dandelion, *Taraxacum*); **disk flowers only** (discoid, as in tansy, *Tanacetum*); and **heads with both ray and disk flowers** (radiate), the ray flowers surrounding the disk flowers (as in brown-eyed Susan, *Rudbeckia laciniata*, below). In addition to flowers, the receptacle may also have scales called chaff; if no scales are present, the receptacle is termed naked.

Each head is surrounded by **involucral bracts** (sometimes called phyllaries); collectively, the bracts are termed the **involucre**, comparable to the group of sepals (calyx) subtending an individual flower. Fertile flowers have 1 pistil tipped by a 2-cleft style (undivided in sterile flowers); 5 stamens; the ovary (and achene) often topped by a pappus composed of several to many scales, awns or hairs. **Fruit** a seedlike achene.

Rudbeckia laciniata
BROWN-EYED SUSAN

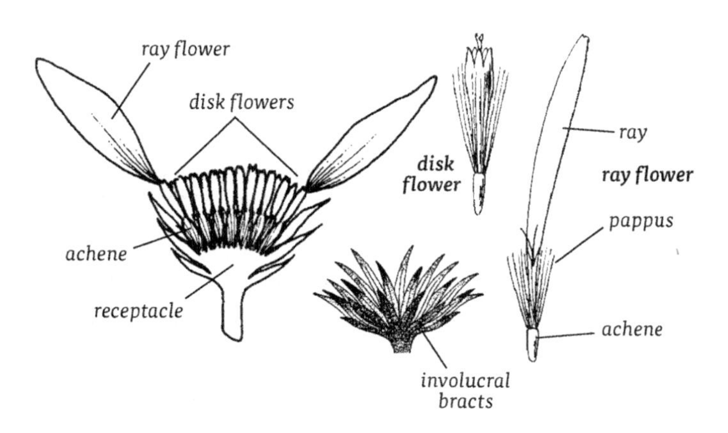

ABOVE Typical composite inflorescence, with both disk and ray flowers.

ASTERACEAE
Aster Family

ASTERACEAE | ASTER FAMILY

1 Plants aquatic, underwater leaves whorled and dissected into narrow segments . . . . *Megalodonta beckii*
. . . . WATER BEGGAR-TICKS; BECK'S WATER-MARIGOLD p. 128

1 Plants not aquatic (sometimes emergent); leaves not as above 2

2 Plants with white milky juice; heads with ray flowers only. 3

2 Plants with watery juice; heads with both ray and disk flowers, or disk flowers only 4

3 Ray flowers yellow; leaves in a basal rosette . . . . *Crepis runcinata*
. . . . DANDELION HAWK'S BEARD p. 119

3 Ray flowers pink, purple, or rarely white; stems leafy . . . .
. . . . *Prenanthes racemosa* | GLAUCOUS RATTLESNAKE-ROOT p. 130

4 Leaves with sharp spines . . . . . . *Cirsium muticum* | MARSH-THISTLE p. 118

4 Leaves without spines 5

5 Leaves opposite or whorled 6

5 Leaves at least in part alternate, or the leaves mostly all at base of plant . .
. . . . 10

6 Receptacle (disk to which flowers are attached) not chaffy. . . . *Eupatorium*
. . . . JOE-PYE-WEED p. 122

6 Receptacle chaffy 7

7 Pappus (bristles or scales atop ovary|achene) of bristly hairs or bristle-tipped scales 8

7 Pappus absent 9

8 Involucral bracts in 2 series, the outer often leaflike and much larger than the inner; pappus of 2–4 awl-shaped bristles, the bristles with sharp, usually downward-pointing barbs . . . . *Bidens*
. . . . BEGGAR-TICKS; BUR-MARIGOLD p. 112

8 Involucral bracts in several series and overlapping, ± equal in size; pappus of 2 lance-shaped, unbarbed awns. . . . . *Helianthus* | SUNFLOWER p. 125

9 Rays yellow, 1 cm or more long; large perennial herb, 1–3 m tall . . *Silphium*
. . . . CUP-PLANT; ROSIN-WEED p. 135

9 Rays small, white; annual herb to 1 m tall. . . . . *Eclipta prostrata*
. . . . YERBA-DE-TAJO p. 119

10 Heads with both disk and ray flowers; the rays yellow. 11

10 Heads with disk flowers only; or with both disk and ray flowers, the rays not yellow 16

11 Pappus of 2–several awns or scales **12**
11 Pappus of many hairlike bristles **14**

12 Receptacle not chaffy; leaves tapered to a stalkless base, continuing downward as wings on the stem. . . . . *Helenium autumnale* . . . . COMMON SNEEZEWEED p. 125
12 Receptacle chaffy; leaves short-stalked, not continuing downward on the stem. **13**

13 Leaves all alternate . . . . *Rudbeckia* | CONEFLOWER p. 131
13 Lower leaves opposite, upper leaves alternate . . . . *Helianthus* . . . . SUNFLOWER p. 125

14 Involucral bracts in 1 series, of similar length and not overlapping in rows . . . . *Senecio* | GROUNDSEL p. 132
14 Involucral bracts in several series, of different lengths and overlapping **15**

15 Flowers in a corymblike (± flat-topped) head; leaves narrow, 2–10 mm wide, entire, dotted with glands. . . . . *Euthamia* . . . . FLAT-TOPPED GOLDENROD p. 124
15 Flowers in a paniclelike head; leaves wider, 1–4 cm wide, toothed, not gland-dotted . . . . *Solidago* | GOLDENROD p. 136

16 Heads either male or female and of different shapes, the male flowers in small heads above the larger female heads; involucral bracts of the female heads with hooked spines, enclosing the flowers to form a bur . . . . *Xanthium strumarium* | COMMON COCKLEBUR p. 139
16 Heads with both male and female flowers, or rarely either male or female; involucral bracts neither spiny nor bur-like. **17**

17 Main leaves large and at base of plant, arrowhead-shaped or palmately lobed, white-woolly at least on underside; plants flowering in spring to early summer before leaves develop. . . . . *Petasites* . . . . SWEET COLTSFOOT p. 128
17 Main leaves along stem, neither arrowhead-shaped nor palmately lobed; plants flowering late-summer or fall (except the earlier flowering *Erigeron philadelphicus*) **18**

18 Heads with disk flowers only **19**
18 Heads with both disk and ray flowers (in *Erigeron lonchophyllus*, the rays narrow and ± equal to the involucral bracts, and therefore inconspicuous) **23**

19 Leaves pinnately dissected; pappus absent. . . . . *Artemisia biennis* . . . . BIENNIAL WORMWOOD p. 104

19 Leaves simple, entire or toothed; pappus of many hairlike bristles **20**

20 Plants annual; leaves linear **21**
20 Plants perennial (sometimes biennial in *Erigeron lonchophyllus*); leaves wider.... **22**

21 Stems with wool-like hairs.... *Gnaphalium uliginosum* .... MARSH CUDWEED p. 124
21 Stems without hairs .... *Aster brachyactis* | RAYLESS ASTER p. 107

22 Flowers white.... *Cacalia* | INDIAN PLANTAIN p. 117
22 Flowers purple or pink-purple **24**

23 Flowers in a long, spikelike head; leaves linear.... *Liatris pycnostachya* .... PRAIRIE BLAZING STAR p. 127
23 Flowers in an open corymb; leaves lance-shaped .... *Vernonia fasciculata* .... SMOOTH IRONWEED p. 138

24 Pappus of 2 bristles about as long as achene and several very small bristles .... *Boltonia asteroides* | WHITE BOLTONIA p. 116
24 Pappus of many hairlike bristles **25**

25 Rays wider than 0.5 mm .... *Aster* | WILD ASTER p. 105
25 Rays very narrow, to 0.5 mm wide .... *Erigeron* | FLEABANE p. 120

ARTEMISIA | *Wormwood, Sage*

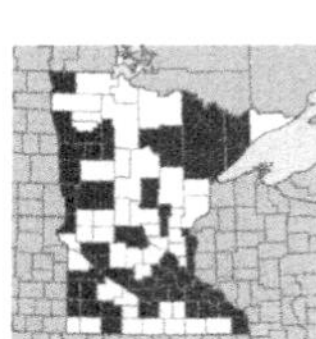

Artemisia biennis Willd.
BIENNIAL WORMWOOD
NC **FACW** | MW **FACW** | GP **FACU**

Taprooted, annual or biennial herb; introduced. **Stems** erect, to 1 m or more long, often branched, smooth, only faintly scented. **Leaves** alternate, pinnately dissected nearly to middle, 5–12 cm long and 2–5 cm wide, the segments linear and toothed. **Flowers** in stalkless heads from upper leaf axils; the heads composed of many small green disk flowers, grouped into spikelike inflorescences, with leafy bracts much longer than the clusters of heads; pappus absent. **Fruit** a small oblong achene. Aug–Sept. **Habitat** Sandy lakeshores, streambanks, ditches, mud flats, disturbed areas; often where seasonally flooded. Native to nw USA, occasional in Minnesota and somewhat weedy.

Artemisia biennis
BIENNIAL WORMWOOD

ASTER | *Wild aster*

Mostly perennial herbs. **Leaves** simple, alternate. **Flower heads** with both ray and disk flowers (disk flowers only in *Aster brachyactis*); **ray flowers** white, pink, blue or purple, usually more than 0.5 mm wide (in contrast to the fleabanes, *Erigeron*); **disk flowers** red, purple or yellow; involucral bracts in 2 or more series, usually overlapping; receptacle naked (not chaffy), ± flat; pappus of numerous hairlike bristles.

ASTER | WILD ASTER

1 Upper stem leaves stalkless, the base of leaf clasping stem; involucral bracts sometimes with gland-tipped hairs 2
1 Upper stem leaves not clasping; involucral bracts without gland-tipped hairs 3

2 Leaf margins entire; involucral bracts and flower stalks with glands . . . . *Aster novae-angliae* | New England aster p. 108
2 Leaf margins usually with at least some teeth; involucral bracts without glands. . . . . . *Aster puniceus* | BRISTLY ASTER; PURPLE-STEM ASTER p. 110

3 Plants annual; rays absent or less than 2 mm long . . . . . . *Aster brachyactis* . . . . RAYLESS ASTER p. 107
3 Plants perennial; rays present and larger 4

4 Heads in a ± flat-topped cluster; midrib of involucral bract ± same width its entire length . . . . *Aster umbellatus* . . . . TALL FLAT-TOPPED WHITE ASTER p. 111
4 Heads in elongate panicles; midvein of involucral bracts expanded to a ± diamond-shape above middle of bract 5

5 Involucral bracts covered with hairs. *Aster ontarionis* . . . . LAKE ONTARIO ASTER; BOTTOMLAND-ASTER p. 109
5 Involucral bracts smooth, or only fringed with hairs on margins 6

6 Involucral bracts tipped with a short sharp spine . . . . *Aster pilosus* . . . . FROST ASTER p. 109
6 Involucral bracts not spine-tipped 7

7 Leaf undersides net-veined, the veins dark green, the spaces enclosed by the veins paler and not elongate; rays blue. . . . . *Aster praealtus* . . . . VEINY LINED ASTER p. 110
7 Leaf undersides not distinctly net-veined with paler spaces, or if somewhat net-veined, the spaces elongate; rays mostly white to pink. 8

8 Leaves hairy, at least on underside midvein; disk flowers deeply lobed (to half or more the length of expanded portion of corolla above the tube) . . 9

8 Leaf undersides ± smooth; disk flowers shallowly lobed (to 1/3 or less the length of expanded portion of corolla above tube) **10**

9 Leaf underside smooth except for hairs on midvein; stems in a clump from a crown . . . . *Aster lateriflorus* | CALICO ASTER; GOBLET ASTER p. 108

9 Leaf underside covered with short hairs; stems single from slender rhizomes . . . . *Aster ontarionis* . . . . LAKE ONTARIO ASTER; BOTTOMLAND-ASTER p. 109

10 Stems less than 2 mm wide; leaves less than 6 mm wide, margins often rolled under; flowers in clusters of 1–15 heads, the flower stalks spreading . . . . *Aster borealis* | NORTHERN BOG-ASTER p. 106

10 Stem 3 mm or more wide; leaves more than 6 mm wide, margins flat; flowers in clusters of 20 heads or more, the flower stalks usually ascending . . . . *Aster lanceolatus* | EASTERN LINED ASTER p. 107

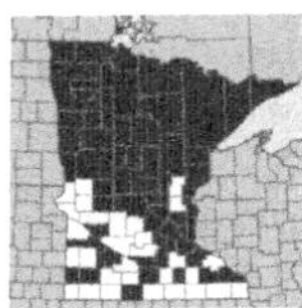

Aster borealis Prov.
NORTHERN BOG-ASTER
NC **OBL** | MW **OBL** | GP **OBL**

Native perennial herb, from rhizomes 1–2 mm wide. **Stems** erect, slender, 3–8 dm tall and to 2 mm wide, unbranched below, usually branched in the head; smooth except for lines of short, appressed hairs below base of upper leaves. **Leaves** alternate, linear, 4–12 cm long and 2–6 mm wide, sometimes slightly clasping at base, margins rough-to-touch, petioles absent. **Flower heads** usually few to rarely many, in an open, broad inflorescence; the heads 1.5–2 cm wide; involucre 5–7 mm high, the involucral bracts overlapping, often purple at tips and on margins; ray flowers 20–50, white to light blue or lavender, 1–1.5 cm long. **Fruit** an achene; pappus of pale hairs. Aug–Sept. **Synonyms** *Aster junciformis, Symphyotrichum boreale*. **Habitat** Conifer swamps, calcareous fens, open bogs, wet meadows, shores and seeps.

Aster borealis
NORTHERN BOG-ASTER

Aster brachyactis S. F. Blake
RAYLESS ASTER
NC **FAC** | MW **FAC** | GP **FACW**

Annual taprooted herb, introduced. **Stems** unbranched and erect, to branched and spreading, 2–6 dm long, smooth. **Leaves** alternate, linear, 2–10 cm long and mostly 2–5 mm wide, margins fringed with scattered hairs, petioles absent. **Flower heads** several to many, in an open inflorescence which forms much of plant; flower heads bell-shaped, 1–2 cm wide, involucre 5–10 mm high, the involucral bracts mostly green, linear, of equal length or slightly overlapping; ray flowers absent. **Fruit** a flattened achene, 1–2 mm long; pappus of many long, soft hairs. Aug–Sept. **Synonyms** *Brachyactis angusta, Brachyactis ciliata, Symphyotrichum ciliatum*. **Habitat** Shores (including along Great Lakes), streambanks, wet meadows, roadside ditches, usually where brackish. In Minnesota, introduced from w North America.

Aster brachyactis
RAYLESS ASTER

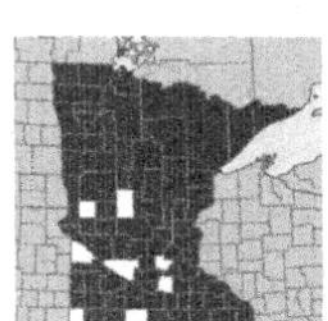

Aster lanceolatus Willd.
EASTERN LINED ASTER
NC **FACW** | MW **FAC** | GP **FACW**

Native perennial herb, forming colonies from long rhizomes. **Stems** 0.5–1.5 m long, upper stem with lines of hairs. **Leaves** alternate, all on stem, lance-shaped to linear, 8–15 cm long and 3–30 mm wide, upper surface smooth or slightly rough-to-touch, margins toothed or sometimes entire; petioles absent or blades tapered to petiolelike base, sometimes slightly clasping stem. **Flower heads** many in an elongate leafy inflorescence; the involucre 3–6 mm high, the involucral bracts tapered to a green tip, smooth or margins fringed with hairs, strongly overlapping; ray flowers 20–40, usually white, sometimes lavender or blue, 4–12 mm long. **Fruit** an achene; pappus white. Aug–Oct. **Synonyms** *Aster hesperius, Aster interior, Aster paniculatus, Aster simplex, Symphyotrichum lanceolatum*. **Habitat** Common; marshes, wet meadows, fens, swamp openings, low prairie, streambanks and shores.

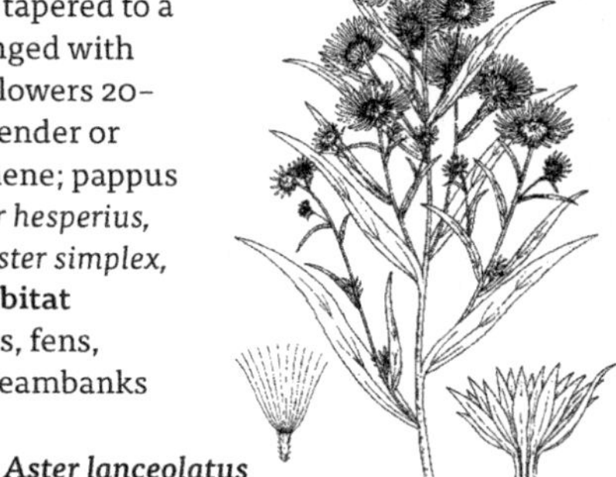

Aster lanceolatus
EASTERN LINED ASTER

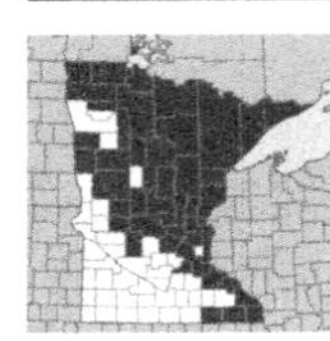

Aster lateriflorus (L.) Britton

CALICO ASTER; GOBLET ASTER

NC **FAC** | MW **FACW** | GP **FACW**

Native perennial herb, from a branched base or short rhizome. **Stems** 3–14 dm long, smooth to finely hairy. **Leaves** from base of plant and along stem; basal leaves ovate, with a long petiole; stem leaves alternate, lance-shaped, 5–15 cm long and 0.5–3 cm wide, branch leaves much smaller; upper surface rough-to-touch or smooth, underside smooth except for finely hairy midvein; margins with forward-pointing teeth or sometimes entire; petioles short or absent; basal and lower stem leaves usually soon deciduous. **Flower heads** many, mostly on 1 side of inflorescence branches; involucre smooth, 4–6 mm long, the involucral bracts overlapping in 3–4 series, with a green-purple tip; ray flowers 9–15, white or pale purple, 4–7 mm long. Aug–Oct. **Synonyms** *Aster agrostifolius, Aster hirsuticaulis, Symphyotrichum lateriflorum*. **Habitat** Swamps, thickets, floodplain forests, shores, streambanks, roadsides, moist hardwood and mixed conifer-hardwood forests; often where shaded.

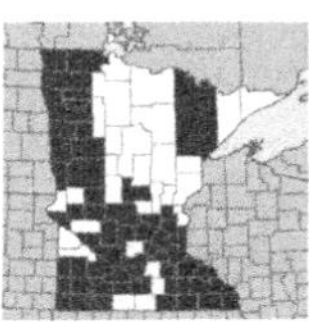

Aster novae-angliae L.

NEW ENGLAND ASTER

NC **FACW** | MW **FACW** | GP **FACW**

Native perennial herb, from a short rhizome or crown. **Stems** stout, erect, 4–10 dm long, with stiff, spreading, sometimes gland-tipped hairs. **Leaves** alternate, lance-shaped, 3–7 cm long and 1–2.5 cm wide, upper surface rough-to-touch or with short hairs, underside soft hairy, base of leaf strongly clasping stem, margins entire, petioles absent. **Flower heads** several to many, in clusters at ends of branches, 1.5–3 cm wide; involucre 7–12 mm high, the involucral bracts awl-shaped, glandular-hairy, sometimes purple; ray flowers 40 or more, blue-violet to less often red or pink, 1–2 cm long. **Fruit** a hairy achene; pappus red-tinged. Aug–Oct. **Synonyms** *Symphyotrichum novae-angliae*. **Habitat** Wet meadows, low prairie, shores, thickets, calcareous fens, roadsides; usually in moist or wet open areas.

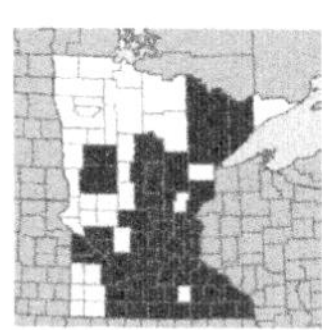

Aster ontarionis Wieg.
LAKE ONTARIO ASTER; BOTTOMLAND-ASTER
NC **FAC** | MW **FAC** | GP **FAC**

Native perennial herb, from long creeping rhizomes. **Stems** branched, 3–8 dm long, upper stems with short spreading hairs. **Leaves** alternate, thin, oblong lance-shaped, 5–10 cm long and 1–3 cm wide (upper leaves smaller), upper surface rough-hairy to nearly smooth, underside finely to densely hairy; margins with sharp, forward-pointing teeth above middle of blade; petioles absent. **Flower heads** 1–2 cm wide, on short stalks from short leafy branches; involucre smooth to finely hairy, 5–7 mm high, the involucral bracts overlapping; ray flowers white, 9 or more. **Fruit** an achene; pappus white. Sept–Oct. **Synonyms** *Symphyotrichum ontarionis*. **Habitat** Floodplain forests, river terraces, thickets.

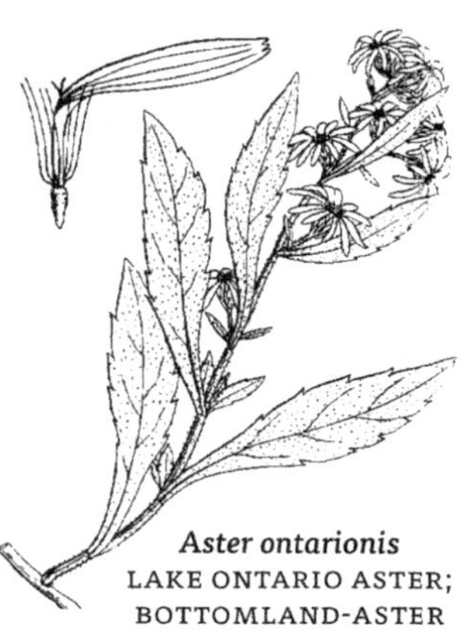

Aster ontarionis
LAKE ONTARIO ASTER; BOTTOMLAND-ASTER

Aster pilosus Willd.
FROST ASTER
NC **FACU** | MW **FACU** | GP **FACU**

Native perennial herb, from a large crown. **Stems** to 1.5 m long, ± smooth (var. *pringlei*) or stems and leaves with spreading hairs (var. *pilosus*). **Leaves** alternate, lower leaves oblong lance-shaped, 5–10 cm long and 1–2 cm wide, stalked; upper leaves smaller, linear, stalkless; margins entire or slightly toothed; petioles fringed with hairs; basal leaves and lower stem leaves soon deciduous (or basal leaves persistent). **Flower heads** at ends of small branches, forming an open inflorescence; involucre urn-shaped, narrowed near middle and flared upward, 3–5 mm high, smooth, involucral bracts overlapping to nearly equal in length, green-tipped; ray flowers 15–35, white. **Fruit** an achene; pappus white. **Synonyms** *Symphyotrichum pilosum*. **Habitat** Sandy and gravelly shores, interdunal swales, wet meadows; often where calcium-rich; sometimes weedy in disturbed fields and roadsides; se Minnesota.

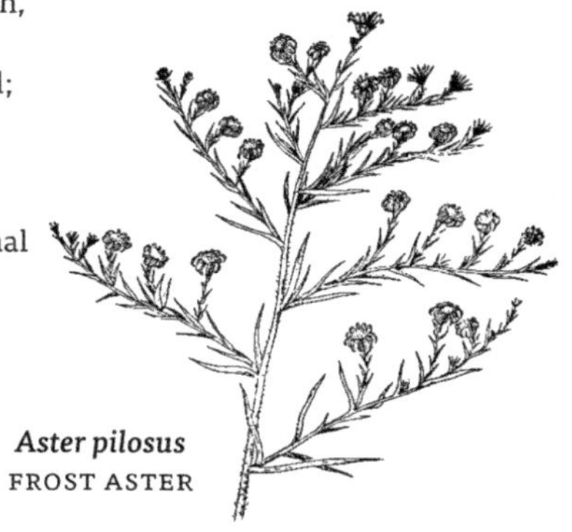

Aster pilosus
FROST ASTER

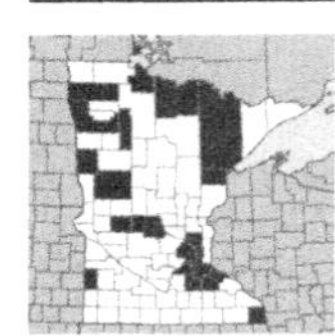

Aster praealtus Poiret
VEINY LINED ASTER
NC **FACW** | MW **FACW** | GP **FACW**

Native perennial herb, spreading by rhizomes and forming colonies. **Stems** to 1 m long, with lines of hairs, especially in upper stem. **Leaves** alternate, firm, lance-shaped, 6–12 cm long and 1–2 cm wide, upper surface rough-to-touch to nearly smooth, underside smooth or finely hairy, with conspicuous netlike veins surrounding lighter colored areas (areole); margins ± entire; petioles absent, the base of leaves often slightly clasping. **Flower heads** at ends of short leafy branches; involucre 5–7 mm high, the involucral bracts overlapping; rays 5–15 mm long, blue-purple (rarely white). **Fruit** an achene; pappus white. Sept–Oct. **Synonyms** *Aster nebraskensis, Aster woldeni, Symphyotrichum praealtum*. **Habitat** Wet meadows, low prairie, moist fields, thickets.

Aster praealtus
VEINY LINED ASTER

Aster puniceus L.
BRISTLY ASTER; PURPLE-STEM ASTER
NC **OBL** | MW **OBL** | GP **OBL**

Large native perennial herb, from a short rhizome or crown, sometimes also with short stolons. **Stems** stout, red-purple, 0.5–2 m long, unbranched, or branched in head, with long stiff hairs or sometimes nearly smooth. **Leaves** alternate, lance-shaped to oblong lance-shaped, 6–18 cm long and 1–4 cm wide, rough-to-touch to nearly smooth above, underside smooth or with long hairs on midvein; margins with scattered sharp teeth or sometimes entire; petioles absent, base of leaf clasping. **Flower heads** numerous, 1.5–2.5 cm wide; involucre 6–10 mm high, involucral bracts about equal in length, smooth or fringed with hairs, green and spreading; ray flowers 20–50, blue (rarely white). **Fruit** a smooth achene; pappus ± white. Aug–Sept. **Synonyms** *Aster firmus, Symphyotrichum puniceum*. **Habitat** Swamps, sedge meadows, thickets, calcareous fens, streambanks, shores, springs.

Aster umbellatus Miller
TALL FLAT-TOPPED WHITE ASTER
NC **FACW** | MW **FACW** | GP **OBL**

Native perennial herb, from thick rhizomes. **Stems** 0.5–2 m long, upper stem with appressed, short hairs. **Leaves** alternate, lance-shaped to oblong lance-shaped, 4–15 cm long and 1–4 cm wide, rough-to-touch above, densely short-hairy below; margins entire; petioles short, or absent on upper leaves. **Flower heads** usually many, 1–1.5 cm wide, in a ± flat-topped inflorescence; involucre 3–5 mm high, the involucral bracts short-hairy and overlapping; rays 5–10, white, 5–8 mm long. **Fruit** a nerved achene; pappus nearly white. July–Sept. **Synonyms** *Doellingeria umbellata*. **Habitat** Openings in swamps and moist forests, thickets, streambanks, sedge meadows, calcareous fens, roadside ditches.

Aster umbellatus (TALL FLAT-TOPPED WHITE ASTER)

BIDENS | *Beggar-ticks*

Weedy annual or biennial herbs. **Leaves** opposite, simple, lobed, or pinnately divided. **Flower heads** with both disk and ray flowers, or with disk flowers only; ray flowers often about 8, yellow; involucral bracts in 2 series, the outer row leaflike and spreading, the inner row much shorter and erect; receptacle ± flat and chaffy. **Fruit** a flattened achene; pappus of 2–5 barbed awns which persist atop the achene; the body of achene barbed or with stiff hairs (at least on the angles), the "stick-tights" facilitating seed dispersal by animals.

BIDENS | BEGGAR-TICKS

1 Leaves simple and toothed, or rarely lobed; achenes 3–4-awned 2
1 Leaves all (or mostly) pinnately divided or compound; achenes 2-awned 5

2 Leaves mostly stalkless 3
2 Leaves with a petiole 1–4 cm long 4

3 Heads nodding when mature; outer involucral bracts widely spreading . . . . *Bidens cernua* | NODDING BEGGAR-TICKS p. 113
3 Heads mostly upright; outer involucral bracts ± erect. . . . . . *Bidens comosa* . . . . STRAWSTEM BEGGAR-TICKS p. 113

4 Stems straw-colored; disk flowers 4-lobed, pale yellow; achenes mostly 3-awned . . . . *Bidens comosa* | STRAWSTEM BEGGAR-TICKS p. 113
4 Stems green-purple; disk flowers 5-lobed, yellow-orange; achenes mostly 4-awned (sometimes 2-awned) . . . . *Bidens connata* . . . . PURPLESTEM BEGGAR-TICKS p. 114

5 Heads with both disk and ray flowers, the rays over 10 mm long. . . . . *Bidens coronata* | NORTHERN TICKSEED-SUNFLOWER p. 114
5 Heads with disk flowers only, or with short rays less than 5 mm long . . . 6

6 Outer involucral bracts 2–5 (usually 4), not fringed with hairs . . . . *Bidens discoidea* | FEW-BRACTED BEGGAR-TICKS p. 115
6 Outer involucral bracts 6 or more, fringed with hairs (at least near base). 7

7 Disk flowers orange; outer involucral bracts mostly 6–8 . . *Bidens frondosa* . . . . DEVIL'S BEGGAR-TICKS p. 115
7 Disk flowers yellow; outer involucral bracts 10 or more . . . . *Bidens vulgata* . . . . TALL BEGGAR-TICKS p. 116

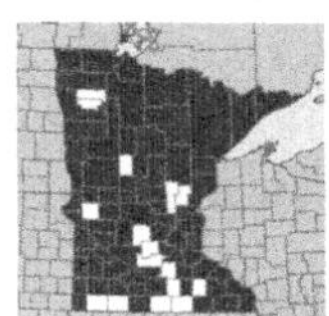

Bidens cernua L.

NODDING BEGGAR-TICKS

NC **OBL** | MW **OBL** | GP **OBL**

Native annual herb. **Stems** often branched, to 1 m long, smooth or with spreading hairs. **Leaves** opposite, smooth, lance-shaped to oblong lance-shaped, 3–16 cm long and 0.5–5 cm wide; margins with sharp, forward-pointing teeth and often rough-to-touch; petioles absent, the leaves usually clasping at base. **Flower heads** many, globe-shaped, 1.5–3 cm wide, usually nodding after flowering; rays yellow, 6–8, to 1.5 cm long, or absent; outer involucral bracts 4–8, unequal in length, the margins often fringed with hairs. **Fruit** a ± straight-sided achene, 5–7 mm long, with downward-pointing barbs on margins; pappus with 4 (sometimes 2) awns, the awns with downward-pointing barbs. July–Oct. **Habitat** Exposed sandy or muddy shores, streambanks, marshes, forest depressions, wet meadows, ditches and other wet places.

Bidens cernua
NODDING BEGGAR-TICKS

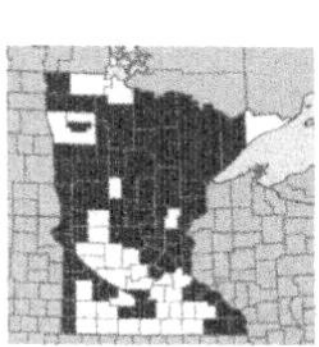

Bidens comosa (A. Gray) Wieg.

STRAWSTEM BEGGAR-TICKS

NC **FACW** | MW **OBL** | GP **FACW**

Native annual herb. **Stems** yellow, 1–12 dm tall, branched, smooth. **Leaves** opposite, lance-shaped to oval, 3–15 cm long and 0.5–5 cm wide, margins with coarse, forward-pointing teeth, rough-to-touch; petioles absent, or leaves tapered to a short, winged petiole. **Flower heads** 1–2.5 cm wide, several to many, remaining erect after flowering; disk flowers yellow-green; rays absent; outer involucral bracts leaflike, 5–10 or more, 2–4x longer than head. **Fruit** an achene, 3–7 mm long, downwardly barbed on the margins; pappus of 3 downwardly barbed awns, the awns shorter than the achenes. Aug–Oct. **Synonyms** *Bidens acuta, Bidens tripartita*. **Habitat** Exposed shores, streambanks, mudflats, forest depressions, pond, wet meadows, ditches and other wet places.

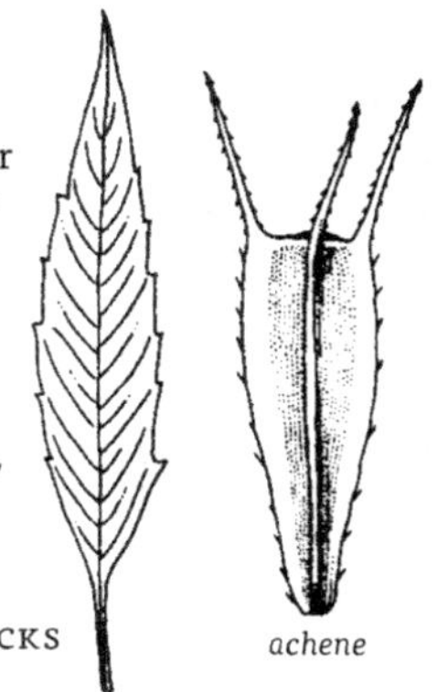

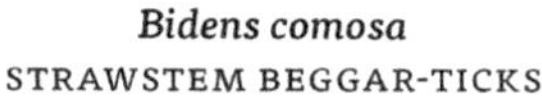

Bidens comosa
STRAWSTEM BEGGAR-TICKS

Bidens connata Muhl.
PURPLESTEM BEGGAR-TICKS
NC **FACW** | MW **OBL** | GP **FACW**

Native annual herb. **Stems** green-purple, to 2 m long, usually branched, smooth. **Leaves** opposite, smooth, the lower leaves sometimes deeply lobed, 3–15 cm long and 1–4 cm wide; margins with coarse, forward-pointing teeth; petioles present on lower leaves, upper leaves short-petioled or stalkless. **Flower heads** several to many, 1–2 cm wide, upright; disk flowers orange-yellow; rays absent, or few and 3–4 mm long; outer involucral bracts 4–9, usually not much longer than the head. **Fruit** an achene, 3–7 mm long; pappus of 2–4 downwardly barbed awns, about half as long as achene. Aug–Oct. **Synonyms** *Bidens tripartita* (in part). **Habitat** Exposed muddy shores, streambanks, marshes, pond, forest depressions, wet meadows, ditches and other wet places.

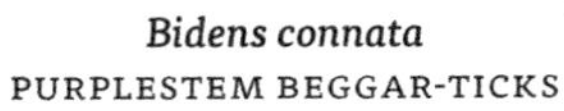

Bidens connata
PURPLESTEM BEGGAR-TICKS

Bidens coronata (L.) Britton
NORTHERN TICKSEED-SUNFLOWER
NC **OBL** | MW **OBL** | GP **OBL**

Native annual or biennial herb. **Stems** branched, 3–15 dm tall, smooth, often purple. **Leaves** opposite, smooth, to 15 cm long, pinnately divided into 3–7 narrow leaflets; margins coarsely toothed or deeply lobed to sometimes entire; petioles 3–15 mm long. **Flower heads** with both disk and ray flowers, large and numerous on slender stalks; rays about 8, gold-yellow, 1–2.5 cm long; outer involucral bracts 6–10, to 1 cm long, short-hairy on margins, inner bracts shorter. **Fruit** a flattened achene, 5–9 mm long, with long, stiff hairs on margins; pappus of 2 short, scalelike awns, 1–2 mm long. July–Oct. **Synonyms** *Bidens trichosperma*. **Habitat** Open bogs, fens, tamarack swamps, shores, streambanks, marshes, sand bars.

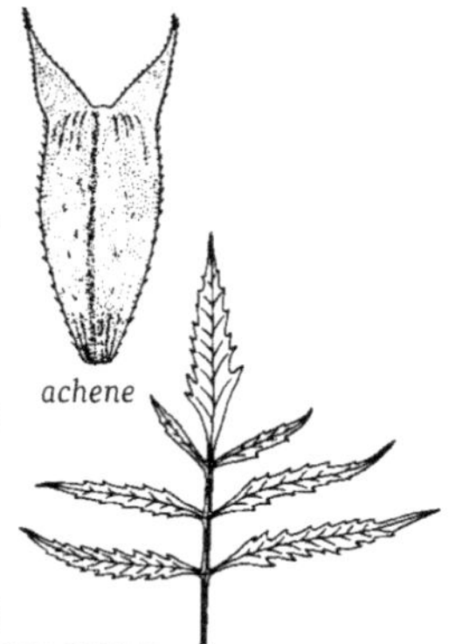

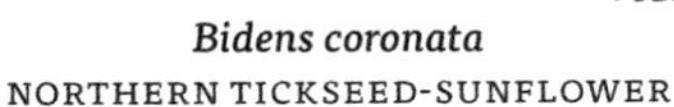

Bidens coronata
NORTHERN TICKSEED-SUNFLOWER

Bidens discoidea (T. & G.) Britton
FEW-BRACTED BEGGAR-TICKS
NC **FACW** | MW **FACW** | GP **FACW**

Native annual herb. **Stems** smooth, 3–10 dm long. **Leaves** opposite, smooth, divided into 3-leaflets, the leaflets lance-shaped, the terminal leaflet largest, to 10 cm long and 4 cm wide; margins with coarse, forward-pointing teeth; petioles slender, 1–6 cm long. **Flower heads** many on slender stalks, the disk to 1 cm wide; rays absent; outer involucral bracts usually 4, leaflike, much longer than disk. **Fruit** a flattened achene, 3–6 mm long; pappus of 2 awns to 2 mm long, with short, upward pointing bristles. Aug–Sept. **Habitat** Hummocks or logs in swamps, exposed muddy shores; usually where shaded.

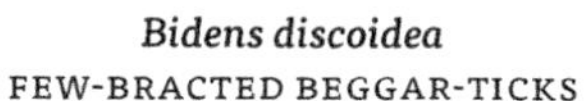

Bidens discoidea
FEW-BRACTED BEGGAR-TICKS

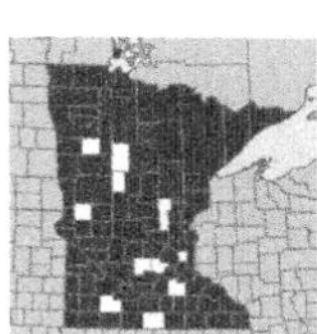

Bidens frondosa L.
DEVIL'S BEGGAR-TICKS
NC **FACW** | MW **FACW** | GP **FACW**

Native annual herb. **Stems** erect, 2–10 dm tall, branched, purple-tinged, ± smooth. **Leaves** pinnately divided into 3–5 segments, the segments lance-shaped, to 10 cm long and 3 cm wide, underside sometimes with short hairs; margins with coarse, forward-pointing teeth; petioles slender, 1–6 cm long. **Flower heads** many on long stalks; disk flowers orange, the disk to 1 cm wide; rays absent or very small; the outer involucral bracts usually 8, green and leaflike, longer than disk, fringed with hairs on margins. **Fruit** a flattened, nearly black achene, 5–10 mm long; pappus of 2 slender awns with downward-pointing barbs. July–Oct. **Habitat** Wet, sandy or gravelly shores, forest depressions, streambanks, pond margins; weedy in wet disturbed areas.

achene

Bidens frondosa
DEVIL'S BEGGAR-TICKS

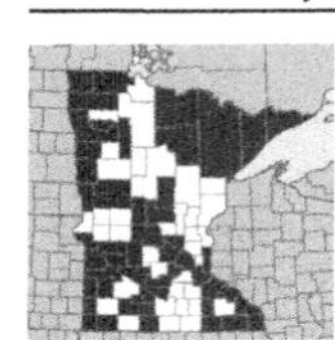

Bidens vulgata Greene
TALL BEGGAR-TICKS
NC **FAC** | MW **FACW** | GP **FAC**

Annual herb. **Stems** to 2 m tall, smooth or upper stem and leaves short-hairy. **Leaves** opposite, pinnately divided into 3–5 segments, the segments lance-shaped, to 15 cm long and 5 cm wide, with prominent veins; margins with sharp, forward-pointing teeth; petioles present. **Flower heads** on stout, leafless stalks, disk flowers yellow; ray flowers usually present, small, yellow; outer involucral bracts about 13, leaflike. **Fruit** a flattened, olive-green or brown achene, 10–12 mm long; pappus of 2 awns with downward-pointing barbs. Aug–Oct. **Synonyms** *Bidens puberula*. **Habitat** Streambanks, wet meadows, wet forests; weedy in moist disturbed areas.

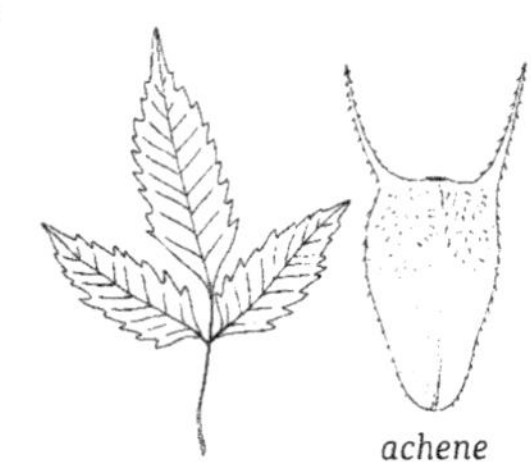

BOLTONIA | *Boltonia*

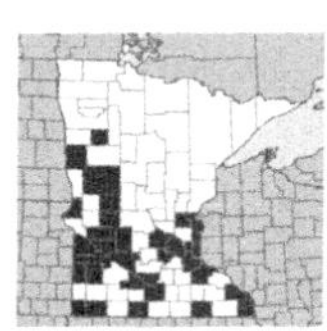

Boltonia asteroides (L.) L'Her.
WHITE BOLTONIA
NC **FACW** | MW **OBL** | GP **FACW**

Native perennial herb, fibrous-rooted, sometimes with shallow rhizomes. **Stems** stout, erect, 3–15 dm long, smooth. **Leaves** alternate, lance-shaped or oval, 5–16 cm long and 0.5–2 cm wide, becoming smaller in the head, narrowed to stalkless or slightly clasping base; margins entire but rough-to-touch. **Flower heads** many, with both disk and ray flowers, 1.5–2.5 cm wide; disk flowers yellow; rays white, pink, or lavender, 5–15 mm long; involucres 2.5–5 mm high, the bracts overlapping, chaffy on outer margins, with a green midvein. **Fruit** a flattened, obovate achene, 1–2 mm long, with a winged margin; pappus of 2 awns and 2–4 shorter bristles. Aug–Sept. **Synonyms** *Boltonia latisquama*. **Habitat** Seasonally flooded muddy shores, wet meadows, marshes, low prairie.

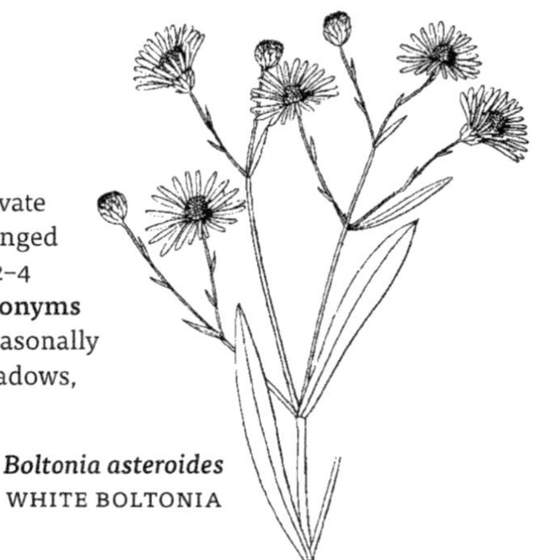
Boltonia asteroides
WHITE BOLTONIA

CACALIA | *Indian plantain*

Large perennial herbs with basal or alternate leaves. **Flower heads** with white disk flowers only, the ray flowers absent. **Fruit** an achene, tipped by a pappus of numerous, slender bristles.

CACALIA | INDIAN PLANTAIN

1 Leaves entire or with a few teeth; involucral bracts 10–15 . . . . *Cacalia plantaginea* | TUBEROUS INDIAN PLANTAIN

1 Leaves sharply toothed; involucral bracts 5 . . . . *Cacalia suaveolens* . . . . SWEET-SCENTED INDIAN PLANTAIN

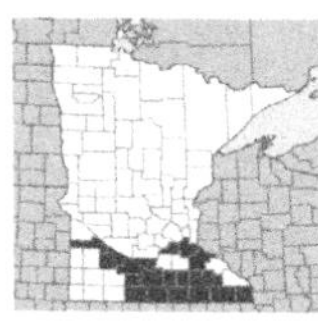

Cacalia plantaginea (Raf.) Shinners

TUBEROUS INDIAN PLANTAIN

NC **FAC** | MW **FAC** | GP **FAC** | *threatened*

Native perennial herb, from a tuberous base and fleshy roots. **Stems** stout, smooth, 5–18 dm long, winged. **Leaves** alternate, mostly at base and lower stem of plant, thick, oval to ovate, strongly 5–9-veined, 5–20 cm long and 2–10 cm wide; margins entire or slightly wavy; petioles long on lower leaves, becoming short or absent on upper leaves. **Flower heads** many, of white disk flowers only, in a branched inflorescence at end of stem; involucral bracts 5, of equal length. **Fruit** a smooth achene; pappus of many, rough white bristles. June–Aug. **Synonyms** *Cacalia tuberosa, Arnoglossum plantagineum, Mesadenia tuberosa*. **Habitat** Marshes, low prairie, fens, sedge meadows, calcareous shores. **Status** Threatened; rare in se Minnesota.

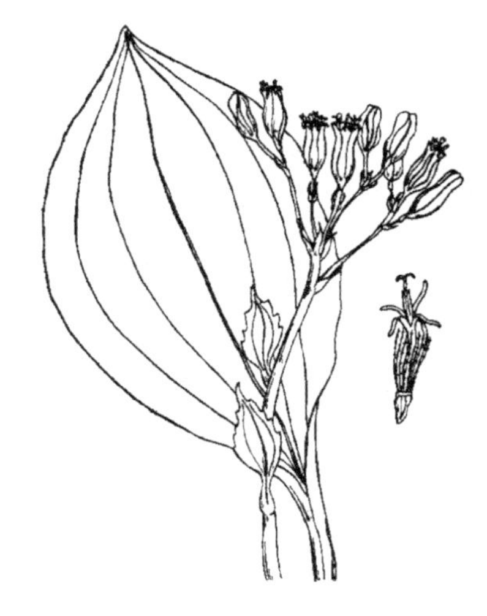

Cacalia plantaginea
TUBEROUS INDIAN PLANTAIN

Cacalia suaveolens L.
SWEET-SCENTED INDIAN PLANTAIN
NC **FACW** | MW **FACW** | *endangered*

Native perennial herb, from fleshy roots. Stems ± smooth, grooved, 1–2.5 m tall, leafy to the inflorescence. Leaves alternate, smooth; lower leaves triangular with a pair of outward-pointing lobes at base, 5–20 cm long and nearly as wide; upper leaves smaller and often not lobed; margins sharply and irregularly toothed; petioles winged. Flower heads of disk flowers only, in a ± flat-topped inflorescence, the disk about 1 cm wide; disk flowers white or light pink; involucre 1 cm long, the main involucral bracts 10–15. Fruit an achene; pappus of many soft, white bristles. July–Sept. Synonyms *Hasteola suaveolens, Synosma suaveolens*. Habitat Riverbanks, shores, calcareous fens, wet low areas. Status Endangered; se Minn.

Cacalia suaveolens
SWEET-SCENTED INDIAN PLANTAIN

CIRSIUM | *Thistle*

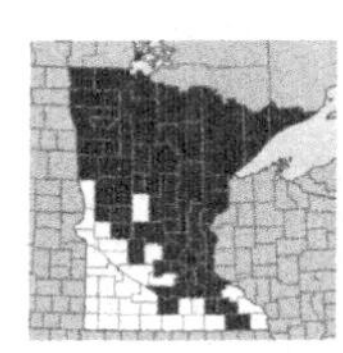

Cirsium muticum Michx.
SWAMP-THISTLE
NC **FAC** | MW **FAC** | GP **FAC**

Stout native biennial herb. **Stems** 0.5–2 m long, branched in head, with long, soft hairs when young, becoming ± smooth. **Leaves** from base of plant or alternate on stem (second year) deeply lobed into pinnate segments, 1–2 dm long, underside often with matted, cobwebby hairs, becoming ± smooth with age; margins toothed and often tipped with spines; petioles present on lower leaves, stem leaves stalkless. **Flower heads** of purple or pink disk flowers only, single on leafless stalks over 1 cm long at ends of stems; involucre 2–3.5 cm high; the involucral bracts overlapping, densely hairy with cottony hairs (especially on margins), sometimes tipped with a short spine 0.5 mm

long. **Fruit** an achene, 5 mm long; pappus of long, slender bristles. Aug–Oct. **Habitat** Swamps, thickets, calcareous fens, sedge meadows, streambanks, shores.

Colonies of **Canada thistle** *(Cirsium arvense, introduced) are sometimes found on wetland margins, especially where the soil is disturbed. It is distinguished from other thistles by its rhizomatous habit and lack of a rosette of basal leaves.*

CREPIS | *Hawk's beard*

Crepis runcinata (James) T. & G.

DANDELION HAWK'S BEARD

NC **FACW** | MW **FACW** | GP **FAC**

Native perennial herb with milky juice. **Stems** 2–6 dm long, smooth or sparsely hairy, the stem leaves small and bractlike. **Leaves** in a rosette at base of plant, oblong lance-shaped to oval, 5–20 cm long and 1–4 cm wide, rounded at tip, tapered to a petiolelike base, margins entire or with widely spaced teeth. **Flower heads** 1–10, 1–2 cm wide, of yellow ray flowers only; involucre 8–15 mm high, with gland-tipped hairs, the involucral bracts in 2 series, the outer bracts shorter than inner. **Fruit** a brown achene, round in section, 4–5 mm long; pappus of many white slender bristles. June–July. **Synonyms** *Crepis glaucella*. **Habitat** Wet meadows, low prairie, swales; especially where alkaline.

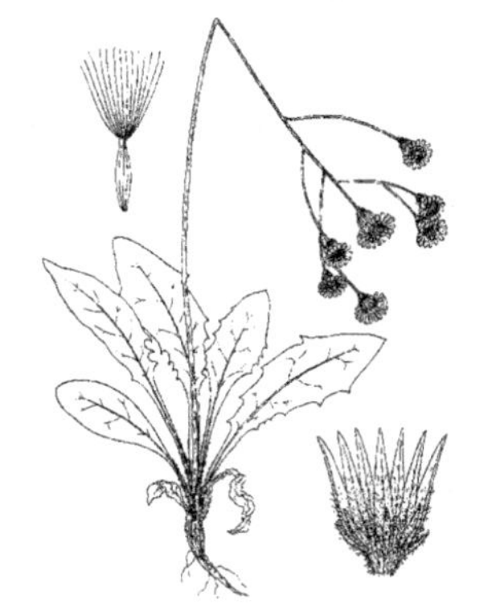

ECLIPTA | *Yerba-de-tajo*

Eclipta prostrata (L.) L.

YERBA-DE-TAJO

NC **FACW** | MW **FACW** | GP **FACW**

Introduced annual herb. **Stems** spreading, branched, 5–8 dm long, with rough, appressed hairs, often rooting at the nodes. **Leaves** opposite, lance-shaped, 2–10 cm long and 0.5–2.5 cm wide, margins with shallow teeth; petioles absent, or short on lower leaves. **Flower heads** with both disk and ray flowers, in clusters of 1–3 at ends of stems or from leaf axils, on stalks or nearly stalkless; the disk 4–6 mm wide; rays short, nearly white. **Fruit** a flat-topped achene, 2–3 mm long; pappus a crown of very short bristles. July–Oct. **Synonyms** *Eclipta alba, Verbesina alba*. **Habitat** Mud flats, muddy stream banks and ditches, where somewhat weedy. In Minnesota, known from a single collection in Washington County.

Eclipta prostrata
YERBA-DE-TAJO

ERIGERON | *Daisy, Fleabane*

Biennial or perennial herbs with simple, alternate **leaves**. **Flower heads** with both disk and ray flowers; disk flowers yellow; rays white to pink, very narrow, only to about 0.5 mm wide; involucral bracts in 1–2 series, linear, about equal in length, green in middle and at base, translucent at tip and on upper margins. **Fruit** a flattened achene; pappus of 20–30 slender, rough bristles.

ERIGERON | DAISY, FLEABANE

1 Leaves clasping stem; common and widespread . . . *Erigeron philadelphicus* . . . PHILADELPHIA DAISY

1 Leaves not clasping stem; uncommon . . . *Erigeron lonchophyllus* . . . LOW MEADOW FLEABANE

Erigeron lonchophyllus Hook.

LOW MEADOW FLEABANE

NC **FACW** | MW **FACW** | GP **FACW** | *special concern*

Native biennial or short-lived perennial herb. **Stems** 1–4 dm tall, with spreading hairs. **Leaves** alternate, lower leaves oblong lance-shaped, 5–15 cm long and 1–5 mm wide, tapered to a short petiolelike base, upper leaves linear and stalkless, not clasping, margins entire, fringed with hairs. **Flower heads** several to many, 1–1.5 cm wide; involucre 5–10 mm high, the involucral bracts coarsely hairy, the outer bracts shorter than inner; rays many, white, turning brown at tip, only to 0.2 mm wide. **Fruit** an achene; pappus of slender, rough bristles. July–Sept. **Synonyms** *Trimorpha lonchophylla*. **Habitat** Wet meadows, low prairie, seeps. **Status** Special concern.

Erigeron philadelphicus L.
PHILADELPHIA DAISY
NC **FAC** | MW **FACW** | GP **FAC**

Native biennial or short-lived perennial herb. **Stems** 1 to several, branched in head, 2–7 dm long, usually long-hairy. **Leaves** alternate, lower leaves spatula-shaped, 5–15 cm long and 1–4 cm wide, tapered to a short petiole; upper leaves smaller, lance-shaped, clasping at base, hairy to nearly smooth, rounded at tip; margins entire or with rounded teeth. **Flower heads** few to many, with both disk and ray flowers, 1.5–2.5 cm wide; involucre 3–6 mm high, the involucral bracts hairy, of ± equal length; rays many, white to deep pink, 5–10 mm long and to 0.5 mm wide. **Fruit** a short-hairy achene; pappus of long rough bristles. May–Aug. **Habitat** Wet meadows, shores, streambanks, wet woods, floodplains, springs; also weedy in open disturbed areas and lawns.

Erigeron lonchophyllus
LOW MEADOW FLEABANE

Erigeron philadelphicus
PHILADELPHIA DAISY

EUPATORIUM | *Joe-pye-weed*

Perennial herbs from a thick rhizome. **Stems** stout, erect. **Leaves** whorled, or opposite and joined at base, the stem passing through the joined leaves; lower leaves smaller; margins toothed. **Flower heads** of pink, purple or white disk flowers only, usually many in a ± flat-topped head at ends of stems; involucral bracts overlapping or nearly equal length. **Fruit** an angled achene; pappus of many slender bristles.

EUPATORIUM | JOE-PYE-WEED

1 Leaves whorled .... *Eupatorium maculatum* | SPOTTED JOE-PYE-WEED
1 Leaves opposite 2

2 Most leaves joined at base and perforated by the stem.................
.... *Eupatorium perfoliatum* | BONESET
2 Leaves on distinct stalks to 5 mm long............. *Eupatorium rugosum*
.... WHITE SNAKEROOT

Eupatorium maculatum L.

SPOTTED JOE-PYE-WEED
NC **OBL** | MW **OBL** | GP **OBL**

Native perennial herb. **Stems** 5–20 dm long, spotted or tinged with purple, short-hairy above, especially on branches of head. **Leaves** in whorls of mostly 4–5, lance-shaped to ovate, 5–20 cm long and 2–7 cm wide, upper surface with sparse short hairs, underside often densely short-hairy; margins with sharp, forward-pointing teeth; petioles to 2 cm long. **Flower heads** of light pink to purple disk flowers only, the inflorescence ± flat-topped; involucres 6–9 mm high, purple-tinged, the involucral bracts overlapping. **Fruit** a black, angled achene, 2–4 mm long; pappus of long, slender bristles. July–Sept. **Synonyms** *Eupatoriadelphus maculatus, Eutrochium maculatum*. **Habitat** Common; wet meadows, marshes, low prairie, shores, streambanks, ditches, cedar swamps, open bogs, calcareous fens.

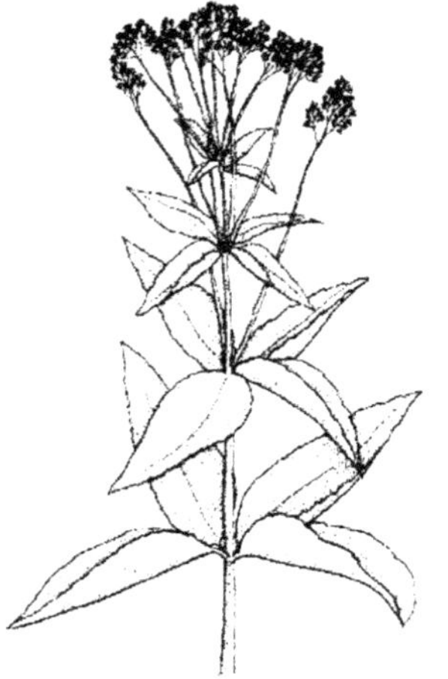

Eupatorium maculatum
SPOTTED JOE-PYE-WEED

Eupatorium perfoliatum L.

BONESET

NC **FACW** | MW **OBL** | GP **FACW**

Native perennial herb. **Stems** 3–15 dm tall, with long, spreading hairs. **Leaves** opposite, mostly joined at the broad base and perforated by the stem (upper leaves sometimes separate), lance-shaped, 6–20 cm long and 1.5–5 cm wide, upper surface sparsely hairy, underside hairy, both sides dotted with yellow glands; margins finely toothed and rough-to-touch; petioles absent. **Flower heads** of dull white disk flowers only, in a flat-topped inflorescence; involucre 3–6 mm high, the involucral bracts green with white margins, hairy, overlapping in 3 series. **Fruit** a black achene, 1–2 mm long; pappus of long slender bristles. July–Sept. **Habitat** Common; marshes, wet meadows, low prairie, shores, streambanks, ditches, cedar swamps, thickets, calcareous fens. Often occurring with spotted joe-pye-weed (*Eupatorium maculatum*).

Eupatorium perfoliatum
BONESET

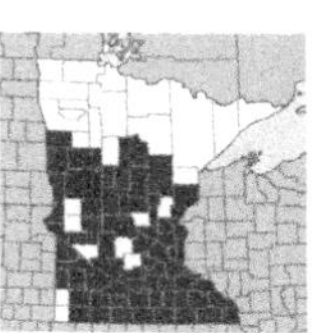

Eupatorium rugosum Houttuyn

WHITE SNAKEROOT

NC **FACU** | MW **FACU** | GP **UPL** | *Caution - toxic*

Native perennial herb. **Stems** 1–3, 3–15 dm long, smooth or with short hairs. **Leaves** opposite, ovate, 5–16 cm long and 3–12 cm wide, smooth or hairy, especially on underside veins, margins coarsely sharp-toothed; petioles 1–3 cm long. **Flower heads** of bright white disk flowers only, in a flat-topped or rounded inflorescence; involucre 3–5 mm high, smooth or short-hairy, the involucral bracts linear, nearly equal, in 1–2 series. **Fruit** a ± smooth achene; pappus of long, slender bristles. July–Oct. **Synonyms** *Eupatorium urticaefolium, Ageratina altissima*. **Habitat** Floodplain forests, cedar swamps, thickets, streambanks, wooded ravines.

White snakeroot *was the cause of "milk sickness" in pioneers who drank the milk from cows which had eaten these plants.*

EUTHAMIA | *Flat-topped goldenrod*

Euthamia graminifolia (L.) Nutt.

COMMON FLAT-TOPPED GOLDENROD

NC **FAC** | MW **FACW** | GP **FACW**

Native perennial herb, spreading by rhizomes. **Stems** erect, 5–15 dm tall, smooth to hairy, usually branched in head. **Leaves** alternate, linear to narrowly lance-shaped or oval, 3–15 cm long and 3–10 mm wide, 3-veined, covered with small glandular dots, smooth or rough-to-touch; margins entire; petioles absent or very short. **Flower heads** small, in flat-topped clusters at ends of stems; with yellow disk and ray flowers, the rays small, to 1 mm long; involucre 3–5 mm high, somewhat sticky, the involucral bracts overlapping in several series, yellow or green-tipped. **Fruit** a finely hairy achene, 1 mm long; pappus of many white, slender bristles. Aug–Sept. **Synonyms** *Solidago graminifolia*. **Habitat** Shores, wet meadows, low prairie, springs, fens, swamps, interdunal wetlands, streambanks, often where sandy or gravelly; also weedy in abandoned fields.

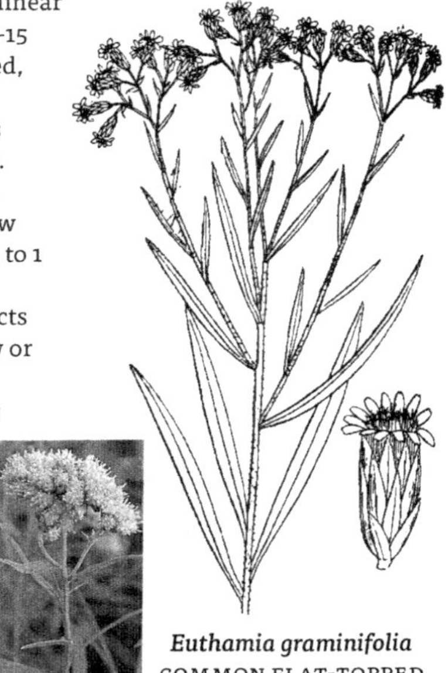

Euthamia graminifolia
COMMON FLAT-TOPPED GOLDENROD

GNAPHALIUM | *Cudweed*

Gnaphalium uliginosum L.

MARSH CUDWEED

NC **FAC** | MW **FAC** | GP **FAC**

Introduced annual herb. **Stems** branched and spreading, 10–25 cm tall, covered with white wool-like hairs. **Leaves** alternate, entire, linear or oblong lance-shaped, 2–4 cm long and to 5 mm wide, with sparse woolly hairs. **Flower heads** of whitish disk flowers only, the rays absent, in numerous clusters from leaf axils and at end of stem-branches, shorter than the subtending leaves. **Fruit** a smooth or bump-covered achene. **Synonyms** *Filaginella uliginosa*. **Habitat** Introduced from Europe and now weedy, especially on streambanks, and in wet to dry disturbed areas.

HELENIUM | *Sneezeweed*

Helenium autumnale L.

COMMON SNEEZEWEED
NC **FACW** | MW **FACW** | GP **FACW**

Native perennial herb. **Stems** single or clustered, erect, 3–13 dm tall, smooth or finely hairy, branched in head. **Leaves** alternate, bright green, lance-shaped to oval, 4–12 cm long and 0.5–3.5 cm wide, usually short-hairy, glandular-dotted; margins entire to shallowly toothed; petioles absent, the blades tapered to a narrow base extending downward as wings on stem. **Flower heads** ± round, 1.5–4 cm wide; few to many on slender stalks in a leafy inflorescence, with both disk and ray flowers, the disk flowers yellow to brown, the rays yellow and 3-lobed, 1.5–2.5 cm long; involucral bracts in 2–3 series, linear, short-hairy, bent downward with age. **Fruit** a finely hairy, 4–5-angled achene, 1–2 mm long; pappus of several translucent, awn-tipped scales. July–Sept. **Habitat** Wet meadows, shores, streambanks, marshes, fens, tamarack swamps.

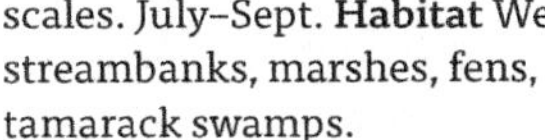

Gnaphalium uliginosum
MARSH CUDWEED

Helenium autumnale
COMMON SNEEZEWEED

HELIANTHUS | *Sunflower*

Large perennial herbs (those included here), with fibrous or fleshy roots and short to long rhizomes. **Stems** unbranched or branched above. **Leaves** usually opposite on lower part of stem and alternate above, lance-shaped, margins entire or with forward-pointing teeth; petioles present. **Flower heads** large, mostly 1 to several (rarely many), at ends of stems and branches, with yellow disk and ray flowers, the rays large and showy; involucre of several series of narrow, overlapping bracts; receptacle chaffy. **Fruit** a flattened achene; pappus of 2 deciduous, awn-tipped scales.

HELIANTHUS | SUNFLOWER

1 Upper side of leaf rough-to-touch; stems hairy ...... *Helianthus giganteus* .... SWAMP-SUNFLOWER

1 Upper side of leaf not (or only slightly) rough-to-touch; stems smooth, often waxy-coated, or sometimes with sparse hairs on upper stem .... *Helianthus grosseserratus* | SAWTOOTH SUNFLOWER

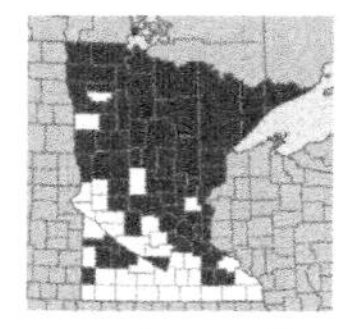

Helianthus giganteus L.

SWAMP-SUNFLOWER

NC **FACW** | MW **FACW** | GP **FAC**

Native perennial herb, with short rhizomes and thick, fleshy roots. **Stems** 1–3 m long, often purple, with coarse hairs or sometimes nearly smooth, often branched in head. **Upper leaves** generally alternate, lower leaves opposite; lance-shaped, 6–20 cm long and 1–4 cm wide, base with 3 main veins, upper surface very rough-to-touch, underside with short, stiff hairs; margins toothed to ± entire; petiole short or absent. **Flower heads** 3–6 cm wide, several to many, on long stalks in an open inflorescence; with yellow disk and ray flowers, the rays 1.5–3 cm long; involucral bracts narrow, awl-shaped, green or dark near base, hairy or margins fringed with hairs. **Fruit** a smooth achene; pappus of 2 awl-shaped scales. July–Sept. **Synonyms** *Helianthus subtuberosus*. **Habitat** Wet meadows, low prairie, sedge meadows, fens, floodplain forests, streambanks.

Helianthus giganteus
SWAMP-SUNFLOWER

Helianthus grosseserratus Martens

SAWTOOTH SUNFLOWER

NC **FACW** | MW **FACW** | GP **FACW**

Native perennial herb, with fleshy roots, spreading by rhizomes and forming colonies. **Stems** 1–3 m tall, short-hairy in head, smooth and often waxy below, purple or blue-green. **Upper leaves** alternate, **lower leaves** opposite; lance-shaped, 10–20 cm long and 2–5 cm wide, rough-to-touch on both sides, also densely short hairy on the paler underside; margins with coarse, forward-pointing teeth, upper leaves often entire; petioles 1–4 cm long. **Flower heads** 3–8 cm wide, several to many at ends of stems and branches; with yellow disk flowers and deep yellow ray flowers, the rays 2.5–4 cm long; involucral bracts narrowly lance-shaped, fringed with hairs and sometimes hairy on back. **Fruit** a smooth achene, 3–4 mm long; pappus of 2 lance-shaped scales. July–Oct. **Habitat** Wet meadows, low prairie, streambanks, swamps, ditches, roadsides.

Helianthus grosseserratus
SAWTOOTH SUNFLOWER

LIATRIS | *Blazing star*

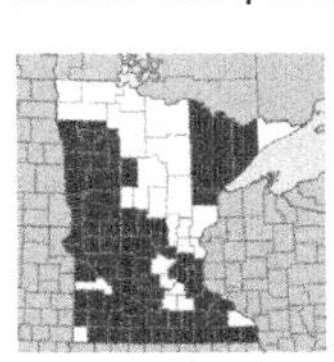

Liatris pycnostachya Michx.

PRAIRIE BLAZING STAR
NC **FAC** | MW **FAC** | GP **FAC**

Native perennial herb. **Stems** arising from a corm, stout, unbranched, to 1.2 m tall, with numerous small longitudinal ridges, stems usually with scattered white hairs. **Leaves** alternate, entire, with prominent central vein; basal leaves to 25 cm long and about 1.5 cm wide; stem leaves linear, upright, progressively smaller upward on stem. **Flowers** in a terminal spike of numerous stalkless heads, each head with 5–10 pink to purple-pink disk flowers; ray flowers absent; styles 2, long and curved; each head subtended by green or reddish bracts, recurved sharply outward at tip. **Fruit** an achene tufted with brown hairs. Late summer. **Habitat** Moist prairie, calcareous fens.

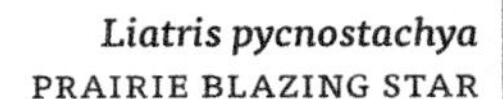

Liatris pycnostachya
PRAIRIE BLAZING STAR

MEGALODONTA | *Water beggar-ticks*

Megalodonta beckii (Torr. ex Spreng.) Greene
WATER BEGGAR-TICKS
NC **OBL** | MW **OBL** | GP **OBL**

Native perennial aquatic herb. **Stems** 0.4–2 m long, little-branched. **Underwater leaves** opposite or whorled, dissected into threadlike segments; **emersed leaves** simple, opposite, lance-shaped to ovate, margins with forward-pointing teeth, petioles absent. **Flower heads** single or few at ends of stems; rays 6–10, gold-yellow, 1–1.5 cm long, notched at tip; involucral bracts smooth. **Fruit** an achene, ± round in section, 10–15 mm long; pappus of 3–6 slender awns, longer than achenes, the upper portion of awn with downward-pointing barbs. June–Sept. **Synonyms** *Bidens beckii*. **Habitat** Quiet, shallow to deep water of lakes, ponds, rivers and streams.

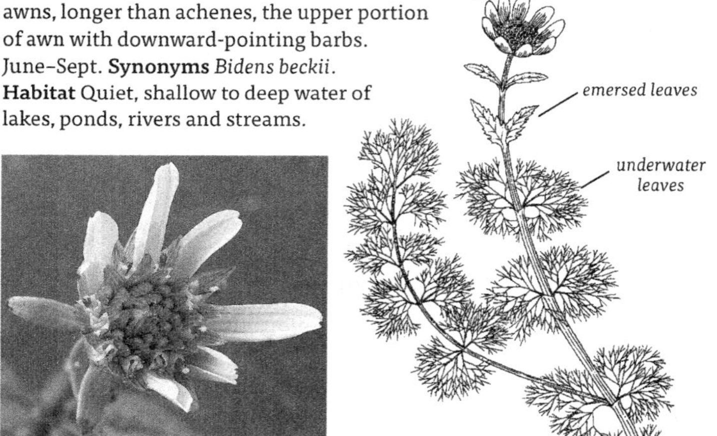

Megalodonta beckii
WATER BEGGAR-TICKS

PETASITES | *Sweet coltsfoot*

Perennial herbs, spreading by rhizomes. **Leaves** mostly from base of plant on long petioles, arrowhead-shaped or palmately lobed, with white woolly hairs on underside; stem leaves alternate, reduced to bracts. Flowering before or as leaves expand in spring, the heads white, the **flowers** mostly either male and female and on different plants; or the heads sometimes with both male and female flowers, the male heads usually with disk flowers only, the female heads with all disk flowers or sometimes with short rays; involucral bracts in a single series; receptacle not chaffy. **Fruit** a linear, ribbed achene; pappus of many white, slender bristles.

PETASITES | SWEET COLTSFOOT

1 Leaf blades palmately lobed . . . . *Petasites frigidus* . . . . NORTHERN SWEET COLTSFOOT

1 Leaf blades arrowhead-shaped, toothed and not lobed . . . . *Petasites sagittatus* | ARROWHEAD SWEET COLTSFOOT

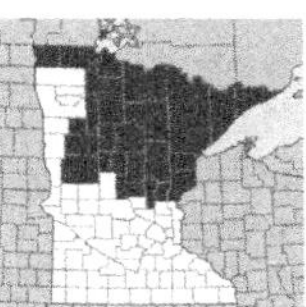

Petasites frigidus (L.) Fries.
NORTHERN SWEET COLTSFOOT
NC **FACW** | MW **FACW** | GP **FAC**

Native perennial herb, spreading by rhizomes. **Stems** 1–6 dm long, smooth or short-hairy in the head. **Leaves** mostly from base of plant, triangular to nearly round in outline, palmately lobed, 5–30 cm wide, upper surface green and smooth, underside densely white-hairy, sometimes becoming smooth with age; margins coarsely toothed; petioles of basal leaves 1–3 dm long; stem leaves small and bractlike, 2–6 cm long. **Flower heads** nearly white, male and female flowers mostly on separate plants; rays of female heads to 7 mm long, involucre 4–9 mm high. **Fruit** a narrow achene; pappus of many slender bristles. May–June. **Habitat** Wet conifer forests and swamps, wet trails and clearings, aspen woods.

Petasites frigidus
NORTHERN SWEET COLTSFOOT

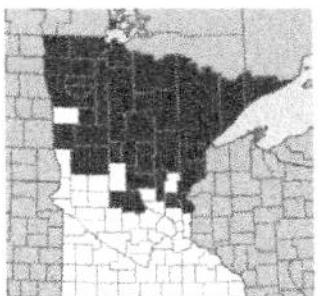

Petasites sagittatus (Banks) A. Gray
ARROWHEAD SWEET COLTSFOOT
NC **FACW** | MW **FACW** | GP **FAC**

Native perennial herb. **Stems** 3–6 dm tall, sparsely covered with woolly white hairs. **Leaves** mostly from base of plant, arrowhead-shaped, 10–40 cm long and 3–30 cm wide, upper surface smooth to sparsely hairy, densely white hairy below; margins wavy with outward-pointing teeth; petioles 1–3 dm long; the stem leaves reduced in size. **Flower heads** ± white; rays of female heads 8–9 mm long. **Fruit** a linear achene; pappus of slender bristles. May–June. **Synonyms** *Petasites frigidus* var. *sagittatus*. **Habitat** Wet meadows, marshes, sedge meadows, open swamps.

PRENANTHES | *Rattlesnake-root*

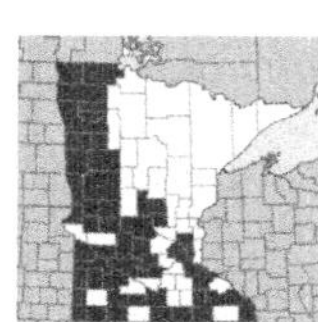

Prenanthes racemosa Michx.

GLAUCOUS RATTLESNAKE-ROOT

NC **FACW** | MW **FACW** | GP **FACU**

Native perennial herb. Stems slender, erect, ridged, 4–18 dm tall, smooth and somewhat waxy, hairy in the head. **Leaves** thick, smooth and waxy; lower leaves oval to obovate, 10–20 cm long and 2–10 cm wide; margins shallowly toothed; petioles long and winged; stem leaves becoming smaller upwards, stalkless and partly clasping the stem. **Flower heads** many in a narrow, elongate inflorescence, of ray flowers only, pink or purplish; involucre 9–14 mm high, purple-black, long-hairy. **Fruit** a linear achene; pappus of straw-colored bristles. Aug–Sept. **Habitat** Sandy or gravelly shores, streambanks, wet meadows, low prairie, fens.

White rattlesnake-root *(Prenanthes alba), typically found in deciduous woods, sometimes occurs in wet woods or along streams in Minnesota. It differs from Prenanthes racemosa by its usually deeply lobed leaves and hairless involucre.*

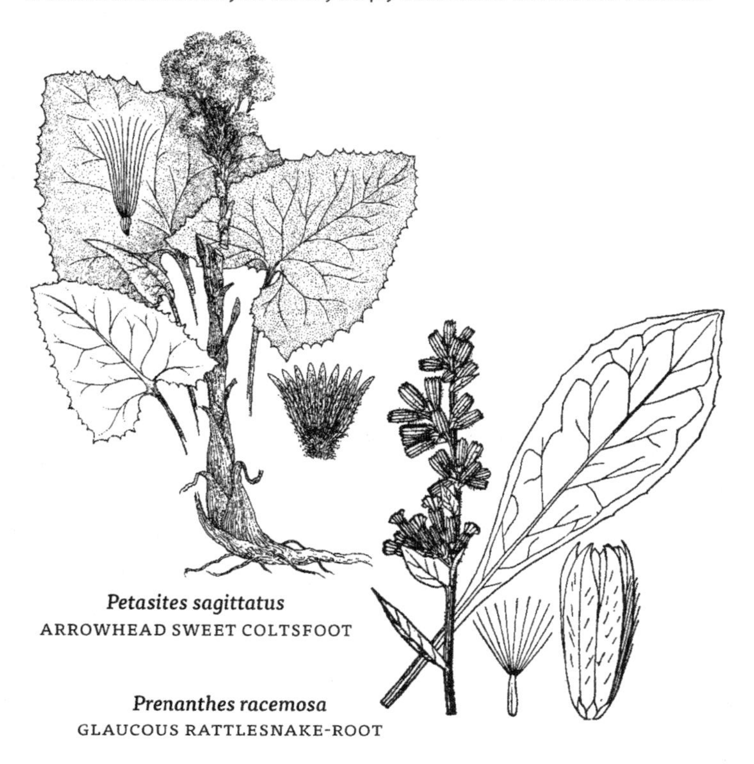

Petasites sagittatus
ARROWHEAD SWEET COLTSFOOT

Prenanthes racemosa
GLAUCOUS RATTLESNAKE-ROOT

RUDBECKIA | *Coneflower*

Perennial herbs. Stems and leaves rough-hairy. **Leaves** alternate, entire to deeply lobed. **Flower heads** with both disk and ray flowers, the rays yellow to orange; involucral bracts green, overlapping; receptacle rounded, chaffy. **Fruit** a smooth, 4-angled achene; pappus absent or a short crown.

RUDBECKIA | CONEFLOWER

1 Largest leaves 3 lobed; disk dark purple .... *Rudbeckia triloba*
.... BROWN-EYED SUSAN

1 Largest leaves 5–7 lobed; disk green-yellow .... *Rudbeckia laciniata*
.... TALL OR CUTLEAF CONEFLOWER

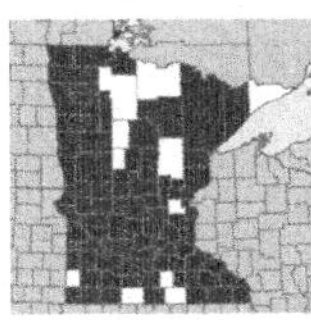

Rudbeckia laciniata L.

TALL OR CUTLEAF CONEFLOWER
NC **FACW** | MW **FACW** | GP **FAC**

Native perennial herb, from a woody base. **Stems** branched, 0.5–3 m long, smooth and often waxy. **Leaves** alternate, to 30 cm wide, deeply lobed, nearly smooth to hairy on underside; margins coarsely toothed as well as lobed, or entire on upper leaves; petioles long on lower leaves, becoming short above. **Flower heads** several to many at ends of stems, with both disk and ray flowers, disk flowers green-yellow, rays lemon-yellow, drooping, 3–6 cm long; involucral bracts of unequal lengths; receptacle round at first, becoming cylindric. **Fruit** a 4-angled achene. July–Sept. **Habitat** Floodplain forests, swamps, streambanks, thickets, ditches; usually in partial or full shade.

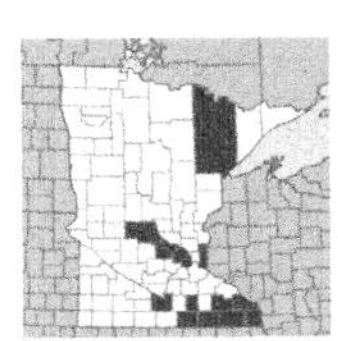

Rudbeckia triloba L.

BROWN-EYED SUSAN
NC **FACU** | MW **FACU** | GP **FACU** | *special concern*

Native perennial herb. **Stems** 0.5–1.5 m long, with coarse spreading hairs or sometimes nearly smooth. **Leaves** alternate, thin, coarsely hairy (or sometimes nearly hairless), the basal leaves broadly ovate to heart-shaped, with long petioles; the stem leaves narrower, with short petioles or the petioles absent, usually with some leaves deeply trilobed; margins sharp-toothed or nearly entire. **Flower heads** several to many at ends of stems, with both disk and ray flowers, the disk dark purple, rays 6–12, yellow

but usually orange at base, 1–3 cm long; involucral bracts of unequal lengths, green and leaflike. **Fruit** a 4-angled achene. July–Sept. Ours are var. *triloba*. **Habitat** Edges of wet forests and marshes, wet prairie, shorelines. **Status** Special concern.

Rudbeckia triloba
BROWN-EYED SUSAN

SENECIO | *Groundsel*

Erect perennial or annual herbs. **Leaves** alternate or from base of plant, stalked near base, stalkless and usually smaller upward. **Flower heads** with both disk and ray flowers, few to many in clusters at ends of stems; disk flowers perfect and yellow, the rays yellow; involucral bracts in 1 series and not overlapping, of equal lengths; receptacle flat or convex, not chaffy. **Fruit** an achene, nearly round in section; pappus of slender bristles.

SENECIO | GROUNDSEL

1 Plants annual or biennial, hollow-stemmed; basal rosette of leaves absent, leaves ± similar in size and shape . . . . . *Senecio congestus* . . . . . NORTHERN SWAMP GROUNDSEL

1 Plants perennial, solid-stemmed; basal leaves crowded and larger than stem leaves 2

2 Heads with disk flowers only . . . . . *Senecio indecorus* . . . . . TALLER DISCOID GROUNDSEL

2 Heads with both ray and disk flowers 3

3 Basal leaves heart-shaped at base . . . . . *Senecio aureus* . . . . . HEART-LEAVED GROUNDSEL

3 Basal leaves ovate . . . . . *Senecio pseudaureus* . . . . . WESTERN HEART-LEAVED GROUNDSEL

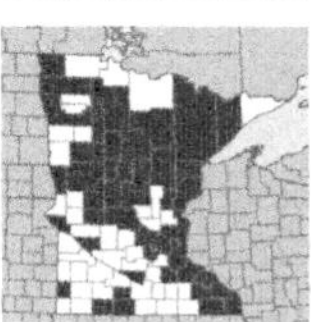

Senecio aureus L.
HEART-LEAVED GROUNDSEL
NC **FACW** | MW **FACW** | GP **FACW**

Native erennial herb, from a spreading crown or rhizome. **Stems** single or clumped, 3–8 dm long, slightly hairy when young, soon becoming smooth. **Basal leaves** heart-shaped, 5–10 cm long and to as wide, often purple-tinged, on long petioles, the margins with rounded teeth; **stem leaves** much smaller and ± pinnately lobed, becoming stalkless. **Flower heads** several to many, the disk 5–10 mm wide, rays gold-yellow, 6–13 mm long involucre 5–8 mm high, the involucral bracts often purple-tipped. **Fruit** a smooth achene; pappus of slender white bristles. May–July. **Synonyms** *Senecio gracilis, Packera aurea*. **Habitat** Floodplain forests, wet forest depressions, swamp openings and hummocks, sedge meadows, thickets, fens, ditches.

Senecio aureus
HEART-LEAVED GROUNDSEL

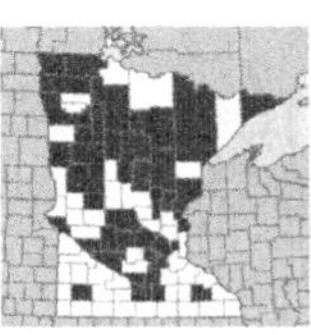

Senecio congestus (R. Br.) DC.
NORTHERN SWAMP GROUNDSEL
[NC **FACW** | MW **FACW** | GP **FACW**]

Native annual or biennial herb. **Stems** stout, single, 2–10 dm long, hollow near base, sparsely to densely hairy. **Leaves** lance-shaped to oblong or the lower spatula-shaped, 5–20 cm long and 0.5–5 cm wide, smooth or hairy, rounded at tip; margins entire to coarsely toothed or cleft; lower leaves stalked, upper leaves stalkless and clasping at base. **Flower heads** 1–1.5 cm wide, usually many in crowded clusters; with both disk and ray flowers, the rays pale yellow, 4–9 mm long; involucre 4–8 mm high, the involucral bracts chaffy near tip. **Fruit** a smooth achene, 1–3 mm long; pappus bristles white, very slender and numerous, lengthening after flowering. May–Aug. **Habitat** Shores and mud flats.

Senecio congestus
NORTHERN SWAMP GROUNDSEL

Senecio indecorus Greene
TALLER DISCOID GROUNDSEL
NC **FACW** | GP **FACW** | *special concern*

Native perennial herb. **Stems** 3–8 dm long, smooth apart from woolly hairs in leaf axils. **Basal leaves** ovate, 3–6 cm long and 2–4 cm wide, margins with coarse, forward-pointing teeth, the petioles longer than blades; **stem leaves** few, much smaller, deeply cleft, ± stalkless. **Flower heads** few to several on slender stalks and forming a rounded cluster; with yellow disk flowers only; involucre 6–10 mm high, the involucral bracts often purple-tipped. **Fruit** a smooth achene; pappus of long slender bristles. July–Aug. **Synonyms** *Senecio discoideus, Packera indecora*. **Habitat** Cedar swamps, moist mixed conifer and deciduous forests, rocky Lake Superior shores, streambanks. **Status** Special concern.

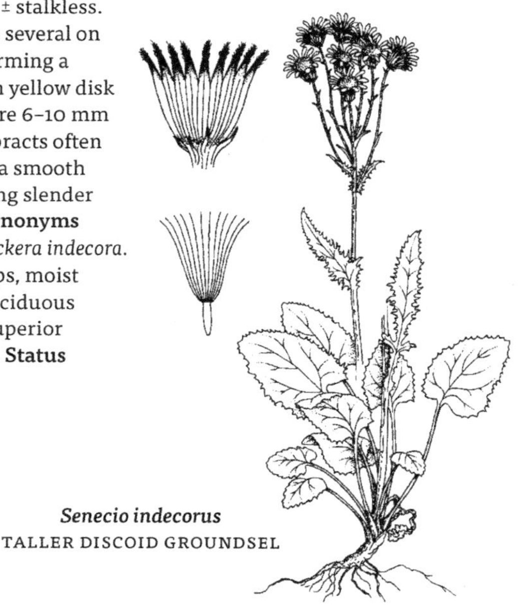

Senecio indecorus
TALLER DISCOID GROUNDSEL

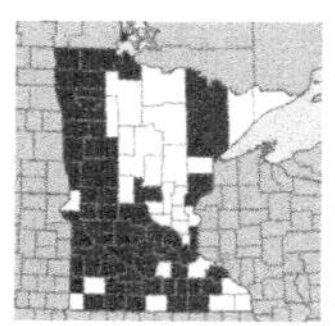

Senecio pseudaureus Rydb.
WESTERN HEART-LEAVED GROUNDSEL
NC **FACW** | MW **FACW** | GP **FACW**

Native perennial herb, from a crown or short rhizome. **Stems** single or few, solid, 2–5 dm long, smooth or with tufts of woolly hairs in leaf axils when young. **Basal leaves** ovate to oval, 2–4 cm long and 1–2 cm wide, underside often purple, margins with rounded teeth, petioles long and slender; **stem leaves** 2–6 cm long and 0.5–2 cm wide, pinnately cleft at least near base, stalkless and often clasping. **Flower heads** 1–1.5 cm wide, few to many in a single cluster; with both disk and ray flowers, the rays pale yellow, 6–10 mm long; involucre 4–7 mm high, the involucral bracts green. **Fruit** a smooth achene, 1–2 mm long; pappus of white bristles. May–July. **Synonyms** *Packera pseudaurea*. **Habitat** Wet meadows, low prairie, fens.

SILPHIUM | *Cup-plant, Rosin-weed*

Silphium perfoliatum L.
CUP-PLANT; INDIAN-CUP
NC **FACW** | MW **FACW** | GP **FAC**

Tall native perennial herb, with resinous juice, spreading by rhizomes. **Stems** erect, 4-angled, smooth, 1–2.5 m long. **Leaves** opposite, broadly ovate, 8–30 cm long and 4–15 cm wide, rough-to-touch, margins coarsely toothed, the lower leaves often short-stalked and joined by wings on the petioles; upper leaves joined at base, forming a cup around stem. **Flower heads** several to many in an open inflorescence, with both disk and ray flowers, the disk 1.5–2.5 cm wide, the rays yellow, 1.5–2.5 cm long; involucre 1–2.5 cm high, the involucral bracts ovate, nearly equal, fringed with hairs on margins; receptacle flat, chaffy. **Fruit** a flat, obovate achene, 8–10 mm long and 5–6 mm wide, the margins narrowly winged; pappus absent. July–Sept. **Habitat** Floodplain forests, streambanks, springs.

Silphium perfoliatum
CUP-PLANT; INDIAN-CUP

Senecio pseudaureus
WESTERN HEART-LEAVED GROUNDSEL

SOLIDAGO | *Goldenrod*

Erect perennials, spreading by rhizomes or from a crown. **Leaves** alternate, margins entire or toothed. **Flower heads** small, many, in flat-topped (corymblike), rounded (paniclelike) or spikelike clusters at ends of stems; the flowers sometimes mostly on 1 side of inflorescence branches (secund) in species with paniclelike heads; the heads with yellow disk and ray flowers; involucral bracts in several overlapping series, papery at base and tipped with green; receptacle flat or convex, not chaffy. **Fruit** an achene, angled or nearly round in cross-section; pappus of many slender white bristles.

SOLIDAGO | GOLDENROD

1 Heads in a ± flat-topped cluster at end of stems . . . . *Solidago riddellii* . . . . RIDDELL'S GOLDENROD
1 Heads in an elongate or pyramid-shaped cluster 2

2 Flower heads spiraled around branches of inflorescence . . . . *Solidago uliginosa* | NORTHERN BOG GOLDENROD
2 Flower heads mostly on upper side of branches of inflorescence. 3

3 Stem leaves with 3 prominent veins . . . . *Solidago gigantea* . . . . SMOOTH GOLDENROD
3 Stem leaves with strong midvein only, not 3-veined . . . . *Solidago uliginosa* . . . . NORTHERN BOG GOLDENROD

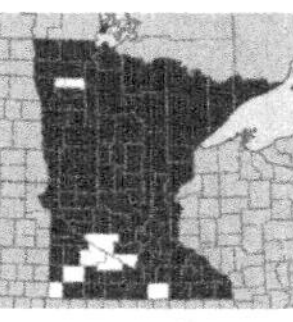

Solidago gigantea Aiton
SMOOTH GOLDENROD
NC **FACW** | MW **FACW** | GP **FAC**

Native perennial herb, from stout rhizomes, often forming colonies. **Stems** 0.5–2 m long, mostly smooth, sometimes waxy, short-hairy on upper branches. **Leaves** alternate, lance-shaped to oval, 6–15 cm long and 1–4 cm wide, prominently 3-veined, tapered to an stalkless or short, petiolelike base, smooth, or sparsely hairy on underside veins; margins with sharp, forward-pointing teeth. **Flower heads** many, in large panicle-like clusters, on 1 side of the spreading branches, with yellow disk and ray flowers, the rays 2–3 mm long; involucre 2–5 mm high, the involucral bracts linear. **Fruit** a short-hairy achene, 1–2 mm long; pappus of slender white bristles. July–Sept. **Synonyms** *Solidago serotina*. **Habitat** Common;

Solidago gigantea
SMOOTH GOLDENROD

wet meadows, streambanks, swamps, floodplain forests, thickets, marshes, calcareous fens, ditches; also in moist to dry open woods and roadsides.

Canada goldenrod *(Solidago canadensis) is a common species that sometimes occurs in moist to wet open areas (as well as drier places). It is similar to smooth goldenrod but generally smaller and densely short-hairy on leaf undersides and upper stems.*

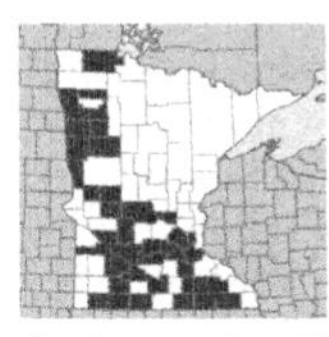

Solidago riddellii Frank
RIDDELL'S GOLDENROD
NC **OBL** | MW **OBL** | GP **OBL**

Native perennial herb, from a crown and sometimes also with rhizomes. **Stems** 2–10 dm long, smooth but sometimes sparsely hairy in head. **Leaves** alternate, smooth, largest at base of plant, these often early-deciduous, lance-shaped to linear, 10–20 cm long and 5–30 mm wide, becoming smaller upward, the upper leaves sickle-shaped and folded along midrib; margins entire; petioles of lower leaves long and winged, upper leaves stalkless and clasping stem. **Flower heads** many, crowded in a branched, rounded to flat-topped inflorescence, the heads not confined to 1 side of the branches, with yellow disk and ray flowers, the rays 1–2 mm long; involucre 5–6 mm high, the involucral bracts rounded at tip. **Fruit** a smooth achene, 1–2 mm long; pappus of slender bristles. Aug–Oct. **Synonyms** *Oligoneuron riddellii*. **Habitat** Wet meadows, calcareous fens, low prairie, lakeshores, streambanks.

Solidago riddellii
RIDDELL'S GOLDENROD

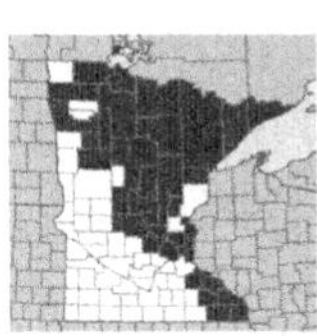

Solidago uliginosa Nutt.
NORTHERN BOG GOLDENROD
NC **OBL** | MW **OBL** | GP **OBL**

Native perennial herb, from a branched crown. **Stems** stout, 5–15 dm long, smooth but finely hairy in the head. **Leaves** alternate, largest at base of plant, 5–35 cm long and 1–5 cm wide, becoming smaller upward, lance-shaped to oblong lance-shaped, smooth, margins finely toothed, or entire on upper leaves, rough-to-touch; lower leaves tapered to long petioles, somewhat clasping stem, upper leaves stalkless. **Flower heads** in a long, crowded spikelike inflorescence, the branches ascending, straight or curved downward at tip, the heads sometimes mostly on 1 side of branches, with yellow disk and ray flowers; involucre 3–5 mm high, the inner involucral bracts rounded at tip, the outer often acute. **Fruit** a ± smooth achene; pappus of slender bristles.

Aug–Sept. **Habitat** Conifer swamps, fens, open bogs, low prairie, wet meadows, interdunal wetlands, Lake Superior rocky shore.

VERNONIA | *Ironweed*

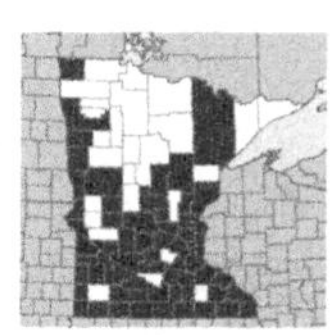

Vernonia fasciculata Michx.

SMOOTH IRONWEED

NC **FACW** | MW **FACW** | GP **FAC**

Stout native perennial herb, from a thick rootstock. **Stems** erect, single or clumped, 5–12 dm long, red or purple, smooth but short-hairy on branches of the head. **Leaves** alternate, lance-shaped, 5–15 cm long and 1–4 cm wide, smooth above, underside finely pitted, margins sharp-toothed, petioles short. **Flower heads** usually many, crowded in flat-topped clusters to 10 cm wide, with purple disk flowers only; involucre 6–9 mm high, the involucral bracts overlapping, green with purple tips; receptacle flat, not chaffy. **Fruit** a ribbed achene, 3–4 mm long; pappus of purple to brown, slender bristles. July–Sept. **Habitat** Marshes, low prairie, streambanks.

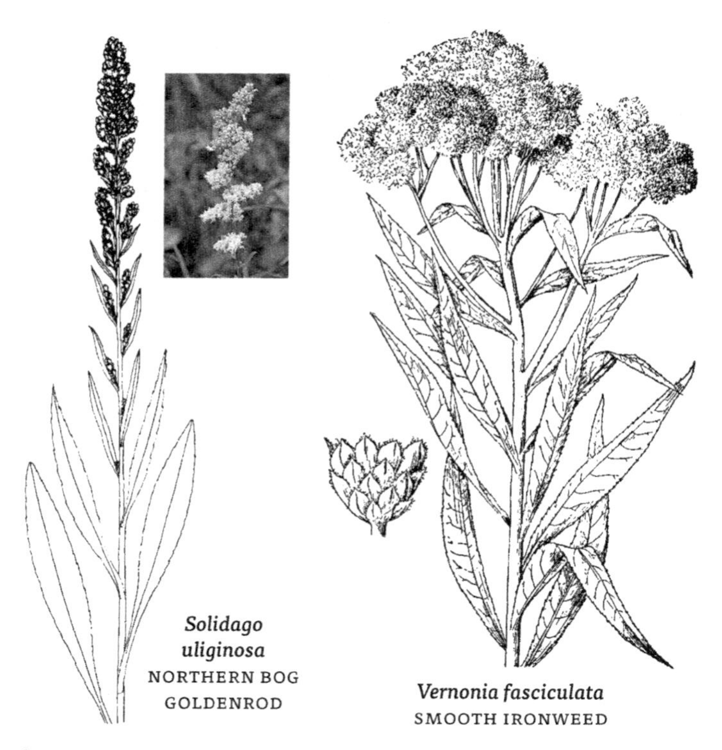

Solidago uliginosa
NORTHERN BOG GOLDENROD

Vernonia fasciculata
SMOOTH IRONWEED

XANTHIUM | *Cocklebur*

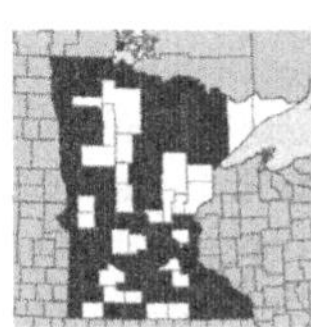

Xanthium strumarium L.

COMMON COCKLEBUR
NC **FAC** | MW **FAC** | GP **FAC**

Weedy taprooted annual herb, native; plants variable in size and habit, rough-to-touch or sometimes nearly smooth. **Stems** 2–15 dm long, often brown-spotted. **Leaves** alternate, ovate to nearly round, sometimes with 3–5 shallow lobes, 3–15 cm long and 2–20 cm wide, margins with blunt teeth; petioles 3–10 cm long. **Flower heads** either male or female, the male flowers brown, in clusters of small round heads at ends of stems above the larger female heads; female heads in several to many clusters from leaf axils, each head with 2 flowers, with a spiny involucre enclosing the head; petals absent. **Fruit** a brown bur formed by the involucre, 1.5–3 cm long, covered with hooked prickles; achenes thick, 1 in each of the 2 chambers of the bur. Aug–Sept. **Habitat** Shores, streambanks, wet meadows, sand bars, dried depressions, often where disturbed; also in cultivated and abandoned fields, roadsides and waste places. Widespread and weedy across North America.

Xanthium strumarium
COMMON COCKLEBUR

IMPATIENS | *Touch-me-not*

Smooth annual herbs with hollow, succulent stems and shallow, weak roots. **Leaves** simple, alternate, the blades shallowly toothed. **Flowers** with both male and female parts, irregular, yellow to orange-yellow, pouchlike and spurred, hanging from the petioles in few- flowered racemes from upper leaf axils; sepals 3, petal-like; petals 3; stamens 5. **Fruit** a 5-valved capsule; the mature capsules splitting when jarred or touched, scattering the seeds away from parent plants.

Small, **cleistogamous** *(self-fertile) flowers lacking petals are sometimes produced in summer and are often the only flowers found on plants growing in shaded situations.*

IMPATIENS | TOUCH-ME-NOT

1 Flowers orange-yellow, usually with red-brown spots . . *Impatiens capensis* . . . . ORANGE TOUCH-ME-NOT; JEWEL-WEED

1 Flowers pale yellow, spots faint or absent . . . . *Impatiens pallida* . . . . YELLOW TOUCH-ME-NOT

Impatiens capensis Meerb.

ORANGE TOUCH-ME-NOT; JEWEL-WEED
NC **FACW** | MW **FACW** | GP **FACW**

Native annual herb. **Stems** 3–10 dm long, usually branched above. **Leaves** ovate to oval, 3–9 cm long and 1.5–4 cm wide, tapered to tip or rounded and tipped with a short slender point, margins shallowly and irregularly toothed; petioles longest on lower leaves, shorter upward, 0.5–5 cm long. **Flowers** orange-yellow, 1.5–3 cm long, usually mottled with red-brown spots, with a spur recurved parallel to the sac and to half its length. **Fruit** a capsule about 2 cm long, splitting when mature to forcefully eject the seeds. July–Sept.

Synonyms *Impatiens biflora*. **Habitat** Swamps, low areas in woods, floodplain forests, thickets, streambanks, shores, marshes, fens, springs; often where disturbed.

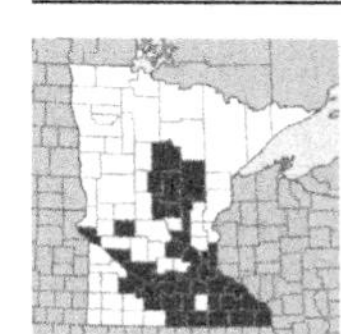

Impatiens pallida Nutt.
YELLOW TOUCH-ME-NOT
NC **FACW** | MW **FACW** | GP **FACW**

Native annual herb, similar to orange touch-me-not (*Impatiens capensis*) but much less common. *Impatiens pallida* is typically larger, the leaves to 12 cm long and 8 cm wide, and more finely toothed than those of *I. capensis*. **Flowers** pale yellow, unspotted or with faint red-brown spots, 2–4 cm long, the spur recurved at a right angle to sac, and to 1/4 length of sac. July–Sept. **Habitat** Floodplain forests, low spots in woods, swamps, streambanks, shores; often where somewhat disturbed.

TREES OR SHRUBS. Leaves deciduous, simple, alternate, with toothed margins and pinnate veins. **Flowers** small, male and female flowers separate on same plant, crowded into catkins (aments) that open in spring before leaves fully open; **male catkins** hang downward; conelike **female catkins** erect or drooping. **Fruit** a small, 1-seeded, winged nutlet.

BETULACEAE | BIRCH FAMILY

1 Female catkins in loose clusters of several catkins; scales of female catkins persistent, becoming hard and stiff .... *Alnus* | ALDER
1 Female catkins single; scales of female catkins soon deciduous .... *Betula* .... BIRCH

ALNUS | *Alder*

Thicket-forming shrubs. **Leaves** deciduous, ovate, toothed on margins. **Male and female flowers** separate on same plant, male flowers in long, drooping catkins which fall after shedding pollen; female flowers in short, persistent conelike clusters. **Fruit** a flattened achene with winged or thin margins.

ALNUS | ALDER

1 Twigs and young leaves sticky, leaves with small, sharp teeth; catkins on long stalks; winter buds not stalked; fruit broadly winged,... *Alnus viridis* .... GREEN OR MOUNTAIN ALDER
1 Twigs and young leaves not sticky, leaves unevenly double-toothed; catkins stalkless or on short stalks; winter bud stalked, blunt at tip; fruit narrowly winged .... *Alnus incana* | SPECKLED ALDER; TAG ALDER

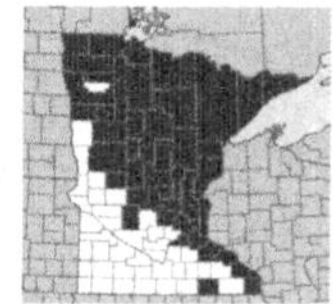

Alnus incana (L.) Moench

SPECKLED ALDER; TAG ALDER
NC **FACW** | MW **FACW** | GP **FACW**

Thicket-forming native shrub to 5 m tall; **twigs** red-brown, waxy, with conspicuous pale lenticels. **Leaves** ovate to oval, broadest near or below middle, 6–14 cm long and 4–7 cm wide, dark green and smooth above, paler and hairy below; margins sharply toothed and shallowly lobed; petioles 1–2.5 cm long. **Flowers** in catkins clustered at ends of branches; male catkins developing in late summer, short-stalked, elongate, 4–9 cm long; female catkins appear in late summer, stalkless, rounded, 1–2 cm long and to 1 cm wide, the scales unlobed, becoming conelike, persistent. **Fruit** a flat nutlet, narrowly winged on margin, 2–4 mm long. April–June. **Synonyms** *Alnus rugosa*. **Habitat** Swamps, thickets, bog margins, shores and streambanks.

Alnus viridis (Villars) Lam.
GREEN OR MOUNTAIN ALDER
NC **FAC** | MW **FAC** | GP **FAC**

Thicket-forming native shrub to 4 m tall; **bark** red-brown to gray; **twigs** brown, sticky, somewhat hairy, lenticels pale and scattered. **Leaves** round-oval, bright green above, slightly paler and shiny below, sticky when young, margins wavy with small, sharp teeth; petioles 6–12 mm long. **Flowers** in catkins; male catkins stalked, slender, developing in late summer and expanding in spring; female catkins appear in spring, becoming long-stalked, blunt and conelike, persistent, 1–2 cm long. **Fruit** a nutlet, 2–3 mm long, with a pale, thin wing. **Synonyms** *Alnus crispa, Alnus mollis*. **Habitat** Lakeshores, wet depressions in woods, rock outcrops, beaches along Lake Superior.

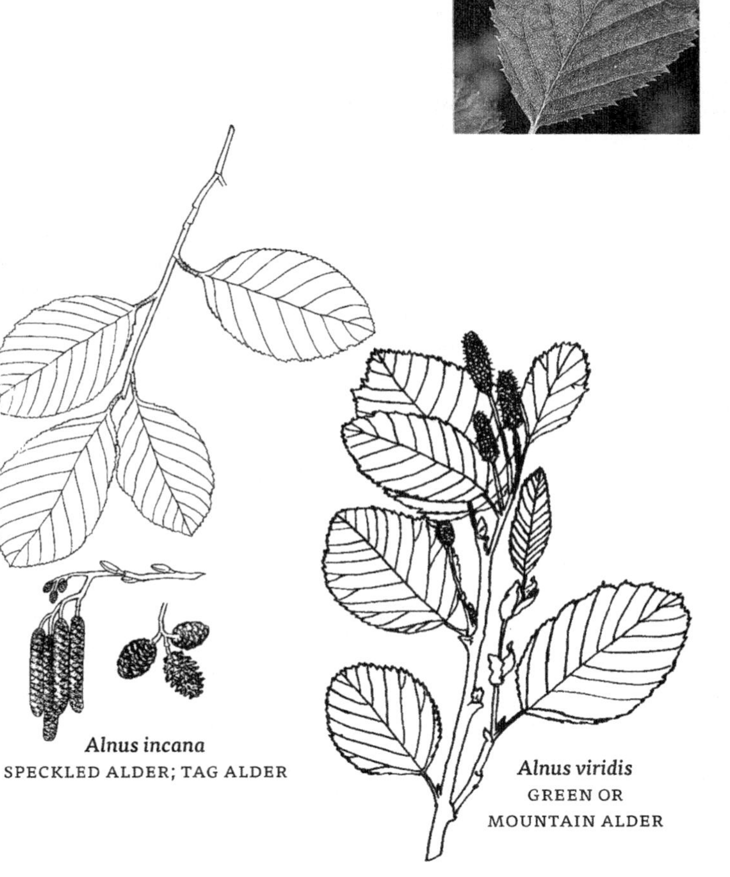

Alnus incana
SPECKLED ALDER; TAG ALDER

Alnus viridis
GREEN OR
MOUNTAIN ALDER

BETULA | *Birch*

Trees or shrubs (often with many stems from base); **bark** sometimes peeling in thin layers. **Leaves** deciduous, alternate, sharply toothed. **Male and female flowers** separate on same plant, catkins appearing in fall, opening the following spring, **male flowers** in drooping slender catkins; **female flowers** in erect conelike catkins. **Fruit** an achene with a winged margin.

BETULA | BIRCH

1 Shrub to 2 m tall; bark not shredding; leaves to 5 cm long . . . *Betula pumila* . . . . BOG BIRCH

1 Small to large trees; bark shredding with age 2

2 Bark yellow-gray; leaves rounded at base, margins not wavy-toothed . . . . *Betula alleghaniensis* | YELLOW BIRCH

2 Bark red-brown; leaves wedge-shaped at base, margins wavy-toothed . . . . *Betula nigra* | RIVER BIRCH

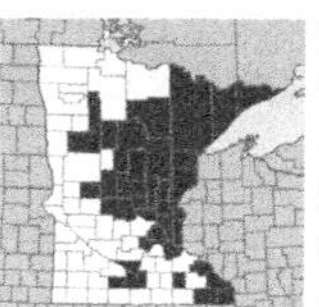

Betula alleghaniensis Britton

YELLOW BIRCH

NC **FAC** | MW **FAC** | GP **FACU**

Medium to large native tree to 25 m tall; **bark** on young trees thin and smooth with conspicuous horizontal lenticels, becoming yellow-gray and shredding into thin, shaggy horizontal strips; bark of old trees breaking into large plates; **twigs** hairy when young, becoming smooth and shiny, wintergreen-scented when crushed. **Leaves** alternate, simple, ovate, tapered to a short, sharp tip, dark green above, paler yellow-green below, 6–12 cm long, margins coarsely double-toothed, petioles grooved and hairy. **Male catkins** drooping, yellow-purple, 7–10 cm long; **female catkins** erect, green, 2–4 cm long, ± stalkless. **Fruit** a winged nutlet, 3–5 mm wide. April–May. **Synonyms** *Betula lutea*. **Habitat** Swamps,

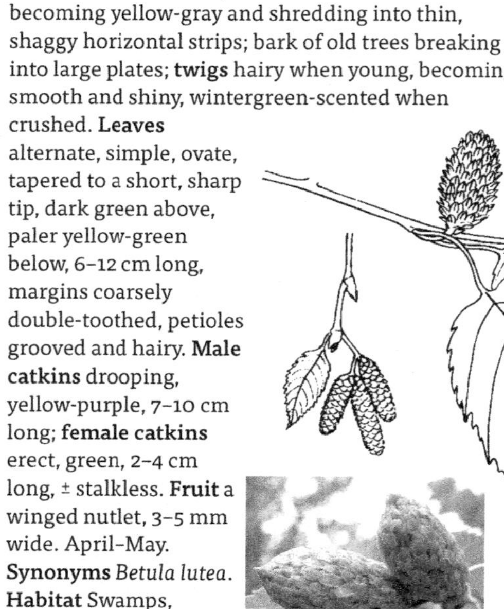

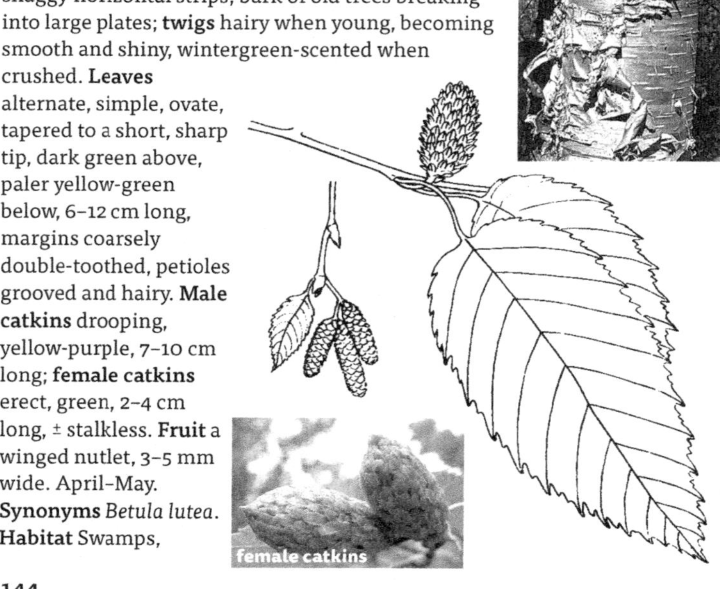

female catkins

thickets, forest depressions and streambanks; more common in moist forests with sugar maple (*Acer saccharum*).

Betula nigra L.

RIVER BIRCH

NC **FACW** | MW **FACW** | GP **FACW**

Small or medium native tree to 20 m tall, trunk to 6 dm wide, crown rounded; **bark** red-brown, shredding and curly; **twigs** slender, red-brown; buds pointed, hairy. **Leaves** alternate, simple, ovate, 4–8 cm long, upper surface smooth, lower surface paler and densely hairy; margins coarsely double-toothed, except untoothed near base; petioles with woolly hairs. Male and female **flowers** small, separate but on same tree; male flowers in slender drooping clusters; female flowers in short, woolly clusters. **Fruit** a small hairy nutlet with a 3-lobed, winged margin, crowded in a cylindrical cone 1.5–3 cm long. May. **Habitat** Floodplain forests, riverbanks, swamps.

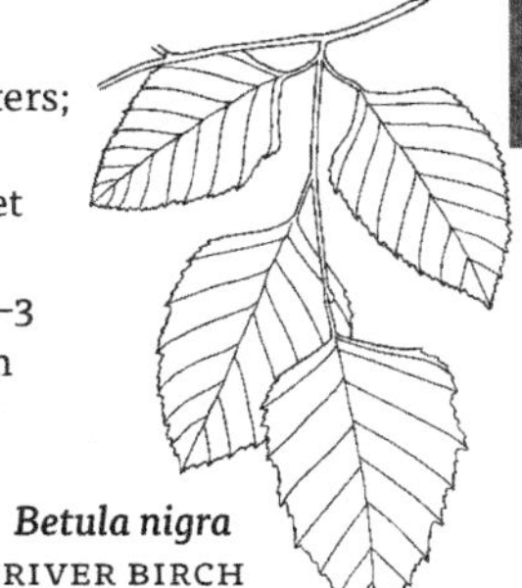

Betula nigra
RIVER BIRCH

Betula pumila L.

BOG BIRCH

NC **OBL** | MW **OBL** | GP **OBL**

Native shrub 1–3 m tall; **bark** dull gray or brown; **twigs** gray, short-hairy and dotted with resin glands, becoming red-brown and waxy with age. **Leaves** leathery, rounded to obovate, 2–4 cm long and 1–3 cm wide, dark green above, paler and often waxy below; margins coarsely toothed, the teeth blunt or sharp; petioles 3–6 mm long. **Flowers** in catkins; male catkins stalkless, cylindric, 15–20 mm long and 2–3 mm wide; female catkins stalked, cylindric, 1–2 cm long and 5 mm wide; scales 3-lobed. **Fruit** a flat, winged, rounded nutlet, 2–3 mm long and 2–4 mm wide. May. **Synonyms** *Betula glandulosa* var. *glandulifera*. **Habitat** Swamps, bogs, fens, seeps; often where calcium-rich.

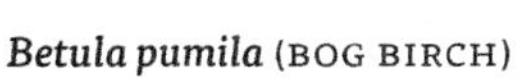

Betula pumila (BOG BIRCH)

ANNUAL OR PERENNIAL HERBS with usually bristly stems and alternate, bristly leaves. **Flowers** typically in a spirally coiled, spikelike head that uncurls as flowers mature; flowers perfect (with both male and female parts), with 5 petals, 4–5 sepals, and 5 stamens. **Fruit** a dry capsule with 4 nutlets.

BORAGINACEAE | BORAGE FAMILY

1 Flowers tubelike, the petal lobes erect or slightly spreading .... *Mertensia* .... BLUEBELL
1 Flowers tubelike, but petal lobes abruptly flared and flattened... *Myosotis* .... FORGET-ME-NOT; SCORPION-GRASS

MERTENSIA | *Bluebell*

Perennial herbs; plants smooth or hairy. **Leaves** alternate and entire. **Flowers** usually blue (pink in bud), tube-, funnel- or bell-shaped, petals widened and shallowly lobed at tip; in small clusters at ends of stems and branches. **Fruit** a smooth or wrinkled nutlet.

MERTENSIA | BLUEBELL

1 Leaves and sepals hairy ... *Mertensia paniculata* | NORTHERN BLUEBELLS
1 Leaves and sepals without hairs .... *Mertensia virginica* .... EASTERN BLUEBELLS

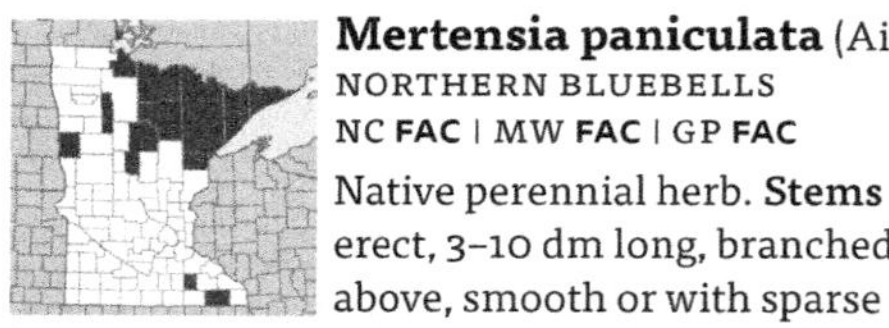

Mertensia paniculata (Aiton) G. Don.

NORTHERN BLUEBELLS
NC **FAC** | MW **FAC** | GP **FAC**

Native perennial herb. **Stems** erect, 3–10 dm long, branched above, smooth or with sparse hairs. **Basal leaves** ovate, rounded at base; **stem leaves** lance-shaped to ovate, 5–15 cm long, tapered to a tip, hairy, entire; petioles short on lower leaves, upper leaves ± stalkless. **Flowers** blue-purple, narrowly bell-shaped, 10–15 mm long, on slender stalks, in few-flowered racemes at ends of stems and branches; sepal lobes lance-shaped, 3–6 mm long, with dense, short hairs. **Fruit** a nutlet. June–July. **Habitat** Conifer swamps, streambanks, seeps.

Mertensia paniculata
NORTHERN BLUEBELLS

Mertensia virginica (L.) Pers.
EASTERN BLUEBELLS
NC **FAC** | MW **FACW**

Native perennial herb; plants smooth. **Stems** upright, 3–7 dm long. **Leaves** oval to obovate, entire, 5–15 cm long, rounded or blunt at the tip; upper leaves stalkless, lower leaves with winged petioles. **Flowers** showy, blue-purple, trumpet-shaped, 5-lobed at tip, 2–3 cm long, stalked, in a cluster at end of stem; sepals rounded at tip, 3 mm long. **Fruit** a nutlet. April–May. **Habitat** Floodplain forests, moist deciduous forests, streambanks; sometimes escaping from gardens where grown as an ornamental.

Mertensia virginica
EASTERN BLUEBELLS

MYOSOTIS | *Forget-me-not, Scorpion-grass*

Perennial (sometimes annual) herbs; plants with short, appressed hairs. **Leaves** alternate and entire. **Flowers** blue, tube-shaped and abruptly flared outward at tip, in a 1-sided raceme. **Fruit** a nutlet.

MYOSOTIS | FORGET-ME-NOT, SCORPION-GRASS

1 Plants without stolons; lobes of sepals as long or longer than corolla tube; flowers up to 6 mm wide; nutlets longer than style .... *Myosotis laxa* .... SMALLER FORGET-ME-NOT

1 Plants creeping and spreading by stolons; lobes of sepals shorter than corolla tube; flowers mostly 6 mm or more wide; nutlets shorter than style .... *Myosotis scorpioides* | WATER SCORPION-GRASS

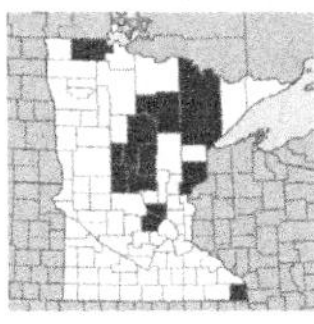

Myosotis laxa Lehm.
SMALLER FORGET-ME-NOT
NC **OBL** | MW **OBL** | GP **OBL**

Short-lived native perennial (or sometimes annual) herb. **Stems** slender, 1–4 dm long, often lying on ground at base, but not creeping, with fine, short, appressed hairs. **Leaves** oblong or spatula-shaped, 2–6 cm long. **Flowers** blue, on stalks usually much longer than the flower, in 1-sided clusters at ends of stems; sepals covered with short hairs, sepal lobes shorter than the tube; petal lobes shorter or slightly longer than the tube. **Fruit** a nutlet distinctly longer than the style. June–Sept. **Habitat** Cedar swamps, wet shores and streambanks.

Myosotis scorpioides L.
WATER SCORPION-GRASS
NC **OBL** | MW **OBL** | GP **OBL**

Colony-forming perennial herb, introduced. **Stems** 2–6 dm long, with short, appressed hairs, often creeping at base and producing stolons. **Leaves** 3–8 cm long and 0.5–2 cm wide, lower leaves oblong lance-shaped, upper leaves oblong or oval; stalkless or the lower leaves on short petioles. **Flowers** blue with a yellow center, tube-shaped, abruptly flared at tip, in a 1-sided raceme at ends of stems; flower stalks spreading in fruit; sepals with short, appressed hairs, sepal lobes equal or shorter than the tube. **Fruit** a nutlet shorter than the style. May–Sept. **Synonyms** *Myosotis palustris*. **Habitat** Streambanks, shores, ditches, swamps, wet depressions in forests. Introduced and naturalized in ne and c USA and s Canada.

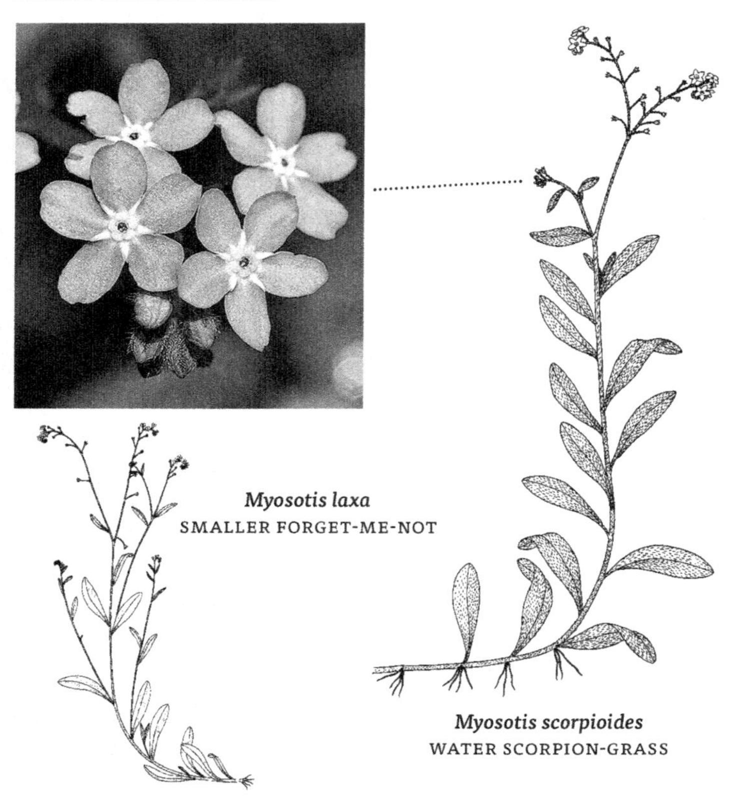

Myosotis laxa
SMALLER FORGET-ME-NOT

Myosotis scorpioides
WATER SCORPION-GRASS

BRASSICACEAE
Mustard Family

ANNUAL, BIENNIAL, OR PERENNIAL HERBS. **Leaves** simple or compound, alternate on stems or basal, smooth or hairy, some species with branched hairs. **Flowers** in terminal or lateral clusters (racemes), the lower portion often fruiting while tip in flower, the stalks elongating in fruit; flowers perfect, cross-shaped, with 4 sepals and 4 yellow, white, pink or purple petals; stamens 6, the outer 2 stamens shorter than the inner 4; pistil 1, style 1, ovary superior. **Fruit** a pod, cylindrical (silique) or round (silicle), with 2 chambers and 1 to many seeds in 1 or 2 rows in each chamber.

BRASSICACEAE | MUSTARD FAMILY

1 Flowers yellow 2
1 Flowers white, pink, or purple 3

2 Leaf segments rounded or broadly oval in outline, the terminal segment much larger than lateral segments; plants without hairs . . . . *Barbarea orthoceras* | NORTHERN WINTER-CRESS
2 Leaf segments tapered to a tip, the terminal segments about same size as lateral segments; plants hairy or smooth . . . . *Rorippa* | WATER-CRESS

3 Petals pink or purple 4
3 Petals white or green 5

4 Plants smooth throughout; rare in se Minnesota . . . . *Iodanthus pinnatifidus* | PURPLE ROCKET
4 Plants hairy, at least on lower stems or leaves . . . . *Cardamine* . . . . BITTER-CRESS; TOOTHWORT

5 Leaves all basal, linear and entire, plants usually flowering and fruiting underwater; uncommon plant of ne Minnesota . . . . *Subularia aquatica* . . . . WATER AWLWORT
5 Leaves basal and on stems, entire (*Alliaria*) or pinnately divided; widespread species 6

6 Plants in water or mud, rooting along underwater nodes of stem; basal rosette absent; fruit often curved, on spreading stalks; seeds netlike on surface . . . . *Rorippa nasturtium-officinale* | WATER-CRESS
6 Plants of moist sites; rooting only at base, basal rosette usually present on young plants; fruit straight and upright 7

7 Plants strongly onion- or garlic-scented . . . . *Alliaria petiolata* . . . . GARLIC-MUSTARD
7 Plants not strongly scented . . . . . *Cardamine* | BITTER-CRESS; TOOTHWORT

ALLIARIA | *Alliaria*

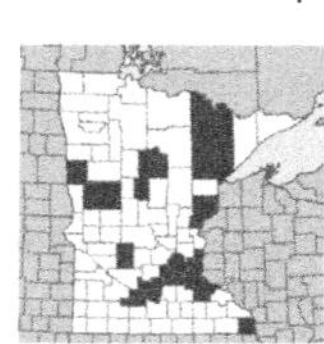

Alliaria petiolata (M. Bieb.) Cavara & Grande
GARLIC-MUSTARD
NC **FACU** | MW **FAC** | GP **FACU**

Biennial introduced invasive herb, strongly onion- or garlic-scented when crushed (especially in spring and early summer); root slender, white, "s"-shaped at top. **Stems** to 1 m long. **Leaves** narrowly oblong to ovate; petioles long on lower leaves; stem leaves alternate, triangular-shaped, margins with large teeth; leaves at base long-petioled; petioles on stem leaves becoming shorter. First-year plants a rosette of 3–4 round, scallop-margined leaves; leaves remain green over winter. Second-year plants with 1–2 flowering stems; **flowers** many; petals white, 4. **Fruit** a slender capsule (silique) 3–6 cm long, with a single row of black seeds. May. **Habitat** Moist, shaded forest openings, streambanks, roadsides, floodplain forests; soils not highly acidic.

Garlic-mustard *is a highly invasive, European biennial.*

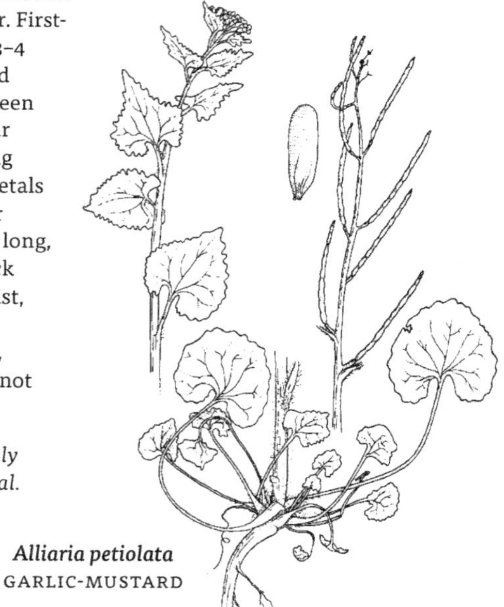

Alliaria petiolata
GARLIC-MUSTARD

BARBAREA | *Winter-cress*

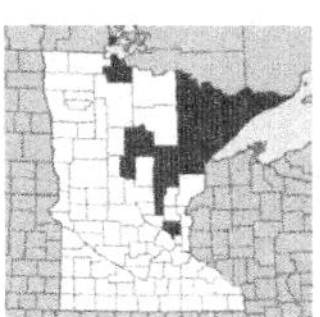

Barbarea orthoceras Ledeb.
NORTHERN WINTER-CRESS
NC **OBL** | MW **OBL** | GP **OBL**

Introduced biennial herb; plants smooth or with sparse covering of unbranched hairs. **Stems** 3–8 dm long, unbranched, or branched above. **Leaves** simple or with 1–4 pairs of lateral lobes, the middle and upper leaves deeply lobed. **Flowers** in racemes; on short stalks to 1 mm long, the stalks clublike at tip; petals yellow, 3–5 mm long. **Fruit** upright, 2–4 cm long, with a beak 0.5–2 mm long. June–July. **Habitat** Rocky shores, swamps and wet woods.

Barbarea orthoceras
NORTHERN WINTER-CRESS

CARDAMINE | *Bitter-cress, Toothwort*

Annual, biennial or perennial herbs, smooth or with short hairs near base of stem. **Leaves** simple to pinnately divided, the basal leaves often different in shape than stem leaves. **Flowers** in racemes or umbel-like clusters; sepals green to yellow, early deciduous; petals white. **Fruit** a 2-chambered, linear pod (silique), the seeds in a single row in each chamber.

CARDAMINE | BITTER-CRESS, TOOTHWORT

1 Stem leaves simple or with 1–2 small lobes only . . . *Cardamine rhomboidea* . . . SPRING-CRESS
1 Stem leaves pinnately dissected, with 2 or more deep lobes . . . 2

2 Plants annual or biennial; petals 2–4 mm long. . . . *Cardamine pensylvanica* . . . PENNSYLVANIA BITTER-CRESS
2 Plants perennial; petals 5 mm or more long . . . *Cardamine pratensis* . . . CUCKOO-FLOWER

Cardamine pensylvanica Muhl.
PENNSYLVANIA BITTER-CRESS
NC **FACW** | MW **FACW** | GP **FACW**

Native annual or biennial herb. **Stems** erect or spreading, to 6 dm long, usually hairy on lower stem. **Leaves** pinnately divided into 2–5 pairs of lateral leaflets and a single terminal segment, 4–8 cm long and 1–4 cm wide, the leaflets entire or with a few teeth or lobes, the terminal leaflet largest, 1–4 cm long and 1–2 cm wide; petioles shorter than blades, becoming shorter upward. **Flowers** in a raceme; sepals 1–2 mm long; petals white, 2–4 mm long. **Fruit** an upright silique, 2–3 cm long and to 1 mm wide, with a style-beak to 2 mm long, on stalks 5–15 mm long. May–Sept. **Synonyms** *Cardamine parviflora*. **Habitat** Streambanks, swamps, and wet forests (often where seasonally flooded); wet, disturbed areas.

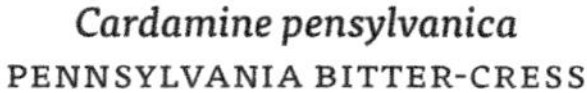

Cardamine pensylvanica
PENNSYLVANIA BITTER-CRESS

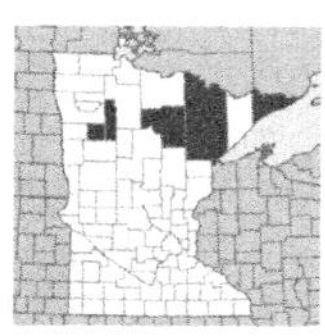

Cardamine pratensis L.
CUCKOO-FLOWER
[NC **OBL** | MW **OBL** | GP **OBL**]

Native perennial upright herb. **Stems** 2–5 dm long. **Basal leaves** on long petioles, divided into 3–8 broad leaflets, 5–20 mm long, the terminal segment largest and ± entire; lower stem leaves similar to basal ones, becoming shorter and with shorter petioles upward on stem; **stem leaves** with 7–17 oval to linear leaflets. **Flowers** in a crowded raceme; petals white, 8–15 mm long. **Fruit** an upright silique, 2.5–4 cm long, with a style-beak 1–2 mm long, on stalks 8–15 mm long. May–June. **Synonyms** Ours are the native *Cardamine pratensis* var. *palustris*. *Cardamine palustris*. **Habitat** Peatlands, tamarack and cedar swamps, wet depressions in forests.

Cardamine pratensis
CUCKOO-FLOWER

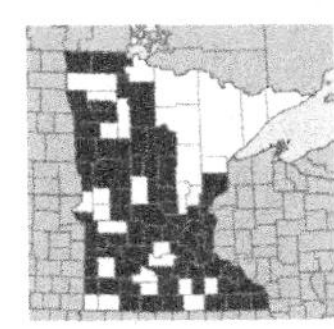

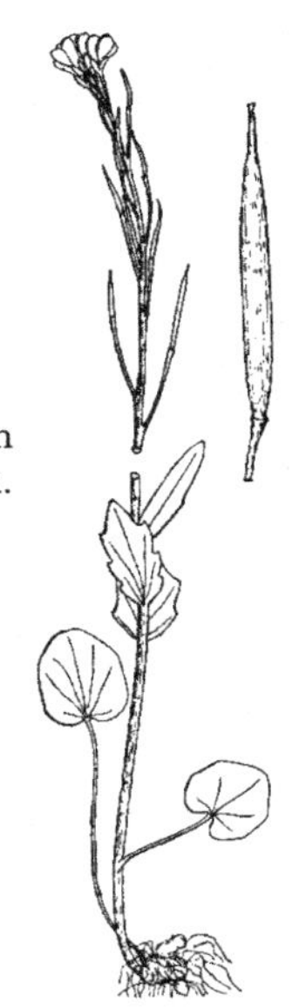

Cardamine rhomboidea (Pers.) DC.
SPRING-CRESS
NC **OBL** | MW **OBL** | GP **OBL**

Native perennial herb. **Stems** 1 to several from a short thick tuber, unbranched or with a few branches above, 2–6 dm long, smooth or with short hairs on lower stems. **Leaves** simple, sparsely to densely covered with short hairs; basal leaves round or heart-shaped, on long petioles, withering before plants in full flower; stem leaves 4–8, oblong to oval, 2–7 cm long and 0.5–2.5 cm wide; petioles shorter upward on stem. **Flowers** in racemes; sepals green, turning yellow after flowering, 2–4 mm long; petals white (rarely pink), 6–15 mm long. **Fruit** a silique, 1–2.5 cm long and 1–2 mm wide, with a style beak 2–4 mm long, on spreading stalks 1–3 cm long, the pod often falling before mature. May–June. **Synonyms** *Cardamine bulbosa*. **Habitat** Wet forest depressions, floodplain forests, streambanks, wet meadows, swamps, calcareous fens.

IODANTHUS | *Purple rocket*

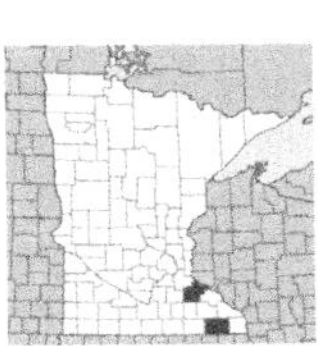

Iodanthus pinnatifidus (Michx.) Steudel
PURPLE ROCKET
NC **FACW** | MW **FACW** | GP **FACW** | *endangered*

Native perennial herb; plants smooth. **Stems** to 1 m long, unbranched except in head. **Leaves** lance-shaped to oval or oblong, leaf base often with lobes which clasp stem; lower leaves often divided at base into 1–4 pairs of small segments; margins deep-toothed; petioles short. **Flowers** in a branched raceme, on stalks 5–10 mm long, pale violet to white; sepals rounded at tip, 3–5 mm long; petals 10–13 mm long. **Fruit** a linear, cylindric silique, 2–4 cm long and 1–2 mm wide, on spreading stalks. June–July. **Habitat** Seasonally flooded floodplain forests (silver maple and American elm common) along the Mississippi and Root Rivers. **Status** Endangered.

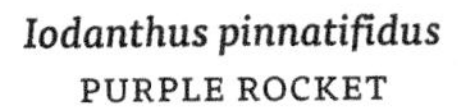

Iodanthus pinnatifidus
PURPLE ROCKET

RORIPPA | *Yellow-cress*

Annual, biennial or perennial herbs; plants smooth or with unbranched hairs. **Leaves** sometimes in a basal rosette in young plants, toothed to pinnately divided, petioles short or absent. **Flowers** small, in racemes at ends of stems or from lateral branches; sepals green to yellow, deciduous by fruiting time; petals yellow or white, shorter to longer than sepals. **Fruit** a short-cylindric to linear pod (silique), mostly 2-chambered, the seeds in 2 rows.

RORIPPA | YELLOW-CRESS

1 Petals white; plants rooting from lower nodes of stem. . . . . *Rorippa nasturtium-officinale* | WATER-CRESS
1 Petals yellow; plants usually not rooting at lower nodes 2

2 Plants annual or biennial, taprooted; petals shorter or equal to sepals. . . 3
2 Plants perennial, roots creeping; petals longer than sepals 4

3 Stalks of fruit 3 mm or more long; fruit to 1.5x longer than its stalk . . . . *Rorippa palustris* | COMMON YELLOW-CRESS
3 Stalks of fruit 1–2 mm long; fruit more than 1.5x longer than its stalk . . . . *Rorippa sessiliflora* | SOUTHERN YELLOW-CRESS

4 Stems sprawling or spreading; lateral leaf segments entire or with few shallow teeth; beak of fruit 1–2 mm long . . . . *Rorippa sinuata* . . . . WESTERN YELLOW-CRESS
4 Stems ± erect; lateral leaf segments with sharp teeth; beak of fruit 0.5–1 mm long . . . . *Rorippa sylvestris* | CREEPING YELLOW-CRESS

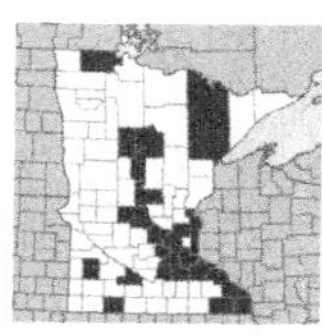

Rorippa nasturtium-officinale (L.) Hayek

WATER-CRESS
NC **OBL** | MW **OBL** | GP **OBL**

Introduced perennial herb; plants smooth. **Stems** underwater, floating, or trailing on mud; rooting from lower nodes. **Leaves** 4–12 cm long and 2–5 cm wide, pinnately divided into 3–9 segments, the lateral segments round to ovate in outline, the terminal segment largest; margins entire or with a few shallow rounded teeth; petioles present. **Flowers** in 1 to several racemes per stem, flat-topped and elongating in fruit; flowers 5 mm wide, sepals green-white, oblong, 1–3 mm long; petals white, sometimes purple-tinged, obovate, 4–5 mm long. **Fruit** a linear, often curved pod (silique), 1–2.5 cm long and 2 mm wide, tipped with a short style beak to 1 mm long. May–Sept. **Synonyms** *Nasturtium officinale*. **Habitat** Springs, slow-moving streams, ditches. Introduced and naturalized across North America.

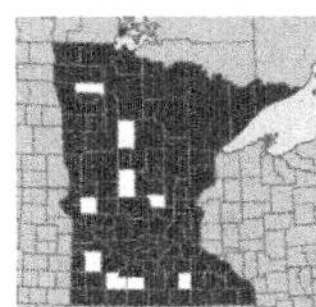

Rorippa palustris (L.) Besser
COMMON YELLOW-CRESS
NC **OBL** | MW **OBL** | GP **OBL**

Native annual or biennial herb. **Stems** erect, usually 1, to 1 m long, unbranched or branched upward. **Leaves** lance-shaped to obovate, mostly pinnately divided; the blades oblong to oblong lance-shaped, 5–30 cm long and 2–6 cm wide, middle stem leaves usually with basal lobes and clasping stem, smooth to densely hairy on lower surface; margins deeply lobed and slightly wavy; petioles short or absent. **Flowers** in racemes at ends of stems and from leaf axils, the terminal raceme flowering and fruiting first, the oldest siliques on lowest portions of raceme; sepals green, 1–3 mm long, early deciduous; petals yellow, drying white, 2–3 mm long. **Fruit** a round to short-cylindric pod, 3–10 mm long and 1–3 mm wide, straight-sided or slightly tapered to tip, on stalks 3–10 mm long. June–Sept. **Synonyms** *Rorippa islandica*. **Habitat** Marshes, wet meadows, shores, streambanks, ditches and other wet places.

Rorippa nasturtium-officinale
WATER-CRESS

Rorippa palustris
COMMON YELLOW-CRESS

Rorippa sessiliflora (Nutt.) A. S. Hitchc.
SOUTHERN YELLOW-CRESS
NC **OBL** | MW **OBL** | GP **OBL** | *special concern*

Native annual (sometimes biennial) herb. **Stems** erect, 2–4 dm long, branched, smooth. Lower leaves oblong, coarsely round-toothed, lower part of blade usually deeply cleft; upper leaves smaller, ovate, entire or toothed. **Flowers** in racemes from ends of branches and upper leaf axils; sepals yellow, petals absent; stamens 3–6. **Fruit** a pod (silique), 6–10 mm long and 3–4 mm wide, often somewhat sickle-shaped, on short, spreading or ascending stalks 1–2 mm long, the style beak very short. June–July. **Synonyms** *Radicula sessiliflora*. **Habitat** Muddy shores along Miss and Minn Rivers, but suitable habitat may no longer be present; species likely absent from the state. **Status** Special concern.

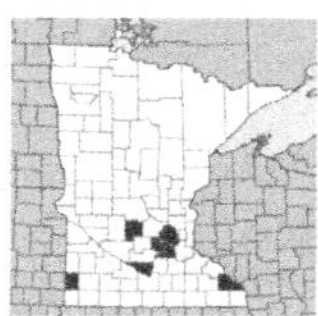

Rorippa sinuata (Nutt.) A. S. Hitchc.
WESTERN YELLOW-CRESS
NC **FACW** | MW **FACW** | GP **FACW**

Native perennial herb, spreading by rhizomes. **Stems** usually several, sprawling, 1–4 dm long, sparsely to densely covered with blunt-tipped hairs. **Leaves** all from stem (basal leaves absent), 2–8 cm long and 0.5–2 cm wide, oblong, pinnately divided into 5–7 pairs, sometimes with basal lobes clasping stem, margins entire or with a few teeth. **Flowers** in racemes at ends of stems and from upper leaf axils, all flowering at about same time or flowers from axils first; sepals yellow-green, 3–5 mm long, early deciduous; petals yellow, 4–6 mm long, longer than sepals. **Fruit** a linear pod (silique), 5–12 mm long and 1–2 mm wide, tapered to the style beak, on upright to spreading stalks, 4–10 mm long. June–Aug. **Synonyms** *Radicula sinuata*. **Habitat** Stream and riverbanks, ditches, and other low places, especially where sandy.

Rorippa sinuata
WESTERN YELLOW-CRESS

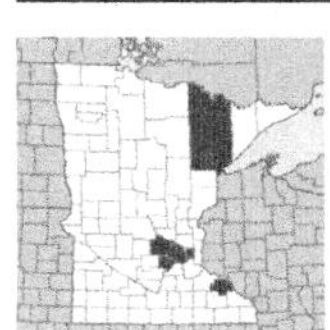

Rorippa sylvestris (L.) Besser
CREEPING YELLOW-CRESS
NC **OBL** | MW **OBL** | GP **FACW**

Introduced perennial herb, spreading by rhizomes and sometimes stolons. **Stems** erect, branched above, 2–6 dm long, smooth or sparsely hairy on lower stem; basal rosettes present on young plants. **Stem leaves** pinnately divided, oblong in outline, 3–15 cm long and 2.5 cm wide, gradually reduced in size upward on stem, margins usually toothed; petioles present on lower leaves, petioles absent on upper leaves. **Flowers** in racemes at ends of stems and from upper leaf axils, all flowering at about same time or the oldest siliques on lower portion of terminal racemes; sepals yellow-green, 2–3 mm long; petals yellow, 3–5 mm long, to 2 mm longer than the sepals. **Fruit** a linear pod (silique), 4–10 mm long and to 1 mm wide, usually upright on spreading stalks 5–10 mm long. June–Aug. **Synonyms** *Radicula sylvestris*. **Habitat** Wet forests, lakeshores, muddy streambanks and ditches; sometimes weedy. Introduced to N America from Europe, now across s Canada and much of USA.

SUBULARIA | *Awlwort*

Subularia aquatica L.
WATER AWLWORT
NC **OBL** | GP **OBL** | *threatened*

Small, annual aquatic herb, native; plants underwater or sometimes on muddy shores. **Stems** 3–10 cm long. **Leaves** all basal, awl-shaped or linear, 1–5 cm long. **Flowers** small, 2–10, widely separated in a raceme; sepals persistent, petals white. **Fruit** a short, oval or oblong pod (silicle), 2–4 mm long. June–Aug. Our plants are var. *americana*. **Habitat** Cold lakes in shallow water to 1 m deep; soils usually sandy. **Status** Threatened; rare in ne Minnesota and on Isle Royale and e Upper Peninsula of Michigan.

BRASENIA | *Water-shield*

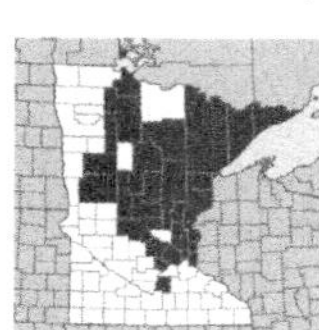

Brasenia schreberi J. F. Gmelin
WATER-SHIELD
NC **OBL** | MW **OBL** | GP **OBL**

Aquatic perennial aquatic herb, native; underwater portions of plant with a slippery jelly-like coating. **Stems** to 2 m long. **Leaf blades** floating, oval, 4–12 cm long and half as wide; petiole attached to center of blade underside. **Flowers** perfect (with both male and female parts), dull-purple, on emergent stalks to 15 cm long from leaf axils; sepals 3, petals 3, 12–15 mm long.
Fruit an oblong capsule, 3–5 mm long. July.
Habitat Quiet ponds and lakes; water usually acidic.

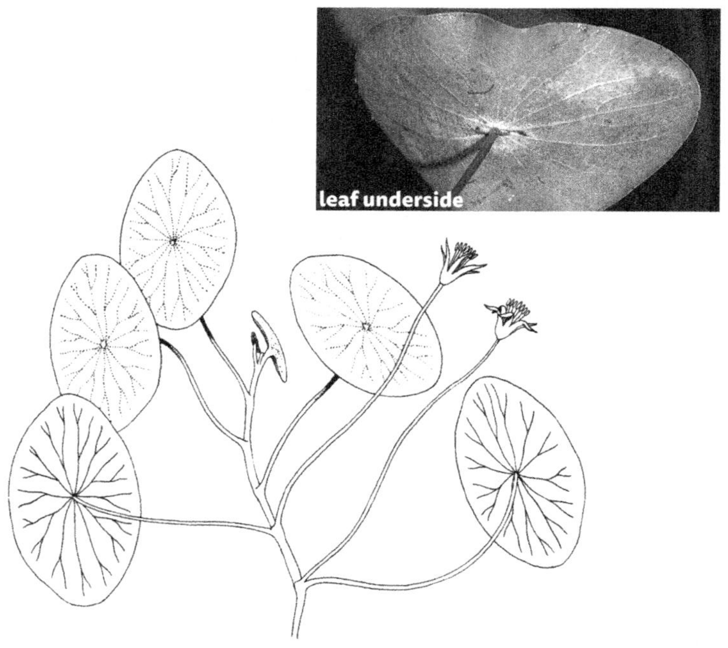

Brasenia schreberi
WATER-SHIELD

CALLITRICHE | *Water-starwort*

Small, annual aquatic herbs with weak, slender stems and fibrous roots. **Leaves** simple, opposite, all underwater or upper leaves floating; underwater leaves linear, 1-nerved, entire except for shallowly notched tip; floating leaves mostly in clusters at ends of stems, obovate to spatula-shaped, 3–5-nerved, rounded at tip. **Flowers** tiny, male and female flowers usually separate on same plant, each flower with 1 stamen or 1 pistil; single and stalkless in middle and upper leaf axils, or 1 male and 1 female flower in each axil, subtended by a pair of thin, translucent, deciduous bracts, or the bracts absent; styles 2, ovary flattened, oval to round, 4-chambered, separating when mature into 4 nutlets.

CALLITRICHE | WATER-STARWORT

1 Leaves all underwater, 1-veined, linear. . . . . *Callitriche hermaphroditica* . . . . AUTUMNAL WATER-STARWORT

1 Leaves both underwater and floating; floating leaves 3-veined, spatula-shaped or obovate 2

2 Fruit as long as wide, rounded at base, pits on fruit not in rows . . . . *Callitriche heterophylla* | LARGER WATER-STARWORT

2 Fruit slightly longer (0.2 mm) than wide, narrowed at base, pitted in rows . . . . *Callitriche palustris* | SPINY WATER-STARWORT

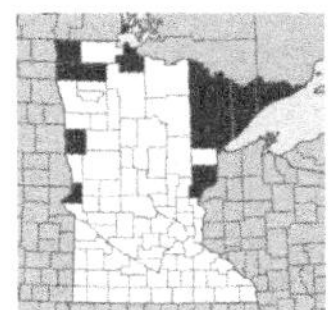

Callitriche hermaphroditica L.

AUTUMNAL WATER-STARWORT

NC **OBL** | MW **OBL** | GP **OBL**

Native annual herb. **Stems** 10–30 cm long. **Leaves** all underwater, alike, linear, 1-nerved, 3–12 mm long and to 1.5 mm wide, shallowly notched at tip, clasping at base, the opposite leaf bases not connected; darker green than our other species. **Flowers** either male or female; single in leaf axils, not subtended by translucent bracts. **Fruit** flattened, rounded, 1–2 mm long, deeply divided into 4 segments. June–Sept. **Synonyms** *Callitriche autumnalis*. **Habitat** Shallow to deep water of lakes, ponds, marshes, ditches and slow-moving streams.

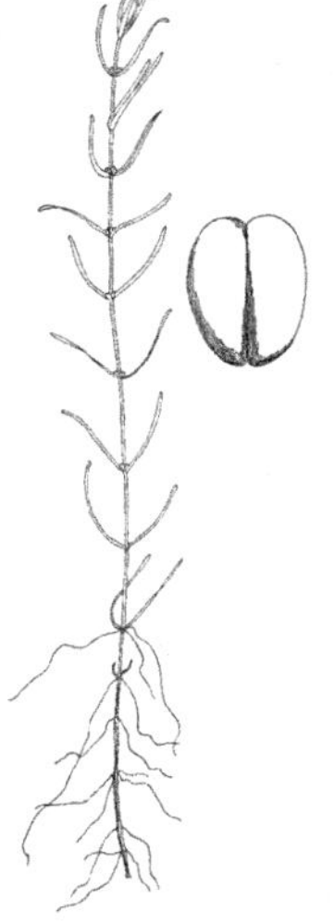

Callitriche hermaphroditica
AUTUMNAL WATER-STARWORT

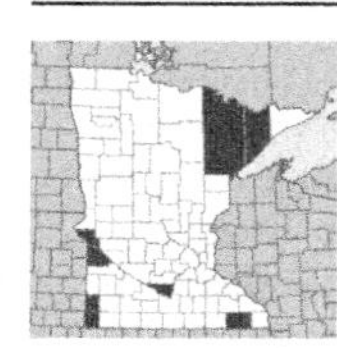

Callitriche heterophylla Pursh

LARGER WATER-STARWORT

NC **OBL** | MW **OBL** | GP **OBL** | *special concern*

Native annual herb. **Stems** 10–20 cm long. **Leaves** of 2 types; **underwater leaves** linear, 1–2 cm long and to 1.5 mm wide, 1-nerved, notched at tip, the leaf pairs connected at base by a narrow wing; **floating leaves** in clusters at ends of stems or opposite along upper stems, 3–5-nerved, obovate to spatula-shaped, rounded at tip, 6–15 mm long and 3–7 mm wide; leaves intermediate between underwater and floating leaves often present. **Flowers** either male or female; usually 1 male and 1 female flower together in leaf axils, subtended by a pair of translucent, deciduous bracts. **Fruit** about 1 mm long and not more than 0.1 mm longer than wide, often broadest above middle, not wing-margined, pits on surface not in rows. May–Aug. **Habitat** Shallow water of rainwater pools on rock outcrops (sw Minn populations); shallow water of protected lake bays (ne Minn occurences). **Status** Special concern.

Similar to **spiny water-starwort** *(Callitriche palustris) but less common, differing mainly in the fruits.*

Callitriche palustris L.

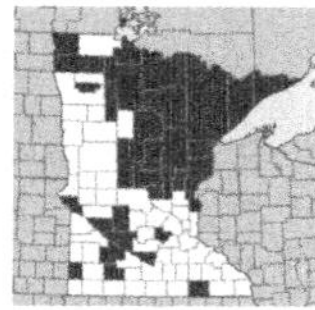

SPINY WATER-STARWORT

NC **OBL** | MW **OBL** | GP **OBL**

Native annual herb. **Stems** 10–20 cm long. **Leaves** of 2 types; **underwater leaves** mostly linear, 1–2 cm long and to 1 mm wide, shallowly notched at tip, the leaf pairs connected at base by a narrow wing; floating leaves in clusters at ends of stems or opposite along upper stems, 3–5-nerved, obovate to spatula-shaped, rounded at tip, 5–15 mm long and 2–5 mm wide; leaves intermediate between underwater and floating leaves usually present. **Flowers** either male or female; usually 1 male and 1 female flower together in leaf axils, subtended by a pair of translucent bracts, these soon deciduous. **Fruit** 1–1.5 mm long and about 0.2 mm longer than wide, broadest above middle, narrowly winged near tip, pitted in vertical rows. June–Sept. **Synonyms** *Callitriche verna*. **Habitat** Shallow water of lakes, ponds, streams; exposed mudflats.

PERENNIAL HERBS. Stems usually with milky sap. **Leaves** simple, alternate. **Flowers** in racemes at ends of stems or single from upper leaf axils, perfect (with both male and female parts), 5-parted, regular and funnel-shaped (*Campanula*) or irregular (*Lobelia*); petals blue, white or scarlet; stamens separate or joined into a tube around style. **Fruit** a many-seeded capsule.

CAMPANULACEAE | BELLFLOWER FAMILY

1 Flowers regular, stamens separate; plants weak and reclining on surrounding plants ...... *Campanula aparinoides* | MARSH-BELLFLOWER
1 Flowers irregular, stamens united to form a tube around the style; plants with upright stems .... *Lobelia* | LOBELIA

CAMPANULA | *Bellflower*

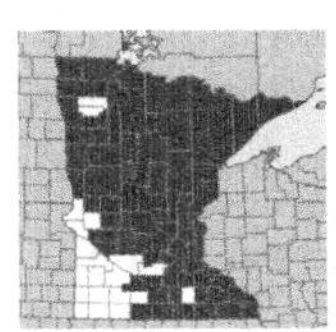

Campanula aparinoides Pursh

MARSH-BELLFLOWER
NC **OBL** | MW **OBL** | GP **OBL**

Native perennial herb, spreading by slender rhizomes. **Stems** slender, weak, usually reclining on other plants, 2–6 dm long, 3-angled, rough-to-touch. **Leaves** linear or narrowly lance-shaped, larger below and smaller upward on stem, 2–8 cm long and 2–8 mm wide, tapered to a sharp tip; margins and midvein on leaf underside often rough; petioles absent. **Flowers** single on long slender stalks from upper leaf axils, funnel-shaped, sepals triangular to lance-shaped, 2–5 mm long; petals pale blue to white, 5–12 mm long. **Fruit** a capsule, opening near its base to release seeds. July–Sept. **Habitat** Sedge meadows, marshes, calcareous fens, conifer swamps (cedar, tamarack), thickets, open bogs; soils often calcium-rich.

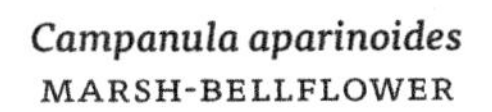

Campanula aparinoides
MARSH-BELLFLOWER

LOBELIA | *Lobelia*

Perennial herbs. **Stems** single, usually with milky juice. **Leaves** alternate. **Flowers** irregular, in racemes at ends of stems; white, bright red, or pale to dark blue, often with white or yellow markings; 2-lipped, the 3 lobes of lower lip spreading, the 2 lobes of upper lip erect or pointing forward, divided to base, the anthers projecting through the split; stamens 5, joined to form a tube around style, the lower 2 anthers hairy at tip and shorter than other 3. **Fruit** a capsule.

LOBELIA | LOBELIA

1 Stem leaves narrow, to 4 mm wide, margins entire or with a few small teeth; or leaves all from base of plant 2
1 Stem leaves broader, 1–5 cm wide, margins toothed 3

2 Leaves all from base of plant, usually underwater . . . . . *Lobelia dortmanna* . . . . WATER-LOBELIA
2 Leaves all from stem . . . . *Lobelia kalmii* | BROOK LOBELIA

3 Flowers small, to 1.5 cm long, lateral slits near base of flower absent . . . . *Lobelia spicata* | SPIKED LOBELIA
3 Flowers larger, 2–4 cm long, base of flower with lateral slits 4

4 Flowers bright red (rarely white), 3 cm or more long . . . . *Lobelia cardinalis* . . . . CARDINAL-FLOWER
4 Flowers blue with white stripes on lower lip (rarely all white), less than 2.5 cm long . . . . *Lobelia siphilitica* | GREAT BLUE LOBELIA

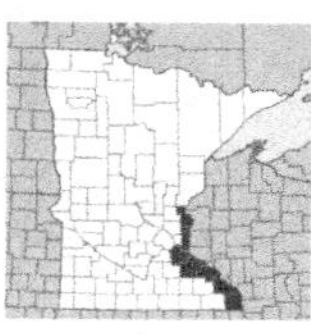

Lobelia cardinalis L.

CARDINAL-FLOWER
NC **OBL** | MW **OBL** | GP **FACW**

Native perennial herb. **Stems** erect, usually unbranched, 5–15 dm long, hairy to smooth. **Leaves** lance-shaped to oblong, 10–15 cm long and 3–5 cm wide, tapered to a point, margins toothed; lower leaves on short petioles, upper leaves ± stalkless. **Flowers** bright scarlet (rarely white), in racemes 1–4 dm long, the racemes with small, leafy, linear bracts; flowers 2–4 cm long, on hairy stalks 5–15 mm long. July–Sept. **Habitat** Floodplain forests, swamps, thickets, streambanks, shores and ditches; sometimes in shallow water.

Lobelia dortmanna L.
WATER-LOBELIA
NC **OBL** | MW **OBL**

Native perennial herb; plants usually underwater or sometimes on exposed sandy shores. **Stems** upright, hollow, smooth, with milky juice. **Leaves** in dense rosettes at base of plants, fleshy, hollow, linear, 3–8 cm long, rounded at tip; stem leaves tiny. **Flowers** pale blue or white, 1–2 cm long, in a few-flowered raceme; sepals 2 mm long. July–Sept. **Habitat** Shallow water of acid lakes and ponds; wet, sandy shores.

Lobelia cardinalis
CARDINAL-FLOWER

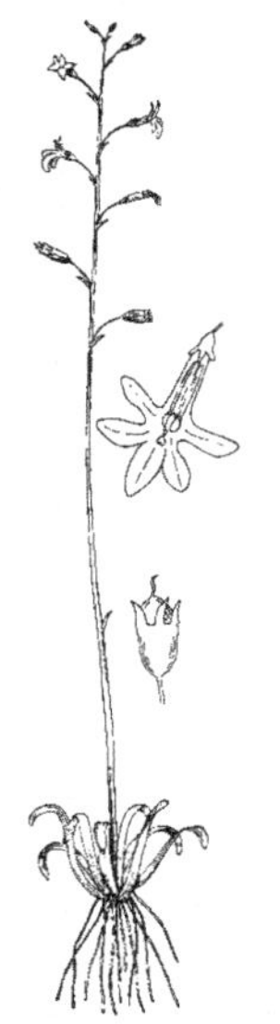

Lobelia dortmanna
WATER-LOBELIA

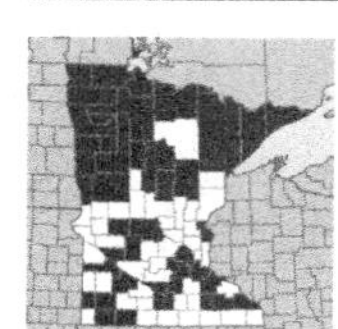

Lobelia kalmii L.

BROOK LOBELIA

NC **OBL** | MW **OBL** | GP **OBL**

Small native perennial herb. **Stems** erect, smooth, 1–4 dm long, unbranched or with a few branches above, sometimes with a rosette of small, obovate leaves at base of plant. **Stem leaves** linear, 1–5 cm long and 1–5 mm wide, blunt to sharp-tipped, margins with a few small teeth. **Flowers** blue with a white center, 6–10 mm long, in an open raceme, the flowers on stalks 4–10 mm long. July–Oct. **Synonyms** *Lobelia strictiflora*. **Habitat** Wet, sandy or gravelly shores, wet meadows, interdunal wetlands, conifer swamps (cedar, tamarack), rock ledges and crevices; usually where calcium-rich.

Lobelia kalmii
BROOK LOBELIA

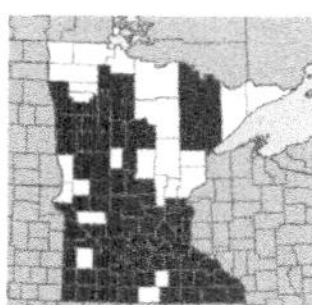

Lobelia siphilitica L.

GREAT BLUE LOBELIA

NC **FACW** | MW **OBL** | GP **OBL**

Native perennial herb. **Stems** stout, erect, 3–12 dm long. **Leaves** oblong or oval, smaller upward, 6–12 cm long and 1–3 cm wide, tip sharp or blunt, margins irregularly toothed, petioles absent. **Flowers** dark blue, in crowded racemes 1–3 dm long; the lower lip blue and white-striped, 1.5–2.5 cm long, on ascending stalks 4–10 mm long; sepals triangular to lance-shaped, 5–20 mm long, usually with narrow lobes near base; petals absent. Aug–Sept. **Habitat** Swamps, floodplain forests, thickets, streambanks, calcareous fens, wet meadows.

Lobelia siphilitica
GREAT BLUE LOBELIA

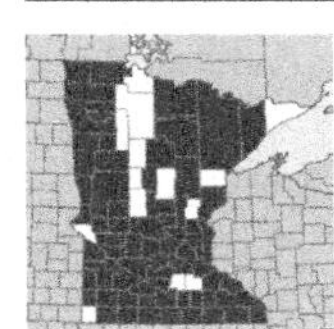

Lobelia spicata Lam.
SPIKED LOBELIA
NC **FAC** | MW **FAC** | GP **FAC**

Native perennial herb. **Stems** unbranched, 3–10 dm long, hairy toward base. **Leaves** obovate to lance-shaped, 5–10 cm long, hairy, becoming smaller above. **Flowers** pale blue to white, 6–10 mm long, on stalks 2–4 mm long, in a slender, crowded raceme; base of sepals often with distinct, curved lobes (auricles), 1–2 mm long. May–Aug. **Habitat** Moist to wet prairie (sometimes where disturbed), wet meadows, swamp margins.

Lobelia spicata
SPIKED LOBELIA

SHRUBS OR VINES, with opposite, mostly simple **leaves** (compound in *Sambucus*). **Flowers** perfect (with both male and female parts), mostly 5-parted. **Fruit** a fleshy berry or dry capsule.

CAPRIFOLIACEAE | HONEYSUCKLE FAMILY

1 Leaves pinnately divided . . . . *Sambucus* | ELDER
1 Leaves simple 2

2 Plants small, creeping, evergreen; flowers paired and nodding at tips of slender stalks . . . . *Linnaea borealis* | TWINFLOWER
2 Plants larger shrubs, upright, deciduous 3

3 Leaf margins lobed and usually toothed . . . . *Viburnum* | VIBURNUM
3 Leaf margins entire 4

4 Flowers in pairs from leaf axils . . . . *Lonicera* | HONEYSUCKLE
4 Flowers in clusters at ends of stems . . . . *Viburnum* | VIBURNUM

LINNAEA | *Twinflower*

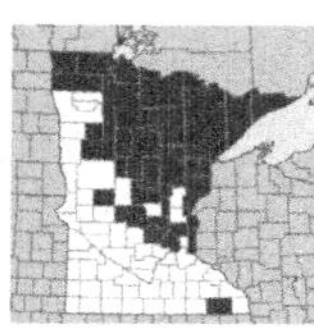

Linnaea borealis L.

TWINFLOWER
NC **FAC** | MW **FAC** | GP **FACU**

Native evergreen, trailing vine. **Stems** slightly woody, to 1–2 m long, with many short, erect, leafy branches to 10 cm long; **branches** green to red-brown, finely hairy; older stems woody, 2–4 mm wide. **Leaves** opposite, simple, evergreen, oval to round, 1–2 cm long, blunt at tip, upper surface and margins with short, straight hairs; margins rolled under, with a few rounded teeth near tip; petiole short, short-hairy. **Flowers** small, pink to white, bell-shaped, shallowly 5-lobed, slightly fragrant, in nodding pairs atop a Y-shaped stalk to 10 cm long, the stalk with gland-tipped hairs and 2 small bracts at the fork and a pair of smaller bracts at base of each flower. **Fruit** a small, dry, 1-seeded capsule. June–Aug. **Synonyms** *Linnaea americana*.

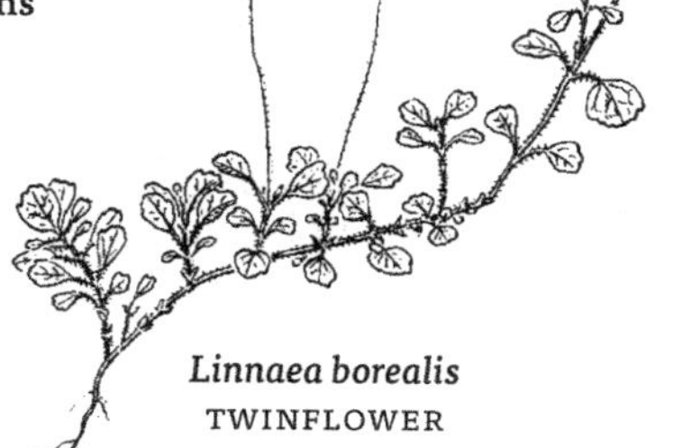

Linnaea borealis
TWINFLOWER

Habitat Hummocks in cedar swamps and thickets, moist conifer woods, on rotten logs and mossy boulders.

***Linnaea** is named for **Carl von Linné**, the 18th century Swedish botanist regarded as the father of modern taxonomy.*

LONICERA | *Honeysuckle*

Shrubs (those included here) or vines. **Leaves** opposite, simple, entire. **Flowers** long and tubular or funnel-shaped, in pairs from leaf axils. **Fruit** a few-seeded, blue or red berry.

LONICERA | HONEYSUCKLE

1 Flowers and fruits on short stalks (1 cm or less), fruit blue; bark red-brown and peeling in papery layers .... *Lonicera caerulea* | WATERBERRY
1 Flowers and fruits on longer stalks (1 cm or more), fruit red; bark gray-brown and shredding .. *Lonicera oblongifolia* | SWAMP FLY-HONEYSUCKLE

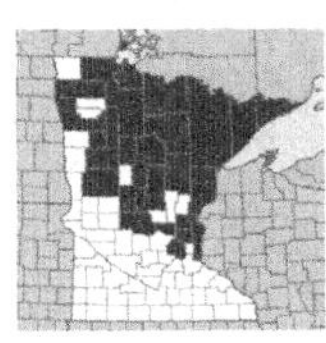

Lonicera caerulea L.

WATERBERRY

NC **FACW** | MW **FACW** | GP **FACW**

Native shrub to 1 m tall; **branches** upright, red-brown to gray, outer thin layers soon peeling to expose red-brown inner layers; **twigs** purple-red, with long, soft hairs. **Leaves** opposite, oval to oblong, 2–6 cm long and 1–3 cm wide, blunt or rounded at tip, upper surface dark green, underside paler and hairy, especially on veins; margins fringed with hairs and often rolled under; petioles absent or to 1–2 mm long. **Flowers** yellow, tubular to funnel-shaped, 10–15 mm long, in pairs on short hairy stalks from axils of lower leaves. **Fruit** an edible dark blue berry consisting of the 2 joined ovaries. May–July. **Synonyms** *Lonicera villosa*. **Habitat** Cedar and tamarack swamps, thickets, fens, shores.

Lonicera caerulea
WATERBERRY

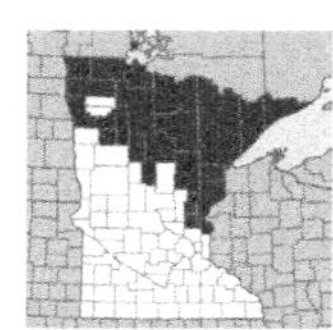

Lonicera oblongifolia (Goldie) Hook.
SWAMP FLY-HONEYSUCKLE
NC **OBL** | MW **OBL** | GP **OBL**

Native thicket-forming shrub 1–1.5 m tall; **branches** upright, with shredding bark and solid pith; **twigs** green to purple, smooth. **Leaves** opposite, oblong or oval, 3–8 cm long and 1–4 cm wide, rounded or blunt at tip, underside hairy when young, becoming smooth; margins entire, not fringed with hairs; petioles absent or to 1–2 mm long. **Flowers** yellow-white, tube-shaped with 2 spreading lips, 10–15 mm long, in pairs at ends of slender stalks up to 4 cm long from leaf axils. **Fruit** an orange-red to red (or sometimes purple), few-seeded berry composed of the 2 joined ovaries. May–June. **Synonyms** *Xylosteon oblongifolia*. **Habitat** Cedar and tamarack swamps, fens, open bogs, wet streambanks and shores; often over limestone.

Lonicera oblongifolia
SWAMP FLY-HONEYSUCKLE

SAMBUCUS | *Elder*

Shrubs or small trees. **Stems** pithy, the bark with wartlike lenticels. **Leaves** pinnately divided. **Flowers** small, white, perfect, 5-lobed; in large, rounded clusters at ends of stems. **Fruit** a red or dark purple, berrylike drupe.

SAMBUCUS | ELDER

1 Flowers opening in summer after leaves developed, in broad, ± flat clusters; fruit purple-black, edible; leaflets usually 7 .... *Sambucus canadensis* .... COMMON ELDER

1 Flowers opening in late spring with unfolding leaves, in pyramid-shaped or rounded clusters; fruit red, inedible; leaflets usually 5 .... *Sambucus racemosa* | RED-BERRIED ELDER

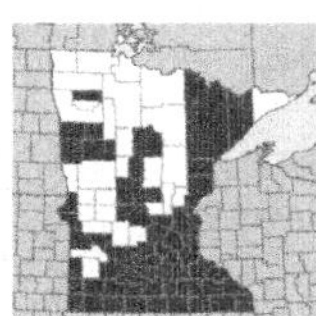

Sambucus canadensis L.

COMMON ELDER

NC **FACW** | MW **FACW** | GP **FAC**

Native shrub to 3 m tall; spreading underground and forming thickets. Young **stems** soft or barely woody, smooth; older stems with warty gray-brown bark; inner pith white. **Leaves** large, opposite, pinnately divided into 5–11 (usually 7) leaflets, the lower pair of leaflets sometimes divided into 2–3 segments; leaflets lance-shaped to oval, tapered to a long sharp tip, base often asymmetrical, smooth or hairy on underside, especially along veins; margins with sharp, forward-pointing teeth. **Flowers** small, white, 5-parted, 3–5 mm wide, numerous, in flat or slightly rounded clusters 10–15 cm wide at ends of stems. **Fruit** a round, purple-black, berrylike drupe, edible. July–Aug (blooming when fruit of *Sambucus racemosa* is about ripe). **Synonyms** *Sambucus nigra*. **Habitat** Floodplain forests, swamps, wet forest depressions, thickets, shores, meadows, roadsides, fencerows.

Sambucus canadensis
COMMON ELDER

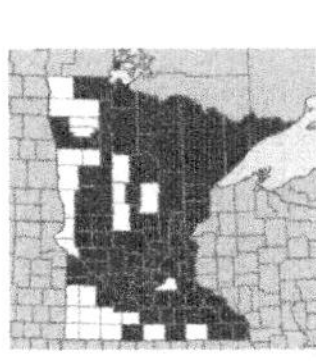

Sambucus racemosa L.

RED-BERRIED ELDER

NC **FACU** | MW **FACU** | GP **FACU**

Native shrub to 3 m tall. **Stems** soft or barely woody; twigs yellow-brown and hairy, branches with warty gray-brown bark; inner pith red-brown. **Leaves** large, opposite, pinnately divided into 5–7 (usually 5) leaflets, the leaflets lance- shaped to ovate, tapered to a long sharp tip, smooth or hairy on underside; margins with small, sharp, forward-pointing teeth. **Flowers** small, white, 5-parted, 3–4 mm wide, many, in elongate, pyramidal clusters at ends of stems, the clusters 5–12 cm long and usually longer than wide. **Fruit** a round, red, berrylike drupe, inedible. May–June (flowers opening with developing leaves). **Synonyms** *Sambucus pubens*. **Habitat** Occasional in swamps and thickets; more common in moist deciduous forests, roadsides and fencerows.

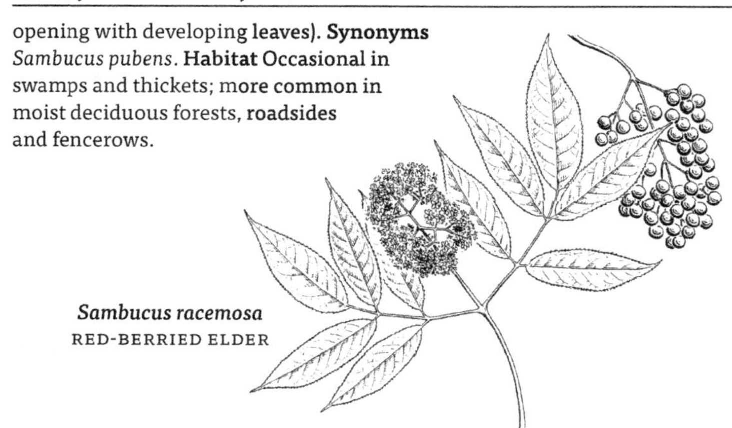

Sambucus racemosa
RED-BERRIED ELDER

VIBURNUM | *Viburnum*

Shrubs or small trees. **Leaves** simple, entire or toothed, often palmately lobed. **Flowers** white or pink, in rounded clusters at ends of stems, sometimes outer florets larger and sterile. **Fruit** a drupe with a large seed; white, yellow, pink, or orange at first, maturing to orange, red, or blue-black.

VIBURNUM I VIBURNUM

1 Flowers on leafy shoots at ends of stems, of 2 types; outer flowers large and showy, inner flowers smaller . . . . *Viburnum opulus* . . . . HIGH-BUSH CRANBERRY

1 Flowers few, all alike, on short branches from lateral buds . . . . *Viburnum edule* | SQUASHBERRY; LOW-BUSH CRANBERRY

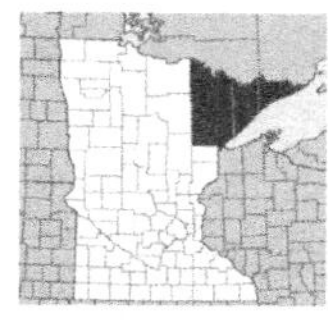

Viburnum edule (Michx.) Raf.

SQUASHBERRY; LOW-BUSH CRANBERRY
NC **FACW** I MW **FACW** I GP **FACW**

Native shrub 1–2 m tall. **Stems** upright or spreading; twigs brown-purple, smooth, often angled or ridged. **Leaves** opposite, mostly shallowly 3-lobed and palmately veined (leaves at ends of stems often unlobed), 5–12 cm long and 3–12 cm wide, tapered to a sharp tip; underside veins hairy; margins coarsely toothed; petioles 1–3 cm long, **Flowers** creamy-white, small, in few-flowered, stalked clusters 1–3 cm wide, on short, 2-leaved branches from lateral buds on last year's shoots. **Fruit** a round, berrylike drupe 6–10 mm long, yellow at first, becoming orange or red. June–July. **Synonyms** *Viburnum pauciflorum*, *Viburnum eradiatum*. **Habitat** Moist conifer forests, thickets, forest openings, talus slopes.

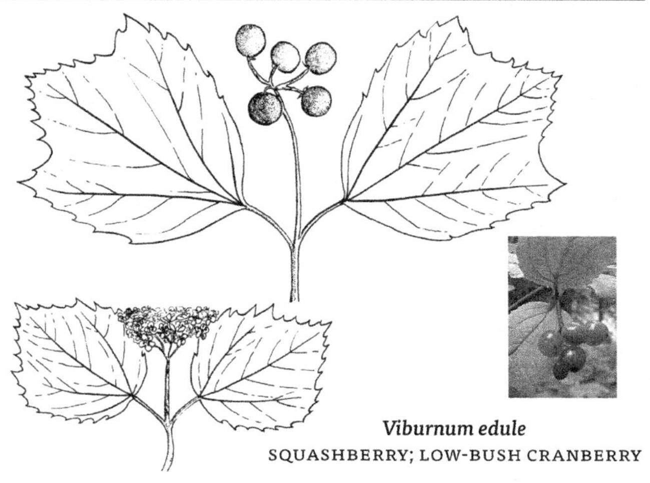

Viburnum edule
SQUASHBERRY; LOW-BUSH CRANBERRY

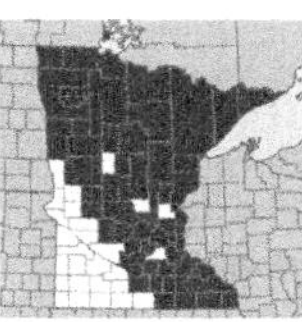

Viburnum opulus L.

HIGH-BUSH CRANBERRY
NC **FACW** | MW **FAC** | GP **FAC**

Native shrub, 3–4 m tall. Young **stems** smooth. **Leaves** opposite, maple-like, sharply 3-lobed and palmately veined, 5–10 cm long and about as wide, the lobes tapered to sharp tips; smooth or hairy beneath, especially on the veins; margins entire or coarsely toothed, petioles grooved, 1–3 cm long, with several club-shaped glands present near base of blade. **Flowers** white, in large, flat-topped clusters 5–15 cm wide at ends of stems; outer flowers sterile with large petals, surrounding the inner, smaller fertile flowers. **Fruit** an orange to red, round or oval drupe, 10–15 mm long. June. **Synonyms** Our plants are var. *americanum* (sometimes considered a separate species —*Viburnum trilobum*). **Habitat** Swamps, fens, streambanks, shores, ditches.

Viburnum opulus
HIGH-BUSH CRANBERRY

ANNUAL OR PERENNIAL HERBS. Leaves simple, entire, mostly opposite but sometimes alternate or whorled. **Stems** often swollen at nodes. **Flowers** perfect (with both male and female parts) or imperfect, in open or compact heads at ends of stems or from leaf axils; sepals usually 5, separate or joined into a tube; petals 5 (sometimes 4), separate, often lobed or toothed, sometimes absent; stamens 3–10, anthers often distinctly colored. **Fruit** a few- to many-seeded capsule.

CARYOPHYLLACEAE | PINK FAMILY

1 Sepals joined to form a toothed or lobed tube . . . . *Silene nivea* . . . . SNOWY CAMPION
1 Sepals free or joined only at base . . . . *Stellaria* | STITCHWORT

SILENE | *Catchfly, Campion*

Silene nivea (Nutt.) Otth
SNOWY CAMPION
NC **FACW** | MW **FACW** | GP **FACW** | *threatened*

Native perennial herb, spreading by rhizomes; plants smooth or with a few short hairs. **Stems** 2–3 dm long. **Leaves** mostly on stem, opposite, lance-shaped or oblong, 5–10 cm long and 1–3 cm wide, stalkless or on short petioles. **Flowers** few, mostly in leaf axils; sepals joined to form a tubelike flower 1.5 cm long; petals white, stamens 10, styles 3. **Fruit** a 1-chambered capsule. June–July. **Synonyms** *Silene alba*. **Habitat** Streambanks, open alluvial situations, margins of calcareous fens. **Status** Threatened.

Silene nivea
SNOWY CAMPION

STELLARIA | *Stitchwort*

Low, spreading or erect perennials (ours), mostly without hairs. **Stems** slender, 4-angled. **Flowers** single in forks of stems or in few-flowered clusters at ends of stems; sepals green with translucent margins; petals white, lobed or deeply cleft (sometimes absent in *Stellaria borealis*); stamens 10 or less; styles 3. **Fruit** an ovate or oblong capsule.

STELLARIA | STITCHWORT

1 Plants large, stems to 8 dm long; styles 5 .... *Stellaria aquatica* .... GIANT CHICKWEED
1 Plants smaller; styles 3–4 2

2 Flowers single in forks of stems, not subtended by membranous bracts 3
2 Flowers in branched terminal clusters, subtended by small membranous bracts 4

3 Stems 25 cm or more long; seeds smooth .... *Stellaria borealis* .... NORTHERN STITCHWORT
3 Stems to 20 cm long; seeds rough .... *Stellaria crassifolia* .... FLESHY STITCHWORT

4 Petals shorter than sepals 5
4 Petals much longer than sepals 6

5 Flowers mostly in clusters from leaf axils; seeds covered with small bumps .... *Stellaria alsine* | BOG-STITCHWORT
5 Flowers in clusters at ends of stems; seeds smooth .... *Stellaria borealis* .... NORTHERN STITCHWORT

6 Heads open and widely branched, the stalks spreading; leaves spreading to ascending, widest at or above middle .... *Stellaria longifolia* .... LONG-LEAVED STITCHWORT
6 Head narrow, the stalks erect to ascending; leaves upright, widest near base .... *Stellaria longipes* | LONG-STALKED STITCHWORT

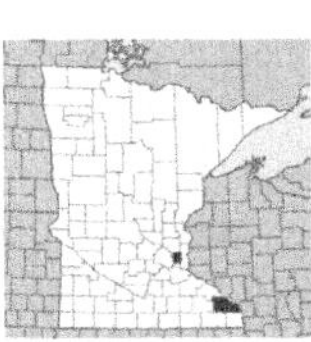

Stellaria alsine Grimm

BOG-STITCHWORT
NC **OBL** | MW **OBL**

Native annual herb. **Stems** sprawling, smooth, angled, rooting at nodes. **Leaves** oval to oblong, 1.5–3 cm long and to 1 cm wide; lower leaves on petioles, upper leaves stalkless. **Flowers** in few-flowered clusters from leaf axils, sepals 3–4 mm long, petals white, shorter than sepals. **Fruit** a capsule, longer than the sepals; seeds 0.5 mm long, covered with small bumps. May–Aug. **Synonyms**

Alsine uliginosa. **Habitat** Marshes, streambanks, seeps; native to ne USA; considered adventive in se Minnesota near Mississippi River.

Stellaria alsine
BOG-STITCHWORT

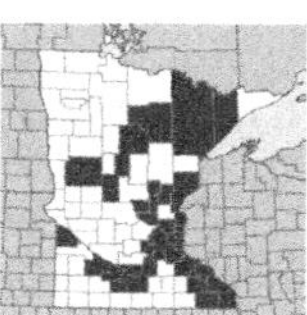

Stellaria aquatica (L.) Scop.
GIANT CHICKWEED
NC **FAC** | MW **FACW** | GP **FAC**
Introduced perennial herb, spreading by rhizomes. **Stems** sprawling and matted, to 8 dm long, rooting at nodes, covered with gland-tipped hairs. **Leaves** ovate to lance-shaped, 2–8 cm long and 1–4 cm wide, petioles short or absent. **Flowers** in open, leafy clusters at ends of stems; sepals 5–9 mm long; petals white, much longer than sepals. **Fruit** a capsule; seeds 0.8 mm long, covered with small bumps. June–Oct. **Synonyms** *Alsine aquatica, Myosoton aquaticum*. **Habitat** Streambanks, ponds, wet or moist disturbed areas, often in partial shade.

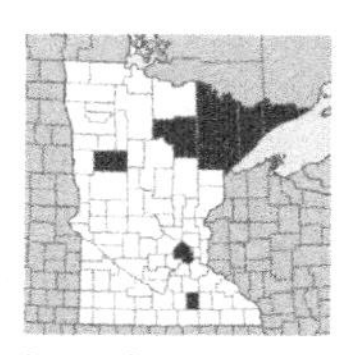

Stellaria borealis Bigelow
NORTHERN STITCHWORT
NC **FACW** | MW **OBL** | GP **FACW**
Native perennial herb, spreading by rhizomes. **Stems** sprawling, to 5 dm long, branched, angled. Leaves lance-shaped, narrowed at base, 1–5 cm long and 2–8 mm wide, margins hairy. **Flowers** in clusters at ends of stems; sepals 2–4 mm long; petals usually absent. **Fruit** a dark capsule, longer than sepals; seeds to 1 mm long, nearly smooth. June–Aug. **Synonyms** *Alsine borealis, Stellaria calycantha*. **Habitat** Openings and hollows in conifer forests, margins of ponds and marshes.

Stellaria borealis
NORTHERN STITCHWORT

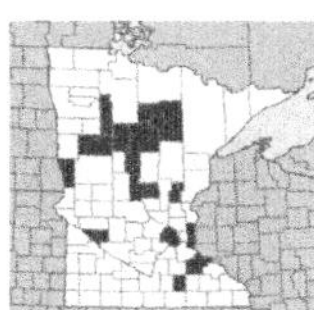

Stellaria crassifolia Ehrh.
FLESHY STITCHWORT
NC **FACW** | MW **FACW** | GP **OBL**

Native perennial herb. **Stems** sprawling and matted to erect, freely branched, 8–30 cm long, fleshy, smooth. **Leaves** soft, oval to lance-shaped, narrowed at base, 1–3 cm long and 1–3 mm wide. **Flowers** single in forks of stem, nodding on stalks 1–3 cm long; sepals 2–4 mm long; petals longer than sepals. **Fruit** an ovate capsule, to 5 mm long and longer than the sepals; seeds red-brown, to 1 mm long. June–July. **Habitat** Streambanks and wet shores.

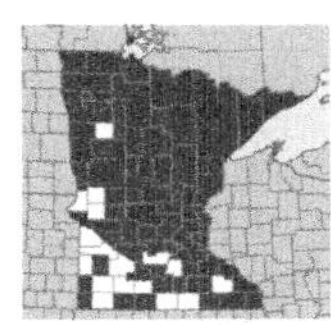

Stellaria longifolia Muhl.
LONG-LEAVED STITCHWORT
NC **FACW** | MW **FACW** | GP **FACW**

Native perennial herb. **Stems** sprawling, prominently 4-angled, usually freely branched, 1–5 dm long. **Leaves** spreading to ascending, linear to lance-shaped, 2–5 cm long and 1–6 mm wide, widest at or above middle, tapered at both ends. **Flowers** in branched clusters at ends of stems; sepals 3–5 mm long; petals longer than sepals. **Fruit** a green-yellow to brown capsule, usually longer than the sepals; seeds light brown, about 1 mm long. May–July. **Habitat** Wet meadows and marshes, shrub thickets, swamps, streambanks, pond margins.

Stellaria longifolia
LONG-LEAVED STITCHWORT

Stellaria crassifolia
FLESHY STITCHWORT

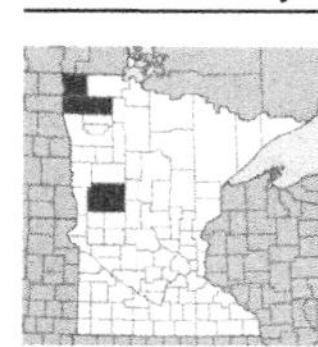

Stellaria longipes Goldie
LONG-STALKED STITCHWORT
NC **FACU** | MW **OBL** | GP **OBL** | *special concern*

Native perennial herb, spreading by rhizomes. **Stems** erect, or sprawling and matted, 5–30 cm long. **Leaves** upright, stiff and shiny-waxy, linear or lance-shaped, 1–4 cm long and 1–4 mm wide, widest near base, tapered to tip. **Flowers** in branched clusters at ends of stems or appearing lateral from stem; sepals 4–5 mm long; petals slightly longer than sepals. **Fruit** a straw-colored to shiny purple capsule, longer than the sepals; seeds red-brown, oblong to oval, about 1 mm long. May–July. Ours are subsp. *longipes*. **Habitat** Wet to moist, sandy shores and meadows. **Status** Special concern.

Stellaria longipes
LONG-STALKED STITCHWORT

CERATOPHYLLUM | *Coontail, Hornwort*

Aquatic perennial herbs, often forming large patches; roots absent, but plants usually anchored to substrate by pale, modified leaves. **Stems** slender, branched. **Leaves** in whorls, with more than 4 leaves per node, whorls crowded at ends of stems (hence the common name of coontail), dissected 2–3x into narrow segments. **Flowers** small, inconspicuous in leaf axils, male and female flowers separate on same plant, male usually above female on stems.

CERATOPHYLLUM | COONTAIL, HORNWORT

1 Leaves usually stiff, forked 1–2 times, margins coarsely toothed; achenes with 2 spines near base . . . . *Ceratophyllum demersum* . . . . COMMON HORNWORT

1 Leaves limp, some larger leaves forked 3–4 times, margins not toothed; achenes with several spines along each margin. . *Ceratophyllum echinatum* . . . . PRICKLY HORNWORT

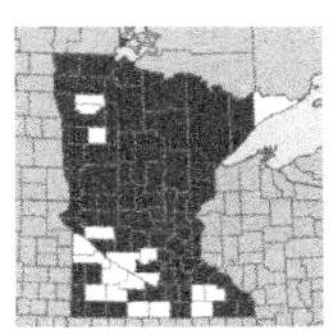

Ceratophyllum demersum L.

COMMON HORNWORT

NC **OBL** | MW **OBL** | GP **OBL**

Native aquatic herb. **Stems** long, branched. **Leaves** in whorls of 5–12 at each node, stiff, 1–3 cm long, 1–2-forked; leaf segments linear, 0.5–1 mm wide, coarsely toothed. **Fruit** an oval achene, 4–6 mm long, with 2 spines at base. **Habitat** Common; shallow to deep water of lakes, ponds, backwater areas, ditches, and may form dense masses; water typically neutral or alkaline.

Ceratophyllum demersum
COMMON HORNWORT

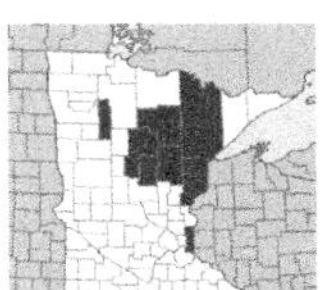

Ceratophyllum echinatum A. Gray

PRICKLY HORNWORT

NC **OBL** | MW **OBL** | GP **OBL**

Native aquatic herb. Similar to *Ceratophyllum demersum*, but **leaves** usually limp, larger leaves usually 3- or sometimes 4-forked, the segments narrower and mostly without teeth. **Fruit** an achene with 2 spines at base and several unequal spines on achene body. **Synonyms** *Ceratophyllum muricatum*. **Habitat** Lakes, ponds and quiet water of rivers and streams; water acidic.

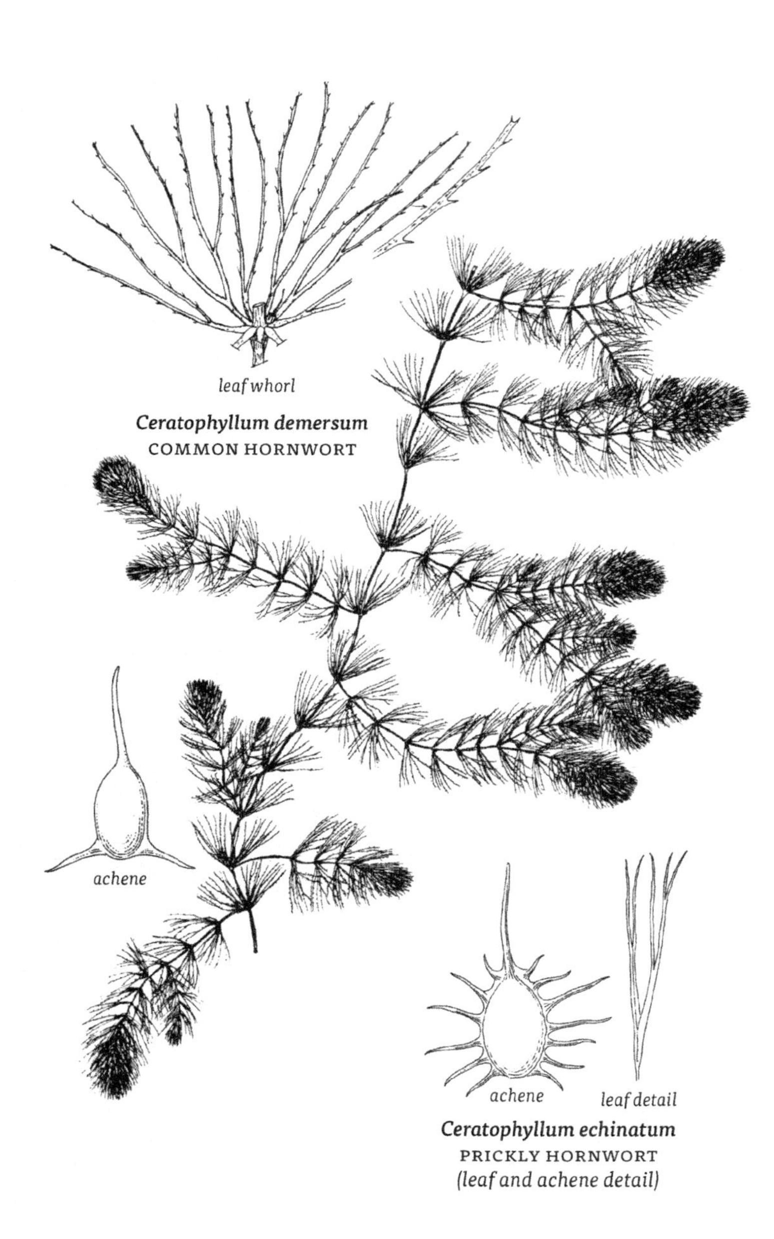

Ceratophyllum demersum
COMMON HORNWORT

Ceratophyllum echinatum
PRICKLY HORNWORT
(leaf and achene detail)

ANNUAL OR PERENNIAL HERBS, often in alkaline soil. **Stems** often angled or jointed, succulent in *Salicornia*. Leaves simple, alternate, or occasionally opposite (*Salicornia*), sometimes covered with thin, flaky scales giving a mealy appearance. **Flowers** 1 to many, small, green or red-tinged, clustered in leaf axils or at ends of stems; perfect, or male and female flowers separate; sepals usually 5; petals absent; ovary superior, 1-chambered. **Fruit** a 1-seeded utricle.

CHENOPODIACEAE | GOOSEFOOT FAMILY

1 Leaves reduced to scales, opposite; stems succulent...... *Salicornia rubra* .... WESTERN GLASSWORT
1 Leaves not scalelike, mostly alternate; stems not succulent **2**

2 Leaves without petioles, linear, round in cross-section .... *Suaeda calceoliformis* | PLAINS SEA-BLITE
2 Leaves mostly with petioles, blades broader **3**

3 Male and female flowers separate, in spikelike heads without leafy bracts or with only a few bracts low on spike; sepals and petals absent in female flowers, fruit enclosed by a pair of sepal-like small bracts............... .... *Atriplex patula* | HALBERD-LEAF SALTBUSH; SPEARSCALE
3 Flowers perfect (with both male and female parts), in spikes which have small leafy bracts throughout; fruit surrounded by the persistent sepals and petals .... *Chenopodium* | GOOSEFOOT

ATRIPLEX | *Saltbush*

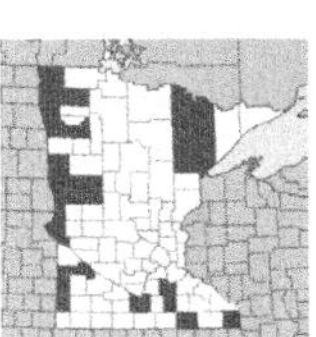

Atriplex patula L.

HALBERD-LEAF SALTBUSH; SPEARSCALE
NC **FACW** | MW **FACW** | GP **FACW**

Taprooted annual herb, introduced. **Stems** erect to sprawling, usually branched, 2–10 dm long. **Leaves** alternate (or the lowest opposite), lance-shaped or triangular, 2–8 cm long and 1–6 cm wide, with outward pointing basal lobes, gray and mealy when young, becoming dull green and smooth with age; petioles present, or absent on upper leaves. **Flowers** tiny, green; either male or female but on same plant, usually intermixed in crowded spikes from leaf axils and at ends of stems, the spikes simple or branched, without bracts or with a few small bracts near base of spikes; male flowers with a 5-lobed group of sepals,

stamens 5; female flowers without sepals or petals, surrounded by 2 sepal-like, small bracts, these expanding and enclosing fruit when mature. **Fruit** lens-shaped, dark brown to black, 1–3 mm wide. Aug–Sept. **Synonyms** *Atriplex acadiensis, Atriplex hastata*. **Habitat** Introduced; shores, streambanks and mud flats, usually where brackish; disturbed places.

CHENOPODIUM | *Goosefoot*

Taprooted annual herbs. **Stems** erect to spreading. **Leaves** alternate, mostly lance-shaped to broadly triangular, somewhat fleshy and often mealy on lower surface. **Flowers** perfect, small and numerous, green or red-tinged, in dense spikelike clusters from leaf axils or at ends of stems, the spikes with small leafy bracts; sepals often curved over the fruit; petals absent; stamens 1–5; styles 2–3. **Fruit** a 1-seeded utricle.

CHENOPODIUM | GOOSEFOOT

1 Leaves persistently white-mealy on underside, dull green above .... *Chenopodium glaucum* | OAK-LEAVED GOOSEFOOT
1 Leaves not white-mealy when mature, green on upper and lower sides, often red-tinged .... *Chenopodium rubrum* | ALKALI-BLITE

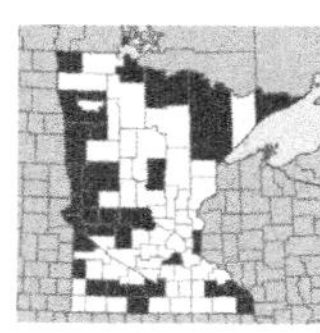

Chenopodium glaucum L.

OAK-LEAVED GOOSEFOOT
NC **FACW** | MW **FACW** | GP **FAC**

Introduced annual herb. **Stems** upright to sprawling, 1–6 dm long, usually branched from base, sometimes red-tinged. **Leaves** lance-shaped to ovate, 1–4 cm long and to 2 cm wide, dull green above, densely white-mealy on underside (especially when young); margins entire, wavy, or with few rounded teeth; petioles slender, shorter on upper leaves. **Flowers** in small, often branched, spikelike clusters from leaf axils, the spikes often shorter than leaves; sepals mostly 3; petals absent. Seeds dark brown, shiny, 1 mm wide. Aug–Oct. **Synonyms** *Chenopodium salinum*. **Habitat** Shores, streambanks, and disturbed areas such as railroad ballast and barnyards; soils often brackish. Introduced from Eurasia and naturalized across much of USA and Canada.

Chenopodium glaucum
OAK-LEAVED GOOSEFOOT

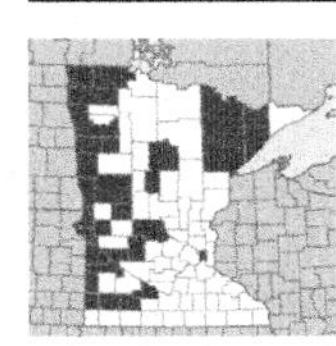

Chenopodium rubrum L.
ALKALI-BLITE
NC **OBL** | MW **OBL** | GP **OBL**

Native annual herb. **Stems** usually erect, sometimes sprawling, 1–8 dm long, often branched from base. **Leaves** lance-shaped to broadly triangular, 2–10 cm long and 1–8 cm wide, green and often red-tinged on both surfaces, smooth, not mealy; margins wavy-toothed or lobed; petioles present. **Flowers** small, in upright, branched spikes from leaf axils and at ends of stems, the spikes often longer than leaves; sepals 3, red; petals absent. Seeds dark brown, shiny, to 1 mm wide. Aug–Oct. **Habitat** Lakeshores, streambanks, wet disturbed areas.

SALICORNIA | *Glasswort*

Salicornia rubra A. Nels.
WESTERN GLASSWORT
NC **OBL** | MW **OBL** | GP **OBL** | *threatened*

Taprooted annual herb, native; plants succulent, green to bright red. **Stems** 0.5–2 dm long; branches opposite, fleshy, jointed at nodes, breaking apart when plants are trampled. Leaves opposite, small and scalelike, 1–2 mm long. **Flowers** in spikes 1–5 cm long at ends of stems; perfect, or some flowers female only; sepals enclosing flower except for a small opening from which stamens and style branches protrude; petals absent. Seeds 1 mm long. Aug–Oct. **Habitat** Shores, swales and ditches; soils brackish. **Status** Threatened; more common in w USA.

Salicornia rubra
WESTERN GLASSWORT

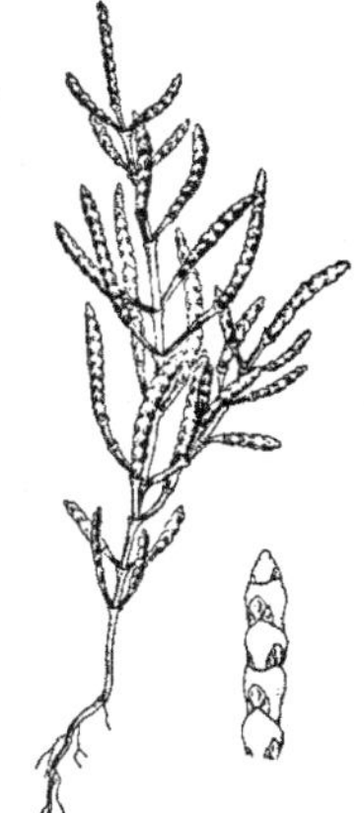

SUAEDA | *Sea-blite*

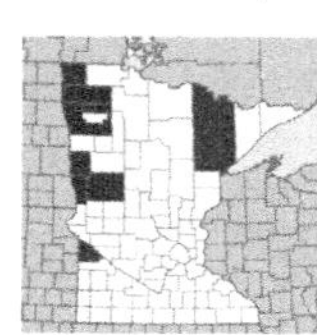

Suaeda calceoliformis (Hook.) Moq.
PLAINS SEA-BLITE
NC **FACW** | MW **FACW** | GP **FACW**

Annual taprooted herb, native. **Stems** upright to sprawling, usually branched, 0.5–6 dm long. **Leaves** alternate, linear, flat on 1 side and convex on other, green, succulent, 5–30 mm long and 1 mm wide, reduced to wider bracts 1–5 mm long in the head; petioles absent. **Flowers** small, perfect, or male or female flowers separate, green or sometimes red-tinged, in dense clusters of 3–7 flowers in bract axils; sepals joined, deeply 5-lobed, the lobes unequal and hooded; stamens 5. **Fruit** a utricle enclosed by the sepals; seeds black, shiny, about 1 mm wide. July–Sept. **Synonyms** *Suaeda depressa*. **Habitat** Brackish wetlands, and along salted highways.

Suaeda calceoliformis
PLAINS SEA-BLITE

CORNUS | *Bunchberry, Dogwood*

Shrubs, or herbaceous shoots from a woody rhizome in bunchberry (*Cornus canadensis*). **Leaves** opposite, simple, entire. **Flowers** 4-parted, perfect, in a rounded or flat-topped cluster; sepals and petals small. **Fruit** a berrylike drupe with 1–2 hard seeds.

The shrubby dogwoods are preferred deer foods and are often reduced in size due to repeated browsing.

CORNUS | BUNCHBERRY, DOGWOOD

1 Plants herbaceous from a woody rhizome, less than 3 dm tall; leaves whorled............. *Cornus canadensis* | BUNCHBERRY; DWARF CORNEL
1 Taller shrubs, 5 dm or more tall; leaves opposite or alternate........... 2

2 Leaves alternate on stems.......................... *Cornus alternifolia* .... PAGODA DOGWOOD; ALTERNATE-LEAVED DOGWOOD
2 Leaves opposite.. 3

3 Fruit white; young twigs densely short-hairy .... *Cornus amomum* .... SILKY DOGWOOD
3 Fruit blue; young twigs ± smooth................................ 4

4 Twigs gray; leaves with fewer than 5 pairs of lateral veins.............. .... *Cornus racemosa* | NORTHERN SWAMP DOGWOOD
4 Twigs red; leaves with 5 or more pairs of lateral veins ...... *Cornus sericea* .... RED-OSIER DOGWOOD

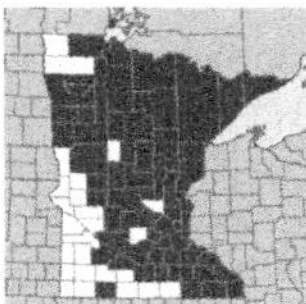

Cornus alternifolia L. f.
PAGODA DOGWOOD;
NC **FACU** | MW **FAC** | GP **FACU**
ALTERNATE-LEAVED DOGWOOD

Native shrub, to 5 m tall; **twigs** red-green or brown, somewhat shiny, alternate on stems, pith white. **Leaves** alternate, sometimes crowded and appearing whorled near ends of stems, oval to ovate, 5–12 cm long and 3–7 cm wide, tapered to a sharp tip, underside finely hairy; lateral veins 4–5 pairs, these curving toward tip of blade; margins entire; petioles to 5 cm long. **Flowers** small, creamy-white, in crowded, flat-topped or rounded clusters at ends of stems. **Fruit** a round, blue, berrylike drupe, 6 mm wide, atop a red stalk. May–July. **Synonyms** *Svida*

alternifolia. **Habitat** Swamps, thickets, streambanks and springs; also in drier deciduous and mixed forests.

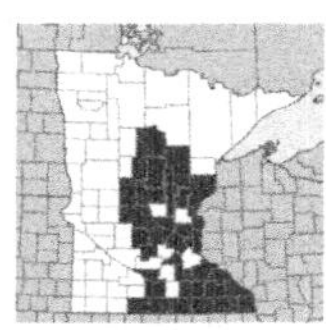

Cornus amomum Miller
SILKY DOGWOOD
NC **FACW** | MW **FACW** | GP **FACW**

Native shrub, 1–3 m tall; older branches red and gray-streaked, young twigs gray, finely hairy; pith brown. Leaves opposite, oval to ovate, 5–12 cm long and 2–5 cm wide, usually less than half as wide as long, tapered to a sharp tip, lateral veins 4–6 on each side, underside finely hairy; margins entire; petioles 1–2 cm long, often curved and causing the leaves to droop. **Flowers** small, creamy-white, in flat-topped or slightly rounded, hairy clusters. **Fruit** a round, blue or blue-white, berrylike drupe, 8 mm wide, atop a long stalk. June–July (our latest flowering dogwood). **Synonyms** *Cornus obliqua*. **Habitat** Conifer swamps, marshes, open bogs, calcareous fens, lakeshores, streambanks, wet dunes.

Cornus amomum
SILKY DOGWOOD

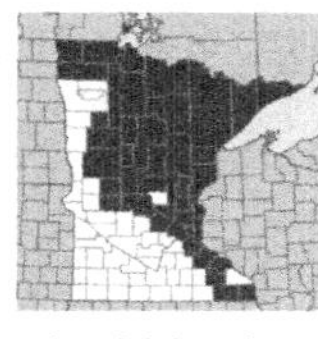

Cornus canadensis L.
BUNCHBERRY; DWARF CORNEL
NC **FAC** | MW **FAC** | GP **FACU**

Native perennial from horizontal, woody rhizomes, often forming large colonies. **Stems** erect, green, 1–2 dm tall, with a pair of small bracts on lower stem, topped with a whorl-like cluster of 4–6 leaves. **Leaves** oval to obovate, 4–7 cm long, tapered at both ends; lateral veins 2–3 pairs, arising from midvein below middle of blade; margins entire; petioles short or absent. **Flowers** small, yellow-green or creamy-white in a single cluster at end of a stalk 1–3 cm long; flowers surrounded by 4 white or pinkish, petal-like showy bracts, 1–2 cm long, these soon deciduous. **Fruit** a cluster of round, bright red berrylike drupes, the drupes 6–8 mm wide. June–July. **Habitat** Cedar swamps, thickets and moist conifer forests, often on hummocks or rotting logs; also in drier, mixed conifer-deciduous forests.

Cornus canadensis
BUNCHBERRY; DWARF CORNEL

Cornus canadensis (BUNCHBERRY; DWARF CORNEL)

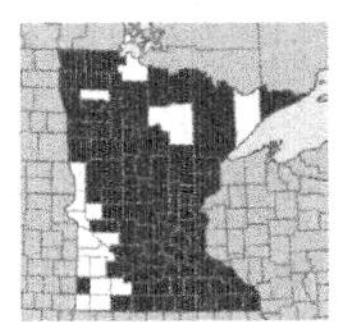

Cornus racemosa Lam.
NORTHERN SWAMP DOGWOOD
NC **FAC** | MW **FAC** | GP **FAC**

Native shrub, 1–3 m tall, often forming dense thickets; **twigs** red, becoming gray or light brown; pith usually brown. **Leaves** opposite, lance-shaped to oval, 4–9 cm long and 2–4 cm wide, abruptly tapered to a rounded tip, underside with short hairs; lateral veins 3 or 4 on each side of midvein; margins entire; petioles to 1 cm long. **Flowers** small, creamy-white, ill-scented, in numerous, open, elongated clusters. **Fruit** a round, berrylike drupe, at first lead-colored, becoming white, 5 mm wide, on red stalks. June–July. **Synonyms** *Cornus foemina*. **Habitat** Lakeshores, streambanks, swamps, thickets, marshes, moist woods, low prairie.

Cornus racemosa
NORTHERN SWAMP DOGWOOD

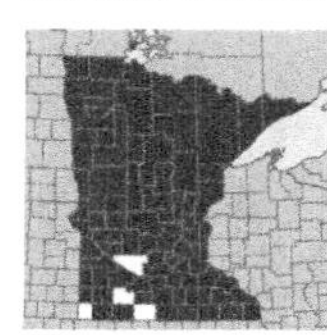

Cornus sericea L.
RED-OSIER DOGWOOD
NC **FACW** | MW **FACW** | GP **FACW**

Many-stemmed native shrub, 1–3 m tall, forming thickets; **branches** upright or prostrate and rooting; twigs and young branches red; pith white. **Leaves** opposite, green, ovate to oval, mostly 5–15 cm long and 2–7 cm wide, tapered to a tip, soft hairy on underside; margins entire; petioles to 2.5 cm long. **Flowers** small, white, many in flat-topped or slightly rounded clusters. **Fruit** a round, white or blue-tinged, berrylike drupe, 6–9 mm wide. May–Aug. **Synonyms** *Cornus alba, Cornus stolonifera*. **Habitat** Common; swamps, marshes, shores, streambanks, floodplain forests, shrub thickets, calcareous fens; also on sand dunes.

Cornus sericea
RED-OSIER DOGWOOD

CRASSULA | *Pygmy-weed*

Crassula aquatica (L.) Schönl
PYGMY-WEED
NC **OBL** | MW **OBL** | GP **OBL** | *threatened*

Small, easily overlooked, native annual herb. **Stems** branched, 2–10 cm long. **Leaves** opposite, linear, succulent, 3–6 mm long, spreading, margins entire, petioles absent. **Flowers** small, 1 mm long, single in leaf axils; petals white or green-white, 4, erect or slightly spreading; stamens 4, alternate with petals; pistils 3–4, with a short style. Aug–Sept. **Synonyms** *Tillaea aquatica*. **Habitat** The several northern Minnesota populations occur submersed in lakes within Voyageurs National Park; the Rock County plants (sw Minnesota) occupy muddy shores and shallow water. **Status** Threatened; species more common in w USA, Louisiana and Texas; also known from New England.

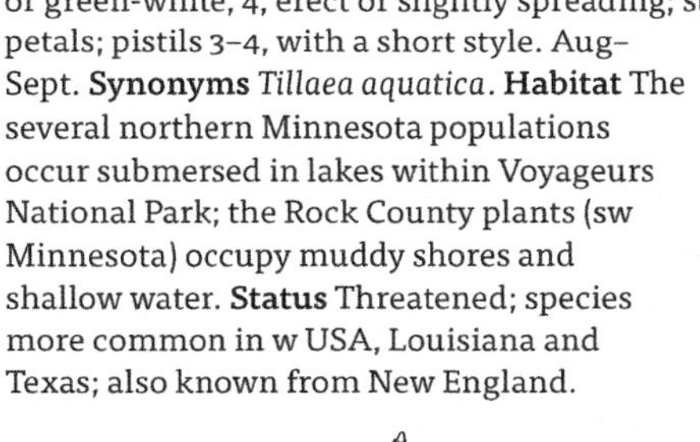

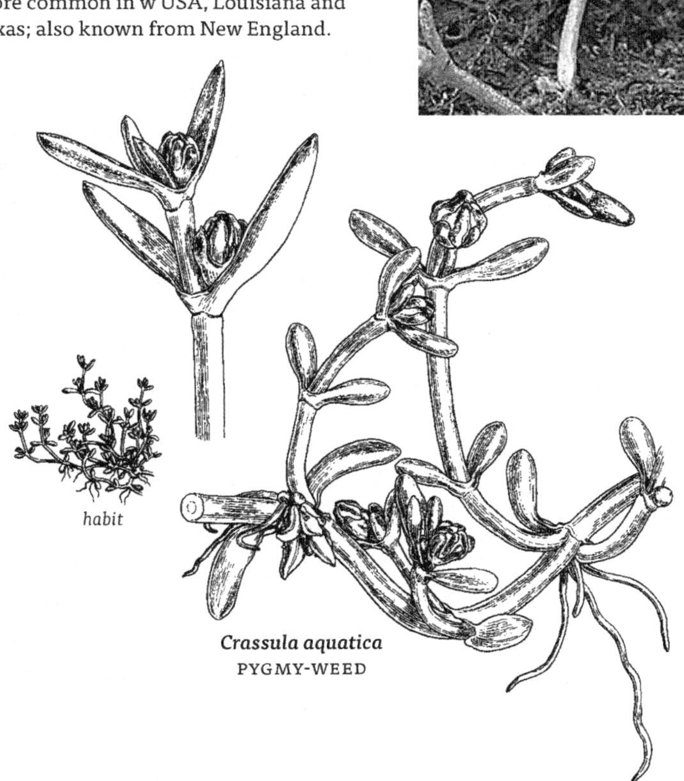

Crassula aquatica
PYGMY-WEED

ANNUAL HERBACEOUS VINES (our species). **Flowers** green or white, either male or female and on same plants (ours). **Fruit** a dry or fleshy, cucumber- or squash-like fruit (pepo).

CURCURBITACEAE | GOURD FAMILY

1 Male flowers 6-lobed; fruit inflated, 3–5 cm long, 4-seeded . . . . *Echinocystis lobata* | WILD CUCUMBER; BALSAM-APPLE

1 Male flowers 5-lobed; fruit not inflated, to 1.5 cm long, 1-seeded . . . . *Sicyos angulatus* | BUR-CUCUMBER

ECHINOCYSTIS | *Wild cucumber*

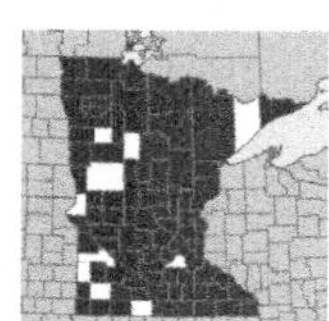

Echinocystis lobata (Michx.) T. & G.
WILD CUCUMBER; BALSAM-APPLE
NC **FACW** | MW **FACW** | GP **FAC**

Native annual vining herb, to 5 m or more long. **Leaves** round in outline, with 3–7 (usually 5) sharp, triangular lobes; petioles 3–8 cm long. **Flowers** white; male flowers 8–10 mm wide, with lance-shaped lobes, in long, upright racemes; female flowers 1 to several on short stalks from leaf axils. **Fruit** green, ovate, inflated, 3–5 cm long, with soft prickles. Aug–Sept. **Habitat** Floodplain forests, wet deciduous forests, streambanks, thickets, and waste ground.

fruit

SICYOS | *Bur-cucumber*

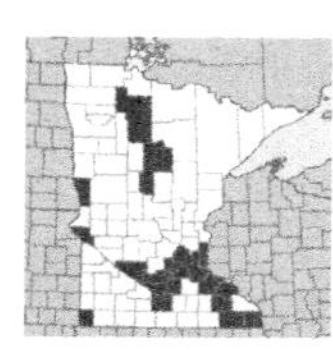

Sicyos angulatus L.
BUR-CUCUMBER
NC **FACW** | MW **FACW** | GP **FACW**

Native annual vining herb, to 2 m long. **Stems** angled, sticky-hairy, with branched tendrils. **Leaves** round in outline, with 3–5 shallow, toothed lobes, rough on both sides; petioles hairy, 3–10 cm long. **Flowers** green or white; male flowers 8–10 mm wide, 5-lobed, on stalks 10 cm or more long; female flowers on stalks to 8 cm long. **Fruit** yellow, ovate, 1.5 cm long, hairy and spine-covered. Aug–Sept. **Habitat** Floodplain forests, wet deciduous forests, streambanks, thickets and waste ground.

flowers

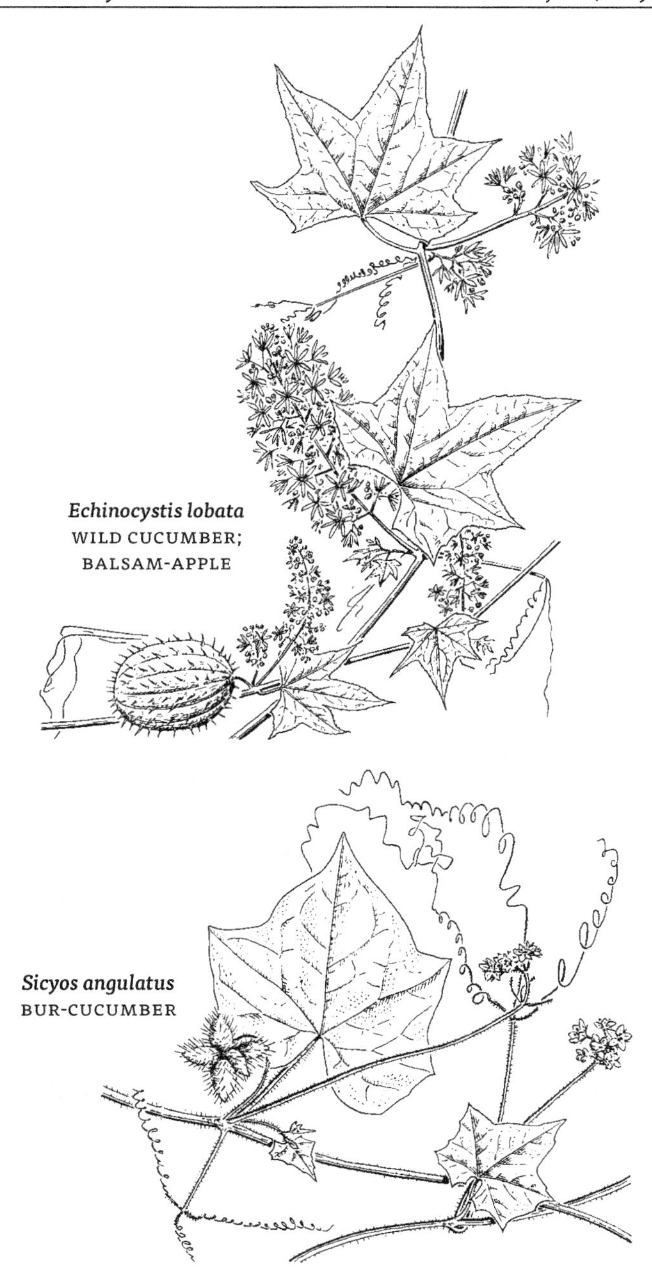

Echinocystis lobata
WILD CUCUMBER;
BALSAM-APPLE

Sicyos angulatus
BUR-CUCUMBER

DROSERA | *Sundew*

Perennial herbs. **Leaves** all from base of plant, covered with stalked, sticky glands that trap and digest insects. **Flowers** white, several, on 1 side of erect, leafless stalks, the stalks nodding at tip; with 5 petals and 5 sepals; stamens mostly 5, styles 3. **Fruit** a dry, many-seeded capsule.

DROSERA | SUNDEW

1 Leaves widely spreading, blades round, wider than long .... *Drosera rotundifolia* | ROUND-LEAVED SUNDEW

1 Leaves upright, blades linear or broad at tip and tapered to base, longer than wide 2

2 Leaf blades linear, 10–20x longer than wide; young petals pink .... *Drosera linearis* | LINEAR-LEAVED SUNDEW

2 Leaf blades broad near tip and narrowed to base, 2–7x as long as wide; young petals white 3

3 Blades 2–3x as long as wide, petioles without hairs; flower stalks from side of plant base and curving upward .... *Drosera intermedia* .... SPOON-LEAVED SUNDEW

3 Blades 5–7x as long as wide, petioles with some hairs; flower stalks erect from center of plant base .... *Drosera anglica* .... ENGLISH OR GREAT SUNDEW

Drosera anglica Hudson

ENGLISH OR GREAT SUNDEW

NC **OBL** | MW **OBL** | *special concern*

Native perennial insectivorous herb. **Leaf blades** obovate to spatula-shaped, 15–35 mm long and 3–4 mm wide, upper surface covered with gland-tipped hairs; petioles 3–6 cm long, smooth or with few glandular hairs. **Flowers** 1–9 in a racemelike cluster atop a stalk 6–25 cm tall; flowers 6–7 mm wide; sepals 5–6 mm long; petals white, 6 mm long, spatula-shaped. **Seeds** black, 1 mm long, with fine lines. June–Aug. **Habitat** Open sedge-dominated fens, often in 'water track' depressions and with *Carex lasiocarpa*. **Status** Special concern.

English sundew *similar to* **spoon-leaved sundew** *(Drosera intermedia) but rarely occur together. Plants of Drosera anglica are generally larger, with shorter petioles (1–3x as long as leaf blades vs. 2.5–3.5x as long in D. intermedia).*

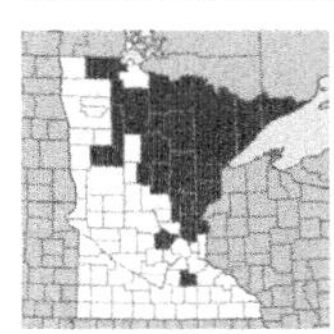

Drosera intermedia Hayne
SPOON-LEAVED SUNDEW
NC **OBL** | MW **OBL** | GP **OBL**

Native perennial insectivorous herb. **Leaves** in a basal rosette and also usually along lower stem; spatula-shaped, 2–4 mm wide, upper surface covered with long, gland-tipped hairs; petioles smooth, 2–5 cm long. **Flowers** on stalks to 20 cm tall; sepals 3–4 mm long; petals white, 4–5 mm long. **Seeds** red-brown, to 1 mm long, covered with small bumps. July–Sept. **Habitat** Low spots in open bogs, sandy shores, often in shallow water.

Drosera intermedia
SPOON-LEAVED SUNDEW

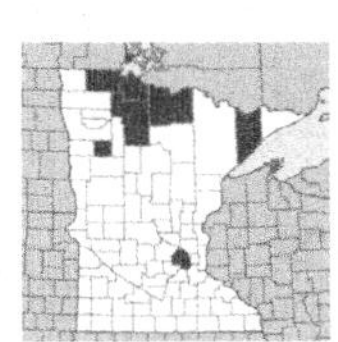

Drosera linearis Goldie
LINEAR-LEAVED SUNDEW
NC **OBL** | MW **OBL** | *special concern*

Native perennial insectivorous herb. **Leaf blades** linear, 2–5 cm long and 2 mm wide; petioles smooth, 3–7 cm long. **Flowers** 1–4 atop stalks 6–15 cm tall; flowers 6–8 mm wide; sepals 4–5 mm long; petals obovate, 6 mm long, white. **Seeds** black, less than 1 mm long, with small craterlike pits on surface. June–Aug. **Habitat** Open sedge-dominated fens, usually in 'water tracks' and with *Carex lasiocarpa*; rarely in sphagnum moss. **Status** Special concern.

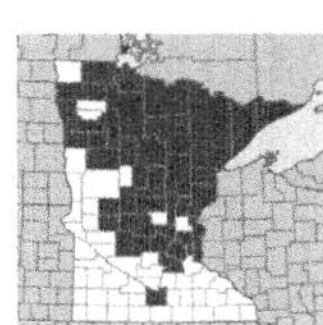

Drosera rotundifolia L.
ROUND-LEAVED SUNDEW
NC **OBL** | MW **OBL** | GP **OBL**

Small, perennial insectivorous herb; native. **Leaf blades** ± round, wider than long, 2–10 mm long and as wide or wider, covered with long, red, gland-tipped hairs; abruptly tapered to a petiole longer than blade; petioles 2–5 cm long covered with gland-tipped hairs. **Flowers** 2–15 in a ± 1-sided, racemelike cluster, on a leafless stalk 10-30 cm tall; flowers 4–7 mm wide, sepals 5, 4–5 mm long; petals white to pink, longer than sepals; stamens 5, shorter than petals. **Seeds** light brown, shiny and with fine lines, 1–1.5 mm long. July–Aug. **Habitat** Swamps and open bogs, usually in sphagnum; wet sandy shores and openings.

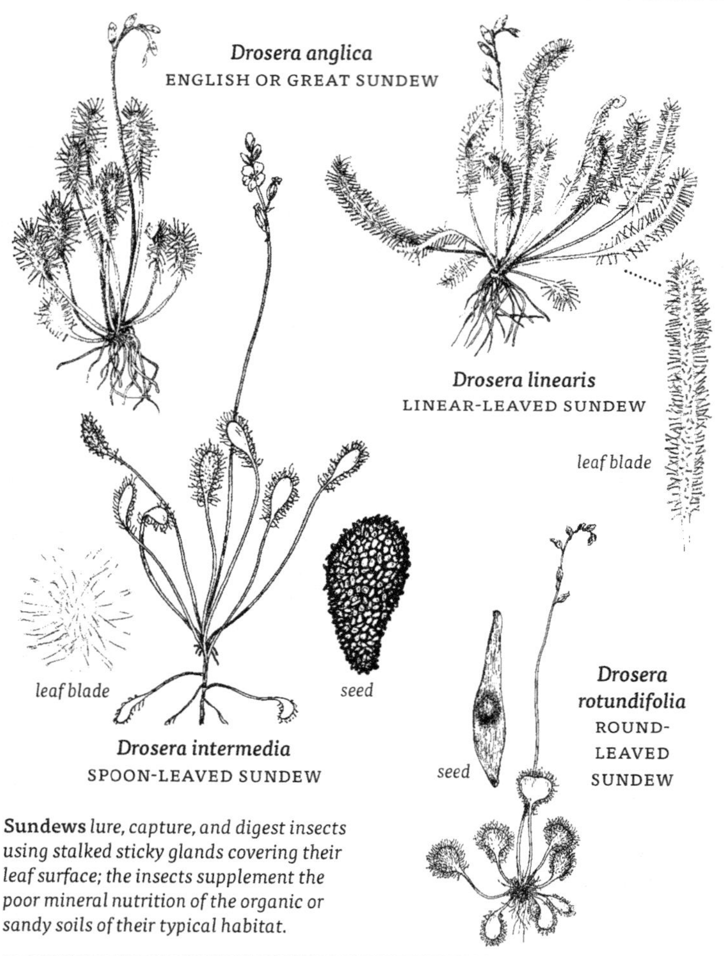

Sundews *lure, capture, and digest insects using stalked sticky glands covering their leaf surface; the insects supplement the poor mineral nutrition of the organic or sandy soils of their typical habitat.*

Drosera rotundifolia
ROUND-LEAVED SUNDEW

ELATINE | *Waterwort*

Small, branched, annual herbs of shallow water, shores and mud flats. **Leaves** simple, opposite, entire or toothed, with small membranous stipules. **Flowers** small, single from leaf axils, perfect; sepals, petals and stamens 2–3 (ours); styles 3; ovary superior, 3–4-chambered. **Fruit** a capsule with many small seeds.

ELATINE | WATERWORT

1 Flowers with 2 sepals and 2 petals, seeds all at base of fruit. . . . . *Elatine minima* | SMALL WATERWORT

1 Flowers with 3 sepals and 3 petals; seeds at differing levels in fruit . . . . *Elatine triandra* | LONG-STEMMED WATERWORT

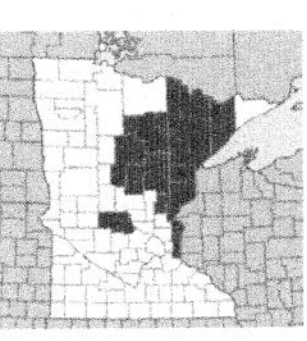

Elatine minima (Nutt.) Fischer & C. A. Meyer

SMALL WATERWORT

NC **OBL** | MW **OBL**

Native annual herb, forming small mosslike mats on mud; plants smooth, with branches to 5 cm long. **Leaves** oblong to obovate, rounded at tip, to 4 mm long, petioles absent. **Flowers** small, single and stalkless in leaf axils, sepals 2, petals 2. **Fruit** a round capsule; seeds with rows of small, rounded pits. **Habitat** Shallow water and wet shores along lakes and ponds, usually where sandy or mucky.

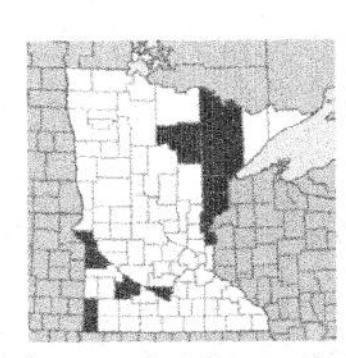

Elatine triandra Schk.

LONG-STEMMED WATERWORT

NC **OBL** | MW **OBL** | GP **OBL**

Native annual herb; plants small, matted, to 15 cm long, somewhat fleshy, smooth, branched from base, the branches sprawling or floating, often rooting at nodes. **Leaves** linear to obovate, 3–10 mm long and 1–3 mm wide, margins entire, petioles absent; stipules very small. **Flowers** small, single and stalkless in leaf axils, 1.5–2 mm wide, sepals 3, petals 3. **Fruit** a round capsule, 1–2 mm wide; seeds 0.5 mm long, ridged and with tiny, angled pits. July–Sept. **Synonyms** *Elatine americana, Elatine brachysperma, Elatine rubella*. **Habitat** Mud flats or in shallow water of lakes and ponds.

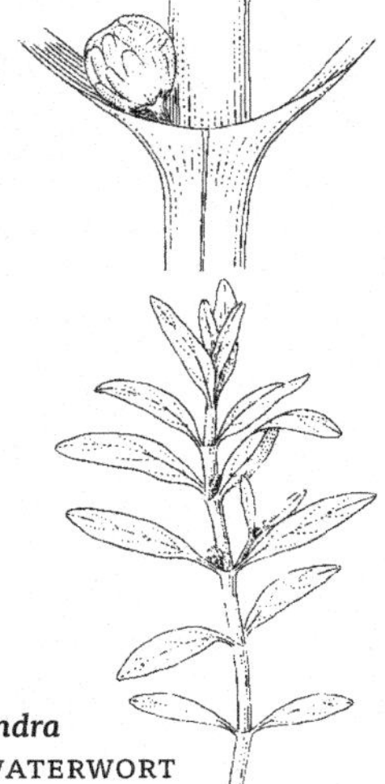

Elatine triandra
LONG-STEMMED WATERWORT

EMPETRUM | *Crowberry*

Empetrum nigrum L.

BLACK CROWBERRY
NC **FAC** | *endangered*

Much-branched low shrub to 3 dm tall, native; sometimes forming mats 1–2 m wide. **Leaves** evergreen and needlelike, dark green and leathery, linear-oblong, only 4–8 mm long, rounded or blunt at tip, narrowed at base to a short stalk, margins rolled under. **Flowers** small, pink to purple, single in axils of upper leaves, either male or female or perfect. **Fruit** purple-black to black, berrylike, 4–6 mm wide, somewhat juicy, with 6–9 hard nutlets. July–Aug. **Habitat** Cedar and black spruce swamps, rocky shorelines; also on drier, sandy, pine-covered ridges. **Status** Endangered; in Minnesota, known only from an island in Lake Superior, but last collection in 1948.

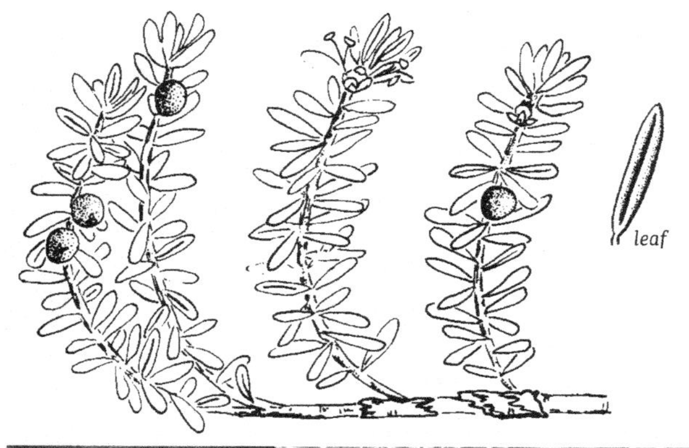

Empetrum nigrum (BLACK CROWBERRY)

SHRUBS OR SCARCELY WOODY SHRUBS. **Leaves** evergreen or deciduous, mostly alternate, simple, with entire or toothed margins. **Flowers** usually perfect (with male and female parts), urn- or vase-shaped, white, pink, or cream-colored; stamens as many (or 2x as many) as petals. **Fruit** a berry or dry capsule.

ERICACEAE | HEATH FAMILY

1 Leaves deciduous 2
1 Leaves evergreen 3

2 Leaves with shiny, orange-yellow resinous dots (especially on underside). .... *Gaylussacia baccata* | BLACK HUCKLEBERRY
2 Leaves without resinous dots .... *Vaccinium* (**blueberries**)

3 Plants creeping on ground surface 4
3 Plants upright 5

4 Stems covered with brown hairs; fruit a white berry .... *Gaultheria hispidula* | CREEPING SNOWBERRY
4 Stems not covered with brown hairs; fruit a red berry .... *Vaccinium* (**cranberries**)

5 Leaf margins distinctly rolled under 6
5 Leaf margins not rolled under .... *Chamaedaphne calyculata* .... LEATHERLEAF

6 Leaf underside densely covered with woolly hairs .... *Ledum groenlandicum* | LABRADOR-TEA
6 Leaf underside white; woolly hairs absent 7

7 Leaves alternate .... *Andromeda glaucophylla* | BOG-ROSEMARY
7 Leaves opposite .... *Kalmia polifolia* | BOG-LAUREL

ANDROMEDA | *Bog-rosemary*

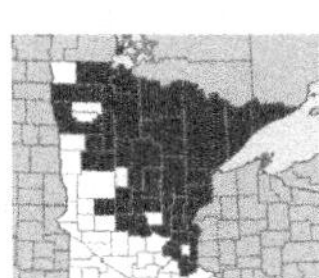

Andromeda glaucophylla Link
BOG-ROSEMARY
NC **OBL** | MW **OBL** | GP **OBL**

Low upright or trailing native shrub, 3–6 dm tall. **Stems** gray to blackish; twigs brown, with hairs in lines running down stems, or sometimes smooth. **Leaves** evergreen and leathery, often blue-green, linear or narrowly oval, 2–5 cm long and 3–10 mm wide, the tip sharp-pointed and tipped with a small spine, the base tapered to the stem or a short petiole, dark green above and whitened below by short stiff hairs; margins entire and distinctly rolled under. **Flowers** in drooping clusters at ends of branches, white or often pink, urn-shaped, 5-parted, 5–6 mm long, on curved stalks to 8 mm long. **Fruit** a rounded capsule to 5 mm wide, the style persistent from indented top of capsule; fruit drooping at first, but erect when mature. May–June. **Synonyms** *Andromeda polifolia*. **Habitat** Sphagnum bogs, black spruce and tamarack swamps.

Andromeda glaucophylla
BOG-ROSEMARY

CHAMAEDAPHNE | *Leatherleaf*

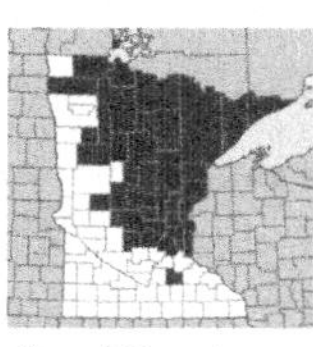

Chamaedaphne calyculata (L.) Moench
LEATHERLEAF
NC **OBL** | MW **OBL** | GP **OBL**

Native upright shrub to 1 m tall. Older **stems** gray, the outer bark shredding to expose the smooth, red inner bark; twigs brown, with fine hairs and covered with small, round scales. **Leaves** evergreen and leathery, becoming smaller toward ends of flowering branches, oval, 1–5 cm long and 3–15 mm wide, the tip rounded or pointed, brown-green and smooth above, pale brown with a covering of small, round scales below; margins entire or with small rounded teeth; petioles short. **Flowers** white, urn-shaped or cylindric, in 1-sided, leafy racemes, hanging from axils of reduced leaves near ends of branches; 5-parted,

Chamaedaphne calyculata (LEATHERLEAF)

5–7 mm long, on stalks 2–5 mm long. **Fruit** a brown, rounded capsule to 6 mm wide, the hairlike style persistent from indented top of capsule; capsules persisting on branches for several years. May–June. **Habitat** Open bogs, black spruce and tamarack swamps, peaty lakeshores and streambanks, often forming low, dense thickets.

GAULTHERIA | *Wintergreen*

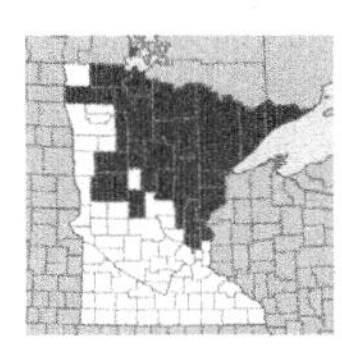

Gaultheria hispidula (L.) Muhl.

CREEPING SNOWBERRY

NC **FACW** | MW **FACW** | GP **FACW**

Low, creeping, matted shrub; native. **Stems** 2–4 dm long, covered with brown hairs. **Leaves** crowded, evergreen, oval to nearly round, 4–10 mm long and to 5 mm wide, abruptly tapered to tip, green above, underside paler, with brown, bristly hairs; margins rolled under; petioles short. **Flowers** few, single in leaf axils, white, bell-shaped, 4-parted, 2–4 mm long, on curved stalks 1 mm long. **Fruit** a translucent, juicy, white berry 5–10 mm wide, slightly wintergreen-flavored. May–June. **Synonyms** *Chiogenes hispidula*. **Habitat** Open bogs, swamps, wet conifer woods, often in moss on hummocks or downed logs.

GAYLUSSACIA | *Huckleberry*

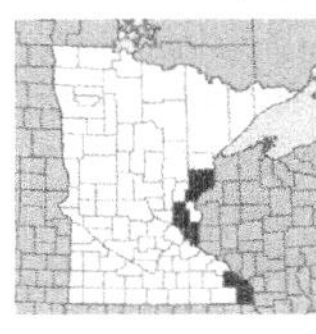

Gaylussacia baccata (Wangenh.) K. Koch.

BLACK HUCKLEBERRY

NC **FACU** | MW **FACU**

Medium-sized native shrub. **Stems** upright, much-branched, 3–10 dm long; branches brown, finely hairy when young, dark and smooth with age. **Leaves** alternate, deciduous, leathery, oval, 2–5 cm long and 1–2.5 cm wide; dark green above, paler below, both sides with shiny, orange-yellow resinous dots; margins entire, often fringed with small hairs; petioles 2–4 mm long. **Flowers** yellow-orange or red-tinged, cylindric, 5-lobed, 4–6 mm long, in ± 1-sided racemes from lateral branches, the flowers on short, gland-dotted stalks 4–5 mm long. **Fruit** a red-purple to black, berrylike drupe, 6–8 mm long, with 10 nutlets; edible but seedy. May–June. **Habitat** Open bogs, usually with tamarack and leatherleaf (*Chamaedaphne calyculata*); more common in dry, acid, sandy or rocky habitats.

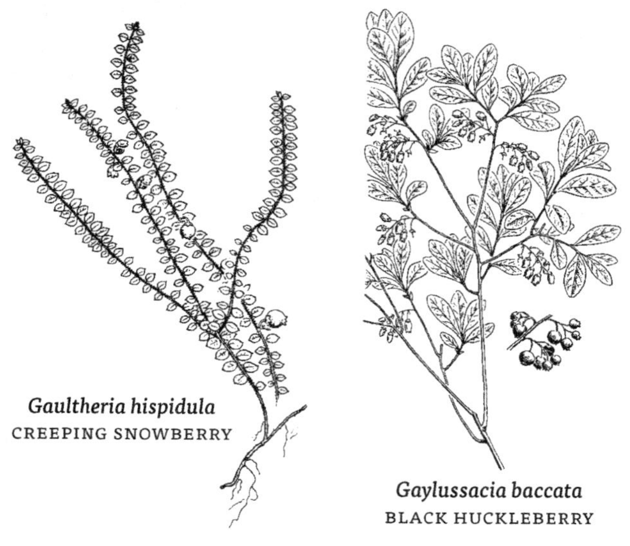

Gaultheria hispidula
CREEPING SNOWBERRY

Gaylussacia baccata
BLACK HUCKLEBERRY

KALMIA | *Laurel*

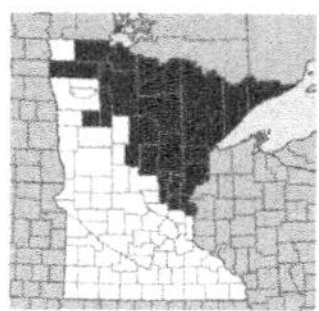

Kalmia polifolia Wangenh.
BOG-LAUREL
NC **OBL** | MW **OBL**

Low evergreen native shrub, to 6 dm tall. Older **stems** dark; twigs swollen at nodes, flattened and 2-edged in section, smooth, pale brown when young. **Leaves** opposite, evergreen and leathery, linear to narrowly oval, 1–4 cm long and 6–12 mm wide, tip blunt or narrowed to an abrupt point; dark green and smooth above, white below with a covering of short, white hairs, midrib on underside with large purple, stalked glands; margins entire and rolled under; petioles absent. **Flowers** showy, pale to rose-pink, in terminal clusters at ends of current year's branches, saucer-shaped, 5-parted, 8–11 mm wide, on stalks to 3 cm long. **Fruit** a round capsule to 6 mm wide, tipped by the persistent style, the capsules in upright clusters. May–June. **Habitat** Sphagnum peatlands, black spruce and tamarack swamps.

Kalmia polifolia
BOG-LAUREL

LEDUM | *Labrador-tea*

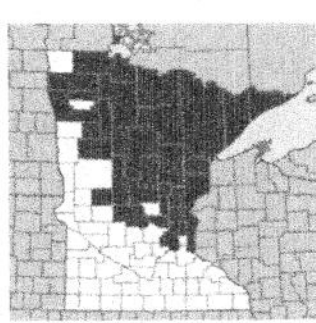

Ledum groenlandicum Oeder
LABRADOR-TEA
NC **OBL** | MW **OBL** | GP **FACW**

Medium-sized native shrub, to 1 m tall. Older **stems** gray or red-brown; twigs covered with woolly, curly brown hairs. **Leaves** alternate, evergreen and leathery, fragrant when rubbed, narrowly oval to oblong, 2.5–5 cm long and 5–20 mm wide, rounded at tip; dark green and smooth above, the midvein sunken; underside covered with tan to rust-colored curly hairs; margins entire and rolled under; petioles short. **Flowers** creamy-white, in rounded clusters at ends of branches, 5-parted, to 1 cm wide, on finely hairy stalks 1–2 cm long. **Fruit** a lance-shaped capsule 5–6 mm long, the style persistent and hairlike; capsules splitting at base to release numerous small seeds, the empty capsules persistent on stems for several years. May–June. **Synonyms** *Rhododendron groenlandicum*. **Habitat** Sphagnum bogs, swamps and wet conifer forests.

Ledum groenlandicum
LABRADOR-TEA

VACCINIUM | *Blueberry, Cranberry*

Deciduous or evergreen shrubs. **Leaves** alternate, simple. **Flowers** 4- or 5-parted, single in leaf axils or in clusters in axils or at ends of branches; ovary inferior. **Fruit** a many-seeded, red, blue, or black berry.

The genus may be divided into subgroups, two of which occur in Minnesota wetlands:

- **Blueberries** - *Vaccinium angustifolium, Vaccinium myrtilloides*
- **Cranberries** - *Vaccinium macrocarpon, Vaccinium oxycoccos, Vaccinium vitis-idaea*

VACCINIUM | BLUEBERRY, CRANBERRY

1 Plants low and trailing; leaves evergreen, less than 5 mm wide; mature fruit a red berry (**cranberries**) 2

1 Plants upright and bushy; leaves deciduous, more than 5 mm wide; mature fruit a blue to black berry (**blueberries**) 4

2 Leaf underside with black, bristly glands . . . . *Vaccinium vitis-idaea* . . . . MOUNTAIN CRANBERRY

2 Leaf underside without black glands. 3

3 Leaves blunt or rounded at tip, pale below; bracts on flower stalk green and leaflike (more than 1 mm wide) . . . . *Vaccinium macrocarpon* . . . . CRANBERRY

3 Leaves tapered to pointed tip, white below; bracts on flower stalk red and narrow (less than 1 mm wide) . . *Vaccinium oxycoccos* | SMALL CRANBERRY

4 Leaf underside hairy; margins with small, bristle-tipped teeth . . . . *Vaccinium angustifolium* | LOWBUSH BLUEBERRY

4 Leaf underside without hairs or only sparsely hairy; margins entire, usually fringed with fine hairs . . . . *Vaccinium myrtilloides* . . . . VELVETLEAF-BLUEBERRY

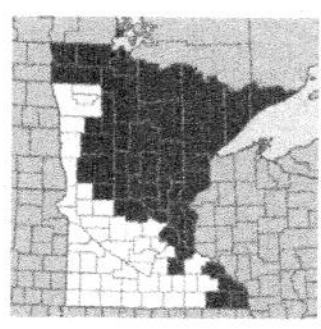

Vaccinium angustifolium Aiton

LOWBUSH BLUEBERRY

NC **FACU** | MW **FACU** | GP **FACU**

Low native shrub 1–6 dm tall, forming colonies from surface runners. Older **stems** red-brown to black; twigs green-brown, with hairs in lines down stems, or sometimes smooth. **Leaves** deciduous, bright green oval, 2–5 cm long and 5–15 mm wide, smooth on both sides or sparsely hairy on veins; margins finely toothed with bristle-tipped teeth; petioles very short. **Flowers** in clusters, opening before or with leaves, white or pale pink, narrowly bell-shaped, 5-parted, 4–6 mm long. **Fruit** blue and wax-covered, 5–12 mm wide, edible and sweet. Flowering April–June, fruit ripening July–Aug. **Habitat** Sphagnum peatlands and wetland margins; also in dry, sandy openings and forests.

Vaccinium angustifolium
LOWBUSH BLUEBERRY

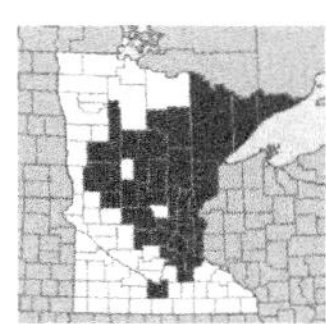

Vaccinium macrocarpon Aiton
CRANBERRY
NC **OBL** | MW **OBL** | GP **OBL**

Evergreen trailing native shrub. **Stems** slender, to 1 m or more long, with branches to 2 dm tall. **Leaves** leathery, oblong-oval, 5–15 mm long and 2–5 mm wide, rounded or blunt at tip, pale on underside; margins flat or slightly rolled under; petioles absent or very short. **Flowers** white to pink, 1 cm wide, 4-lobed, the lobes turned back at tips, single or in clusters of 2–6, on stalks 1–3 cm long, the stalks with 2 bracts, the bracts green, 2–4 mm long and 1–2 mm wide. **Fruit** red, 1–1.5 cm wide, edible but tart, often persisting over-winter. Flowering June–July, fruit ripening Aug–Sept. **Synonyms** *Oxycoccus macrocarpon*. **Habitat** Sphagnum bogs, swamps, peaty pond margins.

***Vaccinium macrocarpon** is the cultivated cranberry; fruit is typically larger than the similar Vaccinium oxycoccos.*

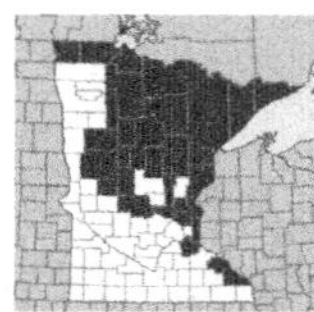

Vaccinium myrtilloides Michx.
VELVETLEAF-BLUEBERRY
NC **FACW** | MW **FACW** | GP **FACW**

Low native shrub, often forming colonies. **Stems** 3–6 dm long, red-brown to black with numerous wartlike lenticels; young twigs green-brown, densely velvety white-hairy. **Leaves** deciduous, thin and soft, oval, 2–5 cm long and 1–2.5 cm wide, dark green above, paler and soft hairy below, not waxy; margins entire and finely hairy; petioles very short. **Flowers** in clusters at ends of short, leafy branches, opening with leaves, creamy or green-white, tinged with pink, bell-shaped or short-cylindric, 5-parted, 4–5 mm long. **Fruit** blue, wax-covered, 6–9 mm wide; edible but tart. Flowering May–July, fruit ripening July–Sept. **Habitat** Sphagnum bogs and swamps; also in dry to moist woods and clearings.

Vaccinium myrtilloides
VELVETLEAF-BLUEBERRY

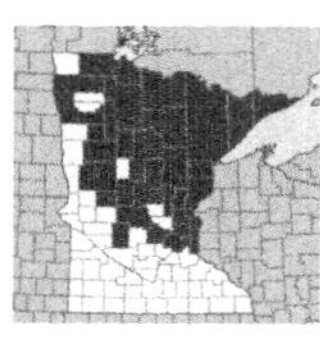

Vaccinium oxycoccos L.
SMALL CRANBERRY
NC **OBL** | MW **OBL** | GP **OBL**

Evergreen trailing native shrub. **Stems** slender, 0.5 m or more long, with upright branches 1–2 dm tall. **Leaves** leathery, ovate to oval or narrowly triangular, 2–10 mm long and 1–3 mm wide, pointed or rounded at tip, strongly whitened on underside; margins flat or strongly rolled under; petioles absent or very short. **Flowers** pale pink, 1 cm wide, 4-lobed, the lobes turned back at tips, single or in clusters of 2–4, on stalks 1–3 cm long, the stalks with 2 bracts usually at or below middle

of stalk, the bracts red, scalelike, to 2 mm long and less than 1 mm wide. **Fruit** pale and red-speckled when young, becoming red, 6–12 mm wide, edible but tart. Flowering June–July, fruit ripening Aug–Sept. **Synonyms** *Oxycoccus oxycoccos*. **Habitat** Wet, acid, sphagnum bogs.

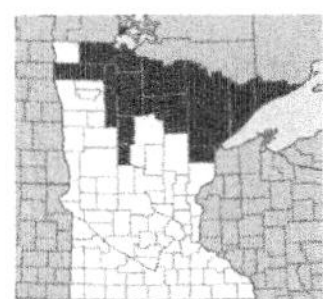

Vaccinium vitis-idaea L.

MOUNTAIN CRANBERRY
NC **FAC** | MW **FAC** | GP **FAC**

Low evergreen, trailing shrub; native. Older **stems** brown-black with peeling bark, branching, the branches upright, slender, 1–2 dm long, often forming mats; twigs green-brown to red, ± smooth. **Leaves** alternate, leathery, oval to oval, 0.5–2 cm long and 4–15 mm wide, rounded or slightly indented at tip; upper surface dark green, shiny and smooth, paler and with dark bristly glands below; margins entire and rolled under; petioles hairy, 1–2 mm long. **Flowers** white to pink, bell-shaped and 4-lobed, style longer than petals, several in 1-sided clusters at ends of branches, the flowers on short glandular stalks, the stalks with 2 small bracts at base. **Fruit** a dark red berry, to 1 cm wide, persisting over winter, tart but edible, especially the following spring. June–July. **Habitat** Sphagnum bogs; also in drier, sandy or rocky places.

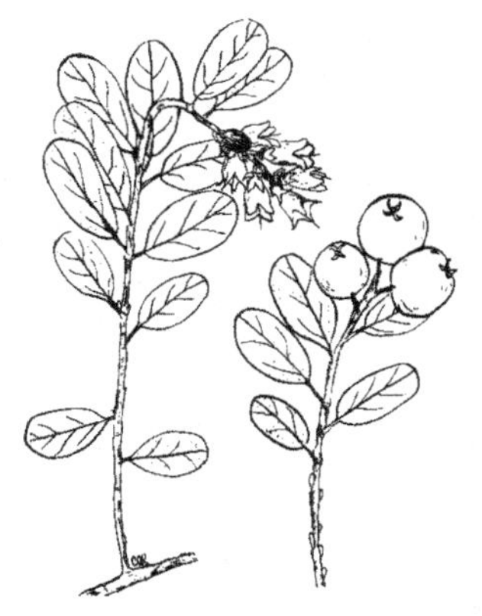

Gathered in Europe, where known as **lingen** *or* **red whortleberry**, *and North America (where available); cooked and eaten like commercial cranberries. Mountain cranberry can be distinguished from the more common cranberries (Vaccinium macrocarpon and Vaccinium oxycoccos) by the black, bristly, glandular dots on the leaf underside.*

Vaccinium vitis-idaea
MOUNTAIN CRANBERRY

PERENNIAL SHRUBS (*Amorpha*) and herbs. **Leaves** alternate, pinnately divided, the terminal leaflet sometimes modified as a tendril (*Lathyrus*). **Flowers** in simple or branched racemes, perfect (with both male and female parts), irregular, 5-lobed (only 1 lobe in *Amorpha*), the upper lobe (banner) larger than the other lobes, with 2 outer lateral petals (wings), and 2 inner petals which are partly joined (the keel), and enclosing the 10 stamens and style; pistil 1, ovary 1-chambered, maturing into a pod.

FABACEAE | BEAN FAMILY

1 Shrub; flowers 1-lobed, only the banner present .... *Amorpha fruticosa* .... FALSE INDIGO
1 Herbs; flowers 5-lobed 2

2 Leaves with tendrils .... *Lathyrus palustris* | MARSH-PEA
2 Leaves without tendrils 3

3 Plants vinelike and sprawling .... *Apios americana* .... COMMON GROUND-NUT; WILD-BEAN; INDIAN-POTATO
3 Plants upright, not vinelike .... *Astragalus* | MILKVETCH

AMORPHA | *False indigo, Lead-plant*

Amorpha fruticosa L.

FALSE INDIGO

NC **FACW** | MW **FACW** | GP **FACW**

Much-branched native shrub, mostly 1–3 m tall; twigs tan to gray. **Leaves** pinnately divided, 5–15 cm long; leaflets 9–27, oval to obovate, 1–4 cm long and 0.5–3 cm wide, upper surface smooth, underside short-hairy, margins entire, petioles 2–5 cm long, stipules absent. **Flowers** dark purple, in dense spikelike racemes 2–15 cm long at ends of stems; petals 1-lobed, only the banner present, 3–5 mm long, folded to enclose the 10 stamens. **Fruit** an oblong pod, curved near tip, 5–7 mm long, spotted with glands, with 1–2 seeds. June–July. **Habitat** Wet meadows, streambanks, shores, ditches.

Amorpha fruticosa
FALSE INDIGO

APIOS | *Ground-nut*

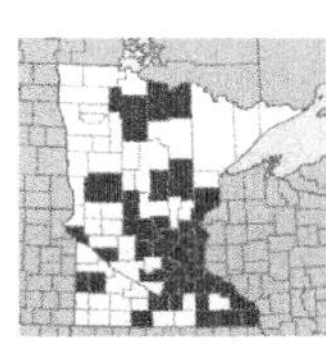

Apios americana Medic.
COMMON GROUND-NUT; WILD-BEAN; INDIAN-POTATO
NC **FACW** | MW **FACW** | GP **FAC**

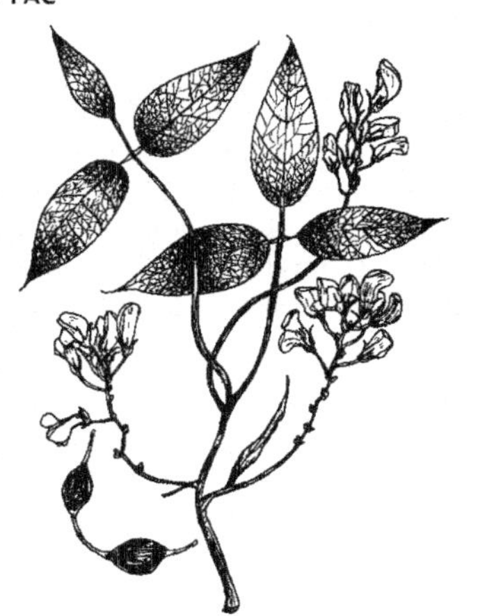

Native perennial herbaceous vine, rhizomes with a necklace-like series of 2 or more tubers; plants with milky juice. **Stems** to 1 m long, climbing over other plants. **Leaves** pinnately divided; main leaves with 5–7 leaflets; leaflets ovate, 4–6 cm long, tapered to a point, smooth to short-hairy beneath, margins entire. **Flowers** brown-purple, 10–13 mm long, single or paired, in crowded racemes from leaf axils. **Fruit** a linear pod, 5–10 cm long. July–Aug. **Habitat** Floodplain forests, thickets, shores, wet meadows, low prairie.

ASTRAGALUS | *Milkvetch*

Perennial herbs. **Leaves** pinnately divided; ours with 13–30 leaflets, margins entire. **Flowers** in crowded racemes. **Fruit** an erect or drooping pod.

ASTRAGALUS | MILKVETCH

1 Flowers and fruit upright; widespread . . . . *Astragalus agrestis* . . . . FIELD MILKVETCH

1 Flowers and fruit drooping; rare in ne Minnesota . . . . . . *Astragalus alpinus* . . . . ALPINE MILKVETCH

Astragalus agrestis Douglas
FIELD MILKVETCH
NC **FACW** | MW **FACW** | GP **FACU**

Native perennial herb, from slender rhizomes. **Stems** slender, 1–3 dm tall, smooth or sparsely hairy. **Leaflets** 13–21, oblong to lance-shaped, sparsely hairy on both sides. **Flowers** purple, upright, 15–20 mm long, in crowded, long-stalked racemes, 2–4 cm long; sepals covered with mix of black and white hairs. **Fruit** an erect, ovate pod, 7–9 mm long, covered with stiff hairs. May. **Synonyms** *Astragalus goniatus*. **Habitat** Wet meadows and low prairie.

Astragalus alpinus L.

ALPINE MILKVETCH

NC **FAC** | GP **FAC** | *endangered*

Native perennial herb, with rhizomes which sprout from the nodes. **Stems** smooth or sparsely hairy, sprawling, to about 70 cm long. **Leaves** pinnately divided; leaflets 15–29 , oblong to ovate. **Flowers** nodding, the keel purple, the wings white. **Fruit** a pendulous, densely pubescent pod, the hairs dark- and light-colored. May–June. Ours are var. *alpinus*. **Habitat** Gravelly shores of shallow groundwater-fed ponds. **Status** Endangered.

In Minnesota, **Astragalus alpinus** *is known from a single locale in Lake County (disjunct from main range of Rocky Mountains); plants occur at several ponds, all of which are less than 1.5 km apart from one another.*

Astragalus agrestis
FIELD MILKVETCH

Astragalus alpinus
ALPINE MILKVETCH

LATHYRUS | *Vetchling, Wild pea*

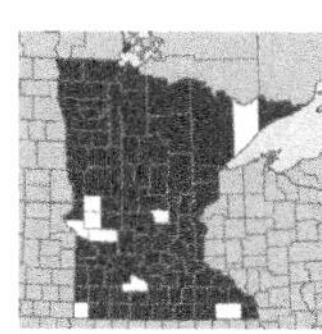

Lathyrus palustris L.

MARSH-PEA

NC **FACW** | MW **FACW** | GP **FACW**

Native perennial vining herb, spreading by rhizomes. **Stems** to 1 m long, strongly winged, climbing and clinging to surrounding plants by tendrils. **Leaves** pinnately divided, with 4–8 leaflets and a terminal leaflet modified into a tendril; leaflets linear to lance-shaped, 2–7 cm long and 3–20 mm wide; stipules prominent, ± arrowhead-shaped, 1–3 cm long; margins entire; petioles absent. **Flowers** in racemes from leaf axils, 2–6 flowers per raceme, red-purple, drying blue to blue-violet; sepals irregular, 7–10 mm long, the lowest lobe longest; petals 12–20 mm long. **Fruit** a flat, many-seeded pod, 3–5 cm long. June–Aug. **Habitat** Conifer swamps, thickets, wet meadows, marshes, streambanks, calcareous fens, low prairie.

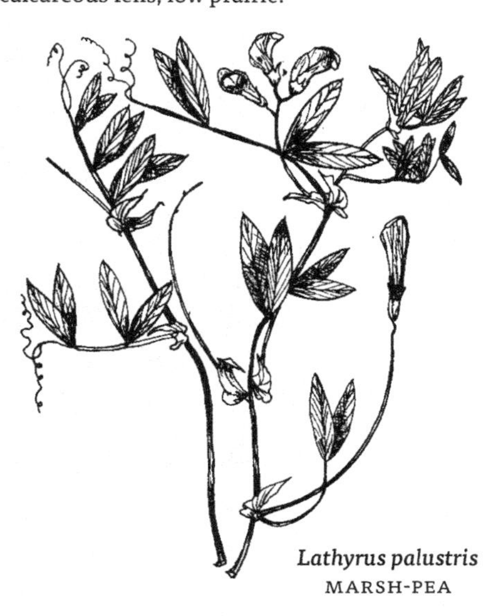

Lathyrus palustris
MARSH-PEA

QUERCUS | *Oak*

Quercus bicolor Willd.
SWAMP WHITE OAK
NC **FACW** | MW **FACW**

Deciduous native tree to 20 m tall; trunk to 1 m wide; crown broad and rounded; **bark** gray-brown, deeply furrowed, becoming flaky; **twigs** gray to yellow-brown; buds clustered at branch tips, yellow-brown, smooth or sparsely hairy. **Leaves** alternate, broadest above middle, to 15 cm long and 10 cm wide, smooth or hairy on upper surface, white and soft hairy on underside; margins with coarse, rounded teeth or shallow lobes; petioles 2–3 cm long. Male and female **flowers** separate but on same tree, appearing with leaves in spring; male flowers in slender, drooping catkins, female flowers in groups of 2–4. **Fruit** a pair of acorns, on stalks 2–3 cm long, the acorns ovate, pale brown, 2.5–4 cm long, the cup rough and hairy, covering about 1/3 of acorn. May. **Habitat** Floodplain forests, low woods and swamps.

Quercus bicolor
SWAMP WHITE OAK

ANNUAL, BIENNIAL, OR PERENNIAL HERBS; plants usually smooth. **Leaves** simple, entire, opposite or whorled, stem leaves without petioles. **Flowers** often showy, perfect (with both male and female parts), regular, single at end of stems or in clusters; petals 4-5, blue, purple, white or green, joined for at least part of their length; stamens 4 or 5. **Fruit** a 2-chambered, many-seeded capsule enclosed by the withered, persistent petals.

GENTIANACEAE | GENTIAN FAMILY

1 Leaves reduced to small, narrow scales less than 3 mm long .... *Bartonia virginica* | YELLOW SCREW-STEM
1 Leaves well-developed, not scalelike **2**

2 Petals 4, spurred at base; flowers green, tinged with purple .... *Halenia deflexa* | SPURRED GENTIAN
2 Petals 4, with fringed lobes; or petals 5 and not spurred; blue, purple or white **3**

3 Petals 4, fringed; flowers on stalks longer than the flowers; seeds covered with small bumps .... *Gentianopsis* | FRINGED GENTIAN
3 Petals 5, not fringed; flower stalks short or absent; seeds smooth **4**

4 Flowers 2.5–4 cm long, on short stalks; seeds flattened and winged .... *Gentiana* | GENTIAN
4 Flowers 1–2 cm long, stalkless; seeds round .... *Gentianella* | GENTIAN

BARTONIA | *Screw-stem*

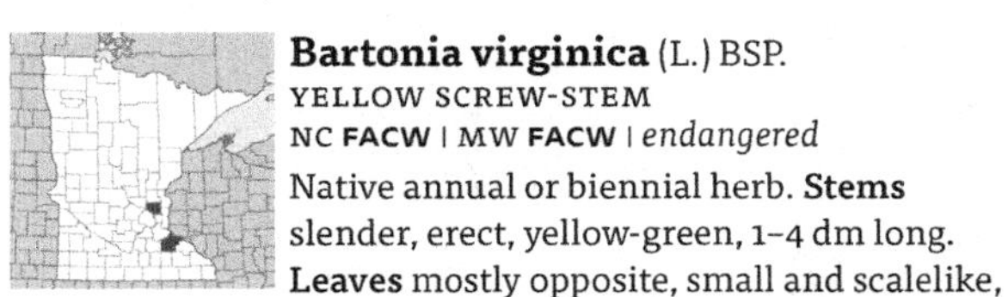

Bartonia virginica (L.) BSP.
YELLOW SCREW-STEM
NC **FACW** | MW **FACW** | *endangered*

Native annual or biennial herb. **Stems** slender, erect, yellow-green, 1–4 dm long. **Leaves** mostly opposite, small and scalelike, 1–2 mm long. **Flowers** 4-parted, bell-shaped, green-yellow or green-white, 3–4 mm long, in a slender raceme or panicle, the branches and flower stalks opposite and upright; sepals awl-shaped; petals oblong, tapered to a rounded tip. **Fruit** a capsule 2–3 mm long. Aug–Sept. **Habitat** Conifer swamps (often in sphagnum moss) and in wet meadows at the edge of conifer swamps; soils sandy. **Status** Endangered.

Bartonia virginica
YELLOW SCREW-STEM

GENTIANA | *Gentian*

Perennial herbs, with thick, fibrous roots. **Leaves** opposite or whorled, simple, margins entire, petioles absent. **Flowers** large, blue, green-white or yellow, 5-parted, in clusters near ends of stems; petals forming a tubelike, shallowly lobed flower, the lobes alternating with a folded membrane as long or longer than petal lobes; stamens 5. **Fruit** a 2-chambered capsule.

GENTIANA | GENTIAN

1 Leaves and petal lobes fringed with tiny hairs (under 10x magnification). .... *Gentiana andrewsii* | BOTTLE GENTIAN

1 Leaves and petal lobes smooth..................... *Gentiana rubricaulis* .... GREAT LAKES GENTIAN

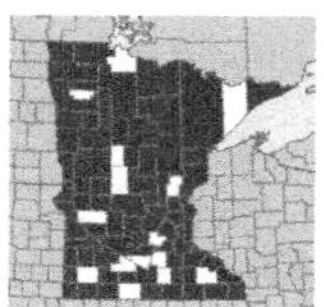

Gentiana andrewsii Griseb.

BOTTLE GENTIAN

NC **FACW** | MW **FACW** | GP **FAC**

Native perennial herb. **Stems** erect, single or few together, 2–8 dm long, unbranched, smooth. **Leaves** opposite, lance-shaped, 4–12 cm long and 1–3 cm wide, margins fringed with hairs. **Flowers** 1 to many, stalkless in upper leaf axils, 3–5 cm long; sepals forming a tube around petals, the sepal lobes unequal, fringed with hairs; petals forming a tubelike flower, usually remaining closed, the folds between petal lobes finely fringed (use hand lens to see this) and longer than the petal lobes. **Fruit** a capsule; seeds winged. Aug–Sept. **Habitat** Wet meadows, swamps and wet woods, thickets, low prairie, shores, ditches.

Gentiana andrewsii
BOTTLE GENTIAN

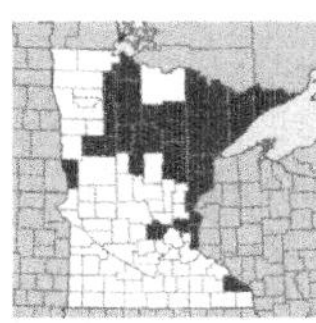

Gentiana rubricaulis Schwein.
GREAT LAKES GENTIAN
NC **OBL** | MW **OBL** | GP **OBL**

Native perennial herb. **Stems** smooth, 3–7 dm long. **Leaves** pale green, lance-shaped, 4–8 cm long and 2–3 cm wide, margins entire. **Flowers** 3–5 cm long, green-blue below, blue above, narrowly open, in a cluster at end of stem; sepal lobes oblong, 4–12 mm long, chaffy and translucent near base. **Fruit** a capsule; seeds winged. **Synonyms** *Gentiana linearis* var. *latifolia*. **Habitat** Wet meadows, peatlands, streambanks, thickets, conifer swamps, Lake Superior rocky shores; soils usually calcium-rich.

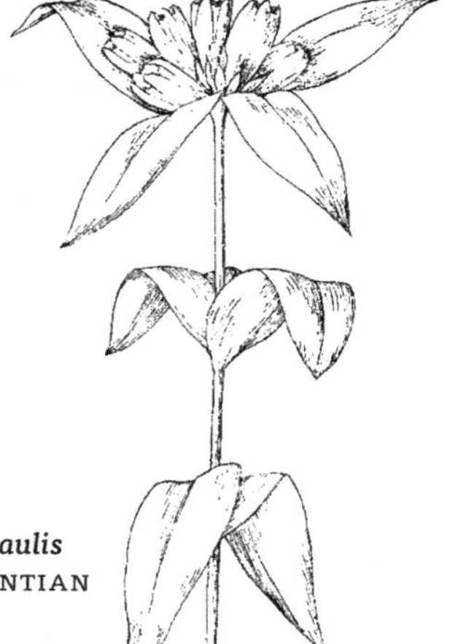

Gentiana rubricaulis
GREAT LAKES GENTIAN

GENTIANELLA | *Gentian*

Annual or biennial herbs. **Leaves** opposite, stalkless, margins entire. **Flowers** 4–5-parted, blue to white, funnel-shaped or tubular; in clusters from ends of stems or upper leaf axils; petals withering and persistent around capsule.

GENTIANELLA | GENTIAN

1 Petal lobes with fringe of hairs at base; nw Minnesota . . . . *Gentianella amarella* | NORTHERN GENTIAN; FELWORT

1 Petal lobes not fringed with hairs at base; se Minnesota . . . . *Gentianella quinquefolia* | STIFF GENTIAN

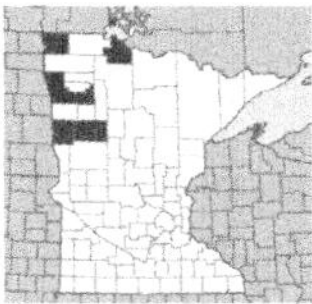

Gentianella amarella (L.) Boerner
NORTHERN GENTIAN; FELWORT
NC **OBL** | MW **OBL** | GP **FACW** | *special concern*

Native annual or biennial herb. **Stems** unbranched, 1–6 dm long. **Leaves** lance-shaped, 2–6 cm long; margins entire. **Flowers** blue, 10–15 mm long, 5-parted (rarely 4-parted); single from middle leaf axils, several in a cluster from upper axils, on stalks less than 1 cm long; sepal lobes linear, 3–5 mm long; petals forming a tubelike to flared, funnel-like flower, petal lobes 3–5 mm long, the base of lobe fringed with hairs 2 mm long. **Fruit** a capsule. July–Aug. **Synonyms** *Gentianella acuta*, *Gentiana amarella*. **Habitat** Low prairie, wet or moist sandy or gravelly soil. **Status** Special concern.

Gentianella quinquefolia (L.) Small
STIFF GENTIAN
NC **FAC** | MW **FAC**

Native annual or biennial herb. **Stems** 4-angled, 2–8 dm long, usually branched. **Leaves** at base spatula-shaped, upper leaves lance-ovate, 2–7 cm long; margins entire. **Flowers** blue (rarely white), 15–25 mm long, in clusters of 1–7 flowers at ends of stems or upper leaf axils, on stalks to 1 cm long; sepal lobes lance-shaped; petals forming a narrowly funnel-shaped flower, petal lobes 4–6 mm long, not fringed with hairs at base. **Fruit** a capsule; seeds round. Aug–Sept. **Synonyms** *Gentianella occidentalis, Amarella occidentalis*. **Habitat** Wet meadows, streambanks, moist woods; often where calcium-rich.

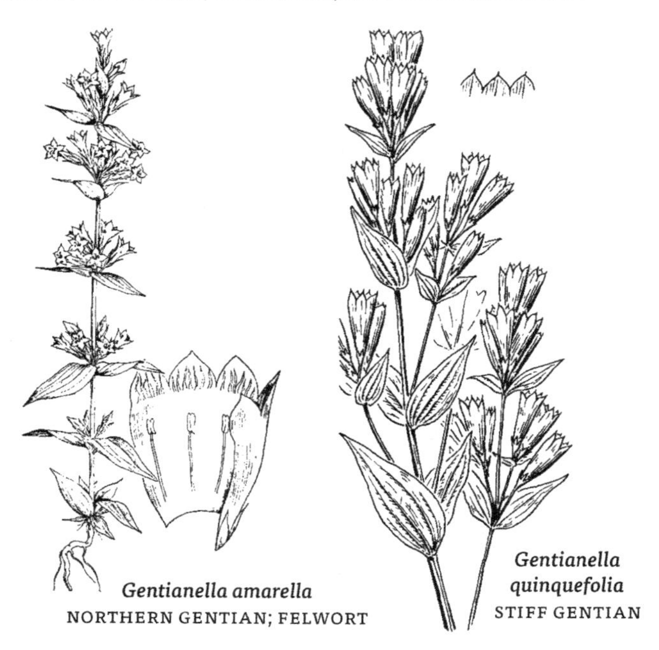

Gentianella amarella
NORTHERN GENTIAN; FELWORT

Gentianella quinquefolia
STIFF GENTIAN

GENTIANOPSIS | *Fringed gentian*

Smooth, taprooted, annual or biennial herbs. **Leaves** opposite, stalkless, margins entire. **Flowers** 1 to several, showy, blue, sometimes tinged with white on outside, long-stalked, at ends of stems and branches, 4-parted; sepals oblong cone-shaped; petals deeply lobed, forming a tubular or bell-shaped flower, the lobes ragged or fringed at tips and sometimes on sides, without a folded membrane between the lobes (present in *Gentiana*); stamens 4. **Fruit** a capsule; seeds covered with bumps.

GENTIANOPSIS | FRINGED GENTIAN

1 Upper leaves lance-shaped to ovate; petal lobes long-fringed across tip and sides, the fringes 2–5 mm long . . . . *Gentianopsis crinita* . . . . FRINGED GENTIAN

1 Upper leaves linear; tips of petal lobes ragged with short, fine teeth, often fringed on sides. . . . . *Gentianopsis procera* | LESSER FRINGED GENTIAN

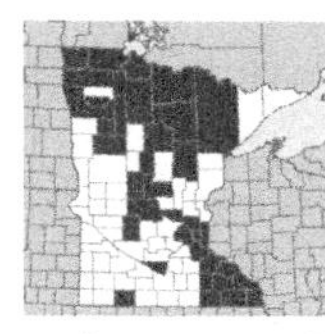

Gentianopsis crinita (Froelich) Ma

FRINGED GENTIAN

NC **FACW** | MW **OBL** | GP **OBL**

Native annual or biennial herb. **Stems** erect, 2–7 dm long, usually branched above. Basal leaves spatula-shaped, smaller than stem leaves; stem leaves ovate, 2–6 cm long and 1–2.5 cm wide, the base usually clasping stem; margins entire. **Flowers** bright blue, 3–6 cm long, 4-parted, single at ends of main stems and branches, on stalks 5–20 cm long; sepals forming a tube, 1–2 cm long; petals joined to form a funnel-like to bell-shaped flower, the petal lobes fringed across tip and part way down sides with linear fringes 2–6 mm long. **Fruit** a capsule, broadest at middle. Aug–Oct. **Synonyms** *Gentiana crinita*. **Habitat** Wet meadows, streambanks, ditches, wet woods; soils usually calcium-rich and sandy or gravelly.

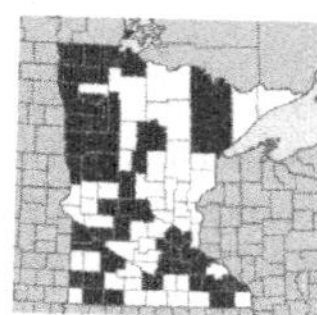

Gentianopsis procera (Holm) Ma
LESSER FRINGED GENTIAN
NC **OBL** | MW **OBL** | GP **OBL**

Native annual herb, similar to ***Gentianopsis crinita*** but smaller. **Stems** simple or few-branched, 1–5 dm long. Basal leaves spatula-shaped; stem leaves linear to linear lance-shaped, 2–5 cm long and 2–7 mm wide, tapered to a blunt tip, the base not clasping stem; margins entire. **Flowers** bright blue, 2–5 cm long, mostly 4-parted, single on stalks at ends of stems; sepal tube 6–15 mm long; petals forming a tubelike flower, flared toward tip, petal lobes ragged toothed across tips, often fringed on sides. **Fruit** a capsule. Sept–Oct. **Synonyms** *Gentiana procera, Gentianopsis virgata*. **Habitat** Sandy and gravelly shores, wet meadows, calcareous fens, wetlands between dunes near Great Lakes; soils usually calcium-rich.

HALENIA | *Spurred gentian*

Halenia deflexa (J. E. Smith) Griseb.
SPURRED GENTIAN
NC **FAC** | MW **FAC** | GP **FAC**

Native annual herb. **Stems** erect, simple or few-branched, rounded 4-angled, 15–40 cm long. **Leaves** opposite, lower leaves spatula-shaped, narrowed to a petiole; stem leaves lance-shaped to ovate, 2–5 cm long and 1–2.5 cm wide, stalkless; margins entire. **Flowers** green, tinged with purple, 10–12 mm long, 4-parted, on stalks to 4 cm long, in loose clusters of 5–9 flowers at ends of stems; petals lance-shaped, usually with downward-pointing spurs at base, the spurs to 5 mm long. **Fruit** an oblong capsule. July–Aug. **Habitat** Cedar swamps, moist conifer woods (especially along shores), old logging roads.

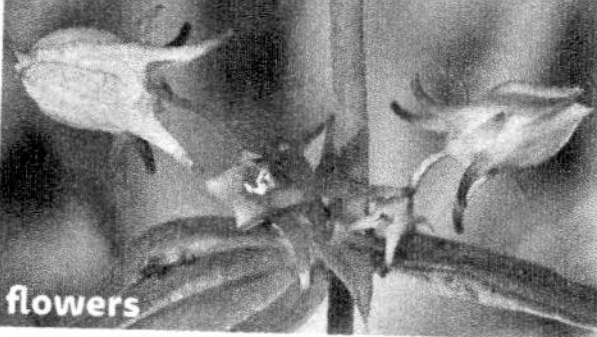

Halenia deflexa
SPURRED GENTIAN

RIBES | *Gooseberry, Currant*

Small to medium shrubs with upright to spreading stems, the **stems** smooth or with spines at nodes and sometimes also with bristles between the nodes. **Leaves** alternate, palmately 3–5-lobed; margins toothed. **Flowers** 1–several in clusters or few to many in racemes; perfect, regular, ovary inferior; sepals 5; petals 5, green to white or yellow, shorter than the sepals; stamens 5, alternate with the petals; styles 2. **Fruit** a many-seeded berry, usually topped by persistent, dry flower parts.

RIBES | GOOSEBERRY, CURRANT

1 Stems with spines or bristles, at least at nodes 2
1 Stems without spines 5

2 Spines and bristles persistent; berries with gland-tipped hairs or bristles 3
2 Spines and bristles deciduous during summer; berries without glands or bristles 4

3 Stems bristly between nodes; leaves with disagreeable scent when rubbed; berries purple-black, with gland-tipped hairs . . . . *Ribes lacustre* . . . . PRICKLY CURRANT
3 Bristles between nodes usually absent; leaves not ill-scented; berries wine-red and spiny . . . . *Ribes cynosbati* | PRICKLY GOOSEBERRY

4 Spines and bristles soft; leaves without glands on underside; bracts below flowers with long hairs . . . . *Ribes hirtellum* | NORTHERN GOOSEBERRY
4 Spines and bristles firm; leaves with scattered glands on underside (at least on veins); bracts below flowers with gland-tipped hairs . . . . *Ribes oxyacanthoides* | NORTHERN OR BRISTLY GOOSEBERRY

5 Stems upright; leaves dotted with resinous glands (at least on underside); berries black. 6
5 Stems spreading and reclining; leaves not resin-dotted; berries red 7

6 Leaves dotted on both sides with yellow to brown resinous glands; flowers in drooping clusters. . . . . *Ribes americanum* . . . . EASTERN BLACK CURRANT
6 Leaves resin-dotted only on underside; flowers in upright clusters. . . . . *Ribes hudsonianum* | HUDSON BAY CURRANT

7 Plants with skunklike odor when rubbed; berries with gland-tipped hairs . . . . *Ribes glandulosum* | SKUNK-CURRANT
7 Plants without skunklike odor; berries smooth . . . . *Ribes triste* . . . . SWAMP RED CURRANT

Ribes americanum Miller
EASTERN BLACK CURRANT
NC **FACW** | MW **FACW** | GP **FACW**

Native shrub, 1–1.2 m tall. **Stems** without spines or bristles, young stems finely hairy; branches upright to spreading; twigs gray-brown and smooth, black with age. **Leaves** 3–8 cm long and 3–10 cm wide, 3-lobed and usually with 2 additional shallow lobes at base, dotted with shiny, yellow to brown resinous glands, especially on underside, smooth or short-hairy above, hairy below; margins coarsely toothed; petioles hairy and resin-dotted, 3–6 cm long. **Flowers** creamy-white to yellow, bell-shaped, 8–12 mm long; 6–15 in drooping racemes 3–8 cm long; each flower with a linear bract longer than the flower stalk, the stalks 2–3 mm long; sepals 4–5 mm long, rounded; petals blunt, 2–3 mm long; stamens about equaling petals. **Fruit** an edible, smooth, black berry, 6–10 mm wide. April–June. **Synonyms** *Ribes floridanum*. **Habitat** Moist to wet forests, swamps, marsh and lake borders, streambanks.

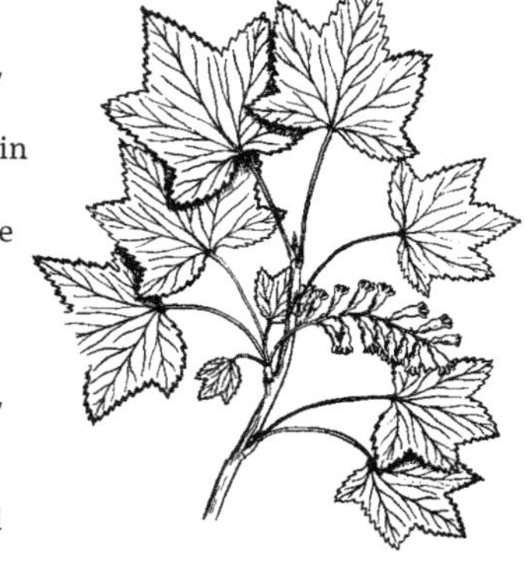

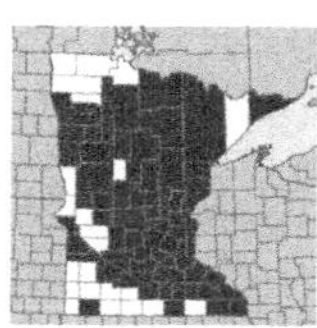

Ribes cynosbati L.
PRICKLY GOOSEBERRY
NC **FACU** | MW **FAC** | GP **FACU**

Native shrub to 6–9 dm tall, branches upright to spreading. **Stems** and branches with 1–3 spines at nodes, outer bark peeling off, inner bark brown-purple to black; young stems brown-gray, finely hairy. **Leaves** 3–8 cm long and 3–7 cm wide, 3–5-lobed, the lobes rounded at tips; upper surface dark green, sparsely hairy, underside paler, finely hairy and with gland-tipped hairs along veins; margins with coarse, round teeth; petioles 2.5–4 cm long, finely hairy and with scattered gland-tipped hairs. **Flowers** green-yellow, bell-shaped, 6–9 mm long, in clusters of 2–3 from spurs on old wood, on stalks with gland-tipped hairs. **Fruit** a red-purple berry, 8–12 mm wide, covered with stiff, brown spines. May–June. **Habitat** Occasional in wet woods, swamps, thickets and streambanks; more typical of moist hardwood forests (where common).

Ribes cynosbati
PRICKLY GOOSEBERRY

Ribes glandulosum Grauer

SKUNK-CURRANT

NC **FACW** | MW **FACW** | GP **OBL**

Native shrub to 8 dm tall. **Stems** sprawling, spines and bristles absent; stems and leaves with skunklike odor when crushed; older stems smooth and dark as outer bark peels off; young stems smooth to finely hairy, brown-gray. **Leaves** 2–8 cm long and 4–8 cm wide, 3–5-lobed, smooth above, paler and finely glandular hairy below (at least along veins); margins toothed or double-toothed; petioles 3–6 cm long, finely hairy. **Flowers** yellow-green to purple, saucer-shaped, in loose upright clusters 3–6 cm long, on slender stalks; bracts very small, the stalks and bracts with gland-tipped hairs; sepals 2 mm long; petals 1–2 mm long. **Fruit** a dark red berry with bristles and gland-tipped hairs, 6 mm wide. June. **Synonyms** *Ribes prostratum*. **Habitat** Cedar and tamarack swamps, cool wet woods, thickets and streambanks.

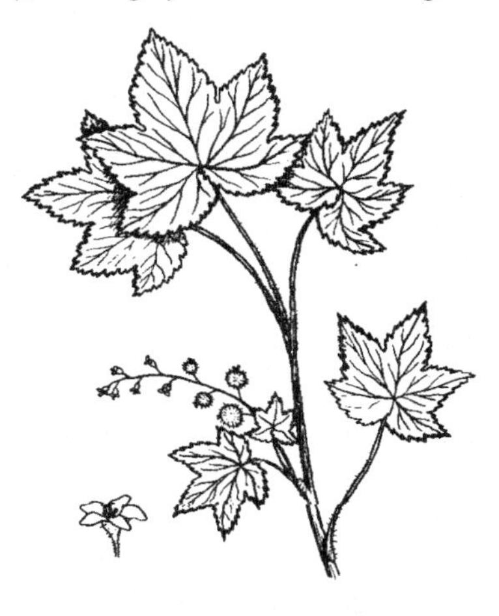

Ribes hirtellum Michx.

NORTHERN GOOSEBERRY

NC **FACW** | MW **FACW** | GP **FAC**

Native shrub to 9 dm tall. **Stems** upright, outer bark pale, soon peeling to expose dark inner layer; young stems gray and smooth, or with 1–3 slender spines at nodes and scattered bristles between nodes. **Leaves** 2.5–5 cm long and 2–5 cm wide, with 3 or 5 pointed lobes, upper surface dark green, smooth to sparsely hairy, lower surface paler, hairy at least along veins, without glands; margins coarsely toothed and fringed with hairs; petioles 1–3 cm long, hairy, some of which are gland-tipped. **Flowers** green-yellow to purple, bell-shaped, 6–9 mm long, in clusters of 2–3 on short, smooth stalks; stamens as long or longer than sepals, the bracts fringed with long hairs. **Fruit** an edible, smooth, dark blue-black berry, 8–12 mm wide. June. **Synonyms** *Ribes huronense*. **Habitat** Cedar and tamarack swamps, thickets, shores, rocky openings.

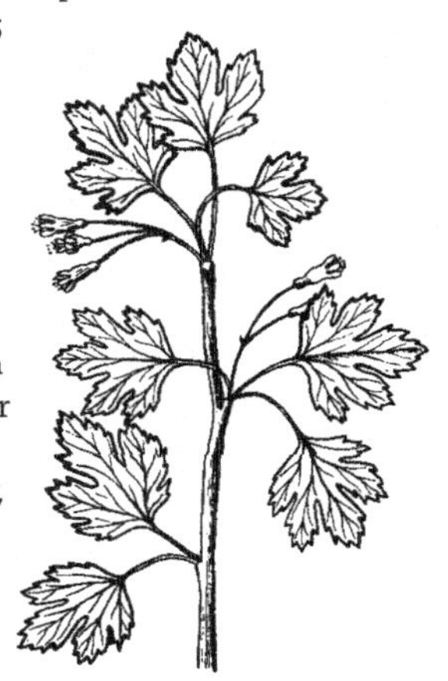

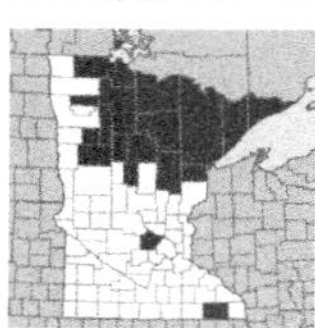

Ribes hudsonianum Richards.
HUDSON BAY CURRANT
NC **OBL** | MW **OBL** | GP **OBL**

Native shrub, 6–9 dm tall. **Stems** upright, spines and bristles absent; bark gray, with scattered yellow resin dots, peeling to expose inner purple-black bark. **Leaves** 5–9 cm long and 6–13 cm wide, 3–5-lobed, with unpleasant odor when rubbed, upper surface dark green and mostly hairless, underside paler, smooth to hairy and with yellow resin dots; margins coarsely toothed, the teeth with a hard tip; petioles 2.5–8 cm long, with fine hairs and resin dots. **Flowers** white, bell-shaped, 4–5 mm long, in small clusters on threadlike stalks. **Fruit** a smooth, blue-black berry, 8–10 mm wide, barely edible. June. **Habitat** Cedar swamps, wet conifer woods and streambanks.

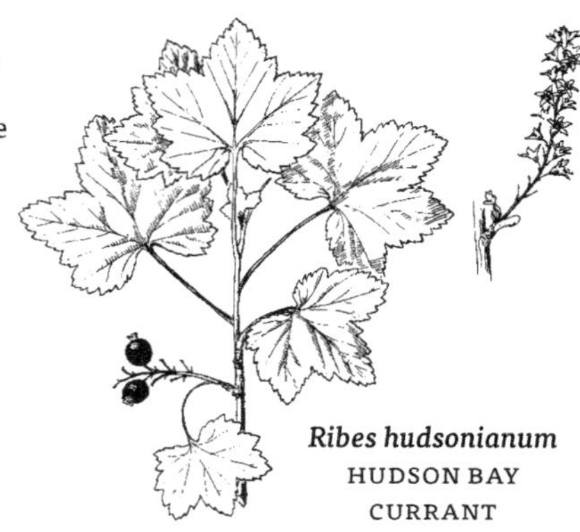

Ribes hudsonianum
HUDSON BAY CURRANT

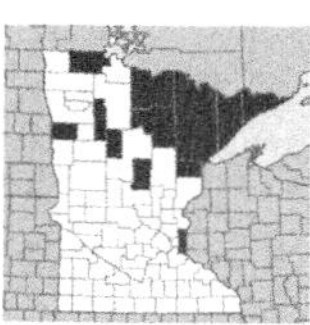

Ribes lacustre (Pers.) Poiret
PRICKLY CURRANT
NC **FACW** | MW **FACW** | GP **FACW**

Native shrub to 1 m tall. **Stems** upright or spreading, densely bristly, long-spiny at nodes; older bark gray, peeling to expose dark inner bark. **Leaves** 4–8 cm long and 4–7 cm wide, with 3–5 deeply parted, pointed lobes, upper surface dark green and mostly smooth, underside paler with scattered gland-tipped hairs; margins cleft into rounded teeth; petioles 2.5–4 cm long, with gland-tipped hairs. **Flowers** yellow-green to pinkish, saucer-shaped, 4–5 mm wide, on stalks with dark, gland-tipped hairs, in arching or drooping clusters. **Fruit** a purple-black berry covered with gland-tipped hairs, 9–12 mm wide. May–June. **Habitat** Moist conifer woods, swamps, thickets, and rock outcrops.

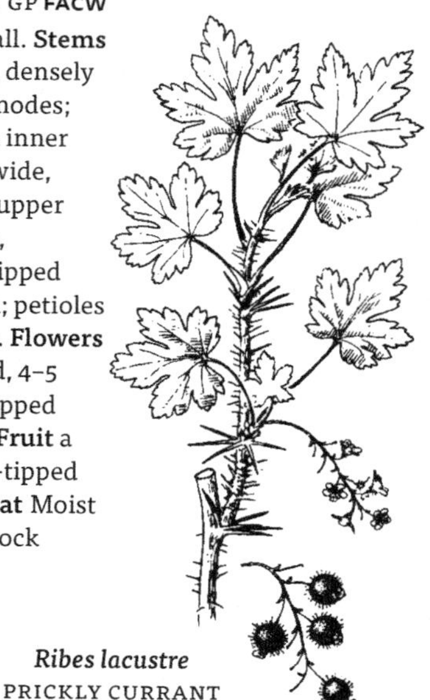

Ribes lacustre
PRICKLY CURRANT

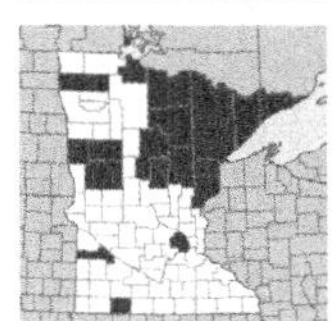

Ribes oxyacanthoides L.

NORTHERN OR BRISTLY GOOSEBERRY

NC **FACU** | MW **FACU** | GP **FACU**

Native shrub to 1 m tall. **Stems** upright with 1–3 spines to 1 cm long at nodes and smaller spines scattered between nodes; young stems gray-brown and finely hairy. **Leaves** 2.5–5 cm long and 2–5 cm wide, with 3–5 blunt or rounded lobes, upper surface sparsely hairy, some hairs tipped with glands, underside resin-dotted, hairy, some gland-tipped, especially along veins; margins coarsely toothed and hairy, some hairs gland-tipped; petioles 0.5–3 cm long, with short hairs and scattered glands. **Flowers** green-yellow, bell-shaped, 6–9 mm long, in clusters of 2–3 on short stalks; stamens shorter than petals. **Fruit** a smooth, edible, blue-black berry, 9–12 mm wide. June. **Synonyms** *Ribes setosum*. **Habitat** Rocky and sandy shores, rocky openings, stabilized dunes, and moist, cold woods.

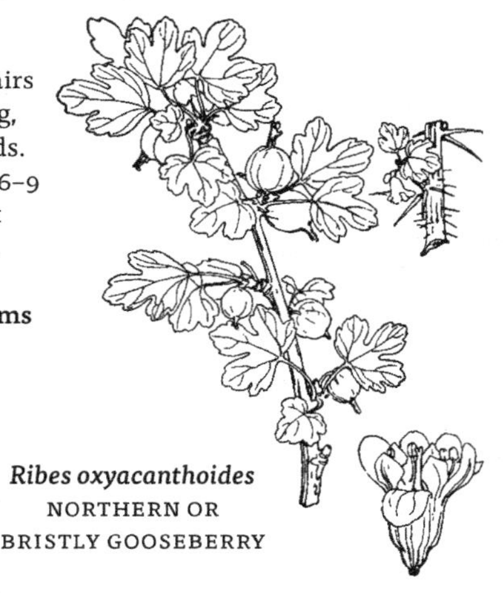

Ribes oxyacanthoides
NORTHERN OR
BRISTLY GOOSEBERRY

Ribes triste Pallas

SWAMP RED CURRANT

NC **OBL** | MW **OBL** | GP **OBL**

Low native shrub, 0.4–1 m tall. **Stems** spreading or lying on ground and rooting at nodes, spines and bristles absent; older stems smooth, purple–black, young stems short-hairy. **Leaves** 4–10 cm long and 4–10 cm wide, with 3–5 broad lobes, dark green and mostly smooth above, paler and usually finely hairy below; margins with both rounded and sharp teeth, the teeth with a hard tip; petioles 2.5–6 cm long, with scattered gland-tipped hairs. **Flowers** green-purple, 4–5 mm wide, on stalks 1–4 mm long, in drooping clusters of 5–12. **Fruit** a smooth, red berry, 6–9 mm wide. May–June. **Habitat** Wet woods and swamps, alder thickets, seeps.

Ribes triste
SWAMP RED CURRANT

PERENNIAL AQUATIC HERBS. **Leaves** alternate or whorled, finely dissected. **Flowers** small, stalkless in axils of leaves or bracts, 3- or 4-parted, regular, perfect (with both male and female parts) or imperfect, petals small or absent. **Fruit** small and nutlike, dividing into 3 or 4 segments.

HALORAGACEAE | WATER-MILFOIL FAMILY

1 Flowers 4-parted; leaves mostly whorled, emersed leaves reduced to small bracts . . . . *Myriophyllum* | WATER-MILFOIL

1 Flowers 3-parted; leaves alternate, emersed leaves not bractlike . . . . *Proserpinaca* | MERMAID-WEED

MYRIOPHYLLUM | *Water-milfoil*

Perennial aquatic herbs. **Stems** submerged, sparsely branched, freely rooting at lower nodes. **Leaves** mostly whorled (alternate in *Myriophyllum farwellii*), pinnately divided into threadlike segments, upper leaves often reduced to bracts. **Flowers** small, mostly imperfect, stalkless in axils of upper emersed leaves (the floral bracts) or axils of underwater leaves; male flowers above female flowers; perfect flowers (if present) in middle portion of spike; sepals inconspicuous; petals 4 or absent; stamens 4 or 8; pistil 4-chambered. **Fruit** nutlike, 4-lobed, each lobe (mericarp) with 1 seed, rounded on back or with a ridge or row of small bumps.

MYRIOPHYLLUM | WATER-MILFOIL

1 Leaves simple, reduced to small, blunt-tipped scales; stems erect from creeping rhizomes . . . *Myriophyllum tenellum* | SLENDER WATER-MILFOIL

1 Leaves dissected into narrow segments 2

2 Leaves alternate, ± opposite, or scattered on stem . . *Myriophyllum farwellii* . . . . FARWELL'S WATER-MILFOIL

2 Foliage leaves all whorled 3

3 Flowers and bracts below flowers alternate on stem. . . . . *Myriophyllum alterniflorum* | ALTERNATE-FLOWER WATER-MILFOIL

3 Flowers and bracts below flowers whorled. 4

4 Bracts surrounding male flowers deeply cleft. . *Myriophyllum verticillatum* . . . . WHORLED WATER-MILFOIL

4 Bracts surrounding male flowers sharply toothed or entire 5

5 Bracts sharply toothed and much longer than flowers. . . . . *Myriophyllum heterophyllum* | TWO-LEAF WATER-MILFOIL

5 Bracts surrounding male flowers entire and not longer than flowers. . . . 6

6 Leaf segments mostly 5–12 on each side of midrib; small bulbs (turions) produced at ends of stems and in upper leaf axils. . . . . *Myriophyllum sibiricum* | COMMON WATER-MILFOIL

6 Leaf segments many, 12–20 on each side of midrib; turions absent . . . . *Myriophyllum spicatum* | EURASIAN WATER-MILFOIL

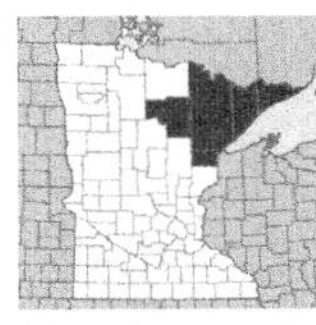

Myriophyllum alterniflorum DC.
ALTERNATE-FLOWER WATER-MILFOIL
NC **OBL**

Native perennial herb. **Stems** very slender. **Leaves** in whorls of 3–5, usually less than 1 cm long and shorter than the stem internodes, pinnately divided. **Flower spikes** raised above water surface, 2–5 cm long; bracts mostly alternate, linear, shorter than the flowers; male flowers with 4 pink petals; stamens 8. **Fruit** segments 1–2 mm long, rounded on back and base. **Habitat** Acidic lakes, Lake Superior shoreline.

Myriophyllum alterniflorum
ALTERNATE-FLOWER WATER-MILFOIL

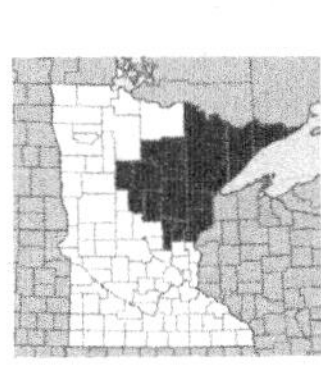

Myriophyllum farwellii Morong
FARWELL'S WATER-MILFOIL
NC **OBL** | MW **OBL**

Native perennial herb; plants entirely underwater, turions present at ends of stems. **Leaves** 1–3 cm long, dissected into threadlike segments, all or most leaves alternate, or ± opposite, or irregularly scattered on stems. **Flowers** underwater, single in axils of foliage leaves; female flowers with 4 purple petals; stamens 4, tiny. **Fruit** 2 mm long, each fruit segment with 2 small, bumpy, longitudinal ridges. **Habitat** Ponds and small lakes.

Myriophyllum farwellii
FARWELL'S WATER-MILFOIL

Myriophyllum heterophyllum Michx.

TWO-LEAF WATER-MILFOIL

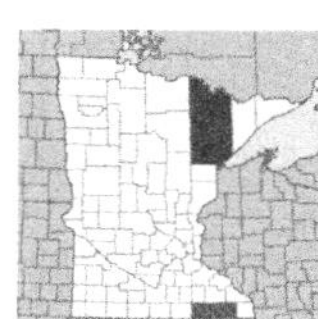

NC **OBL** | MW **OBL** | GP **OBL**

Native perennial herb. **Stems** stout, to 3 mm wide, often red-tinged, to 1 m or more long. **Leaves** whorled, 1.5–4 cm long, divided into threadlike segments. **Flowers** in spikes raised above water surface, 5–30 cm long; floral bracts whorled, smaller than foliage leaves, ovate, sharply toothed, spreading or curved downward. **Flowers** both perfect and imperfect; petals of male and perfect flowers 1–3 mm long; stamens 4. **Fruit** olive, ± round, 2 mm long; fruit segments rounded or with 2 small ridges, beaked by the curved stigma. June–Aug. **Habitat** Lakes, ponds and pools in streams; sometimes where calcium-rich.

Myriophyllum heterophyllum
TWO-LEAF WATER-MILFOIL

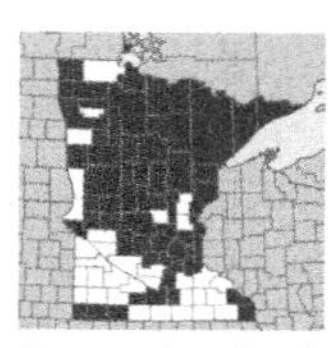

Myriophyllum sibiricum Komarov

COMMON WATER-MILFOIL

NC **OBL** | MW **OBL** | GP **OBL**

Native perennial herb; plants often whitish when dried. **Stems** to 1 m or more long. **Leaves** in whorls of 3–4, 1–4 cm long, with mostly 5–10 threadlike segments on each side of midrib; internodes between whorls about 1 cm long. **Flowers** in spikes with whorled flowers and bracts, raised above water surface, red, clearly different than underwater stems, 4–10 cm long; flowers imperfect, the upper male, the lower female; floral bracts much smaller than the leaves, oblong to obovate; male flowers with pinkish petals (absent in female flowers), 2–3 mm long; stamens 8, the yellow-green anthers conspicuous when flowering. **Fruit** ± round, 2–3 mm long, the segments rounded on back. June–Sept. **Synonyms** *Myriophyllum exalbescens*. **Habitat** Common; shallow to deep water of lakes, ponds, marshes, ditches and slow-moving streams; sometimes where calcium-rich.

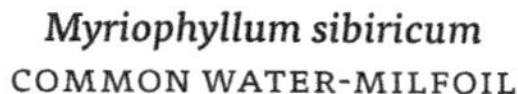

Myriophyllum sibiricum
COMMON WATER-MILFOIL

Eurasian water-milfoil *(Myriophyllum spicatum, below), an introduced species, is similar to Myriophyllum sibiricum but has more finely divided leaves (12–24 threadlike segments on each side of midrib) and larger floral bracts.*

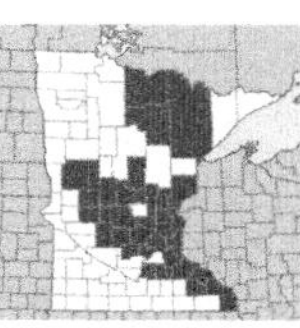

Myriophyllum spicatum L.

EURASIAN WATER-MILFOIL
NC **OBL** | MW **OBL** | GP **OBL**

Introduced, invasive perennial herb, similar to ***Myriophyllum sibiricum***. **Stems** widening below head and curved to a horizontal position, usually many-branched near water surface, internodes between leaves mostly 1–3 cm long, turions absent. **Leaves** with more leaf segments per side (mostly 12–20) than in *Myriophyllum sibiricum*; lower flower bracts often divided into comblike segments and often longer than the flowers. **Fruit** segments 2–3 mm long. Aug–Sept. **Habitat** Lakes and ponds. Introduced, found across most of North America.

Myriophyllum spicatum
EURASIAN WATER-MILFOIL

Eurasian water-milfoil *may form large, dense mats, hindering boating, swimming, and fishing. Map at right shows reported infestations of this species* (MINNESOTA DNR).

Myriophyllum tenellum Bigelow
SLENDER WATER-MILFOIL
NC **OBL** | MW **OBL** | GP **OBL**

Native perennial herb. **Stems** slender, 10–30 cm long, mostly upright and unbranched. **Leaves** absent or reduced to a few spaced scales. **Flowers** in spikes raised above water surface, 2–5 cm long; flower bracts mostly alternate, oblong to obovate, entire, shorter to slightly longer than the flowers. **Fruit** segments rounded on back and at base, 1 mm long. **Habitat** Acidic lakes; often forming large colonies, especially in deep water.

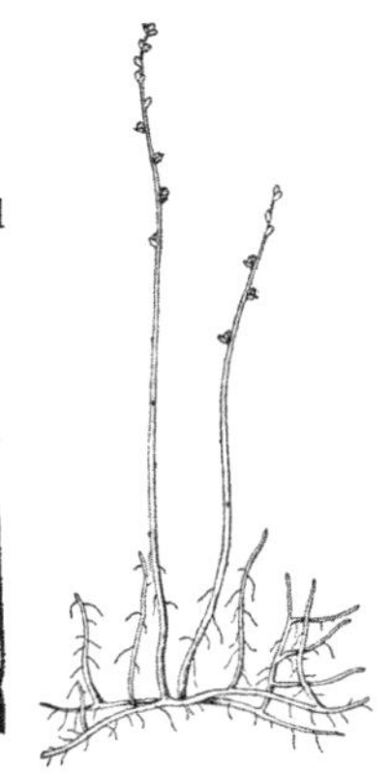

Myriophyllum tenellum
SLENDER WATER-MILFOIL

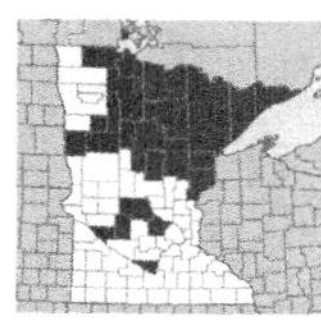

Myriophyllum verticillatum L.
WHORLED WATER-MILFOIL
NC **OBL** | MW **OBL** | GP **OBL**

Native perennial herb, similar to ***Myriophyllum sibiricum***, but plants often larger. Stems 5–25 dm long. **Leaves** in whorls of 4–5, with 9–17 threadlike segments along each side of midrib, 1–5 cm long; lower and middle internodes between whorls mostly less than 1 cm long. **Flowers** perfect, or the lower female and upper male; in spikes 4–12 cm long, the floral bracts much smaller than the leaves, with comb- like segments, mostly longer than the flowers; petals blunt-tipped, 2–3 mm long, smaller in female flowers; stamens 8. **Fruit** ± round, 2–3 mm long, the segments rounded on back. July–Sept. **Habitat** Lakes, ponds and quiet places in rivers.

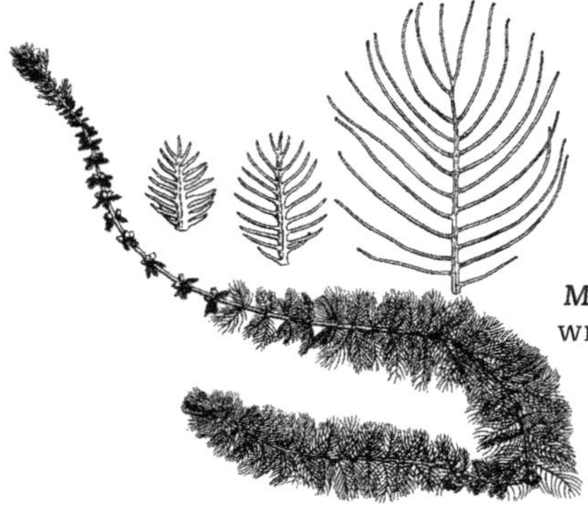

Myriophyllum verticillatum
WHORLED WATER-MILFOIL

PROSERPINACA | *Mermaid-weed*

Proserpinaca palustris L.
MARSH MERMAID-WEED
[NC **OBL** | MW **OBL** | GP **OBL**]

Native perennial aquatic herb, often forming large colonies. **Stems** horizontal at base and often rooting; the flower-bearing branches erect, 1–4 dm tall. **Leaves** alternate; **underwater leaves**, if present, ovate in outline, 2–4 cm long, deeply divided into linear segments; **emersed leaves** narrowly lance-shaped, 2–6 cm long, margins with sharp, forward-pointing teeth. **Flowers** small, perfect, green or purple-tinged, 1–3 in axils of emersed leaves, stalkless; petals absent, stamens 3, stigmas 3. **Fruit** nutlike, 3-angled, with 3 seeds, 2–5 mm long and as wide. June–Aug. **Habitat** Shallow water of ponds, streambanks and ditches, muddy shores, sedge meadows; usually where seasonally flooded.

Reported for Minnesota but distribution unknown; popular as an aquarium plant.

Proserpinaca palustris
MARSH MERMAID-WEED

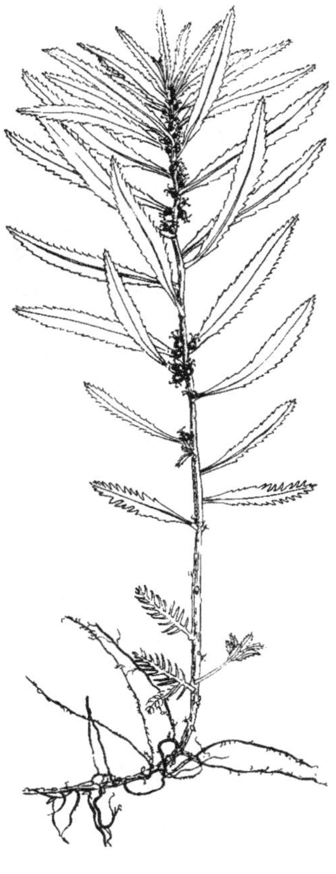

HIPPURIS | *Mare's-tail*

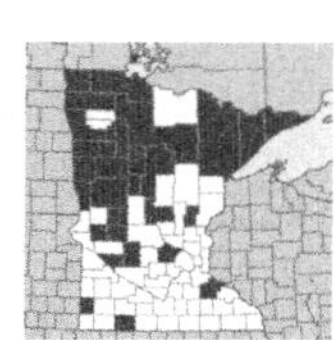

Hippuris vulgaris L.

MARE'S-TAIL

NC **OBL** | MW **OBL** | GP **OBL**

Native perennial herb, from large, spongy rhizomes. **Stems** 2–6 dm long, unbranched, underwater and lax, or emersed and upright, densely covered by the closely spaced whorls of leaves. **Leaves** numerous, in whorls of 6–12, linear, 1–2.5 cm long and 1–3 mm wide, stalkless. **Flowers** very small, perfect, stalkless and single in upper leaf axils, or often absent; sepals and petals lacking; stamen 1, style 1, ovary 1-chambered. **Fruit** nutlike, oval, 2 mm long. June–Aug. **Habitat** Shallow water or mud of marshes, lakes, streams and ditches.

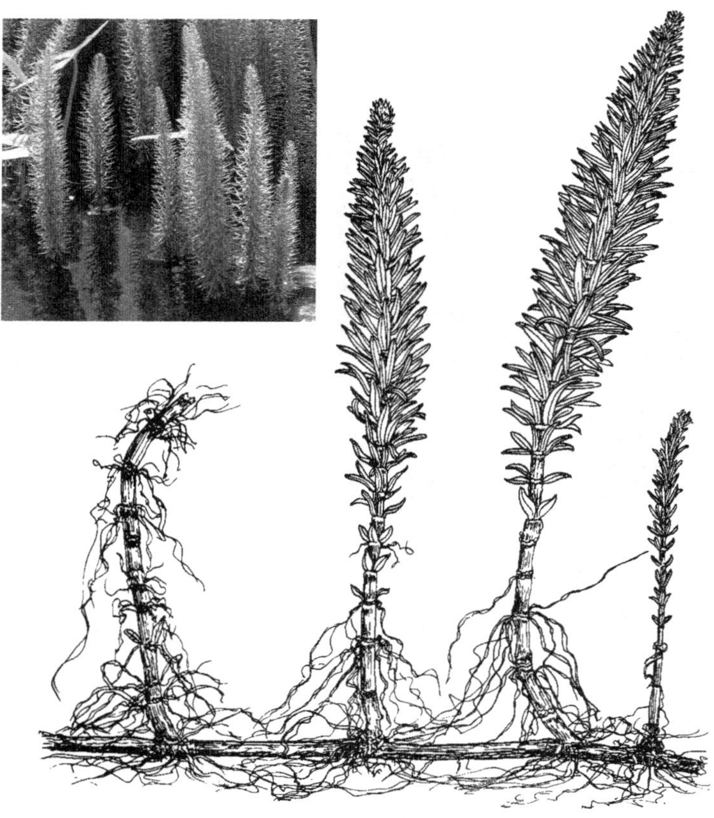

Hippuris vulgaris
MARE'S-TAIL

SMOOTH ANNUAL OR PERENNIAL HERBS. **Stems** usually unbranched below, branched in head. **Leaves** simple, opposite, dotted with dark or translucent glands (visible when held to a light), especially on underside; margins entire; petioles absent. **Flowers** few to many in clusters at ends of stems or from upper leaf axils, perfect, regular, sepals 5, petals 5, yellow or pink to green or purple; stamens 9–35, separate or joined near base into 3 or more groups; styles 3, ovary superior. **Fruit** a 3-chambered, many-seeded capsule.

***Hypericum** and **Triadenum** formerly included within the Clusiaceae (Mangosteen Family).*

HYPERICACEAE | ST. JOHN'S-WORT FAMILY

1 Petals yellow; stamens 15–many .... *Hypericum* | ST. JOHN'S-WORT
1 Petals pink or purple; stamens 9 ... *Triadenum* | MARSH ST. JOHN'S-WORT

HYPERICUM | ST. JOHN'S-WORT

Herbs. **Leaves** opposite, sometimes dotted with black and/or small transparent glands; margins entire. **Flowers** in clusters at ends of stems and upper leaf axils, yellow, perfect, regular, sepals 5, petals 5, stamens 5–many, separate or joined into 3 or 5 bundles. **Fruit** a capsule.

HYPERICUM | ST. JOHN'S-WORT

1 Plants 1–2 m tall; flowers 4 cm or more wide; styles 5..................
.... *Hypericum pyramidatum* | GIANT ST. JOHN'S-WORT
1 Plants smaller; flowers less than 3 cm wide; styles 3 .... **2**

2 Stamens 20 or more; styles joined, persisting on capsule as a straight beak .... *Hypericum ellipticum* | PALE ST. JOHN'S-WORT
2 Stamens less than 20; styles separate to base and often spreading, capsules not beaked **3**

3 Sepals broadest above middle; fruit rounded at tip..... *Hypericum boreale* .... NORTHERN ST. JOHN'S-WORT
3 Sepals lance-shaped, broadest below middle; fruit tapered to tip.........
.... *Hypericum majus* | LARGE CANADIAN ST. JOHN'S-WORT

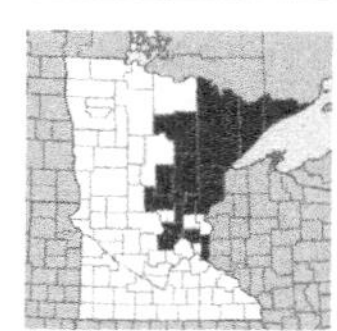

Hypericum boreale (Britton) E. Bickn.

NORTHERN ST. JOHN'S-WORT
NC **OBL** | MW **OBL** | GP **OBL**

Native perennial herb, from slender rhizomes. **Stems** 1–4 dm long, round or slightly 4-angled, branched above. **Leaves** oval or oblong, rounded at ends and nearly clasping stem, 3–5-nerved, larger leaves 1–2 cm long and 0.5–1 cm wide; petioles absent. **Flowers** in clusters at ends of stems and from upper leaf axils; sepals blunt-tipped; petals yellow, 3 mm long; stamens 8–15; styles 3 (sometimes 4), less than 1 mm long. **Fruit** a 1-chambered purple capsule, 3–5 mm long. July–Sept. **Habitat** Pond and marsh margins, low areas between dunes, open bogs.

Hypericum boreale
NORTHERN ST. JOHN'S-WORT

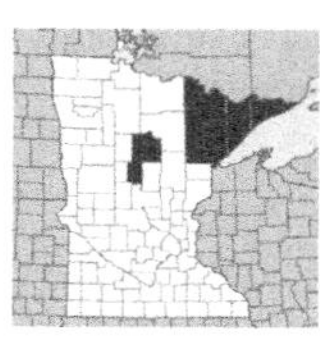

Hypericum ellipticum Hook.

PALE ST. JOHN'S-WORT
NC **OBL** | MW **OBL**

Native perennial herb, spreading by rhizomes. **Stems** 2–5 dm long, branched only in head. Leaves oval, 1–4 cm long and 1–1.5 cm wide, rounded at tip, narrowed at base and sometimes clasping stem; petioles absent. **Flowers** few to many, in clusters at ends of stems; sepals to 6 mm long; petals pale yellow, 6–7 mm long; stigmas 3 (sometimes 4), small. **Fruit** a 1-chambered capsule, 5–6 mm long, rounded to a short beak formed by the persistent styles. July–Aug. **Habitat** Streambanks, sandy shores and flats, thickets, bogs.

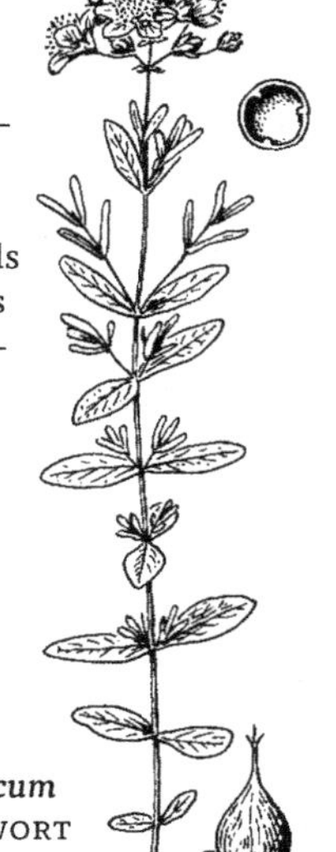

Hypericum ellipticum
PALE ST. JOHN'S-WORT

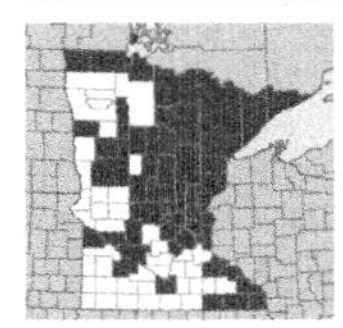

Hypericum majus (A. Gray) Britton
LARGE CANADIAN ST. JOHN'S-WORT
NC **FACW** | MW **FACW** | GP **FACW**

Native perennial herb, spreading from rhizomes or stolons. **Stems** upright, unbranched or branched above, 1–6 dm long. **Leaves** lance-shaped, 2–4 cm long and 3-10 mm wide, dotted with brown sunken glands, 5–7-nerved from base; leaf tip rounded, leaf base rounded or heart-shaped and weakly clasping; petioles absent. **Flowers** few to many in clusters at ends of stems and from upper leaf axils; sepals lance-shaped, 4–6 mm long; petals yellow, equal to sepals but then shriveling to half the length of sepals; stamens 14–21, not joined; styles to 1 mm long. **Fruit** a red-purple ovate capsule, 5–7 mm long. July–Sept. **Habitat** Streambanks, sandy, mucky or calcareous shores, low areas between dunes, marshes, wetland margins.

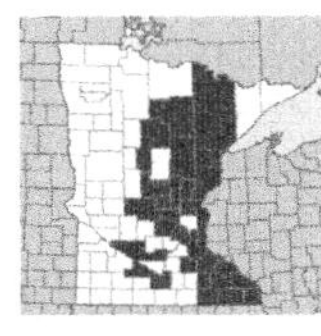

Hypericum pyramidatum Aiton
GIANT ST. JOHN'S-WORT
NC **FAC** | MW **FAC**

Native perennial herb. **Stems** upright, branched, 6–20 dm long. **Leaves** lance-shaped to oval, 4–10 cm long and 1–4 cm wide, base often clasping stem; petioles absent. **Flowers** few, 4–6 cm wide, mostly single on stalks from upper leaf axils; stamens numerous, joined at base into 5 bundles; petals bright yellow; styles 5, not persisting. **Fruit** an ovate, 5-chambered capsule, 15–30 mm long. July–Aug. **Synonyms** *Hypericum ascyron*. **Habitat** Streambanks, ditches, fen and marsh margins.

Hypericum pyramidatum
GIANT ST. JOHN'S-WORT

TRIADENUM | MARSH ST. JOHN'S-WORT

Triadenum fraseri (Spach) Gleason
MARSH ST. JOHN'S-WORT
NC **OBL** | MW **OBL** | GP **OBL**

Smooth perennial native herb, with creeping rhizomes. **Stems** upright, mostly unbranched, red, smooth, 3–6 dm long. **Leaves** opposite, entire, oval or ovate, 3–6 cm long and 1–3 cm wide, pinnately veined, rounded at tip, rounded or heart-shaped and clasping at the base, with dark dots and transparent glands on underside. **Flowers** in clusters at ends of stems and from leaf axils, often remaining closed; sepals 5, 3–5 mm long, rounded at tip; petals 5, pink to green-purple, 5–8 mm long; stamens 9, joined at base into 3 bundles, the bundles alternating with orange glands; styles 3, 1–2 mm long. **Fruit** a purple, cylindric capsule, 7–12 mm long, abruptly narrowed to the 1 mm long persistent style beak. July–Aug. **Synonyms** *Hypericum virginicum* var. *fraseri, Triadenum virginicum* subsp. *fraseri*. **Habitat** Marshes, sedge meadows, open bogs, fens, sandy and calcium-rich shores.

Triadenum fraseri
MARSH ST. JOHN'S-WORT

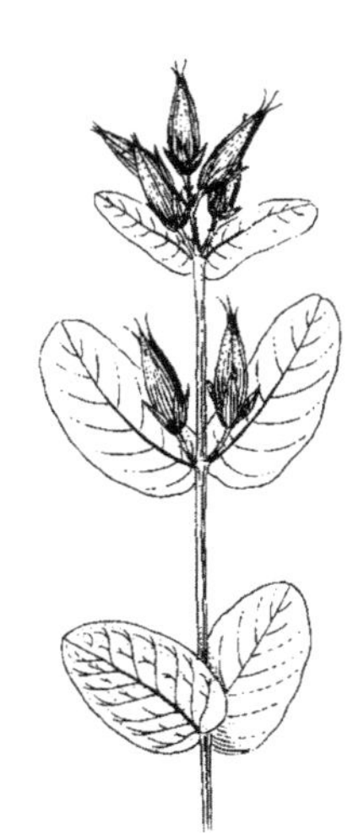

Hardwood swamp, early spring. Black ash (*Fraxinus nigra*) is the dominant tree; red maple (*Acer rubrum*) is also present. Marsh-marigold (*Caltha palustris*) is common in the undergrowth, with scattered plants of several species of violet (*Viola*), three-seeded bog sedge (*Carex trisperma*), rough sedge (*Carex scabrata*), sensitive fern (*Onoclea sensibilis*), and orange touch-me-not or jewelweed (*Impatiens capensis*).

Large marsh dominated by cat-tails (*Typha* spp.) and grasses such as bluejoint (*Calamagrostis canadensis*) and reed canary-grass (*Phalaris arundinacea*). Soils are largely organic and covered with water for all or most of the year, and while productivity is high, overall species diversity is low.

PERENNIAL, OFTEN AROMATIC HERBS. Stems usually 4-angled. **Leaves** simple, opposite, sharply toothed or deeply lobed. **Flowers** in leaf axils or in heads or spikes at ends of stems, perfect (with both male and female parts), nearly regular to irregular; sepals 5-toothed or sometimes 2-lipped; petals white, pink, blue or purple, often 2-lipped; stamens 2 or 4; ovary 4-lobed, splitting into 4, 1-seeded nutlets when mature.

LAMIACEAE | MINT FAMILY

1 Corolla regular or nearly so, with 4–5 lobes of equal length 2
1 Corolla irregular, 1- or 2-lipped 3

2 Stamens 2; plants not strongly scented . . . *Lycopus* | WATER-HOREHOUND
2 Stamens 4; plants strongly mint-scented . . *Mentha arvensis* | FIELD-MINT

3 Upper lip of corolla absent, lower lip large . . . . *Teucrium canadense* . . . . AMERICAN GERMANDER
3 Upper and lower corolla lips well-developed 4

4 Calyx with a rounded bump on upper side; petals blue . . . . *Scutellaria* . . . . SKULLCAP
4 Calyx without bump on upper side; petal colors vary 5

5 Flowers in stalked clusters at branch ends, the clusters forming a flat-topped or rounded inflorescence . . . . *Pycnanthemum virginianum* . . . . VIRGINIA MOUNTAIN-MINT
5 Flowers ± stalkless, in spikes or crowded heads 6

6 Leaf margins ± entire 7
6 Leaf margins regularly toothed 8

7 Leaves linear . . . . *Stachys hyssopifolia* | HYSSOP HEDGE-NETTLE
7 Leaves broader, less than 4x longer than wide . . . . *Prunella vulgaris* . . . . SELF-HEAL

8 Flowers paired in slender spikes; leaves and stems smooth . . . . *Physostegia virginiana* | OBEDIENCE; FALSE DRAGON-HEAD
8 Flowers 3 or more at each node; leaves and stems usually hairy . . . *Stachys* . . . . HEDGE-NETTLE

LYCOPUS | *Water-horehound*

Perennial, ± unscented herbs. **Stems** erect, 4-angled. **Leaves** opposite, coarsely toothed or deeply lobed, smaller on upper stems; petioles short or absent. **Flowers** small, in clusters in middle and upper leaf axils, often appearing whorled; white to pink, the sepals and petals often dotted on outer surface, 4-lobed, stamens 2. **Fruit** a nutlet.

LYCOPUS | WATER-HOREHOUND

1 Sepal lobes broad, triangular to ovate, to 1 mm long, shorter than or nearly as long as nutlets, midvein not prominent 2
1 Sepal lobes slender, 1–3 mm long, longer than nutlets, midvein prominent 3

2 Leaves mostly less than 3 cm wide; stamens and styles visible, longer than petals; outer rim of nutlets taller than the inner rim . . . . *Lycopus uniflorus* . . . . NORTHERN WATER-HOREHOUND
2 Larger leaves 3 cm or more wide; stamens and styles hidden by petals; inner and outer rim of nutlets same height, the 4 nutlets appearing flat-topped across tops . . . . *Lycopus virginicus* | VIRGINIA WATER-HOREHOUND

3 Main leaves stalkless . . . . . *Lycopus asper* | WESTERN WATER-HOREHOUND
3 Leaves stalked . . . . *Lycopus americanus* | AMERICAN WATER-HOREHOUND

Lycopus americanus Muhl.
AMERICAN WATER-HOREHOUND
NC **OBL** | MW **OBL** | GP **OBL**

Native perennial herb, spreading by rhizomes, tubers absent. **Stems** erect, often branched, 2–8 dm long, upper stems smooth or short-hairy. **Leaves** opposite, lance-shaped, 3–8 cm long and 1–4 cm wide, with glandular dots, smooth or rough on upper surface, underside veins short-hairy; margins coarsely and irregularly deeply toothed or lobed, the lowest teeth largest; nearly stalkless or on short petioles. **Flowers** in dense, whorled clusters in leaf axils; sepal lobes narrow, sharp-tipped, 1–3 mm long, longer than fruits; petals white, sometimes pink to purple-dotted, 4-lobed, the upper lobe wider and notched. **Fruit** a nutlet, 1–2 mm long. July–Sept. **Habitat** Marshes, wet meadows, shores, streambanks, ditches, calcareous fens, wetland margins.

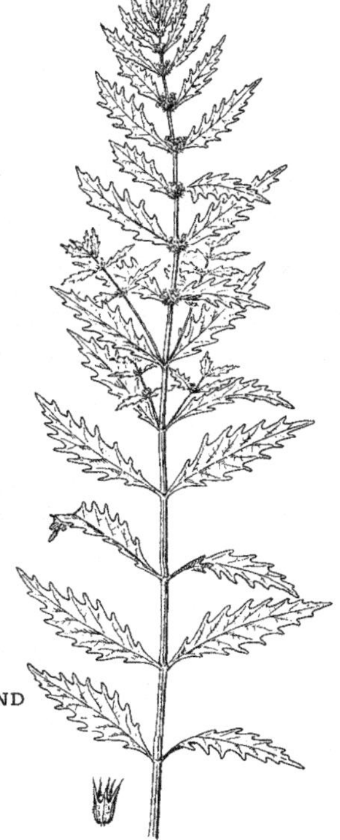

Lycopus americanus
AMERICAN WATER-HOREHOUND

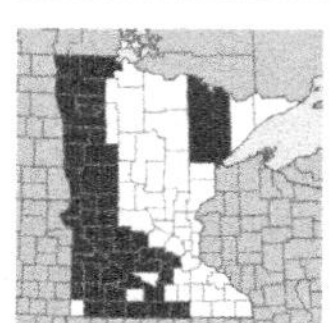

Lycopus asper Greene
WESTERN WATER-HOREHOUND
NC **OBL** | MW **OBL** | GP **OBL**

Native perennial emergent herb, spreading by rhizomes (tubers present) and also usually stolons. **Stems** erect, 2–8 dm long, simple or sometimes branched, hairy, at least on stem angles. **Leaves** opposite, oval to oblong lance-shaped, 3–10 cm long and 0.5–3 cm wide, smooth or rough; margins coarsely toothed; stalkless. **Flowers** in dense, whorled clusters in leaf axils; sepal lobes narrow, firm, sharp-tipped, 1–3 mm long, longer than nutlets; petals white, 4-lobed, only slightly longer than sepals. **Fruit** a nutlet, 1–2 mm long. July–Sept. **Habitat** Shores and ditches, especially where disturbed, often with American water-horehound (*Lycopus americanus*).

Lycopus asper
WESTERN WATER-HOREHOUND

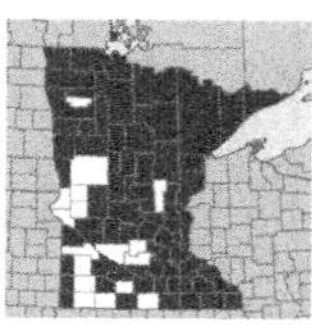

Lycopus uniflorus Michx.
NORTHERN WATER-HOREHOUND
NC **OBL** | MW **OBL** | GP **OBL**

Native perennial herb, similar to **western water-horehound** (*Lycopus asper*). **Stems** smooth or short-hairy, 1–5 dm long. **Leaves** opposite, lance-shaped to oblong, 3–6 cm long and 1–3 cm wide, margins with a few outward-pointing teeth, petioles short or nearly absent. **Flowers** in dense, whorled clusters in leaf axils; sepal lobes broad, triangular to ovate, soft, rounded at tip, to 1 mm long, shorter to as long as nutlets; petals white or pink, 2–3 mm long, 5-lobed, longer than sepals. **Fruit** a nutlet 1–1.5 mm long. Aug–Sept. **Habitat** Swamps, streambanks, thickets, wet meadows, open bogs, calcareous fens, ditches; often with American water-horehound (*Lycopus americanus*).

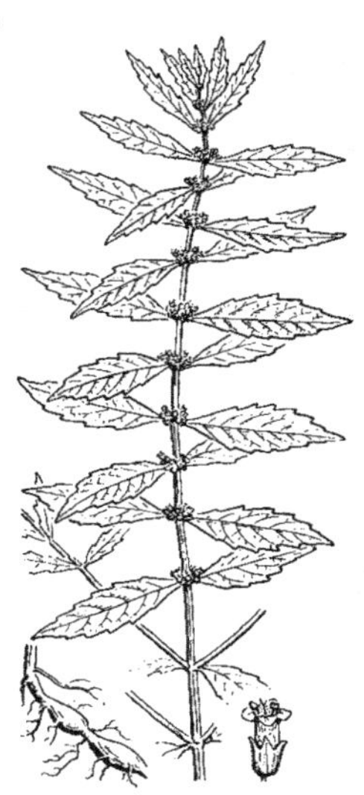

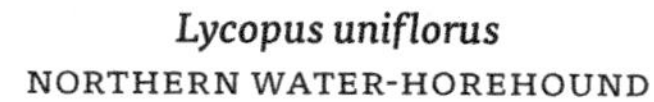

Lycopus uniflorus
NORTHERN WATER-HOREHOUND

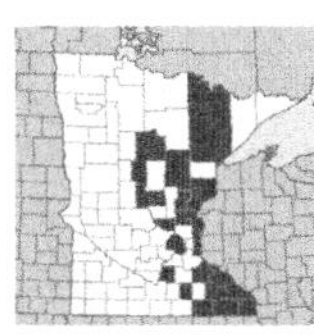

Lycopus virginicus L.
VIRGINIA WATER-HOREHOUND
NC **OBL** | MW **OBL** | GP **OBL**

Native perennial herb, spreading by stolons (tubers usually absent). **Stems** 2–6 dm long, with dense covering of appressed hairs. **Leaves** opposite, lance-shaped to oval, 5–10 cm long and 1.5–5 cm wide, long-hairy, lower surface usually also with short, feltlike hairs; margins coarsely toothed, the lowest tooth just below middle of blade, the margin below tooth concave and petiolelike. **Flowers** in whorled clusters from leaf axils; sepals shorter than nutlets; petals white, 4-lobed, (upper lobe often notched). **Fruit** a nutlet, 1–2 mm long, the group of 4 nutlets ± flat across tips. July–Sept. **Synonyms** *Lycopus membranaceus*. **Habitat** Floodplain forests.

Probably hybridizes with **northern water-horehound** *(Lycopus uniflorus) where their ranges overlap, producing a hybrid swarm called Lycopus x sherardii.*

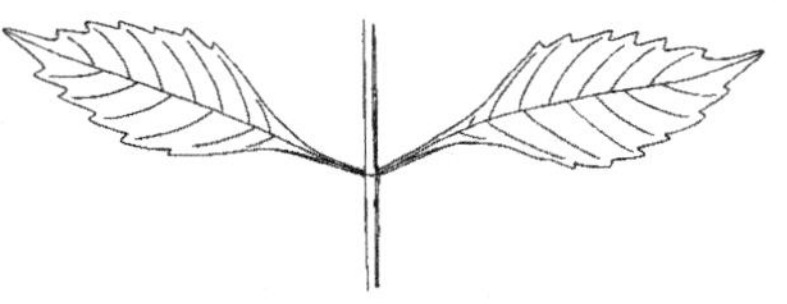

Lycopus virginicus
VIRGINIA WATER-HOREHOUND

MENTHA | *Mint*

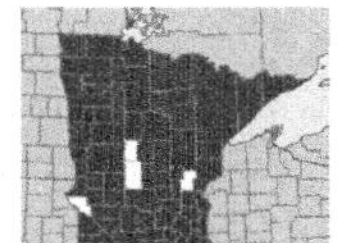

Mentha arvensis L.
FIELD-MINT
NC **FACW** | MW **FACW** | GP **FACW**

Native perennial herb, strongly mint-scented, spreading by rhizomes and often also by stolons. **Stems** 2–8 dm long, 4-angled, hairy at least on stem angles. **Leaves** opposite, ovate to lance-shaped or oval–lance-shaped, 2–7 cm long and 0.5–3 cm wide, smooth or hairy; margins with sharp, forward-pointing teeth; petioles short. **Flowers** small, white or light pink to lavender, hairy, crowded in whorled clusters in middle and upper leaf axils; sepals 2–3 mm long, hairy and glandular; petals ± regular to slightly 2-lipped, 4–6 mm long, glandular on outside, 4- or 5-lobed; stamens and style longer than petals. **Fruit** a smooth nutlet to 1 mm long, enclosed by the persistent sepals. July–Sept. **Habitat** Common; wet meadows, marshes, swamps, thickets, streambanks, ditches, springs and other wet places.

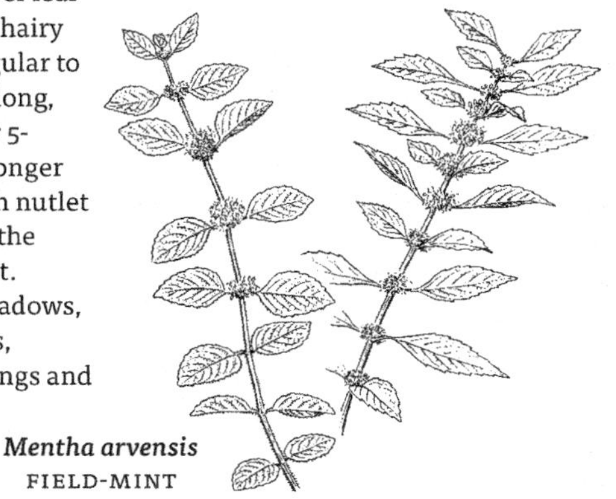

Mentha arvensis
FIELD-MINT

PHYSOSTEGIA | *Obedience*

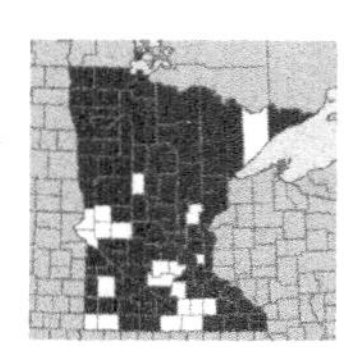

Physostegia virginiana (L.) Benth.
OBEDIENCE; FALSE DRAGON-HEAD
NC **FACW** | MW **FACW** | GP **FACW**

Native perennial herb, spreading by rhizomes. **Stems** erect, 5–15 dm long, often branched near top, 4-angled. **Leaves** opposite, oval to oblong lance-shaped, 2–15 cm long and 1–4 cm wide, sometimes smaller upward; margins with sharp teeth; stalkless, not clasping. **Flowers** in several racemes 5–20 cm long, the stalks short-hairy; sepals 4–8 mm long, often with some gland-tipped hairs; petals pink-purple or white with purple spots, 1.5–3 cm long, short-hairy to smooth. **Fruit** a nutlet, 2–3 mm long. July–Sept. **Synonyms** *Dracocephalum virginianum*. **Habitat** Sedge meadows, low prairie, shores, swamps, floodplain forests, thickets and ditches.

Physostegia virginiana
OBEDIENCE; FALSE DRAGON-HEAD

PRUNELLA | *Self-heal*

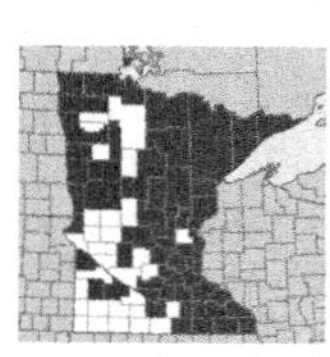

Prunella vulgaris L.
SELF-HEAL
NC **FAC** | MW **FAC** | GP **FAC**

Native perennial herb. **Stems** upright or sometimes spreading, 1–5 dm long, 4-angled. **Leaves** opposite, lance-shaped to oval or ovate, 2–8 cm long and 1–4 cm wide; lower leaves wider than upper; margins entire or with a few small teeth; petioles present. **Flowers** in dense spikes 2–5 cm long and 1–2 cm wide, with obvious bracts; sepals to 1 cm long, green or purple, with spine-tipped teeth; corolla 2-lipped, the upper lip hoodlike and entire, lower lip shorter and 3-lobed; petals blue-violet (rarely pink or white), 1–2 cm long; stamens 4, about as long as petals. **Fruit** a smooth nutlet.June–Oct. **Habitat** Common in many types of wetlands (especially where disturbed): swamps, wet forest depressions, wet trails, streambanks, ditches; also in drier forests, fields and lawns.

PYCNANTHEMUM | *Mountain-mint*

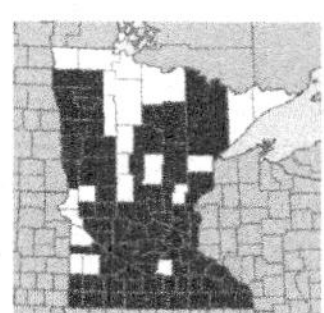

Pycnanthemum virginianum (L.) Durand & B. D. Jackson
VIRGINIA MOUNTAIN-MINT
NC **FACW** | MW **FACW** | GP **FAC**

Native perennial, strongly scented herb. **Stems** to 1 m long, branched above, 4-angled, angles short-hairy. **Leaves** numerous, opposite, narrowly lance-shaped, 3–4 cm long and to 1 cm wide (leaves in heads much smaller), upper surface smooth, with 3–4 pairs of lateral veins, undersides often finely hairy on midvein; margins entire but fringed with short, rough hairs; ± stalkless. **Flowers** small, 2-lipped, in branched, crowded clusters at ends of stems and branches from upper leaf axils; sepals short woolly-hairy; petals white, purple-spotted. **Fruit** a 4-parted nutlet. July–Sept. **Habitat** Wet meadows, marshes, tamarack swamps, calcareous fens, low prairie.

Pycnanthemum virginianum
VIRGINIA MOUNTAIN-MINT

SCUTELLARIA | *Skullcap*

Perennial herbs, spreading by rhizomes. **Stems** erect or spreading, 4-angled. **Leaves** opposite, ovate to lance-shaped, margins toothed, stalked or ± stalkless. **Flowers** blue or blue with white markings, single on short stalks in axils of middle and upper leaves, or in racemes from leaf axils; calyx 2-lipped, with a rounded bump on upper side; corolla 2-lipped, pubescent on outer surface, upper lip hoodlike, lower lip ± flat, 3-lobed; stamens 4, ascending into the upper corolla lip. **Fruit** a 4-parted nutlet.

SCUTELLARIA | SKULLCAP

1 Flowers single in leaf axils; flowers more than 15 mm long. . . . . *Scutellaria galericulata* | MARSH OR HOODED SKULLCAP

1 Flowers in racemes from leaf axils; flowers to 8 mm long . . . . *Scutellaria lateriflora* | BLUE SKULLCAP

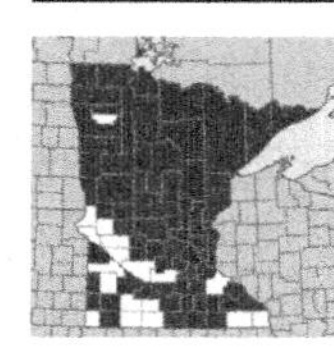

Scutellaria galericulata L.

MARSH OR HOODED SKULLCAP

NC **OBL** | MW **OBL** | GP **OBL**

Native perennial herb, from slender rhizomes. **Stems** erect or spreading, 2–8 dm long, unbranched or branched, 4-angled, short-hairy at least on angles of upper stem. **Leaves** opposite, lance-shaped to narrowly ovate, 2–6 cm long and 0.5–2.5 cm wide, upper surface smooth, underside short-hairy, tapered to a blunt tip; margins with low, rounded, forward-pointing teeth; petioles very short. **Flowers** 2-lipped, single in leaf axils (and paired at nodes), on stalks 1–3 mm long; sepals 3–6 mm long; petals blue, marked with white, 15–25 mm long. **Fruit** a nutlet. June–Sept. **Synonyms** *Scutellaria epilobiifolia*. **Habitat** Shores, streambanks, marshes, wet meadows, swamps, thickets, bogs, ditches.

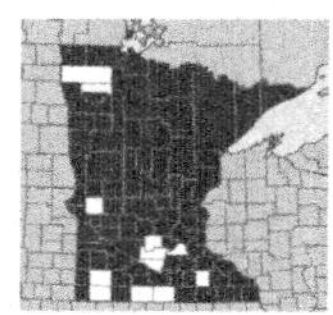

Scutellaria lateriflora L.

BLUE SKULLCAP

NC **OBL** | MW **OBL** | GP **FACW**

Native perennial herb. Stems 2–6 dm long, usually branched, 4-angled, short-hairy on upper stem angles or smooth. **Leaves** opposite, ovate to lance-shaped, 3–8 cm long and 1.5–5 cm wide, smooth; margins coarsely toothed; petioles 0.5–2 cm long. **Flowers** 2-lipped, in elongate racemes from leaf axils; sepals 2–4 mm long; petals blue (rarely pink or white), 5–8 mm long. **Fruit** a nutlet. July–Sept. **Habitat** Shores, streambanks, wet meadows, marshes, swamps, shaded wet areas.

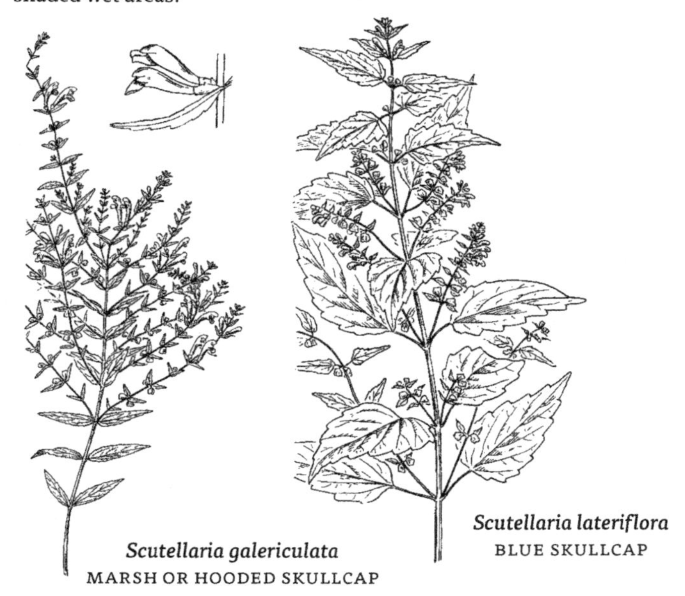

Scutellaria galericulata
MARSH OR HOODED SKULLCAP

Scutellaria lateriflora
BLUE SKULLCAP

STACHYS | *Hedge-nettle*

Erect, perennial herbs, spreading by rhizomes; plants usually hairy. **Stems** 4-angled. **Leaves** opposite, margins entire or toothed, stalkless or with short petioles. **Flowers** in interrupted spikes at ends of stems, appearing whorled in ± evenly spaced clusters; sepals ± regular, with 5 equal teeth; corolla 2-lipped, petals pink, often with purple spots or mottles, upper lip concave, entire, lower lip spreading, 3-lobed; stamens 4, ascending under the upper lip. **Fruit** a dark brown, 4-lobed nutlet, loosely enclosed by the persistent sepals.

STACHYS | HEDGE-NETTLE

1 Stems hairy on angles and sides . . . . *Stachys palustris*
. . . . MARSH HEDGE-NETTLE

1 Stems hairy only on stem angles . . . . *Stachys tenuifolia*
. . . . SMOOTH HEDGE-NETTLE

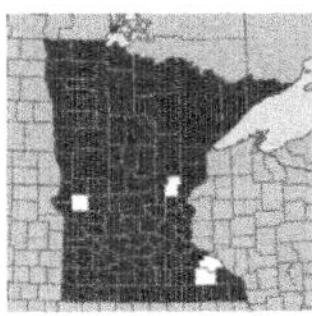

Stachys palustris L.

MARSH HEDGE-NETTLE
NC **FACW** | MW **FACW** | GP **FACW**

Native perennial herb. **Stems** 3–8 dm long, unbranched or branched, 4-angled, stiffly hairy on angles and with short, gland-tipped hairs on sides. **Leaves** opposite, lance-shaped to oblong, 4–12 cm long and 2–5 cm wide, softly hairy on both sides; margins with rounded, forward-pointing teeth; stalkless or with short petioles. **Flowers** in clusters of 6–10 in an interrupted spike at end of stem (sometimes also clustered in upper leaf axils); sepals 5–8 mm long, with long, glandless hairs and shorter gland-tipped hairs; petals purple to pale red with purple spots, 9–13 mm long. **Fruit** a nutlet. June–Aug. **Synonyms** *Stachys pilosa*. **Habitat** Marshes, wet meadows, ditches, thickets, shores, streambanks, openings in swamps.

Stachys tenuifolia Willd.

SMOOTH HEDGE-NETTLE
NC **FACW** | MW **OBL** | GP **FACW**

Native perennial herb. **Stems** 4–10 dm long, 4-angled, smooth, or with downward-pointing, bristly hairs on stem angles. **Leaves** opposite, lance-shaped to ovate, 6–14 cm long and 2–6 cm wide, ± smooth; margins with sharp, forward-pointing teeth; petioles slender, 1–2 cm long or absent. **Flowers** in interrupted spikes at ends of stems or also in upper leaf axils; sepals 5–7 mm long, smooth; petals pale red to purple, 1.5–2.5 cm long. **Fruit** a nutlet. July–Sept.

Synonyms *Stachys hispida*. **Habitat** Floodplain forests, shores, streambanks, thickets, wet meadows.

TEUCRIUM | *Germander, Wood-sage*

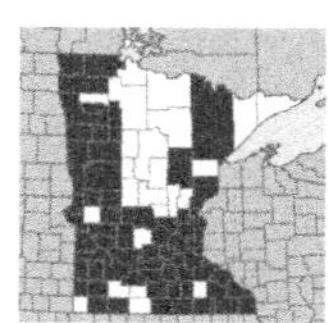

Teucrium canadense L.

AMERICAN GERMANDER

NC **FACW** | MW **FACW** | GP **FACW**

Native perennial herb, spreading by rhizomes. **Stems** 3–10 dm long, mostly unbranched, 4-angled, long-hairy. **Leaves** opposite, lance-shaped or oblong, 4–12 cm long and 1.5–5 cm wide, upper surface smooth or sparsely hairy, underside with dense, matted hairs, margins irregularly finely toothed, petioles 5–15 mm long. **Flowers** in a dense spikelike raceme, 5–20 cm long; bracts present and narrowly lance-shaped; flowers on stalks 1–3 mm long; sepals ± regular, purple or green, 4–7 mm long, covered with long silky hairs and very short glandular ones; corolla irregular, 10–16 mm long, with short gland-tipped hairs, upper lip absent, lower lip large; petals pink to purple; stamens 4, arched over the corolla. **Fruit** a golden nutlet. July–Sept. **Habitat** Marshes, wet meadows, shores, streambanks, thickets, floodplain forests, ditches.

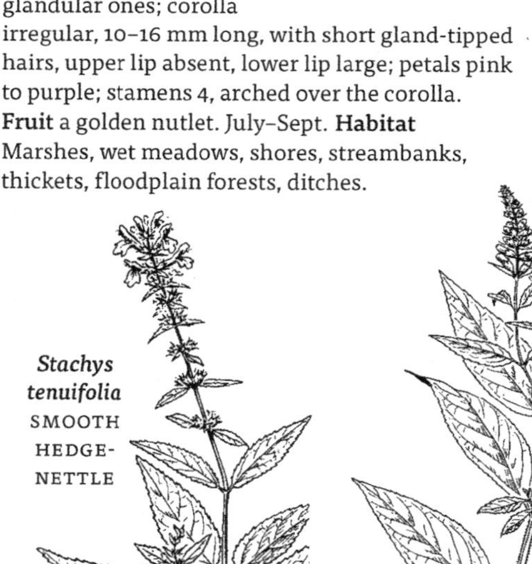

Stachys tenuifolia
SMOOTH HEDGE-NETTLE

Teucrium canadense
AMERICAN GERMANDER

INSECTIVOROUS HERBS. Leaves in a basal rosette (*Pinguicula*); or floating, or in peat, muck, or wet soil (*Utricularia*). **Flowers** perfect (with both male and female parts), irregular, 2-lipped, sometimes with a spur, 1 to several on an erect stem; stamens 2. **Fruit** a capsule.

LENTIBULARIACEAE | BLADDERWORT FAMILY

1 Leaves ovate or oval, in a basal rosette; flowers single on a bractless stalk. . . . . *Pinguicula vulgaris* | COMMON BUTTERWORT

1 Leaves linear or dissected into narrow segments; flowers 1, or several in a raceme, each flower subtended by a bract . . . . *Utricularia* | BLADDERWORT

PINGUICULA | *Butterwort*

Pinguicula vulgaris L.

VIOLET BUTTERWORT; BOG VIOLET

NC **OBL** | *special concern*

Native perennial herb. **Leaves** 3–6 in a basal rosette, ovate or oval, 2–5 cm long, blunt-tipped, narrowed to base, upper surface sticky; margins inrolled. **Flowers** single atop a leafless stalk (scape) 5–15 cm long; corolla violet-purple, spurred, 2-lipped, the upper lip 3-lobed, the lower lip 2-lobed, 1.5–2 cm long (including spur). **Fruit** a 2-chambered capsule. June–July. **Habitat** Rock crevices in sandstone cliffs along Lake Superior, wet areas between dunes, marl flats and calcareous fens; usually with **Mistassini primrose** (*Primula mistassinica*). **Status** Special concern.

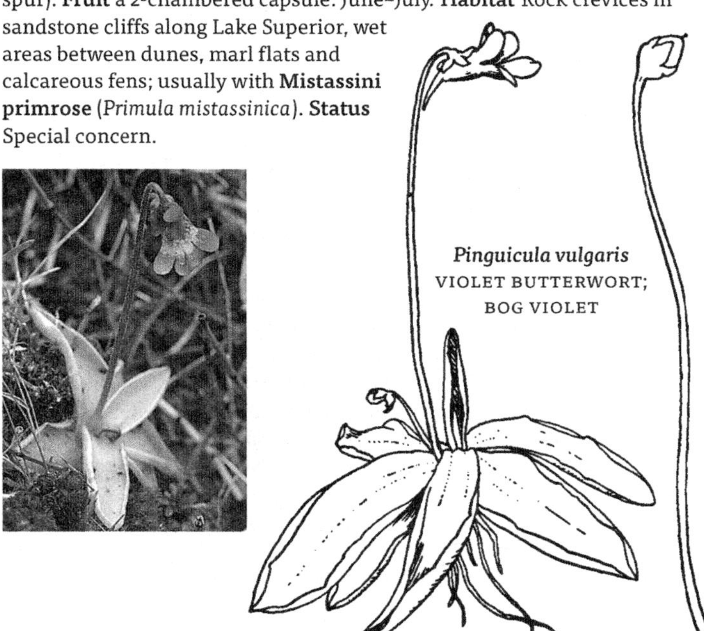

Pinguicula vulgaris
VIOLET BUTTERWORT;
BOG VIOLET

UTRICULARIA | *Bladderwort*

Aquatic or wetland, annual or perennial herbs. **Leaves** underwater, alternate, entire or dissected into many linear segments, some with bladders which trap tiny aquatic invertebrates; or leaves in wet soil and rootlike or absent. **Flowers** perfect, irregular, 1 to several in a raceme atop stalks raised above water or soil surface, each flower subtended by a small bract; corolla yellow or purple, similar to a snapdragon flower, 2-lipped, the upper lip erect, entire or slightly 2-lobed, lower lip entire or 3-lobed, the corolla tube extended backward into a sac or spur, stamens 2. **Fruit** a many-seeded capsule.

***Utricularia** is the largest genus of carnivorous plant, with more than 220 species worldwide. Plants feature small specialized structures (the bladders), which due to negative water pressure, suck in prey when contact is made with the bladder door; enzymes then digest the captured organism.*

UTRICULARIA | BLADDERWORT

1 Flowers purple or pink 2
1 Flowers yellow 3

2 Flowers 2–5 atop a stout stalk; plants floating in water, masses of leaves present . . . . . . *Utricularia purpurea* | SPOTTED OR PURPLE BLADDERWORT
2 Flowers single atop a slender stalk; plants not free-floating, rooted in peat or muck, appearing leafless. . . . . *Utricularia resupinata* . . . . LAVENDER BLADDERWORT

3 Scapes appearing leafless; leaves simple or absent; plants of peat, moist sand, or marl . . . . *Utricularia cornuta* | LEAFLESS BLADDERWORT
3 Scapes with leaves at base, the leaves dissected and with bladderlike traps; plants mostly floating in water. 4

4 Leaf divisions flat in section 5
4 Leaf divisions round or threadlike. 6

5 Bladders borne on leaves; smallest leaf divisions entire (visible with a 10x hand lens); flower with a sac or spur much shorter than lower lip. . . . . *Utricularia minor* | LESSER BLADDERWORT
5 Bladders on branches separate from leaves; smallest leaf divisions finely toothed, the teeth spine-tipped; flower with a spur as long as lower lip . . . . *Utricularia intermedia* | NORTHERN BLADDERWORT

6 Plants large; leaves floating; scapes 1 mm or more wide; flowers 13 mm or more long, 5 or more per head; larger bladders more than 2 mm wide . . . . *Utricularia vulgaris* | COMMON BLADDERWORT

6 Plants smaller; leaves floating or creeping on lake bottom or wet shores; scapes threadlike; flowers to 12 mm long, 1–3 per head; larger bladders mostly less than 2 mm wide *Utricularia gibba* | CREEPING BLADDERWORT

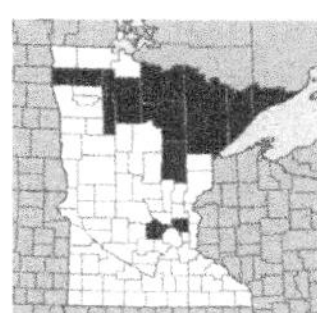

Utricularia cornuta Michx.

LEAFLESS BLADDERWORT
NC **OBL** | MW **OBL** | GP **OBL**

Native annual or perennial herb. **Stems** and leaves underground, roots with tiny bladders. **Flowers** yellow, with a downward-pointing spur 6–15 mm long, on stalks 1–2 mm long, 1–6 atop an erect stalk 10–25 cm long; bracts ovate, 1–2 mm long. **Fruit** a rounded capsule. June–Sept. **Habitat** Acid lakes, shores, peatlands, calcareous pools between dunes, borrow pits.

Utricularia cornuta
LEAFLESS BLADDERWORT

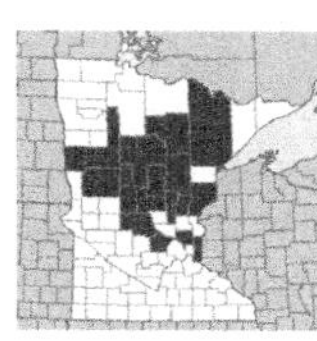

Utricularia gibba L.

CREEPING BLADDERWORT
NC **OBL** | MW **OBL** | GP **OBL**

Native annual or perennial herb. **Stems** creeping on bottom in shallow water, mostly less than 10 cm long, radiating from base of flower stalk (scape) and forming mats. **Leaves** alternate, scattered, to 5 mm long, 1–2-forked into threadlike segments; bladders present. **Flowers** 1–3, yellow, 5–6 mm long, with a thick, blunt spur shorter than lower lip, atop a single stalk 5–10 cm long. **Fruit** a rounded capsule. July–Sept.
Habitat Exposed shores, lakes, ponds, marshes, fens.

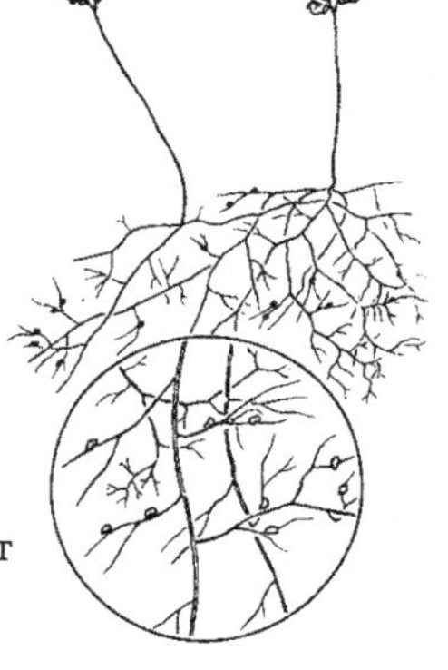

Utricularia gibba
CREEPING BLADDERWORT

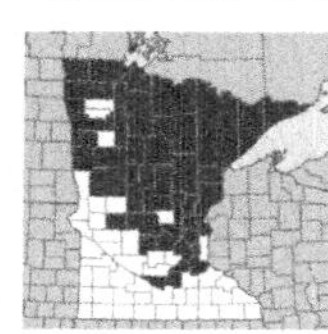

Utricularia intermedia Hayne

NORTHERN BLADDERWORT

NC **OBL** | MW **OBL** | GP **OBL**

Native annual herb. **Stems** very slender, creeping along bottom in shallow water. **Leaves** alternate, 0.5–2 cm long, mostly 3-parted near base, then again divided 1–3x, the segments linear and flat, margins with small, bristly teeth; bladders 2–4 mm wide, borne on branches separate from leaves. **Flowers** yellow, 2–4 atop an emergent stalk 5–20 cm long; individual flower stalks to 15 mm long, remaining erect in fruit; spur nearly as long as lower lip. **Fruit** a capsule. June–Aug. **Habitat** Shallow water (usually alkaline), marly pools between dunes, calcareous fens, marshes, ponds and rivers.

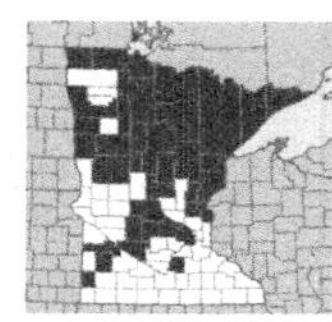

Utricularia minor L.

LESSER BLADDERWORT

NC **OBL** | MW **OBL** | GP **OBL**

Native perennial herb. **Stems** few-branched, 10–30 cm long, creeping on bottom in shallow water or on wet soil. **Leaves** alternate, to 1 cm long, with few divisions, the segments slender, flat, the smallest segments strongly tapered to tip, margins entire; bladders 1–2 mm wide, 1–5 on leaves. **Flowers** pale yellow, 2–8 atop a threadlike stalk 4–15 cm long; individual flower stalks to 1 cm long, curved downward in fruit; lower lip of flower 4–8 mm long, 2x longer than upper lip; spur small, to half length of lower lip. **Fruit** a capsule. June–Aug. **Habitat** Fens, open bogs, sedge meadows and marshes; often in shallow water and where calcium-rich.

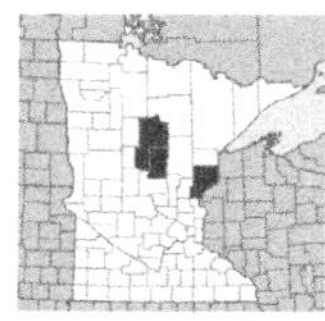

Utricularia purpurea Walter

SPOTTED OR PURPLE BLADDERWORT

NC **OBL** | MW **OBL** | *special concern*

Native annual or perennial herb. **Stems** underwater, to 1 m long. **Leaves** in whorls of 5–7, branched into threadlike segments, many segments tipped by a bladder. **Flowers** red-purple, 1–4 atop a stalk 3–15 cm long; corolla 1 cm long, lower lip 3-lobed, with a yellow spot near base; spur short and appressed to lower lip. **Fruit** a capsule. July–Sept. **Synonyms** *Vesiculina purpurea*. **Habitat** Acid lakes and ponds in water to 1 m deep, peatlands, marshes. **Status** Special concern.

Utricularia purpurea
SPOTTED OR PURPLE BLADDERWORT

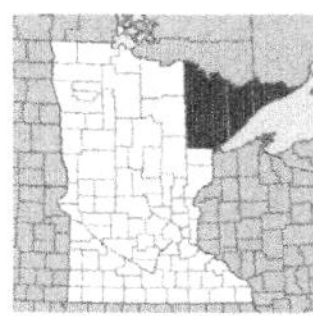

Utricularia resupinata B. D. Greene

LAVENDER BLADDERWORT

NC **OBL** | MW **OBL** | *special concern*

Native annual or perennial herb. **Stems** delicate, on water surface in shallow water or creeping just below soil surface. **Leaves** alternate, 3-parted from base, the middle segment erect and linear, to 3 cm long; the 2 lateral segments slender, rootlike, with bladders. **Flowers** purple, 1 cm long, single atop an erect stalk 2–10 cm long; bract tubelike, surrounding the stem, its margin notched; flower tipped backward on stalk and facing upward; lower lip 3-lobed; spur ± horizontal. **Fruit** a rounded capsule. July–Aug. **Synonyms** *Lecticula resupinata*. **Habitat** Shallow to deep water, wet shores where sandy or mucky. **Status** Special concern.

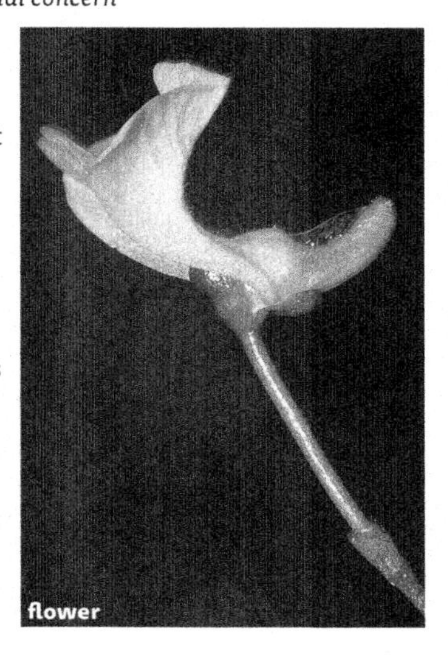

flower

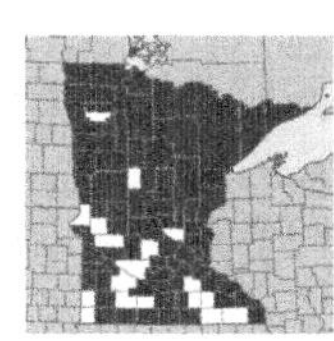

Utricularia vulgaris L.

COMMON BLADDERWORT

NC **OBL** | MW **OBL** | GP **OBL**

Native perennial herb. **Stems** floating below water surface, sparsely branched, often forming large mats. **Leaves** alternate, 1–5 cm long, 2-forked at base and repeatedly 2-forked into segments of unequal length, the segments ± round in section, becoming smaller with each branching, the final segments threadlike; bladders 1–4 mm wide, borne on leaf segments (see photo, right). **Flowers** yellow, 6–20 atop a stout stalk 6–25 cm long; lower flower lip 1–2 cm long, sometimes much smaller on late-season flowers, upper lip ± equal to lower lip; spur about 2/3 as long as lower lip; stalks bearing individual flowers curved downward in fruit. **Fruit** a capsule. June–Aug. **Synonyms** *Utricularia macrorhiza*. **Habitat** Shallow water of lakes, ponds, peatlands, marshes, and rivers.

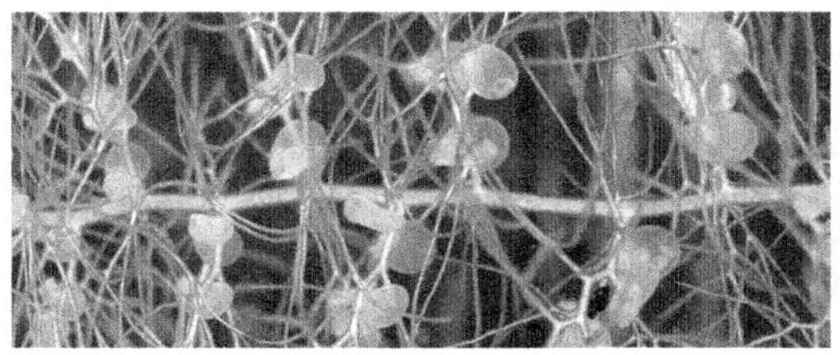

Utricularia vulgaris bladder detail

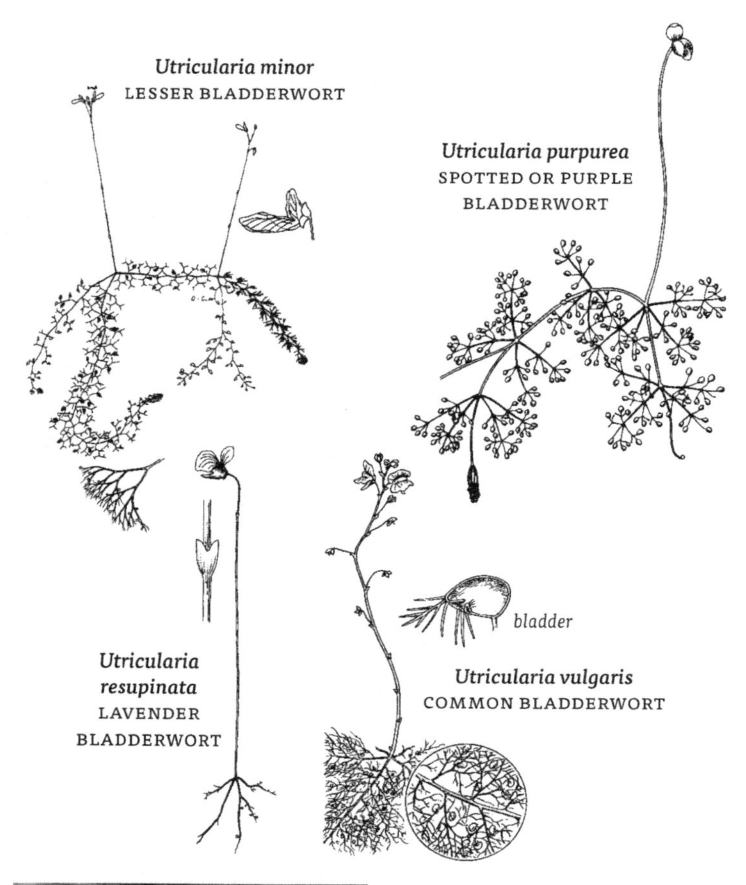

Utricularia intermedia
NORTHERN BLADDERWORT

ANNUAL OR PERENNIAL HERBS, sometimes woody at base (*Decodon*). **Leaves** simple, opposite, or both opposite and alternate, or whorled, margins entire, ± stalkless. **Flowers** 1 or several in leaf axils or in spikelike heads at ends of stems; perfect (with both male and female parts), regular or irregular; sepal lobes 4 or 6; petals 4 or 6, separate, pink or purple, deciduous; stamens usually 2x number of petals. **Fruit** a dry, many-seeded capsule.

LYTHRACEAE | LOOSESTRIFE FAMILY

1 Plants arching, woody near base; leaves with petioles and mostly whorled .... *Decodon verticillatus* | SWAMP OR WHORLED LOOSESTRIFE
1 Plants annual or perennial herbs; leaves opposite, or if whorled, leaves without petioles 2

2 Plants perennial; flowers in spikelike heads at ends of stems; petals and sepals 6.. *Lythrum* | LOOSESTRIFE
2 Plants annual; flowers from leaf axils; petals and sepals 4 or 5 (when present).. 3

3 Flowers mostly 2–5 per leaf axil; flowers purple-tinged *Ammannia robusta* .... TOOTH-CUP; SCARLET LOOSESTRIFE
3 Flowers mostly 1 per axil, not purple.................................... 4

4 Submersed leaves lance-shaped, broadest at base, less than 3 mm wide .. .. *Didiplis diandra* | WATER-PURSLANE
4 Leaves oval, widest near middle, larger leaves 3 mm or more wide .... *Rotala ramosior* | TOOTH-CUP; WHEELWORT

AMMANNIA | *Tooth-cup*

Ammannia robusta Heer & Regel
TOOTH-CUP; SCARLET LOOSESTRIFE
NC **OBL** | MW **OBL** | GP **OBL**

Native annual herb; plants smooth. **Stems** erect, 2–8 dm long, often branched from base. **Leaves** opposite, linear, 2–8 cm long and 3–15 mm wide, heart-shaped and clasping at base; margins entire; petioles absent. **Flowers** stalkless, in clusters of 1–3 per leaf axil; petals 4 (rarely 5), 2–3 mm long, rose-purple, sometimes with a purple midvein at base; stamens 4 or 8. **Fruit** a round, 4-parted capsule, 3–5 mm wide, tipped by the persistent style. July–Oct. **Synonyms** *Ammannia coccinea*. **Habitat** Exposed mud flats and marshes, disturbed open wet areas; sometimes where calcium-rich.

DECODON | *Water-willow*

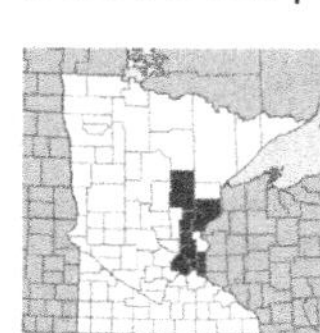

Decodon verticillatus (L.) Elliott
SWAMP OR WHORLED LOOSESTRIFE
NC **OBL** | MW **OBL** | GP **OBL** | *special concern*

Native perennial herb, woody near base. **Stems** slender, angled, smooth or slightly hairy, 1–3 m long, arching downward and rooting at tip when in contact with water or mud. **Leaves** in whorls of 3–4 or opposite, lance-shaped, 5–15 cm long and 1–3 cm wide, smooth above, sparsely hairy below; margins entire; petioles short. **Flowers** in dense clusters in upper leaf axils; sepals 5–7, short, triangular; petals pink-purple, tapered to base, 10–15 mm long; stamens 10 (rarely 8), alternately longer and shorter than petals. **Fruit** a ± round capsule, 5 mm wide. July–Sept. Minnesota plants are var. *laevigatus*. **Habitat** Shallow water and margins of lakes and streams; soils mucky; often with *Typha* or *Scirpus*. **Status** Special concern.

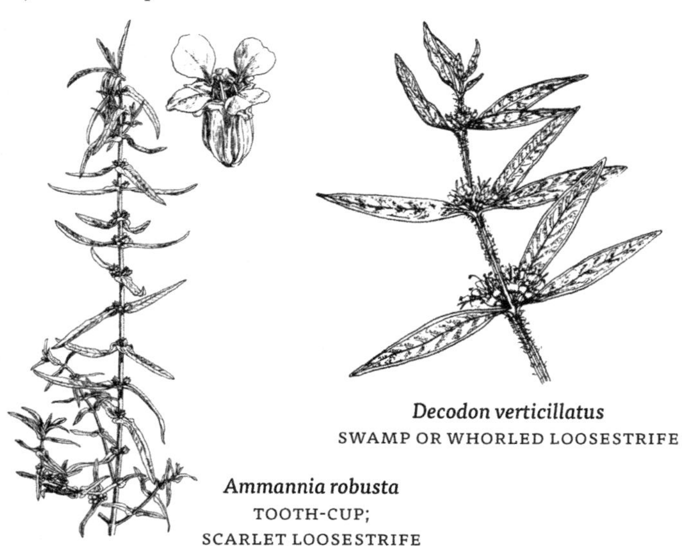

Decodon verticillatus
SWAMP OR WHORLED LOOSESTRIFE

Ammannia robusta
TOOTH-CUP;
SCARLET LOOSESTRIFE

DIDIPLIS | *Water-purslane*

Didiplis diandra (Nutt.) A. Wood
WATER-PURSLANE
NC **OBL** | MW **OBL** | GP **OBL**

Native annual herb; plants underwater or on exposed shores. **Stems** weak, branched, 1–4 dm long. **Leaves** numerous, opposite; underwater leaves linear, straight across base, 1–2.5 cm long; emersed leaves shorter and wider, tapered at base; petioles absent. **Flowers** few, inconspicuous, **green.** **Fruit** a small round capsule. July–Aug. **Synonyms** *Peplis diandra*. **Habitat** Shallow water and muddy pond margins; se Minnesota.

Similar to **water-starwort** *(Callitriche), but in water-starwort the underwater leaves have a shallow notch at tip and the capsule is flattened.*

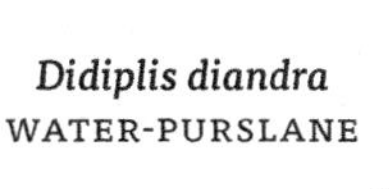

Didiplis diandra
WATER-PURSLANE

LYTHRUM | *Loosestrife*

Perennial herbs. **Stems** erect, sometimes rather woody at base, usually with ascending branches above, upper stems 4-angled. **Leaves** opposite, entire, alternate, or rarely whorled, lance-shaped, stalkless, reduced to bracts in the head. **Flowers** in showy, spikelike heads, 1 to several in axils of upper leaves, regular or somewhat irregular, the stamens and styles of 2 or 3 different lengths. Sepals joined into a tube, the calyx tube cylinder-shaped, green-striped with 8–12 nerves; petals 6, purple, not joined; stamens 6 or 12; ovary 2-chambered. **Fruit** an ovate capsule, enclosed by the calyx tube.

LYTHRUM | LOOSESTRIFE

1 Flowers single in upper leaf axils; stamens usually 6 . . . . . *Lythrum alatum* . . . . . WINGED LOOSESTRIFE

1 Flowers many in spikelike heads at ends of stems; stamens usually 12 (6 long and 6 short). . . . . *Lythrum salicaria* | PURPLE LOOSESTRIFE

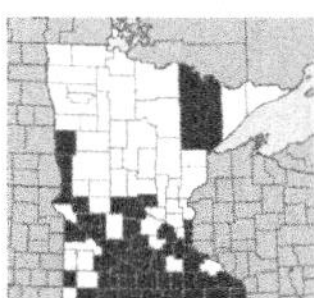

Lythrum alatum Pursh

WINGED LOOSESTRIFE
NC **OBL** | MW **OBL** | GP **OBL**

Smooth native perennial herb, spreading by rhizomes. **Stems** usually branched above, 2–8 dm long, somewhat woody at base. **Lower leaves** usually opposite, **upper leaves** alternate; lance-shaped, 1–4 cm long and 3–10 mm wide, rounded at base; margins entire; petioles absent. **Flowers** single in axils of upper, reduced leaves (bracts), short-stalked; calyx tube 4–6 mm long, smooth; petals 6, deep purple, 3–7 mm long; stamens usually 6. **Fruit** a capsule enclosed by the sepals. June–Aug. **Synonyms** *Lythrum dacotanum*. **Habitat** Lakeshores, wet meadows, marshes, low prairie, calcareous fens, ditches; especially where sandy.

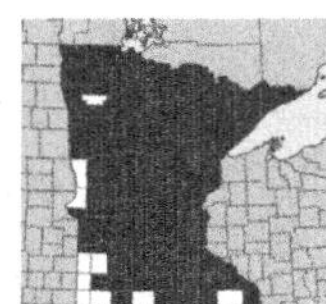

Lythrum salicaria L.
PURPLE LOOSESTRIFE
NC **OBL** | MW **OBL** | GP **OBL**

Introduced and invasive perennial herb, spreading and forming colonies by thick, fleshy roots which send up new shoots. **Stems** erect, 6–15 dm long, 4-angled, with many ascending branches. **Leaves** opposite or sometimes in whorls of 3, becoming alternate and reduced to bracts in the head; lance-shaped, 3–10 cm long and 0.5–2 cm wide, mostly heart-shaped and clasping at base; margins entire; petioles absent. **Flowers** large and showy, 2 or more in axils of reduced upper leaves (bracts), in spikes 1–4 dm long at ends of branches; sepals joined, the calyx tube 4–6 mm long, hairy; petals 6, purple-magenta, 7–10 mm long; stamens usually 12, the stamens and styles of 3 different lengths. **Fruit** a capsule enclosed by the sepals. June–Sept. **Habitat** Introduced from Europe and sometimes misguidedly planted as an ornamental, escaping to marshes, wet ditches, streambanks, cranberry bogs and shores, where a serious threat to native flora and of little value to wildlife. Naturalized over much of e and c USA and s Canada; local in w USA.

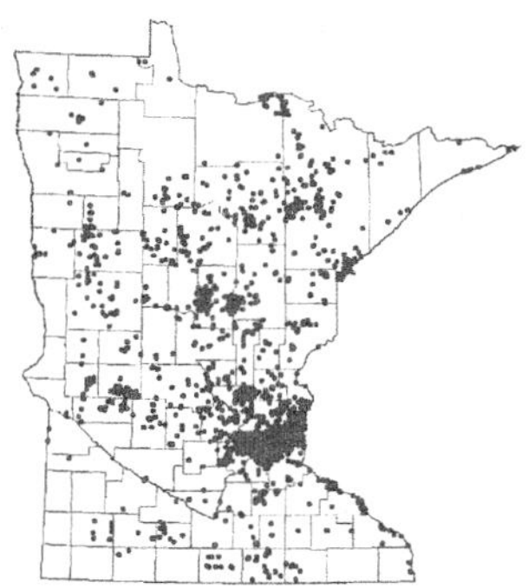

RIGHT *Minnesota DNR reports over 2,000* **purple loosestrife** *infestations from nearly all of the state's 87 counties, with an estimated 58,000 acres (23,400 ha) infested.*

ROTALA | *Tooth-cup*

Rotala ramosior (L.) Koehne
TOOTH-CUP; WHEELWORT
NC **OBL** | MW **OBL** | GP **OBL** | *threatened*

Small native annual herb. **Stems** smooth, 4-angled, to 4 dm long, unbranched or branched from base, the branches spreading to upright. **Leaves** opposite, linear to oblong, 1–5 cm long and 2–12 mm wide; margins entire; stalkless or tapered to a short petiole. **Flowers** single and stalkless in leaf axils, calyx tube bell-shaped to cylindric, 2–5 mm long, not strongly nerved, the lobes alternating with appendages of same length; petals small, white to pink, 4, slightly longer than sepals; stamens 4. **Fruit** a round capsule enclosed by the sepals. July–Oct. **Habitat** Muddy or sandy shores, marshes (especially those that dry during growing season), low spots in fields, ditches and other seasonally flooded places. **Status** Threatened.

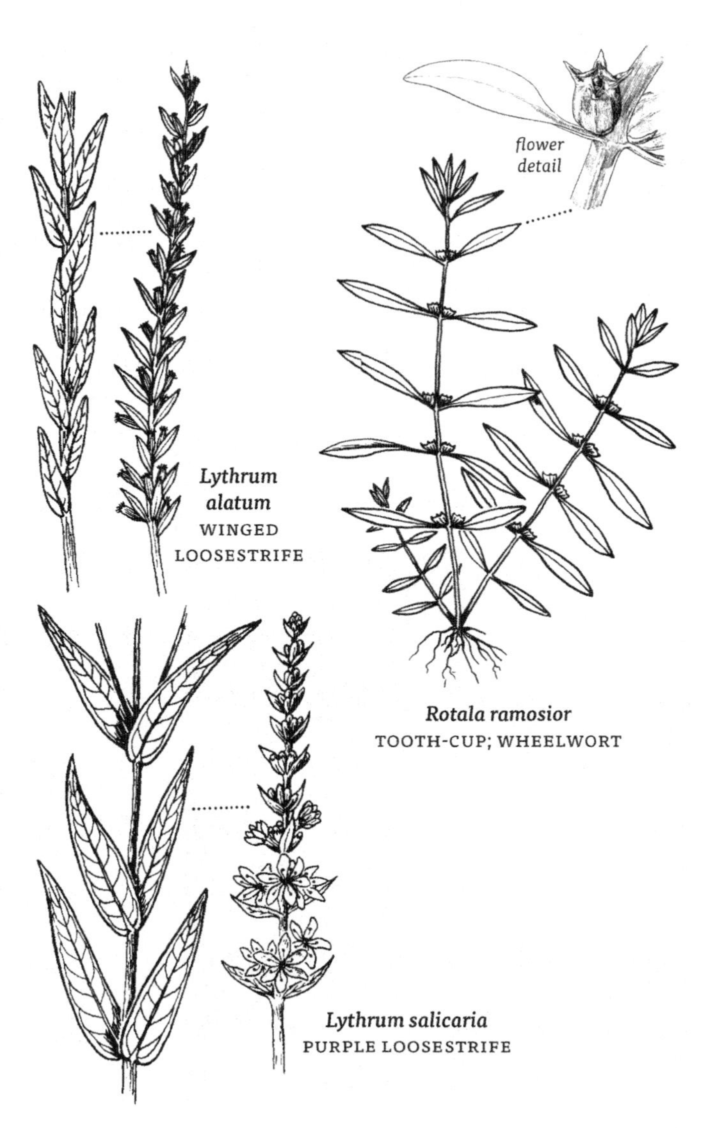

Lythrum alatum
WINGED LOOSESTRIFE

Rotala ramosior
TOOTH-CUP; WHEELWORT

Lythrum salicaria
PURPLE LOOSESTRIFE

NAPAEA | *Glade-mallow*

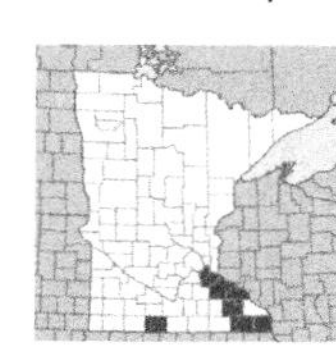

Napaea dioica L.

GLADE-MALLOW

NC **FACW** | MW **FACW** | *threatened*

Large native perennial herb. **Stems** erect, 1–2 m long. **Leaves** alternate, round in outline, 1–3 dm wide, deeply 5–9 lobed, the lobes coarsely toothed, palmately veined, on long petioles; upper leaves smaller, with short petioles. **Flowers** either male or female and on separate plants; numerous in large panicles at ends of stems; petals 5, white, obovate, petals of male flowers 5–9 mm long, petals of female flowers smaller; stamens many and joined near base, forming a tube around the style. **Fruit** a 10-parted capsule, the segments (carpels) 5 mm long, ribbed, and irregularly separating when mature. June–Aug. **Habitat** Moist floodplain forests, riverbanks. **Status** Threatened; se Minnesota.

Napaea dioica
GLADE-MALLOW

MENYANTHES | *Buckbean*

Menyanthes trifoliata L.
BUCKBEAN
NC **OBL** | MW **OBL** | GP **OBL**

Native perennial herb, with thick rhizomes covered with old leaf bases; plants smooth. **Leaves** alternate along rhizomes, palmately divided into 3 leaflets, the leaflets oval to ovate, 3–10 cm long and 1–5 cm wide, entire or sometimes wavy-margined; petioles 5–30 cm long, the base of petiole expanded and sheathing stem. **Flowers** in racemes on leafless stalks 2–4 dm long and longer than the leaves; bracts mostly 3–5 mm long; individual flowers on stalks 5–20 mm long; flowers perfect, regular, 5-parted, often of 2 types, some with flowers with long stamens and a shorter style, others with a long style and shorter stamens; sepal lobes 2–3 mm long; corolla funnel-shaped, 8–12 mm long, petals white, often purple-tinged, bearded with white hairs on inner surface; stamens 5. **Fruit** a rounded capsule, 6–10 mm wide; seeds shiny, yellow-brown. May–July. **Habitat** Open bogs and fens (especially in pools and outer moat), cedar swamps, wet thickets.

Menyanthes trifoliata
BUCKBEAN

MYRICA | *Sweet gale, Bayberry*

Myrica gale L.

SWEET GALE

NC **OBL** | MW **OBL** | GP **OBL**

Much-branched aromatic native shrub, 6–15 dm tall; **bark** dark gray to red-brown with small pale lenticels; **twigs** hairy, dotted with glands. **Leaves** alternate, deciduous, wedge-shaped, tapered to base, broadest above middle, 3–6 cm long and 1–2 cm wide, tip rounded and toothed, dark green on upper surface, paler below, dotted with shiny yellow glands; petioles short, 1–3 mm long. Male and female flowers separate and on different plants, appearing before or with unfolding leaves; **male flowers** in catkins 10–20 mm long, with dark brown, shiny triangular scales; **female flowers** in conelike, brown clusters 10–12 mm long. **Fruit** a flattened, ovate achene, resin-dotted, 2–3 mm long. April–May. **Synonyms** *Gale palustris*. **Habitat** Lakeshores, marshes, swamps and bogs, often at water's edge or in shallow water.

ABOVE *female flowers*
LEFT *male flowers*

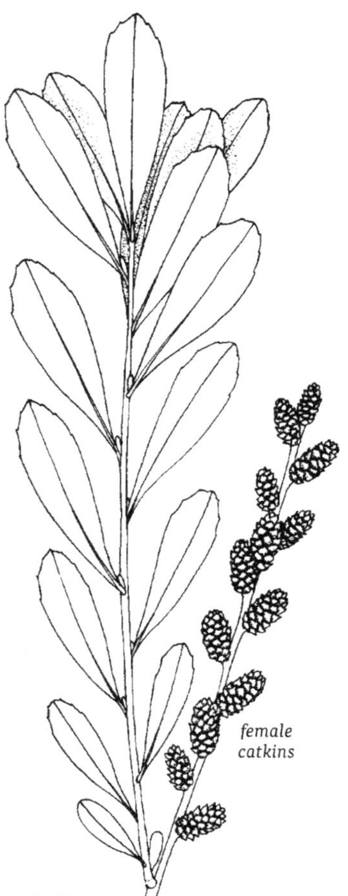

Myrica gale
SWEET GALE

NELUMBO | *Lotus-lily*

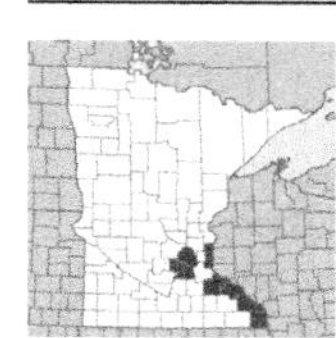

Nelumbo lutea (Willd.) Pers.
AMERICAN LOTUS-LILY
NC **OBL** | MW **OBL** | GP **OBL**

Native perennial aquatic herb, from a large, horizontal rootstock. **Leaves** large and shield-shaped, 3–7 dm wide, ribbed, floating on water surface or held above water surface, smooth above, somewhat hairy below; petioles thick, attached at center of blade. **Flowers** pale yellow, single, 15–25 cm wide; petals obovate, blunt-tipped; receptacle flat-topped, to 1 dm wide. **Seeds** acornlike, 1 cm thick. July–Aug. **Habitat** Lakes, ponds, backwater areas, marshes, where sometimes forming large colonies covering many acres.

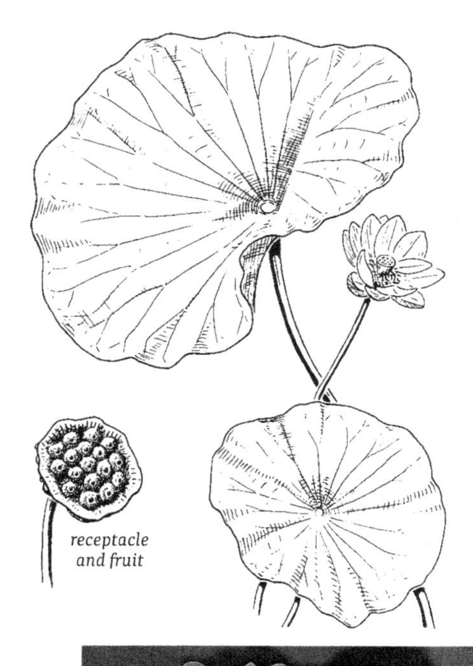

Nelumbo lutea
AMERICAN LOTUS-LILY

AQUATIC PERENNIAL HERBS. **Stems** long and fleshy, from horizontal rhizomes rooted in bottom mud. **Leaves** large, leathery, mostly floating or emergent above water surface, heart-shaped to shield-shaped, notched at base, margins entire. **Flowers** showy, single on long stalks and borne at or above water surface, perfect (with both male and female parts), white or yellow, sepals 4–6, green or yellow; petals numerous, small (*Nuphar*) to large and showy (*Nymphaea*). **Fruit** a many-seeded, berrylike capsule, opening underwater when mature.

NYMPHACEAE | WATER-LILY FAMILY

1 Flowers yellow, often red-tinged, sepals petal-like, true petals small; leaf blades oblong to oval or heart-shaped .... *Nuphar* .... YELLOW WATER-LILY; SPATTERDOCK

1 Flowers white (rarely pink), sepals green, true petals large and showy; leaf blades nearly round .... *Nymphaea* | WATER-LILY

NUPHAR | *Yellow water-lily, Spatterdock*

Aquatic herbs. **Leaves** large and floating or emergent. Sepals 5–6, yellow and petal-like, forming a saucer-shaped flower; petals small and numerous.

NUPHAR | YELLOW WATER-LILY, SPATTERDOCK

1 Leaves small, 5–10 cm wide; anther shorter than its stalk (filament); disk at base of stigma red............ *Nuphar microphylla* | YELLOW WATER LILY

1 Leaves larger; anther longer than its stalk; disk at base of stigma green or yellow.......................... *Nuphar variegata* | BULLHEAD LILY

Nuphar microphylla *(Pers.) Fern.*

YELLOW WATER LILY

NC **OBL** | MW **OBL** | GP **OBL**

Native aquatic perennial herb. **Leaves** both underwater and floating; **floating leaves** 5–10 cm long and 3–8 cm wide, notch at base usually more than half as long as midvein; petioles flattened on upper side; **underwater leaves** membranous, somewhat larger. **Flowers** 1.5–2 cm wide, sepals 5, yellow on inner surface; petals small and many; anthers 1–3 mm long, shorter than the filaments; disk at base of stigma red, 3–6 mm wide, with 6–10 rays. **Fruit** ovate, 15 mm long. July–Aug. **Synonyms** *Nuphar luteum, Nuphar pumila*. **Habitat** Lakes, ponds and slow-moving streams.

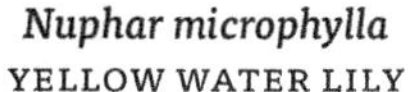

Nuphar microphylla
YELLOW WATER LILY

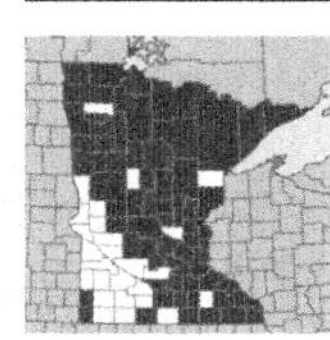

Nuphar variegata Durand
BULLHEAD LILY
NC **OBL** | MW **OBL** | GP **OBL**

Native aquatic perennial herb. **Leaves** mostly floating, 10–25 cm wide, notch usually less than half as long as midvein, petioles flattened on upper side and narrowly winged; underwater leaves absent or few. **Flowers** 2.5–5 cm wide; sepals usually 6, yellow, red-tinged on inner surface; petals small and numerous; anthers 4–7 mm long, longer than filaments; disk at base of stigma green, 1 cm wide, with 10–15 rays. **Fruit** ovate, 2–4 cm long. June–Aug. **Synonyms** *Nuphar fraterna*. **Habitat** Ponds, lakes, quiet streams.

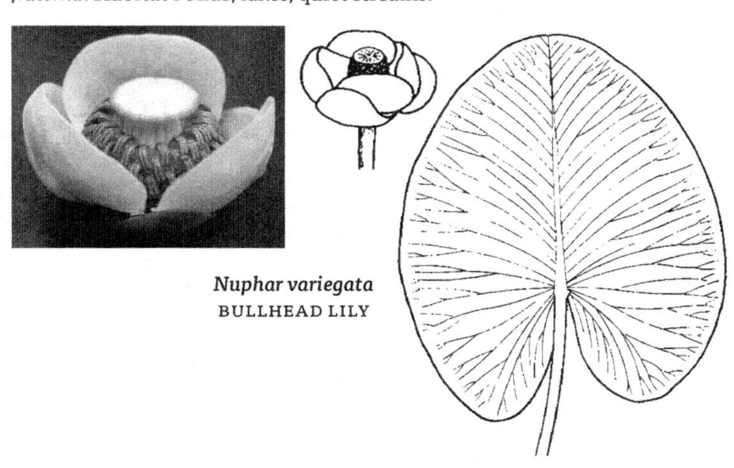

Nuphar variegata
BULLHEAD LILY

NYMPHAEA | *Water-lily*

Large aquatic plants, from stout rhizomes, these sometimes with lateral tubers. **Leaves** floating, round, notched to the petiole, petioles not flattened or winged. **Flowers** white and showy; sepals 4, green; petals white or pink, showy, numerous and overlapping; stamens many, the outer stamens with broadened, petal-like filaments, anthers yellow, ovary depressed at tip with a rounded projection from center, stigmas 10–25. **Fruit** round, covered with persistent petal and stamen bases, maturing under water; seeds numerous, each enclosed within a sac (aril).

NYMPHAEA | WATER-LILY

1 Leaves round in outline, narrowly notched; flowers large and showy, usually fragrant . . . . *Nymphaea odorata* | WHITE WATER-LILY

1 Leaves oval in outline, notch more widely spreading; flowers smaller and scarcely fragrant . . . . *Nymphaea leibergii* | PYGMY WATER-LILY

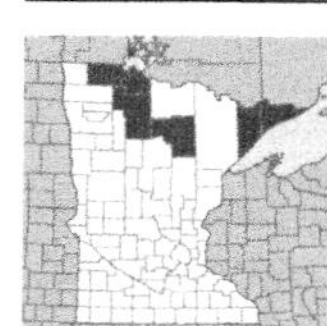

Nymphaea leibergii *Morong*

PYGMY WATER-LILY

NC **OBL** | GP **OBL** | *threatened*

Native aquatic perennial herb, rhizomes ascending. **Stems** arising from tip of rhizomes. **Leaves** 7–12 cm long and to 3/4 as wide, notch fairly wide, upper surface green, green or purple below. **Flowers** white, usually not fragrant, 4–8 cm wide, reported to open in afternoon; sepals 4, green, 2–3 cm long; petals 8–17, about as long as sepals; stamens 20–40. **Fruit** not covered by the erect sepals. Summer. **Synonyms** *Castalia leibergii, Nymphaea tetragona*. **Habitat** Shallow water of ponds and lakes. **Status** Threatened; n Minn.

Nymphaea leibergii
PYGMY WATER-LILY

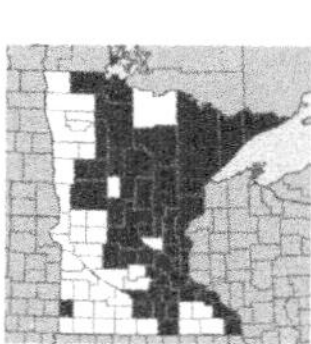

Nymphaea odorata *Aiton*

WHITE WATER-LILY

NC **OBL** | MW **OBL** | GP **OBL**

Native aquatic perennial herb, rhizomes sometimes with knotty tubers. **Leaves** floating, round, 1–3 dm wide, with a narrow notch, green and shiny on upper surface, usually purple or red below. **Flowers** white (rarely pink), usually fragrant, 7–20 cm wide, often opening in morning and closing in late afternoon (or remaining o pen on cool, cloudy days); sepals 4, green, 3–10 cm long; petals 17–25, about as long as sepals, oval, tapered to a rounded tip; stamens 40–100. **Fruit** round, mostly covered by the sepals; seeds 2–4 mm long. June–Aug. **Synonyms** *Castalia tuberosa, Nymphaea tuberosa*. **Habitat** Shallow water of ponds and lakes, quiet water of rivers.

Nymphaea odorata
WHITE WATER-LILY

FRAXINUS | *Ash*

Medium to large trees. **Leaves** deciduous, opposite, pinnately divided into leaflets. **Flowers** in clusters from axils of previous year's twigs, mostly single-sexed, male and female flowers on different trees, rarely perfect, petals absent. **Fruit** a 1-seeded, winged samara.

FRAXINUS | ASH

1 Lateral leaflets ± stalkless . . . . *Fraxinus nigra* | BLACK ASH
1 Lateral leaflets short-stalked . . . . *Fraxinus pennsylvanica* . . . . GREEN ASH; RED ASH

Fraxinus nigra Marshall

BLACK ASH
NC **FACW** | MW **FACW** | GP **FACW**

Native tree to 15 m tall; trunk 30–60 cm wide, crown open and narrow; **bark** gray, thin, flaky; **twigs** smooth, round in section, dark green, becoming gray. **Leaves** opposite, pinnately divided into 7–11 stalkless (except for terminal) leaflets; leaflets lance-shaped to oblong, 7–13 cm long and 2.5–5 cm wide, long-tapered to a tip; margins with sharp, forward-pointing teeth. **Flowers** appear in spring before leaves, in open clusters on twigs of previous year; some perfect, some single-sexed, male and female flowers on different trees. **Fruit** a 1-seeded samara, 2.5–4 cm long and 6–10 mm wide, the wing broad and rounded at tip, deciduous or persisting until following spring. April–May. **Habitat** Floodplain forests, cedar swamps, wet depressions in forests.

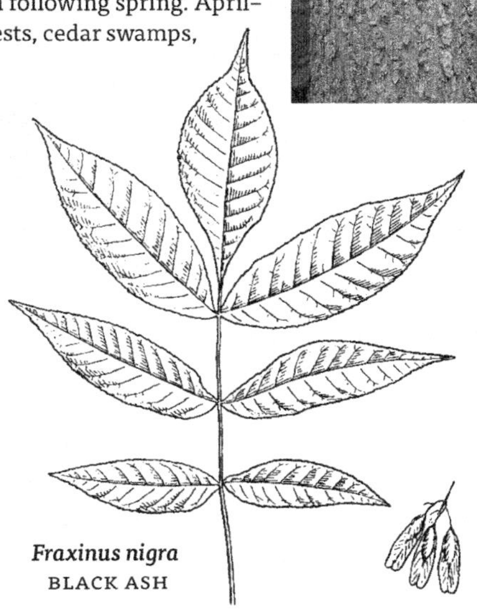

Fraxinus nigra
BLACK ASH

Fraxinus pennsylvanica Marshall
GREEN ASH; RED ASH
NC **FACW** | MW **FACW** | GP **FAC**

Native tree to 15 m tall; trunk 30–60 cm wide; bark dark gray or brown, thick, with shallow furrows and netlike ridges; twigs usually hairy for 1–3 years, becoming light gray or red-brown. **Leaves** opposite, pinnately divided into 7–9 leaflets; leaflets oblong lance-shaped to ovate, 7–13 cm long and 2.5–4 cm wide, upper surface smooth, underside smooth or hairy; margins entire or with few forward-pointing teeth; petioles short, smooth or hairy. **Flowers** appear in spring before or with leaves, in compact, hairy clusters on twigs of previous year; single-sexed, male and female flowers on different trees. **Fruit** a 1-seeded, slender samara, 2.5–5 cm long, in open clusters persisting until following spring. April–May. **Habitat** Floodplain forests, swamps, shores, streambanks.

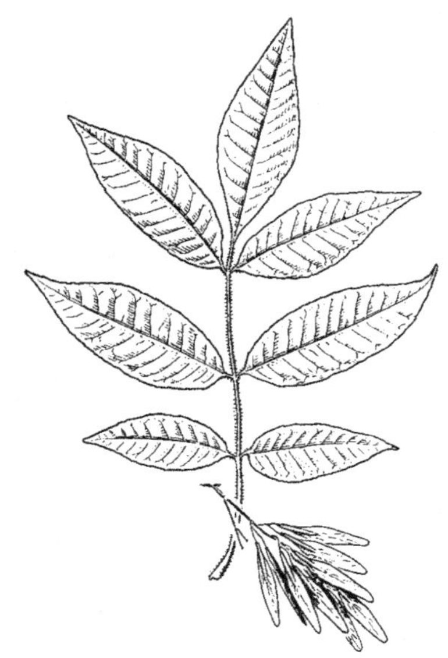

Fraxinus pennsylvanica
GREEN ASH; RED ASH

ANNUAL OR PERENNIAL HERBS. Leaves opposite to alternate, simple to pinnately divided, stalkless or short-petioled. **Flowers** usually large and showy, perfect (with both male and female parts), regular, borne in leaf axils or in heads at ends of stems; sepals 8 or 4; petals 4, white, yellow or pink to rose-purple; ovary 4-chambered. **Fruit** a 4-chambered capsule; seeds many, with or without a tuft of hairs (coma).

ONAGRACEAE | EVENING-PRIMROSE FAMILY

1 Petals 2, white; leaves opposite; fruit with bristly hairs ..... *Circaea alpina* .... SMALL OR ALPINE ENCHANTER'S NIGHTSHADE
1 Petals 4 (rarely absent), white, pink, or yellow; leaves opposite or alternate; fruit without bristly hairs 2

2 Petals pink, white or rose-purple; seeds with a tuft of hairs (coma). .... *Epilobium* | WILLOW-HERB
2 Petals yellow (or absent); seeds without a tuft of hairs .... *Ludwigia* .... WATER-PRIMROSE; PRIMROSE-WILLOW

CIRCAEA | *Enchanter's nightshade*

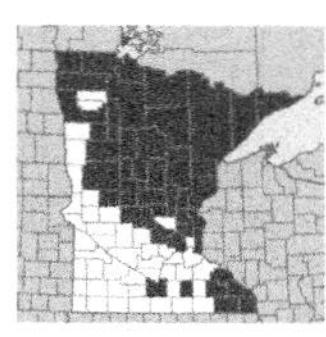

Circaea alpina L.

SMALL ENCHANTER'S NIGHTSHADE
NC **FACW** | MW **FACW** | GP **FACW**

Native perennial herb, spreading from rhizomes thickened and tuberlike at ends. **Stems** weak, 1–3 dm long, mostly smooth. **Leaves** opposite, ovate, 2–5 cm long and 1–3 cm wide; margins coarsely toothed; petioles flat on upper side, underside thin-winged along center. **Flowers** white, in short racemes of 10–15 flowers, becoming 1 dm long in fruit; sepals 1–2 mm long; petals to 2 mm long. **Fruit** a 1-seeded capsule, 2–3 mm long, covered with soft hooked bristles. June–Aug. **Habitat** Cedar swamps (where often on rotting logs), low spots in forests.

Circaea alpina
SMALL ENCHANTER'S NIGHTSHADE

EPILOBIUM | *Willow-herb*

Perennial herbs, often producing leafy rosettes or bulblike offsets (*turions*) at base of stem late in growing season. **Leaves** simple, opposite, alternate, or opposite below and becoming alternate above; stalkless or short-petioled. **Flowers** white to pink, single in axils of upper reduced leaves, or in spike or racemes at ends of stems; sepals 4; petals 4; stamens 8, the inner 4 stamens shorter than outer 4; ovary 4-chambered, maturing into a linear, 4-parted capsule, splitting from tip to release numerous brown seeds which are tipped with a tuft of fine hairs (*coma*).

EPILOBIUM | WILLOW-HERB

1 Leaves more than 1 cm wide, margins toothed, not rolled under; stems usually with lines of hairs 2
1 Leaves less than 1 cm wide, margins ± entire, rolled under; stem hairs not in lines. 3

2 Tuft of hairs attached to tip of seeds (coma) ± white, seeds with a broad, short beak; margins of stem leaves with mostly 10–30 teeth on a side . . . . *Epilobium ciliatum* | AMERICAN WILLOW-HERB
2 Coma brown, seeds beakless; leaf margins with more than 30 teeth on a side . . . . *Epilobium coloratum* | PURPLE-LEAF WILLOW-HERB

3 Hairs spreading . . . . *Epilobium strictum* | DOWNY WILLOW-HERB
3 Hairs flattened against stems and leaves 4

4 Upper surface of leaves finely hairy . . . . *Epilobium leptophyllum* . . . . LINEAR-LEAF WILLOW-HERB
4 Upper surface of leaves ± smooth . . . . *Epilobium palustre* . . . . MARSH WILLOW-HERB

Epilobium ciliatum Raf.

AMERICAN WILLOW-HERB
NC **FACW** | MW **FACW** | GP **FACW**

Native perennial herb, with over-wintering leafy rosettes. **Stems** often branched, 3–10 dm long, smooth below, short-hairy above, especially in the head (where often with gland-tipped hairs). **Leaves** opposite, usually alternate near top; lance-shaped to ovate, 3–10 cm long and 0.5–3 cm wide; margins with few, small, forward-pointing teeth; stalkless or with short, winged petioles to 6 mm long. **Flowers** usually nodding when young, on stalks 3–10 mm long, on branches from upper leaf axils; sepals ovate, 2–5 mm long; petals white (or pink), notched at tip, 2–8 mm long. **Fruit** a linear capsule, 4–8 cm long, with gland-tipped hairs; seeds 1 mm long, the coma white. July–Sept. **Synonyms** *Epilobium adenocaulon, Epilobium glandulosum*. **Habitat** Shores, streambanks, marshes, wet meadows, seeps, ditches and other wet places.

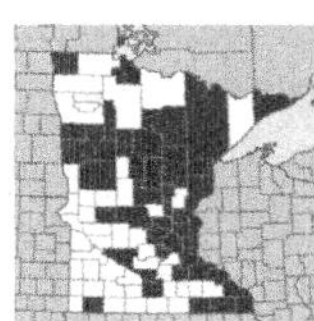

Epilobium coloratum Biehler
PURPLE-LEAF WILLOW-HERB
NC **OBL** | MW **OBL** | GP **OBL**

Native perennial herb, producing basal, leafy rosettes in fall; similar to American willow-herb (*Epilobium ciliatum*) but larger. **Stems** 5–10 dm long, much-branched in the head, smooth below, short-hairy above with hairs often in lines; stems and leaves often purple-tinged. **Leaves** mostly opposite, becoming alternate and smaller above, lance-shaped, 5–15 cm long and 0.5–3 cm wide, long-tapered to a pointed tip; margins finely toothed, with irregular sharp teeth; short-petioled to stalkless. **Flowers** many on branches from upper leaf axils; sepals lance-shaped, 2–3 mm long; petals pink or white, 3–5 mm long, notched at tip; individual flowers on stalks to 10 mm long. **Fruit** a linear capsule, 3–5 cm long; seeds 1.5 mm long, the coma brown when mature. July–Sept. **Habitat** Shores, seeps, swamps and wet woods, wet meadows, fens, ditches.

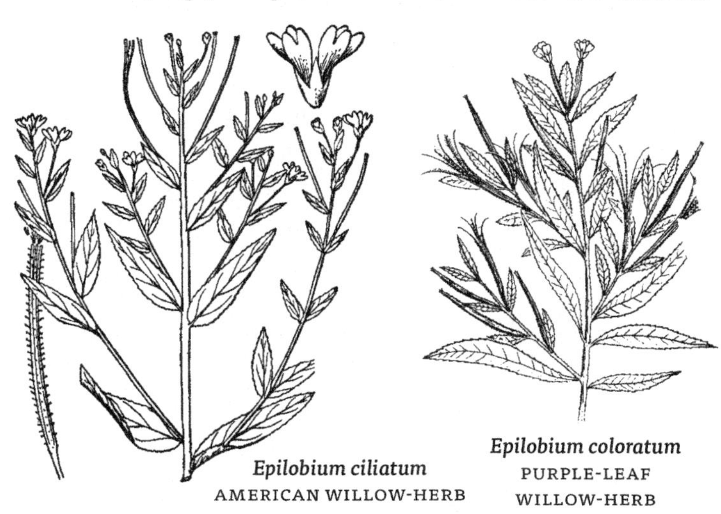

Epilobium ciliatum
AMERICAN WILLOW-HERB

Epilobium coloratum
PURPLE-LEAF
WILLOW-HERB

Epilobium leptophyllum Raf.
LINEAR-LEAF WILLOW-HERB
NC **OBL** | MW **OBL** | GP **OBL**

Native perennial herb, similar to marsh willow-herb (*Epilobium palustre*) but somewhat larger and more hairy. **Stems** simple or branched, 2–10 dm long, with short, incurved hairs. **Leaves** opposite or alternate, linear or linear lance-shaped, 2–7 cm long and 1–6 mm wide, upper surface hairy, underside hairy, at least on midvein, lateral veins indistinct; margins entire and rolled under; petioles short or ± absent. **Flowers** erect in upper leaf axils on short, slender stalks to 1 cm long; petals light pink, 3–5 mm long, entire or slightly notched at tip. **Fruit** a linear, finely hairy capsule, 4–5 cm long; the coma yellow-white. July–Sept. **Synonyms** *Epilobium lineare*. **Habitat** Swamps, marshes, open bogs, sedge meadows, shores, streambanks and springs.

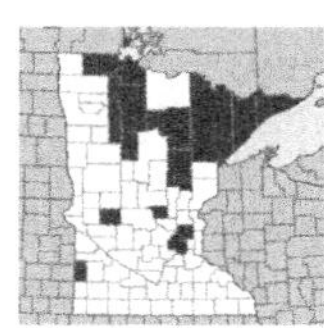

Epilobium palustre L.
MARSH WILLOW-HERB
NC **OBL** | MW **OBL** | GP **OBL**

Native perennial herb, from slender rhizomes or stolons. **Stems** simple or with a few branches above, 1–6 dm long, upper stem hairy with small incurved hairs. **Leaves** mostly opposite, lance-shaped, erect or ascending, 2–6 cm long and 3–15 mm wide, tapered to a rounded tip, upper surface smooth or with sparse hairs along midvein, underside smooth or finely hairy along midvein, lateral veins distinct; margins entire and often rolled under; stalkless. **Flowers** few in upper leaf axils, on short stalks; petals white to pink, 3–5 mm long, notched at tip. **Fruit** a linear, finely hairy capsule; coma pale. July–Aug. **Synonyms** *Epilobium oliganthum*. **Habitat** Open bogs and swamps.

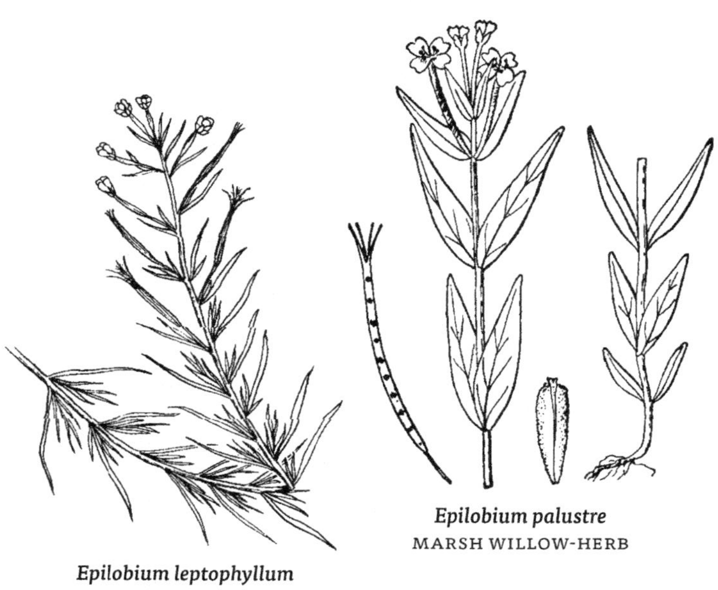

Epilobium palustre
MARSH WILLOW-HERB

Epilobium leptophyllum
LINEAR-LEAF WILLOW-HERB

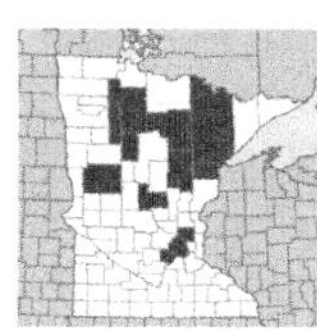

Epilobium strictum Muhl.
DOWNY WILLOW-HERB
NC **OBL** | MW **OBL** | GP **OBL**

Native perennial herb, spreading by slender rhizomes; plants densely soft white-hairy. **Stems** erect, simple or branched above, 3–6 dm long. Lower **leaves** opposite, upper leaves alternate; lance-shaped, ascending, 2–4 cm long and 3–8 mm wide, tapered to a rounded tip; margins mostly entire, rolled under; stalkless. **Flowers** on slender stalks from upper leaf axils; petals pink, 5–8 mm long, notched at tip. **Fruit** a linear, densely hairy capsule; coma pale brown. July–Aug. **Habitat** Conifer swamps, sedge meadows, calcareous fens, marshes.

Epilobium strictum
DOWNY WILLOW-HERB

LUDWIGIA | *Water-primrose, Primrose-willow*

Perennial herbs. **Stems** floating, creeping, or upright. **Leaves** simple, opposite or alternate, entire. **Flowers** single in leaf axils; sepals 4; petals 4 (or absent), yellow or green, large or very small; stamens 4; stigma unlobed. **Fruit** a 4-chambered, many-seeded capsule; seeds without a tuft of hairs at tip (coma).

LUDWIGIA | WATER-PRIMROSE, PRIMROSE-WILLOW

1 Leaves opposite, stalked; stems floating or creeping and rooting at nodes .... *Ludwigia palustris* | COMMON WATER-PURSLANE; MARSH PURSLANE
1 Leaves alternate, ± stalkless; stems mostly erect ...... *Ludwigia polycarpa* .... TOP-POD WATER-PRIMROSE

Ludwigia palustris (L.) Elliott
COMMON WATER-PURSLANE; MARSH PURSLANE
NC **OBL** | MW **OBL** | GP **OBL**

Native perennial herb. **Stems** weak, creeping and rooting at nodes or partly floating, simple to branched, 1–5 dm long, succulent, smooth or with sparse scattered hairs. **Leaves** opposite, lance-shaped to ovate, 0.5–3 cm long and 0.5–2 cm wide, shiny green or red, margins entire, tapered at base to a winged petiole to 2 cm long. **Flowers** single in leaf axils, stalkless; sepals broadly triangular, 1–2 mm long; petals usually absent, or small and red. **Fruit** a capsule, 2–5 mm long and 2–3 mm wide, somewhat 4-angled, with a green stripe on each angle. July–Sept. **Synonyms** *Isnardia palustris*. **Habitat** Shallow water or exposed mud of pond margins, lakeshores, stream-banks, ditches, springs.

Ludwigia polycarpa Short & Peter
TOP-POD WATER-PRIMROSE
NC **OBL** | MW **OBL** | GP **OBL**

Native perennial herb, producing leafy stolons from base in fall; plants smooth. **Stems** erect, 1–9 dm long, often branched, usually 4-angled. **Leaves** alternate, lance-shaped to oblong lance-shaped, 3–12 cm long and 5–15 mm wide; margins entire; ± stalkless. **Flowers** single in leaf axils, stalkless; sepals triangular, 2–4 mm long, usually persistent; petals green and very small or absent. **Fruit** a short-cylindric, rounded 4-angled capsule, 4–7 mm long and 3–5 mm wide. July–Sept. **Habitat** Borders of swamps and marshes, muddy shores, wet depressions.

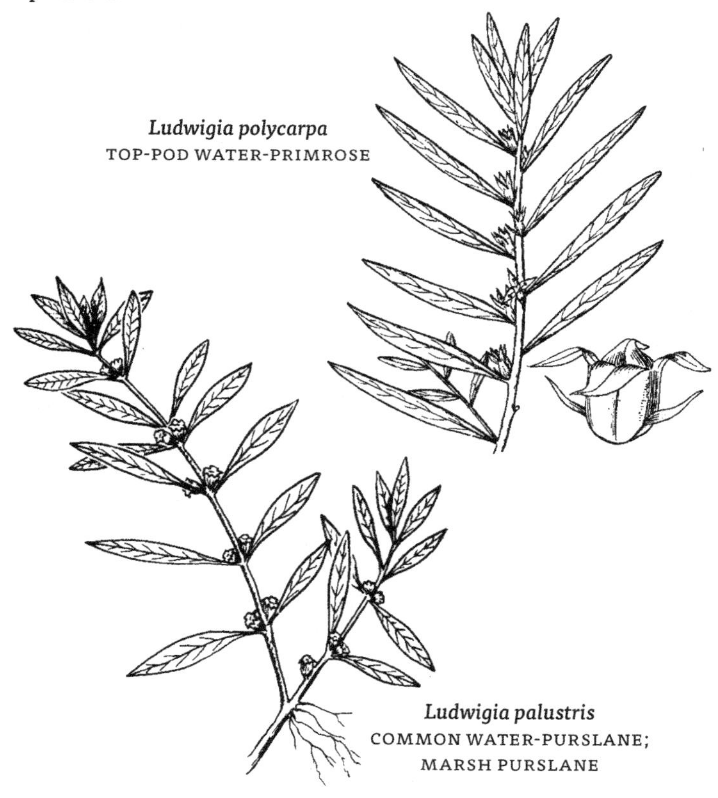

Ludwigia polycarpa
TOP-POD WATER-PRIMROSE

Ludwigia palustris
COMMON WATER-PURSLANE;
MARSH PURSLANE

OXALIS | *Wood-sorrel*

Oxalis montana Raf.
NORTHERN WOOD-SORREL
NC **FACU** | MW **FACU**

Native perennial herb, from slender, scaly rhizomes. **Leaves** single or 3–6 together, all from base of plant, on stalks 4–15 cm long, these joined at base; palmately divided into 3 leaflets, the leaflets notched at tips, sparsely hairy. **Flowers** perfect, broadly bell-shaped, single atop stalks 6–15 cm long (usually slightly taller than leaves), with a pair of small bracts above middle of stalk; sepals 5, much shorter than petals; petals 5, white or pink, with pink veins, 10–15 mm long. **Fruit** a smooth, nearly round capsule. May–July. **Synonyms** *Oxalis acetosella*. **Habitat** Hummocks in swamps, wet depressions in forests, moist wetland margins, usually where shaded.

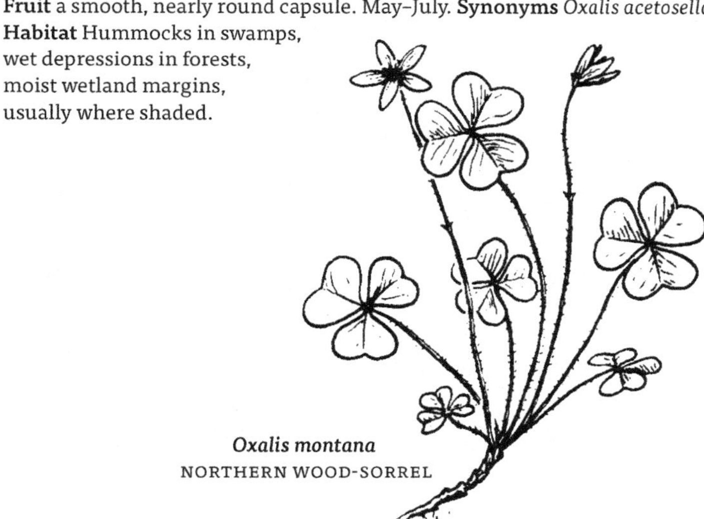

Oxalis montana
NORTHERN WOOD-SORREL

Open bog, late spring. Overall species diversity is low due to the saturated, acidic, organic soil. Common are **Heath Family** (Ericaceae) shrubs such as leatherleaf (*Chamaedaphne calyculata*), bog-rosemary (*Andromeda glaucophylla*). and small cranberry (*Vaccinium oxycoccos*). Herbaceous species within the **Sedge Family** (Cyperaceae) include running bog sedge (*Carex oligosperma*), tussock cotton-grass (*Eriophorum vaginatum*), few-flowered bog sedge (*Carex pauciflora*), and poor sedge (*Carex paupercula*). Other herbaceous species include pod-grass (*Scheuchzeria palustris*), round-leaved sundew (*Drosera rotundifolia*), pitcher-plant (*Sarracenia purpurea*), and leafless bladderwort (*Utriculata cornuta*). Scattered trees of black spruce (*Picea mariana*) have established atop the slightly drier hummocks. **Sphagnum mosses** (examples, lower) form a nearly continuous ground cover.

ANNUAL OR PERENNIAL HERBS. **Leaves** simple, entire, all from base of plant. **Flowers** perfect in a narrow spike (*Plantago*), each flower subtended by bracts, or single-sexed, the male and female flowers on same plant (*Littorella*); flower parts mostly in 4s. **Fruit** a capsule opening at tip.

PLANTAGINACEAE | PLANTAIN FAMILY

1 Flowers single at ends of stalks or stalkless, either male or female; nw Minnesota . . . . *Littorella uniflora* | SHOREWEED

1 Flowers many in spikes, perfect; sw Minnesota. . . . . *Plantago elongata* . . . . SLENDER PLANTAIN

LITTORELLA | *Shoreweed*

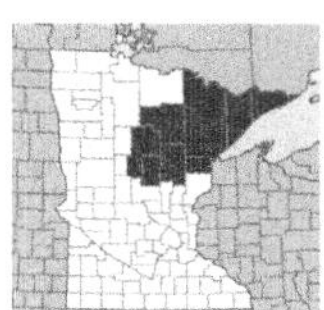

Littorella uniflora (L.) Asch.
SHOREWEED
NC **OBL** | *special concern*

Low native perennial herb; plants clumped, often forming mats from stolons and rhizomes. **Leaves** bright green, linear, to 5 cm long and 2–3 mm wide, succulent; margins entire. **Flowers** only from emersed plants, single-sexed, male and female flowers on same plant; male flowers 1–2 on stalks to 4 cm long; female flowers stalkless among the leaves; sepals 4 (sometimes 3 in female flowers), lance-shaped, 2–4 mm long, with a dark green midrib and lighter margins; petals joined, 4-lobed; stamens 4, longer than the petals. **Fruit** a 1-seeded nutlet, 2 mm long and 1 mm wide. July–Aug. Ours are var. *americana*, or perhaps considered as a valid species, *Littorella americana*. **Habitat** Exposed sandy or gravelly lakeshores, sometimes where mucky; or in water 1 m or more deep. **Status** Special concern.

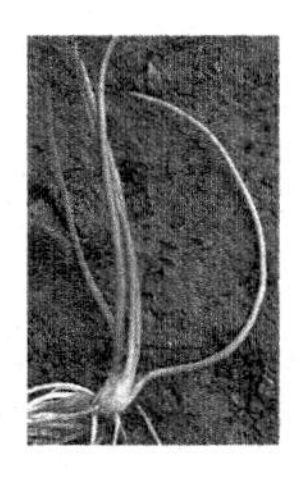

Littorella uniflora
SHOREWEED

PLANTAGO | *Plantain*

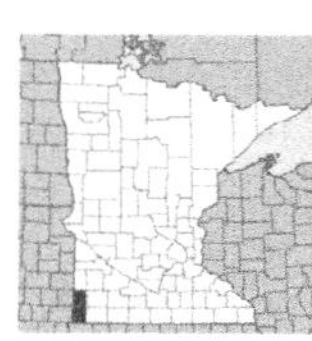

Plantago elongata Pursh

SLENDER PLANTAIN

MW **FACW** | GP **FACW** | *threatened*

Very small native annual herb with a slender taproot. **Stems** to 15 cm long, finely hairy. **Leaves** all from base of plant, narrow and linear, 3–8 cm long, tapered to a blunt tip; margins mostly entire; petioles absent. **Flowers** small, green, in slender spikes 3–8 cm long, with mix of male, female and some perfect flowers; sepals 4, to 2 mm long and as long as bracts; petals 4, joined, the lobes to 1 mm long, usually widely spreading. **Fruit** a capsule 2–3 mm long, opening at or near the middle; seeds 1–2 mm long. May–Aug. **Habitat** Moist, usually alkaline places. **Status** Threatened; rare in sw Minnesota.

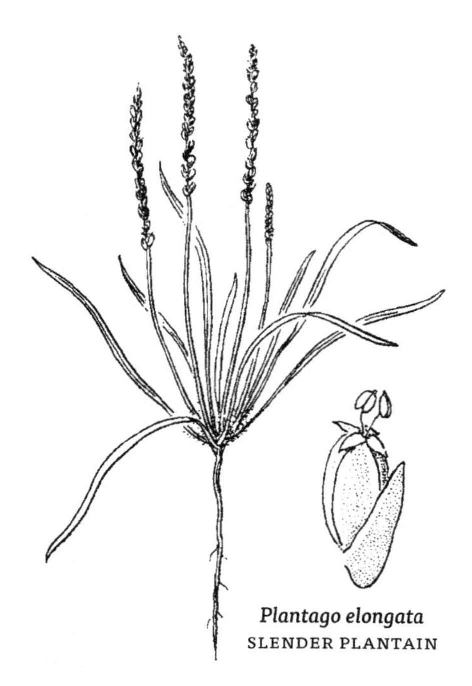

Plantago elongata
SLENDER PLANTAIN

ANNUAL OR PERENNIAL HERBS. Leaves opposite (*Phlox*) or pinnately divided (*Polemonium*). **Flowers** perfect (with both male and female parts), single or in clusters at ends of stems and from leaf axils; sepals and petals 5-parted and joined for part of length. **Fruit** a 3-chambered capsule, with usually 1 seed per chamber.

POLEMONIACEAE | PHLOX FAMILY

1 Leaves undivided . . . . *Phlox maculata*
. . . . SPOTTED PHLOX; WILD SWEET-WILLIAM

1 Leaves pinnately divided into leaflets. . . . . *Polemonium occidentale*
. . . . WESTERN JACOB'S LADDER

PHLOX | *Phlox*

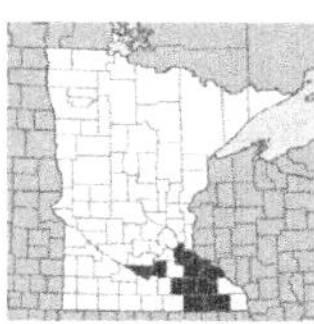

Phlox maculata L.

SPOTTED PHLOX; WILD SWEET-WILLIAM
NC **FACW** | MW **FACW**

Native perennial herb. **Stems** erect, 3–8 dm long, simple or branched above, smooth or finely hairy, usually red-spotted. **Leaves** opposite, smooth, ± firm, lance-shaped, 5–12 cm long and 0.5–1.5 cm wide, long-tapered to a sharp tip; margins entire; petioles absent. **Flowers** 1–2 cm wide, in stalked clusters (cymes) at ends of stems and from several to many upper leaf axils, these short-stalked, forming a long, narrow head 10–20 cm long, the head finely hairy; sepals smooth, joined to form a sharp-tipped tube 6–8 mm long; petals pink or purple; corolla 5-lobed, tubelike but flared outward at tip; stamens 5; style elongated. **Fruit** a 3-chambered capsule. July-Sept. **Synonyms** *Phlox pyramidalis*. **Habitat** Fens, sedge meadows, springs.

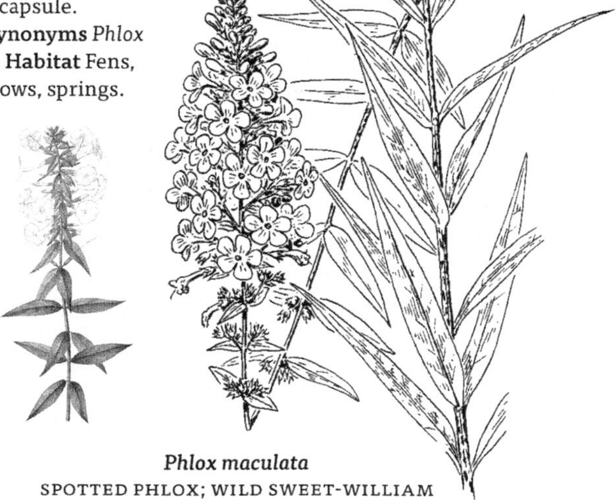

Phlox maculata
SPOTTED PHLOX; WILD SWEET-WILLIAM

POLEMONIUM | *Jacob's ladder*

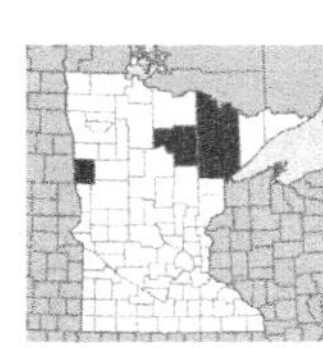

Polemonium occidentale Green

WESTERN JACOB'S LADDER

NC **FACW** | MW **FACW** | GP **FACW** | *endangered*

Native perennial herb. **Stems** erect, to 10 dm long, single from upturned ends of short, unbranched rhizomes. **Leaves** alternate, pinnately divided with up to 27 leaflets, the leaflets 1–4 cm long, smaller upward; margins entire; petioles short or absent. **Flowers** blue, 10–15 mm wide, crowded in a long panicle composed of smaller clusters of flowers; sepals joined to form a tube; petal lobes longer than calyx tube; stamens shorter or equal to corolla; style longer than stamens. **Fruit** a 3-chambered capsule. July. Ours are subsp. lacustre. **Habitat** Forested swamps of black spruce (*Picea mariana*), tamarack (*Larix laricina*), and northern white cedar (*Thuja occidentalis*). Typical associated shrubs include speckled alder (*Alnus incana*) and bog birch (*Betula pumila*). Where it occurs in these habitats, western Jacob's ladder is usually on hummocks of sphagnum moss. **Status** Endangered; disjunct in ne Minnesota and n Wisconsin from main range of Rocky Mountains and western USA.

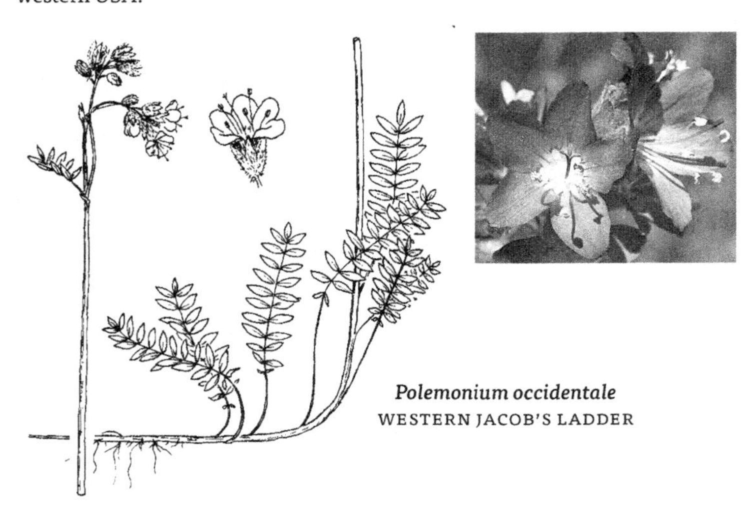

Polemonium occidentale
WESTERN JACOB'S LADDER

POLYGALA | *Milkwort*

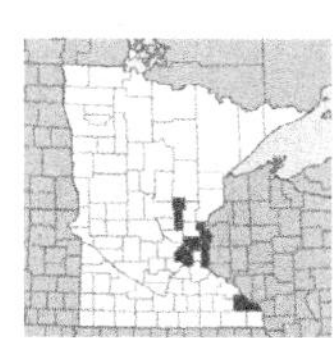

Polygala cruciata L.

DRUM-HEADS; MARSH MILKWORT

NC **FACW** | MW **FACW** | GP **OBL** | *endangered*

Native annual herb. **Stems** erect, 4-angled, 1–4 dm long, usually branched above. **Leaves** mostly in whorls of 4, linear or oblong lance-shaped, 1–4 cm long and 1–5 mm wide, rounded and often with a short, sharp point at tip; margins entire; petioles short or absent. **Flowers** ± stalkless in cylindric, densely-flowered racemes, 1–5 cm long and 1–1.5 cm wide; flowers pale purple or green purple; sepals 5, the 2 lateral sepals (wings) petal-like, 4–6 mm long and 3–4 mm wide at base; petals 3, joined into a tube; stamens 8 (sometimes 6). **Fruit** a 2-chambered capsule, with a single, ± hairy seed in each chamber. July–Aug. **Synonyms** *Polygala ramosior*. **Habitat** Open, sandy or mucky shores; sandy or peaty swales. **Status** Endangered.

Polygala cruciata
DRUM-HEAD; MARSH MILKWORT

ANNUAL OR PERENNIAL HERBS, plants sometimes vining. **Leaves** alternate, simple, sometimes wavy-margined, otherwise entire. Stipules joined to form a membranous or papery sheath (**ocrea**) around stem at each node. **Flowers** in spikelike racemes or small clusters from leaf axils (***Polygonum***), or in crowded panicles at ends of stems (***Rumex***). Flowers small, perfect (with both male and female parts), regular, petals absent; in *Polygonum* the sepals petal-like (*tepals*), white to pink or yellow, mostly 5 (sometimes 4 or 6); in *Rumex* the sepals herbaceous, green to brown, in inner and outer groups, each group with 3 sepals, the 3 inner enlarging after flowering, becoming broadly winged, persisting to enclose the achene; stamens 4–8; ovary 1-chambered, styles 2–3. **Fruit** a 3-angled or lens-shaped achene.

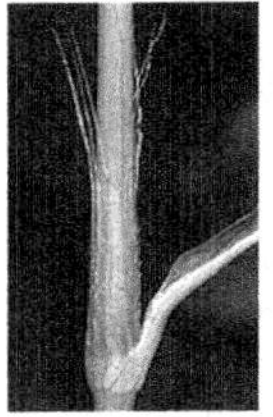

Ocrea
membranous or papery sheath at each node

POLYGONACEAE | SMARTWEED FAMILY

1 Flowers white to pink or green, in spikelike racemes or in groups of 1 to few in leaf axils, the sepals petal-like at least on margins .... *Polygonum* .... SMARTWEED; KNOTWEED; TEARTHUMB
1 Flowers green to brown, in panicles at ends of stems, the sepals not petal-like .... *Rumex* | DOCK

POLYGONUM | *Smartweed, Knotweed, Tearthumb*

Annual or perennial herbs. **Stems** erect to sprawling, often swollen at nodes. **Leaves** arrowhead-shaped to lance-shaped or oval; stipules joined to form a tubular sheath (ocrea) around the stem above each node; the ocreae (plural) membranous or papery, entire or with an irregular, jagged margin or fringed with bristles. **Flowers** small, green, white or pink, usually in slender racemes, the racemes at ends of stems or both terminal and from leaf axils, loosely to densely flowered; sepals usually 5, petal-like, green-white to pink; stamens 8 or less; styles 2–3. **Fruit** a brown to black achene, lens-shaped or 3-angled.

The genus has been subdivided into several new genera (here, ***Persicaria****) based, in part, on features of the sepals and ocreae; new names are listed after the traditional name.*

POLYGONUM | SMARTWEED, KNOTWEED, TEARTHUMB

1 Stems and leaf petioles with downward-curved prickles; leaves ± arrowhead-shaped **2**
1 Plants without prickles; leaves various, mostly lance-shaped or oval ... **3**

2 Basal lobes of leaves pointed downward; achenes 3-sided. . . . . *Polygonum sagittatum* | ARROW-LEAVED TEARTHUMB

2 Basal lobes pointed outward; achenes 2-sided . . . . *Polygonum arifolium* . . . . HALBERD-LEAVED TEARTHUMB

3 Flowers rose-pink, in 1 (sometimes 2) terminal racemes at end of stem; plants perennial. . . . . *Polygonum amphibium* | WATER SMARTWEED

3 Flowers white or pink, in numerous racemes from leaf axils and at ends of stems; plants annual (except *Polygonum hydropiperoides*) 4

4 Ocreae (sheath around stem nodes) entire or with an irregular, jagged margin 5

4 Ocreae fringed with bristles 6

5 Outer sepals strongly 3-nerved, each nerve ending in an anchor shaped fork; racemes nodding to erect . . . . *Polygonum lapathifolium* . . . . DOCK-LEAVED SMARTWEED

5 Outer sepals with faint, irregularly forked nerves; racemes erect . . . . *Polygonum pensylvanicum* | PENNSYLVANIA SMARTWEED

6 Flower stalks and ocreae with spreading hairs . . . . *Polygonum careyi* . . . . CAREY'S HEARTS-EASE

6 Flower stalks and ocreae with flattened hairs or hairs absent 7

7 Sepals covered with shiny yellow glandular dots; plants peppery to taste 8

7 Sepals without yellow glandular dots; plants not peppery 9

8 Ocreae swollen; sepals usually rose-colored at tips; achenes dull brown . . . . *Polygonum hydropiper* | WATER-PEPPER SMARTWEED

8 Ocreae not swollen; sepals green or white-tipped; achenes smooth and shiny-black . . . . *Polygonum punctatum* | DOTTED SMARTWEED

9 Plants perennial; racemes slender, over 3 cm long, loosely flowered and often interrupted; achenes 3-angled . . . . *Polygonum hydropiperoides* . . . . FALSE WATER-PEPPER

9 Plants annual; racemes 1–3 cm long, densely flowered and mostly continuous; achenes mostly lens-shaped (a few sometimes 3-angled) . . . . *Polygonum persicaria* | LADY'S THUMB

Polygonum amphibium L.
(*Persicaria amphibia* (L.) A. Gray)
WATER SMARTWEED
NC **OBL** | MW **OBL** | GP **OBL**

Native perennial herb; plants either aquatic with floating leaves, or emergent and exposed, both types with spreading rhizomes. **Stems** to 1 m or more long, leaves and habit variable. Submerged plants smooth, usually branched, the branches floating, branch tips often upright and raised above water surface; leaves floating, leathery, oval, 4–20 cm long and 1–4 cm wide, rounded at tip; stipules (ocreae) membranous; petioles 1–8 cm long. Exposed plants hairy; leaves stalkless or with short petioles. **Flowers** pink to red, in 1–2 spikelike racemes from branch tips, the racemes 2–15 cm long and 1–2 cm wide; sepals 5-lobed to below middle, 4–5 mm long; stamens 5. **Fruit** a lens-shaped achene, 2–4 mm long, shiny dark brown. June–Sept. **Synonyms** *Polygonum coccineum, Polygonum natans*. **Habitat** Common; ponds, lakes, marshes, bog pools, backwater areas, quiet streams.

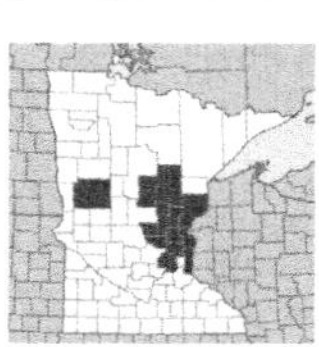

Polygonum arifolium L.
(*Persicaria arifolia* (L.) K. Haraldson)
HALBERD-LEAVED TEARTHUMB
NC **OBL** | MW **OBL** | GP **OBL**

Native annual herb, similar to arrow-leaved tearthumb (*Polygonum sagittatum*). **Leaves** to 20 cm long and 15 cm wide, arrowhead-shaped at base but the triangular-shaped basal lobes pointing outward rather than downward as in *Polygonum sagittatum*. **Flowers** in rounded heads at ends of stems or from leaf axils, flower stalks with glands; sepals pink, 2–3 mm long. **Fruit** a lens-shaped achene, 4–5 mm long. July–Sept. **Habitat** Swamps, wet woods, streambanks and shores.

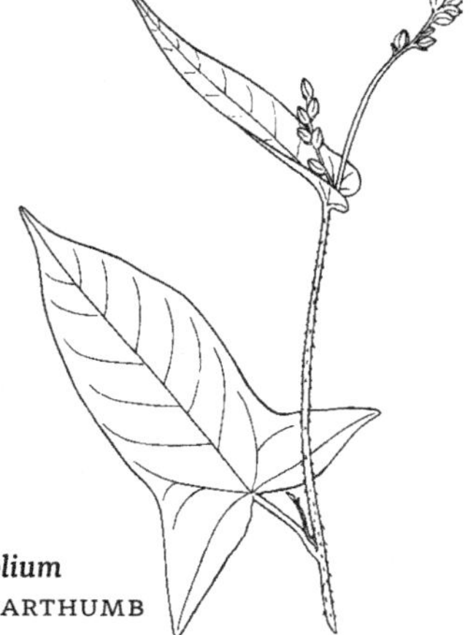

Polygonum arifolium
HALBERD-LEAVED TEARTHUMB

Polygonum careyi Olney
(*Persicaria careyi* (Olney) Greene)
CAREY'S HEARTS-EASE
NC **FACW** | MW **FACW** | *special concern*

Native annual herb. **Stems** upright, branched, to 1 m long, with gland-tipped hairs. **Leaves** lance-shaped; stipules (ocreae) fringed with bristles and covered with stiff, spreading hairs. **Flowers** in cylindric, drooping racemes 3–6 cm long; sepals pink or rose, 3 mm long; stamens 5 (sometimes to 8). **Fruit** a black, smooth, shiny achene, 2 mm wide. July–Aug. **Habitat** Sandy lakeshores and streambanks, marshes. **Status** Special concern; in Minnesota, known from a single 1940 Carlton County collection and not observed in the state since that time.

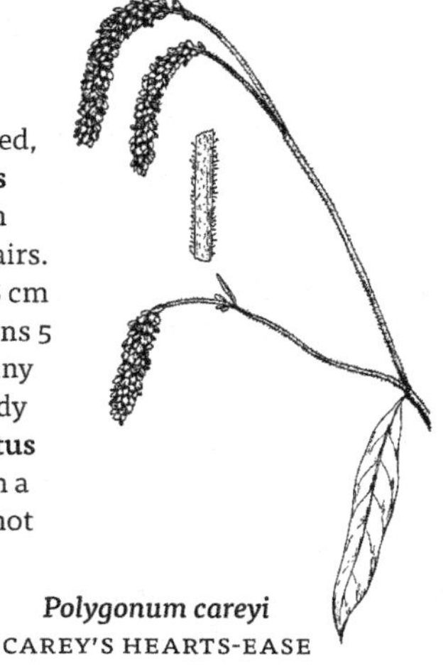

Polygonum careyi
CAREY'S HEARTS-EASE

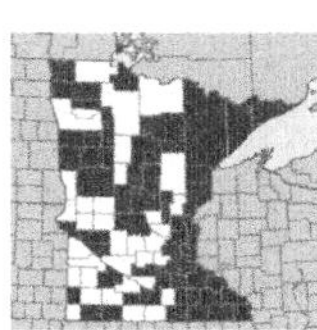

Polygonum hydropiper L.
(*Persicaria hydropiper* (L.) Spach)
WATER-PEPPER SMARTWEED
NC **OBL** | MW **OBL** | GP **OBL**

Introduced annual herb. **Stems** red, erect to sprawling, 2–6 dm long, sometimes rooting at lower nodes, branched or unbranched, peppery-tasting. **Leaves** lance-shaped, 3–8 cm long and to 2 cm wide, hairless except for short hairs on veins and margins, nearly stalkless or with a short petiole; stipules (ocreae) membranous, 5–15 mm long, swollen and fringed with bristles. **Flowers** green and usually white-margined, continuous in slender racemes, often nodding at tip; sepals 5, 3–4 mm long, with glandular dots; stamens 4 or 6. **Fruit** a dull, dark brown achene, 3-angled or lens-shaped, 2–3 mm long. July–Oct. **Habitat** Muddy shores, streambanks, floodplains, marshes, ditches and roadsides.

Polygonum hydropiper
WATER-PEPPER SMARTWEED

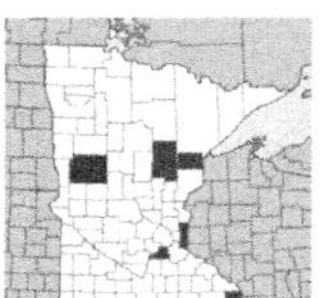

Polygonum hydropiperoides Michx.
(*Persicaria hydropiperoides* (Michx.) Small)
FALSE WATER-PEPPER
NC **OBL** | MW **OBL** | GP **OBL**

Native perennial herb, spreading by rhizomes. **Stems** erect to sprawling with upright tips, to 1 m long, usually branched, nearly smooth or with short hairs. **Leaves** linear to lance-shaped, 4–12 cm long and to 2.5 cm wide, petioles short; stipules (ocreae) membranous, 5–15 mm long, with stiff hairs and fringed with bristles. **Flowers** green, white or pink, in 2 to several slender racemes, 1–6 cm long, often interrupted near base; sepals 2–3 mm long, 5-lobed to just below middle, without glandular dots or only the inner sepals slightly glandular; stamens 8. **Fruit** a black, shiny, 3-angled achene with concave sides, 2–3 mm long. July–Sept. **Habitat** Shallow water or wet soil; ponds, marshes, swamps, bogs and fens, streambanks, lakeshores and ditches.

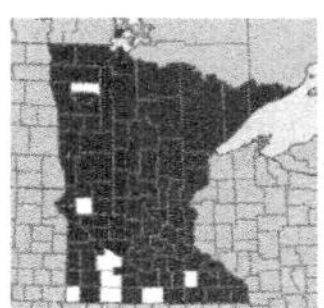

Polygonum lapathifolium L.
(*Persicaria lapathifolia* (L.) A. Gray)
DOCK-LEAVED SMARTWEED
NC **FACW** | MW **FACW** | GP **OBL**

Native annual herb. **Stems** erect to sprawling, unbranched or few-branched, 2–15 dm long. **Leaves** lance-shaped, 4–20 cm long and 0.5–5 cm wide, smooth above, often densely short-hairy on leaf undersides; petioles to 2 cm long, smooth to glandular; stipules (ocreae) 5–20 mm long, entire or with irregular, jagged margins. **Flowers** deep pink, white or green, crowded in erect or nodding racemes 1–5 cm long; sepals 3–4 mm long, 4- or 5-lobed to below middle, the outer 2 sepals strongly 3-nerved; stamens usually 6. **Fruit** a brown, lens-shaped achene, 2–3 mm long. July–Sept. **Habitat** Marshes, wet meadows, shores, streambanks, ditches and cultivated fields. Common and weedy.

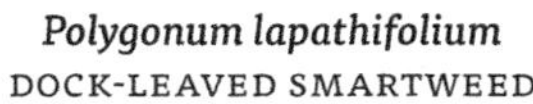

Polygonum lapathifolium
DOCK-LEAVED SMARTWEED

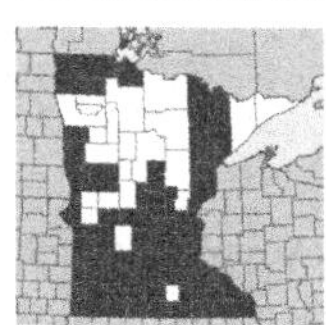

Polygonum pensylvanicum L.

(*Persicaria pensylvanica* (L.) M. Gomez)
PENNSYLVANIA SMARTWEED
NC **FACW** | MW **FACW** | GP **FACW**

Native annual herb. **Stems** erect, 3–20 dm long, unbranched to widely branching. **Leaves** lance-shaped, 3–15 cm long and 1–4 cm wide, smooth except for short hairs on margins; petioles to 2.5 cm long; stipules (ocreae) 0.5–1.5 cm long, entire or with an irregular, jagged margin, hairless, not fringed with bristles. **Flowers** pink to white, in dense racemes 2–3 cm long, the flower stalks with gland-tipped hairs; sepals 3–5 mm long, 5-parted to below middle, the outer sepals faintly nerved; stamens 8 or less. **Fruit** a dark brown to black, shiny achene, lens-shaped, to 3 mm long. June–Sept. **Habitat** Weedy on streambanks, exposed shores, marshes, fens, ditches and cultivated fields.

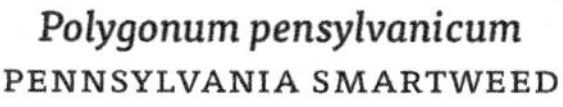

Polygonum pensylvanicum
PENNSYLVANIA SMARTWEED

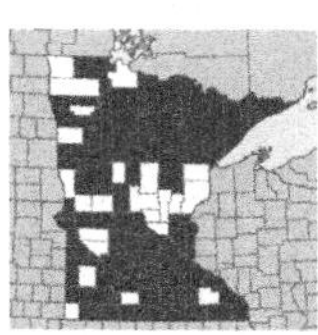

Polygonum persicaria L.

(*Persicaria maculosa* A. Gray)
LADY'S THUMB
NC **FAC** | MW **FACW** | GP **FACW**

Introduced annual herb. **Stems** upright to spreading, 2–8 dm long, unbranched to branched, often red. **Leaves** lance-shaped, 3–15 cm long and 0.5–3 cm wide, smooth or with few hairs, underside usually dotted with small glands, leaves stalkless or on petioles to 1 cm long; ocreae 5–15 mm long, fringed with bristles, with short hairs. **Flowers** pink to rose, crowded in straight, cylindric racemes 1–4 cm long and 0.5–1 cm wide; sepals 2–4 mm long, 5-lobed to near middle; stamens 6. **Fruit** a black, shiny achene, lens-shaped or sometimes 3-angled, 2–3 mm long. July–Sept. **Habitat** Muddy shores, streambanks, ditches and cultivated fields, often weedy. Introduced from Europe and now throughout N America.

Polygonum punctatum Elliott

(*Persicaria punctata* (Elliott) Small)

DOTTED SMARTWEED

NC **OBL** | MW **OBL** | GP **OBL**

Native annual or perennial herb. **Stems** erect to spreading, 4–10 dm long, unbranched to branched. **Leaves** narrowly lance-shaped or oval, 4–15 cm long and 1–2 cm wide, smooth except for small short hairs on margins, underside usually dotted with small glands; petioles short; stipules (ocreae) 5–15 mm long, smooth or with stiff hairs and fringed with bristles. **Flowers** green-white; in numerous slender, loosely flowered racemes, interrupted in lower portion, to 10 cm long; sepals 3–4 mm long, with glandular dots, 5-parted to about middle; stamens 6–8. **Fruit** a dark, shiny achene, lens-shaped or 3-angled, 2–3 mm long. Aug–Sept. **Habitat** Floodplain forests, marshes, shores, streambanks and cultivated fields.

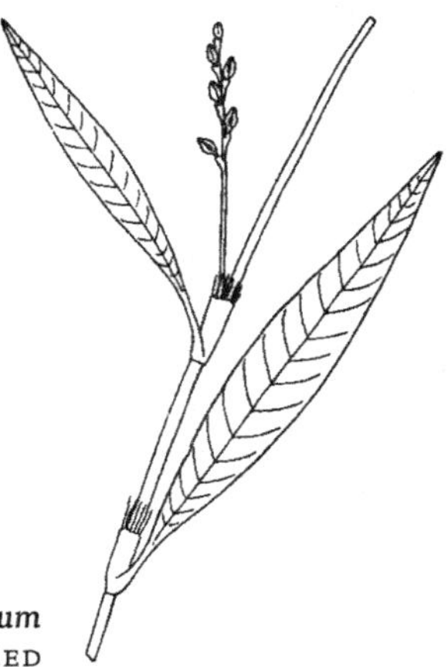

Polygonum punctatum
DOTTED SMARTWEED

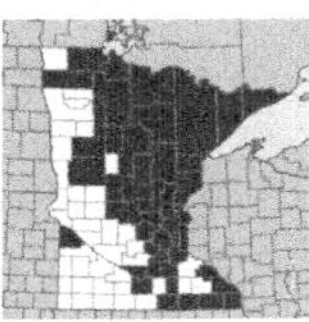

Polygonum sagittatum L.

(*Persicaria sagittata* (L.) H. Gross)

ARROW-LEAVED TEARTHUMB

NC **OBL** | MW **OBL** | GP **OBL**

Slender native annual herb. **Stems** 4-angled, weak, usually supported by other plants, 1–2 m long, with downward pointing prickles on stem angles, petioles, leaf midribs and flower stalks. **Leaves** lance-shaped to oval, arrowhead-shaped at base, 3–10 cm long and to 2.5 cm wide, the basal lobes pointing downward; petioles long on lower leaves, shorter above; stipules (ocreae) 5–10 mm long, with a few hairs on margins. **Flowers** white or pink; in round racemes to 1 cm long, on long slender stalks at ends of stems or from leaf axils; sepals 3 mm long, 5-parted to below middle. **Fruit** a brown to black, shiny achene, 3-angled, 2–3 mm long. July–Sept. **Habitat** Swamps, marshes, wet meadows, burned wetlands.

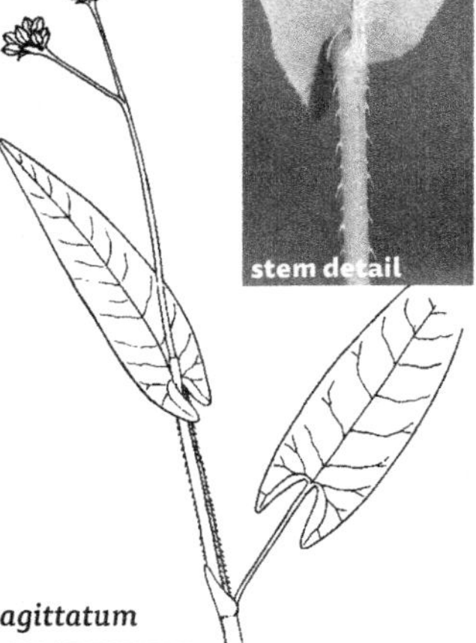

Polygonum sagittatum
ARROW-LEAVED TEARTHUMB

RUMEX | *Dock*

Perennial, sometimes weedy, herbs (annual in *Rumex fueginus*). **Leaves** large and clustered at base of plants, or leafy-stemmed; mostly oblong to lance-shaped, flat to wavy-crisped along margins, usually with petioles. Membranous sheath around stem present at each node (ocrea). **Flowers** in crowded whorls in panicles at ends of stems; flowers small and numerous, green but turning brown; sepals in 2 series of 3, the inner 3 sepals (valves) enlarging, becoming winged and loosely enclosing the achene, giving the appearance of a 3-winged fruit, the midvein of the valve often swollen to produce a grainlike tubercle on the back; stamens 6; styles 3. **Fruit** a brown, 3-angled achene, tipped with a short slender beak.

RUMEX | DOCK

1 Margins of mature valves with coarse or spine-tipped teeth **2**
1 Margins of mature valves entire or shallowly lobed, not toothed **4**

2 Plants annual, fibrous-rooted or with slender taproots; margins of valves dissected into spine-tipped teeth . . . . *Rumex fueginus* | GOLDEN DOCK
2 Plants perennial from stout taproots; valve margins toothed or spine-tipped. **3**

3 Grains 3, margins of valves coarsely toothed.. . . . . *Rumex stenophyllus* . . . . NARROW-LEAF DOCK
3 Grains 1; margins of valves with spine-tipped teeth . . . . *Rumex obtusifolius* . . . . BITTER DOCK

4 Flower stalks without a large swollen joint; grains 3, base of grain distinctly above base of valve . . . . *Rumex orbiculatus* | GREAT WATER-DOCK
4 Flower stalks with a large swollen joint below the middle or near base; grains 1–3, base of grain even with base of valve **5**

5 Fruit with 3 grains, the grains projecting below the valves; flower stalks 2–5x longer than fruit . . . . *Rumex verticillatus* . . . . WATER-DOCK; SWAMP-DOCK
5 Fruit with 1–3 grains, the grains not projecting below the valves; flower stalks 1–2x longer than fruit **6**

6 Leaves crisp-margined (crinkled); grains 2/3 as wide as long . . . . *Rumex crispus* | CURLED DOCK
6 Leaf margins flat; grains narrower, up to half as wide as long **7**

7 Grains usually 1 (sometimes 2–3); leaves mostly less than 4x longer than wide . . . . *Rumex altissimus* | PALE DOCK

7 Grains usually 3; leaves mostly more than 4x longer than wide . . . . *Rumex salicifolius* | WILLOW DOCK

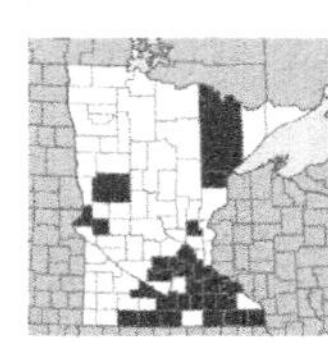

Rumex altissimus A. Wood

PALE DOCK

NC **FACW** | MW **FACW** | GP **FAC**

Native perennial herb, similar to willow dock (*Rumex salicifolius*). **Stems** 3–10 dm long, usually branched from base and with short branches above. **Leaves** all from stem, ovate to lance-shaped, 6–20 cm long and 2–6 cm wide, margins flat or slightly wavy. **Flowers** in panicles 1–3 dm long, the panicle branches short and ± upright; flower stalks short, 3–5 mm long, swollen and jointed near base; valves rounded, 4–6 mm long and as wide, flattened across base, margins smooth or irregularly toothed; grains usually well-developed on only 1 of the 3 valves, although sometimes present on 2–3 valves; the largest grain lance-shaped. **Fruit** a brown achene, 2–3 mm long. May–Aug. **Habitat** Marshes, shores, streambanks, ditches, disturbed areas.

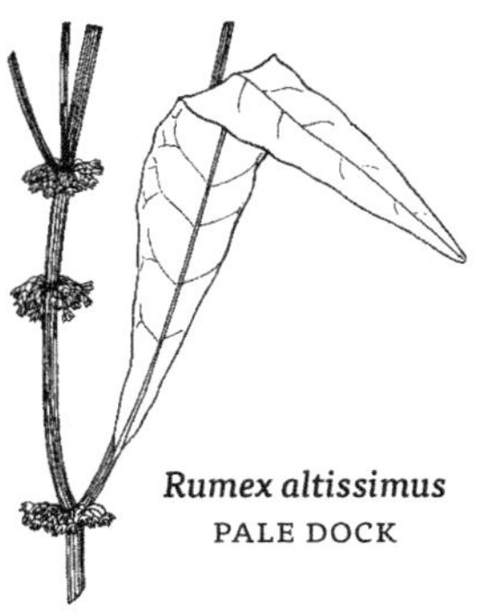

Rumex altissimus
PALE DOCK

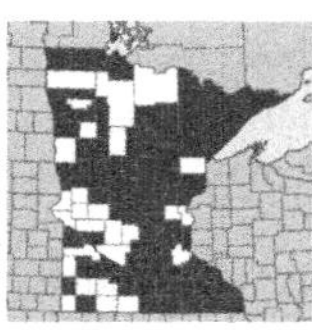

Rumex crispus L.

CURLED DOCK

NC **FAC** | MW **FAC** | GP **FAC**

Introduced perennial herb, from a thick taproot. **Stems** stout, upright, usually single, 5–15 dm long. Basal leaves large, 10–30 cm long and 1–5 cm wide, on long petioles, often drying early in season; stem leaves smaller and with shorter petioles, oval to lance-shaped, margins strongly wavy-crisped (crinkled). **Flowers** in large branched panicles, the panicle branches ± upright; flower stalks drooping at tips, 5–10 mm long, swollen-jointed near base; valves heart-shaped to broadly ovate, 4–5 mm long and as wide, margins ± smooth; grains 3, swollen, often of unequal size, rounded at ends. **Fruit** a brown achene, 2–3 mm long. July–Sept. **Habitat** Wet meadows, shores, ditches, old fields, and other wet and disturbed areas; weedy. Introduced from Eurasia, naturalized throughout USA, s Canada and much of the world.

Rumex crispus
CURLED DOCK

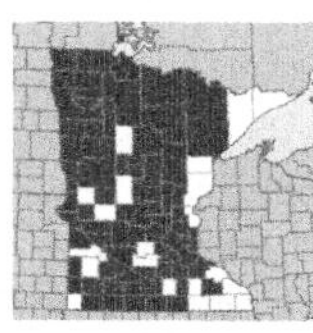

Rumex fueginus Phil.
GOLDEN DOCK
NC **FACW** | MW **FACW** | GP **FACW**

Native annual herb. **Stems** hollow, to 8 dm long, much-branched. **Leaves** mostly on stems, smaller upward, lance-shaped to linear, 5–20 cm long and 0.5–4 cm wide, wedge-shaped or heart-shaped at base, margins flat to wavy-crisped. **Flowers** in large open panicles, the panicle branches ± upright, leafy, the flower stalks jointed near base; valves triangular-ovate, 2–3 mm long, the margins lobed into 2–3 spine-tipped teeth on each side; grains 3. **Fruit** a light brown achene, 1–2 mm long. July–Aug. **Synonyms** *Rumex maritimus*. **Habitat** Marshes, shores, streambanks and ditches, sometimes where brackish.

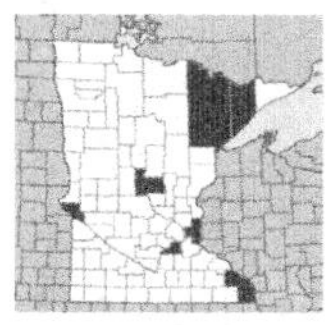

Rumex obtusifolius L.
BITTER DOCK
NC **FAC** | MW **FACW** | GP **FACW**

Introduced perennial herb. **Stems** stout, to 12 dm long, usually unbranched. Lower leaves oblong or ovate, to 30 cm long and 15 cm wide, heart-shaped or rounded at base; upper leaves smaller. **Flowers** in much-branched panicles, flower stalks longer than fruit, jointed near base; valves triangular-ovate, 4–5 mm long, with 2–4 spine-tipped teeth on each side; grains large and with tiny wrinkles. **Fruit** a shiny, red-brown achene. June–Aug. **Habitat** Floodplain forests and openings, cultivated fields and disturbed areas. Introduced from Europe.

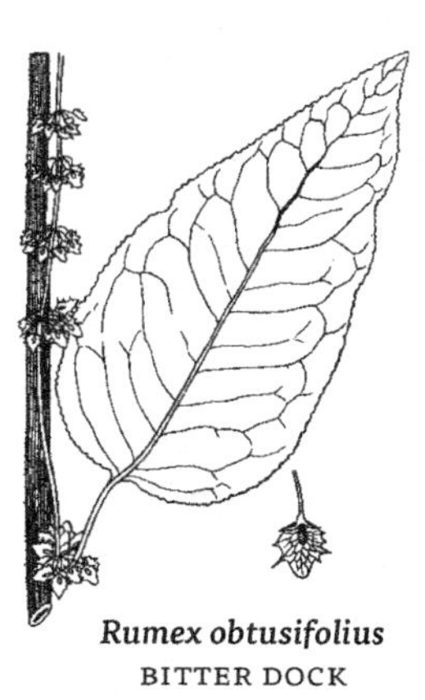

Rumex obtusifolius
BITTER DOCK

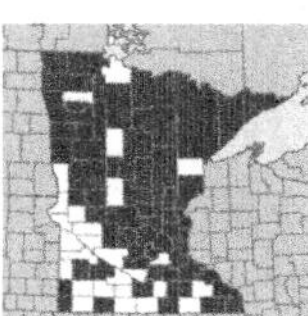

Rumex orbiculatus A. Gray
GREAT WATER-DOCK
NC **OBL** | MW **OBL** | GP **OBL**

Native perennial herb. **Stems** stout, unbranched, 2–2.5 m long. **Leaves** lance-shaped or oblong lance-shaped, lower leaves 30–60 cm long, upper leaves 5–15 cm long; margins flat. **Flowers** in panicles to 5 dm long; valves rounded, flat at base, 5–8 mm long and as wide, smooth or with small teeth; grains 3, narrowly lance-shaped, the base distinctly above base of valve. June–Aug. **Synonyms** *Rumex britannica*. **Habitat** Marshes, fens, streambanks and ditches, often in shallow water.

Rumex salicifolius J. A. Weinm.
WILLOW DOCK
NC **FAC** | MW **FACW** | GP **FACW**

Native perennial taprooted herb. **Stems** smooth, 3–10 dm long, usually branched from base and with short branches on stem. **Leaves** mostly on stems, not much smaller upward, narrowly lance-shaped, tapered at both ends, pale waxy green, 5–16 cm long and 1–3 cm wide, margins mostly flat. **Flowers** in panicles 1–3 dm long, panicle branches few and ± upright, with small linear leaves at base; flower stalks 2–4 mm long, swollen and jointed near base; valves thick, triangular, 3–6 mm long and wide, margins smooth or shallowly toothed; grains usually 3. **Fruit** a brown achene, 2 mm long. June–Aug. **Synonyms** *Rumex mexicanus, Rumex triangulivalvis*. **Habitat** Wet meadows, marshes, shores, streambanks, ditches and other low areas, sometimes where brackish.

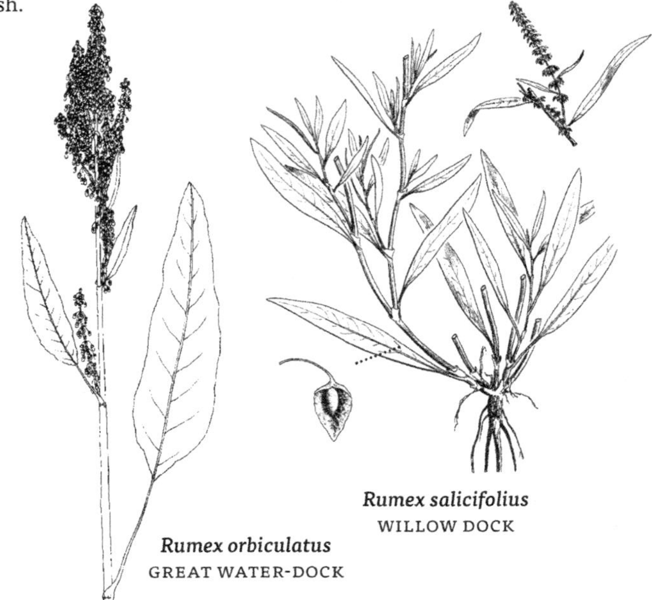

Rumex salicifolius
WILLOW DOCK

Rumex orbiculatus
GREAT WATER-DOCK

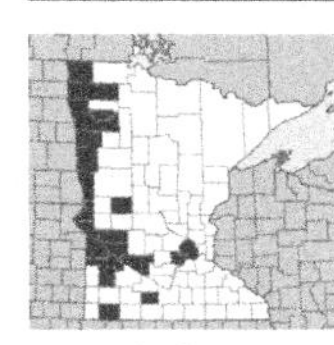

Rumex stenophyllus Ledeb.
NARROW-LEAF DOCK
NC **FACW** | MW **FACW** | GP **FACW**

Introduced perennial herb, similar to **curled dock** (*Rumex crispus*) but leaves less wavy-crisped and the valves with many small teeth on margins. Valves triangular to rounded, 4–6 mm long and as wide; grains 3. **Fruit** a brown achene, 2–3 mm long. July–Sept. **Habitat** Wet meadows, shores, streambanks, ditches, and disturbed places, usually where brackish.

Rumex verticillatus L.
WATER-DOCK; SWAMP-DOCK
NC **OBL** | MW **OBL** | GP **FACW**

Native perennial taprooted herb. **Stems** stout, 1–1.5 m long, with many short branches from leaf axils. **Leaves** narrowly lance-shaped, tapered to base, margins flat. **Flowers** in leafless panicles 2–4 dm long, the panicle branches few and ± upright; flower stalks 10–15 mm long, jointed near base; valves triangular-ovate, 4–6 mm long and wide, thickened at center; grains 3, lance-shaped, the base blunt and projecting 0.5 mm below base of valve. June–Sept. **Synonyms** *Rumex floridanus*. **Habitat** Marshes, swamps, wet forests, backwater areas and muddy shores, often in shallow water.

Rumex stenophyllus
NARROW-LEAF DOCK

Rumex verticillatus
WATER-DOCK; SWAMP-DOCK

MONTIA | *Miner's-lettuce*

Montia chamissoi (Ledeb.) Greene
MINER'S-LETTUCE
NC **OBL** | MW **OBL** | GP **OBL** | *endangered*

Smooth native perennial herb, forming colonies from spreading rhizomes and stolons. **Stems** upright, 5–20 cm long. **Leaves** opposite, spatula-shaped to obovate, 2–5 cm long, margins entire. **Flowers** perfect, 3–10 in drooping racemes from ends of stems or upper leaf axils, flower stalks 1–3 cm long; petals 5, white to pink, 5–8 mm long, stamens 5. June–July. **Habitat** In Minnesota, a single population of miner's-lettuce is known (Winona County). Plants occur in shallow mud on a west aspect, sandstone outcrop, with water seeping over the nearly vertical rock wall. The site is cool and shaded. Associated species are few, but include orange touch-me-not (*Impatiens capensis*) and bulblet fern (*Cystopteris bulbifera*). **Status** Endangered; in midwest USA, miner's-lettuce known only from extreme se Minnesota and adjacent ne Iowa, where disjunct from Rocky Mountains.

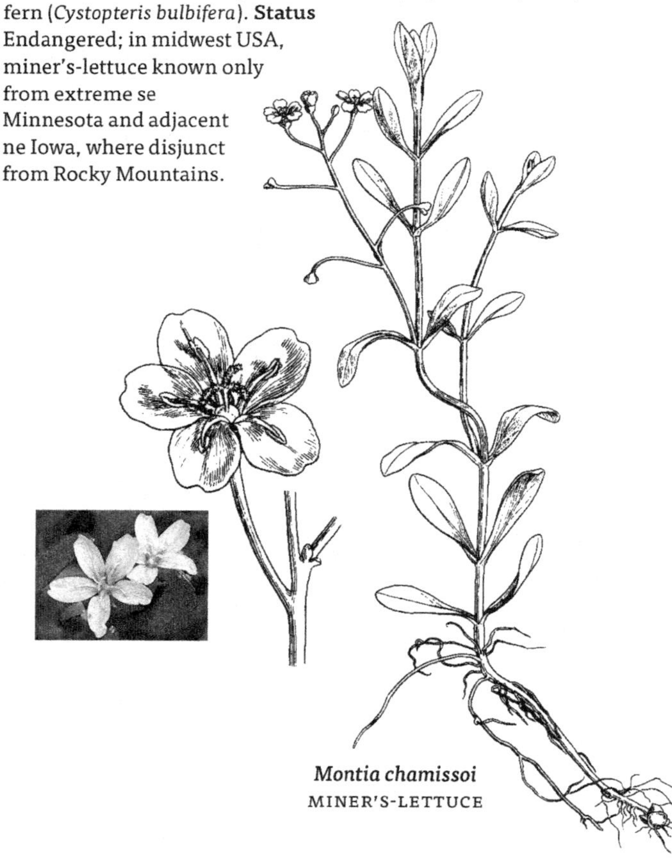

Montia chamissoi
MINER'S-LETTUCE

PERENNIAL HERBS (ours). **Leaves** simple, opposite (sometimes whorled in *Lysimachia*), or leaves all basal. **Flowers** perfect (with both male and female parts), regular, single from leaf axils, or in clusters at ends of stems; sepals 4–5, petals mostly 5 (varying from 4–9, or absent in *Glaux*), joined, tube-shaped below and flared above, deeply cleft to shallowly lobed at tip; ovary superior, style 1; stamens 5. **Fruit** a 5-chambered capsule.

PRIMULACEAE | PRIMROSE FAMILY

1 Leaves all from base of plant, leaf underside strongly whitened .... *Primula mistassinica* | MISTASSINI PRIMROSE
1 Leaves from stem, green on both sides 2

2 Flowers white to pink, single and stalkless in leaf axils; petals absent; sepals petal-like; nw Minnesota .... *Glaux maritima* | SEA-MILKWORT
2 Flowers yellow, single and stalked from leaf axils or in racemes from axils; sepals and petals present; widespread .... *Lysimachia* | LOOSESTRIFE

GLAUX | *Sea-milkwort*

Glaux maritima L.
SEA-MILKWORT
NC **OBL** | MW **OBL** | GP **OBL** | *endangered*

Native perennial herb, with shallow rhizomes; plants smooth, usually waxy blue-green, succulent. **Stems** leafy, 3–25 cm long, unbranched and erect to branched and spreading. **Leaves** fleshy, opposite, sometimes alternate above, oval to linear, 3–20 mm long and 1–5 mm wide, blunt-tipped; margins entire; petioles absent. **Flowers** small, single and stalkless in leaf axils, bell-shaped, 3–4 mm long; petals absent; sepals petal-like, white to pink; stamens equal or slightly longer than sepals. **Fruit** a rounded capsule, 2–3 mm long; seeds several, black, oval and flattened. June–Sept. **Habitat** Alkaline or saline soil of wet meadows, prairies, dry streambeds, and lakeshores. **Status** Endangered.

Glaux maritima
SEA-MILKWORT

LYSIMACHIA | *Loosestrife*

Perennial herbs, spreading by rhizomes. **Stems** erect. **Leaves** opposite (sometimes appearing whorled), ovate or lance-shaped. **Flowers** 5-parted, yellow, single on stalks from leaf axils or in racemes or panicles; sepals green; petals bright to pale yellow. **Fruit** a capsule.

LYSIMACHIA | LOOSESTRIFE

1 Plants creeping; leaves round, on short, smooth petioles . . . . *Lysimachia nummularia* | MONEYWORT; CREEPING JENNIE

1 Plants upright, leaves longer than wide; petioles hairy or absent 2

2 Flowers in dense clusters at ends of stems or from leaf axils 3

2 Flowers single from leaf axils 4

3 Flowers in clusters at ends of stems . . . . *Lysimachia terrestris* . . . . SWAMP CANDLES

3 Flowers in clusters from leaf axils . . . . *Lysimachia thyrsiflora* . . . . SWAMP LOOSESTRIFE

4 Leaves rounded or heart-shaped at base; petioles 1–3 cm long, fringed with hairs. . . . . *Lysimachia ciliata* | FRINGED LOOSESTRIFE

4 Leaves tapered at both ends; petioles absent or short, smooth or fringed with hairs 5

5 Leaves narrowly linear, to 5 mm wide; margins smooth and rolled under . . . . *Lysimachia quadriflora* | SMOOTH LOOSESTRIFE

5 Leaves lance-shaped to oval, usually greater than 9 mm wide; margins finely hairy or rough-to-touch 6

6 Leaves and flowers with red or black dots or streaks; petals entire at tip . . . . *Lysimachia quadriflora* | SMOOTH LOOSESTRIFE

6 Leaves and flowers without red or black dots or streaks; petals ragged-fringed at tip . . . . *Lysimachia hybrida* | LOWLAND LOOSESTRIFE

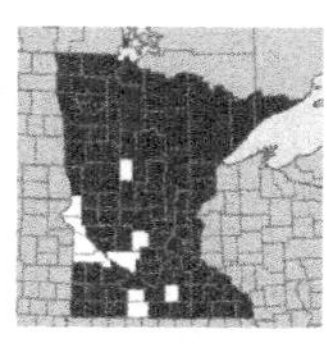

Lysimachia ciliata L.

FRINGED LOOSESTRIFE
NC **FACW** | MW **FACW** | GP **FACW**

Native perennial herb, spreading by rhizomes. **Stems** upright, 3–12 dm long, unbranched or with few branches above. **Leaves** ovate to lance-shaped, 4–15 cm long and 2–6 cm wide, rounded to heart-shaped at base, green above, slightly paler below; margins fringed with short hairs; petioles 0.5–5 cm long, fringed with

hairs. **Flowers** yellow, single from upper leaf axils, on stalks 2–7 cm long; sepal lobes lance-shaped, 4–8 mm long, often with 3–5 parallel red-brown veins; petal lobes rounded and finely ragged at tip, 4–10 mm long and 3–9 mm wide, with a short slender tip. **Fruit** a capsule, 4–7 mm wide. June–Aug. **Synonyms** *Steironema ciliata*. **Habitat** Usually shaded wet areas, such as shores, streambanks, wet meadows, ditches, floodplains, wet woods and thickets.

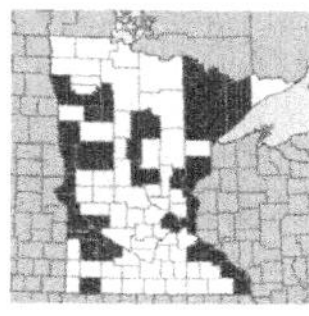

Lysimachia hybrida Michx.
LOWLAND LOOSESTRIFE
NC **OBL** | MW **OBL** | GP **OBL**

Native perennial herb, spreading by rhizomes. **Stems** usually erect, 2–8 dm long, unbranched or sometimes branched from base, usually branched above. **Leaves** narrowly lance-shaped to ovate, 3–10 cm long and 1–2 cm wide, tapered to base, upper surface green, underside green or slightly paler; lower leaves opposite, stalked, withering, petioles fringed with hairs at least near base; upper leaves ± whorled and stalkless, persistent. **Flowers** yellow, single from leaf axils but often appearing crowded, on stalks 1–4 cm long; sepal lobes lance-shaped, 3–6 mm long; petal lobes rounded and finely fringed at tip, 5–10 mm long and 4–10 mm wide, with a short slender tip. **Fruit** a capsule, 4–6 mm wide. July–Aug. **Synonyms** *Lysimachia verticillata*. **Habitat** Wet meadows, marshes, streambanks, ditches and shores, sometimes in shallow water.

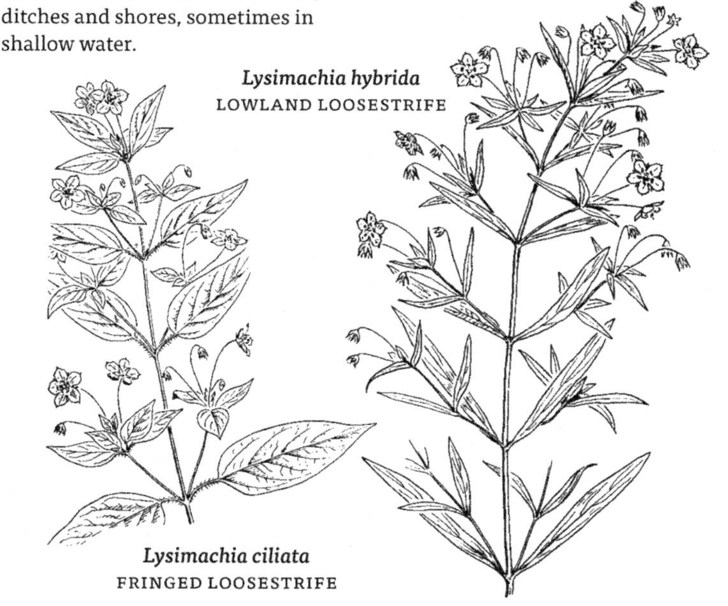

Lysimachia hybrida
LOWLAND LOOSESTRIFE

Lysimachia ciliata
FRINGED LOOSESTRIFE

Lysimachia nummularia L.

MONEYWORT; CREEPING JENNIE

NC **FACW** | MW **FACW** | GP **FACW**

Introduced perennial herb, often forming mats and potentially invasive. **Stems** creeping, to 5–6 dm long. **Leaves** opposite, dotted with black glands, round or broadly oval, 1–2.5 cm long; petioles short. **Flowers** single in leaf axils, on stalks to 2.5 cm long; sepals leaflike, triangular, 6–8 mm long; petals yellow, dotted with dark red, 10–15 mm long. **Fruit** a capsule, shorter than sepals. June–Aug. **Habitat** Swamps, floodplain forests, streambanks, shores, meadows and ditches. Native of Europe; occasional in e Minnesota.

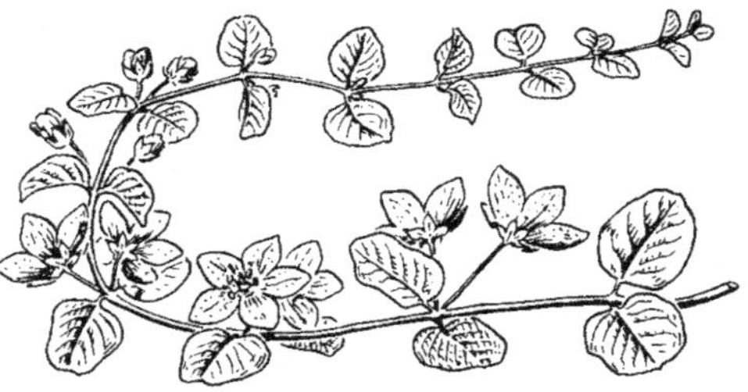

Lysimachia nummularia
MONEYWORT; CREEPING JENNIE

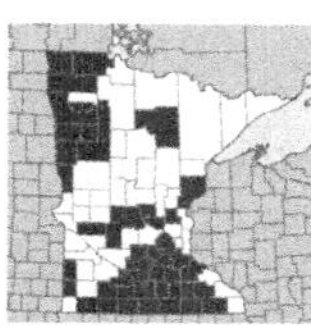

Lysimachia quadriflora Sims

SMOOTH LOOSESTRIFE

NC **OBL** | MW **OBL** | GP **FACW**

Native perennial herb, spreading by rhizomes which form clusters of basal rosettes. **Stems** upright, 3–10 dm long. **Leaves** opposite, sometimes appearing whorled; stem leaves stalkless, often ascending, linear, 3–8 cm long and 2–7 mm wide, margins smooth or rolled under, sometimes fringed with a few hairs near base. **Flowers** yellow, single in clusters at ends of stems and branches, on stalks 1–4 cm long; sepal lobes lance-shaped, 4–6 mm long; petal lobes oval, 7–12 mm long and 5–9 mm wide, entire or finely ragged at tip. **Fruit** a capsule, 3–5 mm wide. July–Aug. **Synonyms** *Lysimachia longifolia*. **Habitat** Wet meadows, pond and marsh margins, low prairie, calcareous fens; often where sandy and calcium-rich.

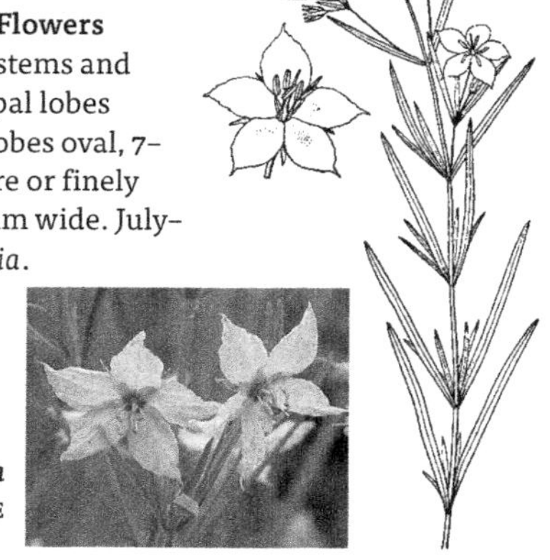

Lysimachia quadriflora
SMOOTH LOOSESTRIFE

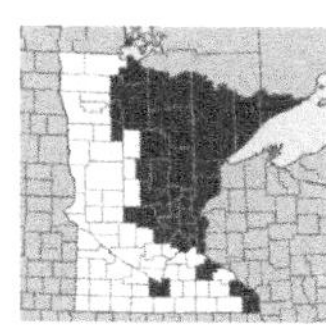

Lysimachia terrestris (L.) BSP.

SWAMP CANDLES

NC **OBL** | MW **OBL** | GP **OBL**

Native perennial herb, spreading by shallow rhizomes. **Stems** smooth, 4–8 dm long, usually branched. **Leaves** opposite, dotted with glands, narrowly lance-shaped, 5–10 cm long and 2–4 cm wide; small bulblike structures produced in leaf axils late in season; bracts awl-like, 3–8 mm long. **Flowers** yellow, in a single, crowded, upright raceme, 1–3 dm long; sepals lance-shaped; petal lobes oval, 5–7 mm long, with dark lines, on stalks 8–15 mm long. **Fruit** a capsule, 2–3 mm wide. June–Aug. **Habitat** Marshes, fens, thickets, muddy shores, and ditches.

Lysimachia terrestris
SWAMP CANDLES

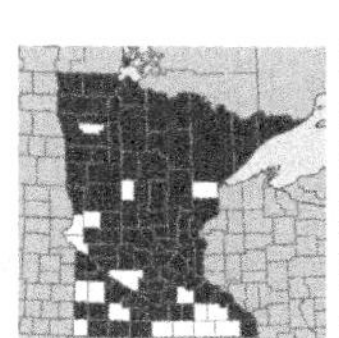

Lysimachia thyrsiflora L.

SWAMP LOOSESTRIFE

NC **OBL** | MW **OBL** | GP **OBL**

Native perennial upright herb, spreading by rhizomes; plants conspicuously dotted with dark glands. **Stems** smooth or with patches of brown hairs, 3–7 dm long, unbranched or branched on lower stem. **Leaves** opposite, linear to lance-shaped, 4–12 cm long and 0.5–4 cm wide, smooth above, smooth or sparsely hairy below; petioles absent. **Flowers** yellow, crowded in dense racemes from leaf axils, on spreading stalks 2–5 cm long; mostly 6-parted; sepal lobes awl-shaped, 1–3 mm long; petal lobes linear, 3 mm long; stamens 2x longer than petals. **Fruit** a capsule, 2–4 mm wide. June–Aug. **Habitat** Many types of wetlands: thickets, shores, fens and bogs, marshes, low places in conifer and deciduous swamps, often in shallow water.

Lysimachia thyrsiflora
SWAMP LOOSESTRIFE

PRIMULA | *Primrose*

Primula mistassinica Michx.
MISTASSINI PRIMROSE
NC **FACW** | MW **FACW** | GP **FACW**

Native perennial herb. **Stems** to 25 cm long. **Leaves** all at base of plant, oblong lance-shaped, 2–7 cm long, long tapered to base, smooth on upper surface, smooth or often white-yellow powdery below; margins with outward pointing teeth; bracts below flowers awl-shaped, 3–6 mm long. **Flowers** 1–2 cm wide, 2–10 in a cluster atop a leafless stalk; sepals joined, shorter than petals; petals joined, tubelike and flared at ends, pink and sometimes with a yellow center. **Fruit** an oblong, upright capsule to 1 cm long. May–June. **Synonyms** *Primula intercedens*. **Habitat** Lake Superior rocky shoreline in nw Minnesota, often with violet butterwort (*Pinguicula vulgaris*); moist, gravelly, calcium- rich soil.

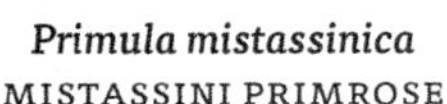

Primula mistassinica
MISTASSINI PRIMROSE

PERENNIAL HERBS or half-shrubs, most dependent on wood-rotting fungi (mycotrophic). **Leaves** alternate to sometimes opposite or nearly whorled, often shiny, evergreen in *Pyrola*, deciduous in *Moneses*. **Flowers** perfect (with both male and female parts), 5-parted, waxy and nodding. **Fruit** a capsule.

PYROLACEAE | SHINLEAF FAMILY

1 Flowers white, single and nodding at ends of stems ..... *Moneses uniflora* .... ONE-FLOWERED SHINLEAF
1 Flowers pink, several in racemes .... *Pyrola asarifolia* | PINK SHINLEAF

MONESES | *One-flowered shinleaf*

Moneses uniflora (L.) A. Gray

ONE-FLOWERED SHINLEAF
NC **FAC** | MW **FAC** | GP **FAC**

Low native perennial herb, roots creeping. **Stems** to 10 cm long. **Leaves** deciduous, mostly at base of plant, opposite or in whorls of 3, nearly round, margins entire or finely toothed. **Flowers** white, single at end of long stalk, nodding, 1–2 cm wide; petals 5. **Fruit** a round capsule, opening from top downward. July–Aug. **Synonyms** *Pyrola uniflora*. **Habitat** Cedar swamps, wet conifer or mixed conifer and deciduous forests.

Moneses uniflora
ONE-FLOWERED SHINLEAF

PYROLA | *Shinleaf, Wintergreen*

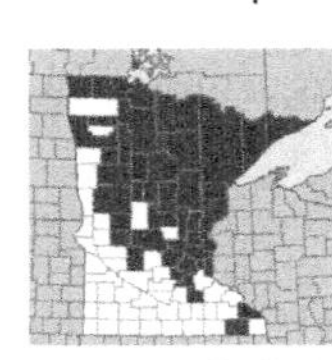

Pyrola asarifolia Michx.
PINK SHINLEAF
NC **FACW** | MW **FACW** | GP **FACU**

Native perennial herb, spreading by rhizomes. **Stems** to 3 dm long. **Leaves** persisting over winter, all near base of plant, kidney-shaped, 3–4 cm long and 3–5 cm wide, margins shallowly rounded-toothed; flower stalk with 1–3 small, scalelike leaves. **Flowers** nodding in a raceme; sepals triangular, 2–3 mm long; petals 5, 5–7 mm long, pink to pale purple. **Fruit** a capsule opening from base upward. June–Aug. **Synonyms** *Pyrola uliginosa*. **Habitat** Cedar swamps, peatlands, marly wetlands, and interdunal wetlands.

inflorescence

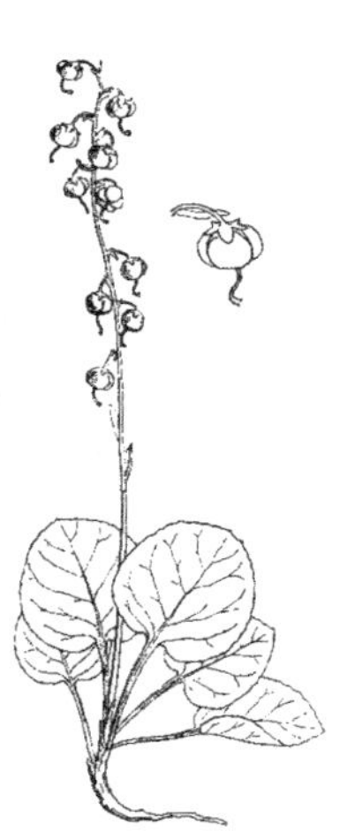
Pyrola asarifolia
PINK SHINLEAF

RANUNCULACEAE
Buttercup Family

ANNUAL OR PERENNIAL, aquatic or terrestrial herbs (or vines in *Clematis*). **Leaves** simple to compound, usually alternate, sometimes opposite or whorled, or all at base of plant. **Flowers** mostly white or yellow, usually with 5 (occasionally more) separate petals and sepals, or petals absent and then with petal-like sepals; sepals leafy and green or petal-like and colored; flowers perfect (with both male and female parts), stamens usually numerous; pistils several to many, ripening into beaked achenes or dry capsules (follicles).

RANUNCULACEAE | BUTTERCUP FAMILY

1 Vines; leaves opposite; fruit with a long, feathery style. . . . . *Clematis* . . . . CLEMATIS

1 Herbs; leaves alternate or from base of plant; fruit not with a long, feathery style 2

2 Leaves linear, 1–2 mm wide, all from base of plant; sepals spurred at base; achenes in a spikelike cluster to 6 cm long. . . . . *Myosurus minimus* . . . . MOUSE-TAIL

2 Leaves not linear; sepals not spurred; achenes in round to short-cylindric heads 3

3 Leaves from base of plant except for 2–3 whorled, leafy bracts below flowers . . . . *Anemone canadensis* | CANADA ANEMONE

3 Stems leafy, the leaves alternate, or leaves all from base of plant. 4

4 Flowers yellow, or leaves simple and unlobed, or plants aquatic 5

4 Flowers not yellow; leaves divided into leaflets; plants not aquatic. 6

5 Leaves all alike, unlobed; sepals yellow, large and petal-like; petals absent . . . . *Caltha* | MARSH-MARIGOLD; COWSLIP

5 Leaves usually of 2 types (stem leaves different from basal leaves), or leaves deeply lobed or divided; sepals green; petals yellow or white . . . . *Ranunculus* | BUTTERCUP; CROWFOOT; SPEARWORT

6 Leaves all from base of plant, flowers single at ends of leafless stalks. . . . . *Coptis trifolia* | ALASKA GOLDTHREAD

6 Stems leafy; flowers many in panicles at ends of stems. . . . . *Thalictrum* . . . . MEADOW-RUE

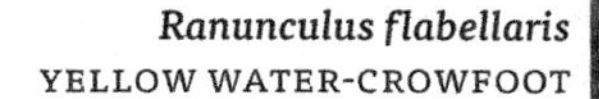

Ranunculus flabellaris
YELLOW WATER-CROWFOOT

ANEMONE | *Anemone*

Anemone canadensis L.

CANADA ANEMONE

NC **FACW** | MW **FACW** | GP **FACW**

Native perennial herb, from slender rhizomes, often forming large patches. **Stems** erect, 1–6 dm long, unbranched below the head. **Leaves** all from base of plant and with long petioles except for 2–3 stalkless leafy bracts below the head; 4–15 cm wide, deeply 3–5-lobed, round to kidney-shaped in outline, underside with long silky hairs, margins sharp-toothed. **Flowers** mostly single at ends of stalks, white and showy, 2–5 cm wide; sepals 5, petal-like, 1–2 cm long; petals absent; stamens and pistils many. **Achenes** clustered in a round, short-hairy head; achene body flat, 3–5 mm long and wide, beak 2–4 mm long. May–Aug. **Habitat** Wet openings, streambanks, thickets, low prairie, ditches, roadsides.

CALTHA | *Marsh-marigold, Cowslip*

Succulent perennial herbs. **Leaves** simple, mostly from base of plant, becoming smaller upward, heart-shaped; margins entire or rounded-toothed. **Flowers** mostly bright yellow (*Caltha palustris*), to pink or white (*Caltha natans*), single at ends of stalks; sepals large and petal-like; petals absent; stamens many. **Fruit** a follicle.

CALTHA | MARSH-MARIGOLD, COWSLIP

1 Flowers bright yellow; common .... *Caltha palustris*
.... COMMON MARSH-MARIGOLD

1 Flowers pink or white; uncommon, ne Minnesota only..... *Caltha natans*
.... FLOATING MARSH-MARIGOLD

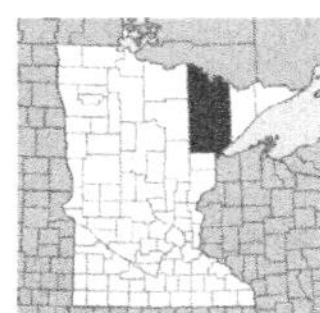

Caltha natans Pallas

FLOATING MARSH-MARIGOLD

NC **OBL** | *endangered*

Native perennial herb. **Stems** floating or creeping, branched, rooting at nodes. **Leaves** heart- or kidney-shaped, 2–5 cm wide, notched at base, upper leaves smaller. **Flowers** pink or white, 1 cm wide; sepals oval; petals absent; stamens 12–25. **Fruit** a follicle, 4–5 mm long, in dense heads of 20–40. July–Aug. **Habitat** Shallow water and shores of ponds and slow-moving streams. In Minnesota, associated species are mannagrass (*Glyceria*), sedges (*Carex*) and pondweeds (*Potamogeton*). **Status** Endangered.

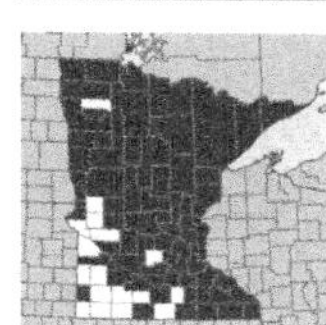

Caltha palustris L.
COMMON MARSH-MARIGOLD
NC **OBL** | MW **OBL** | GP **OBL**

Loosely clumped native perennial herb. **Stems** smooth, 2–6 dm long, hollow. **Leaves** heart-shaped to kidney-shaped, 4–10 cm wide, usually with 2 lobes at base; margins smooth or shallowly toothed; lower leaves with long petioles, stem leaves with shorter petioles. **Flowers** bright yellow, showy at ends of stems or in leaf axils, 2–4 cm wide; sepals 4–9, petal-like, 12–20 mm long; petals absent; stamens many; pistils 4–15, with short styles. **Fruit** a follicle, 10–15 mm long. March–June. **Habitat** Shallow water, swamps, wet woods, thickets, streambanks, calcareous fens, marshes, springs.

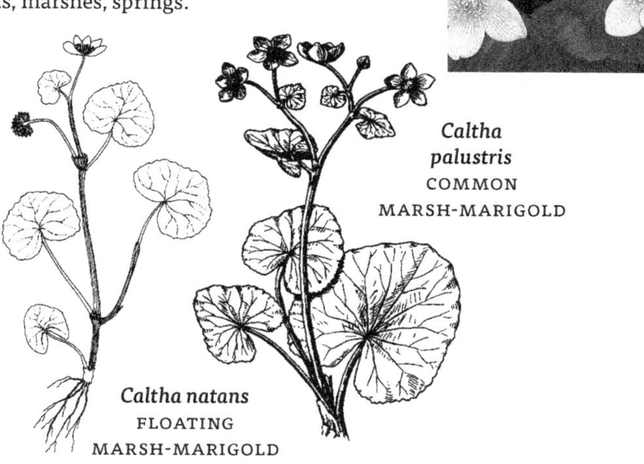

CLEMATIS | *Clematis*

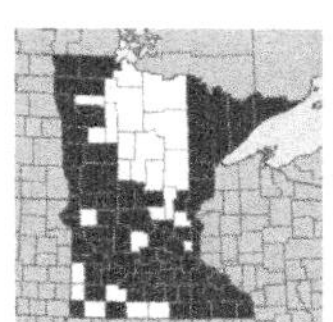

Clematis virginiana L.
VIRGIN'S BOWER
NC **FAC** | MW **FAC** | GP **FAC**

Perennial, woody vine; native. **Stems** slender, to 5 m long or more, trailing on ground or over shrubs, smooth, brown to red-purple. **Leaves** opposite, divided into 3 leaflets, the leaflets ovate, 4–8 cm long and 2.5–5 cm wide; margins sharp-toothed or lobed; petioles 5–9 cm long. **Male and female flowers** separate and on separate plants, in many-flowered, open clusters from leaf axils, on stalks 1–8 cm long, usually shorter than leaf petioles; sepals 4, creamy-white, 6–10 mm long; petals absent. **Fruit** a rounded head of hairy brown achenes tipped with feathery, persistent styles 2.5–4 cm long. July–Sept. **Habitat** Thickets, streambanks, moist to wet woods, rocky slopes.

Clematis virginiana
VIRGIN'S BOWER

COPTIS | *Goldthread*

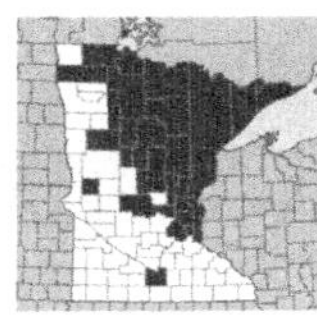

Coptis trifolia (L.) Salisb.
ALASKA GOLDTHREAD
NC **FACW** | MW **FACW** | GP **FACW**

Native perennial herb, with slender, bright yellow rhizomes. **Leaves** from base of plant on long petioles, evergreen, divided into 3-leaflets, the leaflets shallowly lobed, with rounded teeth tipped by an abrupt point. **Flowers** single, white, 10–15 mm wide, on a stalk 5–15 cm long from base of plant; sepals 4–7, petal-like; petals absent; pistils 3–7, narrowed to a short, slender style. **Fruit** a beaked follicle 8–13 mm long. May–June. **Synonyms** *Coptis groenlandica*. **Habitat** Wet conifer woods and swamps, often on mossy hummocks.

Coptis trifolia
ALASKA GOLDTHREAD

MYOSURUS | *Mouse-tail*

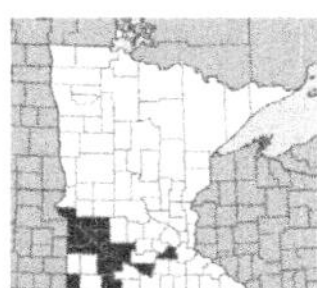

Myosurus minimus L.
MOUSE-TAIL
NC **FAC** | MW **FACW** | GP **FACW**

Small, inconspicuous, annual herb; native. **Stems** 4–15 cm long. **Leaves** in a basal tuft, hairless, linear, mostly less than 1 mm wide. **Flowers** small, few to many, in a spike above leaves on a slender stalk to 6 cm long when mature; sepals 5, green, upright, with a spur at base; petals 5 or sometimes absent, small, white or pink; stamens 5–10; pistils many, in an elongate receptacle. **Fruit** an achene, 2–3 mm long and 1 mm wide, with a small beak. April–June. **Habitat** Wet to moist places such as streambanks and floodplains, sometimes temporarily in shallow water; also in disturbed drier areas. In Minnesota, most common in sw near near Minnesota River.

RANUNCULUS | *Buttercup; Crowfoot; Spearwort*

Aquatic, semi-aquatic, or terrestrial annual and perennial herbs. **Stems** erect to sprawling, sometimes floating in water. **Leaves** simple, or compound and finely dissected, often variable on same plant; alternate on stem or all from base of plant; petioles short to long. **Flowers** borne above water surface in aquatic species; sepals usually 5, green; petals usually 5, yellow or white, often fading to white, usually with a small nectary pit covered by a scale near base of petal; stamens and pistils numerous. **Achenes** many in a round or cylindric head; achene body thick or flattened, tipped with a beak.

RANUNCULUS | BUTTERCUP; CROWFOOT; SPEARWORT

1 Flowers white; leaves divided into linear or threadlike segments; plants typically aquatic .... *Ranunculus aquatilis* | WHITE WATER-CROWFOOT
1 Flowers yellow; leaves simple to deeply lobed or divided into narrow segments; plants aquatic, emergent, or terrestrial 2

2 Sepals 3 (rarely 4); uncommon in northern Minnesota .... *Ranunculus lapponicus* | LAPPLAND BUTTERCUP
2 Sepals 5 (or rarely more).. 3

3 All leaves simple and entire, or shallowly lobed with rounded teeth 4
3 All, or at least stem leaves, deeply lobed, divided or compound......... 5

4 Leaves ovate to round or kidney-shaped, shallowly lobed with rounded teeth; achenes with longitudinal ribs .... *Ranunculus cymbalaria* .... SEASIDE CROWFOOT
4 Leaves oval to lance-shaped or linear, entire to sharp-toothed; achenes not ribbed .... *Ranunculus flammula* | CREEPING SPEARWORT

5 Basal and stem leaves distinctly different in shape, the basal leaves mostly entire or with rounded teeth, the stem leaves deeply divided............ .... *Ranunculus abortivus* | SMALL-FLOWERED CROWFOOT
5 Basal and stem leaves similar, all deeply lobed, divided or compound .. 6

6 Achenes swollen, without a sharp margin 7
6 Achenes flattened, with a sharp or winglike margin 9

7 Petals 2–4 mm long; achenes to 1.2 mm long, nearly beakless; plants terrestrial or in water only part of season .... *Ranunculus sceleratus* .... CURSED CROWFOOT
7 Petals 4–14 mm long; achenes 1.2–2.5 mm long, beaked; plants underwater or exposed later in season 8

8 Petals more than 7 mm long; achene body more than 1.6 mm long, achene margin thickened and white-corky below middle. . . . . *Ranunculus flabellaris* | YELLOW WATER-CROWFOOT

8 Petals less than 7 mm long; achene body less than 1.6 mm long, achene margin rounded but not thickened . . . . *Ranunculus gmelinii* . . . . SMALL YELLOW WATER-CROWFOOT

9 Petals 7–15 mm long; stems often recurved and rooting at nodes **10**

9 Petals 2–5 mm long; stems not rooting at nodes **11**

10 Leaves deeply divided but outer segment not on a petiole; style short, curved; plants weedy . . *Ranunculus acris* | COMMON OR TALL BUTTERCUP

10 Leaves compound, the outermost lobe on a petiole; style long and straight; plants not weedy . . . . *Ranunculus hispidus* . . . . NORTHERN SWAMP BUTTERCUP

11 Beak of achene strongly hooked . . . . *Ranunculus recurvatus* . . . . HOOKED CROWFOOT

11 Beak of achene straight or only slightly curved. **12**

12 Petals shorter than sepals; achenes in cylindric heads longer than wide; widespread species. . . . . . *Ranunculus pensylvanicus* | BRISTLY CROWFOOT

12 Petals equal or longer than sepals; heads ovate or round; n Minnesota . . . . *Ranunculus macounii* | MARSH CROWFOOT

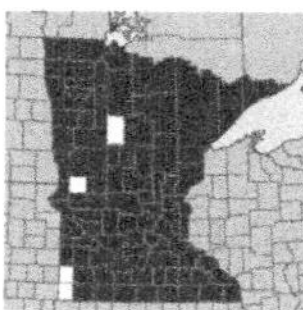

Ranunculus abortivus L.

SMALL-FLOWERED CROWFOOT

NC **FAC** | MW **FACW** | GP **FAC**

Native biennial or perennial herb. **Stems** upright, 2–5 dm long, branched above, smooth or with fine hairs. **Leaves** at base of plant round to kidney-shaped, margins with rounded teeth, some leaves lobed; petioles long; stem leaves 3–5-divided into linear segments, margins entire or broadly toothed, petioles absent. **Flowers** yellow, petals 2–3 mm long, shorter than sepals. **Achenes** in a short, round head; achene body swollen, 1–2 mm long, with a very short, curved beak. April–June. **Habitat** Common; wet to moist woods, floodplains, wet meadows, thickets, ditches; especially where soils disturbed or compacted.

Ranunculus abortivus
SMALL-FLOWERED CROWFOOT

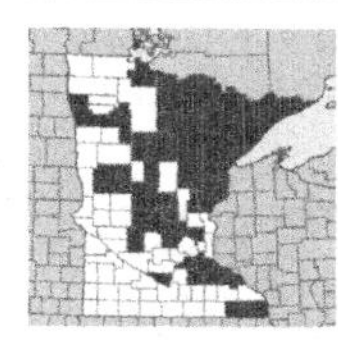

Ranunculus acris L.

COMMON OR TALL BUTTERCUP
NC **FAC** | MW **FAC** | GP **FACW**

Introduced perennial herb, with fibrous roots. **Stems** hairy, to 1 m long, with few branches, most leaves on lower part of stem. **Leaves** kidney-shaped, deeply 3–7-divided, the segments again lobed or dissected; branch leaves much smaller, 3-parted. **Flowers** numerous; sepals 5, half length of petals; petals 5, bright yellow, 6–15 mm long, obovate, often with a rounded notch at tip. **Achenes** in a round head; achene body flat, 2–3 mm long, beak 0.5 mm long. June–Aug. **Habitat** Common weed of fields, thickets, ditches and shores. Introduced from Europe; now throughout e USA.

Ranunculus acris
COMMON OR TALL BUTTERCUP

fruit

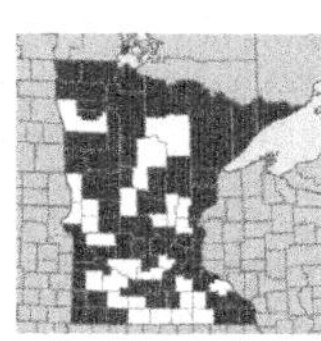

Ranunculus aquatilis L.

WHITE WATER-CROWFOOT
NC **OBL** | MW **OBL** | GP **OBL**

Native perennial aquatic herb; plants mostly smooth. **Stems** underwater or floating (sometimes stranded on muddy shores in late summer), 2–8 dm long, unbranched or with a few branches, rooting from lower nodes. **Leaves** round to kidney-shaped in outline, 2–3x divided into narrow threadlike segments 1–2 cm long; stiff and not collapsing when removed from water; petioles absent or to 4 mm long. **Flowers** at or below water surface, single from upper leaf axils, 1–1.5 cm wide; sepals 5, purple-green, spreading, 2–4 mm long; petals 5, white, yellow at base, 4–9 mm long. **Achenes** many in a round

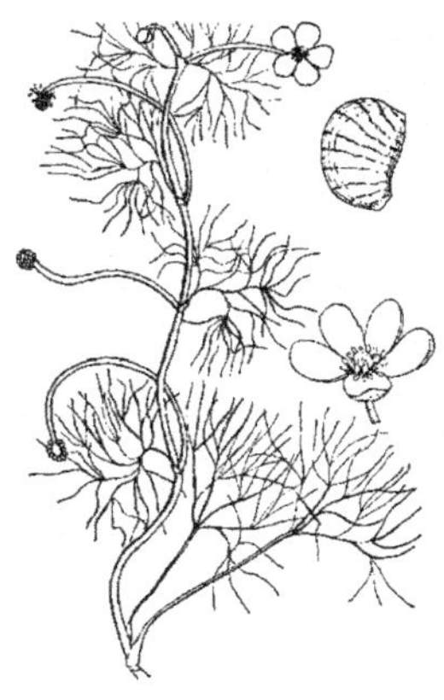

head; achene body obovate, ridged, the beak thin and straight, 1–1.5 mm long. May–Aug. **Synonyms** *Ranunculus trichophyllus*. **Habitat** Ponds, lakes, streams, rivers and ditches.

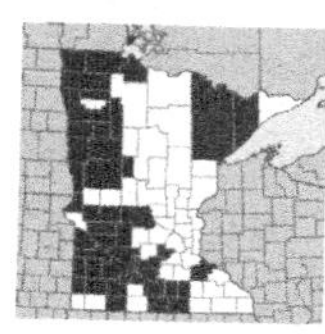

Ranunculus cymbalaria Pursh
SEASIDE CROWFOOT
NC **OBL** | MW **OBL** | GP **OBL**

Native perennial herb, spreading by stolons and forming dense mats. **Stems** 3–20 cm long, smooth. **Leaves** all from base of plant, ovate to kidney-shaped, 5–25 mm long and 4–30 mm wide, heart-shaped at base; margins with rounded teeth, often with 3 prominent lobes at tip; petioles sparsely hairy. **Flower stalks** longer than leaves, unbranched or with a few branches, with 1 to several flowers; sepals 5, green-yellow, 3–5 mm long, deciduous; petals usually 5, yellow, turning white with age, 3–5 mm long; stamens 10–30. **Achenes** numerous in a cylindric head to 10 mm long; achene body 1.5–2 mm long, longitudinally nerved, beak short and straight. June–Sept. **Habitat** Wet meadows, streambanks, shores, ditches and seeps, often in wet mud or sand; often where brackish; found throughout much of North and South America; Eurasia.

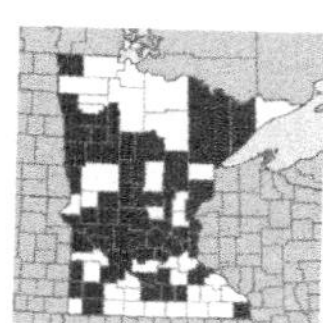

Ranunculus flabellaris Raf.
YELLOW WATER-CROWFOOT
NC **OBL** | MW **OBL** | GP **OBL**

Native perennial herb; plants smooth or sometimes hairy when growing out-of-water. **Stems** floating, or upright from a sprawling base when exposed, branched, rooting at lower nodes, 3–7 dm long. Underwater leaves 3-parted into linear segments 1–2 mm wide, exposed leaves (when present) round to kidney-shaped in outline, 2–10 cm long and 2–12 cm wide, divided into 3 segments, the segments again 3-divided. **Flowers** 1 to several at ends of stems; sepals 5, green-yellow, 4–8 mm long; petals 5–8, bright yellow, 6–15 mm long. **Achenes** 50–75 in a round to ovate head; achene body obovate, to 2 mm long, the margin thickened and corky below middle, beak broad, flat, 1–1.5 mm long. May–July. **Habitat** Shallow water or muddy shores of ponds, quiet streams, swamps, woodland pools, marshes and ditches.

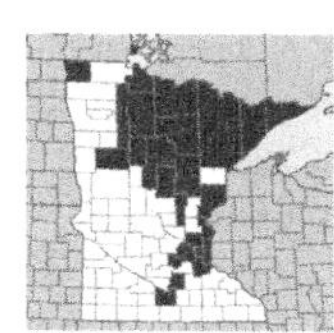

Ranunculus flammula L.
CREEPING SPEARWORT
NC **FACW** | MW **FACW** | GP **FACW**

Native perennial herb, spreading by stolons; plants often covered with appressed hairs. **Stems** sprawling, rooting at nodes, unbranched or few-branched, with upright shoots 4–15 cm long. **Leaves** in small clusters at nodes, simple, linear or threadlike, 1–5 cm long and 1.5 mm wide, margins ± entire; upper leaves smaller and with shorter petioles than lower. **Flowers** single at ends of stems; sepals 5, yellow-green, 2–4 mm long, with stiff hairs; petals 5, yellow, obovate, 3–5 mm long. **Achenes** 10–25 in a round head; achene body swollen, obovate, 1–1.5 mm long, smooth, the beak short, to 0.5 mm long. June–Aug. **Habitat** Sandy, gravelly, or muddy shores; shallow to deep water, water usually acidic.

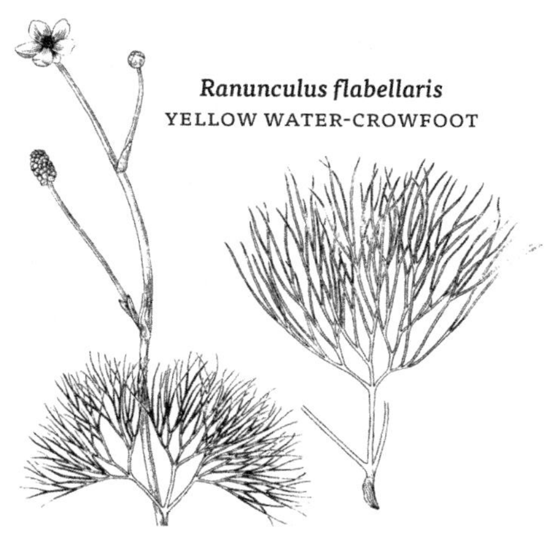

Ranunculus flabellaris
YELLOW WATER-CROWFOOT

Ranunculus flammula
CREEPING SPEARWORT

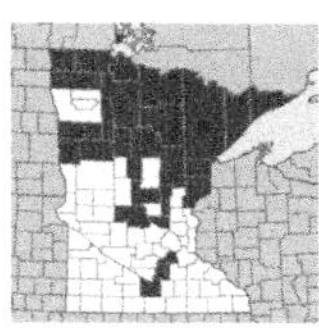

Ranunculus gmelinii DC.
SMALL YELLOW WATER-CROWFOOT
NC **FACW** | MW **FACW** | GP **FACW**

Native perennial herb, similar to **yellow water-crowfoot** (*Ranunculus flabellaris*) but plants aquatic or at least partly underwater; smooth or sometimes with coarse hairs. **Stems** usually sprawling and rooting at nodes, 1–5 dm long, sparsely branched. **Leaves** all on stem or with a few basal leaves on long petioles, deeply 3-lobed or dissected, the segments again forked 2–3 times; underwater leaf segments 2–4 mm wide; exposed leaves to 2 cm long and 1.5–2.5 cm wide. **Flowers** usually 1 to several at ends of stems; sepals 5, green-yellow, 3–6 mm long; petals 5–8, yellow, 4–8 mm long. **Achenes** 50–70 in a round to ovate head; achene body obovate, 1–1.5 mm long, the margin rounded, not corky-thickened, the beak broad and thin, 0.4–0.7 mm long, somewhat curved. May–Aug. **Habitat** Streambanks and lakeshores, springs, pools in swamps and bogs.

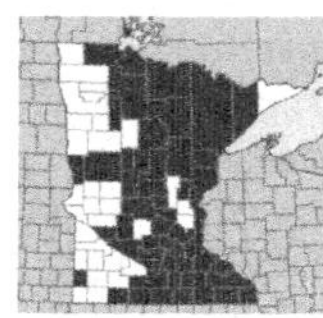

Ranunculus hispidus Michx.
NORTHERN SWAMP BUTTERCUP
NC **FAC** | MW **FAC** | GP **FACW**

Native perennial herb; stems and leaves variable. **Stems** upright, 2–9 dm long, smooth or strongly coarse-hairy. **Leaves** from base of plant and on stems, the basal leaves larger and with longer petioles than stem leaves; 3-lobed, heart-shaped in outline, 3–14 cm long and 4–20 cm wide, with appressed hairs on veins, upper leaves usually strongly toothed. **Flowers** 1 to several; sepals 5, yellow-green, 5–11 mm long, hairy; petals 5–8, yellow, fading to white, 7–15 mm long and 3–10 mm wide. **Achenes** 15–30 or more in a round head; achene body obovate, 2–4 mm long, smooth, winged on margin, the beak straight, 2–3 mm long. May–July. **Synonyms** *Ranunculus septentrionalis*. **Habitat** Wet woods, floodplains and swamps, thickets, lakeshores, wet meadows and fens.

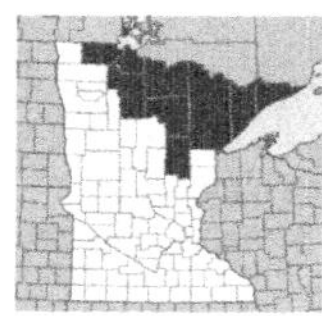

Ranunculus lapponicus L.
LAPPLAND BUTTERCUP
NC **OBL** | GP **OBL** | *special concern*

Native perennial herb, spreading by rhizomes. **Stems** prostrate, 1–2 dm long, sending up 1 shoot from each node, the shoots with 1–2 basal leaves, sometimes with a single smaller leaf above. **Leaves** kidney-shaped, deeply 3-cleft, margins with rounded teeth or shallowly lobed. **Flowers** single at ends of shoots; petals yellow with orange veins, 8–12 mm wide; sepals 3, curved downward. **Achenes** in a round head; achene body 2–3 mm long, swollen near base, flattened above, beak slender, sharply hooked. June–July. **Habitat** Cedar swamps and bogs. **Status** Special concern.

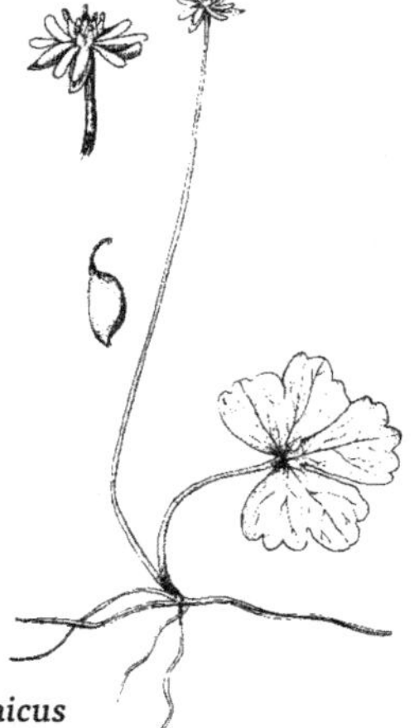

Ranunculus lapponicus
LAPPLAND BUTTERCUP

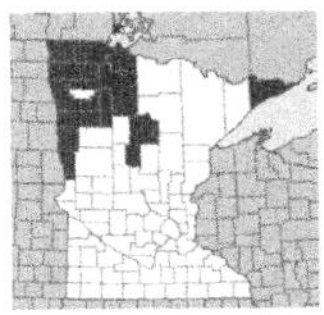

Ranunculus macounii Britton
MARSH CROWFOOT
NC **OBL** | MW **OBL** | GP **OBL**

Native annual or short-lived perennial herb, similar to bristly crowfoot (*Ranunculus pensylvanicus*) but less common; plants smooth to densely hairy. **Stems** erect or reclining, hollow, 2–7 dm long, branched, the branches again branched. **Leaves** from base of plant and on stems, the basal leaves larger and with longer petioles than stem leaves; triangular in outline, 4–14 cm long and 6–16 cm wide, 3-lobed or divided into 3 segments, the segments themselves 3-lobed and coarsely toothed. **Flowers** several at ends of branches; sepals 5, yellow, 3-5 mm long; petals 5, yellow, 3–5 mm long, equal or longer than sepals; stamens 15–35. **Achenes** 30–50 in an ovate to round head; achene body flat, 3 mm long, smooth or shallowly pitted, with a stout, slightly curved or straight beak 1–2 mm long. June–Aug. **Habitat** Wet meadows, marshes, shores, streambanks and ditches.

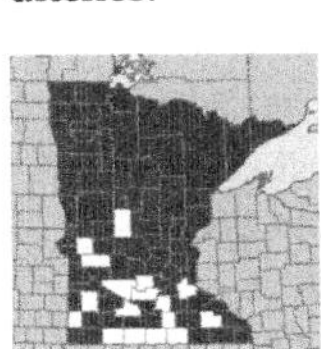

Ranunculus pensylvanicus L. f.
BRISTLY CROWFOOT
NC **OBL** | MW **OBL** | GP **FACW**

Native annual or short-lived perennial herb. **Stems** erect, hollow, 3–8 dm long, branched or unbranched. **Leaves** at base of plant withering early, larger and with longer petioles than the few stem leaves; 4–12 cm long and 4–15 cm wide, with appressed hairs, 3-lobed and coarsely toothed, the terminal leaflet stalked. **Flowers** few, on short stalks; sepals 5, yellow, 4–5 mm long; petals 5, pale yellow, fading to white, shorter than the sepals, 2–4 mm long; stamens 15–20. **Achenes** many, in a rounded cylindric head 10–15 mm long; achene body flattened, 2–3 mm long, smooth, the beak stout, 0.5–1.5 mm long. July–Aug. **Habitat** Marshes, wet meadows, ditches and streambanks, often in muck.

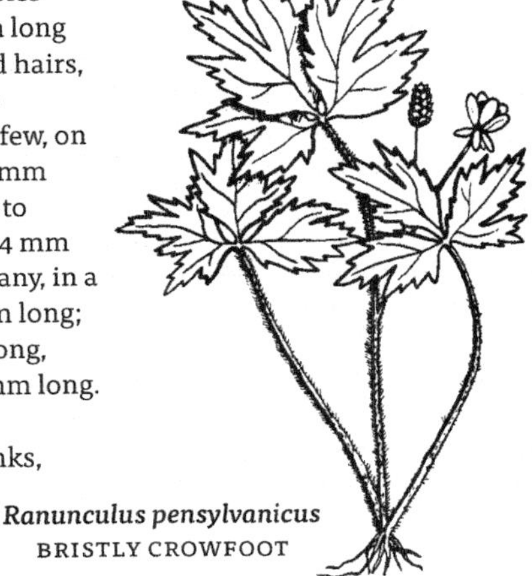

Ranunculus pensylvanicus
BRISTLY CROWFOOT

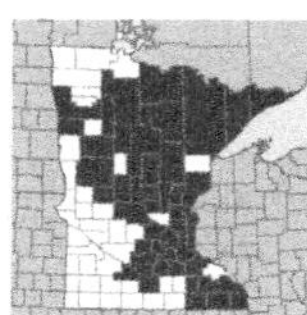

Ranunculus recurvatus Poiret
HOOKED CROWFOOT
NC **FACW** | MW **FACW** | GP **FACW**

Native perennial herb. **Stems** 2–7 dm long, usually hairy, branches few. **Leaves** broadly kidney-shaped or round in outline, 3-parted to below middle, covered with long, soft hairs; petioles present on all but uppermost leaves. **Flowers** on stalks at ends of stems; sepals curved downward, to 6 mm long; petals pale yellow, 4–6 mm long; styles strongly hooked. **Achenes** in a short-cylindric head; achene body flat, round, sharp-margined, to 2 mm long; beak 1 mm long, hooked or coiled. May–June. **Habitat** Moist deciduous forests (especially in openings), swamps; also in drier woods; s in Minnesota, also in partial shade in calcareous fens.

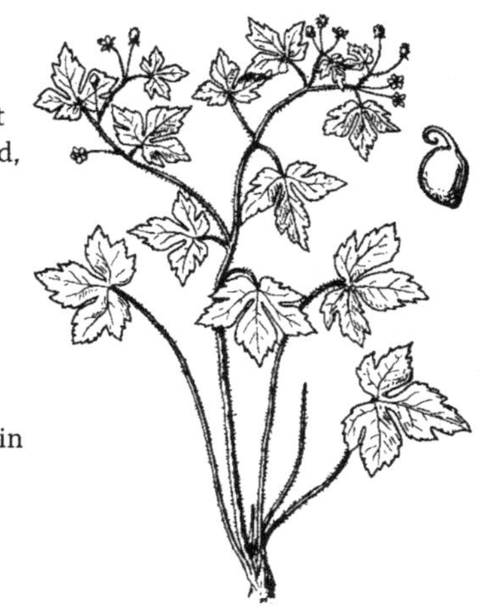

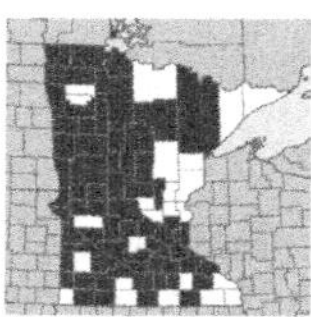

Ranunculus sceleratus L.
CURSED CROWFOOT
NC **OBL** | MW **OBL** | GP **OBL**

Annual herb, native but often weedy; plants smooth, sometimes partly submersed in shallow water. **Stems** upright, hollow, 1–6 dm long, branched above and with many flowers. **Leaves** from base of plant less deeply parted and with longer petioles than stem leaves; upper stem leaves small; leaves deeply 3-parted, the main lobes again lobed, heart-shaped at base, rounded at tip, 1–6 cm long and 3–8 cm wide. **Flowers** numerous at ends of stalks from upper leaf axils and branches; sepals 5, 2–3 mm long, yellow-green, tips curved downward; petals 5, light yellow, fading to white, 3–5 mm long. **Achenes** numerous in a short-cylindric head 4–11 mm long; achene body obovate, 1 mm long, slightly corky-thickened on margins; beak tiny, blunt. May–Sept. **Habitat** Muddy shores, streambanks, wet meadows, ditches, marshes and other wet places.

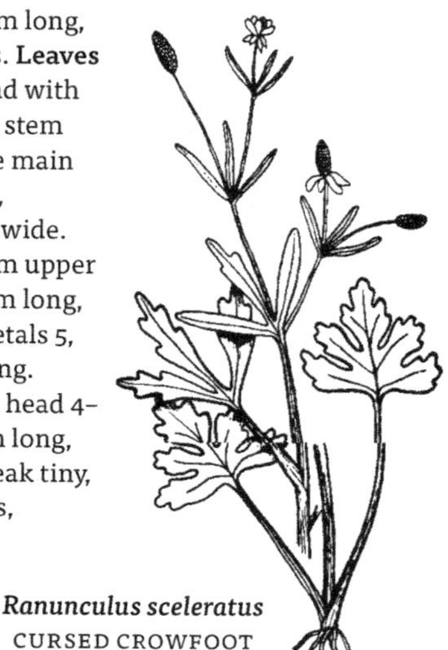

Ranunculus sceleratus
CURSED CROWFOOT

THALICTRUM | *Meadow-rue*

Perennial herbs. **Leaves** alternate, compound. Male and female **flowers** separate, in panicles on separate plants; sepals 4–5, green or petal-like but soon deciduous; petals absent; stamens numerous, the stalks (filaments) long and slender; pistils several to many. **Fruit** a ribbed or nerved achene.

THALICTRUM | MEADOW-RUE

1 Leaflets mostly with 4 or more teeth or lobes, often appearing 3-lobed, each lobe tipped with 1–3 teeth . . . . *Thalictrum venulosum* . . . . NORTHERN MEADOW-RUE

1 Leaflets with 2 or 3 lobes, the lobes usually not toothed 2

2 Underside of leaflets with very short hairs (rarely smooth); not glandular; leaves odorless; common . . . . *Thalictrum dasycarpum* . . . . PURPLE MEADOW-RUE

2 Underside of leaflets with small beads and hairs tipped with gray or amber exudate; leaves with strong odor when crushed; uncommon (more frequent south of Minnesota). . . . . *Thalictrum revolutum* . . . . SKUNK MEADOW-RUE

Thalictrum dasycarpum Fischer & Ave-Lall.

PURPLE MEADOW-RUE
NC **FACW** | MW **FACW** | GP **FAC**

Native perennial herb, from a short rootstock. **Stems** purple-tinged, 1–2 m long, branched above. **Leaves** divided into 3–4 groups of leaflets; leaflets 15 mm or more long, mostly tipped with 3 pointed lobes, dark green above, underside sparsely short-hairy, not waxy and without gland-tipped hairs; margins usually slightly turned under; stem leaves mostly without petioles. **Flowers** in panicles at ends of stems; male and female flowers separate and on different plants (sometimes with some perfect flowers); sepals 3–5 mm long, lance-shaped; anthers linear and sharp-tipped, 2–3 mm long, filaments white; stigmas straight, 2–4 mm long. **Achenes** 4–6 mm long, ribbed, in a round cluster. June–July. **Habitat** Wet to moist meadows, low prairie, swamps, thickets, streambanks.

Thalictrum revolutum DC.
SKUNK MEADOW-RUE
NC **FAC** | MW **FAC** | GP **FACW**

Native perennial herb, from short rootstocks, with strong odor when crushed. **Stems** ± smooth, often purple-tinged, 0.5–1.5 m long. Lowest leaves on petioles, middle and upper leaves stalkless; leaves divided into 3–4 groups of leaflets; leaflets variable in shape and size, usually 3-lobed, some 1–2 lobed, upper surface smooth, underside leathery and conspicuously net-veined, finely hairy with gland-tipped hairs, margins turned under. **Flowers** in panicles at ends of stems; male and female flowers separate and on different plants (sometimes with some perfect flowers); anthers linear, 2–3 mm long, filaments threadlike, 2–5 mm long; pistils 6–12, stigmas 2–3 mm long. **Fruit** an oval or lance-shaped achene, 4–5 mm long, ridged, with tiny gland-tipped hairs. June–July. **Habitat** Streambanks, thickets, moist meadows and prairies.

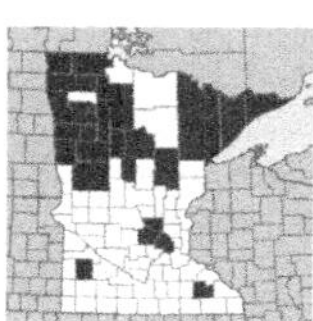

Thalictrum venulosum Trelease
NORTHERN MEADOW-RUE
NC **FACW** | MW **FAC** | GP **FAC**

Native perennial herb, spreading by rhizomes; plants pale green, waxy. **Stems** erect, 3–10 dm long. **Leaves** divided into 3–4 groups of leaflets; leaflets firm, nearly circular or obovate in outline, tipped by 3–5 lobes, underside veiny, appearing wrinkled, usually sparsely covered with gland-tipped hairs; lower leaves on petioles, upper leaves stalkless. **Flowers** in narrow panicles at ends of stems, the panicle branches nearly erect; male and female flowers separate and on different plants; stamens 8–20, anthers linear and pointed at tip, filaments slender. **Fruit** an ovate achene, 4–6 mm long, tapered to a short-beak. June–July. **Synonyms** Ours are var. *confine*. *Thalictrum confine*.
Habitat Streambanks, thickets, shores.

Distinguished from **skunk meadow-rue** *(Thalictrum revolutum) by its less glandular leaflets and its elongate horizontal rhizomes (Thalictrum revolutum has an erect rootstock).*

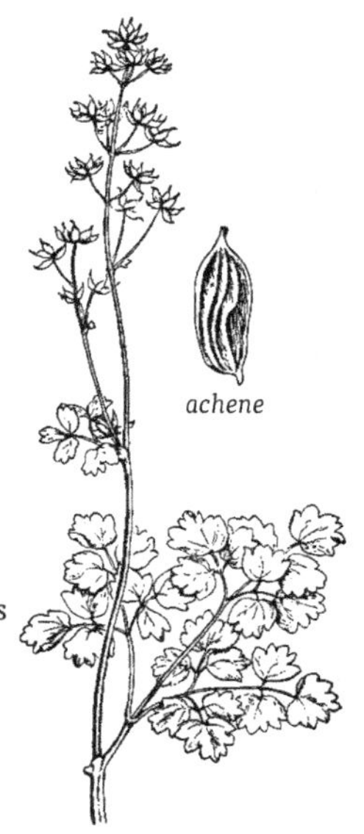

Thalictrum venulosum
NORTHERN MEADOW-RUE

RHAMNUS | *Buckthorn*

Shrubs or small trees. **Leaves** simple, alternate, pinnately veined, usually with stipules. **Flowers** perfect, or male or female, regular, single or few from leaf axils; sepals joined, 4- or 5-parted; petals 4 or 5. **Fruit** a purple-black, berrylike drupe with 2–4, 1-seeded stones.

RHAMNUS | BUCKTHORN FAMILY

1 Leaf margins ± entire; flowers perfect . . . . *Rhamnus frangula* . . . . GLOSSY BUCKTHORN

1 Leaf margins with small, forward-pointing teeth; flowers either male or female . . . . *Rhamnus alnifolia* | ALDER-LEAF BUCKTHORN

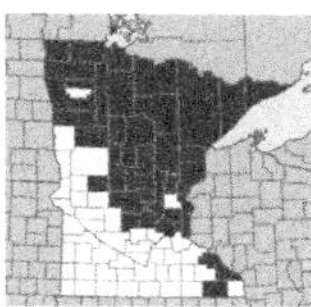

Rhamnus alnifolia L'Her.

ALDER-LEAF BUCKTHORN

NC **OBL** | MW **OBL** | GP **FACW**

Native shrub to 1 m tall, forming low thickets. **Leaves** alternate, oval to ovate, 6–10 cm long and 3–5 cm wide, green above, paler green below; margins with low, rounded teeth; petioles grooved, 5–12 mm long; stipules linear, to 1 cm long, deciduous before fruits mature. **Flowers** appearing with leaves in spring, in clusters of 1–3 flowers from leaf axils; yellow-green, usually 5-parted, 3 mm wide, on short stalks, with both stamens and pistils but one or other is nonfunctional, sepals 1–2 mm long, petals absent. **Fruit** a purple-black, berrylike drupe, 6–8 mm wide, with 1–3 nutletlike stones. May–June. **Habitat** Native, non-invasive species of conifer swamps, thickets, sedge meadows, wet depressions in deciduous forests; usually where calcium-rich.

Rhamnus alnifolia
ALDER-LEAF BUCKTHORN

Rhamnus frangula L.
GLOSSY BUCKTHORN
NC **FAC** | MW **FACW** | GP **FAC**

Introduced shrub or small tree to 5 m tall, often invasive. **Stems** with pale lenticels. **Leaves** mostly alternate but some leaves often nearly opposite, oval or obovate, 5–8 cm long and 3–5 cm wide, tapered to a blunt or sharp tip; margins entire or slightly wavy; petioles stout, 1–2 cm long. **Flowers** appearing after leaves in spring, single or in clusters of 2–8 in leaf axils, perfect, green-yellow, 5-parted, to 5 mm wide; petals 1–2 mm long. **Fruit** a purple-black, berrylike drupe, 7 mm wide, with 2–3 nutlike stones. May–Aug. **Synonyms** *Frangula alnus*. **Habitat** Conifer swamps, thickets, calcareous fens, lakeshores, especially where disturbed or cleared; also invading drier woods. Introduced from Eurasia; escaping from cultivation into disturbed wetlands in ne and c North America.

Rhamnus frangula
GLOSSY BUCKTHORN

ROSACEAE
Rose Family

SHRUBS, OR PERENNIAL, BIENNIAL, OR ANNUAL HERBS. Leaves evergreen or deciduous, mostly alternate and simple or compound. **Flowers** perfect (with both male and female parts), regular, with 5 sepals and petals; stamens numerous. **Fruit** an achene, capsule, or fleshy fruit with numerous embedded seeds (drupe, as in a strawberry), or a fleshy fruit with seeds within (pome, as in apples).

ROSACEAE | ROSE FAMILY

1 Leaves simple 2
1 Leaves divided into leaflets 5

2 Fruit dry; ovary superior 3
2 Fruit fleshy; ovary inferior 4

3 Leaves 3–5 lobed . . . . *Physocarpus opulifolius* | EASTERN NINEBARK
3 Leaves simple . . . . *Spiraea* | SPIRAEA

4 Leaf margins with gland-tipped teeth; flowers less than 1 cm wide . . . . *Aronia melanocarpa* | BLACK CHOKEBERRY
4 Leaves without gland-tipped teeth; flowers 1 cm or more wide . . . . *Amelanchier bartramiana* | MOUNTAIN JUNEBERRY

5 Plants shrubs or brambles 6
5 Plants herbs 7

6 Plants without thorns or bristles; flowers yellow . . . . . . *Potentilla fruticosa* . . . . SHRUBBY CINQUEFOIL
6 Plants with thorns or bristles; flowers white or pink . . . . *Rubus* . . . . RASPBERRY; DEWBERRY

7 Leaves divided into 3 equal leaflets . . . . *Rubus pubescens* . . . . DWARF RASPBERRY
7 Leaves not divided into 3 equal leaflets 8

8 Leaves deeply parted or divided; styles long, jointed near middle . . . *Geum* . . . . AVENS
8 Leaves pinnately or palmately divided; styles short, not jointed . . . . *Potentilla* | CINQUEFOIL

Potentilla fruticosa
SHRUBBY CINQUEFOIL

AMELANCHIER | *Serviceberry, Shadbush*

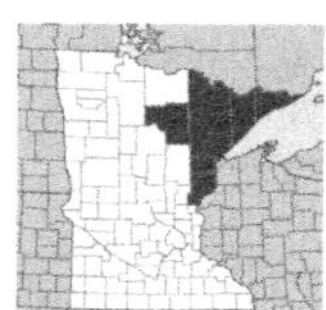

Amelanchier bartramiana (Tausch) Roemer
MOUNTAIN JUNEBERRY
NC **FAC**

Native shrub to 2 m tall, often forming clumps; **twigs** purplish, ± smooth. **Leaves** alternate, ovate to oval, 2–5 cm long and 1–2.5 cm wide, tapered to a blunt or sharp tip and often tipped with a small spine, green above, paler below, often purple-tinged when unfolding; margins with small, sharp, forward-pointing teeth; petioles to 1 cm long. **Flowers** 1 cm or more wide, single or in groups of 2–4 at ends of branches or on stalks 1–2 cm long from leaf axils; sepals lance-shaped; petals white, oval to oblong, 6–10 mm long. **Fruit** a dark purple, edible, berrylike pome, 1 cm long. May–Aug. **Habitat** Conifer swamps, open bogs, thickets.

Hybrids are frequent between various species of ***Amelanchier****, making positive identification difficult. Most other species of serviceberry occur in drier habitats, especially where rocky or sandy.*

ARONIA | *Chokeberry*

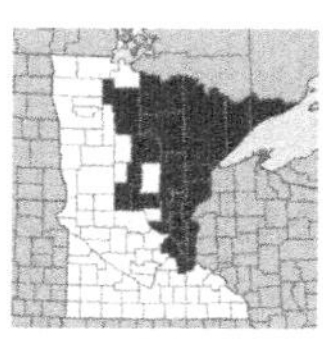

Aronia melanocarpa (Michx.) Elliott
BLACK CHOKEBERRY
NC **FAC** | MW **FACW** | GP **OBL**

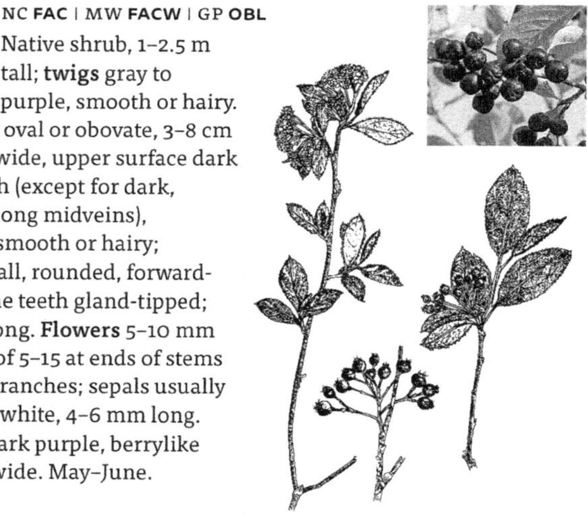

Native shrub, 1–2.5 m tall; **twigs** gray to purple, smooth or hairy. **Leaves** alternate, oval or obovate, 3–8 cm long and 1–4 cm wide, upper surface dark green and smooth (except for dark, hairlike glands along midveins), underside paler, smooth or hairy; margins with small, rounded, forward-pointing teeth, the teeth gland-tipped; petioles to 1 cm long. **Flowers** 5–10 mm wide, in clusters of 5–15 at ends of stems and short, leafy branches; sepals usually glandular; petals white, 4–6 mm long. **Fruit** an edible, dark purple, berrylike pome, 5–10 mm wide. May–June.

Synonyms *Aronia arbutifolia, Aronia prunifolia, Photinia melanocarpa*. **Habitat** Tamarack swamps, open bogs, thickets, marshes and shores.

Aronia melanocarpa
BLACK CHOKEBERRY

GEUM | *Avens*

Perennial herbs. **Lower leaves** pinnately lobed or divided, **upper leaves** smaller, less divided or entire. **Flowers** yellow, white or purple; 1 to many in clusters at ends of stems; petals 5; stamens 10 to many. **Fruit** an achene.

GEUM | AVENS

1 Flowers nodding; sepals red-purple, upright or ascending at flowering time .... *Geum rivale* | PURPLE AVENS
1 Flowers erect; sepals green, curved downward at tip **2**

2 Petals white .... *Geum laciniatum* | ROUGH AVENS
2 Petals yellow **3**

3 Terminal and lateral segment of basal leaves similar in size and shape .... *Geum allepicum* | YELLOW AVENS
3 Terminal segment much larger than lateral segments .... *Geum macrophyllum* | LARGE-LEAF AVENS

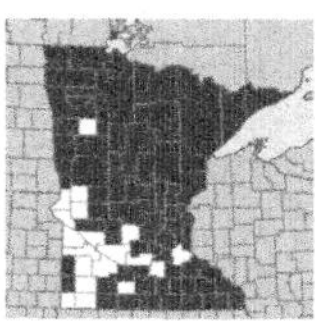

Geum allepicum Jacq.

YELLOW AVENS
NC **FAC** | MW **FACW** | GP **FACU**

Native perennial herb. **Stems** erect or ascending, to 1 m long, branched above, covered with coarse hairs. **Leaves** variable, basal leaves pinnately divided into 5–7 oblong leaflets, wedge-shaped at base, petioles long-hairy; stem leaves divided into 3–5 segments, stalkless or short-petioled; margins coarsely toothed. **Flowers** 1 to several, short-stalked, on branches at ends of stems; sepals lance-shaped; petals 5, yellow; style jointed. **Fruit** an achene, usually with long hairs. June–July. **Habitat** Swamps, wet forests, wet meadows, marshes, calcareous fens, ditches and roadsides.

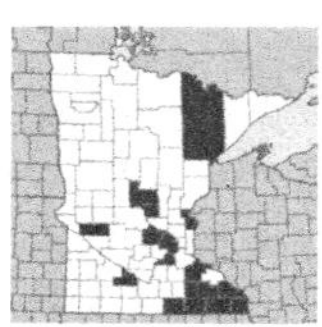

Geum laciniatum Murray
ROUGH AVENS
NC **FACW** | MW **FACW** | GP **FACW**

Native perennial herb. **Stems** 4–10 dm long, covered with long, mostly downward-pointing hairs. Lower leaves pinnately divided, the segments pinnately lobed; upper leaves divided into 3 leaflets or lobes; margins coarsely toothed; petioles hairy. **Flowers** mostly single at ends of densely hairy stalks from ends of stems; sepals triangular, 4–10 mm long; petals 5, white, 3–5 mm long. **Fruit** an achene, 3–5 mm long (excluding style), grouped into round heads 1–2 cm long. May–June. **Habitat** Wet woods, floodplain forests, ditches.

Geum laciniatum
ROUGH AVENS

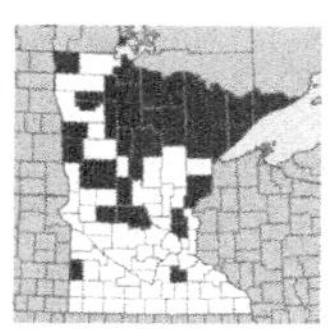

Geum macrophyllum Willd.
LARGE-LEAF AVENS
NC **FACW** | MW **FACW** | GP **FACW**

Native perennial herb. **Stems** to 1 m long, unbranched, or branched above, bristly-hairy. **Leaves** pinnately divided, basal leaves stalked, the terminal segment large, 3–7-lobed, with much smaller segments intermixed; stem leaves smaller, deeply 3-lobed or divided into 3 leaflets, short-stalked or stalkless; margins sharply toothed. **Flowers** 1 to several on branches at ends of stems; sepals triangular, bent backward; petals yellow, obovate, 4–7 mm long; style jointed. **Fruit** a finely hairy achene. May–July. **Habitat** Moist to wet forest openings, streambanks, wet meadows, ditches.

Geum macrophyllum
LARGE-LEAF AVENS

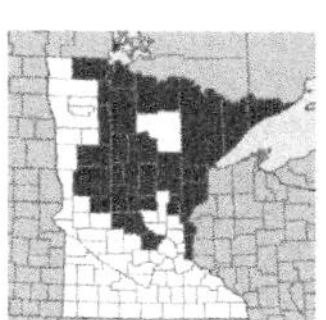

Geum rivale L.
PURPLE AVENS
NC **OBL** | MW **OBL** | GP **FACW**

Native perennial herb. **Stems** erect, 3–8 dm long, mostly unbranched, hairy. **Basal leaves** large, 1–4 dm long, pinnately divided, the terminal 1–3 leaflets much larger than other segments, stalked; stem leaves smaller, 2–5 on stem, pinnately divided or 3-lobed, short-stalked or stalkless; margins shallowly lobed and coarsely toothed. **Flowers** mostly nodding, few on branches at ends of stems, the branches with short, gland-tipped hairs and long, coarse hairs; sepals 5, purple, triangular, 6–10 mm long, ascending; petals 5, yellow to pink with purple veins, tapered to a clawlike base; stamens many; styles jointed above middle, the portion above joint deciduous, lower portion persistent and curved in fruit. **Fruit** a long-beaked, hairy achene, 3–4 mm long, grouped into round heads. May–July. **Habitat** Conifer swamps, wet forests, bogs, fens, wet meadows; often where calcium-rich.

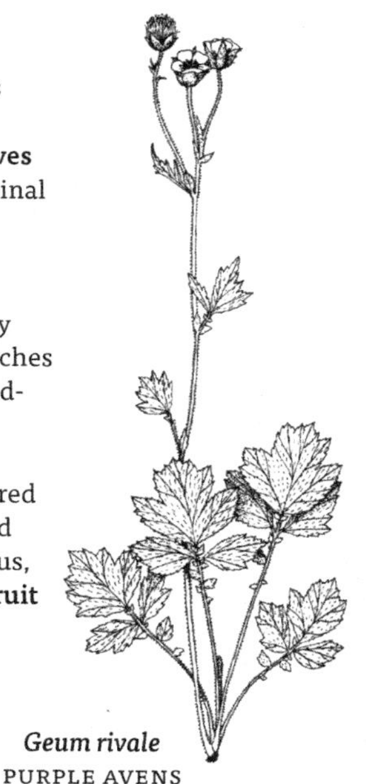

Geum rivale
PURPLE AVENS

PHYSOCARPUS | *Ninebark*

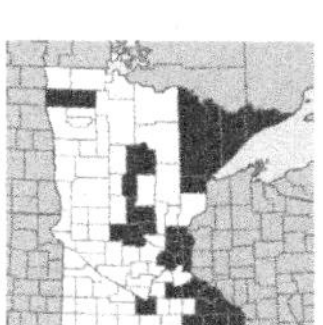

Physocarpus opulifolius (L.) Maxim.
EASTERN NINEBARK
NC **FACW** | MW **FACW** | GP **FACU**

Much-branched native shrub, 2–3 m long; **twigs** greenish, slightly angled or ridged, smooth or finely hairy; bark of older stems shredding in long thin strips. **Leaves** alternate, ovate in outline, mostly 3-lobed, dark green above, paler and often sparsely hairy below; margins irregularly toothed; petioles 1–2 cm long, with a pair of small, deciduous stipules at base. **Flowers** 5-parted, white, 5–10 mm wide, many in stalked, rounded clusters at ends of branches. **Fruit** a red-brown pod, 5–10 mm long, in round clusters; seeds 1–2 mm long, shiny, 3–4 in each pod. June–July. **Habitat** Streambanks, lakeshores, swamps, rocky shores of Lake Superior. In Minnesota, most common near Lake Superior and Miss River.

POTENTILLA | *Cinquefoil*

Shrubs (*Potentilla fruticosa*) and annual or perennial herbs. **Leaves** pinnately or palmately divided, alternate or mostly from base of plant. **Flowers** perfect, regular; sepals 5, alternating with small bracts, the sepals and bractlets joined at base to form a saucer-shaped hypanthium; petals 5, yellow, or dark red in *Potentilla palustris*; stamens many; pistils numerous. **Fruit** a group of many small achenes, surrounded by the persistent hypanthium.

POTENTILLA | CINQUEFOIL

1 Petals dark red-purple; plants often in shallow water, ± woody and creeping at base . . . . *Potentilla palustris* | MARSH-CINQUEFOIL
1 Petals yellow. 2

2 Plants shrubs . . . . *Potentilla fruticosa* | SHRUBBY CINQUEFOIL
2 Plants herbs 3

3 Plants spreading by stolons; leaves densely white-hairy on underside . . . . *Potentilla anserina* | SILVER-WEED
3 Stolons absent; leaves green, smooth to hairy below 4

4 Leaves pinnately divided, with 7–11 leaflets . . . . *Potentilla paradoxa* . . . . BUSHY CINQUEFOIL
4 Leaves palmately divided, with 3–7 leaflets . . . . *Potentilla rivalis* . . . . DIFFUSE CINQUEFOIL

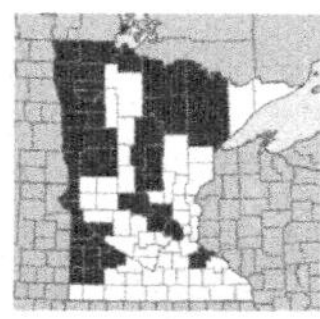

Potentilla anserina L.

SILVER-WEED
NC **FACW** | MW **FACW** | GP **FACW**

Native perennial clumped herb, with a stout rootstock and spreading by stolons to 1 m long. **Leaves** all at base of plant except for a few clustered leaves on stolons, pinnately divided into 7–25 leaflets; leaflets oblong or obovate, 1.5–5 cm long and 0.5–2 cm wide, lower leaflets much smaller; upper surface green and smooth to gray-green and silky-hairy, underside densely white-hairy; margins with deep, sharp, forward-pointing teeth; stipules brown, membranous, at base of petiole. **Flowers** single from leafy axils of stolons, on stalks 5–15 cm long; sepals ovate, white

silky-hairy; petals yellow, oval to obovate, 5–10 mm long; stamens 20–25. **Fruit** a light brown achene. May–Sept. **Synonyms** *Argentina anserina*. **Habitat** Wet meadows, shallow marshes, sandy and gravelly shores and streambanks, ditches; soils often calcium-rich.

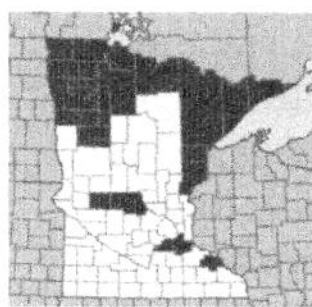

Potentilla fruticosa L.

SHRUBBY CINQUEFOIL

NC **FACW** | MW **FACW** | GP **FACW**

Much-branched native shrub, 0.5–1 m tall; **twigs** brown to red, covered with long, silky- white hairs; bark of older branches shredding. **Leaves** alternate, pinnately divided; leaflets 3–7 (mostly 5), the terminal 3 leaflets often joined at base, oval to oblong, 1–2 cm long and 3–7 mm wide, tapered at each end, upper surface dark green, underside paler, with silky hairs on both sides or at least on underside; margins entire, often rolled under; short-stalked. **Flowers** 5-parted, bright yellow, 1–2.5 cm wide, 1 to few in clusters at ends of branches; bracts lance-shaped, much narrower than the ovate sepals; stamens 15–20. **Fruit** a small head of hairy achenes surrounded by the 10-parted calyx. June–Sept. **Synonyms** *Potentilla floribunda, Dasiphora fruticosa*. **Habitat** Calcareous fens, lakeshores, open bogs, conifer swamps, wet meadows.

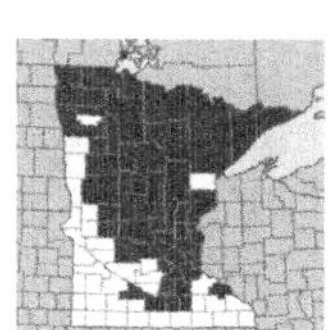

Potentilla palustris (L.) Scop.

MARSH-CINQUEFOIL

NC **OBL** | MW **OBL** | GP **OBL**

Native perennial herb, from long, stout rhizomes. **Stems** 3–8 dm long, ascending to sprawling or floating in shallow water, often rooting at nodes, ± woody at base; lower stems smooth, upper stems sparsely hairy. **Leaves** all from stem, pinnately divided or nearly palmate, with 3–7 leaflets; leaflets oblong to oval, 3–10 cm long and 1–3 cm wide, mostly rounded at tip, underside waxy; margins with sharp, forward-pointing teeth; lower leaves long-petioled, upper leaves nearly stalkless; stipules forming wings around petioles of lower leaves, becoming shorter upward. **Flowers** single or paired from leaf axils, or in open clusters; sepals dark red or purple (at least on inner surface), ovate to lance-shaped, 6–20 mm long; petals 5 (sometimes 10), very dark red, 3–5 mm long, with a short slender tip; stamens about 25, dark red. **Achenes** red to brown, smooth, 1 mm long. June–Aug. **Synonyms** *Comarum palustre*. **Habitat** Open bogs (especially in pools and wet margins), conifer swamps, shores.

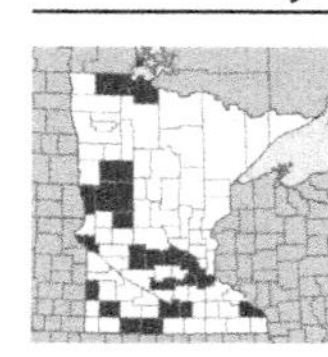

Potentilla paradoxa Nutt.
BUSHY CINQUEFOIL
NC **FACW** | MW **FACW** | GP **FACW**

Native annual or short-lived perennial herb, taprooted. **Stems** erect to reclining on ground with tip upright, 1–6 dm long, unbranched or branched from base, smooth below to long-hairy above. **Leaves** mostly on stem, pinnately divided into 7–11 leaflets; leaflets oval to obovate, 1–4 cm long and 0.5–2 cm wide, finely hairy; margins with rounded teeth; stipules ovate, to 1.5 cm long; lower leaves long-petioled, upper leaves with short petioles. **Flowers** usually numerous in leafy-bracted clusters at ends of stems; sepals ovate, 2–4 mm long; petals yellow, obovate, 2–3 mm long, about as long as sepals; stamens 15–20. **Achenes** brown, 1 mm long. June–Sept. **Habitat** Shores, ditches, floodplains and flats, often where sandy or gravelly.

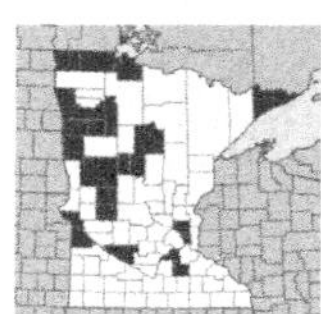

Potentilla rivalis Nutt.
DIFFUSE CINQUEFOIL
NC **FACW** | MW **FACW** | GP **FACW**

Native annual or biennial herb, taprooted. **Stems** upright to spreading, 2–9 dm long, with soft, long hairs, unbranched or branched from base, upper stems branched. **Leaves** mostly on stems, palmately divided, with 3–7 leaflets, or lower leaves pinnately divided; leaflets obovate to oval, 1.5–5 cm long and 0.5–2.5 cm wide, sparsely hairy; margins with coarse, forward-pointing teeth; stipules ovate, to 1.5 cm long; petioles present, but uppermost leaves nearly stalkless. **Flowers** in leafy, branched clusters at ends of stems; sepals triangular, 3–6 mm long; petals yellow, obovate, 1–2 mm long, half as long as sepals (or shorter); stamens 10–15. **Achenes** yellow, less than 1 mm long, smooth. June–Aug. **Synonyms** *Potentilla millegrana, Potentilla pentandra*. **Habitat** Wet meadows, streambanks, shores and ditches.

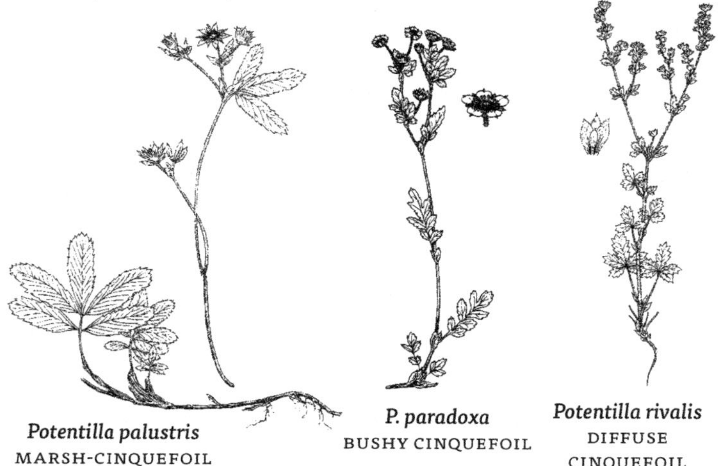

Potentilla palustris
MARSH-CINQUEFOIL

P. paradoxa
BUSHY CINQUEFOIL

Potentilla rivalis
DIFFUSE CINQUEFOIL

RUBUS | *Raspberry, Dewberry, Blackberry*

Perennials, woody at least at base, usually with bristly stems. **Stems** biennial in some species, the first year's canes called primocanes, the second year's growth termed floricanes. **Leaves** alternate, palmately lobed or divided. **Flowers** 5-parted, usually perfect, white to pink or rose-purple; stamens many. **Fruit** a group of small, 1-seeded drupes forming a berry.

RUBUS | RASPBERRY, DEWBERRY, BLACKBERRY

1 Stems without bristles or prickles 2
1 Stems with bristles or prickles 3

2 Flowering stems 1 or several from a short base; petals light to deep pink, 1–2 cm long; northern Minnesota. . . . . *Rubus acaulis* . . . . . NORTHERN DWARF RASPBERRY; ARCTIC RASPBERRY
2 Flowering stems single from a creeping stem; petals green-white, 0.5–1 cm long; widespread . . . . . *Rubus pubescens* | DWARF RASPBERRY

3 Leaves gray-hairy on underside; fruit separating from receptacle when ripe (**raspberries**). . . . . *Rubus idaeus* | WILD RED RASPBERRY
3 Leaves green on both sides, underside veins hairy; fruit detaching with receptacle when ripe 4

4 Plants tall (to 1.5 m); stems erect or arching; flowers mostly more than 10 in a cluster; fruit black (**blackberries**) . . . . . *Rubus setosus* . . . . . BRISTLY BLACKBERRY
4 Plants low and trailing (less than 0.5 m tall); flowers 1 to several in a cluster; fruit red to red-purple (dewberries) 5

5 Stems with prickles, these hooked at tip and broad at base; leaves thin and deciduous; petals more than 1 cm long . . . . . *Rubus flagellaris* . . . . . NORTHERN DEWBERRY
5 Stems with coarse hairs and slender bristles; leaves leathery and often evergreen; petals less than 1 cm long . . . . . *Rubus hispidus* . . . . . BRISTLY BLACKBERRY; SWAMP DEWBERRY

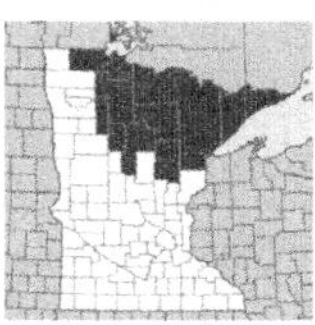

Rubus acaulis Michx.

ARCTIC RASPBERRY
NC **FACW** | MW **FACW** | GP **FACW**

Native perennial, woody at base. **Stems** herbaceous, 5–10 cm long, bristles or prickles absent. **Leaves** alternate, divided into 3 leaflets, 1–4 cm long and 0.5–3 cm wide, terminal leaflet stalked, lateral pair of leaflets nearly stalkless, lateral leaflets often with a shallow lobe, upper surface smooth, underside finely hairy; margins with blunt, forward-pointing teeth; petioles long, finely hairy; stipules small, ovate. **Flowers** single at ends of erect stems; sepals lance-shaped, to 1 cm long; petals 5, light to dark pink, 1–2 cm long. **Fruit** red, nearly round, 1 cm wide, edible. June–Aug. **Synonyms** *Rubus arcticus* var. *grandiflorus*. **Habitat** Conifer swamps, open bogs.

Rubus acaulis
NORTHERN DWARF RASPBERRY;
ARCTIC RASPBERRY

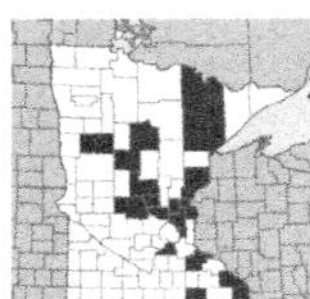

Rubus flagellaris Willd.

NORTHERN DEWBERRY
NC **FACU** | MW **FACU** | GP-UPL

Native shrub. **Stems** long-trailing, 2–4 m long, often rooting at tip, brown to red-purple, with curved, broad-based prickles; bristles absent. **Leaves** alternate; primocane leaves divided into 3–5 leaflets, the terminal leaflet 2–6 cm long and 1–5 cm wide, often with small lobes above middle; floricane leaves smaller, usually divided into 3 leaflets; leaflets ovate to obovate, upper and lower surface ± smooth, or underside veins with appressed hairs; margins with forward-pointing teeth; petioles finely hairy, with scattered, hooked prickles. **Flowers** mostly 2–7 (sometimes 1), on upright, finely hairy stalks, the stalks with scattered prickles; sepals joined, the lobes narrowed to dark tips; petals 5, white, 10–15 mm long. **Fruit** red, ± round, composed of large, juicy drupelets, edible, not easily separating from receptacle. May–June. **Synonyms** *Rubus baileyanus*. **Habitat** Swamps, wetland margins; also in drier sandy woods, prairies and openings.

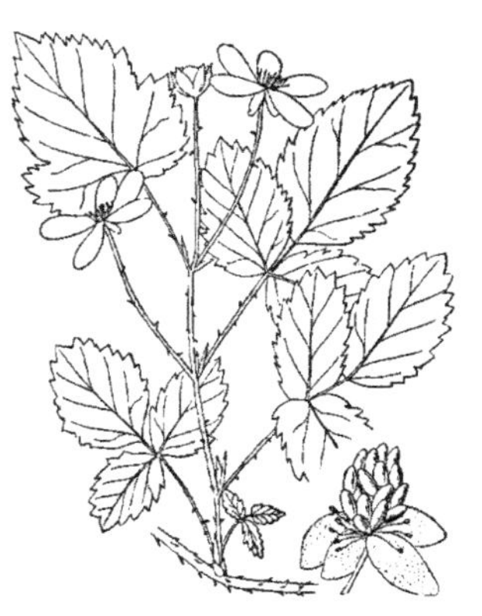

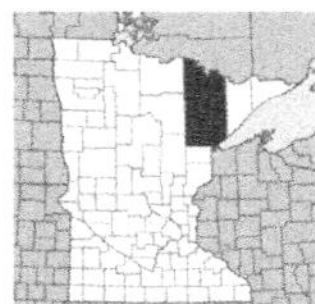

Rubus hispidus L.

BRISTLY BLACKBERRY; SWAMP DEWBERRY
NC **FACW** | MW **FACW**

Native shrub. **Stems** trailing or low-arching, often rooting at tip, with slender bristles or spines 2–5 mm long, these sometimes gland-tipped, not much widened at base. **Leaves** alternate, divided into 3 leaflets (rarely 5); the leaflets ovate to obovate, 2–5 cm long and 1–3 cm wide, upper surface dark green and slightly glossy, slightly paler and ± smooth below, some leaves persisting through winter; margins with rounded teeth; petioles finely hairy and bristly;

stipules linear, persistent. **Flowers** single in upper leaf axils or in open clusters of 2–8 at ends of short branches; sepals joined, the lobes ovate, tipped with a small dark gland; petals 5, white, 5–10 mm long. **Fruit** red-purple, less than 1 cm wide, sour, not easily separated from receptacle. June–Aug. **Habitat** Conifer swamps, wet hardwood forests, thickets, wetland margins, sandy interdunal swales.

Rubus hispidus
BRISTLY BLACKBERRY;
SWAMP DEWBERRY

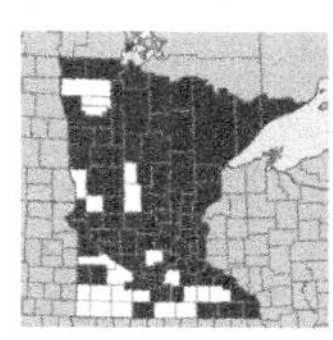

Rubus idaeus L.

WILD RED RASPBERRY
NC **FAC** | MW **FACU** | GP **FACU**

Native shrub. **Stems** erect or spreading, to 1.5 m long, biennial; young stems bristly with slender, often gland-tipped hairs; older stems brown, smooth, **Leaves** alternate, pinnately divided; primocane leaves divided into 3 or 5 leaflets, floricane leaflets usually 3; leaflets ovate to lance-shaped, upper surface dark green and smooth or sparsely hairy, underside gray-hairy; margins with sharp, forward-pointing teeth; petioles with bristly hairs; stipules slender, soon deciduous. **Flowers** in clusters of 2–5 at ends of stems and 1–2 from upper leaf axils; sepals with gland-tipped hairs; petals 5, white, shorter than the sepals. **Fruit** red, about 1 cm wide, edible, separating from receptacle when ripe. May–Aug. **Synonyms** *Rubus strigosus*. **Habitat** Thickets, moist to wet openings, streambanks; often where disturbed.

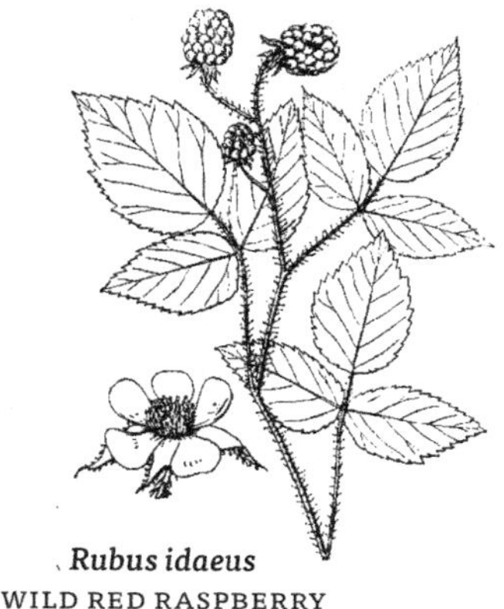

Rubus idaeus
WILD RED RASPBERRY

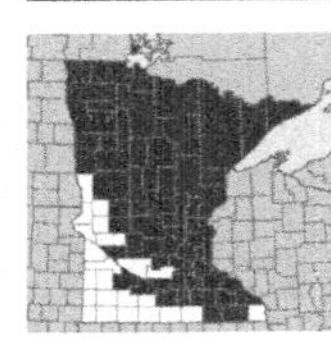

Rubus pubescens Raf.
DWARF RASPBERRY
NC **FACW** | MW **FACW** | GP **FACW**

Low native perennial. **Stems** long-creeping at or near soil surface, with upright, hairy branches 1–3 dm long; the branches herbaceous but woody at base, bristles absent; sterile branches arching to trailing, often rooting at nodes; flowering branches erect, with few leaves. **Leaves** alternate, divided into 3 leaflets; leaflets oval, 2–6 cm long and 1–4 cm wide, tapered to a sharp point; margins with coarse, forward-pointing teeth, often entire near base; petioles hairy; stipules ovate. **Flowers** on glandular-hairy stalks, 1–3 in loose clusters at ends of erect branches, sometimes with 1–2 flowers from leaf axils; petals 5, white or pale pink, to 1 cm long. **Fruit** bright red, round, 5–15 mm wide, the drupelets large, juicy, edible, not separating easily from receptacle. May–July. **Synonyms** *Rubus triflorus*. **Habitat** Conifer swamps, wet deciduous woods, rocky shores.

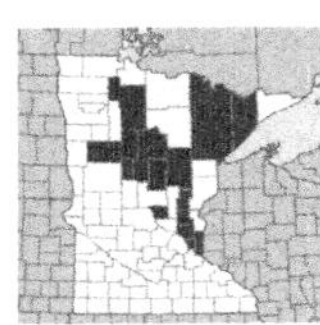

Rubus setosus Bigel.
BRISTLY BLACKBERRY
NC **FACW** | MW **FACW** | GP **FACW**

Native shrub. **Stems** erect to spreading or arching, to 1.5 m long; branches covered with spreading bristles 1–4 mm long; older canes red-brown, ridged, not rooting at tip. **Leaves** alternate; primocane leaves divided into 3–5 leaflets; floricane leaves 3-divided; leaflets ovate to obovate, upper and lower surface ± smooth but often hairy on underside veins; margins with sharp, forward-pointing teeth; petioles bristly; stipules linear, 1–2 cm long. **Flowers** few to many in elongate clusters at ends of stems, with small, leafy bracts throughout the head; petals 5, white, to 1 cm long. **Fruit** red, ripening to black, round, to 1 cm wide, dry, poor eating quality. June–Aug. **Synonyms** *Rubus wheeleri*. **Habitat** Wetland margins, shores, occasional in open bogs; also in drier sandy prairie.

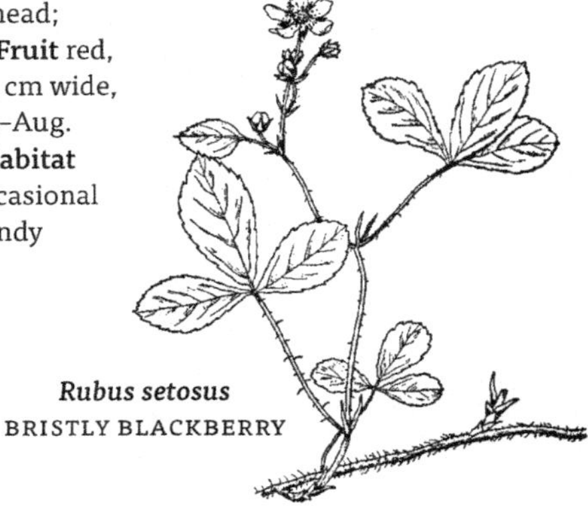

Rubus setosus
BRISTLY BLACKBERRY

SPIRAEA | *Spiraea*

Shrubs with alternate, undivided leaves. **Flowers** 5-parted, white to pink, perfect, numerous in clusters at ends of stems. **Fruit** a cluster of dry, 1-chambered follicles containing small seeds.

SPIRAEA | SPIRAEA

1 Leaves smooth on both sides; flowers white .... *Spiraea alba* .... MEADOWSWEET

1 Leaf underside densely covered with light brown woolly hairs; flowers rose-pink .... *Spiraea tomentosa* | HARDHACK; STEEPLE-BUSH

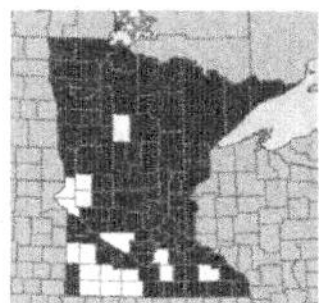

Spiraea alba Duroi

MEADOWSWEET

NC **FACW** | MW **FACW** | GP **FACW**

Much-branched native shrub, often forming colonies. **Stems** somewhat angled or ridged, 0.5–1.5 m long, smooth or short-hairy when young, becoming red-brown and smooth. **Leaves** alternate, often crowded on stems, oval to oblong lance-shaped, 3–7 cm long and 1–2 cm wide, smooth on both sides; margins with sharp, forward-pointing teeth; petioles 2–8 mm long; stipules absent. **Flowers** small, 6–8 mm wide, many in a narrow, pyramid-shaped panicle 5–25 cm long at ends of branches; sepals 5, triangular; petals 5, white. **Fruit** a group of 5–8 small follicles, each with several seeds; the fruiting branches often persistent over winter. June–Aug. **Habitat** Wet meadows, streambanks, lakeshores, conifer swamps; soils often sandy.

Spiraea alba
MEADOWSWEET

Spiraea tomentosa L.
HARDHACK; STEEPLE-BUSH
NC **FACW** | MW **FACW** | GP **FACW**

Sparsely branched native shrub to 1 m tall. Young **stems** covered with brown woolly hairs, becoming smooth and red-brown. **Leaves** alternate, lance-shaped to ovate, 2–5 cm long and 0.5–2 cm wide; ± smooth above, underside gray-green to tan, densely covered with feltlike hairs, the veins prominent; tapered to a pointed or blunt tip; margins with coarse, forward-pointing teeth; petioles 1–4 mm long or absent. **Flowers** small, 3–4 mm wide, in narrow panicles 5–15 cm long at ends of stems, the panicle branches covered with reddish woolly hairs; petals 5, pink or rose (rarely white). **Fruit** a cluster of small, hairy follicles, often persisting over winter. July–Sept. **Habitat** Open bogs, conifer swamps, thickets, lakeshores, wet meadows; soils usually sandy.

Spiraea tomentosa
HARDHACK; STEEPLE-BUSH

Potentilla palustris (MARSH-CINQUEFOIL, page 317)

SHRUBS (*Cephalanthus*), or herbs (*Galium*). **Leaves** simple, opposite or whorled. **Flowers** small, perfect (with both male and female parts), white to green, single or in loose or round clusters; petals joined, 3–4-lobed; stamens 3–4; ovary 2-chambered. **Fruit** a round head of cone-shaped nutlets (*Cephalanthus*), or a bristly to smooth capsule (*Galium*).

RUBIACEAE | MADDER FAMILY

1 Shrub .... *Cephalanthus occidentalis* | BUTTONBUSH
1 Herbs .... *Galium* | BEDSTRAW

CEPHALANTHUS | *Buttonbush*

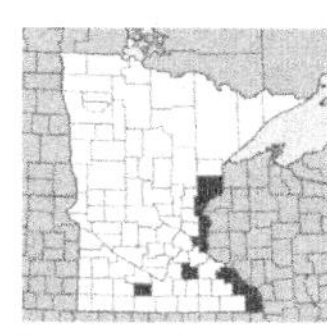

Cephalanthus occidentalis L.

BUTTONBUSH
NC **OBL** | MW **OBL** | GP **OBL**

Native shrub or small tree, 1-4 m tall. Young **stems** green-brown, with lighter lenticels; older stems gray-brown. **Leaves** opposite or in whorls of 3, oval to ovate, 8–20 cm long and to 7 cm wide, upper surface bright green and shiny, paler or finely hairy below; margins entire or slightly wavy; petioles grooved, to 2 cm long. **Flowers** small, perfect, in round, many-flowered heads 2–4 cm wide, on long stalks at ends of stems or from upper leaf axils; petals 4, creamy white, 5–8 mm long; styles longer than petals and swollen at tip. **Fruit** a round head of brown, cone-shaped nutlets, tipped by 4 teeth of persistent sepals. June–Aug. **Habitat** Hardwood swamps, floodplain forests, thickets, streambanks, marshes, open bogs; often in standing water or muck.

Cephalanthus occidentalis
BUTTONBUSH

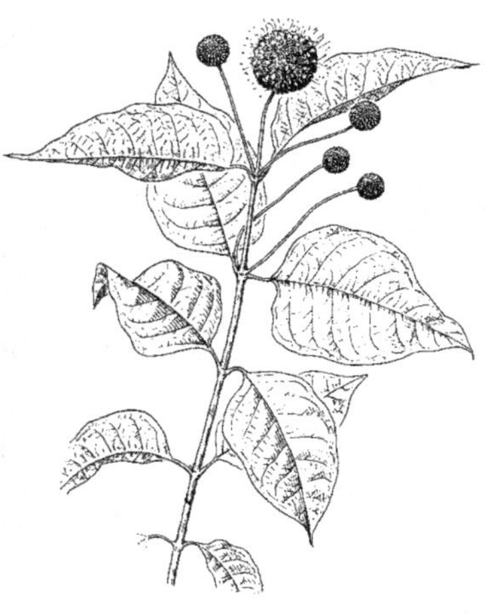

GALIUM | *Bedstraw*

Perennial herbs (our species), from slender rhizomes. **Stems** 4-angled, ascending to reclining, smooth or bristly. **Leaves** entire, in whorls of 4–6. **Flowers** small, perfect, regular, 1 to several from leaf axils or in clusters at ends of stems; sepals absent; petals joined, 3–4-lobed, white; stamens 3–4; styles 2, ovary 2-chambered and 2-lobed, maturing as 2 dry, round fruit segments which separate when mature.

GALIUM | BEDSTRAW

1 Fruit bristly. 2
1 Fruit ± smooth 3

2 Leaves in whorls of 4 . . . . *Galium boreale* | NORTHERN BEDSTRAW
2 Leaves in whorls of 6. . . . . *Galium triflorum* | SWEET-SCENTED BEDSTRAW

3 Leaves sharp-tipped . . . . *Galium asprellum* | ROUGH BEDSTRAW
3 Leaves rounded or blunt at tip 4

4 Corolla lobes mostly 3, the lobes wider than long 5
4 Corolla lobes 4, the lobes longer than wide 6

5 Leaves in whorls of 4–6; fruit on smooth, straight stalks . . . . *Galium tinctorium* | SOUTHERN THREE-LOBED BEDSTRAW
5 Leaves in whorls of 4; fruit on rough, curved stalks . . . . . . *Galium trifidum* . . . . NORTHERN THREE-LOBED BEDSTRAW

6 Leaves bent strongly downward, less than 1.5 cm long and 1–2 mm wide; fruits 1–1.5 mm wide . . . . . . *Galium labradoricum* | LABRADOR-BEDSTRAW
6 Leaves spreading, more than 2 cm long and usually more than 2 mm wide; fruits 2 mm or more wide 7

7 Larger leaves linear to lance-shaped, 3 cm or more long; flowers many in panicles at ends of stems . . . . *Galium boreale* | NORTHERN BEDSTRAW
7 Leaves narrowly oval, up to 3 cm long; flowers 1–3 on stalks at ends of stems . . . . *Galium obtusum* | BLUNTLEAF-BEDSTRAW

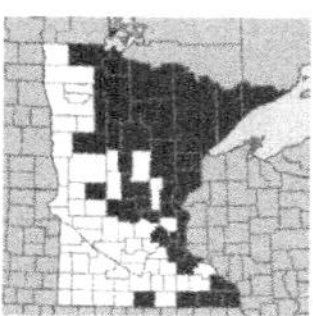

Galium asprellum Michx.

ROUGH BEDSTRAW
NC **OBL** | MW **OBL** | GP **OBL**

Native perennial herb. **Stems** spreading or reclining on other plants, much-branched, to 2 m long, 4-angled, with rough, downward-pointing hairs on stem angles (which cling tightly to clothing). **Leaves** 6 in a whorl or 5-whorled on branches, narrowly oval, usually widest above middle, 1–2 cm long and 4–6 mm wide, tapered to a sharp tip; underside midvein and margins with rough hairs;

petioles absent. **Flowers** in loose, few-flowered clusters at ends of stems and from upper leaf axils; corolla 4-lobed, white, 3 mm wide. **Fruit** smooth. July–Sept. **Habitat** Swamps, streambanks, thickets, marshes, wet meadows, calcareous fens.

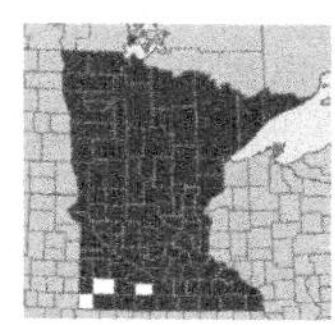

Galium boreale L.

NORTHERN BEDSTRAW

NC **FAC** | MW **FAC** | GP **FACU**

Native perennial herb. **Stems** erect, 2–8 dm long, 4-angled, smooth or with short hairs at leaf nodes, sometimes slightly rough-to-touch. **Leaves** in whorls of 4, linear to lance-shaped, 1.5–4 cm long and 3–8 mm wide, 3-nerved, tapered to a small rounded tip; margins sometimes fringed with hairs; petioles absent. **Flowers** many, 3–6 mm wide, in branched clusters at ends of stems; corolla lobes 4, white. **Fruit** with short, bristly hairs, or smooth when mature. June–Aug. **Habitat** Streambanks, shores, thickets, swamps, moist meadows; also in drier woods and fields.

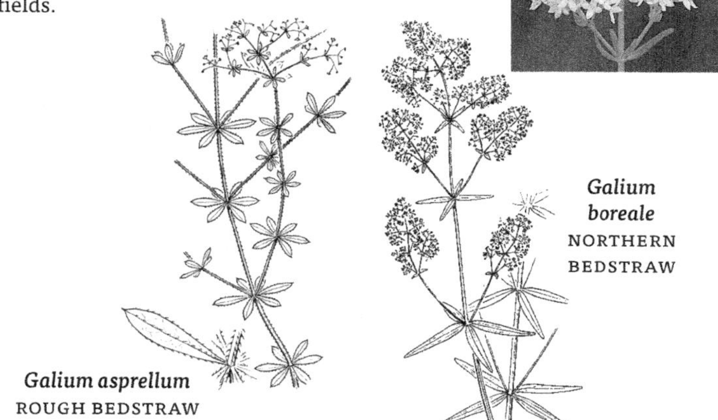

Galium boreale
NORTHERN BEDSTRAW

Galium asprellum
ROUGH BEDSTRAW

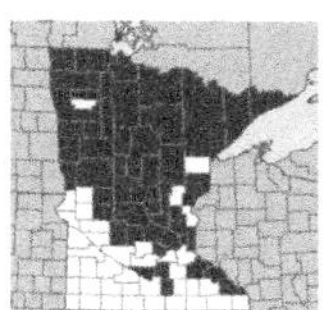

Galium labradoricum (Wieg.) Wieg.

LABRADOR-BEDSTRAW

NC **OBL** | MW **OBL** | GP **OBL**

Native perennial herb. **Stems** simple or branched, 1–3 dm long, 4-angled, hairy at leaf nodes, smooth on stem angles. **Leaves** in whorls of 4, soon curved downward, oblong lance-shaped, 1–1.5 cm long and 1–2 mm wide, blunt-tipped; underside midvein and margins with short, bristly hairs; petioles absent. **Flowers** single or in small groups on stalks from leaf axils; corolla lobes 4, white. **Fruit** smooth, dark. June–July. **Habitat** Conifer swamps, sphagnum bogs, fens, sedge meadows.

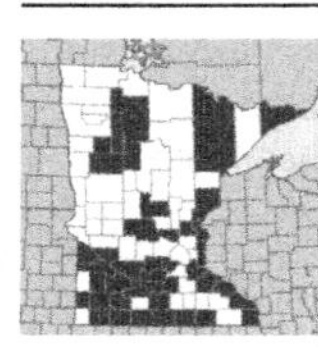

Galium obtusum Bigel.
BLUNTLEAF-BEDSTRAW
NC **FACW** | MW **FACW** | GP **FACW**

Native perennial herb. **Stems** branched, 2–6 dm long, 4-angled, hairy at leaf nodes, otherwise smooth. **Leaves** mostly in whorls of 4 (sometimes 5 or 6), ascending to spreading, linear to lance-shaped or oval, 1–3 cm long and 3–5 mm wide, blunt-tipped; margins with short, bristly hairs and often somewhat rolled under; petioles absent. **Flowers** in clusters at ends of stems; corolla lobes 4, white. **Fruit** smooth, dark, often with only 1 segment maturing. May–July. **Habitat** Wet deciduous forests, wet meadows, streambanks, thickets, floodplains, moist prairie.

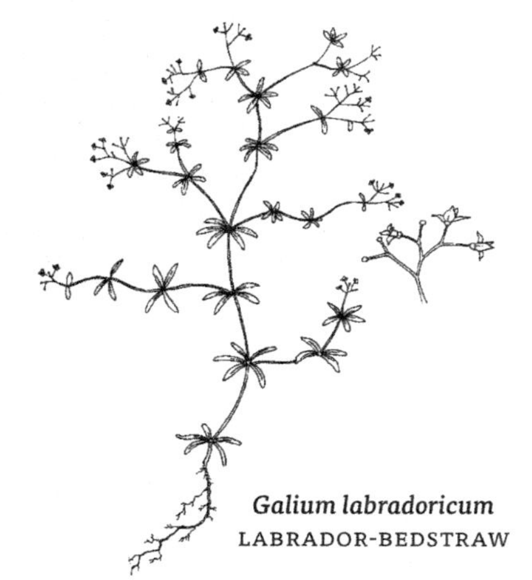

Galium labradoricum
LABRADOR-BEDSTRAW

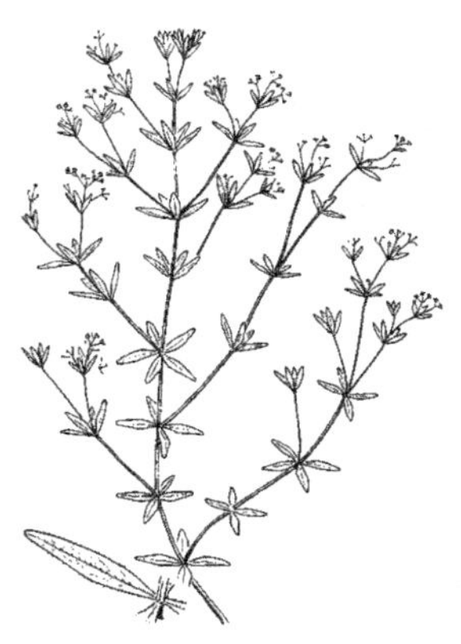

Galium obtusum
BLUNTLEAF-BEDSTRAW

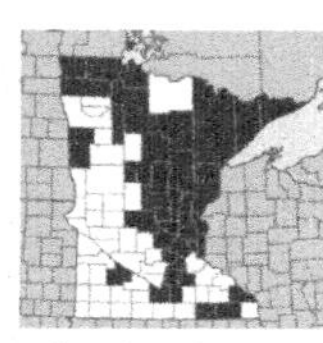

Galium tinctorium L.
SOUTHERN THREE-LOBED BEDSTRAW
NC **OBL** | MW **OBL** | GP **OBL**

Native perennial herb. **Stems** slender, weak, 4-angled, with rough hairs on angles. **Leaves** in whorls of 4 or sometimes 5–6, linear to oblong lance-shaped, 1–2.5 cm long, tapered to a narrow base, dark green and dull; underside midvein and margins with rough hairs; petioles absent. **Flowers** in clusters of 2–3, on slender, smooth, straight stalks at ends of stems; corolla lobes 3, white. **Fruit** smooth. July–Sept. **Synonyms** *Galium claytonii*. **Habitat** Conifer swamps, open bogs, fens, thickets, wet shores and marshes.

Galium trifidum L.

NORTHERN THREE-LOBED BEDSTRAW

NC **FACW** | MW **FACW** | GP **OBL**

Native perennial herb. **Stems** slender, weak, 2–6 dm long, much-branched, sharply 4-angled, with rough, downward-pointing hairs on stem angles. **Leaves** in whorls of 4, linear to oblong lance-shaped, 5–20 mm long and 1–3 mm wide, blunt-tipped, dark green and dull on both sides; underside midvein and margins often rough-hairy; petioles absent. **Flowers** small, on 2–3 slender stalks from leaf axils or at ends of stems, the stalks much longer than the leaves; corolla lobes 3, white. **Fruit** dark, smooth. June–Sept. **Synonyms** Includes plants sometimes considered a separate species (*Galium brevipes*). **Habitat** Lakeshores, streambanks, swamps, marshes, bogs, springs.

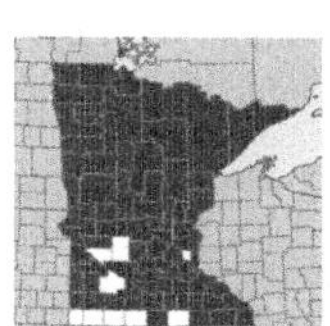

Galium triflorum Michx.

SWEET-SCENTED BEDSTRAW

NC **FACU** | MW **FACU** | GP **FACU**

Native perennial herb. **Stems** prostrate or scrambling, 2–8 dm long, 4-angled, smooth or with rough, downward-pointing hairs on stem angles. **Leaves** shiny, in whorls of 6 (or 4 on smaller branches), narrowly oval to oblong lance-shaped, 2–5 cm long and to 1 cm wide, l-nerved, tipped with a short, sharp point, slightly vanilla-scented, underside midvein with rough hairs, margins with rough, forward-pointing hairs; petioles absent. **Flowers** 2–3 mm wide, on slender stalks from leaf axils and at ends of stems, the stalks with 3 flowers or branched into 3 short stalks, each with 1–3 flowers; corolla lobes 4, green-white. **Fruit** 2-lobed, covered with hooked bristles. June–Aug. **Habitat** Moist to wet woods, hummocks in cedar swamps, wetland margins and shores, clearings.

DECIDUOUS TREES OR SHRUBS. **Leaves** alternate, margins entire or toothed; stipules often present at base of leaf petiole, these usually soon falling. **Flowers** borne in catkins near ends of branches, imperfect (the male and female flowers on separate plants), usually appearing before leaves open, or in a few species after leaves open; flowers without petals or sepals, each flower with either 1 or 2 enlarged basal glands (*Salix*) or a cup-shaped disk (*Populus*). **Fruit** a dry, many-seeded capsule; seeds small, covered with long, silky hairs.

SALICACEAE | WILLOW FAMILY

1 Large trees; leaves heart-shaped to ovate, mostly less than 2x as long as wide; buds often sticky and covered by 2 or more overlapping scales; catkins drooping, flowers subtended at base by a cup-shaped disk; stamens many, 12–80 .... *Populus* | POPLAR; COTTONWOOD

1 Shrubs and trees; leaves ovate, lance-shaped or linear, 2x or more longer than wide; buds covered by 1 scale; catkins upright or drooping, flowers subtended by 1 or 2 enlarged glands; stamens 2–8 .... *Salix* | WILLOW

POPULUS | *Poplar, Cottonwood*

Trees with deciduous, ovate to triangular leaves. **Flowers** in drooping catkins that develop and mature before and with leaves in spring; male and female flowers on separate trees; base of flower with a cup-shaped disk; stamens 10–80. **Fruit** a 2–4 chambered capsule with many small seeds, these covered with long white hairs which aid in dispersal by the wind.

POPULUS | POPLAR, COTTONWOOD

1 Leaf petioles strongly flattened....................... *Populus deltoides* .... EASTERN COTTONWOOD

1 Leaf petioles round in section..... *Populus balsamifera* | BALSAM-POPLAR

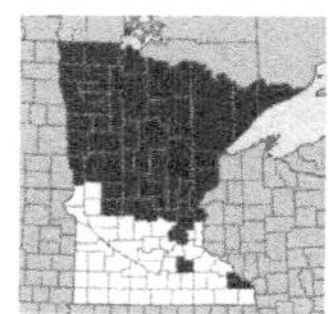

Populus balsamifera L.

BALSAM-POPLAR

NC **FACW** | MW **FACW** | GP **FACW**

Medium to large native tree to 20 m or more tall, trunk 30–60 cm wide, crown open, somewhat narrow; **bark** smooth when young, becoming dark gray and furrowed; **twigs** red-brown when young, becoming gray; **leaf buds** fragrant, very resinous and sticky. **Leaves** resinous, ovate to broadly lance-shaped, 8–13 cm long and 4–7 cm wide, tapered to a long tip, rounded or somewhat heart-shaped at base, dark green and somewhat shiny above, white-green or silvery and often stained with rusty brown resin

below; margins with small, rounded teeth; petioles round in section, 3–4 cm long. **Catkins** densely flowered, drooping, appearing before leaves; scales fringed with long hairs, early deciduous; female catkins 10–13 cm long; female flowers with 2 spreading stigmas; stamens 20–30. **Capsules** ovate, 6–8 mm long, crowded on short stalks. April–May. **Habitat** Swamps, floodplain forests, shores, streambanks, forest depressions, moist dunes.

Populus balsamifera
BALSAM-POPLAR

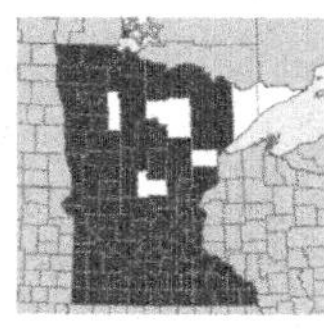

Populus deltoides Marsh.
EASTERN COTTONWOOD
NC **FAC** | MW **FAC** | GP **FAC**

Large native tree to 30 m or more tall, with a large trunk (often 1 m or more wide) and a broad, rounded crown; **bark** gray to nearly black, deeply furrowed; **twigs** olive-brown to yellow, turning gray with age; **leaf buds** very resinous and sticky, shiny, covered by several tan bud scales. **Leaves** smooth, broadly triangular, 8–14 cm long and 6–12 cm wide, short-tapered to tip, heart-shaped or squared-off at base; margins with forward-pointing, incurved teeth, 2–5 large glands usually present at base of blade near petiole; petioles strongly flattened, 3–10 cm long; stipules tiny, early deciduous. **Catkins** loosely flowered, drooping, appearing before leaves; scales fringed, soon falling; flowers subtended by a cup-shaped disk 2–4 mm wide; female catkins green, 7–12 cm long in

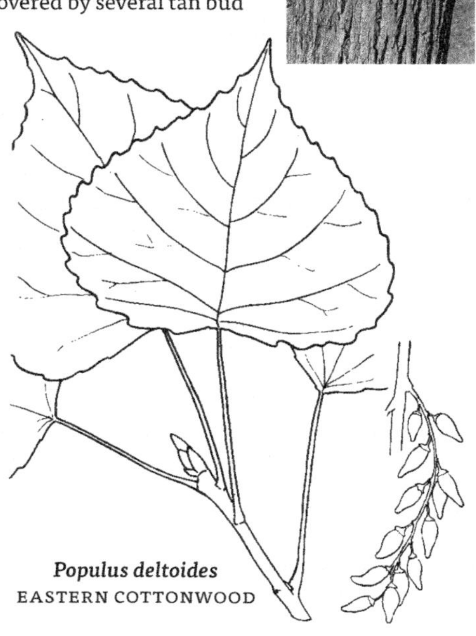

Populus deltoides
EASTERN COTTONWOOD

flower, to 20 cm long in fruit; female flowers with 3–4 spreading stigmas; male catkins dark red, soon deciduous; stamens 30–80. **Capsules** ovate, 6–12 mm long, on stalks 3–10 mm long. April–May. **Habitat** Floodplains, streambanks and bars, shores, wet meadows, ditches.

SALIX | *Willow*

Shrubs and trees. **Leaves** alternate, variable in shape, petioles glandular in some species; stipules early deciduous or persistent, sometimes absent. **Catkins** stalkless or on leafy branchlets, usually shed early in season. Male and female **flowers** on separate plants; male flowers with mostly 2–3 stamens (to 8 in some species). **Fruit** a 2-chambered, stalked or stalkless capsule.

SALIX | WILLOW

1 Leaf petioles with glands at or near base of leaf blade **2**
1 Petioles without glands **7**

2 Trees, usually with a single trunk; leaves narrow **3**
2 Small trees or shrubs, usually with several to many stems; leaves broader **4**

3 Leaves often curved sideways (scythe-shaped), tapered to a long, slender tip; stipules large on vigorous shoots . . *Salix nigra* | BLACK WILLOW p. 340
3 Leaves not curved sideways, tapered to a short tip; stipules small and early deciduous . . . . *Salix alba* | WHITE WILLOW p. 335

4 Leaves not waxy on underside . . . . . *Salix lucida* | SHINING WILLOW p. 339
4 Leaves waxy-coated on underside **5**

5 Leaf tips rounded or with a short point; leaf base heart-shaped or rounded; young leaves thin and translucent; buds and leaves with a balsam-like scent . . . . *Salix pyrifolia* | BALSAM-WILLOW p. 343
5 Leaves tapered to tip; leaf base blunt or rounded; young leaves not translucent; buds and leaves not balsam-scented. **6**

6 Young leaves sparsely hairy; margins with small forward-pointing teeth; flowering in early summer . . . . *Salix amygdaloides* . . . . PEACH-LEAF WILLOW p. 335
6 Young leaves without hairs; margins with small, gland-tipped, forward-pointing teeth; flowering summer or fall . . . . *Salix serissima* . . . . AUTUMN-WILLOW p. 344

7 Mature leaves hairy, at least on underside **8**
7 Mature leaves without hairs (sometimes hairy on petiole and midvein) . **15**

8 Leaves linear or narrowly lance-shaped 9
8 Leaves broadly lance-shaped, oblong or oval 12

9 Underside of leaves with feltlike covering of white tangled hairs; young twigs white-hairy; plant of peatlands, often where calcium-rich .... *Salix candida* | HOARY OR SAGE-LEAVED WILLOW p. 337
9 Leaves not with feltlike hairs; twigs smooth or sparsely hairy 10

10 Leaf margins smooth and turned under; leaf underside with shiny white hairs; uncommon in n Minnesota .... *Salix pellita* .... SATINY WILLOW p. 341
10 Leaf margins with gland-tipped teeth; leaf underside with few long hairs 11

11 Leaf margins with widely spaced sharp teeth; petioles 1–5 mm long; colony-forming shrub of sandy banks .... *Salix exigua* .... SANDBAR-WILLOW p. 338
11 Leaf margins with small teeth at least above middle of blade; petioles 3–10 mm long; stems clustered but not forming large colonies............... .... *Salix petiolaris* | MEADOW-WILLOW p. 342

12 Leaves rounded or heart-shaped at base; margins with gland-tipped, forward-pointing teeth; stipules present and persistent .... *Salix eriocephala* | DIAMOND WILLOW p. 338
12 Leaves tapered to base; margins smooth or toothed; stipules usually deciduous early in season 13

13 Leaves linear to lance-shaped, more than 5x longer than wide, underside velvety with shiny white hairs; uncommon in n Minnesota .... *Salix pellita* | SATINY WILLOW p. 341
13 Leaves oval or obovate, less than 5x longer than wide, underside hairs not shiny 14

14 Small branches widely spreading; young leaves with white hairs; catkins appearing with leaves in spring; catkin bracts yellow or straw-colored ... . *Salix bebbiana* | BEBB'S OR BEAKED WILLOW p. 336
14 Small branches not widely spreading; young leaves with some red or copper-colored hairs; catkins appearing before leaves in spring; catkin bracts dark brown to black .... *Salix discolor* | PUSSY-WILLOW p. 337

15 Leaves green on both sides or slightly paler on underside, not waxy or white below 16
15 Leaves waxy or white-hairy on underside 18

16 Single-stemmed tree; stipules large... *Salix nigra* | BLACK WILLOW p. 340
16 Shrub; stipules small or absent 17

17 Many-stemmed, colony-forming shrub; upper leaf surface not conspicuously shiny; common . . . *Salix exigua* | SANDBAR-WILLOW p. 338
17 Clumped shrub, not forming colonies; upper leaf surface very shiny; rare in nw Minnesota . . . . *Salix maccalliana* | MACCALLA'S WILLOW p. 340

18 Leaf margins entire to shallowly lobed or with irregular teeth, sometimes rolled under **19**
18 Leaf margins distinctly and regularly toothed **22**

19 Leaf margins entire and somewhat rolled under **20**
19 Leaf margins irregularly toothed **21**

20 Stems to 1 m tall, or creeping and rooting in moss; upper surface of leaves with raised, netlike veins; catkins appearing with leaves; capsules not hairy . . . . *Salix pedicellaris* | BOG-WILLOW p. 341
20 Stems 1–4 m tall; leaf veins not netlike; catkins appearing before leaves; capsules hairy . . . . *Salix planifolia* | TEA-LEAF WILLOW p. 342

21 Leaves dull green above, wrinkled below; catkins appearing with leaves; bracts of female catkins green-yellow to straw-colored . . . . *Salix bebbiana* . . . . BEBB'S OR BEAKED WILLOW p. 336
21 Leaves dark green and shiny above; catkins appearing before leaves; bracts of female catkins dark brown to black **23**

22 Stipules large on vigorous shoots; capsules on distinct stalks 2 mm or more long; statewide . . . . *Salix discolor* | PUSSY-WILLOW p. 337
22 Stipules lance-shaped and soon deciduous, or absent or very small; capsules stalkless or on short stalks less than 2 mm long; n Minnesota . . . . *Salix planifolia* | TEA-LEAF WILLOW p. 342

23 Leaves broadly oval, ovate, or oblong lance-shaped . . . . *Salix pyrifolia* . . . . BALSAM-WILLOW p. 343
23 Leaves lance-shaped. **24**

24 Leaves ± equally tapered at tip and base . . . . *Salix petiolaris* . . . . MEADOW-WILLOW p. 342
24 Leaves unequally tapered, the tip tapered to a point, the base usually rounded or heart-shaped **25**

25 Young twigs hairless; stipules small or absent; bracts of female catkins pale yellow and soon deciduous. . . . . *Salix amygdaloides* . . . . PEACH-LEAF WILLOW p. 335
25 Young twigs gray-hairy; stipules large; bracts of female catkins dark brown to black, persistent . . . . *Salix eriocephala* | DIAMOND WILLOW p. 314

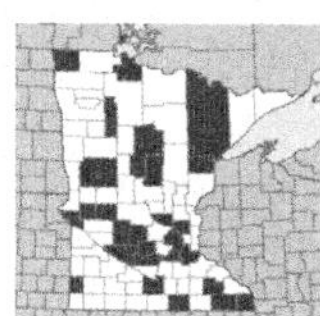

Salix alba L.
WHITE WILLOW
NC **FACW** | MW **FACW** | GP **FACW**

Introduced tree to 20 m tall; **twigs** golden-yellow, often with long, silky hairs. **Leaves** lance-shaped, 4–10 cm long and 1–2.5 cm wide, dark green and shiny above, waxy white below, smooth to sparsely hairy on both sides, margins with small gland-tipped teeth; petioles 2–8 mm long, with silky hairs; stipules lance-shaped, 2–4 mm long, early deciduous. **Catkins** appearing with leaves in spring; female catkins 3–6 cm long, on leafy branches 1–4 cm long; male catkins 3–5 cm long, stamens 2; catkin bracts pale yellow, hairy near base, early deciduous. **Capsules** ovate, 3–5 mm long, without hairs, stalkless or on stalks to 1 mm long. May–June.
Habitat Introduced from Europe and sometimes escaping to streambanks and other wet areas; widely established in temperate North America.

Hybrids between Salix alba and two other introduced species— **weeping willow** *(Salix babylonica) and* **crack-willow** *(Salix fragilis)—are common.*

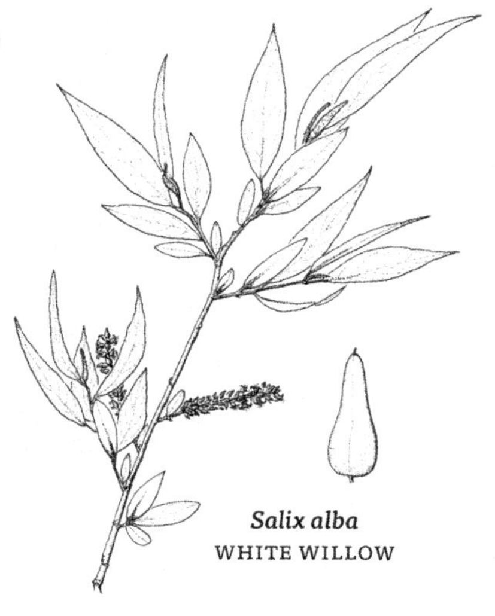

Salix alba
WHITE WILLOW

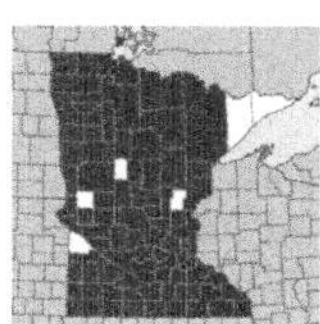

Salix amygdaloides Andersson
PEACH-LEAF WILLOW
NC **FACW** | MW **FACW** | GP **FACW**

Native shrub or tree to 15 m tall, often with several trunks; **twigs** gray-brown to light yellow, shiny and flexible. **Leaves** smooth, lance-shaped, long-tapered to tip, 5–12 cm long and 1–3 cm wide, yellow-green above, waxy-white below, margins finely toothed; petioles 5–20 mm long and often twisted; stipules small and early deciduous. **Catkins** appearing with leaves, linear and loosely flowered; female catkins 3–12 cm long, on leafy branches 1–4 cm long; catkin bracts deciduous, pale yellow, long hairy especially on inner surface; stamens 3–7 (usually 5). **Capsules** smooth, ovate, 3–7 mm long, on stalks 1–3 mm long. May–June. **Habitat** Floodplains, streambanks, lake and pond borders.

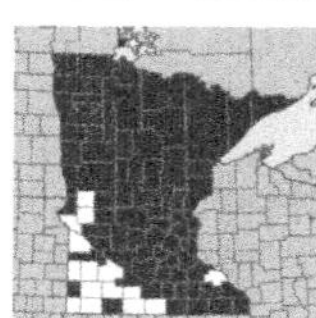

Salix bebbiana Sarg.
BEBB'S OR BEAKED WILLOW
NC **FACW** | MW **FACW** | GP **FACW**

Native shrub or small tree to 8 m tall, stems 1 to several; **twigs** yellow-brown to dark brown, usually with short hairs. **Leaves** oval to ovate or obovate, tapered to tip, 4–8 cm long and 1–3 cm wide, dull gray-green, hairy or sometimes smooth on upper surface, waxy-gray, hairy and wrinkled below, the veins distinctly raised on lower surface; margins entire to shallowly toothed; petioles 5–15 mm long; stipules deciduous or persistent on vigorous shoots. **Catkins** appearing before leaves in spring; female catkins loose, 2–6 cm long, on short leafy branches to 2 cm long; catkin bracts persistent, red-tipped when young, turning brown, long hairy; stamens 2. **Capsules** ovate, 5–8 mm long, finely hairy, on stalks 2–6 mm long. May–June. **Habitat** Common; swamps, thickets, wet meadows, streambanks, marsh borders.

Salix bebbiana
BEBB'S OR BEAKED WILLOW

Salix amygdaloides
PEACH-LEAF WILLOW

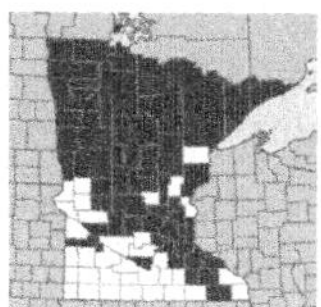

Salix candida Fluegge
HOARY OR SAGE-LEAVED WILLOW
NC **OBL** | MW **OBL** | GP **OBL**

Low native shrub to 1.5 m tall; **twigs** much-branched, covered with dense, matted white hairs. **Leaves** linear-oblong, tapered at tip, 4–10 cm long and 0.5–2 cm wide, dull, dark green and sparsely hairy above, veins sunken, densely white-hairy below; margins entire and rolled under; petioles 3–10 mm long; stipules persistent, 2–10 mm long, white-hairy. **Catkins** appearing with leaves in spring; female catkins 1–5 cm long, on leafy branches 0.5–2 cm long; catkin bracts persistent, brown, hairy; stamens 2. **Capsules** ovate, 4–8 mm long, white-hairy, on stalks to 1 mm long. May–June. **Habitat** Fens, bogs, open swamps, streambanks, often where calcium-rich.

Salix discolor Muhl.
PUSSY-WILLOW
NC **FACW** | MW **FACW** | GP **FACW**

Native shrub or small tree to 5 m tall; twigs yellow-brown to red-brown, dull, smooth with age or with patches of fine hairs. **Leaves** oval and short-tapered to tip, 3–10 cm long and 1–4 cm wide, dark green and smooth above, underside red-hairy when young, becoming white-waxy, smooth and not wrinkled; margins entire or with few rounded teeth; petioles without glands; stipules deciduous, or often persistent on vigorous shoots. **Catkins** appearing and maturing before leaves in spring; female catkins 4–8 cm long, stalkless, sometimes with 2 or 3

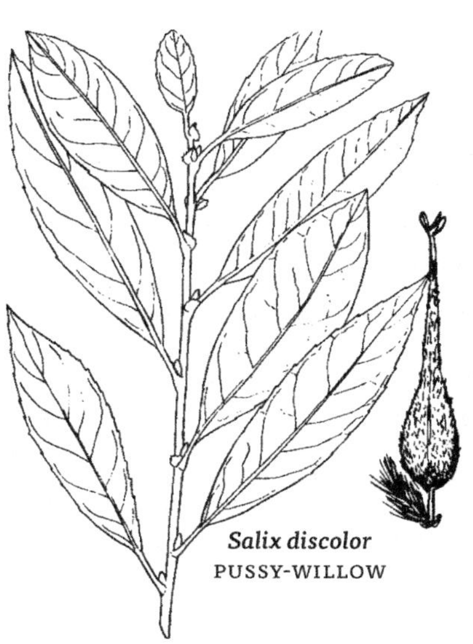

Salix discolor
PUSSY-WILLOW

small, brown, bractlike leaves at the base; stamens 2. **Capsules** ovate with a long neck, 6–10 mm long, densely gray-hairy, on stalks 2–3 mm long. April–May. **Habitat** Common; swamps, fens, streambanks, floodplains, marsh borders.

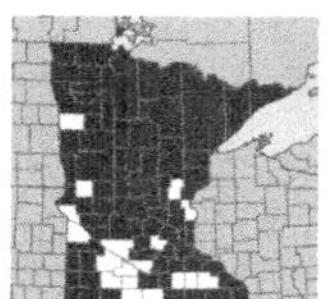

Salix eriocephala Michx.

DIAMOND WILLOW

NC **FACW** | MW **FACW** | GP **FACW**

Native shrub or small tree to 6 m tall; **twigs** red-brown to dark brown, hairy when young. **Leaves** lance-shaped or oblong lance-shaped, 5–12 cm long and 1–3 cm wide, red-purple and hairy when young, upper surface becoming smooth and dark green, underside becoming pale-waxy; margins finely toothed; petioles without glands, 3–15 mm long; stipules persistent (especially on vigorous shoots), ovate or kidney-shaped, to 12 mm long, hairless, toothed. **Catkins** appearing with or slightly before leaves in spring; female catkins 2–6 cm long, on short leafy branches to 1 cm long; catkin bracts persistent, brown to black, hairy; stamens 2. **Capsules** ovate with a long neck, 4–6 mm long, without hairs, on stalks 1–2 mm long. April–May. **Synonyms** *Salix lutea, Salix rigida*. **Habitat** Shores, streambanks, floodplains, ditches and wet meadows, especially along major rivers.

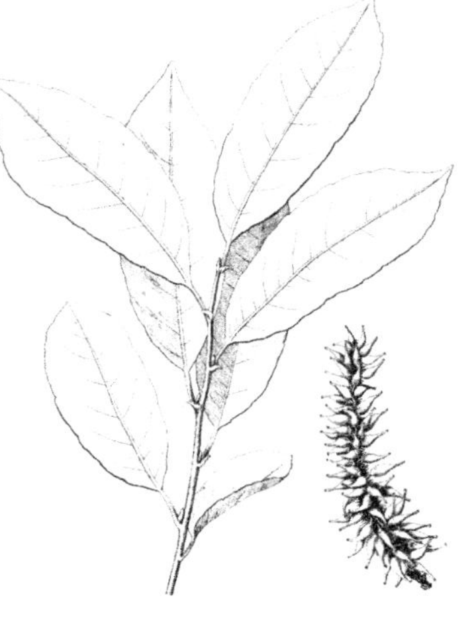

Salix exigua Nutt.

SANDBAR-WILLOW

NC **FACW** | MW **FACW** | GP **FACW**

Native shrub to 4 m tall, spreading by rhizomes and often forming dense thickets; **twigs** yellow-orange to brown, smooth. **Leaves** linear to lance-shaped, tapered at tip and base, 5–14 cm long and 5–15 mm wide, green on both sides but paler below, at first hairy but soon usually smooth; margins with widely spaced, large teeth; petioles without glands, 1–5 mm long; stipules tiny or absent. **Catkins** appearing with leaves in spring on short leafy branches (and plants sometimes again flowering in summer); female catkins loosely flowered, 2–8 cm long; catkin bracts deciduous, yellow; stamens 2. **Capsules** narrowly ovate, 5–8 mm long, hairy when young, smooth when mature, on stalks to 2 mm long. May–June. **Synonyms** *Salix interior*. **Habitat** Common; shores, streambanks, sand and mud bars, ditches and other wet places; often colonizing exposed banks.

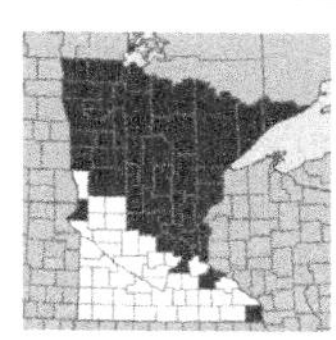

Salix lucida Muhl.
SHINING WILLOW
NC **FACW** | MW **FACW** | GP **FACW**

Native shrub or small tree to 5 m tall; **twigs** yellow-brown or dark brown, smooth and shiny. **Leaves** lance-shaped to ovate, long-tapered and asymmetric at tip, 4–12 cm long and 1–4 cm wide, shiny green above, pale below, red-hairy when young, but soon smooth; margins with small, gland-tipped teeth; petioles with glands near base of leaf; stipules often persistent, strongly glandular. **Catkins** appearing with leaves in spring; female catkins 2–5 cm long, on leafy branches 1–3 cm long; catkin bracts deciduous, yellow, sparsely hairy; stamens 3–6. **Capsules** ovate with a long neck, 4–7 mm long, not hairy, on short stalks to 1 mm long. May. **Habitat** Swamps, shores, wet meadows, moist sandy areas.

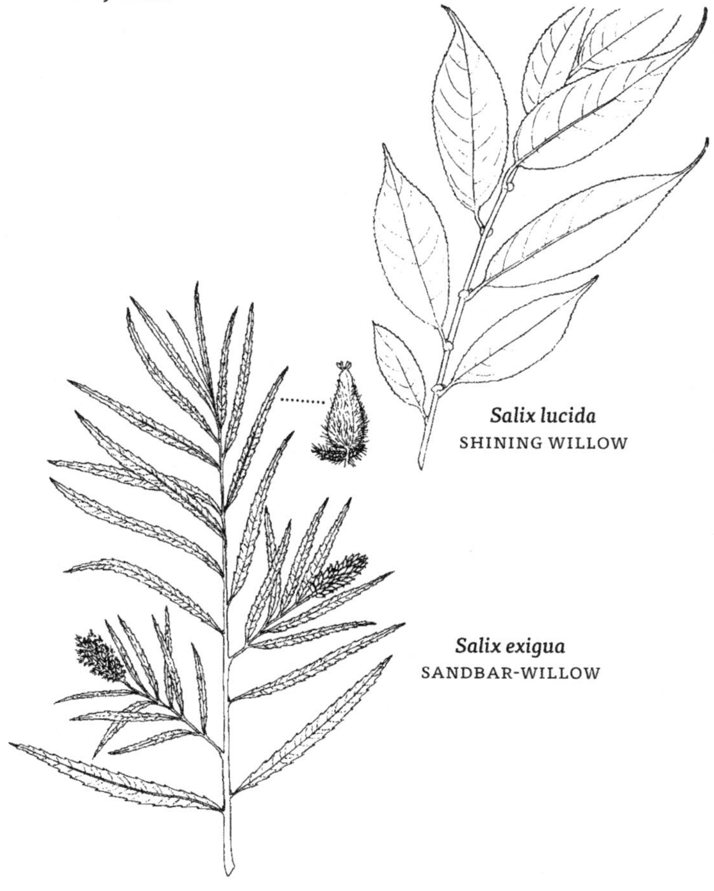

Salix lucida
SHINING WILLOW

Salix exigua
SANDBAR-WILLOW

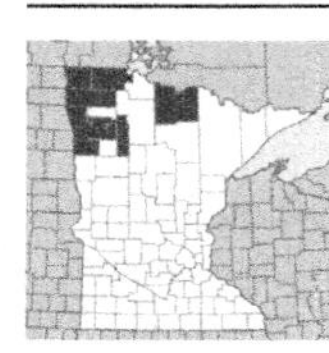

Salix maccalliana Rowlee
MCCALLA'S WILLOW
NC **OBL** | MW **OBL** | GP **OBL** | *special concern*
Native shrub 2–4 m tall; **twigs** upright, red- to yellow-brown, sparsely hairy or smooth, glossy. **Leaves** strap-shaped to narrowly oblong, to 8 cm long and 2.5 cm wide, dark green, upper surface glossy; leaf underside not waxy-coated; margins usually finely toothed; young leaves smooth or with white or reddish hairs; petioles sparsely hairy, without glandular dots; stipules small or leaflike. **Catkins** appearing with leaves in spring; female catkins 2–6 cm long, on leafy branches 1–2.5 cm long. **Capsules** densely hairy when mature. **Habitat** In Minnesota, uncommon in shrubby wetlands, fens, sedge meadows, soils usually sedge-derived peat and not strongly acid; typically found with other willows, dogwoods (*Cornus* spp.), and bog birch (*Betula pumila*). **Status** Special concern.

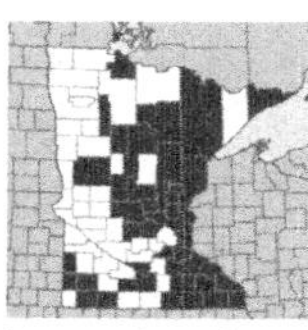

Salix nigra Marshall
BLACK WILLOW
NC **OBL** | MW **OBL** | GP **FACW**
Medium-sized native tree to 15 m tall, trunks 1 or several, crown rounded and open; bark dark brown, furrowed, becoming shaggy; **twigs** bright red-brown, often hairy when young. **Leaves** commonly drooping, linear lance-shaped, 6–15 cm long and 0.5–2 cm wide, long tapered to an often curved tip, green on both sides but satiny above and paler below, lateral veins upturned at tip to form a ± continuous vein near leaf margin; margins finely toothed; petioles 3–8 mm long, hairy, usually glandular near base of blade; stipules to 12 mm long, heart-shaped, usually deciduous. **Catkins** appearing with leaves in spring; female catkins 3–8 cm long, on leafy branches 1–3 cm long; stamens usually 6 (varying from 3–7); catkin bracts yellow, hairy, deciduous. **Capsules** ovate, 3–5 mm long, without hairs, on a short stalk to 2 mm long. May. **Habitat** Streambanks, lakeshores and wet depressions; not tolerant of shade.

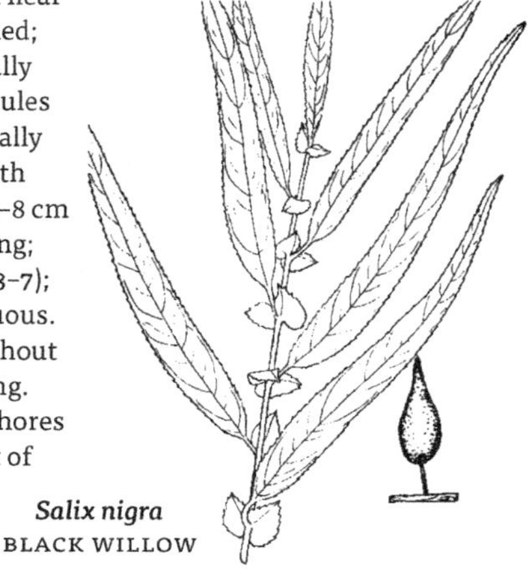

Salix nigra
BLACK WILLOW

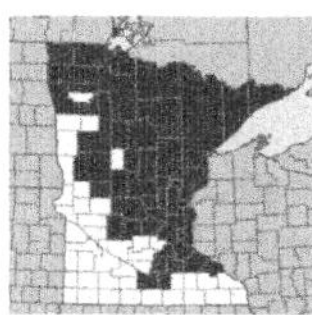

Salix pedicellaris Pursh

BOG-WILLOW

NC **OBL** | MW **OBL** | GP **OBL**

Short, sparsely branched native shrub 4–15 dm tall; **twigs** dark brown and smooth. **Leaves** oblong-lance-shaped to obovate, tapered to tip or blunt and often with a short point, 3–6 cm long and 0.5–2 cm wide, silky hairy when young, becoming hairless and thick and leathery with age, green on upper surface, white-waxy below, veins pale-colored and slightly raised on both sides; margins entire, often slightly rolled under; petioles without glands, 2–8 mm long; stipules absent. **Catkins** appearing with leaves in spring; female catkins 2–4 cm long, on leafy branches 1–3 cm long; catkin bracts persistent, yellow-brown, hairy on inner surface near tip; stamens 2. **Capsules** lance-shaped, 4–7 mm long, without hairs, on stalks 2–3 mm long. May–June. **Habitat** Bogs, fens, sedge meadows, interdunal wetlands.

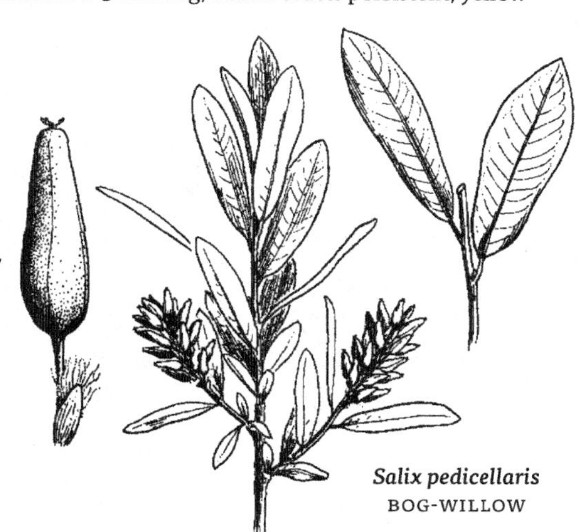

Salix pedicellaris
BOG-WILLOW

Salix pellita Andersson

SATINY WILLOW

NC **FACW** | GP **FACW** | *special concern*

Native shrub, 3–5 m tall; **twigs** easily broken, yellow to olive-brown or red-brown, smooth or sparsely hairy when young, becoming waxy. **Leaves** lance-shaped, 4–12 cm long and 1–2 cm wide, short-tapered to a tip, upper surface without hairs, veins sunken, underside waxy and satiny hairy but becoming smooth with age, with numerous, parallel lateral veins; margins rolled under, entire or with rounded teeth; petioles to 1 cm long; stipules absent. **Catkins** appearing and maturing before leaves in spring; female catkins 2–5 cm long, stalkless or on short branches to 1 cm long; catkin bracts black, long hairy; male catkins uncommon. **Capsules** lance-shaped, 4–6 mm long, silky hairy, ± stalkless. May. **Habitat** Streambanks, sandy shores and rocky shorelines. **Status** Special concern; in the midwest, uncommon in ne Minnesota and n Michigan.

Salix petiolaris J. E. Smith
MEADOW-WILLOW
NC **FACW** | MW **OBL** | GP **OBL**

Native shrub to 5 m tall; **twigs** red-brown to dark brown, sometimes with short, matted hairs when young, smooth with age. **Leaves** narrowly lance-shaped, 4–10 cm long and 1–2.5 cm wide, hairy when young, becoming smooth, dark green above, white-waxy below,; margins entire or with small, gland-tipped teeth; petioles without glands, 3–10 mm long; stipules absent. **Catkins** appearing with leaves in spring; female catkins 1–4 cm long, stalkless or on short branches to 2 cm long; catkin bracts persistent, brown, with a few long, soft hairs; stamens 2. **Capsules** narrowly lance-shaped, 4–8 mm long, finely hairy, on stalks 2–4 mm long. May. **Synonyms** *Salix gracilis*. **Habitat** Common; wet meadows, fens, streambanks, shores, open bogs, floating sedge mats, ditches.

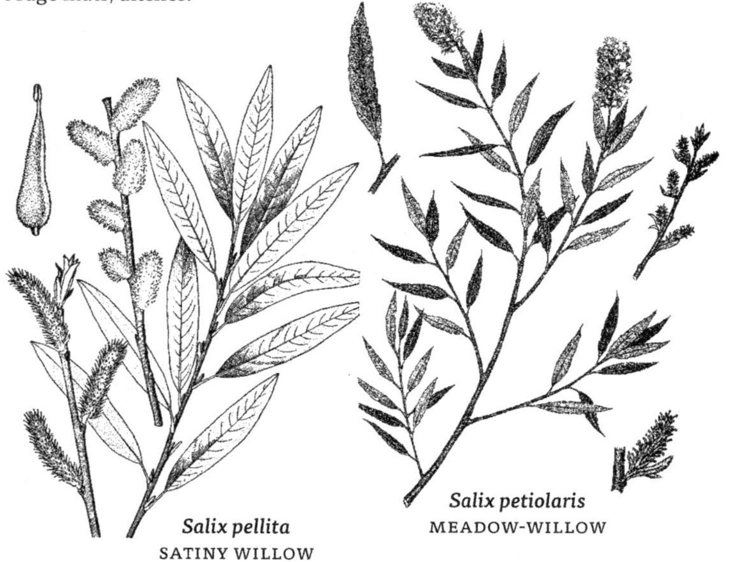

Salix pellita
SATINY WILLOW

Salix petiolaris
MEADOW-WILLOW

Salix planifolia Pursh
TEA-LEAF WILLOW
NC **OBL** | MW **OBL** | GP **OBL**

Native shrub to 3 m tall; **twigs** dark red-brown, short hairy when young, soon smooth and shiny. **Leaves** oval to oblong-lance-shaped, 3–6 cm long and 1–3 cm wide, short-hairy when young, becoming smooth, green above, paler or waxy below; margins entire or with a few small, rounded teeth; petioles 3–6 mm long, without glands; stipules small and deciduous. **Catkins** appearing before leaves in spring; female catkins 2–5 cm long, stalkless; catkin bracts 2–3 mm long, persistent, black, with long, soft hairs; stamens 2. **Capsules** lance-shaped, 4–8 mm long, finely hairy, stalkless or on short stalks to 0.5 mm long. May. **Synonyms** *Salix phylicifolia* subsp. *planifolia*. **Habitat** Rocky lakeshores, cedar swamps, black spruce bogs, streambanks, and margins of sedge meadows.

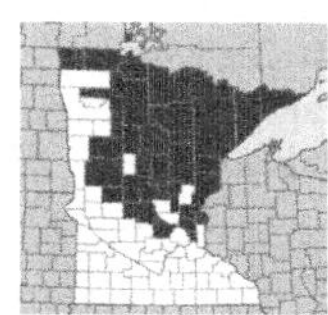

Salix pyrifolia Andersson
BALSAM-WILLOW
NC **FACW** | MW **FACW** | GP **OBL**

Native shrub or small tree to 5 m tall; **twigs** smooth, yellow when young, becoming shiny red. **Leaves** smooth, ovate to lance-shaped, often rounded at tip, rounded to heart-shaped at base, 4–12 cm long and 2–4 cm wide, red-tinged and translucent when unfolding; green on upper surface, waxy and finely net-veined below; with balsam fragrance (especially when dried); margins with small gland-tipped teeth; petioles 1–2 cm long; stipules absent or small and 1–2 mm long. **Catkins** appearing with or after leaves in spring; female catkins loosely flowered, 2–6 cm long, on leafy branches 1–3 cm long; catkin bracts red-brown, white-hairy, 2 mm long; stamens 2. **Capsules** lance-shaped, beaked at tip, 6–8 mm long, smooth, on stalks 2–4 mm long. May–June. **Synonyms** *Salix balsamifera*. **Habitat** Conifer swamps, bogs, rocky shores, wet depressions in boreal forests.

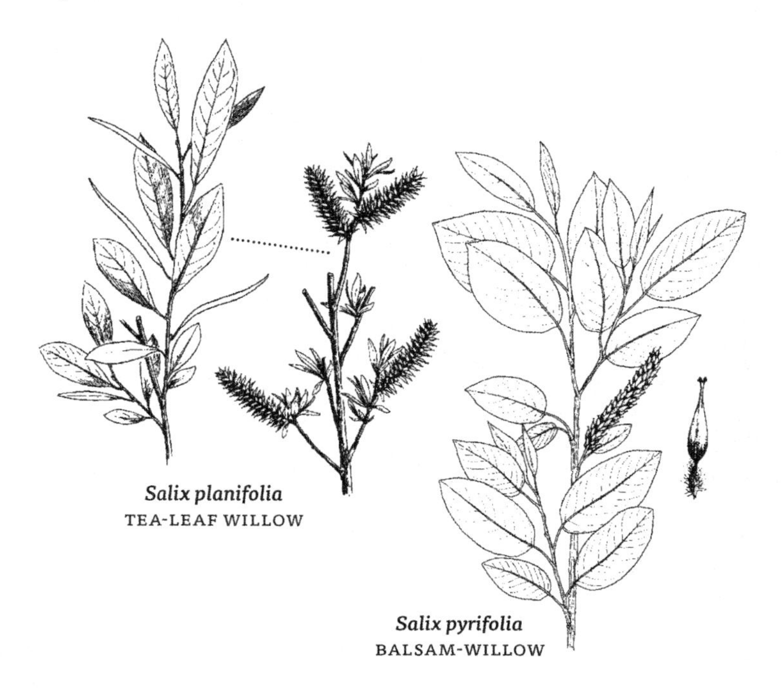

Salix planifolia
TEA-LEAF WILLOW

Salix pyrifolia
BALSAM-WILLOW

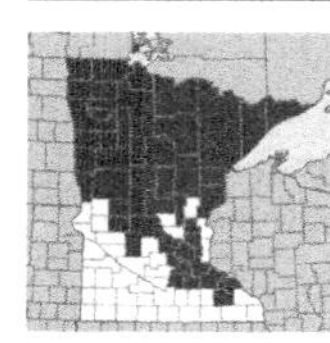

Salix serissima (L. H. Bailey) Fernald
AUTUMN-WILLOW
NC **OBL** | MW **OBL** | GP **OBL**

Native shrub to 4 m tall; **twigs** gray, yellow or dark brown, shiny and smooth. **Leaves** smooth, oval to lance-shaped, 4–10 cm long and 1–3 cm wide, red and hairless when young; green and shiny above, usually white-waxy below; margins with small gland-tipped teeth; petioles with glands near base of leaf; stipules usually absent. **Catkins** appearing with or after leaves in spring; female catkins 2–4 cm long, on leafy branches 1–4 cm long; catkin bracts deciduous, light yellow, long hairy; stamens 3–7. **Capsules** narrowly cone-shaped, 7–10 mm long, smooth, on stalks to 2 mm long. Late May–July; our latest blooming willow. **Habitat** Fens, cedar and tamarack swamps, marshes, floating sedge mats, streambanks and shores, often where calcium-rich.

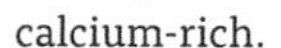

Salix serissima
AUTUMN-WILLOW

Salix discolor
PUSSY-WILLOW

GEOCAULON | *Toadflax*

Geocaulon lividum (Richardson) Fern.
NORTHERN RED-FRUIT TOADFLAX
NC **FAC**

Native perennial herb, from a slender rhizome; at least partially parasitic on other plants. **Stems** smooth, 1–3 dm long. **Leaves** alternate, oval or ovate, 1–3 cm long and 1–1.5 cm wide, rounded at tip; margins entire; petioles short. **Flowers** usually 3 on slender stalks from leaf axils, the lateral 2 flowers typically male, the middle flower perfect; sepals 4–5, triangular, 1–2 mm long; petals absent; style very short. **Fruit** a round, orange or red drupe, about 6 mm wide. June–Aug. **Synonyms** *Comandra livida*. **Habitat** Cedar swamps, open bogs; more commonly in sandy conifer woods and forested dune edges.

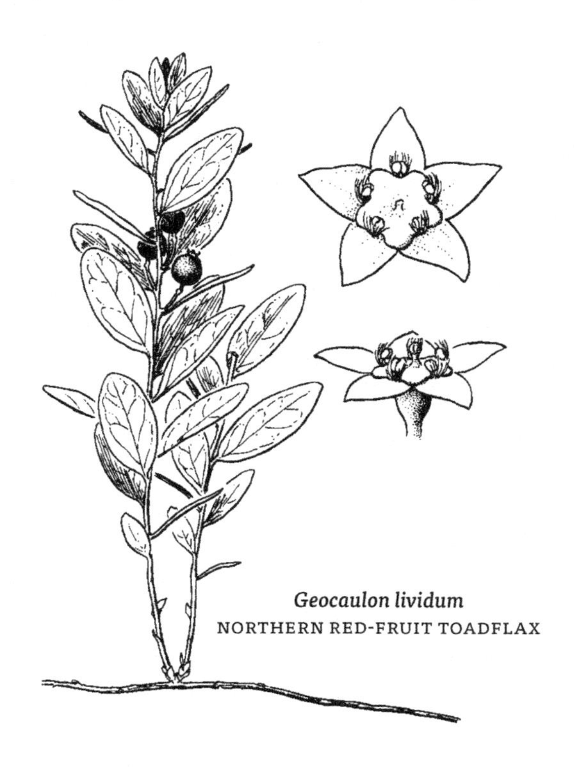

Geocaulon lividum
NORTHERN RED-FRUIT TOADFLAX

SARRACENIA | *Pitcher-plant*

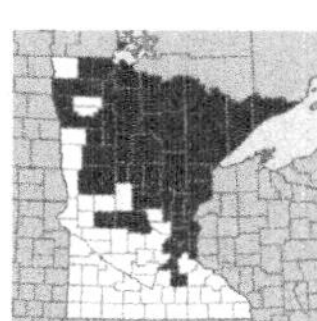

Sarracenia purpurea L.

PITCHER-PLANT
NC **OBL** | MW **OBL** | GP **OBL**

Native perennial insectivorous herb. Flower stalks leafless, 3–6 dm long. **Leaves** clumped, hollow and vaselike, curved and upright from base of plant, 1–2 dm long and 1–5 cm wide, green or veined with red-purple, winged, smooth on outside, upper portion of inside with downward-pointing hairs, tapered to a short petiole at base. **Flowers** large and nodding, 5–6 cm wide, single at ends of stalks, perfect; sepals 5; petals 5, obovate, dark red-purple, curved inward over yellow style; ovary large and round. **Fruit** a 5-chambered capsule; seeds small and numerous. May–July. **Habitat** Sphagnum bogs, floating bog mats, occasional in calcium- rich wetlands.

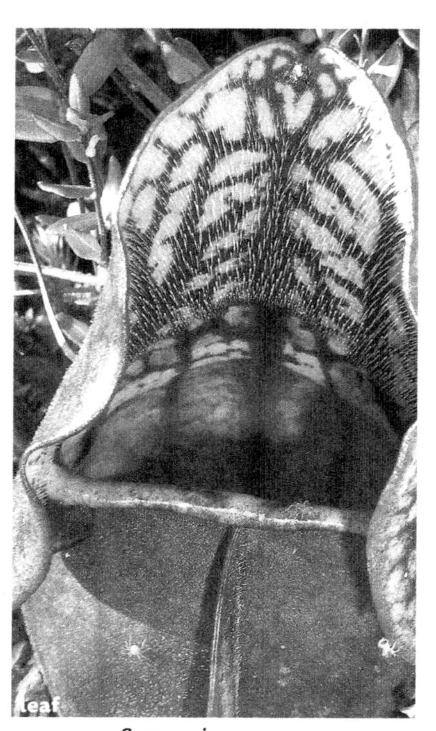

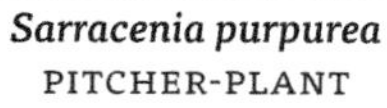

Sarracenia purpurea
PITCHER-PLANT

PERENNIAL HERBS with alternate, opposite or basal leaves. **Flowers** perfect (with both male and female parts), regular, single on stalks or in narrow heads; sepals 5; petals 5 or absent (*Penthorum*); stamens 5 or 10, stigmas 2 or 4. **Fruit** mostly a 2-parted capsule.

SAXIFRAGACEAE | SAXIFRAGE FAMILY

1 Leaves all on stem 2
1 Leaves all (or nearly all) from base of plant 3

2 Leaves mostly opposite; plants low and trailing . . . . *Chrysosplenium americanum* | AMERICAN GOLDEN-SAXIFRAGE
2 Leaves alternate; plants upright . . . . *Penthorum sedoides* . . . . DITCH-STONECROP

3 Leaves with rounded teeth. . . . . *Mitella nuda* | SMALL BISHOP'S-CAP
3 Leaves entire 4

4 Flowers single atop a smooth stalk; leaves more than 2x longer than wide . . . . *Parnassia* | GRASS OF-PARNASSUS
4 Flowers many in a panicle, the stalk hairy; leaves less than 2x longer than wide. . . . . *Saxifraga pensylvanica* | SWAMP-SAXIFRAGE

CHRYSOSPLENIUM | *Golden-saxifrage*

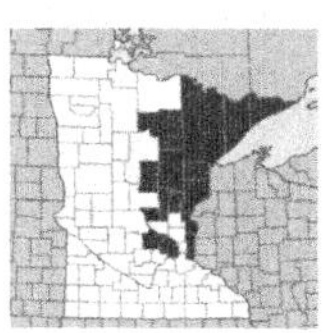

Chrysosplenium americanum Schwein.
AMERICAN GOLDEN-SAXIFRAGE
NC **OBL** | MW **OBL**

Small native perennial herb, often forming large mats. **Stems** creeping, branched, 5–20 cm long. Lower **leaves** opposite, the upper often alternate, broadly ovate, 5–15 mm long and as wide, margins entire or with rounded teeth or lobes; petioles short. **Flowers** single and stalkless from leaf axils, 4–5 mm wide; sepals 4, green-yellow or purple-tinged; petals absent; stamens usually 8 from a red or green disk, anthers red. **Fruit** a 2-lobed capsule. April–June. **Habitat** Springs, shallow streams, shady wet depressions; soils mucky.

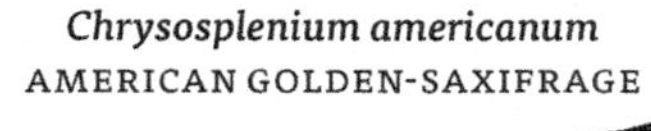

Chrysosplenium americanum
AMERICAN GOLDEN-SAXIFRAGE

RIGHT ***Chrysosplenium iowense***, *more common in arctic regions, is known from several locations in driftless area of ne Iowa and se Minnesota (state endangered). Plants occur on cool, moist, mossy, talus slopes underlain by dolomitic limestone. It differs from Chrysosplenium americanum by having leaf margins with 5–7 large rounded teeth.*

MITELLA | *Mitrewort, Bishop's-cap*

Mitella nuda L.

SMALL BISHOP'S-CAP
NC **FACW** | MW **FACW** | GP **OBL**

Small native perennial herb, spreading by rhizomes or stolons. **Leaves** all from base of plant, or with 1 small leaf on flower stalk, rounded heart-shaped, 1–3.5 cm wide, both sides with sparse coarse hairs; margins with rounded teeth; petioles 2–8 cm long. **Flowers** small, green, on short stalks, in racemes of 3–12 flowers, on a glandular-hairy stalk 10–25 cm tall; calyx lobes 5, 1–2 mm long; petals green, pinnately divided into usually 4 pairs of threadlike segments, the segments 2–4 mm long; stamens 10. **Fruit** a capsule, splitting open to reveal the black, shiny, 1 mm long seeds. June–July. **Habitat** Hummocks in swamps and alder thickets, ravines, seeps, moist mixed conifer and deciduous forests.

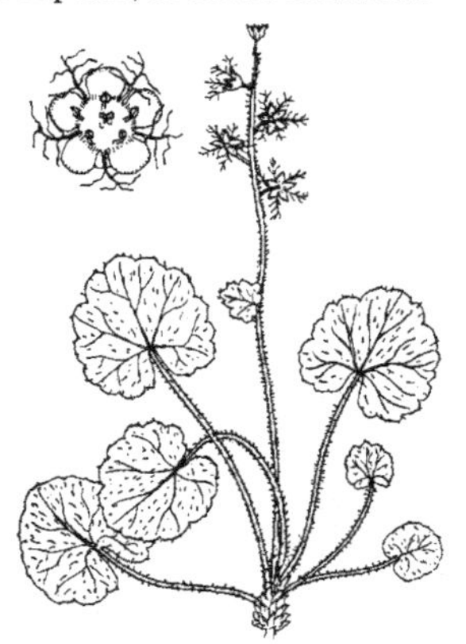

PARNASSIA | *Grass-of-parnassus*

Smooth perennial herbs. **Leaves** all from base of plant but often with 1 stalkless leaf near middle of stalk, margins entire; petioles present. **Flowers** large, white, single at ends of stalks; calyx 5-lobed; petals white, veined, spreading; fertile stamens 5, alternating with petals; staminodes (infertile stamens) attached to base of petals and divided into threadlike segments tipped with glandular knobs; stigmas 4. **Fruit** a 4-chambered capsule with numerous seeds.

PARNASSIA | GRASS-OF-PARNASSUS

1 Sepals with narrow translucent margins; staminodes (sterile stamens) 3-parted, not widened at base .... *Parnassia glauca*
.... AMERICAN-GRASS OF-PARNASSUS

1 Sepal margins green; staminodes 5 to many-parted . . . . *Parnassia palustris*
. . . . NORTHERN GRASS OF-PARNASSUS

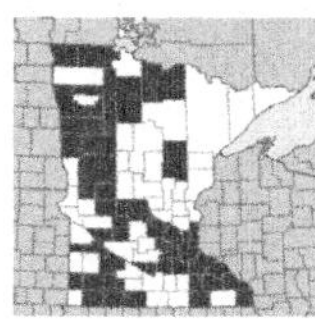

Parnassia glauca Raf.
AMERICAN GRASS-OF-PARNASSUS
NC **OBL** | MW **OBL** | GP **OBL**

Smooth native perennial herb. **Leaves** from base of plant and usually with 1 ± stalkless stem leaf; broadly ovate to nearly round, 2–7 cm long and 1–5 cm wide, rounded to a blunt or somewhat pointed tip; margins entire; petioles long. **Flowers** single atop a stalk 1–4 dm long; sepals ovate, 2–5 mm long, with a narrow, translucent margin; petals white with green veins, 1–2 cm long; staminodes 3-parted from near base, shorter than to equal to stamens. **Fruit** a capsule about 1 cm long. Aug–Sept. **Habitat** Calcareous fens and wet meadows.

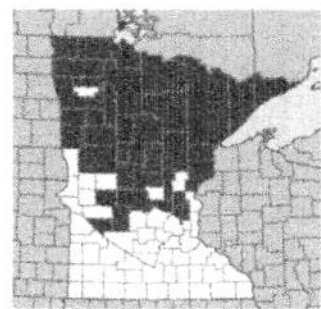

Parnassia palustris L.
NORTHERN GRASS-OF-PARNASSUS
NC **OBL** | MW **OBL** | GP **OBL**

Smooth native perennial herb. **Leaves** from base of plant and usually with 1clasping, heart-shaped leaf below middle of stalk; ovate to nearly round, 1–3 cm long, rounded or blunt at tip; margins entire; petioles long and slender. **Flowers** single atop a stalk 1.5–4 dm long; sepals lance-shaped, to 1 cm long, green throughout; petals white with green veins, ovate, 1–1.5 cm long, longer than sepals; staminodes many-parted from the widened tip, 5–9 mm long. **Fruit** a capsule. July–Sept. **Habitat** Calcareous fens, shores, streambanks, wet meadows.

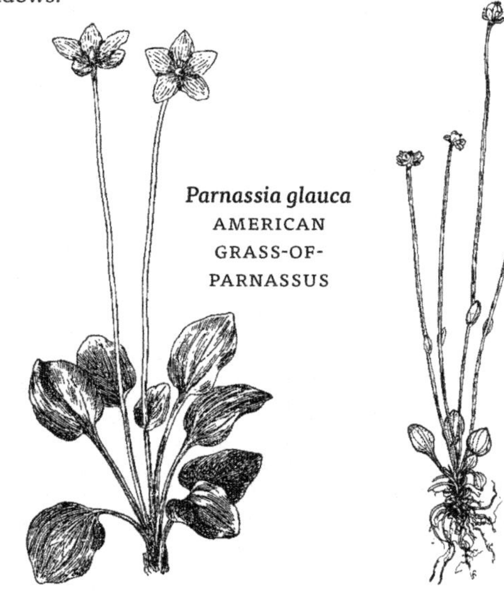

Parnassia glauca
AMERICAN GRASS-OF-PARNASSUS

Parnassia palustris
NORTHERN GRASS-OF-PARNASSUS

PENTHORUM | *Ditch-stonecrop*

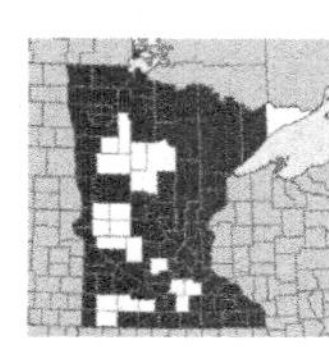

Penthorum sedoides L.

DITCH-STONECROP

NC **OBL** | MW **OBL** | GP **OBL**

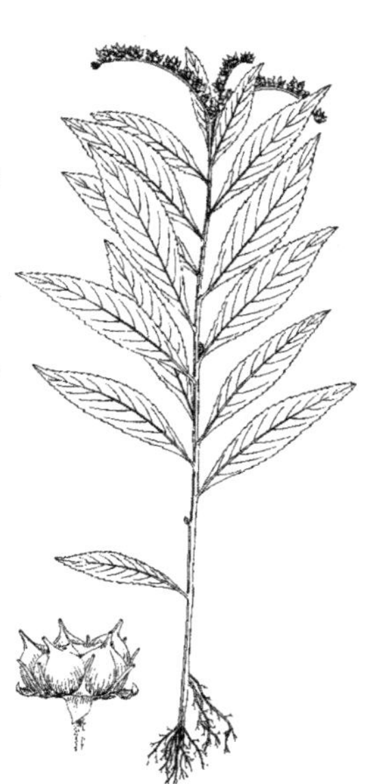

Native perennial herb, spreading by rhizomes; plants often red-tinged. **Stems** 1–6 dm long, smooth and round in section below, upper stem often angled and with gland-tipped hairs. **Leaves** alternate, lance-shaped to narrowly oval, 2–10 cm long and 0.5–3 cm wide, tapered to tip and base; margins with small, forward-pointing teeth; stalkless or on petioles to 1 cm long. **Flowers** star-shaped, perfect, 3–6 mm wide, on short stalks, in branched racemes at ends of stems; sepals 5, green, triangular, 1–2 mm long; petals usually absent; stamens 10; pistils 5, joined at base and sides to form a ring. **Fruit** a many-seeded capsule, the seeds about 0.5 mm long. July–Sept. **Habitat** Streambanks, muddy shores and ditches.

SAXIFRAGA | *Saxifrage*

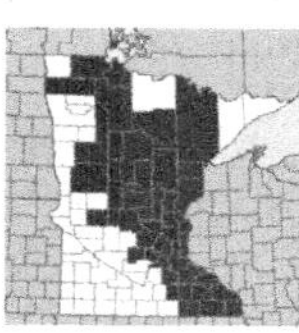

Saxifraga pensylvanica L.

SWAMP-SAXIFRAGE

NC **OBL** | MW **OBL** | GP **FACW**

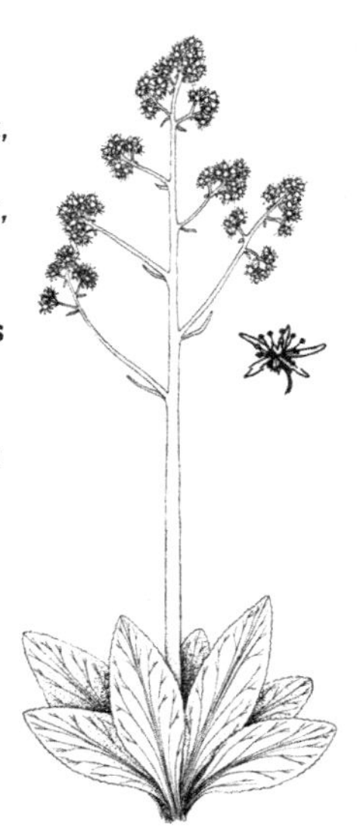

Native perennial herb. **Stems** stout, erect, 3–10 dm long, with sticky hairs. **Leaves** all from base of plant, ovate to oblong ovate, 1–2 dm long and 4–8 cm wide, smooth or hairy; margins entire to slightly wavy or with irregular rounded teeth; petioles wide. **Flowers** small, in clusters atop stem, the head elongating with age; sepals bent backward, 1–2 mm long; petals green-white or purple-tinged, lance-shaped, 2–3 mm long; stamens 10, the filaments threadlike. **Fruit** a follicle. May–June. **Synonyms** *Micranthes pensylvanica*. **Habitat** Swamps, wet deciduous forests, marshes, moist meadows and low prairie; often where calcium-rich.

Saxifraga pensylvanica
SWAMP-SAXIFRAGE

SCROPHULARIACEAE
Figwort Family

ANNUAL, BIENNIAL, OR PERENNIAL HERBS. Leaves opposite, alternate, or (rarely) all from base of plant (*Limosella*). **Flowers** single or few from leaf axils, or numerous in clusters at ends of stems or leaf axils, perfect, usually with a distinct upper and lower lip; sepals and petals 4–5 (petals sometimes absent); stamens usually 4. **Fruit** a capsule.

SCROPHULARIACEAE | FIGWORT FAMILY

1 Leaves all clustered at base of plant . . . . *Limosella aquatica* . . . . NORTHERN MUDWORT
1 Leaves all or mostly from stem 2

2 Sepals (or calyx lobes) 4 or less 3
2 Sepals (or calyx lobes) 5 6

3 Most leaves pinnately lobed . *Pedicularis lanceolata* | SWAMP-LOUSEWORT
3 Leaves entire or toothed, not lobed 4

4 Leaves whorled . . . . *Veronicastrum virginicum* | CULVER'S ROOT
4 Leaves opposite 5

5 Stamens 2, flowers blue (occasionally white) . . . . . . *Veronica* | SPEEDWELL
5 Stamens 4, flowers white or pale yellow.. . . . . *Melampyrum lineare* . . . . AMERICAN COW-WHEAT

6 Flowers in spikes at ends of stems 7
6 Flowers single from leaf axils 8

7 Leaves whorled . . . . *Veronicastrum virginicum* | CULVER'S ROOT
7 Leaves opposite . . . . *Chelone* | TURTLEHEAD

8 Sepals joined to form a tube 9
8 Sepals not joined, free to their base 10

9 Leaves linear, usually rough-to-touch; corolla ± regular . . . . *Agalinis* . . . . AGALINIS
9 Leaves broader, smooth; corolla 2-lipped. . . . *Mimulus* | MONKEY-FLOWER

10 Leaves about as long as wide; corolla regular; upper sepal much wider than other 4 sepals . . . . *Bacopa rotundifolia* | WATER-HYSSOP
10 Leaves longer than wide; corolla 2-lipped; sepals all about same width. . 11

11 Petals white; stamens 2; plants with gland-tipped hairs (at least in upper portion) . . . . *Gratiola* | HEDGE HYSSOP
11 Petals blue-violet; stamens 4; plants smooth throughout. . *Lindernia dubia* . . . . FALSE PIMPERNEL

AGALINIS | *Agalinis*

Annual herbs. **Stems** slender, erect, branched, 4-angled. **Leaves** opposite, linear, stalkless, smooth, or rough-to-touch on upper surface. **Flowers** showy, in clusters at ends of branches; sepals joined, the calyx 5-lobed, bell-shaped; petals united, corolla 5-lobed, bell-shaped, and slightly 2-lipped, pink to purple; stamens 4, of 2 different lengths. **Fruit** a ± round, many-seeded capsule.

AGALINIS | AGALINIS

1 Flowers on short stalks 2–5 mm long .... *Agalinis purpurea* .... SMOOTH AGALINIS

1 Flower stalks 1 cm or more long . . *Agalinis tenuifolia* | COMMON AGALINIS

Agalinis purpurea (L.) Pennell

SMOOTH AGALINIS

NC **FACW** | MW **FACW** | GP **FACW**

Native annual herb. **Stems** slender, 2–8 dm long, 4-angled, smooth to slightly rough, branched and spreading above. **Leaves** opposite, spreading, linear, 1–5 cm long and 1–3 mm wide; margins entire; petioles absent. **Flowers** on spreading stalks 2–5 mm long, in racemes on the branches; calyx 4–6 mm long; corolla purple, 2–3 cm long, the lobes spreading, 5–10 mm long. **Fruit** a round capsule, 4–6 mm wide. Aug–Sept. **Synonyms** *Gerardia purpurea*. **Habitat** Wet meadows, fens, shores of lakes and ponds, ditches; usually where sandy, sometimes where calcium-rich.

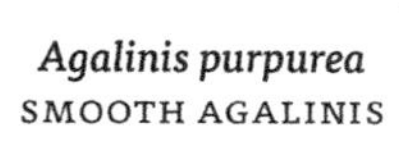

Agalinis purpurea
SMOOTH AGALINIS

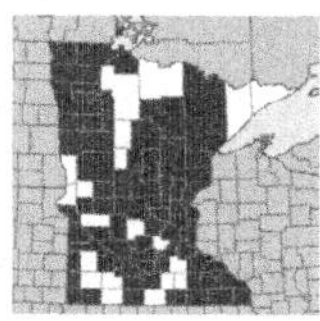

Agalinis tenuifolia (M. Vahl) Raf.
COMMON AGALINIS
NC **FACW** | MW **FACW** | GP **FAC**

Native annual herb. **Stems** slender, erect, 2–6 dm tall, smooth, usually with many branches. **Leaves** opposite, spreading, linear, 1–5 cm long and 1–3 mm wide, upper surface slightly rough; margins entire; petioles absent. **Flowers** on slender, ascending stalks, 1–2 cm long; calyx 3–5 mm long, with short teeth; corolla purple (rarely white), often spotted, 10–15 mm long, the lobes 3–5 mm long. **Fruit** a round capsule, 4–6 mm wide. Aug–Sept. **Synonyms** *Agalinis besseyana*, *Gerardia tenuifolia*. **Habitat** Wet meadows, low prairie, fens, shores, streambanks and ditches, usually where sandy.

Agalinis tenuifolia
COMMON AGALINIS

BACOPA | *Water-hyssop*

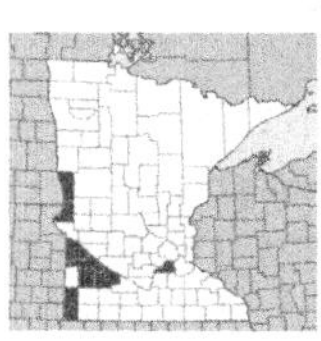

Bacopa rotundifolia (Michx.) Wettst.
WATER-HYSSOP
NC **OBL** | MW **OBL** | GP **OBL** | *special concern*

Native perennial succulent herb, spreading by stolons. **Stems** creeping, 0.5–4 dm long, rooting at leaf nodes, smooth when underwater, usually hairy when emersed. **Leaves** opposite, obovate to nearly round, 1–3.5 cm long and 1–2.5 cm wide, smooth, palmately veined, rounded at tip, clasping at base; margins entire or slightly wavy; petioles absent. **Flowers** 1–2 from leaf axils, on hairy stalks to 1.5 cm long and shorter than the leaves; sepals 5, unequal, 4–5 mm long; corolla white with a yellow throat, 2-lipped, bell-shaped, 5–10 mm long, the 5 lobes shorter than the tube; stamens 4. **Fruit** a round capsule, about as long as sepals. Aug–Sept. **Synonyms** *Bacopa simulans*, *Hydranthelium*

Bacopa rotundifolia
WATER-HYSSOP

roundifolium. **Habitat** Mud flats and shallow water of ponds and marshes. **Status** *Special concern.*

CHELONE | *Turtlehead*

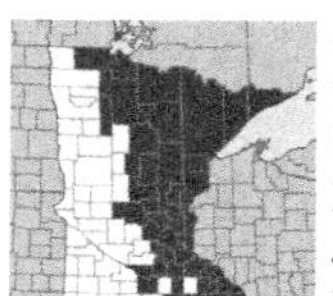

Chelone glabra L.

WHITE TURTLEHEAD

NC **OBL** | MW **OBL** | GP **OBL**

Native perennial herb. **Stems** erect, 5–10 dm long, rounded 4-angled, unbranched or sometimes branched above. **Leaves** opposite, lance-shaped, to 15 cm long and 1–3 cm wide, tapered to a sharp tip; margins with sharp, forward-pointing teeth; petioles very short or absent. **Flowers** large, 5-parted, 2-lipped, the lower lip 3-lobed and woolly on inner surface; perfect, in dense spikes at ends of stems, sepals 5; corolla white or light pink, 2.5–3.5 cm long; stamens 5 (4 fertile and 1 sterile and smaller), anthers woolly; style threadlike. **Fruit** an ovate, many-seeded capsule. Aug–Sept. **Habitat** Swamp openings, thickets, streambanks, shores, wet meadows, marshes, calcareous fens.

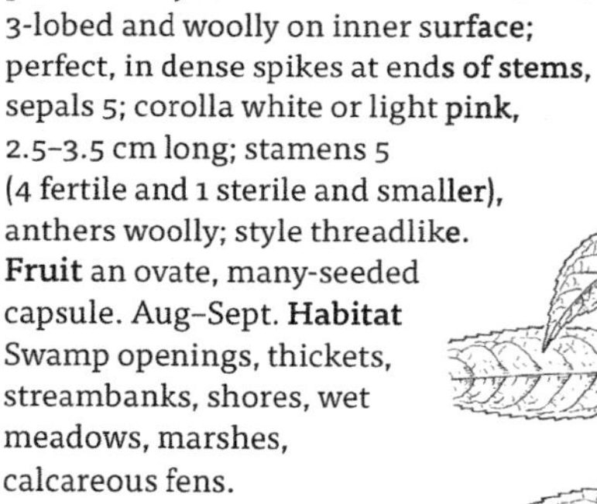

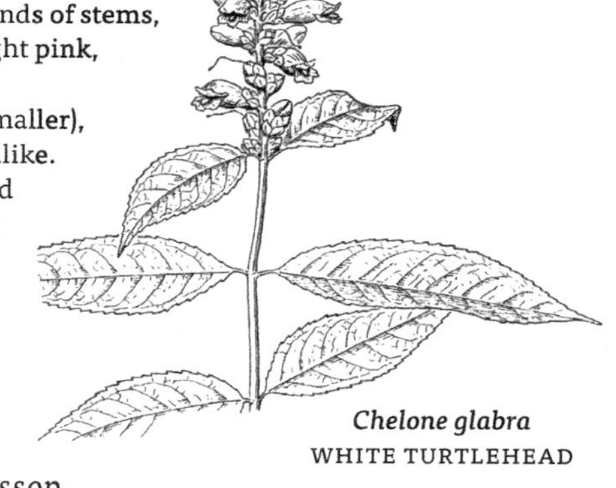

Chelone glabra
WHITE TURTLEHEAD

GRATIOLA | *Hedge hyssop*

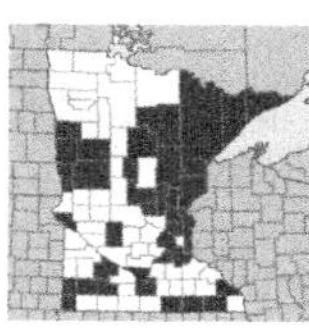

Gratiola neglecta Torr.

CLAMMY HEDGE HYSSOP

NC **OBL** | MW **OBL** | GP **OBL**

Native annual herb. **Stems** erect to horizontal, 5–25 cm long, usually branched, glandular-hairy above. **Leaves** opposite, linear to lance-shaped, 5–25 mm long and 1–10 mm wide, clasping at base; margins entire to wavy-toothed; petioles absent. **Flowers** single in the leaf axils, on slender stalks 1–2 cm long, subtended by a pair of small narrow bracts; sepals 5, unequal, 3–6 mm long, enlarging after flowering; corolla white, tube-shaped, slightly 2-lipped, the lower lip 3-lobed , 6–10 mm long; fertile

stamens 2. **Fruit** an ovate, 4-chambered capsule, 3–5 mm long. June–Sept. **Habitat** Mud flats, shores of ponds and marshes.

LIMOSELLA | *Mudwort*

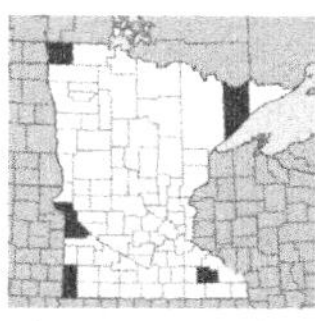

Limosella aquatica L.

NORTHERN MUDWORT

NC **OBL** | MW **OBL** | GP **OBL** | *special concern*

Small native annual herb (sometimes perennial by stolons); plants smooth, succulent. **Stems** 3–10 cm long. **Leaves** from base of plant, linear but wider at tip, the tip emersed or floating, 1–2.5 cm long and 2–8 mm wide, tapered to a long petiole; margins entire. **Flowers** small, single on stalks from base of plant, the stalks shorter than the leaves; calyx lobes triangular, corolla white or pink, 1–3 mm wide, slightly longer than sepals; stamens 4. **Fruit** an ovate capsule, 2–3 mm long. June–Sept. **Habitat** Streambanks, shores and mud flats of temporary ponds and marshy areas. **Status** *Special concern.*

Limosella aquatica
NORTHERN MUDWORT

LINDERNIA | *False pimpernel*

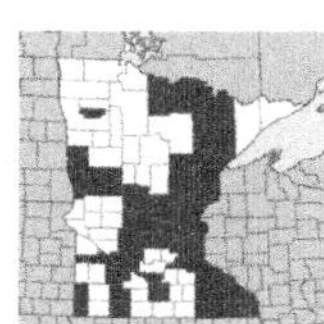

Lindernia dubia (L.) Pennell

FALSE PIMPERNEL

NC **OBL** | MW **OBL** | GP **OBL**

Native annual herb. **Stems** smooth, 1–2 dm long, widely branched. **Leaves** opposite, ovate to obovate, 5–30 mm long and 3–10 mm wide, the upper leaves smaller; margins entire or with small, widely spaced teeth; petioles absent. **Flowers** single, on slender stalks 0.5–2.5 cm long from leaf axils; sepals 5, linear; corolla light blue-purple, 5–10 mm long, 2-lipped, the upper lip 2-lobed, the lower lip 3-lobed and wider than upper lip; fertile stamens 2, staminodes (sterile stamens) 2. **Fruit** an ovate capsule, 4–6 mm long. June–Sept. **Synonyms**

Lindernia anagallidea. **Habitat** Mud flats, sandbars, shores of temporary ponds and marshes, streambanks.

MELAMPYRUM | *False pimpernel*

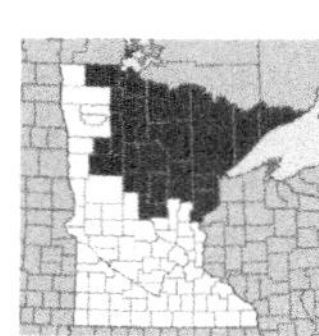

Melampyrum lineare Desr.
AMERICAN COW-WHEAT
NC **FACU** | MW **FAC**

Native annual herb, partially parasitic on other plants, often red-tinged when in open habitats. **Stems** usually branched, 1–4 dm long. **Leaves** opposite, lower leaves oblong lance-shaped, upper leaves linear or lance-shaped, often tootherd near base; petioles short or absent. **Flowers** from upper leaf axils; sepals joined, calyx lobes 5 and awl-shaped; corolla about 1 cm long, 2-lipped, the upper lip white, the lower pale yellow. **Fruit** a capsule to 1 cm long. Summer. **Habitat** Common in a wide variety of habitats, ranging from wet to dry forests and openings; in wetlands occasional in swamps and on hummocks in open fens.

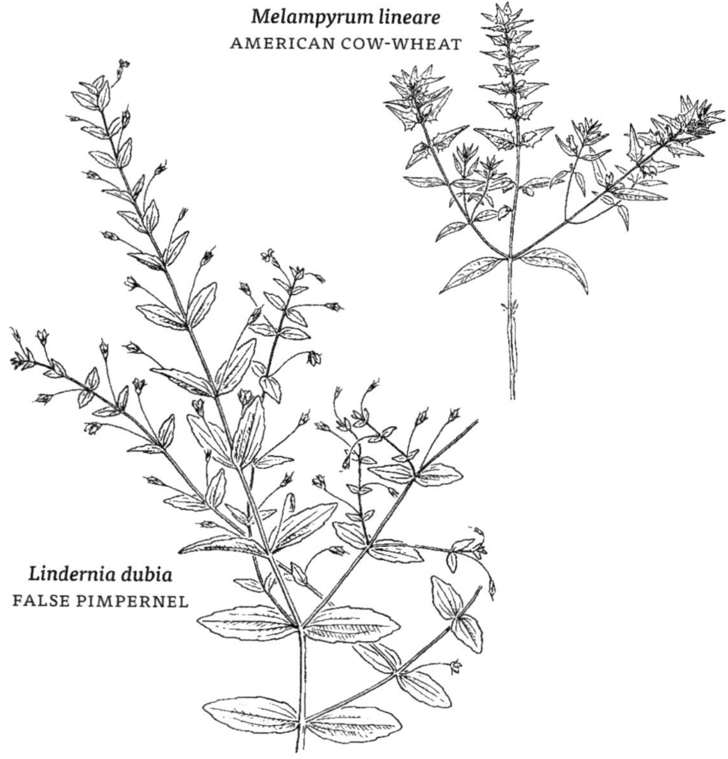

Melampyrum lineare
AMERICAN COW-WHEAT

Lindernia dubia
FALSE PIMPERNEL

MIMULUS | *Monkey-flower*

Perennial herbs. **Leaves** opposite, margins shallowly toothed. **Flowers** often large and showy, single on stalks from leaf axils or in leafy racemes at ends of stems; sepals joined, the calyx tube-shaped; corolla 2-lipped, the upper lip 2-lobed, the lower lip 3-lobed, yellow or blue-violet; stamens 4, of 2 different lengths; stigmas 2. **Fruit** a cylindric capsule.

MIMULUS | MONKEY-FLOWER

1 Flowers blue to violet . . . . *Mimulus ringens* . . . . ALLEGHENY OR SQUARE-STEMMED MONKEY-FLOWER

1 Flowers yellow . . . . . . *Mimulus glabratus* | ROUND-LEAF MONKEY-FLOWER

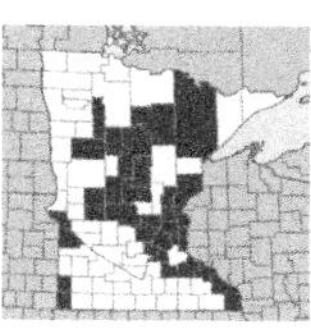

Mimulus glabratus HBK.

ROUND-LEAF MONKEY-FLOWER

NC **OBL** | MW **OBL** | GP **OBL**

Native perennial herb, spreading by stolons and often forming large mats. **Stems** succulent, smooth, 0.5–5 dm long, creeping and rooting at nodes, the stem ends angled upward. **Leaves** opposite, nearly round to broadly ovate, 1–2.5 cm wide, palmately veined, rounded at tip, hairy when young, becoming smooth; margins shallowly toothed or entire; petioles short and winged, or the upper leaves stalkless. **Flowers** yellow, on stalks from leaf axils and at ends of stems; calyx 5–9 mm long, barely toothed, irregular, the upper lobe large, the other lobes smaller; corolla 2-lipped, 9–15 mm long, the throat open and bearded on inner surface. **Fruit** an ovate capsule, 5–6 mm long. June–Aug. **Habitat** Cold springs, seeps, and banks of spring-fed streams; usually where calcium-rich.

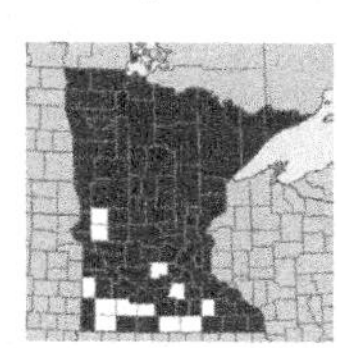

Mimulus ringens L.

ALLEGHENY OR SQUARE-STEMMED MONKEY-FLOWER

NC **OBL** | MW **OBL** | GP **OBL**

Smooth native perennial herb, from stout rhizomes. **Stems** usually erect, 3–8 dm long, 4-angled and sometimes winged. **Leaves** opposite, oblong to lance-shaped, 4–12 cm long and 1–3.5 cm wide, upper leaves smaller; margins with forward-pointing teeth; petioles absent, the base of leaf clasping stem. **Flowers** single from upper leaf axils, on slender stalks 1–5 cm long and longer than the sepals; calyx regular, angled, 1–2 cm long, the lobes awl-shaped, 3–5 mm long; corolla blue-violet, 2-lipped, 2–3 cm long, the throat nearly closed, the upper lip erect and bent upward, lower lip longer and bent backward. **Fruit** a capsule, about as long as calyx tube. July–Aug. **Habitat** Streambanks, oxbow marshes, swamp openings, floodplain forests, muddy shores, ditches; sometimes where disturbed.

PEDICULARIS | *Lousewort*

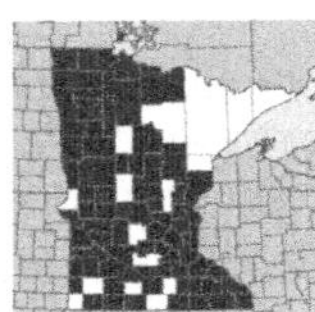

Pedicularis lanceolata Michx.
SWAMP-LOUSEWORT
NC **FACW** | MW **OBL** | GP **OBL**

Native perennial herb; plants at least partially parasitic on other plants. **Stems** 3–8 dm long, ± smooth, unbranched or few-branched. **Leaves** opposite, or in part alternate, mostly lance-shaped, 4–9 cm long and 1–2 cm wide, pinnately lobed; margins with small rounded teeth; lower leaves short-petioled, upper leaves stalkless. **Flowers** ± stalkless, in spikes at ends of stems and from upper leaf axils; the spikes 2–10 cm long; calyx 2-lobed; corolla yellow, about 2 cm long, the upper lip entire and arched, lower lip upright. **Fruit** an unequally ovate capsule, mostly shorter than the sepals. July–Sept. **Habitat** Wet meadows, calcareous fens, wetland margins, springs, streambanks.

Mimulus ringens
ALLEGHENY OR SQUARE-STEMMED MONKEY-FLOWER

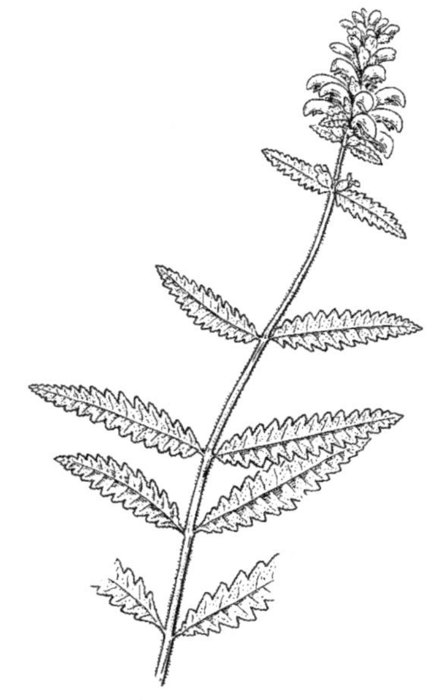

Pedicularis lanceolata
SWAMP-LOUSEWORT

VERONICA | *Speedwell*

Annual or perennial herbs. **Leaves** opposite, or becoming alternate in the head. **Flowers** single or in racemes from leaf axils or at ends of stems; sepals deeply 4-parted, enlarging after flowering; corolla blue or white, 4-lobed, somewhat 2-lipped, the tube shorter than the lobes; stamens 2. **Fruit** a flattened capsule, lobed at tip; styles usually persistent on fruit.

VERONICA | SPEEDWELL

1 Flowers in racemes at ends of stems; flower stalks 1–2 mm long . . . . *Veronica peregrina* | PURSLANE-SPEEDWELL
1 Flowers in racemes from leaf axils; flower stalks longer than 2 mm **2**

2 Leaves with short petioles . . *Veronica americana* | AMERICAN SPEEDWELL
2 Petioles absent. **3**

3 Leaves with sharp forward-pointing teeth, or entire; upper leaves clasping stem; capsules swollen. . . . . *Veronica anagallis–aquatica* . . . . WATER SPEEDWELL
3 Leaves with small, gland-tipped, outward-pointing teeth; leaves narrowed to base, not clasping stem; capsules strongly flattened . . . . *Veronica scutellata* | NARROW-LEAVED SPEEDWELL

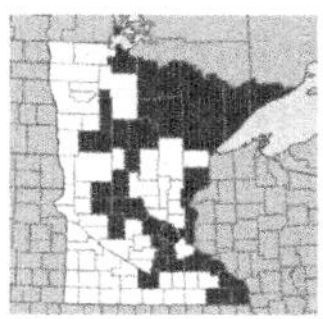

Veronica americana L.

AMERICAN SPEEDWELL
NC **OBL** | MW **OBL** | GP **OBL**

Native perennial herb, spreading by rhizomes; plants smooth and succulent. **Stems** erect to creeping, 1–6 dm long. **Leaves** opposite, ovate to lance-shaped (or lower leaves oval), 2–8 cm long and 0.5–3 cm wide, upper leaves tapered to a tip, lower leaves often rounded; margins with forward-pointing teeth; petioles short. **Flowers** in stalked racemes from leaf axils; the racemes with 10–25 flowers and to 15 cm long; corolla 4-lobed, blue (sometimes white), often with purple stripes. **Fruit** a ± round, compressed capsule, 3–4 mm long, slightly notched at tip, the styles persistent, 2–4 mm long. July–Sept. **Synonyms** *Veronica beccabunga*. **Habitat** Streambanks and wet shores, hummocks in swamps, springs.

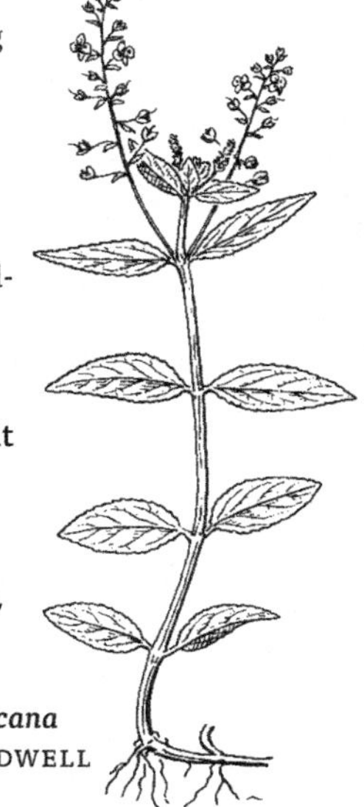

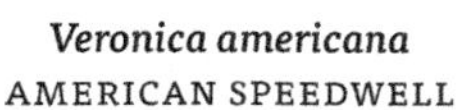

Veronica americana
AMERICAN SPEEDWELL

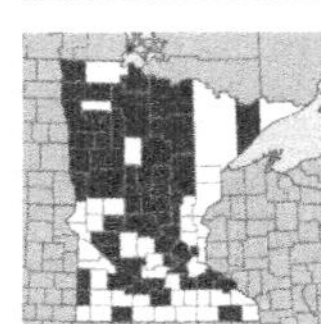

Veronica anagallis-aquatica L.
WATER SPEEDWELL
NC **OBL** | MW **OBL** | GP **OBL**

Biennial or short-lived perennial herb, spreading by stolons or leafy shoots produced in fall; plants ± smooth. **Stems** erect to spreading, 1–6 dm long, often rooting at lower nodes. **Leaves** opposite, lance-shaped to ovate, 2–10 cm long and 0.5–5 cm wide, tapered to a blunt or rounded tip; margins entire or with fine, forward-pointing teeth; petioles absent, the leaves often clasping. **Flowers** in many-flowered racemes from leaf axils, the racemes 5–12 cm long; corolla 4-lobed, blue or striped with purple, about 5 mm wide. **Fruit** a round, compressed capsule, 2–4 mm long, notched at tip, the styles persistent, 1–2 mm long. June–Sept. **Synonyms** *Veronica catenata, Veronica comosa*. **Habitat** Wet, sandy or muddy streambanks and ditches; often in shallow water. Evidently introduced from Eurasia and naturalized throughout most of North America.

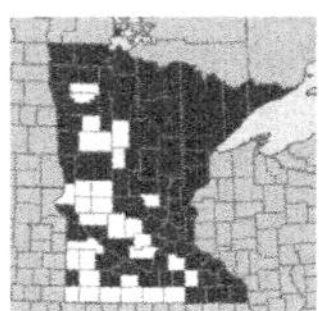

Veronica peregrina L.
PURSLANE-SPEEDWELL
NC **FAC** | MW **FACW** | GP **FACW**

Small native annual herb. **Stems** upright, 0.5–3 dm long, unbranched or with spreading branches, usually glandular-hairy. Lower leaves opposite, becoming alternate and smaller in the head, oval to linear, 5–25 mm long and 1–5 mm wide, rounded at tip; margins of lower leaves sparsely toothed, upper leaves entire; petioles short or absent. **Flowers** small, on short stalks from upper leaf

Veronica anagallis-aquatica
WATER SPEEDWELL

Veronica peregrina
PURSLANE-SPEEDWELL

axils; corolla 4-lobed, ± white, about 2 mm wide. **Fruit** an oblong heart-shaped capsule, 2–4 mm long, notched at tip, the styles not persistent. May–July. **Habitat** Mud flats, shores, ditches, temporary ponds, swales; also weedy in cultivated fields, lawns and moist disturbed areas.

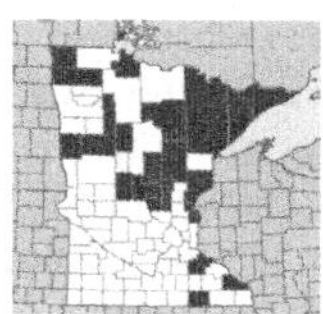

Veronica scutellata L.
NARROW-LEAVED SPEEDWELL
NC **OBL** | MW **OBL** | GP **OBL**

Native perennial herb, spreading by rhizomes or leafy shoots produced in fall; plants smooth (or sometimes with sparse hairs). **Stems** slender, erect to reclining, 1–4 dm long, often rooting at lower nodes. **Leaves** opposite, linear to narrowly lance-shaped, 3–8 cm long and 2–10 mm wide, tapered to a sharp tip; margins entire or with small, irregularly spaced teeth; petioles absent. **Flowers** in racemes from leaf axils, the racemes with 5–20 flowers, as long or longer than the leaves; corolla 4-lobed, blue, 6–10 mm wide. **Fruit** a strongly flattened capsule, 3–4 mm long, notched at tip, the styles persistent, 3–5 mm long. June–Sept.
Habitat Marshes, pond margins, hardwood swamps, thickets, springs, streambanks, wet swales and depressions.

VERONICASTRUM | *Culver's root*

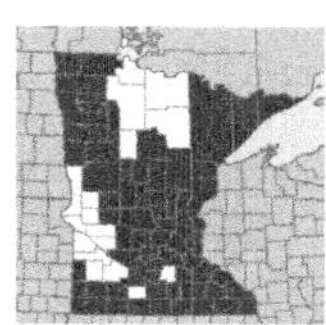

Veronicastrum virginicum (L.) Farw.
CULVER'S ROOT
NC **FAC** | MW **FAC** | GP **FAC**

Erect native perennial herb, 1–2 m tall, usually with several upright branches. **Leaves** in whorls of 3–6, lance-shaped; margins with fine, forward-pointing teeth; petioles to 1 cm long. **Flowers** in erect, spikelike racemes to 15 cm long, the flowers crowded and spreading; corolla white, nearly regular, 4–5-parted, the lobes shorter than the tube; stamens 2, long-exserted from the corolla mouth. **Fruit** a capsule, 4–5 mm long. June–Aug. **Synonyms** *Veronica virginicum, Leptandra virginicum*. **Habitat** Moist to wet prairie, fens and streambanks; also in drier deciduous woods and sandy grasslands.

ULMUS | *Elm*

Ulmus americana L.

AMERICAN ELM

NC **FACW** | MW **FACW** | GP **FAC**

Native tree to 25 m tall, trunk to 1 m wide, crown broadly rounded or flat-topped, smaller branches usually drooping; **bark** gray, furrowed, breaking into thin plates with age; **twigs** brown, smooth or with sparse hairs, often zigzagged; buds red-brown. **Leaves** alternate, simple, to 15 cm long and 7–8 cm wide, oval, pointed at tip, base strongly asymmetrical, upper surface dark green and smooth, lower surface pale and smooth or soft-hairy; margins coarsely double-toothed; petioles short, usually yellow. **Flowers** small, green-red, hairy, in drooping clusters of 3–4; appearing before leaves unfold in spring. **Fruit** 1-seeded, oval, 1 cm wide, with a winged, hairy margin, notched at tip. **Habitat** Floodplain forests, streambanks and moist, rich woods; less common now than formerly due to losses from Dutch elm disease.

bark

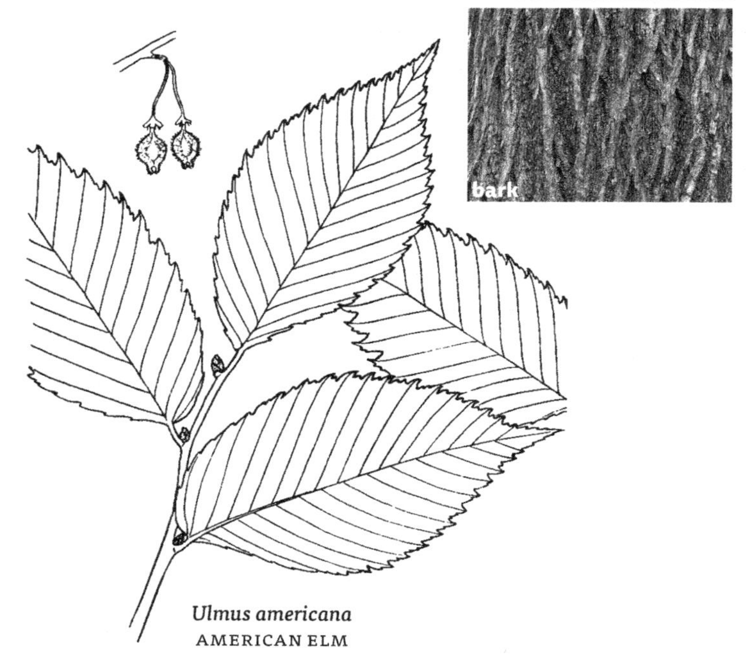

Ulmus americana
AMERICAN ELM

ANNUAL OR PERENNIAL HERBS with watery juice, sometimes with stinging hairs. **Leaves** alternate or opposite, simple, with petioles. **Flowers** small, green, in simple or branched clusters from leaf axils, male and female flowers usually separate, on same or separate plants; sepals joined, 3–5-lobed; petals absent; ovary superior, 1-chambered. **Fruit** an achene, often enclosed by the sepals which enlarge after flowering.

URTICACEAE | NETTLE FAMILY

1 Plants with stiff stinging hairs 2
1 Plants without stinging hairs; smooth or with sparse small hairs 3

2 Leaves alternate .... *Laportea canadensis* | WOOD-NETTLE
2 Leaves opposite .... *Urtica dioica* | STINGING NETTLE

3 Flowers in cylindric spikes from leaf axils; achene shorter than and hidden by sepals .... *Boehmeria cylindrica* | FALSE NETTLE
3 Flowers in dense short clusters from leaf axils; achene equal or longer than sepals .... *Pilea* | CLEARWEED

BOEHMERIA | *False nettle*

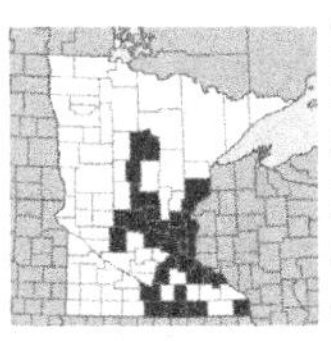

Boehmeria cylindrica (L.) Swartz

FALSE NETTLE
NC **OBL** | MW **OBL** | GP **FACW**

Perennial, nettle-like native herb, stinging hairs absent. **Stems** upright, 4–10 dm long, usually unbranched. **Leaves** opposite, rough-textured, ovate to broadly lance-shaped, narrowed to a pointed tip, with 3 main veins; margins coarsely toothed; petioles shorter than blades. **Flowers** tiny, green, male and female flowers usually on separate plants, in small clusters along unbranched stalks from upper leaf axils, forming cylindric, interrupted spikes of male flowers or continuous spikes of female flowers. **Fruit** an achene, enclosed by the enlarged bristly sepals and petals, ovate and narrowly winged. July–Aug. **Habitat** Floodplain forests, swamps, marshes and bogs.

Boehmeria cylindrica
FALSE NETTLE

LAPORTEA | *Wood-nettle*

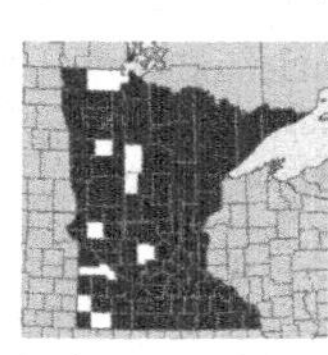

Laportea canadensis (L.) Wedd.
WOOD-NETTLE
NC **FACW** | MW **FACW** | GP **FAC** | *Caution - stinging hairs*
Native perennial herb, spreading by rhizomes. **Stems** somewhat zigzagged, 5–10 dm long. **Leaves** alternate, 8–15 cm long, ovate and narrowed to a tip, with small stinging hairs, margins coarsely toothed. **Flowers** small, green, male and female flowers separate but borne on same plant; male flowers in branched clusters from lower leaf axils, shorter than leaf petioles; female flowers in open, spreading clusters from upper axils, usually much longer than petioles. **Fruit** a flattened achene, longer than the 2 persistent sepals. July–Sept. **Habitat** Common; floodplain forests, rich moist woods, low places in hardwood forests, streambanks.

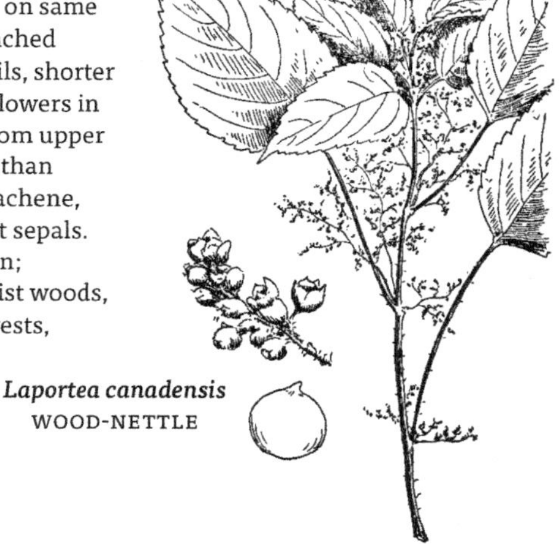

Laportea canadensis
WOOD-NETTLE

PILEA | *Clearweed*

Annual herb, sometimes forming colonies from seeds of previous year. **Stems** erect to sprawling, smooth, translucent and watery. **Leaves** opposite, stinging hairs absent, thin and translucent, ovate, with 3 major veins from base of leaf, margins toothed. **Flowers** green, male and female flowers separate, borne on same or different plants, in clusters from leaf axils; male flowers with 4 sepals and 4 stamens; female flowers with 3 sepals, ovary superior. **Fruit** a flattened, ovate achene.

PILEA | CLEARWEED

1 Achenes olive-green to dark purple with a narrow pale margin, about as long as wide, covered with low bumps; leaf petioles 1/5–1/2 length of blade . . . . *Pilea fontana* | BOG CLEARWEED

1 Achenes to 1 mm wide, green to yellow, longer than wide, often marked with purple spots, smooth; petioles 1/3 to as long as blade . . . . *Pilea pumila* | CLEARWEED

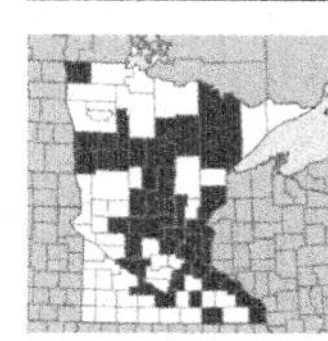

Pilea fontana (Lunell) Rydb.
BOG CLEARWEED
NC **FACW** | MW **FACW** | GP **OBL**

Native annual herb. **Stems** 1–4 dm long, often sprawling. **Leaves** opposite, 2–6 cm long and 1–4 cm wide; petioles 0.5–5 cm long. **Flowers** in clusters, male flowers usually innermost when mixed with female flowers. **Fruit** a dark olive-green to purple-black achene, 1–1.5 mm wide, with a narrow pale margin; sepals persistent, shorter to slightly longer than achene. Aug–Sept. **Habitat** Lakeshores, riverbanks, swamps, marshes and springs.

achene

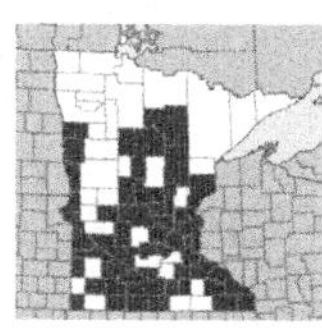

Pilea pumila (L.) A. Gray
CLEARWEED
NC **FACW** | MW **FACW** | GP **FAC**

Native annual herb; plants similar to *Pilea fontana*, but sometimes taller (to 5 dm). **Leaves** opposite, usually larger (to 12 cm long and 8 cm wide), thinner and more translucent than in *Pilea fontana*; petioles to 8 cm long. The green achenes (rather than nearly black), often marked with purple and to 1 mm wide, and more translucent stems, distinguish Pilea *pumila* from *Pilea fontana*. July–Sept. **Habitat** Swampy woods (often on logs), wooded streambanks, floodplain forests, wet depressions, rocky hollows; usually in partial shade.

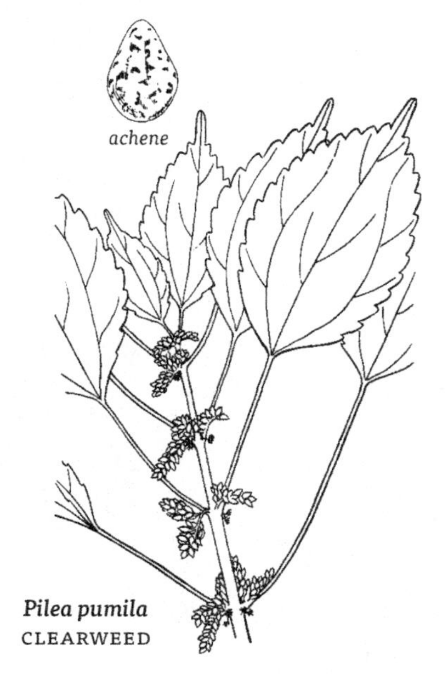

Pilea pumila
CLEARWEED

URTICA | *Nettle*

Urtica dioica L.

STINGING NETTLE

NC **FAC** | MW **FACW** | GP **FAC** | *Caution - stinging hairs*

Stout native perennial herb, often forming dense patches from spreading rhizomes. **Stems** 8–20 dm tall, usually unbranched, with stinging hairs on stems and leaves, the hairs irritating to skin. **Leaves** opposite, ovate to lance-shaped, 5–15 cm long and 2–8 cm wide; margins coarsely toothed; petioles 1–6 cm long; stipules lance-shaped, 5–15 mm long. **Flowers** small, green, male and female flowers separate but mostly on same plants; flower clusters branched and spreading from leaf axils, the clusters usually longer than petioles, all of one sex or a mix of male and female flowers, the female clusters usually above the male clusters when both are present. **Fruit** an ovate achene, 1–2 mm long, enclosed by the inner pair of sepals. July–Sept. **Synonyms** *Urtica procera*. **Habitat** Common; moist woods, thickets, ditches, streambanks, disturbed areas.

Urtica dioica
STINGING NETTLE

VALERIANA | *Valerian*

Valeriana edulis Nutt.
COMMON VALERIAN; TOBACCO-ROOT
NC **FACW** | MW **FACW** | GP **FAC** | *threatened*

Perennial, strongly scented native herb, from a stout taproot. **Stems** smooth, 3–12 dm long. **Leaves** from base of plant and opposite along stem, thick, ± parallel-veined, often hairy when young, becoming smooth or with the margins fringed with hairs when mature; basal leaves oblong lance-shaped, 1–3 dm long and 1–2 cm wide, margins entire or with several lobes, tapered to a winged petiole; stem leaves stalkless, pinnately divided into lance-shaped segments. **Flowers** typically single-sexed; the male flowers 2–4 mm wide, female flowers to 1 mm wide, in widely branched panicles at ends of stems; corolla tube-shaped, 5-lobed, yellow-white; stamens 3. **Fruit** an ovate, 1-chambered achene, 3–4 mm long. May–June. **Synonyms** Ours are var. *ciliata*. *Valeriana ciliata*. **Habitat** Wet meadows, calcareous fens, low prairie. **Status** *Threatened.*

*In Minnesota, most **Valeriana edulis** populations are located in moist prairie remnants found within railroad rights-of way.*

Valeriana edulis
COMMON VALERIAN; TOBACCO-ROOT

PERENNIAL HERBS with 4-angled, erect or prostrate stems. **Leaves** opposite, toothed. **Flowers** small, numerous, perfect (with both male and female parts), in branched or unbranched spikes or heads at ends of stems or from upper leaf axils, the spikes elongating as flowers open upward from the base. Calyx 5-toothed (*Verbena*) or 2-parted (*Phyla*); corolla 5-lobed (*Verbena*) or 4-lobed (*Phyla*), somewhat 2-lipped; stamens 4, of 2 lengths. **Fruit** dry, enclosed by the sepals, splitting lengthwise into 2 or 4 nutlets when mature.

VERBENACEAE | VERVAIN FAMILY

1 Flowers in round heads or short-cylindric spikes on leafless stalks from leaf axils . . . . *Phyla lanceolata* | FOGFRUIT

1 Flowers in long spikes at ends of stems and from upper leaf axils . . . . *Verbena hastata* | COMMON VERVAIN; WILD HYSSOP

PHYLA | *Fogfruit*

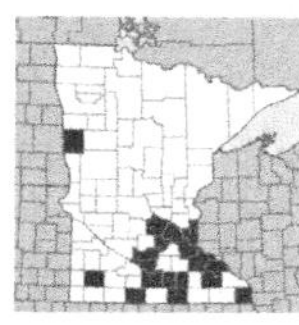

Phyla lanceolata (Michx.) Greene
FOGFRUIT
NC **OBL** | MW **OBL** | GP **FACW**

Native perennial herb, sometimes forming mats; plants smooth or with sparse, short, forked hairs. **Stems** slender, weak, 4-angled, creeping to ascending, often rooting at nodes, the stem tips and lateral branches upright. **Leaves** opposite, ovate to oblong lance-shaped, 2–7 cm long and 0.5–3 cm wide, bright green, tapered to a sharp tip; margins with coarse, forward-pointing teeth to below middle of blade; tapered to a short petiole. **Flowers** small, crowded in spikes from leaf axils, the spikes single, at first round, becoming short-cylindric, 0.5–2 cm long and 5–7 mm wide, on slender stalks 2–9 cm long; calyx 2-parted and flattened, about as long as corolla tube; corolla pale blue or white, 3–4 mm long, 4-lobed and 2-lipped, the lower lip larger than upper lip; withering but persistent in fruit. **Fruit** round, enclosed by the sepals, separating into 2 nutlets. June–Sept.
Synonyms *Lippia lanceolata*.
Habitat Margins of lakes, ponds, streams, ditches, mud flats; often where seasonally flooded.

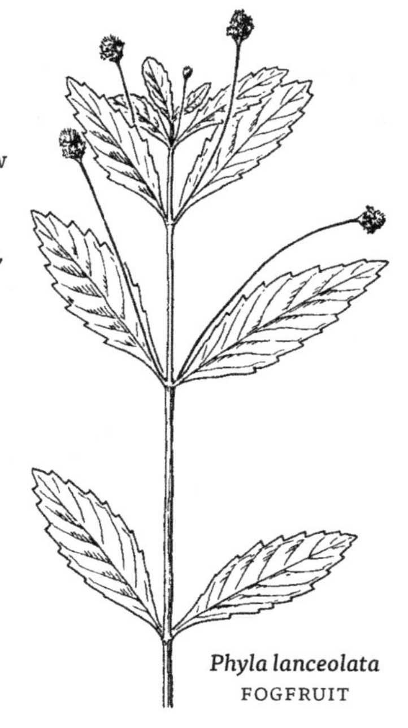

Phyla lanceolata
FOGFRUIT

VERBENA | *Vervain*

Verbena hastata L.
COMMON VERVAIN; WILD HYSSOP
NC **FACW** | MW **FACW** | GP **FACW**

Native perennial herb; plants with short, rough hairs. **Stems** stout, erect, 4–12 dm tall, 4-angled, sometimes branched above. **Leaves** opposite, lance-shaped to oblong lance-shaped, 4–12 cm long and 1–5 cm wide; margins with coarse, forward-pointing teeth and sometimes lobed near base; petioles short. **Flowers** small, numerous, slightly irregular, in long, narrow spikes 5–15 cm long at ends of stems, the spikes elongating as flowers open upward from base; calyx unequally 5-toothed, 1–3 mm long; corolla dark blue to purple, 5-lobed, trumpet-shaped, slightly 2-lipped, 2–4 mm wide. **Fruit** 4-angled, splitting into 4 nutlets. July–Sept. **Habitat** Marshes, wet meadows, shores, streambanks, openings in swamps, ditches.

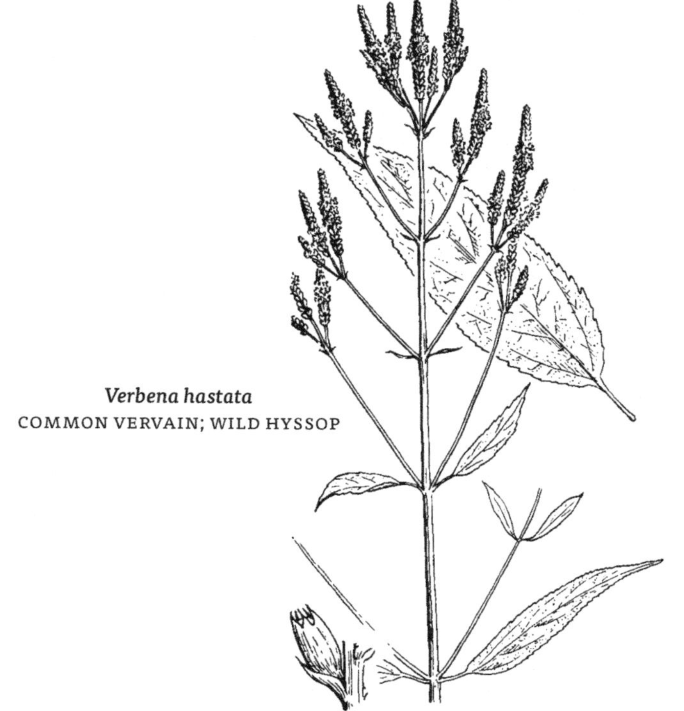

Verbena hastata
COMMON VERVAIN; WILD HYSSOP

VIOLA | *Violet*

Perennial herbs, with or without leafy stems. **Leaves** all at base of plant or alternate on stems; petioles with membranous stipules. **Flowers** perfect, nodding and single at ends of stems, with 5 unequal sepals, 2 upper petals, 2 lateral, bearded petals, and 1 lower petal prolonged into a nectar-holding spur at the petal base. **Fruit** an ovate capsule which splits to eject the seeds.

VIOLA | VIOLET

1 Plants with stems; leaves and flowers borne on the upright stems .... *Viola labradorica* | LABRADOR VIOLET

1 Plants without stems; leaves and flowers directly from rootstock 2

2 Flowers white 3

2 Flowers purple 6

3 Leaves lance-shaped, tapered to a narrow base , more than 1.5x longer than wide .... *Viola lanceolata* | LANCE-LEAVED VIOLET

3 Leaves often wider than long, less than 1.5x longer than wide 4

4 Leaves dull, not shiny, upper and lower surface without hairs, lower surface not paler than upper; margins ± entire or with rounded teeth; petioles often with long soft hairs .... *Viola macloskeyi* .... WILD WHITE VIOLET

4 Leaves shiny and smooth on upper surface, or dull and hairy on either upper or lower surface; underside paler than upper surface; margins with sharp, forward-pointing teeth 5

5 Plants with stolons and horizontal rhizomes; upper and lower surface of leaves sparsely to densely hairy with short hairs less than 1 mm long .... *Viola blanda* | SWEET WHITE VIOLET

5 Plants without stolons, rhizomes upright; leaves often shiny and smooth on upper surface, or densely hairy on upperside with hairs about 1–2 mm long and smooth below .... *Viola renifolia* | KIDNEY-LEAVED VIOLET

6 Leaves longer than wide .... *Viola cucullata* | BLUE MARSH VIOLET

6 Leaves as wide or wider than long 7

7 Sepals long-tapered to a sharp tip; lateral petals with short, knob-tipped hairs on inner surface; spurred petal without hairs .... *Viola cucullata* .... BLUE MARSH VIOLET

7 Sepals oblong to broadly lance-shaped, rounded at tip; lateral petals with long, threadlike hairs on inner surface 8

8 Flowers held above leaves; leaves and stems without hairs, leaves rounded at tip, margins with rounded teeth; spurred petal densely hairy within;

plant of open wetlands and peatlands. . . . . *Viola nephrophylla*
. . . . NORTHERN BOG VIOLET

8 Flowers overtopped by leaves; leaves and stems usually hairy, leaves tapered to a pointed tip, margins with sharp, forward-pointing teeth; spurred petal smooth to slightly hairy within; plant of moist forests. . . . .
. . . . *Viola sororia* | COMMON BLUE VIOLET

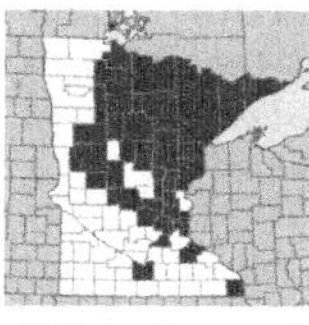

Viola blanda Willd.

SWEET WHITE VIOLET
NC **FACW** | MW **FACW** | GP **FACW**

Native perennial herb, spreading by short rhizomes (and stolons later in season). **Stems** smooth. **Leaves** all from base of plant, heart-shaped, dark green and satiny, 2–5 cm wide, upper surface near base of blade usually with short, stiff white hairs; petioles usually red. **Flowers** white, fragrant, on stalks shorter than longer than leaves; lower 3 petals with purple veins near base, all ± beardless; upper 2 petals narrow, twisted backward, 2 side petals forward-pointing. **Fruit** a purple capsule 4–6 mm long, seeds dark brown. April–May. **Synonyms** *Viola incognita*. **Habitat** Hummocks in swamps and bogs, low wet areas in deciduous and conifer forests.

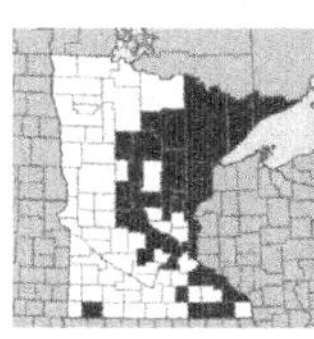

Viola cucullata Aiton

BLUE MARSH VIOLET
NC **OBL** | MW **OBL**

Native perennial herb, spreading by short, branched rhizomes; plants smooth. **Leaves** all from base of plant, ovate to kidney-shaped, to 10 cm wide, heart-shaped at base; margins coarsely toothed; blade angled from the upright petioles. **Flowers** light purple or white, dark at center, on slender stalks longer than leaves; the 2 side petals densely bearded with short hairs, the hairs mostly knobbed or club-tipped. **Fruit** a cylinder-shaped capsule, seeds dark. April–June. **Synonyms** *Viola obliqua*. **Habitat** Swamps, sedge meadows, shady seeps; occasionally in bogs and low areas in forests.

Viola cucullata
BLUE MARSH VIOLET

Viola labradorica Schrank
LABRADOR VIOLET
NC **FAC** | MW **FACW** | GP **FAC**

Native perennial herb; plants smooth. **Leaves** in clumps from rhizomes, at first all from base of plants, later with leafy, horizontal stems to 15 cm long; light green, ovate to kidney-shaped, 1–2.5 cm wide; margins with rounded teeth; petioles 2–6 cm long. **Flowers** pale blue, side petals bearded on inner surface. **Fruit** 4–5 mm long, seeds dark brown. April–June. **Synonyms** *Viola adunca* var. *minor*, *Viola consperma*. **Habitat** Swamps, streambanks, moist hardwood forests.

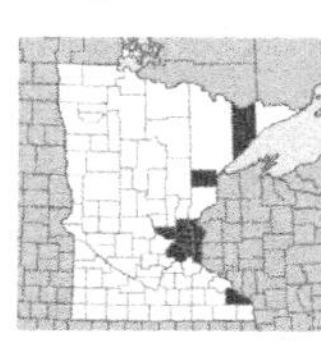

Viola lanceolata L.
LANCE-LEAVED VIOLET
NC **OBL** | MW **OBL** | GP **OBL** | *threatened*

Native perennial herb, spreading by rhizomes and stolons. **Leaves** from base of plant, narrowly lance-shaped, more than 2x longer than wide, tapered to base; margins toothed. **Flowers** white, all beardless; lower 3 petals purple-veined near base. **Fruit** a green capsule 5–8 mm long, seeds brown. April–June. **Habitat** Wet meadows, moist swales in dunes and other sandy areas, sandy lakeshores. **Status** *Threatened.*

The narrow, toothed leaves of ***Viola lanceolata*** *help identify this species.*

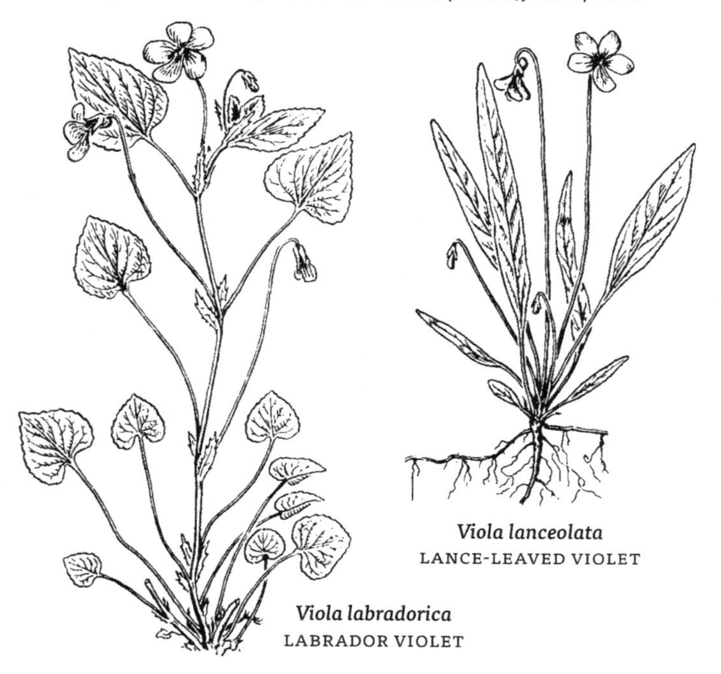

Viola lanceolata
LANCE-LEAVED VIOLET

Viola labradorica
LABRADOR VIOLET

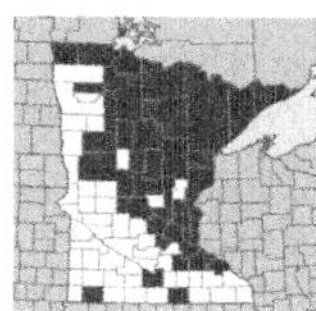

Viola macloskeyi F. Lloyd
WILD WHITE VIOLET
NC **OBL** | MW **OBL** | GP **FACW**

Small native perennial herb (our smallest violet), spreading by rhizomes and stolons. **Leaves** all from base of plant, heart-shaped to kidney-shaped, 1–3 cm wide at flowering, later to 8 cm wide, underside orange-tinged; margins with rounded teeth. **Flowers** white, on upright stalks equal or longer than leaves, 3 lower petals purple-veined near base, 2 side petals beardless or with sparse hairs. **Fruit** a green capsule 4–6 mm long, seeds olive-black. April–July. **Synonyms** *Viola pallens*. **Habitat** Marshes, sedge meadows, open bogs and swamps, alder thickets; sometimes in shallow water.

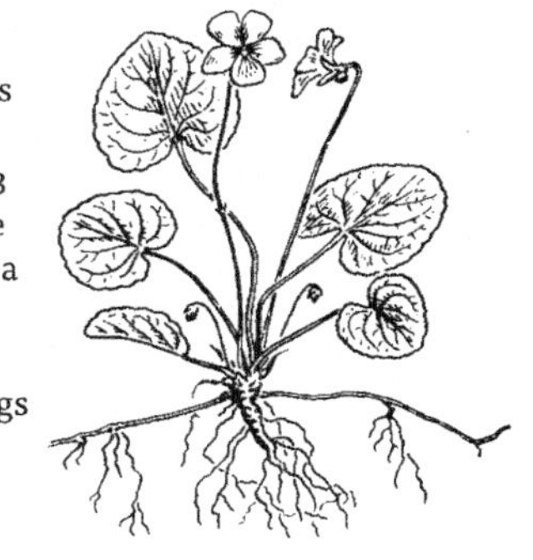

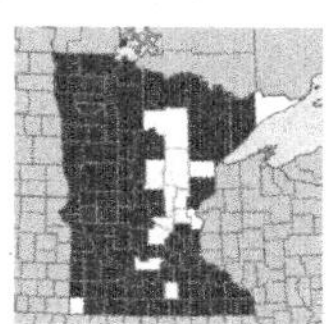

Viola nephrophylla Greene
NORTHERN BOG VIOLET
NC **FACW** | MW **FACW** | GP **FACW**

Low native perennial herb, spreading by short rhizomes. **Leaves** all from base of plant, smooth, heart-shaped to kidney-shaped, 1–4 cm long and 2–6 cm wide, rounded at tip; margins with rounded teeth; petioles slender, 2–16 cm long. **Flowers** single, nodding on slender stalks, the stalks longer than leaves. **Flowers**violet, bearded near base on inside, or upper pair of petals not bearded. **Fruit** a capsule 5–10 mm long. May, sometimes again flowering in Aug or Sept. **Habitat** Wet meadows, fens, calcium-rich wetlands, low areas between dunes, streambanks, rocky shores.

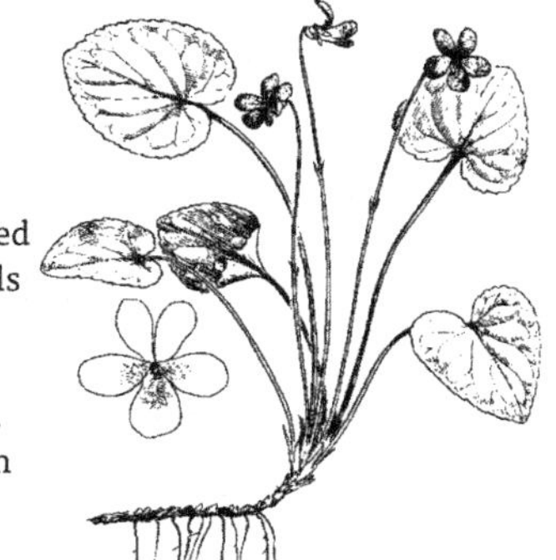

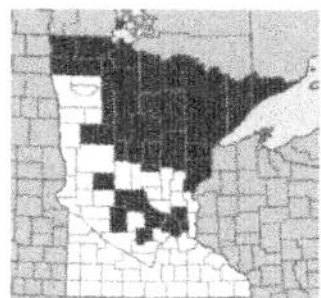

Viola renifolia A. Gray
KIDNEY-LEAVED VIOLET
NC **FACW** | MW **FACW** | GP **FACW**

Native perennial herb, spreading by long rhizomes. **Leaves** all from base of plant, mostly kidney-shaped, rounded at tip, varying from smooth and shiny above to hairy on lower surface only; margins with few rounded teeth. **Flowers** white, all bearded or beardless, 3 lower petals purple-veined at base. **Fruit** a capsule 4–5 mm long, seeds brown and dark-flecked. May–July. **Habitat** Cedar swamps, sphagnum hummocks in peatlands.

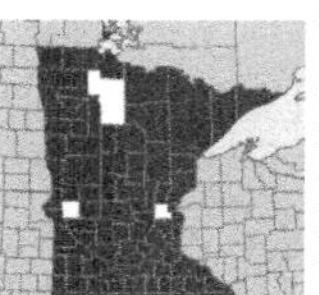

Viola sororia Willd.
COMMON BLUE VIOLET
NC **FAC** | MW **FAC** | GP **FAC**

Native perennial herb, spreading by short rhizomes. **Leaves** all from base of plant, ovate to heart-shaped, sometimes expanding to 10 cm wide in summer, with long hairs; margins with rounded teeth; blades angled from the upright petioles. **Flowers** blue-violet, on stalks about as high as leaves, the 2 side petals densely bearded with hairs 1 mm long and not club-tipped. **Fruit** a purple-flecked capsule, seeds dark brown. April–June. **Habitat** Moist hardwood forests; occasionally in swamps, floodplain forests and along rocky streambanks.

Viola sororia
COMMON BLUE VIOLET

Viola renifolia
KIDNEY-LEAVED VIOLET

VITIS | *Grape*

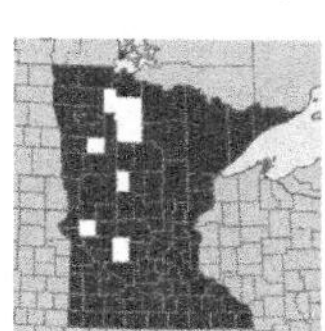

Vitis vulpina L.

RIVERBANK GRAPE

NC **FAC** | MW **FACW** | GP **FAC**

Perennial, woody, climbing native vine to 5 m or more long; young **branches** green or red, hairy, becoming smooth. **Leaves** alternate, heart-shaped in outline, 1–2 dm long and as wide, with a triangular tip and 2 smaller lateral lobes, the lobes variable; leaf base with a U-shaped indentation, upper surface smooth, bright green, underside paler and sparsely hairy along veins; margins with coarse, forward-pointing teeth; petioles shorter than blades. **Flowers** small, sweet-scented, green-white to creamy, in stalked clusters 5–10 cm long. **Fruit** a dark blue to black berry, 6–12 mm wide, with a waxy bloom, sour when young, becoming sweeter when ripe in fall. May–July. **Synonyms** *Vitis riparia*. **Habitat** Floodplain forests, moist sandy woods, streambanks, thickets; also on sand dunes.

Vitis vulpina
RIVERBANK GRAPE

Sedge Meadow community, early summer. Typical growth-form of **common tussock sedge** (*Carex stricta*), a dominant component of sedge meadows. The large hummocks, up to 1 m tall, are composed of plant roots and the previous years' leaves.

Monocots

MONOCOTYLEDONEAE (or Liliopsida, the Monocots) include families with showy flowers such as members of the **Liliaceae** and **Orchidaceae** as well as families lacking petals and sepals such as the **Cyperaceae** (Sedge Family), **Poaceae** (Grass Family), and **Juncaceae** (Rush Family, see examples below). Identification of species within these three families is frequently dependent on characters of small floral parts, and a hand lens or dissecting microscope is often useful.

Orchidaceae
Small yellow lady's-slipper
Cypripedium parviflorum var. makasin

Cyperaceae
Beaked sedge
Carex utriculata

Poaceae
Bluejoint
Calamagrostis canadensis

Juncaceae
Soft rush
Juncus effusus

ACORUS | *Sweet flag*

Acorus americanus (Raf.) Raf.
SWEET FLAG
NC **OBL** | MW **OBL** | GP **OBL**

Native or introduced perennial herb, from stout, aromatic rhizomes. **Leaves** linear, long and swordlike, bright-green, leathery, 2-ranked, 5–15 dm long and 1–2 cm wide, sweet-scented when crushed; margins entire, sharp-edged, translucent near base. **Flowers** small, in a cylindric, yellow-green spadix, appearing lateral from a leaflike, tapered stalk; the spadix upright, 5–10 cm long and 1–2 cm wide; flowers perfect, yellow or brown, composed of 6 papery tepals and 6 stamens. **Fruit** a 1–3-seeded berry, dry outside and jellylike on inside. June–July. **Habitat** Marshes (often with cat-tails), shores, moist depressions, streambanks. Widely established in North America; Asia.

Native plants are separated as **Acorus americanus** *and may be distinguished from Acorus calamus by having a raised leaf midvein plus 1-5 other veins raised above the leaf surface. In Acorus calamus, only the midvein is prominently raised. However, the distinctions are not always clear, and populations of both species may be present in the same location.*

Acorus americanus *Acorus calamus*

Acorus
SWEET FLAG

PERENNIAL, AQUATIC OR EMERGENT HERBS; plants swollen and tuberlike at base. **Leaves** all from base of plant and clasping an erect stem; underwater leaves often ribbonlike; emergent leaves broader. **Flowers** perfect (with both male and female parts) or imperfect, in racemes or panicles at ends of stems, with 3 sepals and 3 petals; stamens 6 or more. **Fruit** a compressed achene, usually tipped by the persistent style.

ALISMATACEAE | WATER-PLANTAIN FAMILY

1 Leaves never arrowhead-shaped; flowers perfect, in a panicle-like head; pistils or achenes in a single whorl on a small, flat receptacle ..... *Alisma* .... WATER-PLANTAIN

1 Leaves often arrowhead-shaped; flowers mostly imperfect, whorled on stem; male flowers above female in the head; pistils or achenes in several series around a large, round receptacle, and forming a dense, round head .... *Sagittaria* | ARROWHEAD

ALISMA | *Water-plantain*

Perennial herbs, from cormlike rootstocks. **Leaves** emersed or floating, ovate to lance-shaped, never arrowhead-shaped; underwater leaves sometimes ribbonlike (in *Alisma gramineum*). **Flowers** perfect, in whorled panicles, sepals 3, green; petals 3, white or light pink; stamens 6. **Fruit** a flattened achene in a single whorl on a flat receptacle, style beak small or absent.

ALISMA | WATER-PLANTAIN

1 Leaves lance-shaped to oval, or if underwater, leaves long and ribbonlike; flower stalks rarely longer than leaves; petals usually pink .... *Alisma gramineum* | NARROW-LEAVED WATER-PLANTAIN

1 Leaves ovate; flower stalks much longer than leaves; petals white .... *Alisma subcordatum* | WATER-PLANTAIN

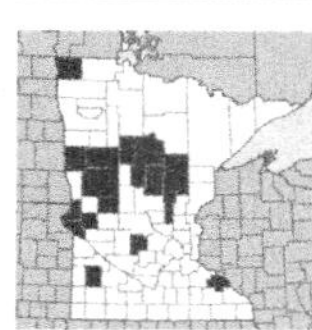

Alisma gramineum Lej.
NARROW-LEAVED WATER-PLANTAIN
NC **OBL** | MW **OBL** | GP **OBL**

Native perennial herb. **Stems** upright to spreading, 0.5–3 dm tall. **Emersed leaves** lance-shaped to oval, 2–10 cm long and 1–3 cm wide, petioles mostly longer than blades; **underwater leaves** reduced to linear, ribbonlike petioles, to 6 dm long and 1 cm wide. **Flowers** many on spreading stalks, the stalks shorter to somewhat longer than the leaves; sepals 3; petals pink, 1–3 mm long. **Fruit** an achene, 2–3 mm long, with a central ridge and 2 lateral ridges. July–Sept. **Synonyms** *Alisma geyeri*. **Habitat** Shallow, often brackish water, muddy shores, streambanks.

Alisma gramineum
NARROW-LEAVED WATER-PLANTAIN

Alisma subcordatum Raf.
WATER-PLANTAIN
NC **OBL** | MW **OBL** | GP **OBL**

Native perennial herb. **Leaves** ovate to oval, 3–15 cm long and 2–12 cm wide, rounded to nearly heart-shaped at base; petioles long. **Flowers** clustered on slender stalks 1–10 dm long, in whorls of 3–10; sepals 3; petals white, 3–5 mm long. **Fruit** an achene, 2–3 mm long, with a central groove. July–Sept. **Synonyms** *Alisma brevipes*, *Alisma plantago-aquatica*, *Alisma triviale*. **Habitat** Shallow water marshes, shores, ditches.

Alisma subcordatum
WATER-PLANTAIN

SAGITTARIA | *Arrowhead*

Perennial or annual herbs, with fleshy rootstocks. **Leaves** sheathing, all from base of plant, variable in shape and size. Emersed and floating leaves usually arrowhead-shaped with large lobes at base, or sometimes ovate to oval and without lobes; **underwater leaves** often linear in a basal rosette, normally absent by flowering time. **Flowers** in a raceme of mostly 3-flowered whorls; upper flowers usually male, lower flowers usually female or sometimes perfect; sepals 3, green, persistent; petals 3, white, deciduous; stamens 7 to many; pistils crowded on a rounded receptacle. **Fruit** a crowded cluster of achenes in ± round heads, the achenes flattened and winged, beaked with a persistent style.

SAGITTARIA | ARROWHEAD

1 Emersed leaves not arrowhead-shaped, basal lobes absent 2
1 Emersed leaves all or mostly arrowhead-shaped, with large basal lobes 3

2 Female flowers and fruiting heads ± stalkless . . . . *Sagittaria rigida* . . . . SESSILE-FRUITED ARROWHEAD
2 Female flowers and fruiting heads obviously stalked . . . . *Sagittaria graminea* | GRASS-LEAVED ARROWHEAD

3 Plants annual, rhizomes absent; sepals appressed to fruiting heads; stalks of fruiting heads stout . . . . *Sagittaria calycina* | MISSISSIPPI ARROWHEAD
3 Plants perennial, with rhizomes; sepals reflexed on fruiting heads; stalks of fruiting heads slender 4

4 Bracts below flowers mostly less than 1 cm long; achene beak projecting horizontally from tip of achene . . . . *Sagittaria latifolia* . . . . COMMON ARROWHEAD; WAPATO
4 Bracts below flowers usually more than 1 cm long; achene beak upright 5

5 Achene beak short, erect, to 0.4 mm long; basal lobes of leaves mostly shorter than terminal lobe . . . . *Sagittaria cuneata* . . . . NORTHERN ARROWHEAD; WAPATO
5 Achene beak larger, curved and ascending, 0.5 mm long or more; basal lobes of leaves usually equal or longer than terminal lobe . . . . *Sagittaria brevirostra* | MIDWESTERN ARROWHEAD

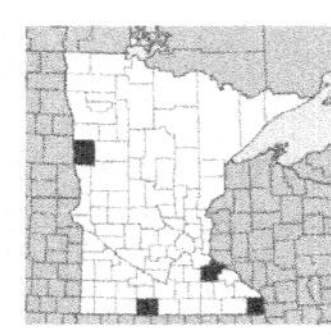

Sagittaria brevirostra Mackenzie & Bush
MIDWESTERN ARROWHEAD
NC **OBL** | MW **OBL** | GP **OBL**

Native perennial herb. **Leaves** arrowhead-shaped, mostly 10–30 cm long and to 20 cm wide; basal lobes lance-shaped and usually equal or longer than terminal lobe; petioles long. **Flowers** in heads 1–2 cm wide, on a mostly unbranched stalk usually longer than leaves, the male flowers above the female flowers; bracts below flowers 1–5 cm long; female flowers on ascending stalks 0.5–2 cm long; sepals bent backward in fruit; petals white, 1–2 cm long. **Fruit** a winged achene, 2–3 mm long, separated from style beak by a saddlelike depression, the beak usually curved-ascending. July–Sept. **Synonyms** *Sagittaria engelmanniana* var. *brevirostra*. **Habitat** Shallow water and muddy shores, marshes.

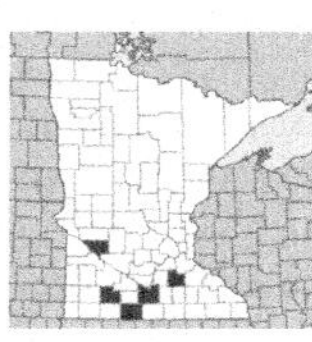

Sagittaria calycina Engelm.
MISSISSIPPI OR LONG-LOBED ARROWHEAD
NC **OBL** | MW **OBL** | GP **OBL**

Annual or perennial native herb, rhizomes and tubers absent. **Leaves** erect to spreading; emersed blades arrowhead-shaped (the lobes sometimes outward spreading) or oval to ovate, 3–40 cm long and 2–25 cm wide, the basal lobes usually longer than terminal lobe; petioles and flower stalks spongy, round in cross-section. Lower flowers usually perfect; upper flowers usually male, in heads 1–2 cm wide **Flowers** from a stalk 1–10 dm tall, heads leaning when fruiting; bracts membranous, short and rounded at lower nodes, upper bracts longer (to 1 cm long)

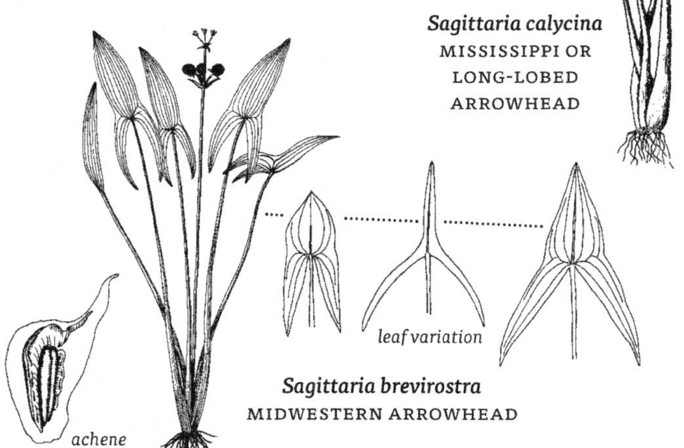

Sagittaria calycina
MISSISSIPPI OR LONG-LOBED ARROWHEAD

Sagittaria brevirostra
MIDWESTERN ARROWHEAD

and tapered to a tip; sepals blunt-tipped, bent backward in flower but appressed to the head in fruit; petals white with a yellow base. **Fruit** an achene, 2–3 mm long, beak ± horizontal from top of achene. July–Sept. **Synonyms** *Lophotocarpus calycinus, Sagittaria montevidensis*. **Habitat** Muddy shores.

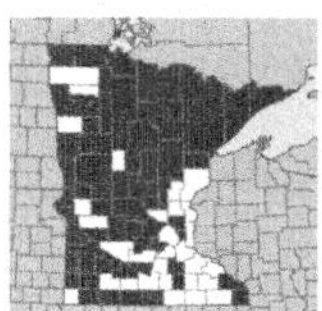

Sagittaria cuneata Sheldon
NORTHERN ARROWHEAD; WAPATO
NC **OBL** | MW **OBL** | GP **OBL**

Native perennial herb, with rhizomes and large, edible tubers. **Submerged leaves** (if present) often awl-shaped or reduced to bladeless, expanded petioles (phyllodes); **emersed leaves** long-stalked, usually arrowhead-shaped, 5–20 cm long and 2–15 cm wide, the basal lobes much shorter than terminal lobe; floating leaves often heart-shaped (unlike our other species of *Sagittaria*). **Flowers** imperfect, the male flowers above the female, in ± round heads 5–12 mm wide, with 2–10 whorls of heads on a stalk 1–6 dm tall, the stalks often branched at lowest node; bracts tapered to tip, 1–4 cm long; sepals ovate, bent backward in flower and fruit; petals white, 7–15 mm long. **Fruit** an achene, 2–3 mm long; beak erect, small, 0.1–0.4 mm long. June–Sept. **Habitat** Shallow water, lakeshores and streambanks.

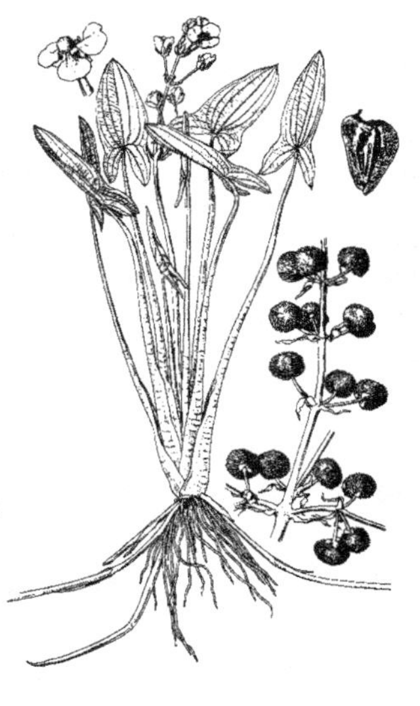

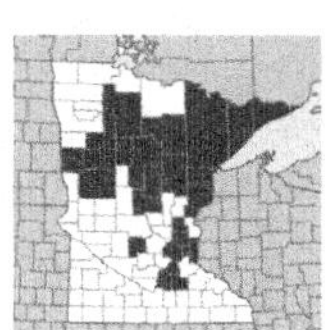

Sagittaria graminea Michx.
GRASS-LEAVED ARROWHEAD
NC **OBL** | MW **OBL** | GP **OBL**

Native perennial herb, with rhizomes. **Underwater plants** sometimes only a rosette of bladeless, ribbonlike petioles (phyllodes) to 1 cm wide; **emergent leaves** lance-shaped to oval, never arrowhead-shaped, 3–20 cm long and 0.5–3 cm wide, tapered to a blunt tip. **Flowers** imperfect, the male flowers usually above the female, clustered in ± round heads, 5–12 mm wide, the heads on spreading stalks 1–4 cm long; with 2–10 whorls of flowers along an unbranched stalk mostly shorter than leaves; bracts broadly ovate, joined in their lower portion, 2–8 mm long; sepals ovate, bent backward in fruit; petals white, equal or longer than sepals. **Fruit** a winged achene, 1–2 mm long, beak small or absent. June–Sept. **Synonyms** *Sagittaria cristata*. **Habitat** Shallow water and shores.

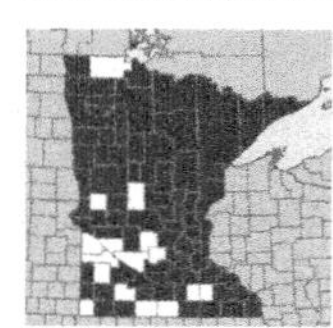

Sagittaria latifolia Willd.
COMMON ARROWHEAD; WAPATO
NC **OBL** | MW **OBL** | GP **OBL**

Native perennial herb, with rhizomes and edible tubers in fall. **Leaves** variable; **emersed leaves** arrowhead-shaped, mostly 8–40 cm long and 1–15 cm wide, lobes typically narrow on plants in deep water to broad on emersed plants; plants sometimes with bladeless, expanded petioles (phyllodes). **Flowers** male above and female below, clustered in ± round heads 1–2.5 cm wide, at ends of slender, spreading stalks 0.5–3 cm long, in whorls of 2–15 along a stalk 2–10 dm tall; bracts tapered to a tip or blunt, 0.5–1 cm long; sepals ovate, bent backward by fruiting time; petals white, 7–20 mm long. **Fruit** a winged achene, 2–4 mm long, the beak projecting horizontally, 1–2 mm long. July–Sept. **Habitat** Shallow water, shores, marshes, and pools in bogs.

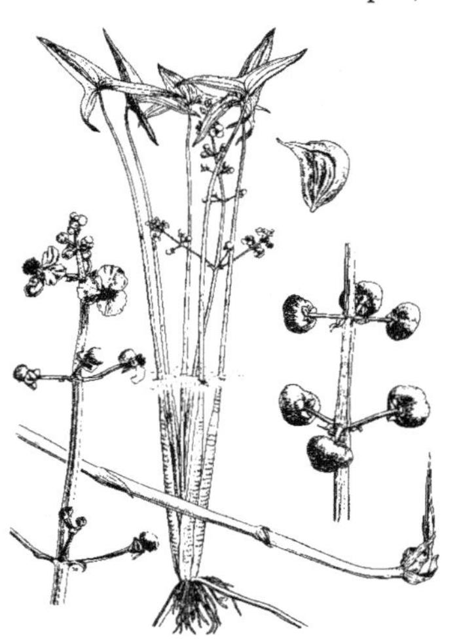

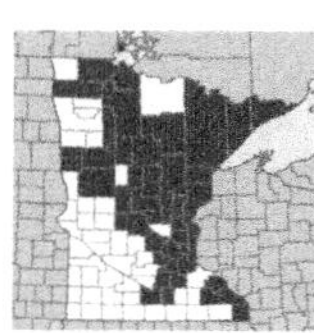

Sagittaria rigida Pursh
SESSILE-FRUITED ARROWHEAD
NC **OBL** | MW **OBL** | GP **OBL**

Native perennial herb, rhizomes present. **Stems** erect or lax. **Emersed leaves** lance-shaped to ovate, rarely with short, narrow basal lobes, (but not arrowhead-shaped), 4–15 cm long and to 7 cm wide; petioles sometimes bent near junction with blades; deep water plants often with only linear, bladeless, expanded petioles (phyllodes). **Flowers** in ± round heads to 1.5 cm wide, the heads stalkless and bristly when mature due to achene beaks; in 2–8 whorls on a stalk 1–8 dm tall, the stalk often bent near lowest node; flowers imperfect, male flowers above the female, male flowers on threadlike stalks 1–3 cm long, female flowers ± stalkless; bracts ovate, 5 mm long, joined at base; sepals ovate, 4–7 mm long, bent backward when in fruit; petals white, 1–3 cm long. **Fruit** a narrowly winged achene, 2–4 mm long; beak ascending, 1–1.5 mm long. June–Sept. **Habitat** Shallow water, shores and streambanks.

PERENNIAL HERBS; **leaves** alternate, simple or compound, often fleshy. **Flowers** small and numerous, mostly single-sexed, male flowers usually above female, crowded in a cylindric or rounded spadix subtended by a leaflike spathe; sepals 4–6 or absent; petals absent; stamens mostly 2–6; pistils 1–3-chambered. **Fruit** a usually fleshy berry, containing 1 to few seeds, or the entire spadix ripening as a fruit.

ARACEAE | ARUM FAMILY

1 Leaves compound .... *Arisaema* | JACK-IN-THE-PULPIT
1 Leaves simple, not divided 2

2 Leaves arrowhead-shaped, lobes 1/3 or more length of blade .... *Peltandra virginica* | TUCKAHOE; ARROW-ARUM
2 Leaves heart-shaped or rounded at base 3

3 Leaves broadly heart-shaped, abruptly tapered to a tip; spathe white, long-stalked; flowering late spring .... *Calla palustris* | WATER-ARUM
3 Leaves large, ovate, tapered to a rounded tip; spathe green-yellow to purple-brown, short-stalked or stalkless; flowering late winter to early spring ... . *Symplocarpus foetidus* | SKUNK-CABBAGE

ARISAEMA | *Jack-in-the-pulpit*

Perennial herbs. **Leaves** compound. **Flowers** either male or female, on same or different plants; male flowers with 2–5, ± stalkless stamens, above the female flowers on a fleshy spadix, the spadix subtended by a green or purple-brown spathe; sepals and petals absent. **Fruit** a cluster of round, red berries, each berry with 1–3 seeds.

ARISAEMA | JACK-IN-THE-PULPIT

1 Leaflets 7–13; spadix longer than spathe .... *Arisaema dracontium* .... GREEN DRAGON
1 Leaflets usually 3; spathe arching over spadix. .... *Arisaema triphyllum* .... JACK-IN-THE-PULPIT

Arisaema dracontium (L.) Schott

GREEN DRAGON
NC **FACW** | MW **FACW** | GP **FACW**

Native perennial herb, from corms. **Leaf** usually single, palmately branched into 7–15 leaflets, the leaflets oval to oblong lance-shaped, tapered to a point and narrowed at base, the central leaflets 1–2 dm long and to 8 cm wide, the outer leaflets progressively smaller; petioles 2–10 dm long. **Flowers** male or female and on different plants, or plants with both male and female flowers, the male above female, on a long, slender spadix exserted 5–10 cm beyond spathe; the spathe green, slender, rolled inward, 3–6 cm long; the flower stalk shorter than the leaf petiole. **Fruit** a cluster of red-orange berries. May–July. **Habitat** Wet woods and floodplain forests.

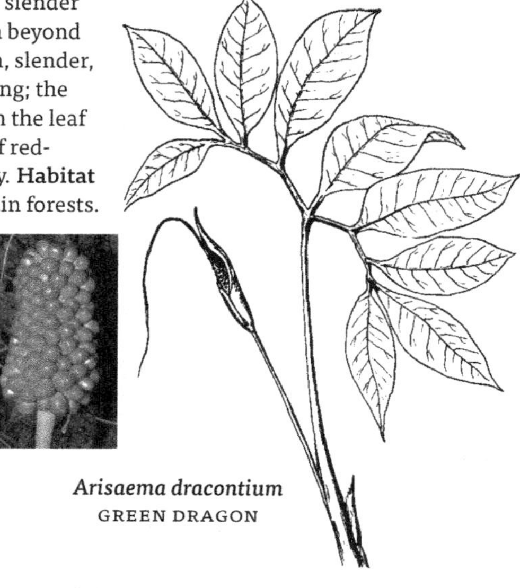

Arisaema dracontium
GREEN DRAGON

Arisaema triphyllum (L.) Schott
JACK-IN-THE-PULPIT
NC **FAC** | MW **FACW** | GP **FAC**

Native perennial herb, from bitter-tasting corms. **Stems** 3–12 dm long. **Leaves** usually longer than the flower stalk, mostly 2, divided into 3 leaflets, the terminal leaflet oval to ovate, the lateral leaflets often asymmetrical at base. **Flowers** male or female and usually on separate plants, borne near base of a cylindric, blunt-tipped spadix, subtended by a green, purple-striped spathe, rolled inward below, expanded and arched over the spadix above, abruptly tapered to a tip. **Fruit** a cluster of shiny red berries. April–July. **Synonyms** *Arisaema atrorubens*. **Habitat** Moist forests, cedar swamps.

Arisaema triphyllum
JACK-IN-THE-PULPIT
spathe and spadix

CALLA | *Water-arum*

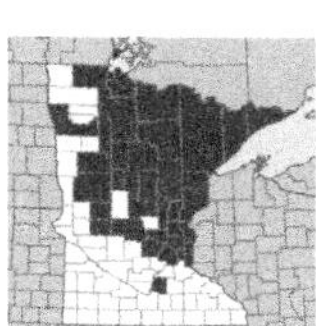

Calla palustris L.

WATER-ARUM

NC **OBL** | MW **OBL** | GP **OBL**

Native perennial herb, from rhizomes, the rhizomes creeping in mud or floating. **Leaves** broadly heart-shaped, abruptly tapered to a tip, 5–15 cm long and about as wide; petioles stout, 1–2 dm long (or longer when underwater). **Flowers** perfect or the uppermost male, on a short-cylindric spadix, 1.5–3 cm long, shorter than the spathe; the spathe white, ovate, tipped with a short, sharp point to 1 cm long; sepals and petals absent; stamens 6. **Fruit** a fleshy, few-seeded berry, turning red when ripe. May–July. **Habitat** Bog pools, swamps, shores and wet ditches.

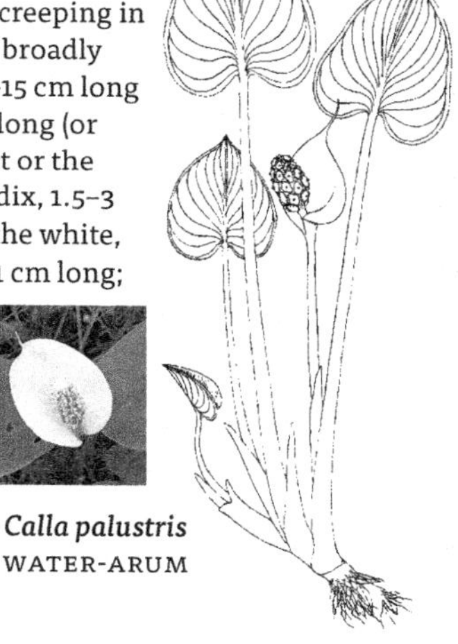

Calla palustris
WATER-ARUM

PELTANDRA | *Arrow-arum*

Peltandra virginica (L.) Schott & Endl.

ARROW-ARUM; TUCKAHOE

NC **OBL** | MW **OBL** | GP **OBL**

Native perennial herb, with thick fibrous roots. **Leaves** all from base of plant on long petioles, bright green, oblong to triangular in outline, 1–3 dm long and 8–15 cm wide at flowering, to 8 dm long later; leaf base with pair of lobes. **Flowers** in white to orange spadix about as long as the spathe, atop a curved stalk 2–4 dm long; flowers either male or female, the male flowers covering upper 3/4 of the spadix, the female flowers on lower portion; spathe green with a pale margin, 1–2 dm long, the lower portion covering the fruit. **Fruit** a head of green-brown berries, the berries with 1–3 seeds surrounded by a jellylike material. June–July. **Synonyms** *Peltandra luteospadix*.

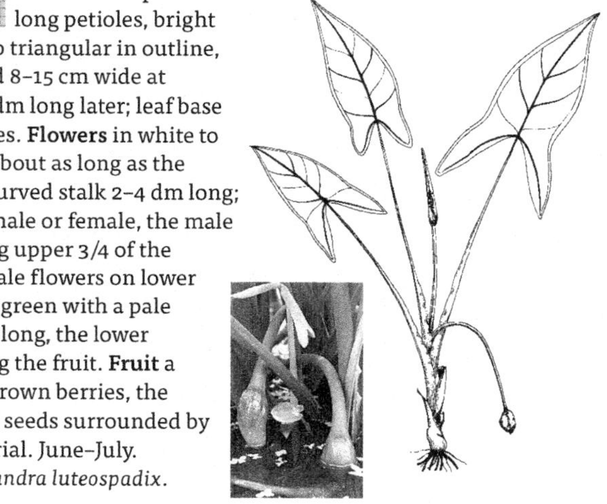

Habitat Shallow water, shores, bog pools; often where shaded. In Minnesota, known from one location only; disjunct from main range further south in USA.

SYMPLOCARPUS | *Skunk-cabbage*

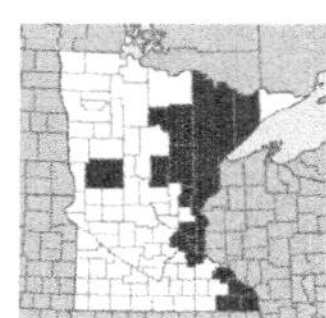

Symplocarpus foetidus (L.) Nutt.

SKUNK-CABBAGE

NC **OBL** | MW **OBL** | GP **OBL** | *Caution - toxic*

Perennial, foul-smelling native herb, from thick rootstocks. **Leaves** all from base of plant, ovate to heart-shaped, 3–8 dm long and to 3 dm wide, strongly nerved; petioles short, channeled. **Flowers** appearing before leaves in late winter or early spring, perfect; the spathe ovate, curved over spadix, 8–15 cm long, green-purple and often mottled; sepals 4. **Fruit** round, 8–12 cm wide; seeds 1 cm thick. Feb–May. **Habitat** Floodplain forests, swamps, streambanks, calcareous fens, moist wooded slopes.

Skunk-cabbage *is the earliest flowering plant in the state, the flowers often appearing while still partially covered with snow. The plant's roots are toxic and the leaves can burn the mouth if tasted.*

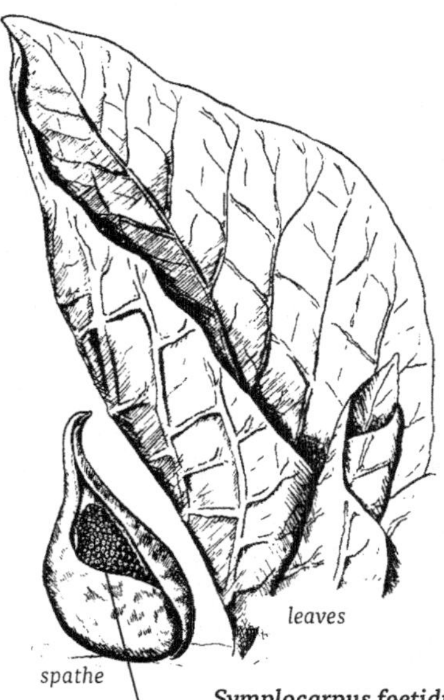

spathe, spadix

Symplocarpus foetidus (SKUNK-CABBAGE)

BUTOMUS | *Flowering rush*

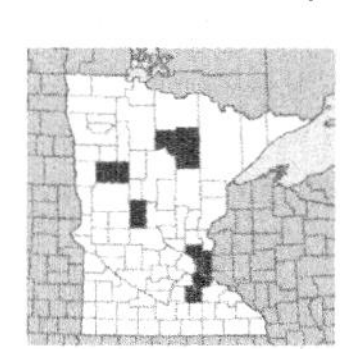

Butomus umbellatus L.

FLOWERING RUSH

NC **OBL** | MW **OBL** | GP **OBL**

Introduced, invasive perennial herb, from creeping rhizomes. **Leaves** all from base of plant, erect when emersed, or floating when in deep water, linear, to 1 m long and 5–10 mm wide, parallel-veined; petioles absent. **Flowers** pink, perfect, 2–3 cm wide, stalked, in a many-flowered umbel, borne on a round stalk 1–1.2 m tall; with 4 lance-shaped bracts subtending the umbel; sepals 3, petal-like; petals 3; stamens 9; pistils 6. **Fruit** a dry, many-seeded capsule, splitting open on inner side. June–Aug. **Habitat** Marshes and lakeshores.

Flowering rush *was introduced to North America from Eurasia, and is now established (and often invasive) across much of n USA and s Canada.*

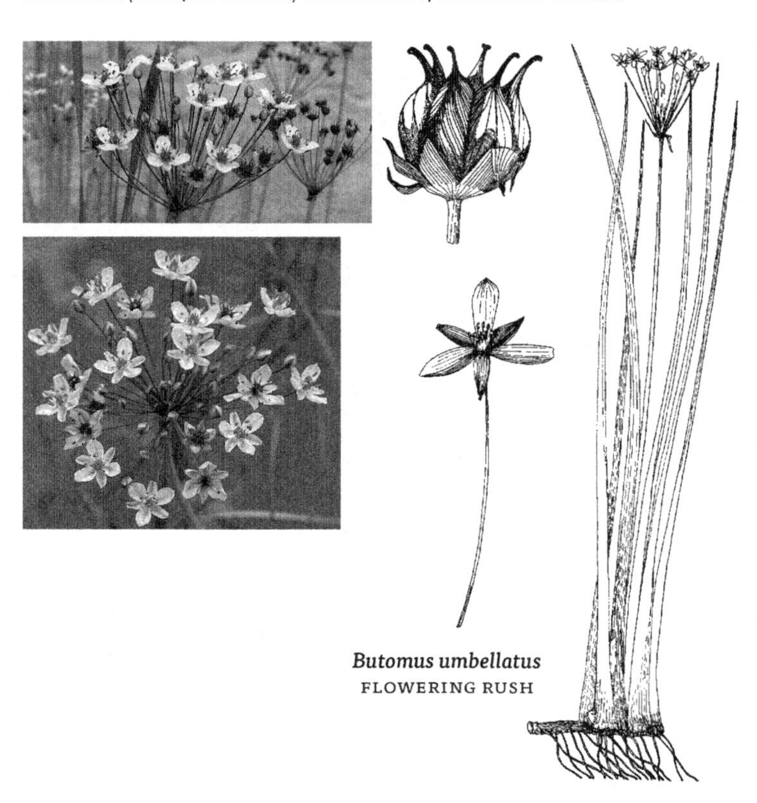

Butomus umbellatus
FLOWERING RUSH

CYPERACEAE
Sedge Family

MOSTLY PERENNIAL, grasslike, rushlike or reedlike plants. **Stems** 3-angled or ± round in section, solid or pithy. **Leaves** 3-ranked or reduced to sheaths at base of stem; leaf blades, when present, grasslike, parallel-veined, often keeled; sheaths mostly closed around the stem. **Flowers** small, perfect (with both male and female parts), or single-sexed, each flower subtended by a bract (scale); perianth of 1 to many (often 6) small bristles, or a single perianth scale, or absent; stamens usually 3; ovary 2–3-chambered, contained in a saclike covering (**perigynium**) in *Carex*, maturing into an **achene**, stigmas 3 or 2. Flowers arranged in **spikelets** (termed **spikes** in *Carex*), the spikelets single as a terminal or lateral spike, or several to many in various types of heads, the head often subtended by 1 to several bracts.

CYPERACEAE | SEDGE FAMILY

1 Flowers either male or female; achene enclosed in a saclike perigynium . . . . . *Carex* | SEDGE p. 391
1 Flowers perfect (except in *Scleria*); achene not in a sac. 2

2 Scales in 2 rows in the spikelets 3
2 Scales overlapping in a spiral in the spikelets, or (rarely) spikelets not in spikes or heads 4

3 Heads terminal at ends of stems; achene not subtended by bristles . . . . . . *Cyperus* | FLATSEDGE p. 454
3 Heads from leaf axils; achene subtended by bristles . . . . . . *Dulichium arundinaceum* | THREE-WAY SEDGE; DULICHIUM p. 460

4 Spikelets of 2 kinds on each plant; achenes exposed, ± round, white to gray . . . . . . *Scleria verticillata* | LOW NUT-RUSH p. 487
4 Spikelets ± all alike; achenes rarely white. 5

5 Base of style persistent as a swelling (tubercle) atop the achene, different in color and texture from achene body, or if style base not different from achene body, then the spikelets single at end of leafless stems 6
5 Base of style persistent as a small beak atop achene, not different from achene body, or the style base not persistent; spikelets usually few to many, or if single, then on leafy stems 7

6 Spikelets single at ends of leafless stems; leaves reduced to sheaths at base of plant . . . . . . *Eleocharis* | SPIKE-RUSH p. 460
6 Spikelets 1 to several, stems leafy at base of plant, leaves with narrow blades . . . . . . *Rhynchospora* | BEAK-RUSH p. 474

7 Sepals and petals of many long, white or coppery bristles, longer than the scales; spikelets with a cottony look .... *Eriophorum* .... COTTON-GRASS p. 467
7 Sepals and petals of 1–6 small bristles or scales, or the sepals and petals absent; spikelets not long-cottony 8

8 Sepals and petals of 1 or 3 scales, sometimes absent; plants annual .... *Hemicarpha micrantha* | DWARF-BULRUSH p. 473
8 Sepals and petals of 1–6 small bristles or absent; plants mostly perennial 9

9 Sepals and petals absent; base of style deciduous and swollen above the attachment to achene .... *Fimbristylis* | FIMBRY p. 472
9 Sepals and petals of 1–6 bristles (or rarely absent); base of style not swollen, often persistent as a beak on achene 10

10 Bracts 2 or more, leaflike .... *Cladium mariscoides* | TWIG-RUSH p. 454
10 Bracts 1, erect, appearing to be a continuation of stem .... *Scirpus* .... BULRUSH p. 477

CAREX | *Sedge*

Perennial grasslike plants. **Stems** mostly 3-angled. **Leaves** 3-ranked, margins often finely toothed. **Flowers** either male or female, with both sexes in same spike, or in separate spikes on same plant, or the male and female flowers on different plants. Male flowers with 3 or rarely 2 stamens; female flowers with style divided into 2 or 3 stigmas. **Achenes** lens-shaped or flat on 1 side and convex on other (in species with 2 stigmas), or achenes 3-angled or nearly round (in species with 3 stigmas), enclosed in a sac called the **perigynium** (singular) or **perigynia** (plural). Sedges are the largest genus of wetland plant in Minnesota.

For unknown specimens, start with the **Group Key** (next page), followed by keys to Sections containing closely related *Carex* species. The Section Keys and species descriptions are arranged alphabetically after the Group Keys.

As evident in the following keys, identification of Carex species is often based on characteristics of the **perigynium**—*the sac around the achene. These are illustrated in the sidebar next to each species' description. For small plants, a hand lens or dissecting scope may be needed to see diagnostic features of the perigynium.*

perigynium
(Carex oligosperma)

CAREX GROUPS (SEDGES)

1 One terminal spike on each stem .... **Group 1** p. 392
1 Two or more spikes on each stem 2

2 Achenes lens-shaped, or flat on 1 side and convex on other; stigmas 2.... .... **Group 2** p. 393
2 Achenes 3-angled or nearly round; stigmas 3 3

3 Style persistent on achene and becoming firm .... **Group 3** p. 394
3 Style withering and usually deciduous as the achene matures 4

4 Perigynia beaked, the beak prolonged into well-developed, usually stiff, sharp teeth.................................... **Group 4** p. 395
4 Perigynia beaked or beakless, the teeth small or soft and blunt 5

5 Terminal spike with some female flowers or perigynia ... **Group 5** p. 395
5 Terminal spike entirely of male flowers 6

6 Perigynia covered with hairs, at least around base of beak....... **Group 6** .... p. 396
6 Perigynia without hairs 7

7 Leafy bract of lowest female spike with a well-developed sheath enclosing the stem .... **Group 7** p. 395
7 Bract of lowest female spike without a well-developed sheath ... **Group 8** .... p. 397

CAREX GROUP KEYS

CAREX GROUP 1

1 Achenes lens-shaped or flat on 1 side and convex on other; stigmas 2 ... 2
1 Achenes 3-angled or round in cross-section; stigmas 3 3

2 Plants spreading by long rhizomes .... **Dioiceae** (*Carex dioica*) p. 408
2 Plants forming clumps, long rhizomes absent......... **Stellulatae** p. 441

3 Spikes with both male and female flowers, male below female .... **Squarrosae** p. 440
3 Spikes with both male and female flowers, male above female 4

4 Perigynia long and narrow or awl-shaped, spreading or angled downward .... **Orthocerates** (*Carex pauciflora*) p. 424
4 Perigynia wider and upright in spike ..... **Polytrichoidae** (*Carex leptalea*) .... p. 438

CAREX GROUP 2

1 Spikes all similar, usually less than 1.5 cm long, and not on stalks 2
1 Spikes not alike, the top and bottom spikes distinctly different; side spikes with or without stalks; if without stalks, then the spikes usually long and slender 14

2 Stems single or few together from rhizomes, stolons, or stems lying along ground 3
2 Stems in clumps, rhizomes absent or very short 6

3 Stems becoming horizontal, new stems upright from old leaf axils; plants in shallow water and wet peatlands .... **Chordorrhizae** .... (*Carex chordorrhiza*) p. 405
3 Stems from long rhizomes, not reclining; plants mostly not of peatlands 4

4 Spikes separated from one another .... **Bracteosae** p. 403
4 Spikes crowded into dense heads 5

5 Leaves well-developed and mostly clustered near base of plant; perigynia usually without nerves, often sharp-edged but not thin-margined .... **Divisae** (*Carex praegracilis*) p. 409
5 Lowest leaves scalelike, leaves with blades located above base of plant; perigynia with nerves and usually thin-margined .... **Intermediae** .... (*Carex sartwellii*) p. 418

6 Male flowers at top of some or all spikes.............................. 7
6 Male flowers at base of some or all spikes 11

7 Spikes single at each node; heads usually with less than 10 spikes 8
7 Spikes two or more (at least at lowermost node)................... 9

8 Side spikes with a few male flowers at tip; perigynia usually green even when mature.................................. **Bracteosae** p. 403
8 Side spikes either all male or all female; perigynia usually brown when mature .... **Stellulatae** p. 441

9 Perigynia firm, not spongy, the tip abruptly narrowed to a beak, olive-green to dark brown or nearly black .... **Paniculatae** p. 437
9 Perigynia often spongy and thickened at base, usually gradually tapered to a beak, pale green to light golden-brown 10

10 Female scales awn-tipped, the awn 1–5 mm long; perigynia base slightly or not at all spongy-thickened .... **Multiflorae** (*Carex vulpinoidea*) p. 423
10 Female scales unawned, or if awned, then perigynia base broad and spongy-thickened .... **Vulpinae** p. 451

11 Perigynia margins thin-winged at least along upper part of body and lower beak. . . . . **Ovales** p. 425
11 Perigynia mostly filled by achene, sometimes sharp edged but never thin-winged. **12**

12 Perigynia nearly filled by the achene . . . . **Heleonastes** p. 414
12 Perigynia spongy at base; achene filling only upper half of perigynia . . **13**

13 Perigynia (especially the lowest in spike) spreading or curved downward; perigynia often smaller or broader or shorter-beaked than in Deweyanae . . . . **Stellulatae** p. 441
13 Perigynia upright, 4–6 mm long and 3–5x as long as wide . . . . **Deweyanae** . . . . p. 407

14 Lower female spikes spreading or drooping on slender stalks . . . . **Cryptocarpae** p. 406
14 Lower female spikes erect, stalkless or on short stalks **15**

15 Perigynia round and swollen, orange when mature (white-tinged when dried) . . . . **Bicolores** (*Carex aurea, Carex garberi*) p. 403
15 Perigynia flattened or lens-shaped, green or brown. 16

16 Perigynia dull; achene jointed with the style . . . . **Acutae** p. 397
16 Perigynia shiny; achene continuous with the persistent style . . . . **Vesicariae** p. 446

CAREX GROUP 3

1 Perigynia lance-shaped, 8–15 mm long and 1.5–3 mm wide, gradually tapered to a beak . . . . **Folliculatae** p. 411
1 Perigynia wider, ovate or oval **2**

2 Lowest female spikes drooping on long stalks . . . **Pseudocypereae** p. 438
2 Female spikes upright, stalked or stalkless **3**

3 Perigynia thick-walled, usually not inflated . . . . **Paludosae** p. 432
3 Perigynia thin-walled, usually inflated **4**

4 Perigynia broadest above middle, rounded or blunt at tip and abruptly narrowed to a beak . . . . **Squarrosae** p. 440
4 Perigynia broadest below middle, ovate or oval, tapered or rounded to a beak **5**

5 Bract of lowest female spike with an obvious sheath around stem; perigynia large, 10–20 mm long . . . . **Lupulinae** p. 420

5 Bract of lowest female spike without an obvious sheath, or if sheath present, perigynia less than 10 mm long .... **Vesicariae** p. 446

CAREX GROUP 4

1 Perigynia 10–20 mm long .... **Lupulinae** p. 420
1 Perigynia less than 8 mm long **2**

2 Female scales awn-tipped; top third of terminal spike female, lower two-thirds male .... **Gracillimae** p. 412
2 Female scales unawned or with only a small point; terminal spike all male or sometimes with a few perigynia at top or bottom .... **Sylvaticae** p. 445

CAREX GROUP 5

1 Perigynia at bottom of terminal spike .... **Sylvaticae** p. 445
1 Perigynia at top or middle of terminal spike **2**

2 Perigynia with an obvious beak **3**
2 Perigynia beakless, or with a short beak not much different than body . . **4**

3 Side spikes short-cylindric, upright, mostly without stalks; plants mostly of calcium-rich wetlands .... **Extensae** p. 409
3 Side spikes long-cylindric, spreading or drooping on slender stalks; habitats various .... **Sylvaticae** p. 445

4 Female spikes slender, spreading or drooping, 2–6 cm long; perigynia in several open rows; perigynia usually more than 4 mm long . . **Gracillimae** .... p. 412
4 Female spikes short-cylindric, mostly 1–2 cm long; perigynia in dense clusters; perigynia less than 4 mm long **5**

5 Female scales longer than perigynia; spikes spreading or drooping on slender stalks .... **Limosae** p. 419
5 Female scales shorter than perigynia, or if longer, then the spikes stalkless **6**

6 Terminal spike about half female, half male .... **Atratae** p. 401
6 Terminal spike with only one or several perigynia; plants mostly of calcium-rich wetlands .... **Paniceae** p. 435

CAREX GROUP 6

1 Perigynia with 2 prominent ribs but without nerves .... **Montanae** .... (*Carex deflexa*) p. 423
1 Perigynia with 2 prominent ribs and with several distinct nerves 2

2 Female spikes short-cylindric, upright and ± stalkless ...... **Hirtae** p. 417
2 Female spikes slender, spreading or drooping on stalks .... **Sylvaticae** .... p. 445

CAREX GROUP 7

1 Perigynia with 2 prominent ribs, otherwise ± without nerves 2
1 Perigynia with 2 ribs and usually many conspicuous nerves 5

2 Perigynia beakless .... **Paniceae** p. 435
2 Perigynia beaked 3

3 Female spikes ± upright, perigynia abruptly narrowed to a beak .... **Paniceae** 435
3 Female spikes spreading or drooping, mostly on slender stalks, perigynia various 4

4 Female spikes 1–6 cm long, if only 1 cm long, then the perigynia 3–6 mm long .... **Sylvaticae** p. 445
4 Female spikes 5–15 cm long, perigynia 2–3 mm long .... **Capillares** .... (*Carex capillaris*) p. 404

5 Female spikes ± stalkless, in dense short-cylindric to round heads; perigynia spreading (at least at their tips) and with conspicuous beaks .... **Extensae** p. 409
5 Lower female spikes on short stalks, short to long cylindric; spikes densely or loosely clustered, the perigynia mostly upright, not curved downward 6

6 Perigynia ovate or lance-shaped, distinctly narrowed to their base 7
6 Perigynia ovate or oval, rounded at base............................ 8

7 Female spikes short, cylinder-shaped, upright .... **Oligocarpae** .... (*Carex conoidea*) p. 424
7 Female spikes elongated, at least the lower spikes spreading or drooping .... **Sylvaticae** p. 445

8 Perigynia beakless 9
8 Perigynia with a short but distinct beak 10

9 Female spikes short, cylinder-shaped, 1–3 cm long, ± erect on short stalks .... **Oligocarpae** (*Carex conoidea*) p. 424
9 Female spikes very slender, 3–6 cm long, spreading or drooping on slender stalks .... **Gracillimae** p. 412

10 Leaf blades flat; plants of freshwater habitats.......... **Granulares** p. 413
10 Leaf blades rolled inward; plants of brackish or calcium-rich habitats ... . **Extensae** p. 409

CAREX GROUP 8

1 Perigynia beakless or with only a short, soft tip 2
1 Perigynia with a distinct, short or long beak 4

2 Female spikes short-cylindric on short stalks or stalkless, ± upright...... .... **Paniceae** p. 435
2 Female spikes short-cylindric on slender stalks, at least the lower spikes spreading or drooping 3

3 Female spikes usually less than 2x as long as wide; perigynia flattened... .... **Limosae** p. 419
3 Female spikes at least 3x as long as thick; perigynia 3-angled in section .. .. **Gracillimae** p. 412

4 Female spikes upright, stalkless or on short stalks ...... **Extensae** p. 409
4 Female spikes spreading or drooping on slender stalks 5

5 Female spikes 1–2 cm long and 6–8 mm wide .... **Sylvaticae** p. 445
5 Female spikes slender, 2–5 cm long and about 5 mm wide.... **Gracillimae** .... p. 412

CAREX SECTION ACUTAE

ACUTAE

1 Longest bract much longer than spikes 2
1 Longest bract shorter or equal to spikes 3

2 Plants forming large clumps or patches; leaves 3–8 mm wide; female spikes 3–6 cm long; perigynia without nerves .... *Carex aquatilis* .... WATER SEDGE
2 Plants densely clumped, long rhizomes absent; leaves 2–3 mm wide; female spikes 1.5–3 cm long; perigynia with a few prominent raised nerves .... *Carex lenticularis* | SHORE SEDGE

3 Lower sheaths rough-to-touch, red-brown, splitting into fibers .... *Carex stricta* | COMMON TUSSOCK SEDGE

3 Lower sheaths smooth, not splitting into fibers **4**

4 Perigynia olive-green with red dots, without nerves, or with 1–3 faint nerves .... *Carex haydenii* | LONG-SCALED TUSSOCK SEDGE

4 Perigynia green, becoming straw-colored, without red dots, usually with several nerves .... *Carex emoryi* | RIVERBANK SEDGE

Carex aquatilis Wahlenb. | Acutae
WATER SEDGE
NC **OBL** | MW **OBL** | GP **OBL**

Large native perennial, forming clumps or turfs; spreading by many slender rhizomes. **Stems** 3–12 dm long and longer than the leaves, 3-angled, usually rough-to-touch below the spikes. **Leaves** waxy blue-green, 2–7 mm wide; sheaths white or purple-dotted. **Spikes** 3–5, the upper spikes male, the middle and lower spikes female or often with male flowers borne above female, 2–5 cm long; **female scales** tapered to tip. **Perigynia** pale green to yellow-brown or red-brown, obovate, broadest near tip, not inflated, 2–3 mm long; beak tiny. **Achenes** lens-shaped, 1–2 mm long; stigmas 2. May–Aug. **Habitat** Wet meadows, marshes, shores, streambanks, kettle lakes, ditches and fens.

Carex aquatilis
WATER SEDGE

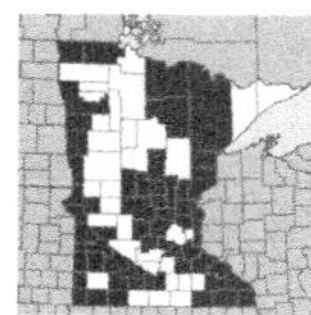

Carex emoryi Dewey | Acutae
RIVERBANK SEDGE
NC **OBL** | MW **OBL** | GP **OBL**

Loosely clumped native perennial from scaly rhizomes. **Stems** 3-angled, 4–12 dm long. 3–7 mm wide, lowest **leaves** reduced to red-brown sheaths; upper sheaths white or yellow-tinged and translucent, lower sheaths red-brown. **Spikes** 3–7, the terminal 1 or 2 all male, 2–4.5 cm long, the lateral spikes all female or with male flowers above the female, 2–10 cm long; lowest bract leaflike; **female scales** blunt or tapered to tip, narrower than the perigynia. **Perigynia** light green, becoming straw-colored at maturity, convex on both sides, oval or ovate,

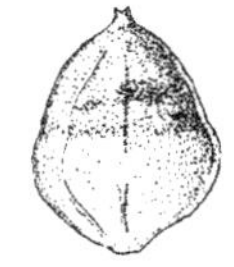

1.5–3 mm long, stigmas 2. May–July. **Habitat** Shores, streambanks, wet meadows and floodplain forests, sometimes forming pure stands.

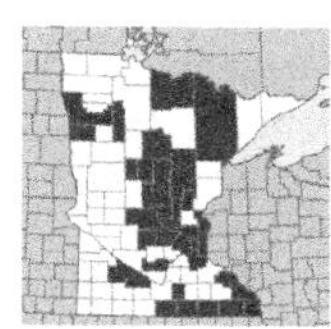

Carex haydenii Dewey | Acutae
LONG-SCALED TUSSOCK SEDGE
NC **OBL** | MW **OBL** | GP **OBL**

Loosely clumped native perennial, from short rhizomes. **Stems** arising from previous year's clumps of leaves (which are persistent at the base of the new leaves), 3-angled, 3–10 dm long, usually longer than the leaves, rough-to-touch above. **Leaves** green, 2–5 mm wide, the lower leaves bladeless and sheathlike; sheaths white to yellow on front, green on back, translucent. **Spikes** 3–6, the upper 1–3 male, the terminal one largest, 2–5 cm long, the others smaller; the lower 2–3 female or with male flowers above female, 1–3 cm long; lowest bract leaflike, usually shorter than the head; **female scales** tapered to tip, longer than the perigynia. **Perigynia** pale brown when mature, often with dark brown spots, convex on both sides, obovate, inflated at tip, 2–3 mm long; beak tiny. **Achenes** lens-shaped, 1 mm long; stigmas 2. May–July. **Habitat** Wet to moist meadows and swales, marshes and streambanks; often with dark-scaled sedge (*Carex buxbaumii*).

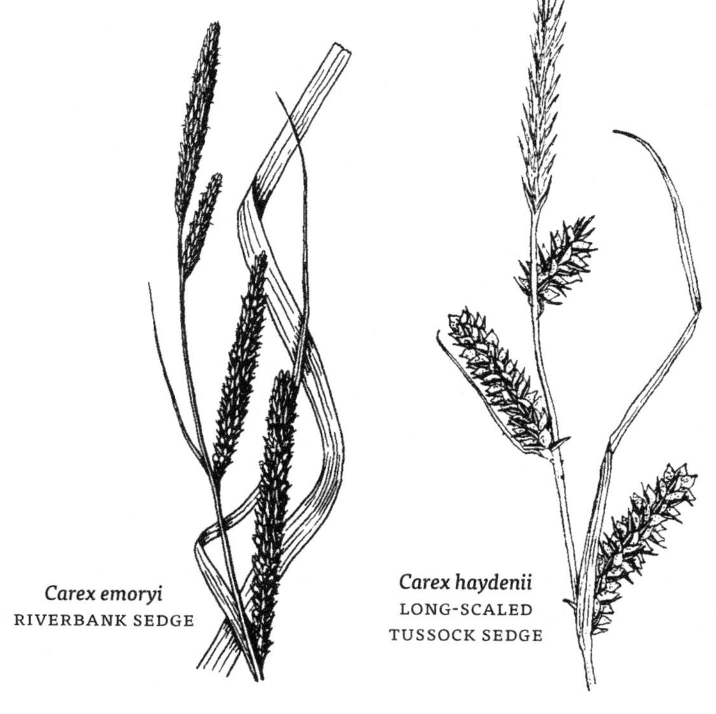

Carex emoryi
RIVERBANK SEDGE

Carex haydenii
LONG-SCALED TUSSOCK SEDGE

Carex lenticularis Michx. | Acutae
SHORE SEDGE
NC **OBL** | MW **OBL** | GP **OBL**

Densely clumped native perennial. **Stems** 1–6 dm long, upright, slender, usually shorter than leaves, brown at base. **Leaves** clustered on lower one-third of stem, upright, long-tapered to tip, 1–2 mm wide; sheaths dotted with yellow-brown on front. **Male spike** single, sometimes with a few female flowers, stalked, linear, 8–30 mm long and 2–3 mm wide; female spikes 3–5, upright, the upper stalkless, the lower stalked, the upper grouped, the lower separate, linear, 1–5 cm long and 3–4 mm wide; lowest bract leaflike, erect, much longer than the head, the upper bracts shorter; **female scales** ovate, red or red-brown, with a 3-veined, green center, the margins translucent near tip, narrower and usually shorter than the perigynia. **Perigynia** upright, soon deciduous, obovate, flattened, convex on both sides and sharply two-edged, 2–3 mm long and 1–1.5 mm wide, waxy blue-green, with a few yellow glandular dots or bumps, tapered or rounded at the abruptly pointed tip; the beak small, to 0.2 mm long. **Achenes** lens-shaped, 2 mm long, brown; stigmas 2. June–Sept. **Habitat** Rocky and sandy lakeshores, rock pools along Lake Superior, shallow ponds, sedge mats.

Carex lenticularis
SHORE SEDGE

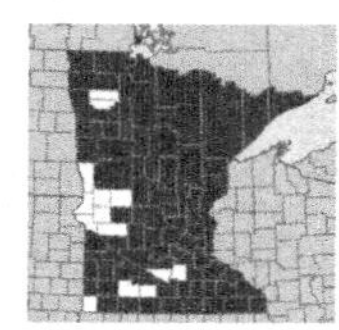

Carex stricta Lam. | Acutae
COMMON TUSSOCK SEDGE
NC **OBL** | MW **OBL** | GP **OBL**

Densely clumped native perennial from long scaly rhizomes, forming large, raised hummocks to 1 m tall. **Stems** 3-angled, 3–10 dm long, rough-to-touch, longer than leaves. **Leaves** 2–6 mm wide, the lower leaves reduced to sheaths around the base of stem; sheaths white to red-brown on front, green on back, the lower sheaths breaking into ladderlike thin strands. **Spikes** mostly all male or female (sometimes mixed), the upper 1–3 spikes male, the terminal spike 1.5–5 cm long, the lower 2–5 spikes female or some with male flowers borne above female, 2–8 cm long; lowest bract leaflike; **female scales** rounded or tapered to tip, equal or longer than the perigynia but narrower. **Perigynia** green at tip and margins, golden to yellow-brown in middle, with white or brown bumps, convex on both sides to nearly

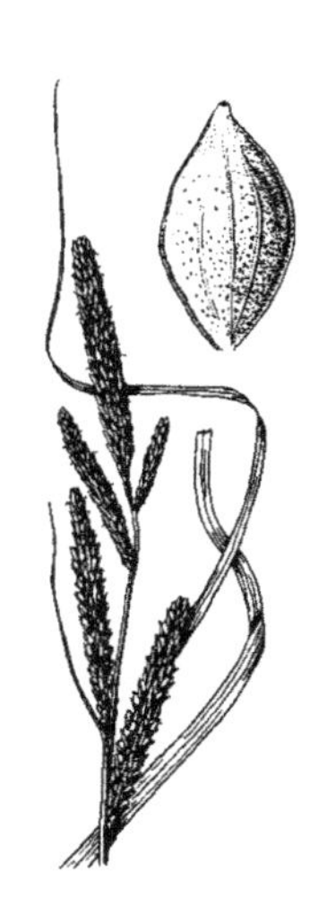

flat, ovate to oval, 2–3 mm long and 0.5–2 mm wide, 2-ribbed with a few faint nerves on both sides; beak short and tubelike, to only 0.3 mm long. **Achenes** lens-shaped, 1.5 mm long; stigmas 2. May–July. **Habitat** Sedge meadows, marshes, fens, shores, streambanks, ditches. Common, and a dominant component of sedge meadow communities (see photos, page 376).

CAREX SECTION ATRATAE

ATRATAE

1 Lowest leaves reduced to scales; stems not surrounded by dried sheaths of previous years; perigynia light gray-green, covered with many small bumps; female spikes 1–3 cm long; female scales brown to nearly black with a lighter midrib . . . . *Carex buxbaumii* | DARK-SCALED SEDGE
1 Lowest leaves well-developed or slightly scalelike; stems surrounded by dried sheaths of previous years; perigynia usually straw-colored but sometimes green or purple, without small bumps; side spikes usually less than 1 cm long; female scales black with white translucent margins **2**

2 Spikes short, the terminal spike 6–14 mm long, female scales dark purple . . . . *Carex norvegica* | SCANDINAVIAN SEDGE
2 Spikes longer, the terminal spike 15–30 mm long, female scales straw-colored to brown . . . . *Carex parryana* | PARRY'S SEDGE

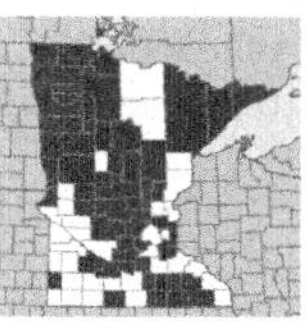

Carex buxbaumii Wahlenb. | Atratae

DARK-SCALED SEDGE
NC **OBL** | MW **OBL** | GP **OBL**

Loosely clumped native perennial, from long rhizomes. **Stems** single or few together, 3-angled, 3–10 dm long, rough-to-touch above, red-tinged near base. **Leaves** 1–3 mm wide, the lowest leaves without blades; lower sheaths shredding into thin strands, the upper sheaths membranous and purple-dotted. **Spikes** 2–5, 1–3 cm long, terminal spike with female flowers above male and larger than the lateral spikes, lateral spikes female, short-cylindric, stalkless or nearly so; bracts leaflike, the lowest shorter than the head; **female scales** dark brown, tapered to an awn at tip. **Perigynia** light green, golden brown near base, oval, 2.5–3.5 mm long, 2-ribbed, with 6–8 faint nerves on each side; beak tiny, notched. **Achenes** 3-angled, 1–2 mm long; stigmas 3. May–Aug. **Habitat** Wet meadows and fens, shallow marshes, low prairie, hollows in patterned peatlands.

Carex norvegica Retz. | Atratae
SCANDINAVIAN SEDGE
NC **FACW** | MW **FACW** | GP **FACW**

Loosely clumped native perennial, from short rhizomes. **Stems** slender, not stiff, 2–8 dm long, smooth or slightly rough-to-touch above, sharply triangular above, much longer than the leaves, red-tinged at base. **Leaves** 7–15 and mostly near base of stem, pale-green, flat or margins slightly rolled under, 2–3 mm wide, rough-to-touch on margins, the dried leaves of previous year conspicuous; sheaths translucent. **Spikes** usually 3, densely flowered, the terminal with both male and female flowers, the male below the female, clustered, upright, oblong to nearly round in outline per spike, 4–8 mm long and 3–5 mm wide, stalkless; the lateral spikes female, on short stalks; lowest bract usually shorter than the head; **female scales** ovate, 2–3 mm long, purple-black, acute to rounded, margins white-translucent, nearly as wide as perigynia but much shorter. **Perigynia** obovate, 2–4 mm long and 1.5 mm wide, rounded 3-angled, slightly inflated, yellow-green to brown, two-ribbed, otherwise without nerves, tip rounded and abruptly beaked, the beak short (0.5 mm long), red-tinged, with a small notch. **Achenes** obovate, 1–2 mm long, 3-angled, yellow-brown; stigmas 3. July–Aug. **Synonyms** *Carex media*. **Habitat** Rocky streambanks, rocky Lake Superior shores, talus slopes.

Carex parryana Dewey | Atratae
PARRY'S SEDGE
NC **FACW** | MW **FACW** | GP **FAC**

Loosely clumped native perennial, from short scaly rhizomes. **Stems** 2–6 dm long. **Leaves** clustered near base of plant, 2–4 mm wide, mostly less than 20 cm long. **Spikes** 1–5, upright, all female or the terminal spike male (or with both male and female flowers, the male usually below the female), 1–3 cm long; bract usually shorter than the head; **female scales** brown with translucent margins, shorter or longer than the perigynia. **Perigynia** obovate, 2–3 mm long, tapered to a short beak up to 0.5 mm long, often short-hairy near tip, 2-ribbed. **Achenes** 3-angled; stigmas 3. June–Aug. **Synonyms** Includes *Carex hallii*, sometimes considered a separate species (and of special concern in Minnesota). **Habitat** Prairie swales and wet meadows.

CAREX SECTION BICOLORES

Two wetland species in Minnesota . . . . *Carex aurea, Carex garberi*

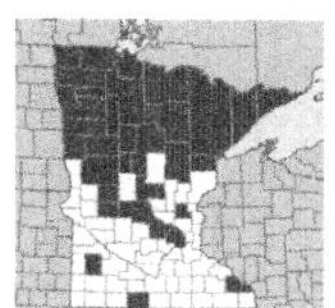

Carex aurea Nutt. | Bicolores

GOLDEN SEDGE
NC **FACW** | MW **FACW** | GP **OBL**

Small, loosely clumped native perennial. **Stems** upright, 3-angled, 5–30 cm long. **Leaves** 1–4 mm wide. **Spikes** 2–5 per stem, the lower spikes stalked; terminal spike male-flowered, 3–18 mm long; lateral spikes female, 8–20 mm long; bract of lowest spike longer than the head; **female scales** white-tinged to yellow-brown, with a green midvein, tipped with a short, sharp point, shorter than the perigynia. **Perigynia** with short white hairs when young, becoming a gold-orange when mature (drying paler), round to obovate, beakless or with a very short beak, several-ribbed, 2–3 mm long. **Achenes** dark brown to black, lens-shaped,1–1.5 mm long; stigmas 2. May–July. **Habitat** Wet meadows, low prairie, swales, wet woods and along sandy or gravelly shores; often where calcium-rich.

achene

RIGHT *The similar* **Carex garberi** *(threatened) is known from two sites in northern Minnesota, one a sedge fen, the other along Lake Superior. One distinction between the 2 species is terminal spike of Carex garberi tipped with female flowers (with male flowers below); in Carex aurea, terminal spike is of male flowers only.*

Carex garberi

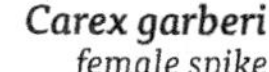

Carex garberi
female spike

CAREX SECTION BRACTEOSAE

BRACTEOSAE

1 Female scales brown or red-purple; perigynia 3 mm or more long; spikes many and crowded into a head; leaves 3–6 mm wide . . . *Carex alopecoidea*
. . . . BROWN-HEADED FOX SEDGE

1 Female scales green or green-white with translucent margin (or drying to pale brown); perigynia less than 3 mm long; spikes 2–5 and widely separated; leaves narrow, 1–2 mm wide . . . . *Carex disperma*
. . . . TWO-SEEDED SEDGE

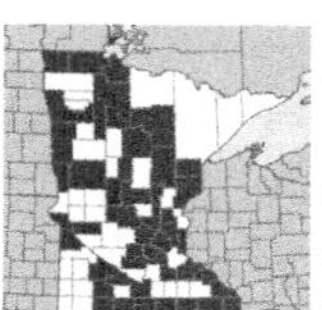

Carex alopecoidea Tuckerman | Bracteosae
BROWN-HEADED FOX SEDGE
NC **FACW** | MW **FACW** | GP **FACW**

Clumped native perennial. **Stems** soft, 4–10 dm long, 3-angled and sharply winged. **Leaves** 3–8 mm wide; sheaths purple-dotted, not cross-wrinkled. **Spikes** with both male and female flowers, male flowers above female, in heads 1.5–5 cm long; **female scales** tapered to tip or with a short sharp tip. **Perigynia** yellow-brown when mature, ovate, flat on 1 side and convex on other, 3–5 mm long, spongy-thickened at base, narrowed to a beak half to as long as the body. **Achenes** lens-shaped, 1–2 mm long; stigmas 2. May–July. **Habitat** Swamps and floodplain forests, streambanks, swales and moist fields.

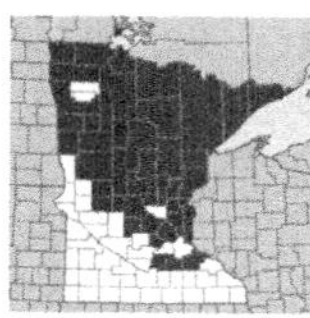

Carex disperma Dewey | Bracteosae
TWO-SEEDED SEDGE
NC **OBL** | MW **OBL** | GP **FACW**

Small, loosely clumped native perennial, from slender rhizomes. **Stems** slender, weak, 3-angled, 1–4 dm long, shorter to longer than leaves. **Leaves** soft and spreading, 1–2 mm wide; sheaths tight, translucent. **Spikes** 2–5, with both male and female flowers, male flowers borne above female, few flowered and small, with 1–6 perigynia and 1–2 male flowers, to 5 mm long, stalkless, separate or upper spikes grouped in interrupted heads 1.5–2.5 cm long; bracts sheathlike and resembling the female scales, or threadlike and to 2 cm long; **female scales** white, translucent except for the darker midrib, tapered to tip or short-awned, 1–2 mm long. **Perigynia** convex on both sides to nearly round in section, oval, 2–3 mm long, strongly nerved and rounded on the margins, beak tiny. **Achenes** lens-shaped, oval, 1–2 mm long; stigmas 2. May–July. **Habitat** Hummocks in conifer swamps and alder thickets, usually where shaded, wetland margins.

Carex disperma
TWO-SEEDED SEDGE

CAREX SECTION CAPILLARES

One wetland species in Minnesota. . . . . *Carex capillaris* | HAIRLIKE SEDGE

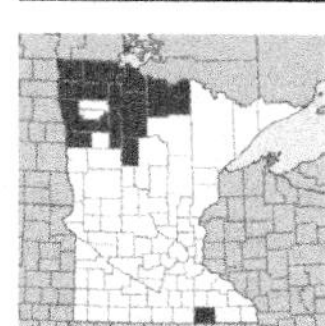

Carex capillaris L. | Capillares
HAIRLIKE SEDGE
NC **FACW** | MW **FACW** | GP **FACW**

Small, densely clumped native perennial. **Stems** slender, 3-angled, 1.5–4 dm long. **Leaves** mostly at base of plant and much shorter than stems, 1–3 mm wide; sheaths tight. **Spikes** either male or female; terminal spike male, 4–8 mm long; lateral spikes 1–4, separated on stem, loosely flowered, short-cylindric, 5–15 mm long, on threadlike, spreading to drooping stalks 5–15 mm long; **female scales** white, translucent on outer edges, green or light brown in middle, blunt or acute at tip, shorter but usually wider than perigynia, deciduous. **Perigynia** shiny brown to olive-green, ovate, round in section, 2–4 mm long, 2-ribbed, otherwise without nerves, tapered to a translucent-tipped beak 0.5 mm or more long. **Achenes** 3-angled with concave sides, 1–1.5 mm long; stigmas 3. June–July. **Habitat** Alder thickets, wetland margins, usually in shade.

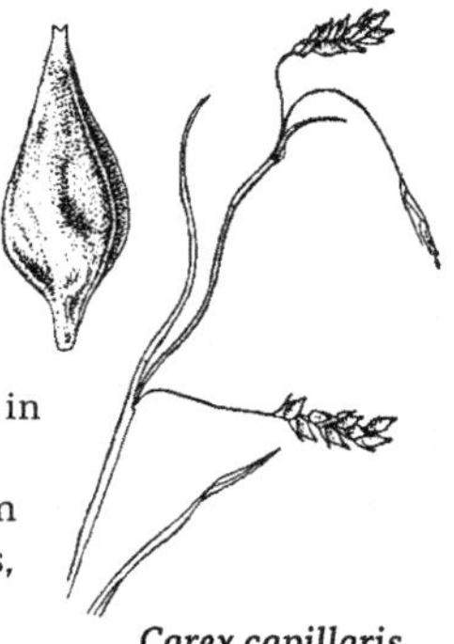

Carex capillaris
HAIRLIKE SEDGE

CAREX SECTION CHORDORRHIZAE

One wetland species in Minnesota .... *Carex chordorrhiza* .... CORDROOT SEDGE

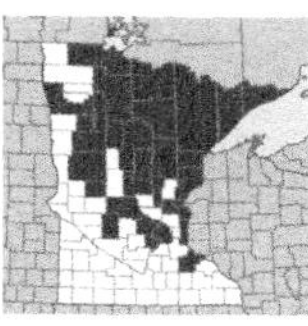

Carex chordorrhiza L. f. | Chordorrhizae
CORDROOT SEDGE
NC **OBL** | MW **OBL** | GP **OBL**

Native perennial from long, creeping stems. Flowering **stems** upright, rounded 3-angled in section, 1–3 dm tall, single or several together, arising from axils of dried leaves on older, reclining sterile stems. **Leaves** several on stem, the lower ones often bladeless, 1–2 mm wide; sheaths translucent. **Spikes** 3–8, with both male and female flowers, male flowers borne above female, crowded in an ovate head 5–15 mm long; bracts absent; **female scales** dark brown, ovate, about equaling the perigynia. **Perigynia** brown, compressed, ovate, 2–3.5 mm long, leathery, with many nerves on both sides; beak short. **Achenes** lens-shaped; stigmas 2. May–Aug. **Habitat** Open floating mats around lakes and ponds, fens, conifer swamps, interdunal hollows.

The creeping stems which root at each flowering stem are distinctive in **cordroot sedge**.

CAREX SECTION CRYPTOCARPAE

CRYPTOCARPAE

1 Stem sheath smooth; perigynia inflated, obovate, rounded at tip but abruptly tapered to a beak . . . . *Carex crinita* | FRINGED SEDGE
1 Stem sheath rough-to-touch; perigynia flattened, oval, tapered from near middle to a small beak . . . . *Carex gynandra* | FRINGED SEDGE

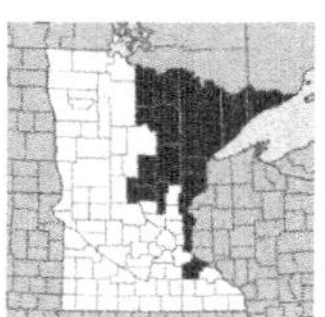

Carex crinita Lam. | Cryptocarpae

FRINGED SEDGE

NC **OBL** | MW **OBL** | GP **OBL**

Large, densely clumped native perennial. **Stems** 5–15 dm long and longer than leaves. **Leaves** 7–13 mm wide, lowest stem leaves reduced to scales; sheaths smooth. **Spikes** male or female, drooping on slender stalks; male spikes 1–3, above female spikes, 4–10 cm long; female spikes 2–5, narrow cylindric, 4–12 cm long; bract leaflike, without a sheath; **female scales** rounded and notched at tip with pale midvein prolonged into a toothed awn to 10 mm long, scale edges copper-brown. **Perigynia** green, 2-ribbed, nerves faint or absent, round in cross-section, abruptly tapered to a tiny beak. **Achenes** lens-shaped; stigmas 2. May–July. **Habitat** Swamps and alder thickets, wet openings, ditches and potholes.

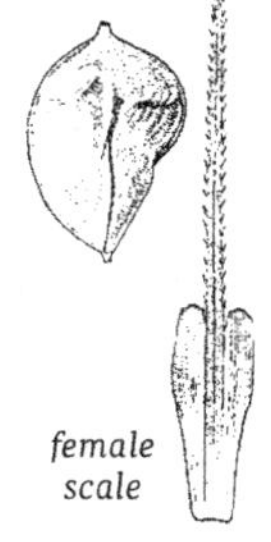

Similar to **Carex gynandra** *but with smooth sheaths, lower female scales rounded at tip, and perigynia inflated.*

Carex gynandra Schwein. | Cryptocarpae

FRINGED SEDGE

NC **OBL** | MW **FACW**

Large, clumped native perennial. **Stems** 5–15 dm long, longer than leaves. **Leaves** 7–14 mm wide, lowest leaves reduced to scales; sheaths finely hairy; bracts leaflike, lowest bract 1–3.5 dm long. **Spikes** either male or female, spreading or drooping and often curved, stalked; male spikes 1–3, above female, 5–9 cm long; female spikes 2–5, long-cylindric, 5–12 cm long; lower **female scales** 5–6 mm long, with a pale midrib, tapered to an awned tip about 5 mm long. **Perigynia** green, ovate to oval, not inflated, 3–4 mm long. **Achenes** lens-shaped, stigmas 2. June–July. **Synonyms** *Carex crinita* var. *gynandra*. **Habitat** Wet openings and swamps.

Carex gynandra *similar to Carex crinita, but has finely hairy sheaths, lower female scales tapered to an awned tip, and perigynia somewhat flattened and not inflated.*

female scale

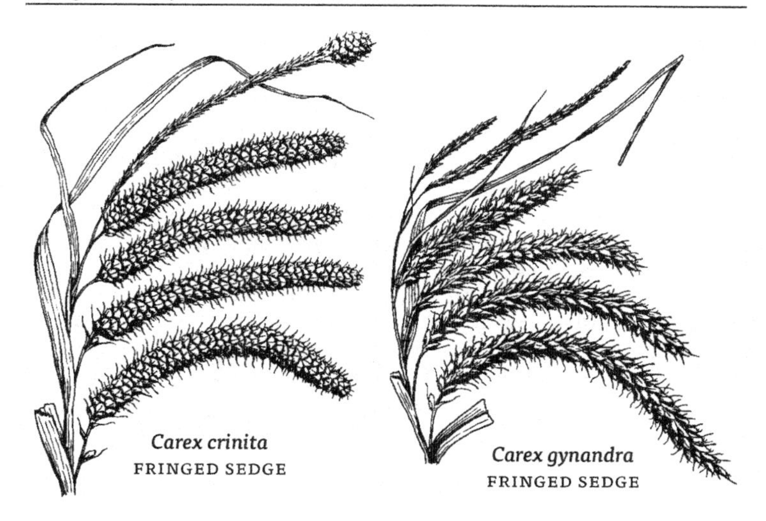

Carex crinita
FRINGED SEDGE

Carex gynandra
FRINGED SEDGE

CAREX SECTION DEWEYANAE

DEWEYANAE

1 Perigynia to 1.2 mm wide and 4–5x as long as wide, usually strongly nerved on both sides .... *Carex bromoides* | BROME HUMMOCK SEDGE
1 Perigynia 1.3–1.6 mm wide and 3–4x as long as wide, nerves faint or absent .... *Carex deweyana* | DEWEY'S HUMMOCK SEDGE

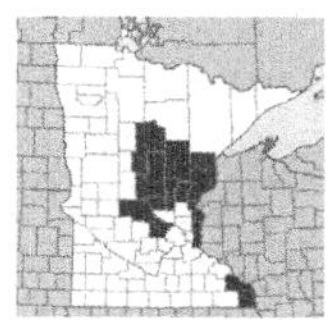

Carex bromoides Willd. | Deweyanae

BROME HUMMOCK SEDGE
NC **FACW** | MW **FACW**

Densely clumped native perennial. **Stems** very slender, 3–8 dm long. **Leaves** 1–2 mm wide. **Spikes** 3–7, narrowly oblong, 1–2 cm long, terminal spike with both male and female flowers, the male below female; lateral spikes all female or with a few male flowers at base, the spikes clustered or overlapping; **female scales** obovate, about as long as perigynia body, pale brown or orange-tinged with translucent margins, tapered to tip or short-awned. **Perigynia** lance-shaped, flat on 1 side and convex on other, light green, 4–6 mm long and 1–1.5 mm wide, nerved on both sides, gradually tapered to a finely sharp-toothed beak, the beak 1/2–2/3 as long as body. **Achenes** lens-shaped, to 2 mm long, in upper part of perigynium body; stigmas 2. April–July. **Habitat** Floodplain forests, old river channels, swamps.

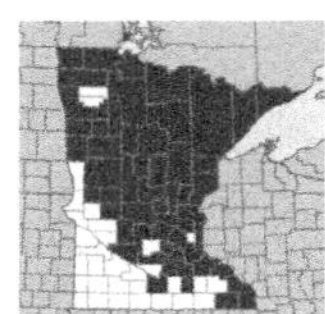

Carex deweyana Schwein. | Deweyanae
DEWEY'S HUMMOCK SEDGE
NC **FACU** | MW **FACU** | GP **FACU**

Large, loosely clumped native perennial from short rhizomes. **Stems** weak and spreading, 2–12 dm long, rough-to-touch below the head. **Leaves** shorter than stems, yellow-green to waxy blue-green, soft, flat, 2–5 mm wide; sheath tight. **Spikes** 2–6, the lower separate, the upper grouped, forming a head 2–6 cm long and often drooping near tip; terminal spike with male flowers at base, lateral spikes usually female, the perigynia upright; **female scales** ovate, blunt to short-awned at tip, thin and translucent with green center, slightly shorter than perigynia. **Perigynia** flat on 1 side and convex on other, 4–6 mm long and 1–2 mm wide, oblong lance-shaped, pale-green, very spongy at base, the beak 2–3 mm long, finely toothed and weakly notched. **Achenes** lens-shaped, nearly round, yellow-brown, 2 mm long; stigmas 2. May–Aug. **Habitat** Thickets, swamps and moist to dry woods.

CAREX SECTION DIOICAE

One wetland species in Minnesota. *Carex dioica* | NORTHERN BOG SEDGE

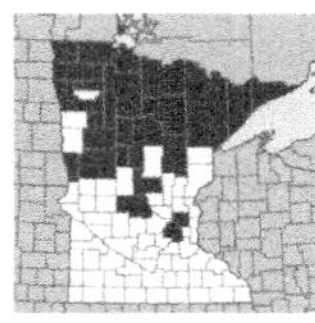

Carex dioica L. | Dioicae
NORTHERN BOG SEDGE
NC **OBL** | MW **FACW**

Small native perennial, from long, slender rhizomes. **Stems** single or few together, 0.3–3 dm long, smooth, usually longer than the leaves, brown at base. **Leaves** clustered near base of plant, blades inrolled and threadlike, to 1 mm wide. **Spikes** only 1 per stem, all male or all female, or with both male and female flowers and with the male flowers borne above the female, 0.5–2 cm long; the male spike or portion of spike narrowly cylindric, the female spike or portion short-cylindric; bract absent; **female scales** brown or red-brown, obovate, tapered to tip, shorter but wider than perigynia. **Perigynia** 4–10, widely spreading, yellow to dark brown, shiny, plump, obovate, 2–4 mm long and 1–2 mm wide, spongy at base, abruptly contracted to the beak; beak nearly entire to unequally notched, 0.5 mm long. **Achenes** lens-shaped, 1–2 mm long; stigmas 2. June–July. **Synonyms** *Carex gynocrates*. **Habitat** Conifer swamps and open peatlands, usually in sphagnum and wet, peaty soils.

CAREX SECTION DIVISAE

One wetland species in Minnesota . . . . *Carex praegracilis* . . . . EXPRESSWAY SEDGE

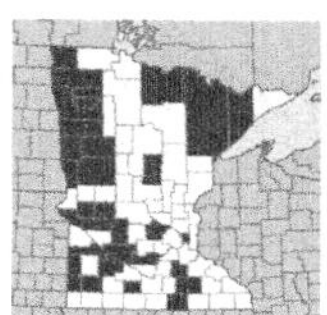

Carex praegracilis W. Boott | Divisae
EXPRESSWAY SEDGE
NC **FACW** | MW **FACW** | GP **FACW**

Colony-forming native perennial, from long black rhizomes. **Stems** single or few together, 3-angled, 1–7 dm long, longer than the leaves. **Leaves** on lower part of stems, 2–3 mm wide; sheaths white-translucent. **Spikes** with both male and female flowers, male flowers above female, or spikes nearly all male or female, 4–8 mm long, upper spikes crowded, lower spikes separated, in narrowly ovate heads 1–4 cm long; bracts absent; **female scales** brown, shiny, shorter or equal to perigynia. **Perigynia** green-brown, turning dark brown, flat on 1 side and convex on other, ovate to lance-shaped, 3–4 mm long and 1 mm wide, sharp-edged, spongy at base, tapered to a finely toothed beak 2 mm long, unequally notched. **Achenes** lens-shaped, 1–2 mm long; stigmas 2. May–June. **Habitat** Wet to moist meadows, shores, streambanks, ditches, along salted highways.

CAREX SECTION EXTENSAE

EXTENSAE

1 Perigynia 2–3.3 mm long, achene filling most of perigynium. . . . . *Carex viridula* | GREEN YELLOW SEDGE
1 Perigynia 3.5–6 mm long, achene filling only lower half of perigynium. . 2

2 Female scales conspicuous copper-brown, beak of perigynia rough-margined . . . . *Carex flava* | LARGE YELLOW SEDGE
2 Female scales inconspicuous, about same color as perigynia; beak smooth . . . . *Carex cryptolepis* | SMALL YELLOW SEDGE

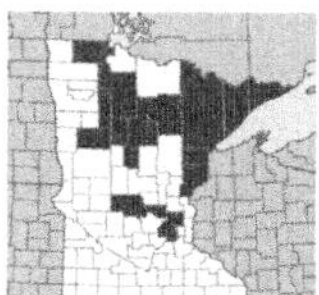

Carex cryptolepis Mackenzie | Extensae
SMALL YELLOW SEDGE
NC **OBL** | MW **OBL** | GP **OBL**

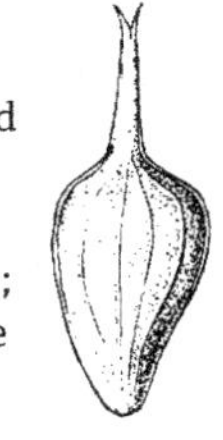

Clumped native perennial. **Stems** 2–6 dm long and longer than leaves. **Leaves** 2–4 mm wide. **Spikes** male or female; male spikes short-stalked or stalkless, the stalk shorter than the female spikes; female spikes 3–4, the upper 2 spikes grouped, the third separate, the fourth spike lower on stem, short-cylindric, 1–2 cm long, stalkless; bracts leaflike and spreading; **female scales** narrowly ovate, same color as perigynia and as long as perigynia body. **Perigynia**

yellow-brown when mature, lower ones curved outward and downward, body obovate, 3–5 mm long, 2-ribbed and several nerved, contracted into a smooth beak 1–1.5 mm long. **Achenes** 3-angled; stigmas 3. June–Aug. **Habitat** Wet meadows and marshy areas, peatlands, swamp margins; often where calcium-rich.

Carex cryptolepis
SMALL YELLOW SEDGE

Carex flava L. | Extensae
LARGE YELLOW SEDGE
NC **OBL** | MW **OBL** | GP **OBL** | *special concern*

Densely clumped native perennial from short rootstocks. **Stems** stiff, 1–7 dm high, usually longer than the leaves. **Leaves** 4–8 to a stem, mostly near base, 3–5 mm wide. Terminal spike male (or rarely partly female), stalkless or short-stalked, linear, 0.5–2 cm long and 2–3 mm wide; female spikes 2–5, sometimes with male flowers at tip, the uppermost spikes nearly stalkless, the lower stalked, short-oblong to nearly round, 6–18 mm long and 7–12 mm wide, perigynia 15–35, crowded in several to many rows, their beaks turned downward; bracts conspicuous, leaflike, spreading outward, much longer than the head; **female scales** ovate, narrower and much shorter than the perigynia, red-tinged except for the pale, three-nerved middle and the narrow translucent margins. **Perigynia** 4–6 mm long and 1–2 mm wide, obovate, yellow-green becoming yellow with age, conspicuously ribbed, tapered to a slender, finely toothed beak about as long as the body, the tip notched. **Achenes** 1.5 mm long, obovate, 3-angled, yellow-brown; stigmas 3. May–Aug. **Synonyms** *Carex laxior*. **Habitat** Wet, peaty meadows, often where calcium-rich. **Status** Special concern.

Carex flava
LARGE YELLOW SEDGE

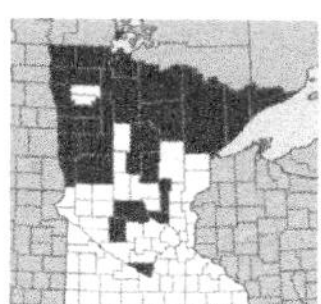

Carex viridula Michx. | Extensae
GREEN YELLOW SEDGE
NC **OBL** | MW **OBL** | GP **OBL**

Clumped native perennial. **Stems** stiff, slightly 3-angled, 0.5–4 dm long, longer than leaves. **Leaves** 1–3 mm wide; sheaths white-translucent. **Spikes** either male or female (or sometimes mixed), the terminal spike male or with a few female flowers at tip or middle, 3–15 mm long, short-stalked or stalkless, longer than the female spikes or clustered with them; lateral spikes female, 2–6, ovate to short-cylindric, 5–10 mm long, clustered and stalkless above, the lower spikes often separate and on short stalks; bracts leaflike, usually upright, much longer than the heads; **female scales** brown on sides, rounded or with a short, sharp point, about equal to perigynia. **Perigynia** yellow-green to brown, rounded 3-angled, obovate, 2–4 mm long, 2-ribbed, tapered to a slightly notched beak 0.5–1 mm long. **Achenes** 3-angled, to 1.5 mm long; stigmas 3. May–Aug. **Habitat** Wet meadows, sandy lake margins, fens and seeps; often where calcium-rich.

Carex viridula
GREEN YELLOW SEDGE

CAREX SECTION FOLLICULATAE

One wetland species in Minnesota .... *Carex michauxiana*
.... MICHAUX'S SEDGE

Carex michauxiana Boeckeler | Folliculatae
MICHAUX'S SEDGE
NC **OBL** | *special concern*

Clumped native perennial. **Stems** 2–6 dm long. **Leaves** 2–4 mm wide. **Spikes** either male or female; terminal spike male, 0.5–1.5 cm long, stalkless or short stalked; female spikes 2–4, broadly ovate, 1.5–2.5 cm long, upright, the lower spikes stalked, the upper on shorter stalks; bracts leaflike, 1–3 mm wide, longer than the stems; **female scales** ovate, shorter than the perigynia, margins translucent or brown, with a green midrib, tapered to a tip. **Perigynia** narrowly lance-shaped, 8–13 long and to 2 mm wide, round in section, long-tapered to a beak with upright teeth 1 mm long. **Achenes** rounded 3-angled; stigmas 3. June–Aug. **Habitat** Wet meadows, sphagnum peatlands, ditches and swales. **Status** Special concern.

CAREX SECTION GRACILLIMAE

GRACILLIMAE

1 Perigynia sharply angled; plants brown or green at base .... *Carex prasina* .... DROOPING SEDGE

1 Perigynia somewhat 3-angled; plants purple at base .... *Carex davisii* .... AWNED GRACEFUL SEDGE

Carex davisii Schwein. & Torr. | Gracillimae
AWNED GRACEFUL SEDGE
NC **FAC** | MW **FAC** | GP **FAC**

Clumped native perennial. **Stems** 3–10 dm long, purple at base. **Leaves** 4–8 mm wide, hairy on underside; sheaths hairy. Terminal spike male with female flowers near tip; female spikes 2–3, the upper 2 overlapping, cylindric, 2–4 cm long and 5–6 mm wide, upright to nodding on short stalks; **female scales** obovate, white or translucent with green center, tipped with a long awn, shorter or longer than perigynia. **Perigynia** ovate, dull orange when mature, 4–6 mm long and 2–3 mm wide, somewhat 3-angled, tapered to a notched beak to 1 mm long. **Achenes** 3-angled; stigmas 3. May–June. **Habitat** Floodplain forest and moist woods.

Carex prasina Wahlenb. | Gracillimae
DROOPING SEDGE
[NC **OBL** | MW **OBL**]

Clumped native perennial. **Stems** 3–8 dm long, brown or green at base. **Leaves** 3–5 mm wide. Terminal spike male or with a few female flowers at tip; female spikes 2–4, widely separated, cylindric, 2–5 cm long and 5 mm wide, curved or nodding, lower spikes on long stalks, the upper stalks much shorter; upper bract ± sheathless; **female scales** ovate to obovate, shorter than the perigynia, tipped with a short point. **Perigynia** 3–4 mm long, ovate, 3-angled, tapered to beak. **Achenes** 3-angled; stigmas 3. May–June. **Habitat** Low areas and seeps in deciduous woods, streambanks.

Carex prasina *is not known from Minnesota but may occur in St. Croix Valley as this species is present in adjacent Wisc.*

CAREX SECTION GRANULARES

GRANULARES

1 Male spike stalkless, or on a stalk shorter than uppermost female spike . . . . *Carex granularis* | PALE SEDGE

1 Stalk of male spike longer than uppermost female spike . . . . . *Carex crawei* . . . . EARLY FEN SEDGE

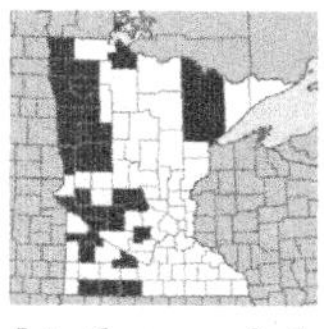

Carex crawei Dewey | Granulares
EARLY FEN SEDGE
NC **FACW** | MW **FACW** | GP **FACW**

Native perennial, from long-creeping rhizomes. **Stems** single or several together, faintly 3-angled, 0.5–4 dm long. **Leaves** 1–4 mm wide; sheaths tight, translucent. **Spikes** either male or female, cylindric, densely flowered, 1–3 cm long, terminal spike male; lateral spikes female, 2–5, separate, the lowest spike near base of plant; bract leaflike with well-developed sheath, its blade shorter than terminal spike; **female scales** red-brown with a pale or green midrib, shorter and narrower than the perigynia. **Perigynia** green to brown, ovate, 2–3.5 mm long, many-nerved; beak absent or very short, entire to notched. **Achenes** 3-angled, 1–2 mm long; stigmas 3. May–July. **Habitat** Wet to moist meadows and prairies, marly lakeshores, ditches, especially where calcium-rich.

Carex crawei
EARLY FEN SEDGE

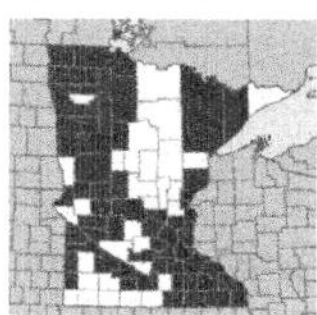

Carex granularis Muhl. | Granulares
PALE SEDGE
NC **FACW** | MW **FACW** | GP **OBL**

Clumped native perennial, from short rhizomes. **Stems** rounded 3-angled, 1–5 dm long. **Leaves** often longer than stems, 3–13 mm wide; sheaths membranous on front, divided-with small swollen joints on back. **Spikes** either all male or female, the terminal spike male, stalkless; the lateral spikes female, clustered around the male spike; bracts longer than the head; **female scales** brown, tapered to tip or with a short, sharp point, half as long as perigynia. **Perigynia** crowded in several rows, green or olive to brown, oval to obovate, 2–3 mm long and 1–2 mm wide, 2-ribbed, strongly nerved; beak tiny or absent, entire to slightly notched. **Achenes** 3-angled, 1–2 mm long; stigmas 3. May–July. **Habitat** Wet to moist meadows and swales, streambanks and pond margins, especially where calcium-rich.

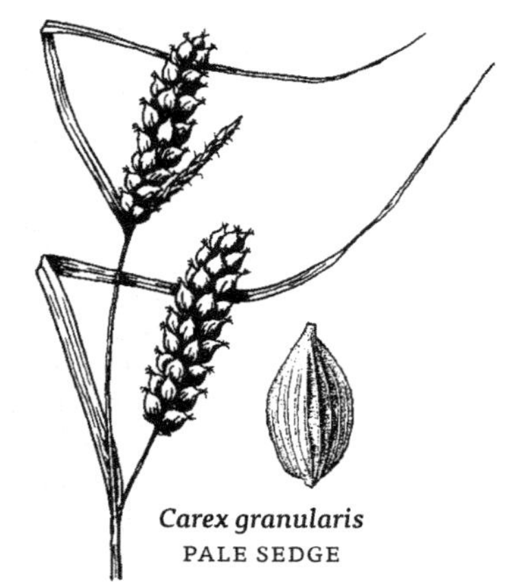

Carex granularis
PALE SEDGE

CAREX SECTION HELEONASTES

HELEONASTES

1 Spikes 2–4, in a short, overlapping cluster .... *Carex tenuiflora* .... SMALL-HEADED BOG SEDGE
1 Spikes single or 2–9, the lower spikes generally separate and not overlapping 2

2 Spikes 1–3 (usually 2), perigynia 2.5–4 mm long, 1–5 in each spike........ .... *Carex trisperma* | THREE-SEEDED BOG SEDGE
2 Spikes 4–9; perigynia 1.5–2.5 mm long, 5–30 in each spike 3

3 Perigynia mostly 5–10 in each spike, nerveless or with faint nerves, beak rough-margined; perigynia loosely spreading, the perigynia tips noticeable and interrupting the outline of spike; spikes green; leaves 1–2 mm wide . ... *Carex brunnescens* | GREEN BOG SEDGE
3 Perigynia usually 15–30, with evident nerves (visible under a hand lens), beak smooth-margined; perigynia more upright, the perigynia beaks not interrupting the outline of spike; spikes silvery green; leaves waxy blue-green, 2–4 mm wide................. *Carex canescens* | GRAY BOG SEDGE

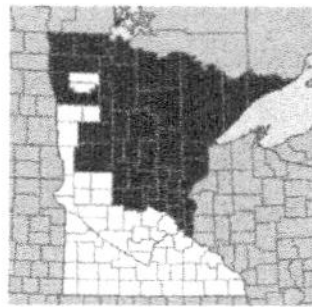

Carex brunnescens (Pers.) Poiret | Heleonastes
GREEN BOG SEDGE
NC **FACW** | MW **FACW** | GP **FAC**

Densely clumped native perennial from a short, fibrous rootstock. **Stems** slender, sharply 3-angled, 0.5–5 dm long, smooth or slightly rough-to-touch below the head, mostly longer than the leaves. **Leaves** 1–3 mm wide; sheaths tight, thin and translucent. **Spikes** 5–10 in a head 2–5 cm long, all with female flowers borne above male, 4–8 mm long, each spike with 5–15 perigynia, lower spikes separated; lowermost bract bristlelike, shorter or longer than lowermost spike; **female scales** ovate, rounded or acute at tip, shorter than the perigynia. **Perigynia** 3-angled, not winged or sharp-edged, 2–3 mm long, faintly nerved on both sides, not spongy-thickened at base, tapered at tip to a short, minutely notched beak, the beak and upper body finely toothed and white-dotted. **Achenes** lens-shaped, 1–1.5 mm long; stigmas 2. June–Aug. **Habitat** Wet forests and swamps, peatland margins. Circumboreal.

Carex brunnescens
GREEN BOG SEDGE

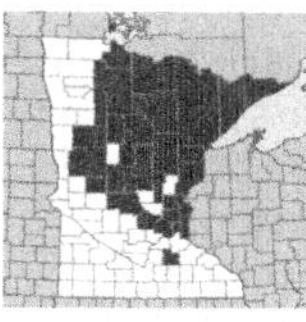

Carex canescens L. | Heleonastes
GRAY BOG SEDGE
NC **OBL** | MW **OBL** | GP **OBL**

Clumped native perennial. **Stems** 2–6 dm long. **Leaves** waxy blue- or gray-green, 2–4 mm wide, mostly near base of plant and shorter than stems. **Spikes** 4–8, silvery green or grayish, with both male and female flowers, the male below the female, ovate to cylindric, 5–10 mm long, the lower spikes ± separate, each spike with 10–30 perigynia. **Perigynia** flat on one side and convex on other, 2–3 mm long and 1–2 mm wide, with a beak to 0.5 mm long, not noticeably finely toothed on the margins; **female scales** shorter than perigynia. **Achenes** lens-shaped; stigmas 2. May–July. **Habitat** Peatlands (including hummocks in patterned fens), tamarack swamps, floating mats, alder thickets, wet forest depressions.

*Similar to **Carex brunnescens** (page 414) but leaves waxy blue-green rather than green and spikes somewhat larger and silver-green vs. brown in Carex brunnescens.*

Carex canescens
GRAY BOG SEDGE

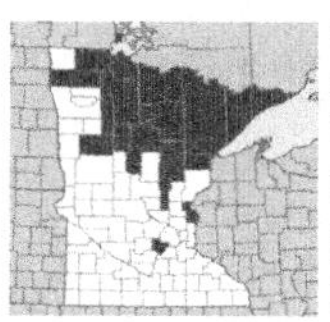

Carex tenuiflora Wahlenb. | Heleonastes
SMALL-HEADED BOG SEDGE
NC **OBL** | MW **OBL** | GP **OBL**

Loosely clumped, delicate native perennial, spreading from long, slender rhizomes. **Stems** very slender, 2–6 dm long. **Leaves** 1–2 mm wide. **Spikes** 2–4, with both male and female flowers, the male below the female, stalkless, clustered into a head 8–

15 mm long; **female scales** white-translucent with green center, covering most of the perigynium. **Perigynia** 3–15, oval, flat on 1 side and convex on other, 3–4 mm long, dotted with small white depressions, sharp-edged, beakless. **Achenes** lens-shaped, nearly filling the perigynia; stigmas 2. June–Aug. **Habitat** Hummocks in peatlands, floating mats, conifer swamps; s in Minnesota, mostly confined to tamarack swamps.

Easily overlooked as plants similar to **Carex disperma** *(page 404) and* **Carex trisperma** *and often growing with these more common species.*

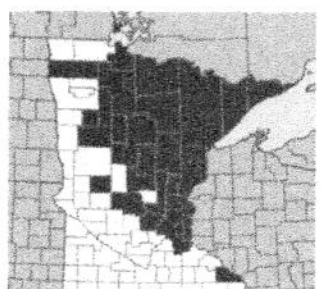

Carex trisperma Dewey | Heleonastes
THREE-SEEDED BOG SEDGE
NC **OBL** | MW **OBL** | GP **OBL**

Loosely clumped native perennial, with short, slender rhizomes. **Stems** very slender and weak, 2–7 dm long. **Leaves** 1–2 mm wide. **Spikes** 1–3 (usually 2), stalkless, 1–4 cm apart in a slender, often zigzagged head, each spike with 2–5 perigynia and a few male flowers at the base; lowest spike subtended by a bristlelike bract 2–6 cm long; **female scales** ovate, translucent with a green center, shorter or equal to the perigynia. **Perigynia** flat on 1 side and convex on other, oval, 3–4 mm long, finely many-nerved, tapered near tip to a short, smooth beak 0.5 mm long. **Achenes** oval-oblong, filling the perigynia; stigmas 2. May–Aug. **Habitat** Forested wetlands and conifer swamps, alder thickets, true bogs; a dominant species of forested wetlands in the Red Lake peatland.

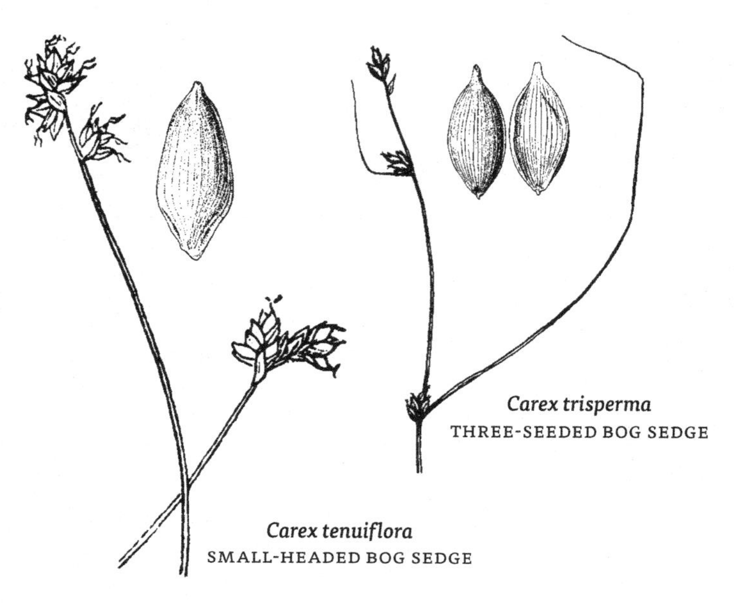

Carex trisperma
THREE-SEEDED BOG SEDGE

Carex tenuiflora
SMALL-HEADED BOG SEDGE

CAREX SECTION HIRTAE

HIRTAE

1 Leaves ± flat, mostly 2–5 mm wide . . . . . *Carex lanuginosa* | WOOLLY SEDGE
1 Leaves folded along midrib and inrolled, only 1–2 mm wide . . . . *Carex lasiocarpa* | SLENDER SEDGE

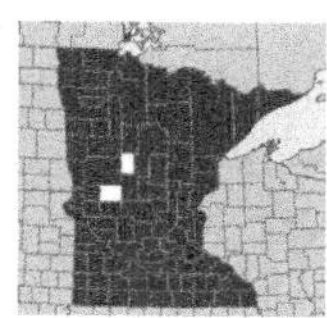

Carex lanuginosa Michx. | Hirtae
WOOLLY SEDGE
NC **OBL** | MW **OBL** | GP **OBL**

Colony-forming native perennial, from scaly rhizomes. **Stems** 3-angled, 2–10 dm long. **Leaves** 2–5 mm wide; sheaths thin and translucent, lower sheaths often purple-tinged on back, shredding into a loose network of fibers. **Spikes** either all male or female, the upper 1–3 male, 2–6 cm long; the lower 1–3 spikes female, separate, stalkless or nearly so, cylindric, 1–4 cm long; bracts leaflike, the lowest usually longer than the head; **female scales** brown to purple-brown, tapered to a tip to awned, shorter or longer than the perigynia. **Perigynia** brown to yellow-green to gray-brown, nearly round in section, obovate, 2.5–5 mm long, densely hairy, many-nerved, contracted to a finely toothed beak 1–2 mm long, the beak teeth spreading. **Achenes** 3-angled with concave sides, 1.5–2 mm long; stigmas 3. June–Aug. **Synonyms** *Carex lasiocarpa* var. *latifolia*, *Carex pellita*. **Habitat** Common; wet to moist meadows and swales, marshes, shores, streambanks and other wet places.

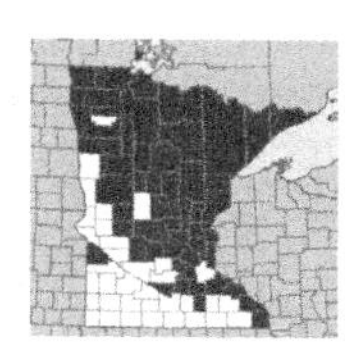

Carex lasiocarpa Ehrh. | Hirtae
SLENDER SEDGE
NC **OBL** | MW **OBL** | GP **OBL**

Colony-forming native perennial, from long, scaly rhizomes. **Stems** loosely clumped, 3-angled, 3–10 dm long. **Leaves** elongate and inrolled, 1–2 mm wide; sheaths tinged with yellow-brown. **Spikes** either all male or female, usually the upper 2 male; the male spikes slender, on a long stalk; the lower 1–3 spikes female, widely separate, ± stalkless, cylindric, 1–4 cm long; bracts leaflike, the lowest usually longer than the stem; **female scales** purple-brown with a green center, narrowly ovate, narrower and shorter or longer than the perigynia. **Perigynia** dull brown green, obovate, nearly round in section, 3–5 mm long, densely soft hairy, contracted to a beak about 1 mm long, the beak teeth erect. **Achenes** yellow-brown, 3-angled with concave sides, to 2 mm long; stigmas 3. June–Aug. North American plants are var.

americana. **Habitat** Peatlands and wet peaty soils, open bogs, pond margins (where a pioneer mat-former), hollows in Red Lake peatlands.

Carex lasiocarpa
SLENDER SEDGE

CAREX SECTION INTERMEDIAE

One wetland species in Minnesota .... *Carex sartwellii* .... RUNNING MARSH SEDGE

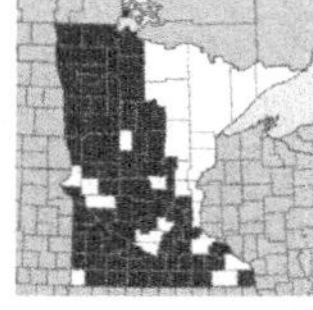

Carex sartwellii Dewey | Intermediae
RUNNING MARSH SEDGE
NC **OBL** | MW **FACW** | GP **FACW**

Colony-forming native perennial, from long black rhizomes. **Stems** single or few together, stiff, sharply 3-angled, 3–8 dm long, longer than the leaves. **Leaves** 2–4 mm wide, few per stem, the lowest leaves small and without blades; sheaths with green lines on front, and a translucent ligule around stem. **Spikes** with both male and female flowers, male flowers above female, or upper spikes male; clustered or lower spikes separate, 5–10 mm long, in cone-shaped heads, 3–6 cm long; bracts small, the lower bracts sometimes bristlelike and longer than the spike; **female scales** brown with a prominent green midvein, acute or with a short sharp point,

about equal to perigynia. **Perigynia** tan to brown, flat on 1 side and convex on other, ovate, 2.5–3.5 mm long and 1–2 mm wide, finely nerved on both sides, sharp-edged, tapered to a short, finely toothed beak. **Achenes** lens-shaped, 1–1.5 mm long; stigmas 2. May–July.
Habitat Wet to moist meadows, marshes, fens and shores, often where calcium-rich.

CAREX SECTION LIMOSAE

LIMOSAE

1 Female scales oval, about as wide and long as the perigynia. . . . . *Carex limosa* | MUD SEDGE
1 Female scales lance-shaped, narrower and longer than perigynia . . . . *Carex paupercula* | POOR SEDGE

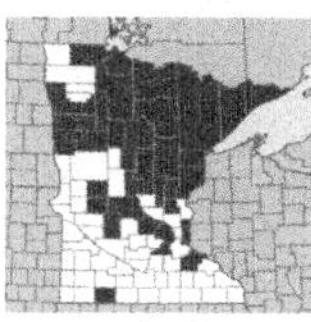

Carex limosa L. | Limosae

MUD SEDGE
NC **OBL** | MW **OBL** | GP **OBL**

Loosely clumped native perennial, from long scaly rhizomes. **Stems** sharply 3-angled, 3–5 dm long, longer than leaves, usually rough-to-touch above. **Leaves** 1–3 mm wide; sheaths translucent, shredding into threadlike fibers near base. **Spikes** either all male or female, the terminal spike male, 1–3 cm long; the lower 1–3 spikes female, drooping on lax, threadlike stalks 1–3 cm long, ovate to short-cylindric, 1–2 cm long; **female scales** brown, rounded or with a short, sharp point, about same size as perigynia. **Perigynia** waxy blue-green, ovate, flattened except where filled by achene, 2.5–4 mm long and 1–2 mm wide, strongly 2-ribbed with a few faint nerves on each side; beak tiny. **Achenes** 3-angled, 2 mm long; stigmas 3. May–July. **Habitat** Open bogs and floating mats. Common northward, less common in s Minnesota where mostly confined to calcareous fens and tamarack swamps.

Carex paupercula *is similar but has scales much narrower than the perigynia.*

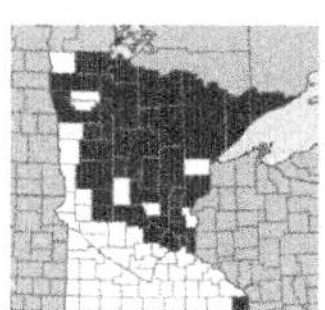

Carex paupercula Michx. | Limosae

POOR SEDGE
NC **OBL** | MW **OBL** | GP **OBL**

Loosely clumped native perennial, from slender branching rhizomes. **Stems** slender, 1–8 dm high, longer than the leaves, red-brown at base. **Leaves** 3–12 on lower half of stem, flat but with slightly rolled under margins, 2–4 mm wide, the dried leaves of previous year conspicuous; sheaths red-dotted. **Terminal spike** male (or sometimes with a few female flowers at tip), on a long stalk, linear, 4–12 mm long and 2–4 mm wide, usually

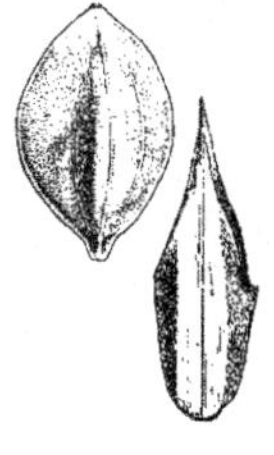

upright; **female spikes** 1–4 (rarely with several male flowers at base), clustered, usually drooping on slender stalks, 4–20 mm long and 4–8 mm wide, nearly round to oblong; lowest bract leaflike, equal or longer than the head. **Female scales** lance-shaped to ovate, tapered to a tip, narrower but usually longer than the perigynia, brown or green in center, margins brown. **Perigynia** broadly ovate or oval, 2–3 mm long and 1.5–2.5 mm wide, flattened and 2-ribbed, with several evident nerves, pale or somewhat waxy blue-green, covered with many small bumps, the tip rounded and barely beaked. **Achenes** 3-angled, obovate, 2 mm long; stigmas 3. July–Aug. **Synonyms** *Carex magellanica*. **Habitat** Open bogs, partly shaded peatlands, floating mats, swamps and thickets, usually in sphagnum moss.

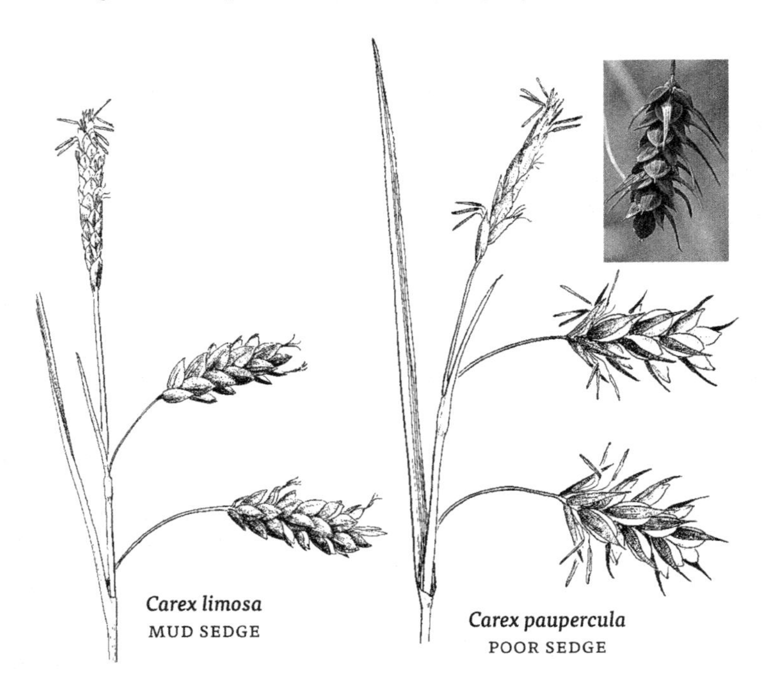

Carex limosa
MUD SEDGE

Carex paupercula
POOR SEDGE

CAREX SECTION LUPULINAE

LUPULINAE

1 Plants clumped; sheath of uppermost stem leaf (not bract) absent or less than 15 cm long; perigynium beak 1–4 mm long 2
1 Plants usually with long rhizomes; sheath of uppermost leaf usually longer than 15 cm; perigynium beak 4–10 mm long . . . . *Carex lupulina* . . . . COMMON HOP SEDGE

2 Perigynia dull, tapered to the base; spikes with mostly 8–35 perigynia, these radiating in all directions .... *Carex grayi* | COMMON BUR SEDGE

2 Perigynia shiny, convexly rounded to the base; spikes with 1–12 perigynia, these ascending or spreading ... *Carex intumescens* | SHINING BUR SEDGE

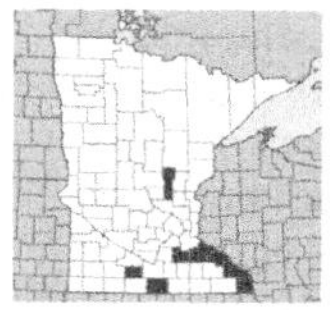

Carex grayi Carey | Lupulinae
COMMON BUR SEDGE
NC **FACW** | MW **FACW** | GP **FACW**

Native perennial, rhizomes absent. **Stems** single or forming small clumps, 3–9 dm long, rough on upper stem angles, sheaths at base of stem persistent, red-purple. **Leaves** 5–12 mm wide. **Spikes** either male or female; terminal spike male, 1–6 cm long, stalked; female spikes 1–2, rounded, stalked; bracts leaflike; **female scales** ovate, body shorter than perigynia but sometimes tipped with an awn to 7 mm long. **Perigynia** 10–30 per spike, spreading in all directions, not shiny, 10–20 mm long, strongly nerved, tapered from widest point to a notched beak 2–3 mm long. **Achenes** 3–5 mm long, with a persistent, withered style; stigmas 3. June–Sept. **Habitat** Floodplain forests and backwater areas (as along Mississippi River).

Carex grayi
COMMON BUR SEDGE

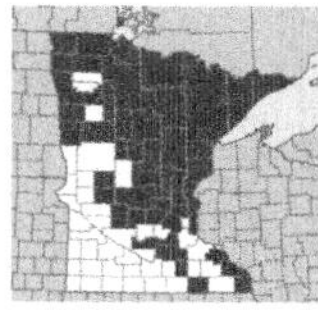

Carex intumescens Rudge | Lupulinae
SHINING BUR SEDGE
NC **FACW** | MW **FACW** | GP **OBL**

Native perennial, rhizomes absent. **Stems** single or in small clumps, 3–9 dm long, rough on upper stem angles; sheaths at base of stem persistent, red-purple. **Leaves** 4–12 mm wide, bracts leaflike. **Spikes** either male or female, or sometimes male spikes with a few female flowers; terminal spike male, 1–6 cm long, stalked; female spikes 1–4, grouped, 1–3 cm long and wide, rounded, on stalks to 1.5 cm long; **female scales** narrowly ovate, shorter and narrower than perigynia. **Perigynia** 1–12 per spike, satiny (not dull), 10–17 mm long, tapered to a beak 2–4 mm long. **Achenes**

3–6 mm long, flattened; stigmas 3. May–Aug. **Habitat** Mixed and deciduous moist forests, kettle wetlands in woods, alder thickets.

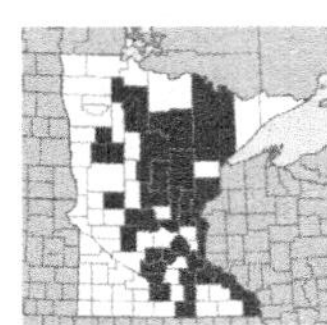

Carex lupulina Muhl. | Lupulinae
COMMON HOP SEDGE
NC **OBL** | MW **OBL** | GP **OBL**

Loosely clumped native perennial, from rhizomes. **Stems** stout, 3-angled, 3–12 dm long. **Leaves** much longer than head, 4–15 mm wide; upper sheaths white and translucent, the lower sheaths brown. **Spikes** either all male or female, the upper spike male, short-stalked, 2–5 cm long; female spikes 2–6, clustered or overlapping, the lowermost sometimes separate, 2.5–6 cm long and 1.5–3 cm wide; bracts leaflike and spreading, much longer than head; **female scales** narrowly ovate, tapered to tip or with a short awn, much shorter than the perigynia. **Perigynia** many, upright, dull green-brown, lance-shaped, inflated, 10–20 mm long and 4–7 mm wide, many-nerved, tapered to a finely toothed beak 5–10 mm long, the beak teeth 1–2 mm long. **Achenes** 3-angled, 3–4 mm long; stigmas 3. June–Aug. **Habitat** Wet woods, swamps, wet meadows and marshes, ditches and shores. In Minnesota most common in floodplain forests of Mississippi and St. Croix Rivers.

Shining bur sedge *(Carex intumescens) is similar but differs from common hop sedge by having fewer, uncrowded perigynia which are olive-green and glossy.*

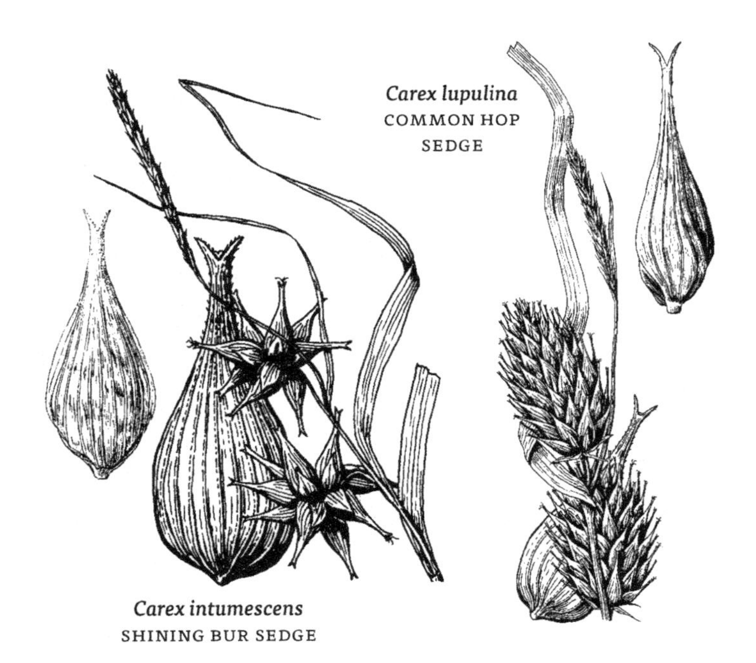

Carex lupulina
COMMON HOP SEDGE

Carex intumescens
SHINING BUR SEDGE

CAREX SECTION MONTANAE

One wetland species in Minnesota ..... *Carex deflexa* | NORTHERN SEDGE

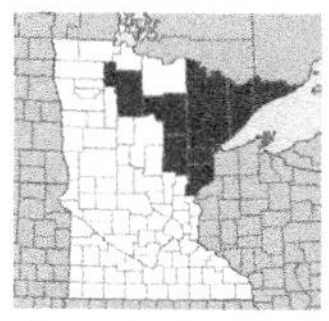

Carex deflexa Hornem. | Montanae
NORTHERN SEDGE
[NC **FACW** | MW **FACW** | GP **FACW**]

Small, loosely clumped native perennial. **Stems** 1–2 dm long, purple-tinged at base, shorter than the leaves. **Leaves** soft, 1–3 mm wide. **Spikes** either male or female; male spike short, to 5 mm long; female spikes on long, slender stalks near base of plant and also 2–4 spikes on stem near male spike; bract leaflike, to 2 cm long; **female scales** ovate, shorter than perigynia. **Perigynia** green, oblong-ovate, 2–3 mm long, covered with short hairs, abruptly tapered to a small beak about 0.5 mm long. **Achenes** 3-angled; stigmas 3. June–Aug. **Habitat** Moist woods and swamps, wetland margins.

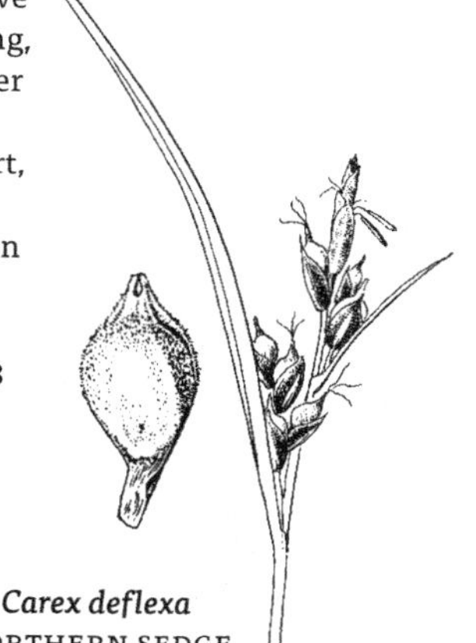

Carex deflexa
NORTHERN SEDGE

CAREX SECTION MULTIFLORAE

One wetland species in Minnesota .... *Carex vulpinoidea* .... BROWN FOX SEDGE

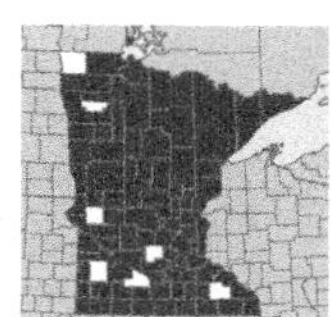

Carex vulpinoidea Michx. | Multiflorae
BROWN FOX SEDGE
NC **OBL** | MW **FACW** | GP **FACW**

Densely clumped native perennial, from short rootstocks. **Stems** stiff, sharply 3-angled, 3–9 dm long, shorter to longer than the leaves. 2–4 mm wide; sheaths tight, cross-wrinkled and translucent on front, mottled green and white on back. **Spikes** with both male and female flowers, male flowers borne above female; heads oblong to cylindric, 3–9 cm long, with several spikes per branch at lower nodes; bracts small and bristlelike, longer than the spikes; **female scales** awn-tipped, the awns equal or longer than the perigynia. **Perigynia** yellow-green, becoming straw-colored or brown when mature, flat on 1 side and convex on other, ovate to nearly round, 2–3 mm long and 1–2 mm wide, abruptly contracted to a notched, finely toothed beak 1 mm long. **Achenes** lens-shaped, 1–2 mm long; stigmas 2. May–Aug. **Synonyms** *Carex annectens* (sometimes considered a separate species). **Habitat** Wet to moist meadows, marshes, lakeshores, streambanks, roadside ditches.

CAREX SECTION OLIGOCARPAE

One wetland species in Minnesota . . . . *Carex conoidea*
. . . . PRAIRIE GRAY SEDGE

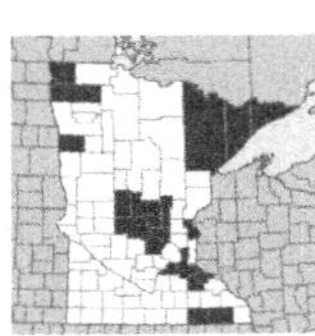

Carex conoidea Schk. | Oligocarpae
PRAIRIE GRAY SEDGE
NC **FACW** | MW **FACW** | GP **FAC**

Clumped native perennial. **Stems** 1–7 dm long, much longer than leaves. **Leaves** 2–4 mm wide. **Spikes** either male or female; male spike on a long stalk and overtopping female spikes, linear, 1–2 cm long; female spikes 2–4, widely spaced or upper 2 grouped, short cylindric, 1–2 cm long, on short, rough stalks; bract leaflike with a rough sheath; **female scales** ovate and much shorter than perigynia, with a green midvein prolonged into an awn. **Perigynia** oval, 3–4 mm long and 1–2 mm wide. **Achenes** 3-angled; stigmas 3. May–July. **Habitat** Wet meadows and prairies.

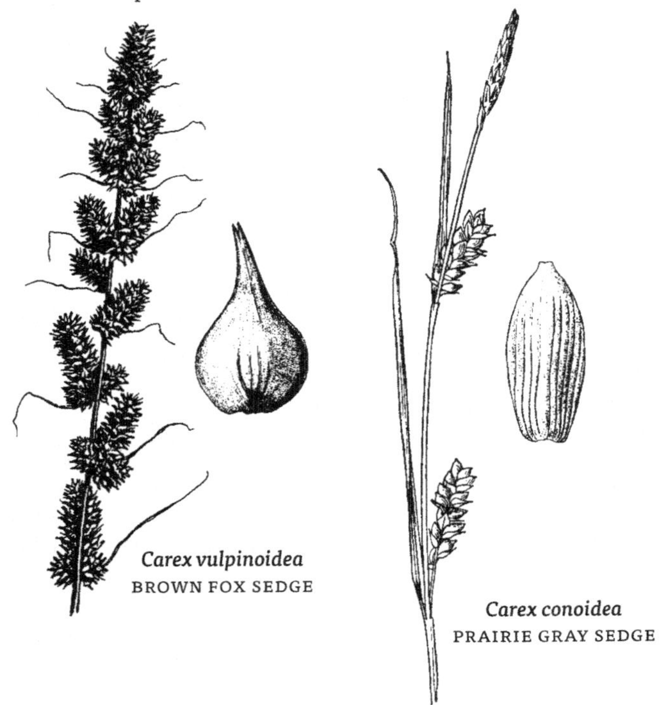

Carex vulpinoidea
BROWN FOX SEDGE

Carex conoidea
PRAIRIE GRAY SEDGE

CAREX SECTION ORTHOCERATES

One wetland species in Minnesota . . . . *Carex pauciflora*
. . . . FEW-FLOWERED BOG SEDGE

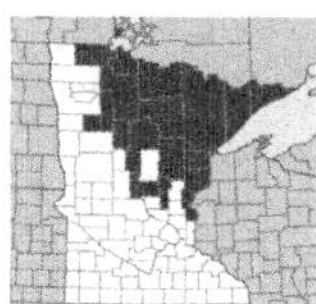

Carex pauciflora Lightf. | Orthocerates
FEW-FLOWERED BOG SEDGE
NC **OBL** | MW **OBL** | GP **OBL**

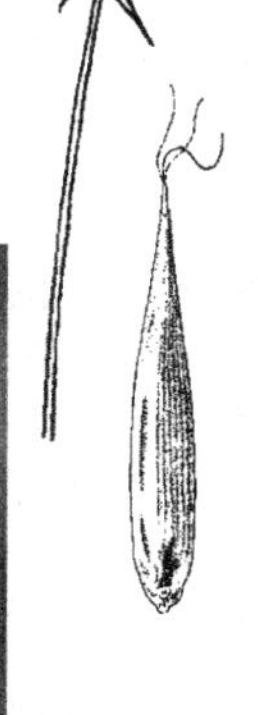

Native perennial, from long slender rhizomes. **Stems** single or several togethedm long, longer than leaves. **Leaves** 1–2 mm wide, lower stem leaves reduced to scales; bract absent. Spike single, to 1 cm long, with both male and female flowers, the male above the female; male scales infolded to form a slender terminal cone; **female scales** lance-shaped, 4–6 mm long, pale brown, soon deciduous. **Perigynia** 1–6, soon turned downward, slender, spongy at base, nearly round in section, straw-colored or pale brown, deciduous when mature, 6–8 mm long. **Achenes** 3-angled, not filling the perigynium; stigmas 3. June–July. **Habitat** Open peatlands and floating mats in sphagnum moss, true bogs.

Carex pauciflora
FEW-FLOWERED BOG SEDGE

spike

CAREX SECTION OVALES

OVALES *Species within Section Ovales are notoriously difficult to accurately identify; for most species, mature perigynia are needed.*

1 Bracts leaflike and many times longer than head .... *Carex sychnocephala* .... MANY-HEAD SEDGE
1 Bracts short and bristlelike, usually less than 2x as long as head 2

2 Achenes narrow, mostly 0.5–0.8 mm wide; perigynia usually more than 2.5x as long as wide, not oval-shaped or broadest above the middle 3
2 Achenes wider, 0.9–1.5 mm wide (or if a little narrower, then the perigynium body obovate), perigynia to about 2.5x as long as wide 9

3 Perigynia 7–10 mm long .... *Carex muskingumensis* | SWAMP OVAL SEDGE
3 Perigynia 2–6 mm long 4

4 Perigynia 2–4 mm long and 1–1.5 mm wide, to 3x as long as wide 5
4 Perigynia either at least 4 mm long, or more than 3x as long as wide, or both 6

5 Perigynia beak somewhat spreading, the perigynia often not winged to their base .... *Carex cristatella* | CRESTED OVAL SEDGE

5 Perigynia with a stiff, upright beak, the perigynia winged to their base . . . . *Carex bebbii* | BEBB'S OVAL SEDGE

6 Leaves 1–3 mm wide **7**
6 Leaves 7 mm or more wide **8**

7 Perigynia 3–5 mm long and 1.5–2 mm wide, strongly flattened . . . . *Carex crawfordii* | CRAWFORD'S OVAL SEDGE
7 Perigynia 4–6 mm long and to 1 mm wide, flat on 1 side, convex on other . . . . *Carex scoparia* | LANCE-FRUITED OVAL SEDGE

8 Spikes overlapping and crowded, 8–12 mm long, with more than 30 perigynia in each spike . . . . *Carex tribuloides* | AWL-FRUITED OVAL SEDGE
8 Lowermost spikes separate, spikes 5–8 mm long, with 15–30 perigynia in each spike . . . . *Carex projecta* | NECKLACE SEDGE

9 Perigynia less than 4 mm long and less than 2 mm wide. **10**
9 Perigynia more than 4 mm long or more than 2 mm wide, or both **11**

10 Heads compact; leaves mostly 2.5–6 mm wide. . . . . *Carex normalis* . . . . SPREADING OVAL SEDGE
10 Heads loose, the lower spikelets widely spaced; leaves mostly 1.5–2.5 mm wide . . . . *Carex tenera* | NARROW-LEAVED OVAL SEDGE

11 Portion of leaf sheath long-translucent; perigynia oval, widest at about one-third of total length . . . . *Carex normalis* | SPREADING OVAL SEDGE
11 Leaf sheath green-veined almost to its top, with only a short translucent area; perigynia round to obovate, widest at or above middle . . . . *Carex suberecta* | WEDGE-FRUITED OVAL SEDGE

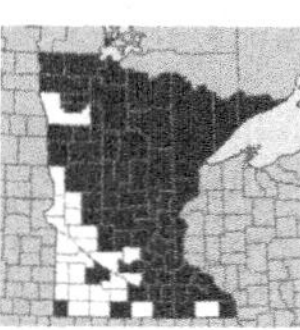

Carex bebbii (L. H. Bailey) Fernald | Ovales
BEBB'S OVAL SEDGE
NC **OBL** | MW **OBL** | GP **OBL**

Clumped native perennial. **Stems** sharply 3-angled, 2–8 dm long. **Leaves** shorter to slightly longer than the stems, 2–5 mm wide; sheaths white, thin and translucent. **Spikes** 5–10, with both male and female flowers, female flowers above male, 5–8 mm long, clustered in an ovate head 1.5–3 cm long; **female scales** tapered to tip, narrower and slightly shorter than the perigynia. **Perigynia** green to brown, flat on 1 side and convex on other, ovate, 2.5–3.5 mm long, finely nerved on back, nerveless on front, wing-margined, with a finely toothed beak 1/3–1/2 the length of the body, shallowly notched at tip. **Achenes** lens-shaped, 1–1.5 mm long; stigmas 2. June–Aug. **Habitat** Wet to moist meadows, marshes, streambanks, ditches and other wet places; calcareous fens.

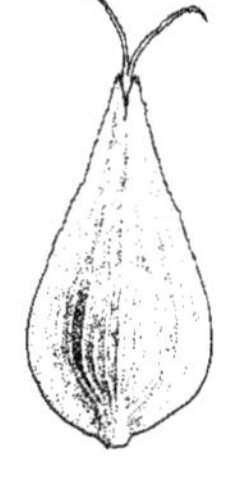

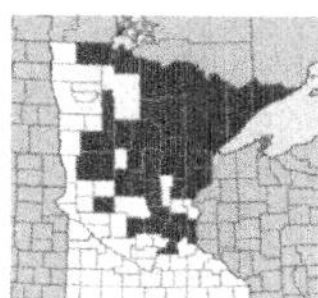

Carex crawfordii Fernald | Ovales
CRAWFORD'S OVAL SEDGE
NC **FACW** | MW **FAC** | GP **OBL**

Densely clumped native perennial. **Stems** 1–8 dm long, stiff. **Leaves** 3–4 on each stem, longer or shorter than the stems, 1–4 mm wide. **Spikes** 3–15, with both male and female flowers, the male below the female, grouped into a narrowly oblong, sometimes drooping head, 1–3 cm long and 4–15 mm wide; **female scales** ovate, light brown with green center, shorter and about as wide as perigynia. **Perigynia** flattened except where enlarged by the achenes, lance-shaped, 3–4 mm long and 0.5–1 mm wide, brown, narrowly winged nearly to the base, finely toothed above the middle, tapered to a long, slender, toothed, notched beak. **Achenes** brown, lens-shaped, 1 mm long; stigmas 2. July–Sept. **Habitat** Moist openings and wetland margins, sandy shorelines.

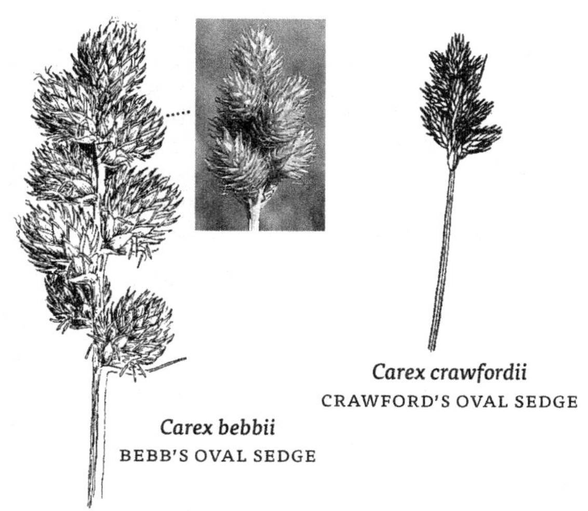

Carex crawfordii
CRAWFORD'S OVAL SEDGE

Carex bebbii
BEBB'S OVAL SEDGE

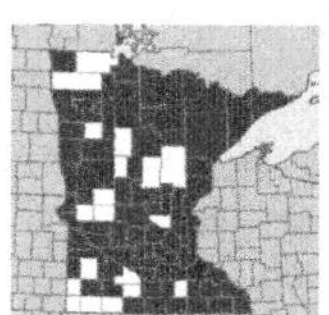

Carex cristatella Britton | Ovales
CRESTED OVAL SEDGE
NC **FACW** | MW **FACW** | GP **FACW**

Clumped native perennial, from short rhizomes. **Stems** sharply 3-angled, 3–10 dm long, slightly shorter to longer than the leaves. **Leaves** 3–7 mm wide; sheaths loose, with fine green lines. **Spikes** with both male and female flowers, female flowers borne above male; spikes 5–12, 4–8 mm long, crowded in an ovate to oblong head 2–3.5 cm long; bracts much reduced; **female scales** tapered to tip, shorter than the perigynia. **Perigynia** widely spreading when mature, green to pale brown, flat on 1 side and convex on other, ovate to lance-shaped,

2.5–4 mm long and 1–2 mm wide, faintly nerved on both sides, strongly winged above the middle, tapered to a finely toothed, notched beak 1–2 mm long. **Achenes** lens-shaped, 1–1.5 mm long; stigmas 2. June–Aug. **Habitat** Wet meadows, ditches, floodplains, marshy shores and streambanks.

Carex cristatella
CRESTED OVAL SEDGE

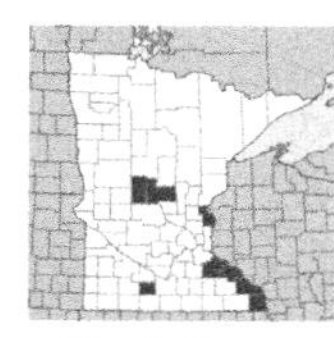

Carex muskingumensis Schwein. | Ovales
SWAMP OVAL SEDGE
NC **OBL** | MW **OBL**

Clumped native perennial. **Stems** stout, 5–10 dm long, with many leafy sterile stems present. **Leaves** 3–5 mm wide. **Spikes** 5–10, with both male and female flowers, the male below the female, pointed at both ends, 15–25 mm long and 4–6 mm wide, in a dense head 4–8 cm long; **female scales** lance-shaped, pale brown with translucent margins, half as long as the perigynia. **Perigynia** upright, lance-shaped, 6–10 mm long, finely nerved on both sides, tapered to a finely toothed, deeply notched beak half as long as the body. **Achenes** lens-shaped, 2 mm long; stigmas 2. June–Aug. **Habitat** Floodplain forests (as along Miss River), wet woods.

Carex muskingumensis
SWAMP OVAL SEDGE

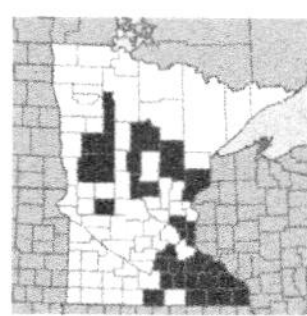

Carex normalis Mackenzie | Ovales
SPREADING OVAL SEDGE
NC **FACW** | MW **FACW** | GP **OBL**

Clumped native perennial. **Stems** 3–8 dm long, longer than the leaves. **Leaves** 2–6 mm wide, lower stem leaves reduced to scales. **Spikes** 5–10, with both male and female flowers, the male below the female, round in outline, 6–9 mm long, stalkless, loosely grouped in heads 3–5 cm long; **female scales** ovate, translucent, lightly brown-tinged, with green midvein, tapered to a point or blunt-

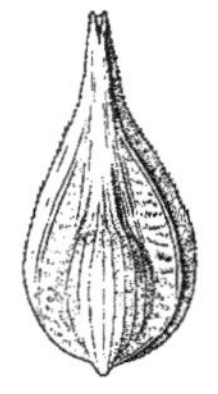

tipped, shorter than the perigynia. **Perigynia** upright, ovate, flat on 1 side and convex on other, green or pale green-brown, 3–4 mm long, finely nerved, tapered to a finely toothed beak. **Achenes** lens-shaped, 2 mm long; stigmas 2. June–Aug. **Habitat** Moist to wet deciduous woods, floodplain forests, alder thickets, marshes, pond margins.

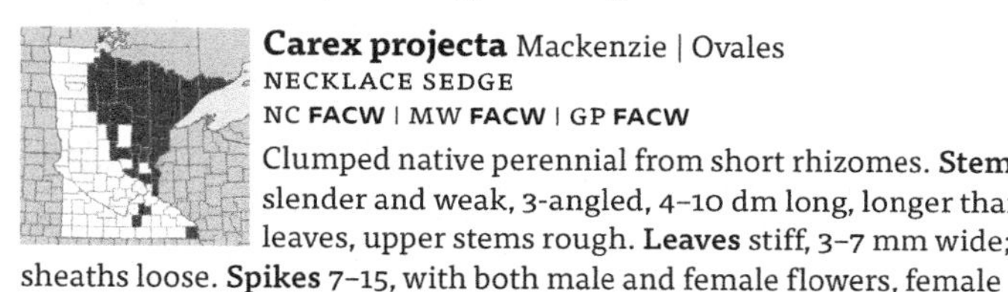

Carex projecta Mackenzie | Ovales
NECKLACE SEDGE
NC **FACW** | MW **FACW** | GP **FACW**

Clumped native perennial from short rhizomes. **Stems** slender and weak, 3-angled, 4–10 dm long, longer than leaves, upper stems rough. **Leaves** stiff, 3–7 mm wide; sheaths loose. **Spikes** 7–15, with both male and female flowers, female flowers above male in each spike, obovate to nearly round, 5–10 mm long and about as wide, straw-colored, distinct and ± separated (at least the lower spikes) in a somewhat lax and zigzagged inflorescence 3–5 cm long; bracts inconspicuous; **female scales** narrowly ovate, straw-colored, narrower and shorter than the perigynia. **Perigynia** ascending to spreading when mature, lance-shaped, 3–5 mm long and 1–2 mm wide, dull brown, flattened except where filled by the achene, winged on margin, the wing gradually narrowing from middle to base, tapered to a notched, finely toothed beak 1–2 mm long. **Achenes** lens-shaped, 1–2 mm long; stigmas 2. June–Aug. **Habitat** Floodplain forests, swamps, thickets, wet openings, shaded slopes.

*Similar to **Carex tribuloides** but the perigynia tips spreading rather than erect as in Carex tribuloides.*

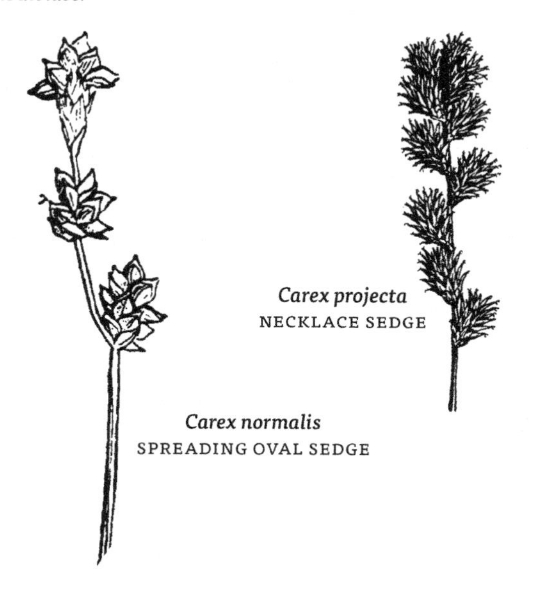

Carex projecta
NECKLACE SEDGE

Carex normalis
SPREADING OVAL SEDGE

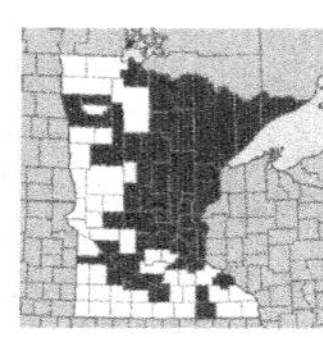

Carex scoparia Schk. | Ovales
LANCE-FRUITED OVAL SEDGE
NC **FACW** | MW **FACW** | GP **FACW**

Densely clumped native perennial, sometimes spreading by surface runners. **Stems** 2–10 dm long, sharply 3-angled, usually longer than the leaves. **Leaves** 1–3 mm wide; sheaths tight, white-translucent. **Spikes** 4–10, with both male and female flowers, female flowers borne above male, ovate to broadest at middle, 6–12 mm long, clustered or separate, in a narrowly ovate head 1–5 cm long; bracts small, the lowest often bristlelike; **female scales** tapered to tip, slightly shorter than perigynia. **Perigynia** green-white, flat, narrowly lance-shaped, 3–7 mm long and 1–2 mm wide, margins narrowly winged, tapered to a finely toothed, slightly notched beak 1–2 mm long. **Achenes** lens-shaped, 1–1.5 mm long; stigmas 2. May–July. **Habitat** Wet meadows and openings, low prairie, swamps and sandy lakeshores.

Carex suberecta (Olney) Britton | Ovales
WEDGE-FRUITED OVAL SEDGE
[NC **OBL** | MW **OBL**]

Clumped native perennial. **Stems** 3–7 dm long, shorter or longer than the leaves. **Leaves** 2–3 mm wide, lower stem leaves reduced to scales. **Spikes** 2–5, stalkless, loosely grouped into a head, with both male and female flowers, the male below the female, 7–12 mm long; **female scales** shorter and narrower than the perigynia, yellow-brown with a pale midvein and narrow translucent margins, tapered to a tip. **Perigynia** numerous, conspicuously swollen over the achene, 4–5 mm long, abruptly contracted to the flat, finely toothed beak. **Achenes** lens-shaped, 1.5 mm long; stigmas 2. May–July. **Habitat** Calcareous swamps, marshes, wet meadows, low prairie, calcareous fens and shores.

Carex suberecta *reported for Minnesota but not verified; known to occur in Wisconsin and Iowa.*

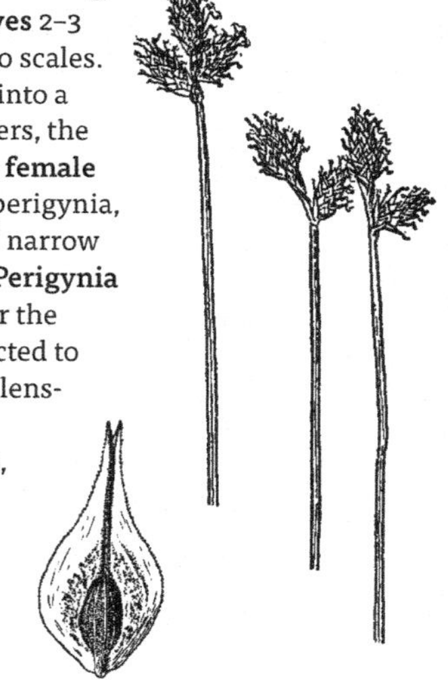

Carex suberecta
WEDGE-FRUITED OVAL SEDGE

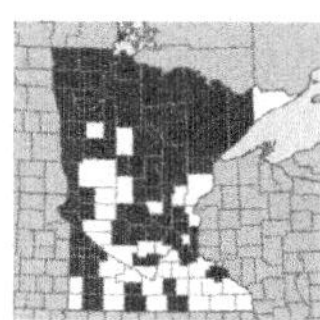

Carex sychnocephala Carey | Ovales
MANY-HEAD SEDGE
NC **FACW** | MW **FACW** | GP **FACW**

Clumped native perennial (sometimes annual), from fibrous roots. **Stems** many and crowded, rounded 3-angled, 0.5–6 dm long. **Leaves** 1.5–4 mm wide; sheaths tight, white-translucent. **Spikes** with both male and female flowers, female flowers borne above male, densely clustered in ovate heads 1.5–3 cm long; bracts leaflike, 2–4 per head, the longest bracts much longer than the heads; **female scales** thin and translucent with a green midvein, 2|3 length of perigynia, tapered to a tip or with a short sharp point. **Perigynia** green to straw-colored, flat, lance-shaped, 5–7 mm long and to 1 mm wide, narrowly wing-margined, spongy at base when mature, tapered to a finely toothed, notched beak 3–5 mm long. **Achenes** lens-shaped, 1–1.5 mm long; stigmas 2. June–Aug. **Habitat** Wet meadows, sandy lake and stream shores, marshes.

Carex sychnocephala
MANY-HEAD SEDGE

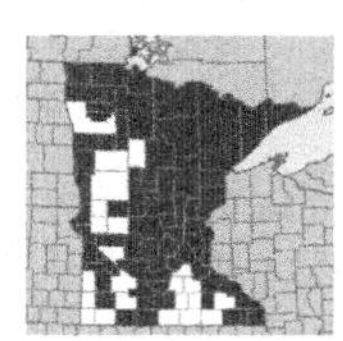

Carex tenera Dewey | Ovales
NARROW-LEAVED OVAL SEDGE
NC **FAC** | MW **FACW** | GP **FACW**

Clumped native perennial, from short rhizomes. **Stems** slender, sharply 3-angled, 3–8 cm long, rough-to-touch above, longer than leaves. **Leaves** 0.5–3 mm wide; sheaths white-translucent on front, mottled green and white on back. **Spikes** 4–8, with both male and female flowers, female flowers borne above male, ovate to round, 4–10 mm long, loose in nodding heads 2.5–5 cm long; bracts small, sometimes bristlelike, longer than the spike; **female scales** tapered to tip, slightly shorter than perigynia. **Perigynia** ovate, flat on 1 side and convex on other, straw-colored when mature, 2.5–4 mm long and 1–2 mm wide, wing-margined, tapered to a notched, finely toothed beak 1–2 mm long. **Achenes** lens-shaped, 1–2 mm long; stigmas 2. June–Aug. **Habitat** Wet to moist meadows, streambanks, floodplains and moist woods.

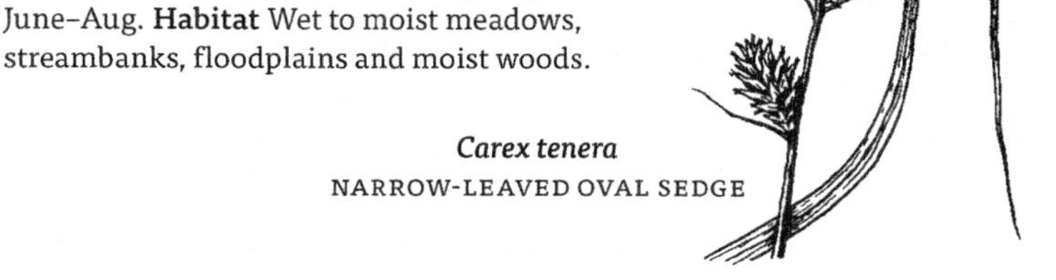

Carex tenera
NARROW-LEAVED OVAL SEDGE

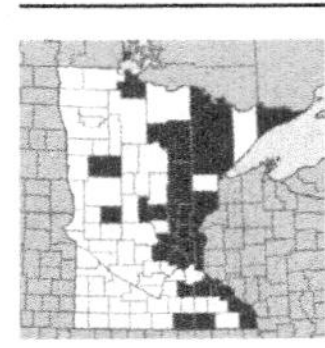

Carex tribuloides Wahlenb. | Ovales
AWL-FRUITED OVAL SEDGE
NC **FACW** | MW **OBL** | GP **OBL**

Clumped native perennial from short rhizomes. **Stems** sharply 3-angled, 3–9 dm long, longer than leaves. **Leaves** stiff, 3–7 mm wide; sheaths loose, with green lines. **Spikes** 5–15, with both male and female flowers, female flowers borne above male, obovate, 6–13 mm long, densely to loosely clustered into an ovate or oblong head 2–5 cm long; bracts inconspicuous; **female scales** tapered to tip, shorter than the perigynia. **Perigynia** light green to pale brown, flattened except where filled by the achenes, lance-shaped, 3–6 mm long and 1–2 mm wide, broadly winged near middle, tapered to a notched, finely toothed beak 1–2 mm long. **Achenes** lens-shaped, 1.5 mm long; stigmas 2. June–July. **Habitat** Floodplain forests, shady low areas in woods, pond and lake margins, marshes, low prairie.

Carex tribuloides
AWL-FRUITED OVAL SEDGE

CAREX SECTION PALUDOSAE

PALUDOSAE

1 Perigynia covered with hairs . . . . *Carex trichocarpa* . . . . HAIRY-FRUIT LAKE SEDGE

1 Perigynia smooth and hairless 2

2 Teeth of perigynia beak 1 mm or less in length . . . . *Carex lacustris* . . . . COMMON LAKESHORE SEDGE

2 Teeth of beak 1–3 mm long 3

3 Sheaths and blades smooth and without hairs . . . . *Carex laeviconica* . . . . LONG-TOOTHED LAKE SEDGE

3 Sheaths hairy, underside of blades usually hairy near their base . . . . *Carex atherodes* | SLOUGH SEDGE

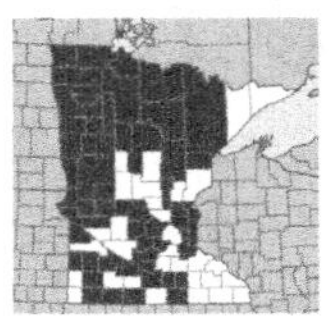

Carex atherodes Sprengel | Paludosae
SLOUGH SEDGE
NC **OBL** | MW **OBL** | GP **OBL**

Loosely clumped native perennial from long scale-covered rhizomes. **Stems** 3-angled, 5–12 dm long. **Leaves** 3–12 mm wide; sheaths hairy on back, brown to purple-tinged at the mouth, the lower sheaths shredding into narrow strands. **Spikes** either male or female; male spikes 2–6 at ends of stems; female spikes 2–4, widely

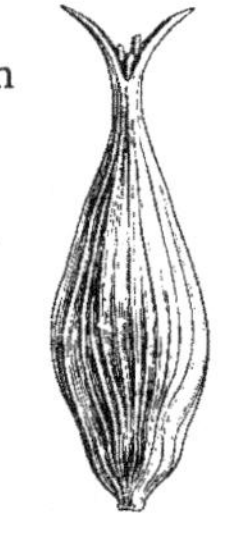

spaced, cylindrical, 2–11 cm long; bracts leaflike, longer than the stems; **female scales** thin, translucent or pale brown, shorter than the perigynia, tipped with a slender awn. **Perigynia** ovate, 6–11 mm long, long-tapered to a smooth beak, with many distinct nerves, the beak with spreading teeth 1.5–3 mm long. **Achenes** 3-angled, 2–2.5 mm long; stigmas 3. June–Aug. **Habitat** Marshes, wet meadows, prairie swales, stream and pond margins, usually in shallow water where may form dense colonies.

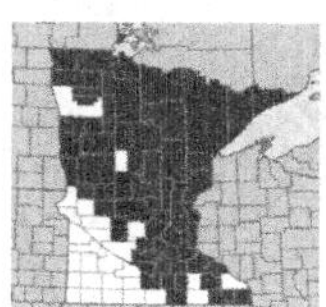

Carex lacustris Willd. | Paludosae
COMMON LAKESHORE SEDGE
NC **OBL** | MW **OBL** | GP **OBL**

Large clumped native perennial, from scaly rhizomes. **Stems** erect, 3-angled, 6–13 dm long, rough-to-touch. **Leaves** equaling or slightly longer than the stem, 6–15 mm wide; sheaths often red-tinged, the lower ones disintegrating into a network of fibers. **Spikes** either male or female, the upper 2–4 male, stalkless, 4–7 cm long; the lower 2–4 spikes female, erect, usually separate, stalkless or short-stalked, cylindric, 3–10 cm long and 3–15 mm wide; bracts leaflike, some or all longer than the head; **female scales** awned or tapered to tip, the body shorter than the perigynia, the sides thin and translucent to pale brown. **Perigynia** olive, flattened to nearly round in section, narrowly ovate, 5–7 mm long, with more than 10 raised nerves, tapered to a smooth beak about 1 mm long. **Achenes** 3-angled, 2.5 mm long; stigmas 3. May–Aug. **Habitat** Common; swamps, marshes, kettle wetlands, wetland margins, usually in shallow water; low areas in tamarack swamps.

Carex atherodes
SLOUGH SEDGE

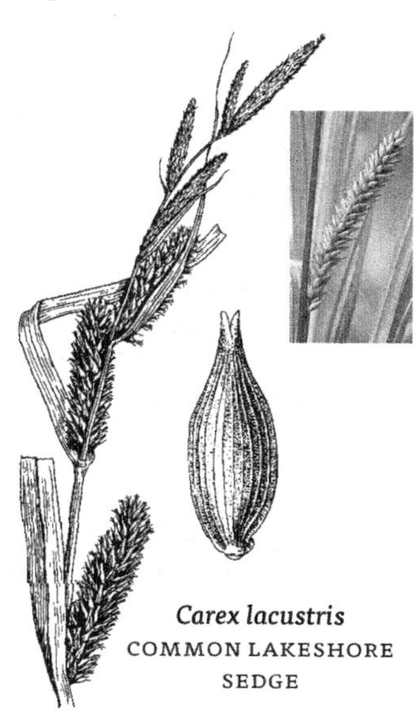

Carex lacustris
COMMON LAKESHORE SEDGE

Carex laeviconica Dewey | Paludosae
LONG-TOOTHED LAKE SEDGE
NC **OBL** | MW **OBL** | GP **OBL**

Loosely clumped native perennial, from scaly rhizomes. **Stems** stout, 3-angled, 3–12 dm long. **Leaves** shorter to longer than the stem, 2–8 mm wide; sheaths smooth, often purple-tinged below and splitting into fibers. **Spikes** either all male or female, the upper 2–6 male, 1–4 cm long; the lower 2–4 spikes female, erect, separate, stalkless or short-stalked, cylindric, 3–10 cm long and 6–10 mm wide; bracts leaflike, equal or longer than the head; **female scales** acute or awn-tipped, the scale body shorter than the perigynium, translucent or brown on the sides. **Perigynia** green-yellow, broadly ovate, inflated, round in section, 4–9 mm long, strongly many-nerved, tapered to a slender beak 1.5–2 mm long. **Habitat** Wet meadows, marshes, lakeshores and streambanks.

Carex laeviconica
LONG-TOOTHED LAKE SEDGE

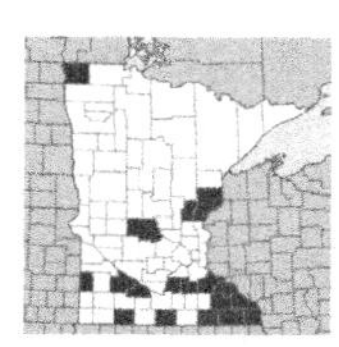

Carex trichocarpa Muhl. | Paludosae
HAIRY-FRUIT LAKE SEDGE
NC **OBL** | MW **OBL** | GP **OBL**

Loosely clumped native perennial, with short rhizomes. **Stems** stout, 6–12 dm long, smooth below, rough-to-touch above. **Leaves** 2–6 mm wide, rough-to-touch on margins, upper leaves and bracts often longer than stems. **Spikes** either all male or female, the upper 2–6 spikes male, long-stalked; female spikes2–4, cylindric, 4–10 cm long, the upper spikes ± stalkless, the lower spikes on slender stalks; **female scales** ovate, with white translucent margins, about half as long as perigynia. **Perigynia** ovate, usually covered with short white hairs, prominently ribbed, gradually tapered to a 2-toothed beak. **Achenes** 3-angled; stigmas 3. May–Aug. **Habitat** Riverbanks and old river channels, marshes, wet meadows, low prairie.

Similar to **slough sedge** *(Carex atherodes, page 432) but sheaths strongly purple-tinged at tip, the leaf blades not hairy on underside, and the perigynia with short white hairs (vs. smooth in Carex atherodes).*

CAREX SECTION PANICEAE

PANICAE

1 Perigynia with a distinct beak 1–2 mm long .... *Carex vaginata* .... SHEATHED SEDGE
1 Perigynia beakless or with a very small beak less than 0.5 mm long 2

2 Plants waxy blue-green; female scales white-translucent on margins .... *Carex livida* | LIVID SEDGE
2 Plants green; female scales purple-brown on margins ..... *Carex tetanica* .... COMMON STIFF SEDGE

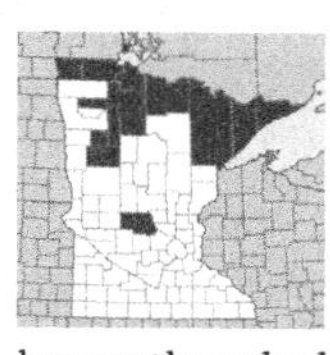

Carex livida (Wahlenb.) Willd. | Paniceae
LIVID SEDGE
NC **OBL** | MW **OBL** | GP **OBL**

Native perennial forming small clumps, from long slender rhizomes. **Stems** slender, erect, 0.5–6 dm long, shorter or longer than the leaves, light brown at base. **Leaves** 6–12 on lower third of stem, strongly waxy blue-green, channeled, 0.5–4 mm wide, dried leaves of the previous year conspicuous; sheaths very thin. **Terminal spike** male (or rarely with both male and female flowers, the male below the female), linear, 1–3 cm long and 3–4 mm wide; **female spikes** 1–3, the lowest ± separate, sometimes long-stalked, the upper grouped, stalkless or short-stalked, oblong, 1–2 cm long and 5 mm wide, with 5–15 upright perigynia; bracts leaflike, sometimes longer than the head; **female scales** ovate, rounded to somewhat acute, shorter than the perigynia, light purple with broad green center and white translucent margins. **Perigynia** obovate, slightly flattened and rounded 3-angled, 2–5 mm long and 1–2 mm wide, strongly waxy blue-green, with small dots, tapered to a beakless tip. **Achenes** ovate, 3-angled with prominent ribs, 2.5 mm long, brown-black; stigmas 3. July–Aug. **Synonyms** *Carex grayana*. **Habitat** Wet meadows and fens, especially where calcium-rich.

achene

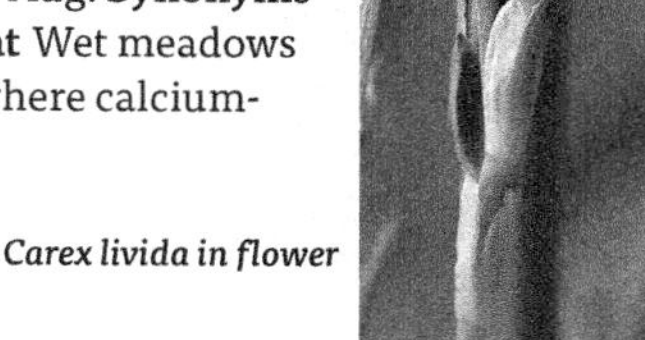

Carex livida in flower

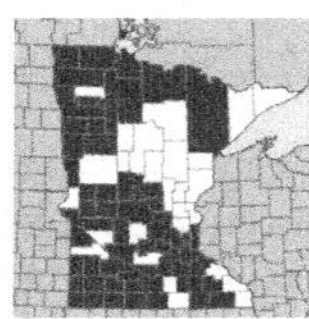

Carex tetanica Schk. | Paniceae
COMMON STIFF SEDGE
NC **FACW** | MW **FACW** | GP **FACW**

Clumped native perennial from slender rhizomes. **Stems** 3-angled, 1–6 dm long, rough-to-touch above. **Leaves** 1–5 mm wide; sheaths tight, white or yellow and translucent. **Spikes** either all male or female, terminal spike male, 1–3 cm long; lateral spikes female, usually widely separated, the lower spikes short-cylindric, stalked, 6–30 mm long and 3–5 mm wide, loosely flowered with perigynia in 3 rows; bracts shorter than the head; **female scales** purple-brown on margins, rounded to acute or short-awned, as wide as but shorter than the perigynia. **Perigynia** green, faintly 3-angled, obovate, 2–4 mm long and about 1–2 mm wide, 2-ribbed; beak tiny, bent. **Achenes** 3-angled with concave sides, 2 mm long; stigmas 3. May–July. **Habitat** Wet meadows and openings, low prairies, marshy areas.

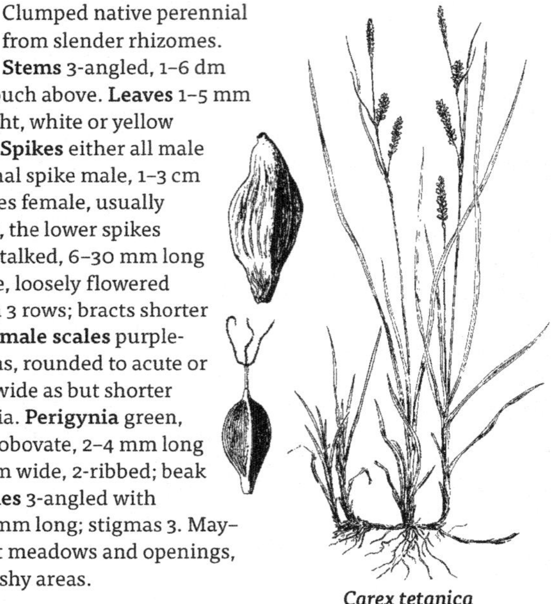

Carex tetanica
COMMON STIFF SEDGE

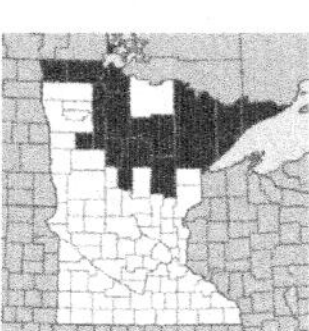

Carex vaginata Tausch | Paniceae
SHEATHED SEDGE
NC **OBL** | MW **OBL** | GP **OBL**

Native perennial, from long rhizomes. **Stems** 2–6 dm long, several together. **Leaves** 2–5 mm wide, leaves not scalelike at base of stem. Terminal spike male, 1–2 cm long; female spikes 1–3, sometimes male at tip, loosely spreading, widely separated, the lower stalks long, the upper shorter; bracts with loose sheaths and blades shorter than the spikes; **female scales** shorter and narrower than the perigynia, purple-brown, sometimes with a narrow green center, tapered to a tip. **Perigynia** usually in 2 rows, the lower separate, the upper overlapping, 3–5 mm long, narrowly obovate, with a curved beak 1 mm long. **Achenes** 3-angled, nearly filling the perigynia; stigmas 3. June–Aug. **Synonyms** *Carex saltuensis*. **Habitat** Swamps and thickets, especially where calcium-rich.

CAREX SECTION PANICULATAE

PANICULATAE

1 Perigynia with a thin, sunken strip up its middle; inner side of sheaths pale and dotted with red. . . . . *Carex diandra* | LESSER PANICLED SEDGE
1 Perigynia without a sunken strip; inner side of upper sheath strongly copper-colored . . . . *Carex prairea* | PRAIRIE SEDGE

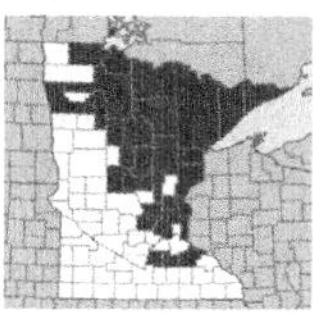

Carex diandra Schrank | Paniculatae
LESSER PANICLED SEDGE
NC **OBL** | MW **OBL** | GP **OBL**

Densely clumped native perennial. **Stems** sharply 3-angled, 3–8 dm long, usually longer than leaves. **Leaves** 1–3 mm wide; sheaths white with fine pale lines, translucent on front or slightly copper-colored at mouth. **Spikes** with both male and female flowers, male flowers borne above female, clustered in ovate heads 1–4 cm long; bracts small and inconspicuous, shorter than the spikes; **female scales** brown, tapered to tip or with a short sharp point, about equaling the perigynia. **Perigynia** brown, shiny, unequally convex on both sides, broadly ovate, 2–3 mm long and 1–2 mm wide, beak finely toothed, entire to notched, 1–2 mm long. **Achenes** lens-shaped, 1 mm long; stigmas 2. May–July. **Habitat** Wet meadows, ditches, peatlands (especially calcareous fens), floating mats.

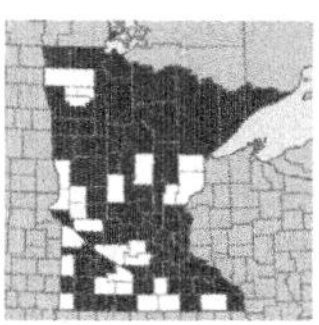

Carex prairea Dewey | Paniculatae
PRAIRIE SEDGE
NC **FACW** | MW **OBL** | GP **OBL**

Densely clumped native perennial, from short rootstocks. **Stems** sharply 3-angled, 5–10 dm long, longer than the leaves. **Leaves** 2–3 mm wide; sheaths translucent, yellow-brown or bronze-colored. **Spikes** with both male and female flowers, male flowers borne above female, ovate, 4–7 mm long, lower spikes usually separate, in linear-oblong heads 3–8 cm long; bracts small; **female scales** red-brown, tapered to tip, as long as and covering most of perigynia. **Perigynia** dull brown, flat on 1 side and convex on other, lance-shaped to ovate, 2–3 mm long and 1–2 mm wide, tapered to a finely toothed, unequally notched beak 1–2 mm long. **Achenes** lens-shaped, 1 mm long; stigmas 2. May–July. **Habitat** Wet meadows, calcareous fens, marshes, tamarack swamps and peaty lakeshores.

Carex prairea
PRAIRIE SEDGE

CAREX SECTION POLYTRICHOIDAE

One wetland species in Minnesota . . . . . . *Carex leptalea* | SLENDER SEDGE

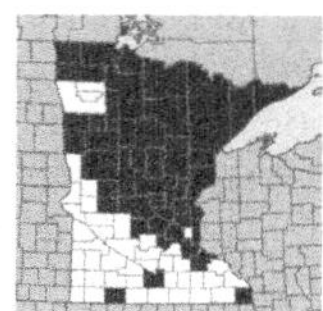

Carex leptalea Wahlenb. | Polytrichoidae
SLENDER SEDGE
NC **OBL** | MW **OBL** | GP **OBL**

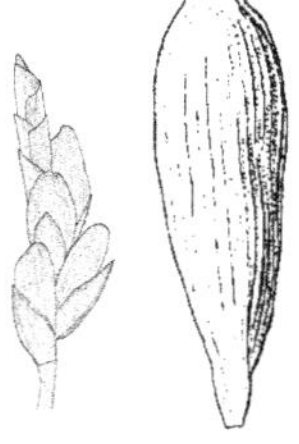

Densely clumped native perennial. **Stems** slender, rounded 3-angled, 1–7 dm long, equal or longer than leaves. **Leaves** narrow, 0.5–1.5 mm wide; sheaths tight, white, translucent on front. **Spikes** single on the stems, few-flowered, 5–15 mm long, with both male and female flowers, the male flowers borne above female; bracts absent; **female scales** rounded or with a short sharp point, shorter than the perigynia (or the tip of lowest scale sometimes longer than the perigynium). **Perigynia** yellow-green, nearly round in section to slightly flattened, oblong to oval, 3–5 mm long, finely many-nerved, beakless or with a short beak. **Achenes** 3-angled, obovate, 1–2 mm long; stigmas 3. May–July. **Habitat** Swamps, alder thickets, open bogs, calcareous fens; usually in partial shade.

CAREX SECTION PSEUDOCYPEREAE

PSEUDOCYPEREAE

1 Mature perigynia spreading or upright in spike, thin-textured, ± round in cross-section, abruptly narrowed to the beak . . . . *Carex hystericina* . . . . PORCUPINE SEDGE

1 Mature perigynia curved downward, leathery, somewhat three-angled, gradually tapered to the beak 2

2 Teeth of perigynium beak curved and spreading, 1–2 mm long . . . . *Carex comosa* | BRISTLY SEDGE

2 Teeth of perigynium beak nearly straight and parallel, 0.5–1 mm long. . . . . *Carex pseudocyperus* | FALSE BRISTLY SEDGE

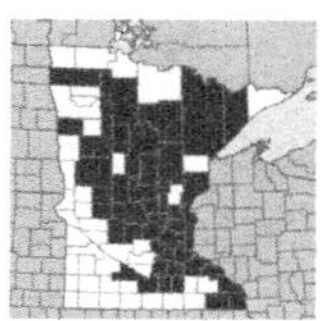

Carex comosa F. Boott | Pseudocypereae
BRISTLY SEDGE
NC **OBL** | MW **OBL** | GP **OBL**

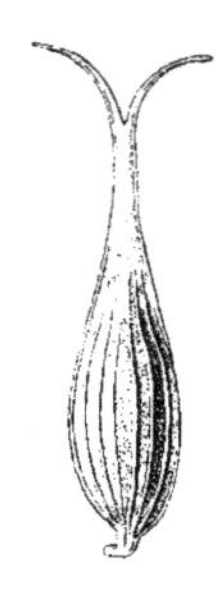

Large native perennial, often forming large clumps. **Stems** stout, sharply 3-angled, 5–15 dm long. **Leaves** 5–12 mm wide; sheaths translucent on front, with small swollen joints on back. **Spikes** either male or female; terminal spike male, 3–7 cm long; lateral spikes female, 3–5, cylindric, 3–8 cm long and 9–12 mm wide, the lower spikes longer stalked and drooping when mature; bracts leaflike, much longer than the head; **female scales** with translucent margins, tapered into a long, rough awn. **Perigynia** numerous, spreading outward when ripe, flattened 3-angled, lance-shaped, 5–8 mm long, shiny, strongly nerved, gradually tapered to the 2–3 mm long beak, the beak with curved teeth 1–2 mm long. **Achenes** 3-angled, 1.5–2 mm long; stigmas 3. June–Aug. **Habitat** Marshes, wetland margins, floating mats, ditches.

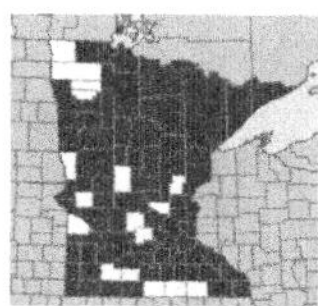

Carex hystericina Muhl. | Pseudocypereae
PORCUPINE SEDGE
NC **OBL** | MW **OBL** | GP **OBL**

Native perennial from short rhizomes, often forming large clumps. **Stems** upright or leaning, 3-angled, 2–10 dm long, usually longer than the leaves. **Leaves** yellow-green, 3–8 mm wide; sheaths white, thin and translucent on front, green to yellow or red on back, the lower sheaths breaking into threadlike fibers. **Spikes** either all male or female, the terminal spike male, 1–5 cm long, usually short-stalked and often with a bract; lateral spikes female or occasionally with male flowers above female, 1–4, short-cylindric, 1–5 cm long and 1–1.5 cm wide, separate or clustered, the lower spikes usually nodding on slender stalks, the upper spikes short-stalked and upright; **female scales** small, narrow and much shorter than the perigynia, tipped with a rough awn. **Perigynia** spreading or upright, green to straw-colored, ovate, round in section when mature, 5–8 mm long, strongly nerved, abruptly tapered to a slender, toothed beak 3–4 mm long; the beak teeth to 1 mm long. **Achenes** 3-angled with concave sides, 1.5 mm long; stigmas 3. May–July. **Habitat** Common; swamps, alder thickets, wet meadows and ditches; calcareous fens.

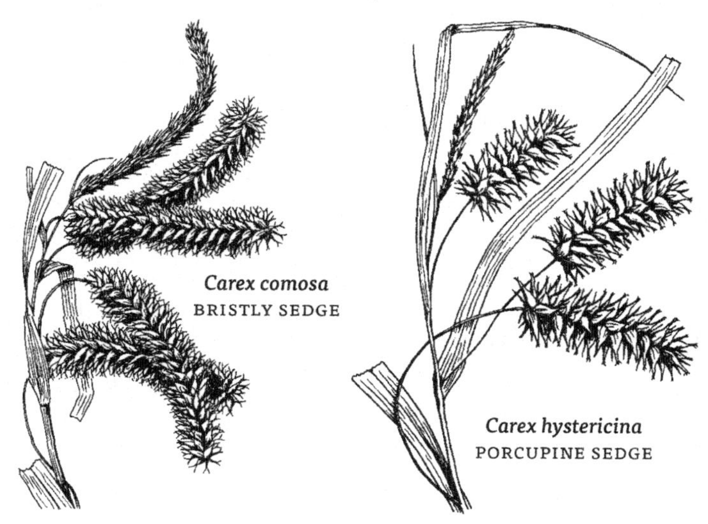

Carex comosa
BRISTLY SEDGE

Carex hystericina
PORCUPINE SEDGE

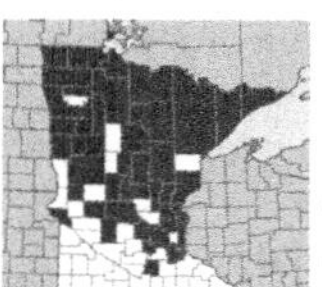

Carex pseudocyperus L. | Pseudocypereae
FALSE BRISTLY SEDGE
NC **OBL** | MW **OBL** | GP **OBL**

Large, clumped native perennial. **Stems** stout, 3–10 dm long, 3-angled, rough-to-touch. **Leaves** 5–15 mm wide; sheaths translucent, yellow-tinged on back. **Spikes** either all male or female, the terminal spike male, 1.5–7 cm long; lateral spikes female, 2–6, cylindric, 3–8 cm long and 1 cm wide, lower spikes drooping on slender stalks; bracts much longer than the head; **female scales** tipped by an awn, the awn shorter or longer than the perigynia. **Perigynia** spreading, ovate, 3-angled, 4–6 mm long, shiny, strongly nerved, tapered to a toothed beak, the beak teeth 0.5–1 mm long. **Achenes** 3-angled, 1.5 mm long; stigmas 3. June–Aug. **Habitat** Marshy lake margins, swamps, fens, wet ditches; Red Lake peatland where indicator of calcium-rich fens.

Carex pseudocyperus
FALSE BRISTLY SEDGE

CAREX SECTION SQUARROSAE

SQUARROSAE

1 Achene slightly less than half as wide as long; style curved near base . . . . *Carex squarrosa* | NARROW-LEAVED CAT-TAIL SEDGE
1 Achene half to 3/5 as wide as long; style ± straight . . . . *Carex typhina* . . . . COMMON CAT-TAIL SEDGE

Carex squarrosa L. | Squarrosae
NARROW-LEAVED CAT-TAIL SEDGE
NC **OBL** | MW **OBL**

Densely clumped native perennial. **Stems** 3–9 dm long. **Leaves** 3–6 mm wide, lower stem leaves reduced to scales. **Spikes** 1 (or sometimes 2–3), with both male and female flowers, the male below the female; female portion oval, 1–3 cm long and 1–2 cm wide; lateral spikes (if present) female, on upright stalks; bract of the terminal spike short and narrow; **female scales** tapered to a tip or short-awned, smaller than the perigynia. **Perigynia** numerous and crowded, spreading, obovate, inflated, the body 3–6 mm long, abruptly tapered to a long notched beak 2–3 mm long. **Achenes** 2–3 mm long, with a persistent, strongly bent style; stigmas 3. June–Aug. **Habitat** Swamps, floodplain forests; rare in Minnesota ; more common in se USA.

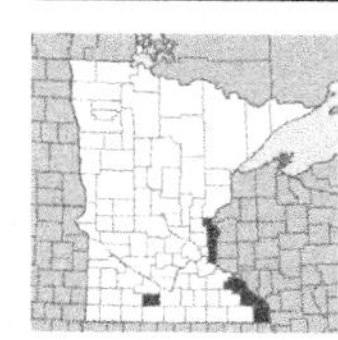

Carex typhina Michx. | Squarrosae
COMMON CAT-TAIL SEDGE
NC **OBL** | MW **OBL** | GP **OBL** | *special concern*

Clumped native perennial. **Stems** 3–8 dm long and usually shorter than upper leaves. **Leaves** 5–10 mm wide. **Spikes** 1–6, the terminal spike mostly female with a short male base, the female portion cylindric, 2–4 cm long and 1–1.5 cm wide, subtended by a short narrow bract; lateral spikes female, smaller, upright or spreading on short stalks; **female scales** hidden by the perigynia, blunt or tapered to a tip. **Perigynia** obovate, crowded, body 3–5 mm long, abruptly narrowed to a notched beak 2–3 mm long. **Achenes** 2–3 mm long; stigmas 3. June–Sept. **Habitat** Floodplain forests of large rivers (especially Mississippi and St. Croix), often occurring with swamp oval sedge (*Carex muskingumensis*) and common bur sedge (*Carex grayi*); marshy areas. **Status** Special concern.

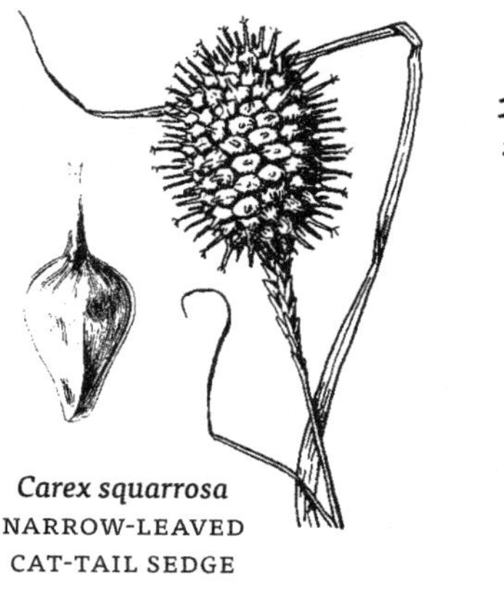

Carex squarrosa
NARROW-LEAVED
CAT-TAIL SEDGE

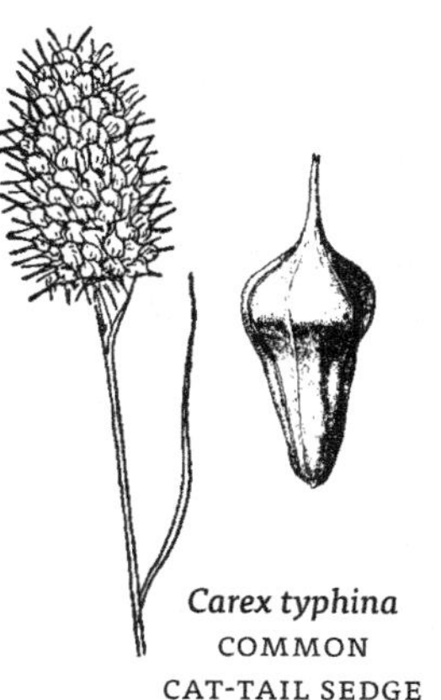

Carex typhina
COMMON
CAT-TAIL SEDGE

CAREX SECTION STELLULATAE

STELLULATAE

1 Spike usually single; leaves inrolled; anthers large, 2–4 mm long .... *Carex exilis* | COAST SEDGE
1 Spikes 2 to many; leaves flat or folded like a fan; anthers smaller, 2 mm or less long 2

2 Spikes 7–15, crowded into long heads; spikes with 20–40 perigynia .... *Carex arcta* | NORTHERN CLUSTERED SEDGE
2 Spikes 2–8, often less crowded or with fewer flowers 3

3 Plants mostly with male and female flowers on different plants .... *Carex sterilis* | FEN STAR SEDGE
3 Plants with male and female flowers on same plant, the terminal spike with male flowers below female 4

4 Perigynia long and narrow and tipped by a long beak, the body mostly 3–3.5 mm long, 2–3x as long as wide, the beak half to almost as long as body . . . . *Carex echinata* | LARGE-FRUITED STAR-SEDGE

4 Perigynia shorter or relatively wider, and with a shorter beak to half as long as body . . . . *Carex interior* | INLAND SEDGE

Carex arcta F. Boott | Stellulatae

NORTHERN CLUSTERED SEDGE

NC **OBL** | MW **OBL**

Loosely to densely clumped native perennial from very short, thick rhizomes. **Stems** 2–8 dm long, soft, sharply triangular, very rough-to-touch above, usually shorter than the leaves. **Leaves** clustered near base, light green, flat, 2–4 mm wide, long-tapered to a tip, very rough; sheaths loose, purple-dotted. **Spikes** 5–15, each with both male and female flowers, the male small and below the female; flowers crowded in oblong heads, 1.5–3 cm long, upper spikes densely packed, lower spikes slightly separate; **female scales** ovate, acute, translucent with a brown-tinged center, shorter than the perigynia. **Perigynia** flat on 1 side and convex on other, ovate, 2–3 mm long and 1–1.5 mm wide, green to straw-colored or brown when mature, covered with white dots, widest near the broad base, tapered to a sharp-toothed, notched beak 0.5–1.5 mm long. **Achenes** lens-shaped, brown, 1–2 mm long; stigmas 2. June–Aug. **Habitat** Floodplain forests, old river channels, swamps and wetland margins.

Carex arcta
NORTHERN CLUSTERED SEDGE

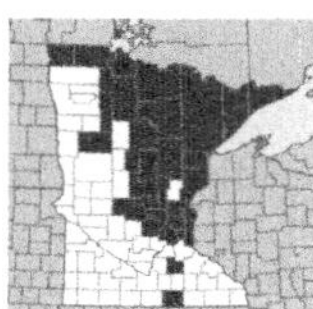

Carex echinata Murray | Stellulatae

LARGE-FRUITED STAR SEDGE

NC **OBL** | MW **OBL** | GP **OBL**

Clumped native perennial. **Stems** 1–6 dm long, rough above. **Leaves** scalelike at base of stem; leaves with blades 3–6 on lower stem, 1–3 mm wide, shorter to as long as the stems. **Spikes** 3–7, stalkless, few-flowered; terminal spike with a slender male portion near its base; lateral spikes usually all female; bract small; **female scales** ovate, shorter than perigynia, yellow-tinged with green midvein. **Perigynia** 5–15 and crowded in each spike, spreading or curved downward, green or light brown, flat on 1 side and convex on other, narrowly ovate, spongy-thickened at base, 3–4 mm long, tapered to a toothed, notched beak 1–2 mm long. **Achenes** lens-shaped; stigmas 2. July–Sept. **Synonyms** *Carex angustior, Carex cephalantha*. **Habitat** Swamp margins, wet sandy lakeshores, hummocks in peatlands (as at Red Lake peatland).

Carex exilis Dewey | Stellulatae
COAST SEDGE
NC **OBL** | GP **OBL** | *special concern*

Densely clumped native perennial. **Stems** stiff, 2–7 dm long and longer than the leaves. **Leaves** narrow and rolled inward. **Spike** usually 1, either male or female, or with both male and female flowers, the male below the female, 1–3 cm long; lateral spikes (if present) 1 or 2 and much smaller than terminal spike; lower 2 scales empty and upright; **female scales** ovate, red-brown with translucent margins, about as long as perigynia. **Perigynia** spreading or drooping, ovate, flat on 1 side and convex on other, 3–5 mm long, spongy-thickened at base, tapered to a toothed beak to 2 mm long. **Achenes** lens-shaped; stigmas 2. June–Aug. **Habitat** Sphagnum moss peatlands, interdunal wetlands near Great Lakes. **Status** Special concern. Coastal disjunct; locally common in Red Lake peatland, uncommon elsewhere in Minnesota.

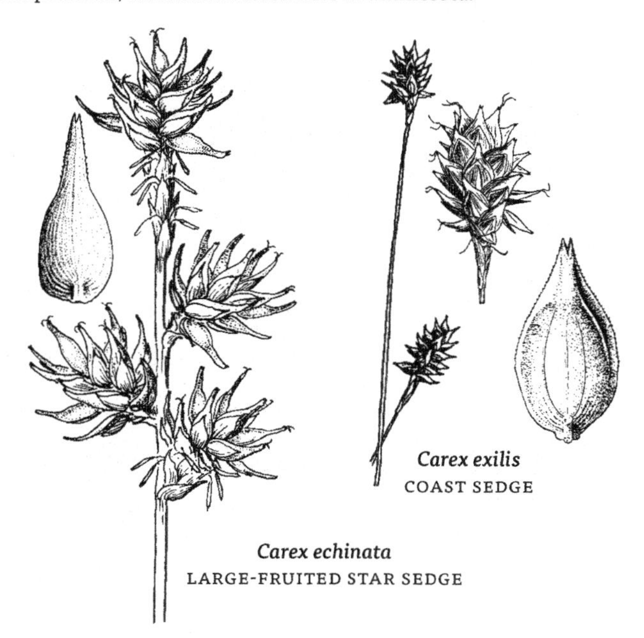

Carex exilis
COAST SEDGE

Carex echinata
LARGE-FRUITED STAR SEDGE

Carex interior L. Bailey | Stellulatae
INLAND SEDGE
NC **OBL** | MW **OBL** | GP **OBL**

Densely clumped native perennial. **Stems** slender, sharply 3-angled, 1–6 dm long, equal or longer than the leaves. **Leaves** 1–2 mm wide; sheaths tight, thin and translucent on front. **Spikes** 2–4, the terminal spike with female flowers borne above male (or rarely all male), the lateral spikes female (or rarely with female flowers borne above male), round in outline, about 5 mm wide, ± overlapping in heads 1–2.5 cm long; bracts small or absent; **female scales** blunt-tipped, much shorter than the perigynia. **Perigynia** green-brown to brown, ovate, filled to margins by the achenes, sharp-edged but not wing-margined, 2–3 mm long and 1–2 mm wide, the base spongy so that achene fills upper perigynium body, tapered to a finely toothed beak to 1 mm long; the beak teeth small, not longer than 0.3 mm. **Achenes** lens-shaped, 1–1.5 mm long; stigmas 2. May–Aug. **Habitat** Swamps, tamarack bogs, alder thickets, wet meadows and wetland margins.

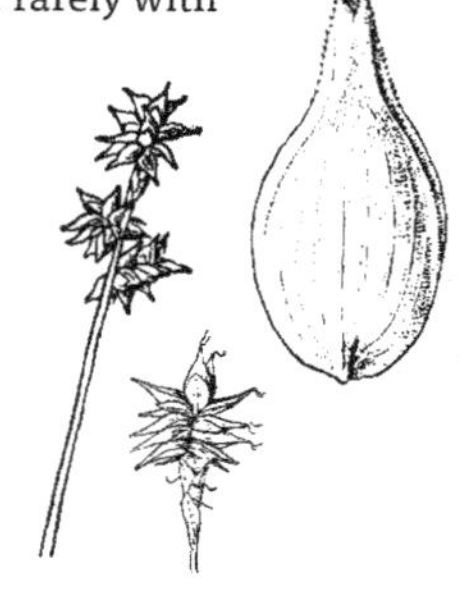

Carex interior
INLAND SEDGE

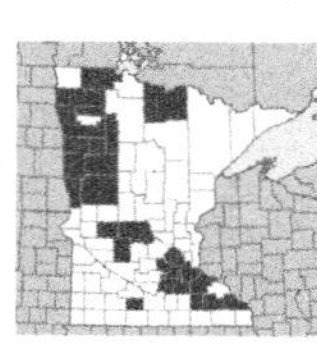

Carex sterilis Willd. | Stellulatae
FEN STAR SEDGE
NC **OBL** | MW **OBL** | GP **OBL** | *threatened*

Clumped native perennial. **Stems** stiff, 1–7 dm long, longer than the leaves, rough-to-touch on the upper stem angles. **Leaves** 3–5 from lower part of stem, 1–4 mm wide, rough, lower stem leaves reduced to scales. **Spikes** 3–8, 3–12 mm long, stalkless, clustered or the lower separate; male and female flowers mostly on separate plants; **female scales** red-brown with green midvein and translucent margins, tapered to a tip or short point, about as long as body of perigynia. **Perigynia** 5–25, the lower spreading, red-brown, flat on 1 side and convex on other, broadly ovate, spongy-thickened at base, 2–4 mm long tapered to a finely toothed, notched beak 0.5–1.5 mm long, the beak teeth sharp, to 0.5 mm long. **Achenes** lens-shaped; stigmas 2. April–June. **Synonyms** *Carex muricata* var. *sterilis*. **Habitat** Uncommon in spring-fed calcareous fens. **Status** Threatened.

Similar to the more common **inland sedge** *(Carex interior, page 443).*

Carex sterilis
FEN STAR SEDGE

CAREX SECTION SYLVATICAE

SYLVATICAE

1 Female spikes to 2.5 cm long; achene stalkless from base of perigynium . . . . *Carex castanea* | CHESTNUT-COLOR SEDGE

1 Female spikes to 3–6 cm long; achene elevated above perigynium base on a slender stalk 0.5–1.5 mm long . . . . *Carex debilis* . . . . SOUTHERN WEAK SEDGE

Carex castanea Wahl. | Sylvaticae

CHESTNUT-COLOR SEDGE

NC **FACW** | MW **FACW** | GP **FAC**

Clumped native perennial. **Stems** 3–10 dm long, purple-tinged at base, longer than leaves. **Leaves** 3–6 mm wide, softly hairy. **Spikes** either male or female; the terminal spike male, upright atop a long stalk; lateral spikes female, usually 3, on slender, drooping stalks, short cylindric, 1–2.5 cm long and 6–8 mm wide; **female scales** ovate, brown-tinged, about as long as perigynia. **Perigynia** ovate, 4–6 mm long, somewhat 3-angled, strongly 2-ribbed with several faint nerves, tapered to a notched beak. **Achenes** 3-angled; stigmas 3. June–July. **Habitat** Swamps, moist openings, wetland margins and ditches.

Carex castanea
CHESTNUT-COLOR SEDGE

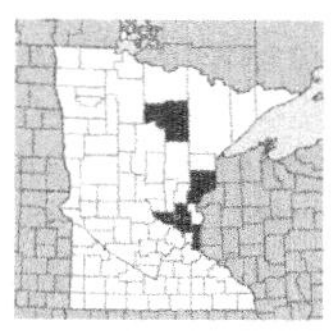

Carex debilis Michx. | Sylvaticae

SOUTHERN WEAK SEDGE

NC **FACW** | MW **FACW** | GP **OBL**

Clumped native perennial. **Stems** 6–10 dm long, purple-tinged at base. 2–4 mm wide. Male spike linear, 2–4 cm long, sometimes with a few female flowers near tip; female spikes 2–4, separate along stem, spreading or nodding, 3–6 cm long and 3–5 mm wide, flowers loose in spikes; **female scales** oblong, half the length of perigynia with translucent or brown margins and a green midrib. **Perigynia** lance-shaped, somewhat 3-angled, 2-ribbed, 5–8 mm long, narrowed to a beak. **Achenes** 3-angled; stigmas 3. May–Aug. **Habitat** Wet woods, swamp margins, wet sandy ditches.

Carex debilis
SOUTHERN WEAK SEDGE

CAREX SECTION VESICARIAE

VESICARIAE

1 Leaves narrow, 1–4 mm wide, perigynium beak with only a small notch at tip .... *Carex oligosperma* | RUNNING BOG SEDGE
1 Leaves usually wider; perigynium beak with two slender, sharp teeth . . . 2

2 Female scales with a slender awn equal or longer than scale body .... *Carex lurida* | BOTTLEBRUSH SEDGE
2 Female scales tapered to tip or with a short, sharp point 3

3 Perigynia ascending when mature, arranged into mostly 6 vertical rows 4
3 Perigynia spreading or bent downward when mature, in mostly 8–12 vertical rows.. 5

4 Perigynia to about 3 mm wide, achenes symmetrical ...... *Carex vesicaria* .... INFLATED SEDGE
4 Perigynia 4–7 mm wide, achenes deeply indented in middle of one angle . ... *Carex tuckermanii* | TUCKERMAN'S SEDGE

5 Bract of lowest female spike 2–several times as long as the head; plants clumped, without long rhizomes .... *Carex retrorsa* .... DEFLEXED BOTTLEBRUSH SEDGE
5 Bract of lowest female spike 1–2x as long as the head; plants with long rhizomes 6

6 Leaves flat, yellow-green, not waxy blue-green, the larger ones 5–12 mm wide; common .... *Carex utriculata* | BEAKED SEDGE
6 Leaves rolled inward, waxy blue-green on upper surface, 2–4 mm wide; uncommon .... *Carex rostrata* | BEAKED SEDGE

Carex lurida Wahlenb. | Vesicariae
BOTTLEBRUSH SEDGE
NC **OBL** | MW **OBL** | GP **OBL**

Clumped native perennial. **Stems** 3–10 dm long, shorter than the leaves, rounded 3-angled and ± smooth, purple-tinged at base. **Leaves** flat, 3–7 mm wide, lower stem leaves reduced to scales. **Spikes** either male or female, terminal spike male, 1–7 cm long; female spikes 1–4 (usually 2), many-flowered, grouped or the lower separate, stalkless and erect or the lower short-stalked and sometimes drooping, 2–7 cm long and 15–20 mm wide; bracts leafy, longer than the head; male scales with the midrib prolonged into an awn, **female scales** awned or sharp-pointed. **Perigynia** in many rows, broadly ovate, somewhat inflated, 6–9 mm long, pale, smooth and shining, strongly nerved, tapered to a notched beak half to as long as the body. **Achenes** 3-angled, loosely enclosed in the lower part of the perigynium, the style persistent and twisted; stigmas 3. June–Aug. **Habitat** River floodplains, swamps, fens and wet meadows; single Minnesota collection from Mille Lacs County.

Carex oligosperma Michx. | Vesicariae
RUNNING BOG SEDGE
NC **OBL** | MW **OBL** | GP **OBL**

Native perennial, forming colonies from creeping rhizomes. **Stems** slender, 4–10 dm long, purple-tinged at base. **Leaves** stiff, rolled inward, 1–3 mm wide. **Spikes** either all male or female; male spike usually single; female spikes l (or 2–3 and widely separated), stalkless or nearly so, ovate to short-cylindric, 1–2 cm long, lowest bract leaflike. **Perigynia** 3–15, ovate, somewhat inflated, compressed, 4–7 mm long, strongly several-nerved, abruptly tapered to a beak 1–2 mm long. **Achenes** 3-angled, 2–3 mm long; stigmas 3. June–Aug. **Habitat** Open bogs and swamps, floating mats, pioneer mat-former along pond margins. Common or dominant sedge of northern peatlands, less common southward.

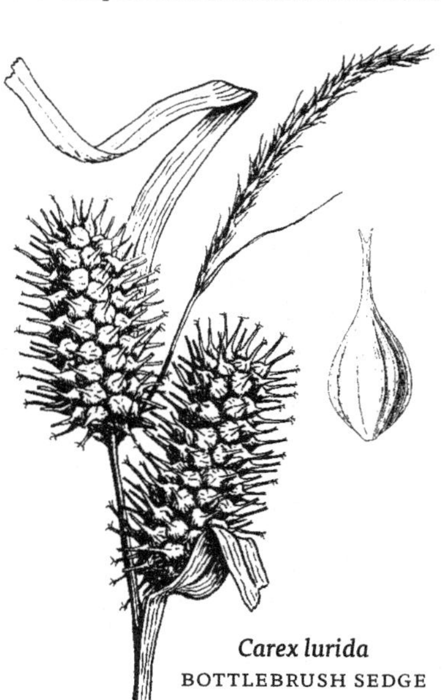

Carex lurida
BOTTLEBRUSH SEDGE

Carex oligosperma
RUNNING BOG SEDGE

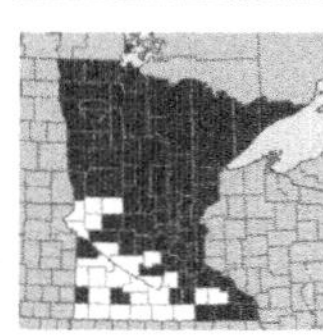

Carex retrorsa Schwein. | Vesicariae
DEFLEXED BOTTLEBRUSH SEDGE
NC **OBL** | MW **OBL** | GP **OBL**

Densely clustered native perennial. **Stems** 4–10 dm long. **Leaves** 3–4 dm long and 4–10 mm wide, flat and soft; sheaths dotted with small bumps. **Spikes** either all male or female, or the terminal 1–2 spikes with both male and female flowers, the male above the female, stalkless or lowest spike on a slender stalk; lower spikes 3–8, female, 1.5–5 cm long and 1.5–2 cm wide; **female scales** conspicuous, shorter and narrower than the perigynia. **Perigynia** crowded in rows, spreading or the lowest perigynia angled downward, smooth and shiny, 6–13-nerved, 7–10 mm long, somewhat inflated, tapered to a long, smooth beak 2–4 mm long, the beak teeth short, to 1 mm long. **Achenes** dark brown, 3-angled, 2 mm long, loose in the lower part of the perigynium; stigmas 3. June–Aug. **Habitat** Floodplain forests, swamps, thickets and marshes.

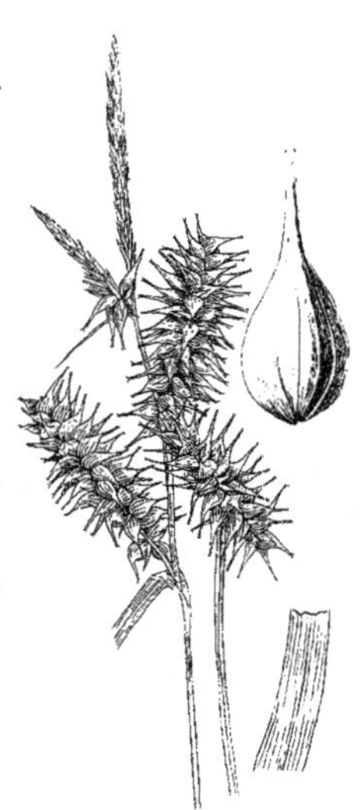

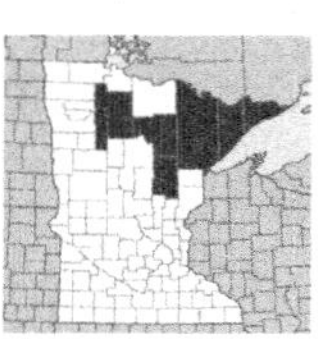

Carex rostrata J. Stokes | Vesicariae
BEAKED SEDGE
NC **OBL** | MW **OBL** | GP **OBL**

Native perennial with short to long-creeping rhizomes. **Stems** round or bluntly 3-angled, 3–10 dm long, smooth below inflorescence. **Leaves** waxy blue, with many fine bumps on upper surface, to 4 mm wide, inrolled or channeled in section. **Spikes** either male or female, the upper 2–5 male; lower 2–5 spikes female or sometimes 1 or 2 with male flowers above the female, cylindric, 2–10 cm long and 8–12 mm wide. **Perigynia** upright when young, becoming widely spreading when mature, yellow-green to brown, shiny, ovate, nearly round in section, inflated, 2–6 mm long, narrowed to a beak about 1 mm long. **Achenes** 3-angled, 1–2 mm long; stigmas 3. July–Sept. **Habitat** Peat mats or shallow water.

Similar to ***Carex utriculata*** *but much less common in Minnesota, and with the leaves waxy blue and dotted with fine bumps on upper surface, v-shaped in section or inrolled, and only 2–4 mm wide.*

Carex rostrata
BEAKED SEDGE

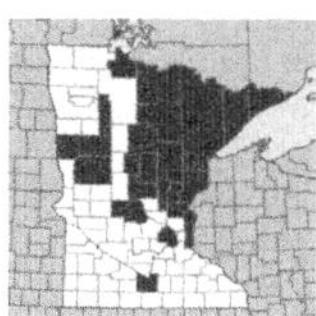

Carex tuckermanii F. Boott | Vesicariae
TUCKERMAN'S SEDGE
NC **OBL** | MW **OBL** | GP **OBL**

Clumped native perennial, from short rhizomes. **Stems** 4–8 dm long. **Leaves** 3–6 mm wide and 2–4 dm long, soft and flat. **Spikes** either male or female; male spikes usually 2, separated, raised above female spikes; female spikes 2–4, separated, cylindric, 2–5 cm long. **Perigynia** overlapping and ascending in 6 rows, broadly ovate, 7–10 mm long and 4–7 mm wide, inflated, tapered to a notched beak 2 mm long. **Achenes** 3-angled, obovate, 3–4 mm long, with a deep indentation near the middle of 1 angle; stigmas 3. June–Aug. **Habitat** Swamps, alder thickets, low areas in forests, pond margins.

Carex tuckermanii
TUCKERMAN'S SEDGE

Carex utriculata F. Boott | Vesicariae
BEAKED SEDGE
NC **OBL** | MW **OBL** | GP **OBL**

Large, densely clumped native perennial from short rootstocks, also forming turfs from long rhizomes. **Stems** bluntly 3-angled, 3–12 dm long, spongy at base. **Leaves** strongly divided with swollen joints 4–12 mm wide; sheaths white-translucent on front, divided with swollen joints on back. **Spikes** either male or female, the upper 2–5 male, held well above the female spikes, the terminal spike 3–6 cm long; lower 2–5 spikes female or sometimes 1 or 2 with male flowers above the female, usually separate, cylindric, 2–10 cm long and 8–12 mm wide, the upper spikes stalkless or short-stalked, lower spikes stalked, upright; bracts shorter to slightly longer than the head; **female scales** acute to awn-tipped, body of scale shorter than perigynia. **Perigynia** upright at first to widely spreading when mature, in many rows, yellow-green to brown, shiny, ovate, nearly round in section, inflated, 3–8 mm long and 2–4 mm wide, strongly 7–9-nerved, contracted to a toothed beak 1–2 mm long, the teeth mostly straight, 0.5 mm long. **Achenes** 3-angled, 2 mm long; stigmas 3. June–Aug. **Habitat** Wet meadows, marshes, fens, swamps and lakeshores.

*Long confused with **Carex rostrata** (a boreal species apparently uncommon Minnesota) which has waxy blue leaves to only 4 mm wide and with numerous small bumps on upper leaf surface.*

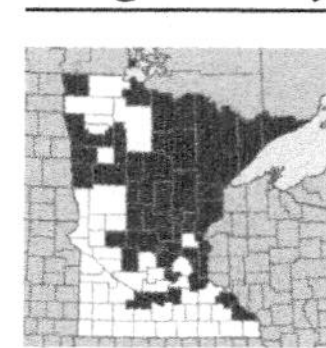

Carex vesicaria L. | Vesicariae
INFLATED SEDGE
NC **OBL** | MW **OBL** | GP **OBL**

Clumped native perennial, from stout, short rhizomes. **Stems** sharply 3-angled and rough-to-touch below the head, 3–10 dm long, not spongy at base (as in *Carex utriculata*). **Leaves** 2–7 mm wide; sheaths white-translucent on front, not conspicuously divided-with small swollen joints on back, the lowest sheaths often shredding into ladderlike fibers. **Spikes** either all male or female, the upper 2–4 male, held well above the female, 2–4 cm long; lower 1–3 spikes female, separate, cylindric, 2–8 cm long and 4–15 mm wide, stalkless or short-stalked, erect; lowest bract usually longer than the head; **female scales** acute to awn-tipped, shorter to as long as perigynia. **Perigynia** upright and overlapping in rows, dull yellow-green to brown, ovate to round, inflated, 3–8 mm long and 3 mm wide, strongly nerved, abruptly tapered to a toothed beak 1–2 mm long, the teeth 0.5–1 mm long. **Achenes** 3-angled, 2–3 mm long; stigmas 3. June–Aug. **Habitat** Wet meadows, marshes, forest depressions and shores.

Carex utriculata
BEAKED SEDGE

Carex vesicaria
INFLATED SEDGE

CAREX SECTION VULPINAE

VULPINAE

1 Perigynia ovate, rounded at base, the beak shorter than the body . . . . *Carex conjuncta* | GREEN-HEADED FOX SEDGE

1 Perigynia lance-shaped to narrowly ovate, broadest near base, the beak as long or longer than the body 2

2 Perigynia spreading at base into a disk-shaped spongy structure wider than the short, somewhat cone-shaped body . . . . *Carex crus-corvi* . . . . CROWFOOT FOX SEDGE

2 Perigynia with a spongy base continuous with body 3

3 Sheaths prolonged beyond base of blade, cross-corrugated (wrinkled) on inner side . . . . *Carex stipata* | COMMON FOX SEDGE

3 Sheaths with thick-margined mouth, not prolonged beyond base of blade, not cross-corrugated . . . . *Carex laevivaginata* . . . . SMOOTH-SHEATHED FOX SEDGE

Carex conjuncta Boott | Vulpinae

GREEN-HEADED FOX SEDGE

NC **FACW** | MW **FACW** | GP **FAC**

Clumped native perennial; plants light green. **Stems** slender, 3-angled, 4–9 dm long, as long or longer than leaves, somewhat roughened above. **Leaves** soft, 5–10 mm wide; margins rough; sheaths somewhat cross-wrinkled on inner side. **Spikes** 6–10, either male or female, the male above the female, in a narrow head 2–5 cm long; bract small and bristlelike or absent; **female scales** ovate, about as long as perigynia, tapered to a sharp tip or short awn. **Perigynia** ovate, green or yellow-green, 3–4 mm long and 1–2 mm wide, slightly spongy at base, tapered to a rough, 2-toothed beak. **Achenes** flattened; stigmas 2. June–July. **Habitat** Low prairie, streambanks, wet meadows and thickets; local in southeastern Minnesota.

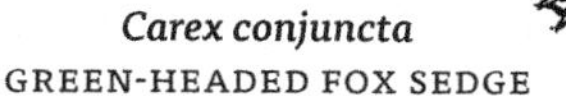

Carex conjuncta
GREEN-HEADED FOX SEDGE

Carex crus-corvi Shuttlew. | Vulpinae
CROWFOOT FOX SEDGE
NC **OBL** | MW **OBL** | GP **OBL** | *special concern*

Clumped native perennial. **Stems** stout, 4–8 dm long, sharply 3-angled, shorter than leaves. **Leaves** 5–10 mm wide; sheaths not cross-wrinkled. **Spikes** in large head, 8–20 cm long, upper spikes grouped, lower spikes separate. **Spikes** with both male and female flowers, male above female; **female scales** triangular-ovate, shorter or equal to perigynia. **Perigynia** ovate, flat on 1 side and convex on other, 5–9 mm long, with a broad, spongy base to 2.5 mm wide, tapered to a notched beak much longer than body. **Achenes** lens-shaped; stigmas 2. June–July. **Habitat** Floodplains, marshes, edges of seasonally wet woodland depressions. **Status** Special concern; historical records from several se Minn locations.

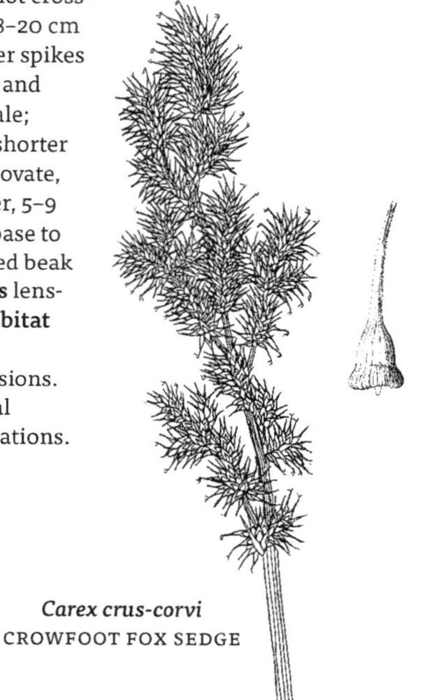

Carex crus-corvi
CROWFOOT FOX SEDGE

Carex laevivaginata (Kuk.) Mack.| Vulpinae
SMOOTH-SHEATHED FOX SEDGE
NC **OBL** | MW **OBL**

Densely clumped native perennial. **Stems** stout, 3-angled, 3–10 dm long. **Leaves** 3–10 mm wide; sheaths not cross-corrugated (as in *Carex stipata*). **Spikes** with both male and female flowers, the male above the female, numerous, grouped into a dense head 2–5 cm long and 1–1.5 cm wide, green or straw-colored when mature; bracts short or reduced to bristles sometimes longer than the spikes; **female scales** shorter than the perigynia, tapered to a tip or short awn. **Perigynia** green or straw-colored, spreading, broadly lance-shaped, 5–6 mm long, long-tapered to the tip, flat on 1 side and convex on other, spongy-thickened at base. **Achenes** lens-shaped; stigmas 2. May–July. **Habitat** Swamps, marshy areas and streambanks.

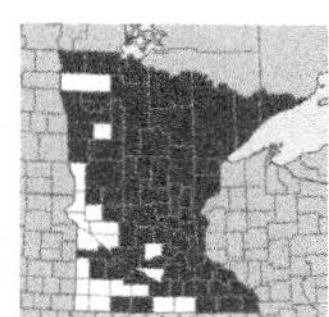

Carex stipata Muhl. | Vulpinae
COMMON FOX SEDGE
NC **OBL** | MW **OBL** | GP **OBL**

Densely clumped native perennial. **Stems** 3-angled and slightly winged, 2–12 dm long. **Leaves** 4–8 mm wide; sheaths cross-wrinkled on front, divided with small swollen joints on back. **Spikes** with both male and female flowers, male flowers borne above female, clustered or the lowest spikes often separate, in oblong heads 3–10 cm long; bracts small and sometimes bristlelike, longer than the spike; **female scales** tapered to a tip or with a short, sharp point, half to 3|4 as long as the perigynia. **Perigynia** yellow-green to dull brown, flat on 1 side and convex on other, narrowly ovate, 3–5 mm long and 1–2 mm wide, strongly several-nerved on both sides, tapered to a finely toothed, notched beak 1–3 mm long. **Achenes** lens-shaped, 1.5–2 mm long; stigmas 2. May–July. **Habitat** Common; floodplain forests and swamps, thickets, wet meadows, wetland margins and ditches; usually not in sphagnum bogs.

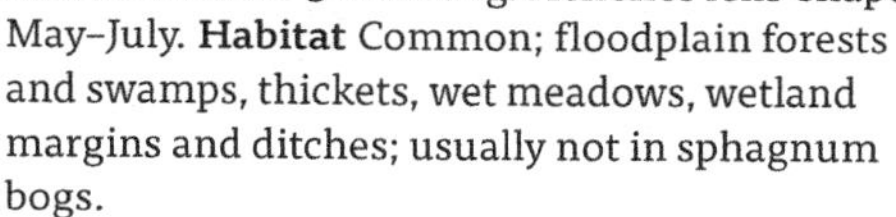

Carex laevivaginata
SMOOTH-SHEATHED FOX SEDGE

Carex stipata
COMMON FOX SEDGE

CLADIUM | *Twig-rush*

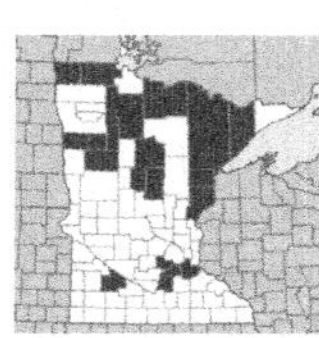

Cladium mariscoides (Muhl.) Torr.
TWIG-RUSH
NC **OBL** | MW **OBL** | GP **OBL** | *special concern*

Grasslike native perennial, spreading by rhizomes and forming colonies. **Stems** single or in small groups, stiff, slender, smooth, 0.3–1 m tall. **Leaves** 1–3 mm wide, upper portion round in section, middle portion flattened. **Flowers** in lance-shaped spikelets, 3–5 mm long, in branched clusters (umbels) at end of stem and also with 1–2 clusters on slender stalks from leaf axils; uppermost flower perfect, the style 3-parted; middle flowers male; lowest scale of each spikelet empty; scales overlapping, ovate, brown; bristles absent. **Achenes** dull brown, 2–3 mm long, pointed at tip; tubercle absent. June–Aug. **Synonyms** *Mariscus mariscoides*. **Habitat** Shallow water, sandy or mucky shores, floating bog mats, calcium-rich wet meadows, seeps, fens and low prairie. **Status** Special concern.

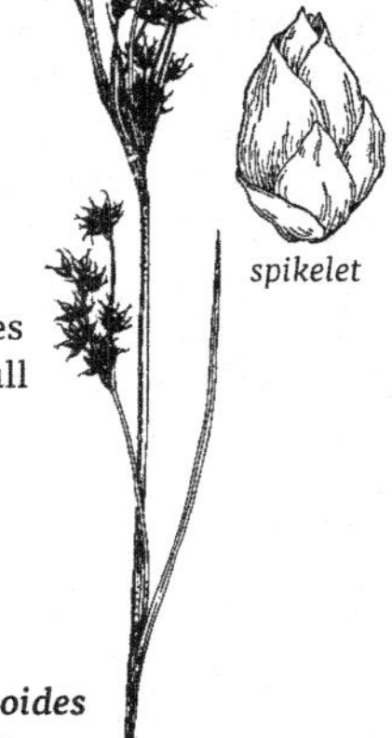

Cladium mariscoides
TWIG-RUSH

CYPERUS | *Flatsedge*

Small to medium, annual or perennial, grasslike plants. **Stems** often clumped, unbranched, sharply 3-angled. **Leaves** mostly from base of plants, with 1 or more leaflike bracts near top of stems, the blades flat or folded along midvein. **Flower heads** in umbels at ends of stems; the spikelets many, grouped in 1 to several rounded or cylindric spikes. **Flowers** perfect; bristlelike sepals and petals absent; stamens 1–3; styles 2–3-parted. **Achenes** lens-shaped or 3-angled, beakless.

CYPERUS | FLATSEDGE

1 Achenes lens-shaped; stigmas 2 2
1 Achenes 3-angled; stigmas 3 3

2 Style cleft to slightly below middle . . . . *Cyperus bipartitus* . . . . SHINING FLATSEDGE
2 Style cleft almost to base . . . . . . *Cyperus diandrus* | UMBRELLA FLATSEDGE

3 Plants perennial 4
3 Plants annual 5

4 Scales 2–3 mm long, only slightly keeled . . . . *Cyperus esculentus* . . . . CHUFA; YELLOW NUTSEDGE
4 Scales 3–5 mm long, keeled . . . . *Cyperus strigosus* . . . . FALSE NUTSEDGE; STRAW-COLORED CYPERUS

5 Scales curved outward at tip; stamens 1 . . . . **6**
5 Scales not curved outward at tip; stamens 3. . . . . **7**

6 Scales 3-nerved, not awn-tipped . . . . *Cyperus acuminatus* . . . . SHORT-POINT FLATSEDGE
6 Scales 7–9-nerved, tipped with a short awn. . . . . *Cyperus squarrosus* . . . . AWNED FLATSEDGE

7 Scales to 2 mm long, with 3–5 veins near center of scale . . . . *Cyperus erythrorhizos* | RED-ROOT FLATSEDGE
7 Scales 2–5 mm long, with 7 or more well-spaced veins. . . . . *Cyperus odoratus* | RUSTY FLATSEDGE

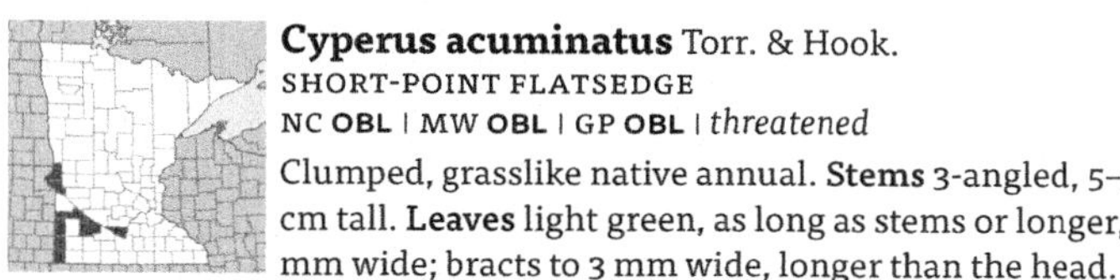

Cyperus acuminatus Torr. & Hook.
SHORT-POINT FLATSEDGE
NC **OBL** | MW **OBL** | GP **OBL** | *threatened*

Clumped, grasslike native annual. **Stems** 3-angled, 5–30 cm tall. **Leaves** light green, as long as stems or longer, to 2 mm wide; bracts to 3 mm wide, longer than the head. **Spikelets** flat, 3–7 mm long, crowded in 1–5 round clusters (spikes), 1 spike stalkless, the other spikes on stalks 1–3 cm long; scales 1–2 mm long, ovate, pale green, becoming tan when mature, strongly 3-nerved; stamens 1; style 3-parted. **Achenes** tan to pale brown, 3-angled, 0.5–1 mm long. Aug–Sept. **Habitat** Muddy or sandy shores, streambanks and flats. **Status** Threatened.

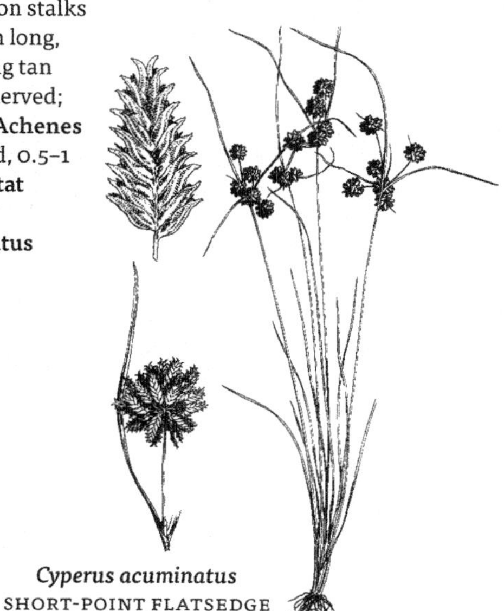

Cyperus acuminatus
SHORT-POINT FLATSEDGE

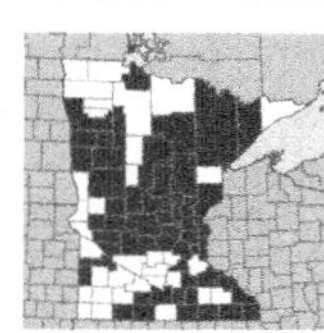

Cyperus bipartitus Torr.
SHINING FLATSEDGE
NC **FACW** | MW **OBL** | GP **FACW**

Clumped, grasslike native annual. **Stems** 3-angled, 1–3 dm tall. usually shorter than stems; leaves and bracts 0.5–2 mm wide, the bracts usually 3, longer than the spikes. **Spikelets** linear, 10–15 mm long and 2–3 mm wide, in clusters (spikes) of 3–10, the spikes stalkless or on stalks to 10 cm long; scales overlapping, ovate, shiny, purple-brown on margins; stamens 2 or 3; style 2-parted, lower third not divided. **Achenes** lens-shaped, 1–2 mm long, hidden by the scales. July–Sept. **Synonyms** *Cyperus rivularis*. **Habitat** Wet, sandy, gravelly or muddy shores, streambanks, wet meadows, ditches.

Cyperus bipartitus
SHINING FLATSEDGE

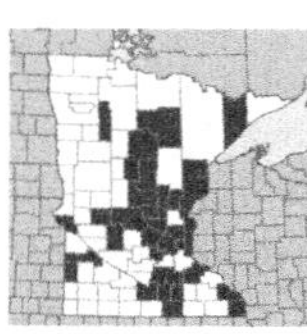

Cyperus diandrus Torr.
UMBRELLA FLATSEDGE
NC **OBL** | MW **FACW** | GP **FACW**

Clumped, grasslike native annual. **Stems** 3-angled, 5–30 cm tall. **Leaves** about as long as stems, 1–3 mm wide; bracts usually 3, longer than the spikes. **Spikelets** 5–10, linear, 5–20 mm long and 2–3 mm wide; in 1–3 loose, rounded spikes, the spikes on stalks to 6 cm long; scales loosely overlapping, ovate, 2–3 mm long, not shiny, purple-brown on margins; stamens 2; style 2-parted, divided nearly to the base, persistent. **Achenes** lens-shaped, pale brown, 1 mm long, visible between the scales. July–Sept. **Habitat** Sandy or muddy shores, streambanks, wet meadows.

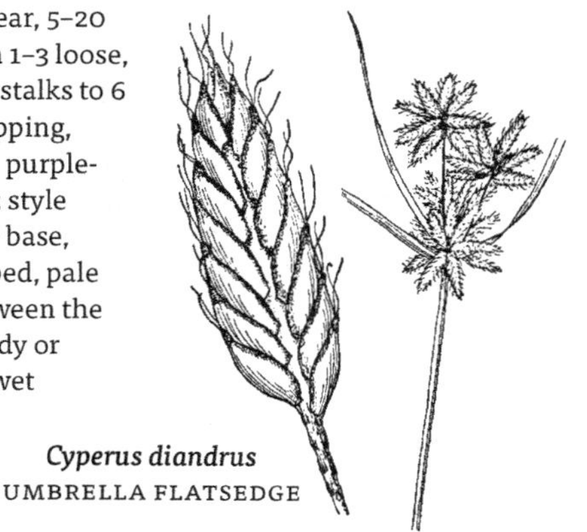

Cyperus diandrus
UMBRELLA FLATSEDGE

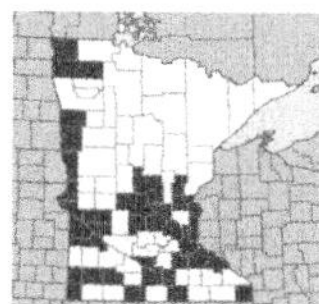

Cyperus erythrorhizos Muhl.

RED-ROOT FLATSEDGE

NC **OBL** | MW **OBL** | GP **OBL**

Clumped, stout or slender native annual, roots red. **Stems** 3-angled, 1–7 dm long. mostly near base of plant, shorter to longer than stems, 2–8 mm wide; bracts 3–7, to 9 mm wide, usually much longer than the spikes. **Spikelets** linear, 2–10 mm long and 1–2 mm wide; grouped in a pinnate manner along a stalk (rachilla), in cylindric clusters, the terminal cluster stalkless, the others on stalks to 8 cm long; scales ovate, satiny brown, 1–2 mm long, overlapping; stamens 3; style 3-parted. **Achenes** ivory white, sharply 3-angled, ovate, 0.5–1 mm long. July–Sept. **Habitat** Sandy or muddy shores, streambanks, exposed mud flats, ditches; often with rusty flatsedge (*Cyperus odoratus*). A widespread species across most of temperate North America.

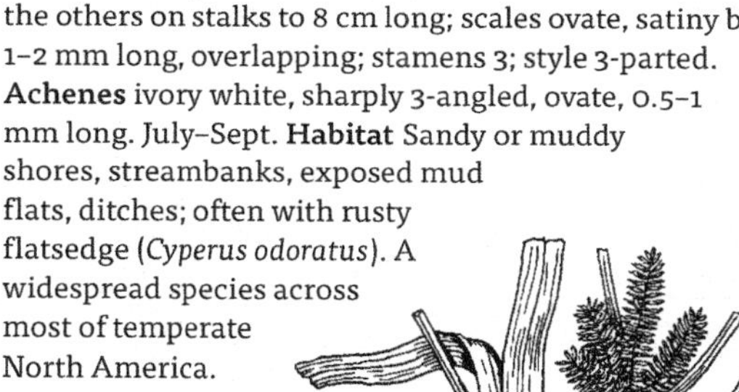

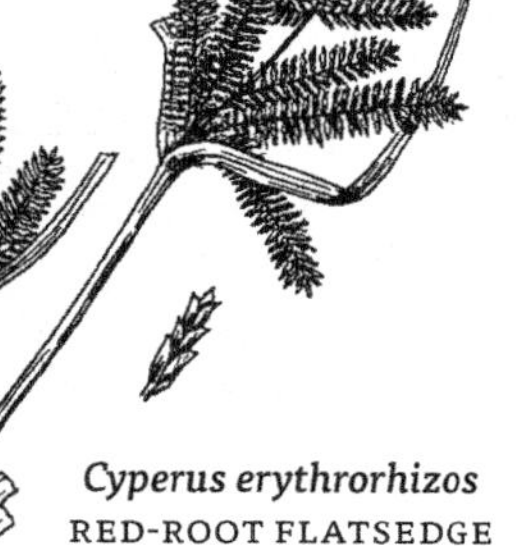

Cyperus erythrorhizos
RED-ROOT FLATSEDGE

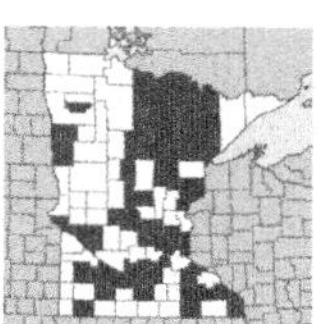

Cyperus esculentus L.

CHUFA; YELLOW NUTSEDGE

NC **FACW** | MW **FACW** | GP **FACW**

Grasslike introduced perennial, with rhizomes ending in small tubers. **Stems** single, 3-angled, erect, 2–7 dm long. **Leaves** light green, mostly from base of plant, about as long as stems, 3–10 mm wide, with a prominent midvein; the bracts 3–6, usually much longer than the spikes. **Spikelets** linear, 3–12 cm long and 1–2 mm wide; pinnately arranged on a stalk, forming loose cylindrical spikes, the spikes to 5 cm long and 1–2 mm wide; scales straw-colored, 2–3 mm long, overlapping; stamens 3; style 3-parted. **Achenes** pale brown, 3-angled, 1–2 mm long. July–Sept. **Habitat** Sandy or muddy shores, streambanks, marshes, ditches and other wet places; weedy in wet or moist cultivated fields; common as a lawn weed in se USA.

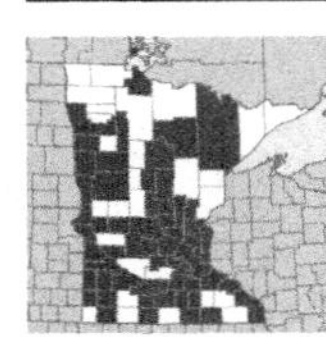

Cyperus odoratus L.
RUSTY FLATSEDGE
NC **OBL** | MW **FACW** | GP **FACW**

Stout, grasslike, fibrous-rooted native annual. **Stems** clumped or single, 3-angled, 2–7 dm long. **Leaves** mostly from base of plant, shorter to longer than flowering stems, the blades 2–8 mm wide; the involucral bracts much longer than the spikes. **Spikelets** linear, 1–2 cm long, pinnately arranged along a stalk, forming several to many cylindrical spikes, the spikes stalkless or stalked; scales red-brown, 2–3 mm long, overlapping; stamens 3; style 3-parted. **Achenes** brown, 3-angled, 1–2 mm long. July–Sept. 2. **Habitat** Sandy or muddy shores, floating mats, ditches, wet cultivated fields.

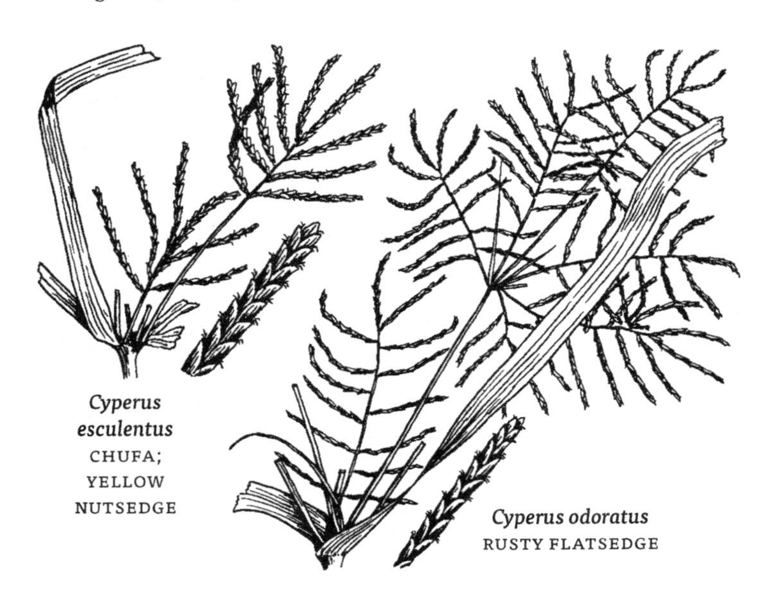

Cyperus esculentus CHUFA; YELLOW NUTSEDGE

Cyperus odoratus RUSTY FLATSEDGE

Cyperus squarrosus L.
AWNED FLATSEDGE
NC **OBL** | MW **OBL** | GP **OBL**

Small, clumped, sweet-scented, grasslike annual; native. **Stems** very slender, 3-angled, 3–15 cm long. **Leaves** few, all at base of plant, 1–2 mm wide; bracts 2–3, longer than the spikes. **Spikelets** linear, flattened, 3–10 mm long, in 1–4 dense, rounded spikes, 1 spike stalkless, the other spikes on stalks to 3 cm long; scales 1–2 mm long, tipped by an awn to 1 mm long, pale brown; stamens 1; style 3-parted. **Achenes** brown, 3-angled, 0.5–1 mm long. July–Sept. **Synonyms** *Cyperus aristatus, Cyperus inflexus*.
Habitat Wet, sandy or muddy lake-shores, streambanks, mud and

gravel bars, wet meadows. Widespread, throughout most of USA and much of world.

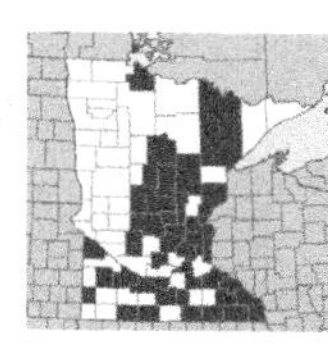

Cyperus strigosus L.
FALSE NUTSEDGE;
NC **FACW** | MW **FACW** | GP **FACW**
STRAW-COLORED CYPERUS

Grasslike native perennial, from tuberlike corms. **Stems** single or few, slender, sharply 3-angled, 1–8 dm long. **Leaves** mostly at base of plants, the blades 2–12 mm wide, margins rough-to-touch; the bracts mostly longer than the spikes. **Spikelets** flat, linear, 6–20 mm long and 1–2 mm wide, golden-brown, pinnately arranged and spreading, in several to many cylindric spikes, the spikes often bent downward, on stalks 1–12 cm long, the stalks sometimes branched; scales straw-colored, 3–5 mm long; stamens 3; style 3-parted. **Achenes** brown, 3-angled, 1–2 mm long. July–Sept. **Habitat** Wet, sandy or muddy shores, streambanks, marshes, wet meadows, ditches, cultivated fields.

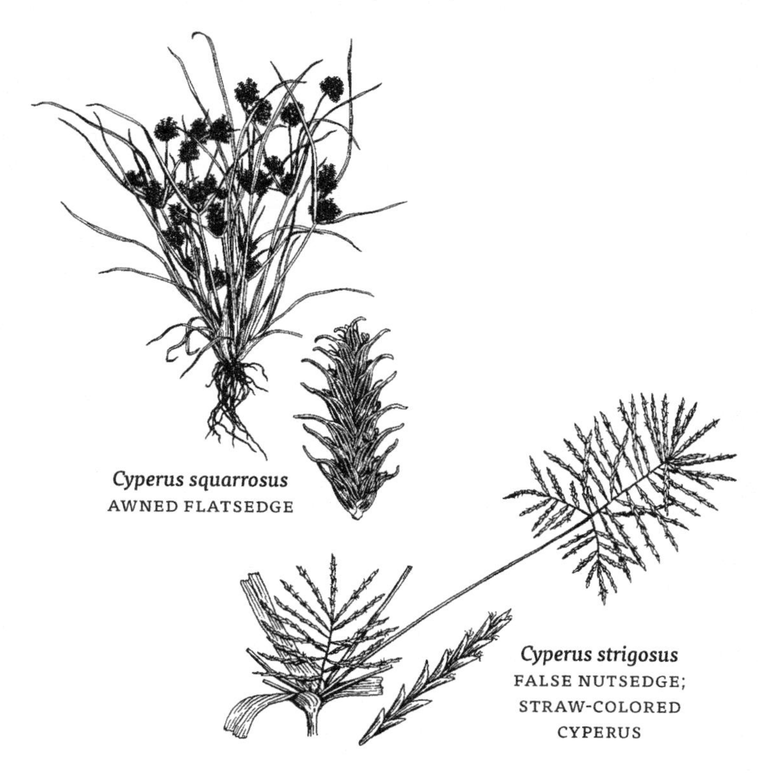

Cyperus squarrosus
AWNED FLATSEDGE

Cyperus strigosus
FALSE NUTSEDGE;
STRAW-COLORED CYPERUS

DULICHIUM | *Three-way sedge*

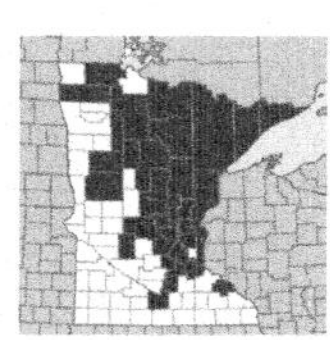

Dulichium arundinaceum (L.) Britton
THREE-WAY SEDGE; DULICHIUM
NC **OBL** | MW **OBL** | GP **OBL**

Grasslike native perennial, spreading by rhizomes and often forming large colonies. **Stems** stout, erect, 3–10 dm long, jointed, hollow, rounded in section. **Leaves** 3-ranked, flat, short, 4–15 cm long and 3–8 mm wide; lower leaves reduced to sheaths. Flower heads from leaf axils, in linear clusters of 5–10 spikelets, the clusters 1–2.5 cm long; scales lance-shaped, green to brown, 5–8 mm long. **Flowers** perfect; sepals and petals reduced to 6–9 downwardly barbed bristles; stamens 3; style 2-parted. **Achenes** light brown, oblong, 2–4 mm long, beaked by the persistent, slender style. July–Sept. **Habitat** Shallow marshes, wet meadows, shores, bog margins.

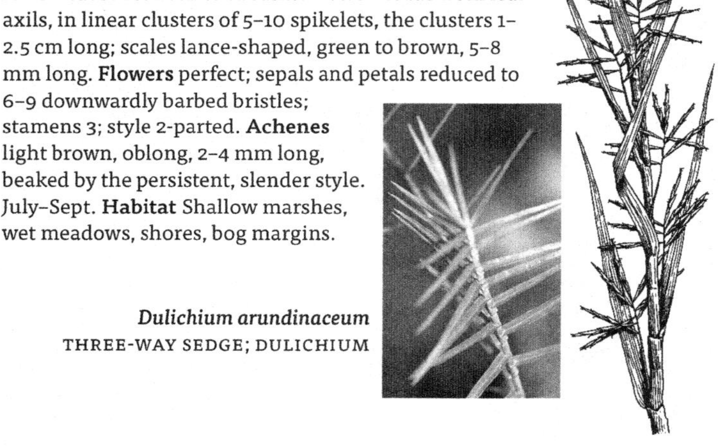

Dulichium arundinaceum
THREE-WAY SEDGE; DULICHIUM

ELEOCHARIS | *Spike-rush*

Small to medium rushlike plants, mostly perennial from rhizomes (annual in *E. ovata*), often forming large, matlike colonies. **Stems** round, flattened, or angled in section. **Leaves** reduced to sheaths at base of stems. Flower head a single spikelet at tip of stem; scales of the spikelets spirally arranged and overlapping. **Flowers** perfect; sepals and petals bristlelike or absent, the bristles usually 6 if present; stamens 3; styles 2–3-parted, the base of style swollen and persistent as a projection (tubercle) atop the achene, or sometimes joined with the achene body. **Achenes** rounded on both sides or 3-angled.

To aid identification, **achenes** *as well as the plant's habit are shown for each species of Eleocharis.*

Dwarf spike-rush *(Eleocharis parvula) has been collected from sw Minnesota; however the last collection is dated 1945 and repeated survey efforts have failed to relocate this species in the state.*

ELEOCHARIS | SPIKE-RUSH

1 Tubercle joined with achene and not forming a distinct cap 2
1 Tubercle forming a distinct cap atop the achene 3

2 Stems 1–4 dm long, not rooting at tips; spikelets with 3–9 flowers . . . . *Eleocharis quinqueflora* | FEW-FLOWER SPIKE-RUSH
2 Stems 4 dm or more long, often rooting at tips; spikelets with 10 or more flowers . . . . *Eleocharis rostellata* | BEAKED SPIKE-RUSH

3 Stems 3-angled in section; spikelets slender, about as wide as stems; the scales persistent . . . . *Eleocharis robbinsii* | ROBBINS' SPIKE-RUSH
3 Stems round or flat i section; spikelets wider than stems; scales deciduous 4

4 Styles 3-parted; achenes 3-angled to ± round 5
4 Styles 2-parted; achenes lens-shaped 9

5 Achene round in section, 2x longer than wide 6
5 Achene 3-angled, nearly as wide as long 8

6 Stems flattened, 1 mm or more wide . . . . *Eleocharis wolfii* . . . . WOLF'S SPIKE-RUSH
6 Stems ± round in section, less than 0.5 mm wide 7

7 Scales 1.5–3 mm long; achene grayish or yellow-white; common . . . . *Eleocharis acicularis* | LEAST SPIKE-RUSH
7 Scales to 1.3 mm long; achene dark-yellow to orange; rare in ne Minn . . . . *Eleocharis nitida* | NEAT SPIKE-RUSH

8 Achenes pitted or rough (use hand lens) . . . . *Eleocharis compressa* . . . . FLAT-STEM SPIKE-RUSH
8 Achenes smooth . . . . *Eleocharis intermedia* | MATTED SPIKE-RUSH

9 Plants annual, clumped . . . . *Eleocharis ovata* | BLUNT SPIKE-RUSH
9 Plants perennial, spreading by short to long rhizomes 10

10 Plants small, 3–15 cm tall; stems clumped . . . . *Eleocharis flavescens* . . . . BRIGHT GREEN SPIKE-RUSH
10 Plants larger, 2–10 dm tall; stems scattered and single, or in small clusters . . . . *Eleocharis palustris* | CREEPING SPIKE-RUSH

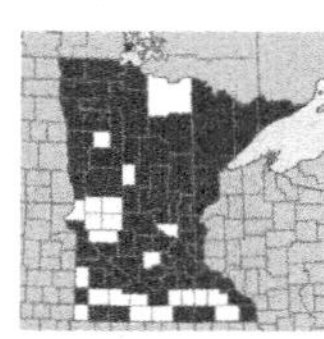

Eleocharis acicularis (L.) Roemer & Schultes
LEAST SPIKE-RUSH
NC **OBL** | MW **OBL** | GP **OBL**

achene

Small, clumped, mat-forming native perennial, from slender rhizomes. **Stems** threadlike, 3–15 cm long and to 0.5 mm wide, somewhat 4-angled and grooved; sheaths membranous, usually red at base. **Spikelets** narrowly ovate, 3–6 mm long and 1–1.5 mm wide; scales with a green midvein and chaffy margins; sepals and petals reduced to 3–4 bristles or absent; style 3-parted. **Achenes** gray, rounded 3-angled, ridged, to 1 mm long; tubercle cone-shaped, constricted at base. May–Sept. **Habitat** Common species of shallow water, exposed muddy or sandy shores, marshes and streambanks; sometimes used as an aquarium plant.

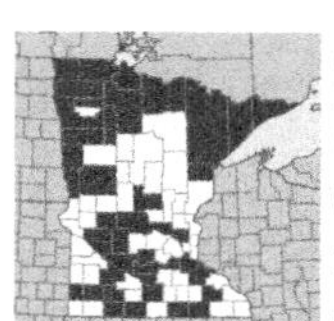

Eleocharis compressa Sullivant
FLAT-STEM SPIKE-RUSH
NC **FACW** | MW **FACW** | GP **FACW**

Clumped native perennial, from stout black rhizomes. **Stems** flattened and often twisted, 1.5–4 dm long and 0.5–1 mm wide, shallowly grooved; sheaths red or purple at base. **Spikelets** ovate, 4–10 mm long and 3–4 mm wide; lowest scale sterile and encircling the stem; fertile scales with a green midvein, purple-brown on sides, and white translucent margins; sepals and petals absent or reduced to 1–5 bristles; style 3-parted. **Achenes** yellow-brown, covered with small bumps, somewhat 3-angled, 1–1.5 mm long; tubercle small, constricted at base. May–Aug. **Synonyms** *Eleocharis acuminata*. **Habitat** Low calcareous prairie, wet meadows, swamps, ditches.

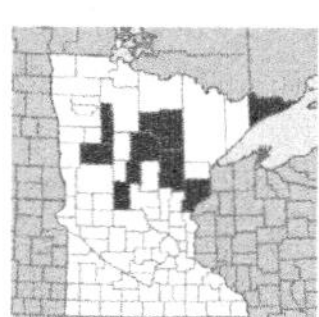

Eleocharis flavescens (Poiret) Urban
BRIGHT GREEN SPIKE-RUSH
NC **OBL** | MW **OBL** | GP **OBL** | *threatened*

Small, clumped, mat-forming native perennial, spreading by slender rhizomes. **Stems** bright green, flattened, 3–15 cm long. **Spikelets** ovate, 2–7 mm long and much wider than stem; scales ovate, red-brown, with a green midvein; sepals and petals reduced to 6–8 barbed bristles; style 2-parted (rarely 3-parted). **Achenes** lens-shaped, brown, 1 mm long; tubercle pale, cone-shaped, constricted at base. **Synonyms** *Eleocharis olivacea*. **Habitat** Shallow water, sandy or muddy lakeshores, mud flats; sometimes where calcium-rich. **Status** Threatened.

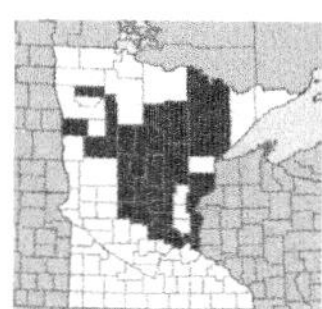

Eleocharis intermedia (Muhl.) Schultes
MATTED SPIKE-RUSH
NC **OBL** | MW **OBL** | GP **OBL**

Small, densely clumped native annual. **Stems** threadlike, grooved, of unequal lengths, 5–20 cm long; sheaths toothed on 1 side. **Spikelets** long-ovate, wider than stem; scales oblong lance-shaped, purple-brown, with a green midvein and white, translucent margins; sepals and petals reduced to barbed bristles or sometimes absent; style 3-parted. **Achenes** light brown to olive, 3-angled, 1 mm long; tubercle cone-shaped, constricted at base. June–Sept. **Synonyms** *Eleocharis macounii.* **Habitat** Wet, sandy or mucky shores, streambanks, mud flats.

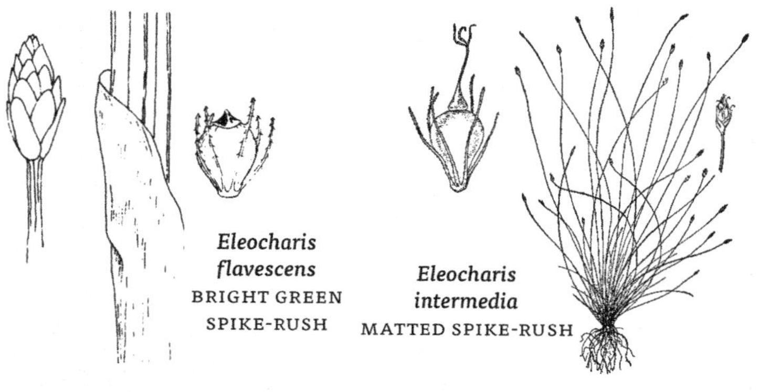

Eleocharis flavescens
BRIGHT GREEN SPIKE-RUSH

Eleocharis intermedia
MATTED SPIKE-RUSH

Eleocharis nitida Fernald
NEAT SPIKE-RUSH
NC **OBL** | *threatened*

Small, mat forming native perennial, from matted or creeping purplish rhizomes. **Stems** round in section to somewhat 4-angled, to 15 cm long, thin and delicate, to only 0.3 mm wide. **Spikelets** ovoid, small, to 4 mm long and 2 mm wide; **fertile scales** 1–1.3 mm long, brown to dark brown, midrib usually pale or greenish, the scales often early-deciduous; bristles absent; style 3-parted. **Achenes** persistent after scales fall, dark yellow-orange or brown, 3-angled (the angles evident), covered with small bumps (under magnification); tubercle brown, flattened and saucer-like, with a tiny central tip. May–June. **Habitat** Wet soil in openings in alder thickets and marshes, sometimes in shallow water, usually where little competing vegetation; disturbed moist places such as ditches and wheel ruts; crevices in rocks along Lake Superior shoreline. **Status** Threatened.

Eleocharis ovata (Roth) Roemer & Schultes
BLUNT SPIKE-RUSH
NC **OBL** | MW **OBL** | GP **OBL**

Clumped, fibrous-rooted native annual. **Stems** slender, round in section, ribbed, 0.5–5 dm long and 1–2 mm wide; sheaths green. **Spikelets** ovate to cylindric, 4–15 mm long and 2–4 mm wide; scales purple-brown, with a green midvein and pale margins; sepals and petals reduced to 6–7 brown bristles, or absent; styles 2- or 3-parted. **Achenes** lens-shaped, light to dark brown or olive, shiny, 1–1.5 mm long; tubercle flattened-triangular, about as wide as the broad top of achene. June–Sept. **Synonyms** *Eleocharis engelmannii, Eleocharis obtusa*. **Habitat** Wet, sandy or muddy shores, marshes, ditches, mud flats, temporary ponds.

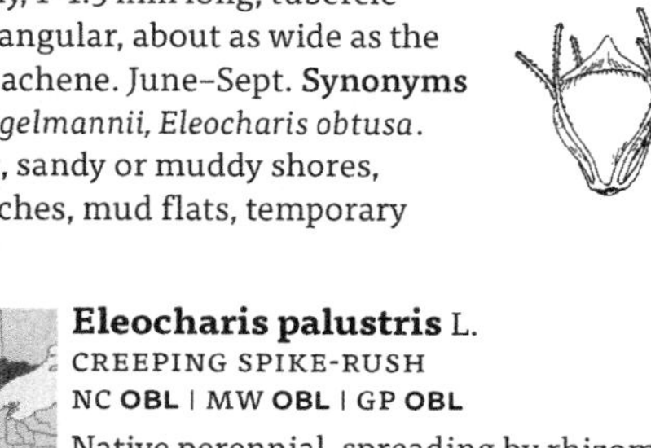

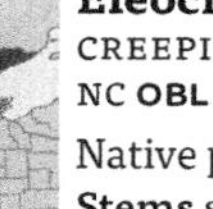

Eleocharis palustris L.
CREEPING SPIKE-RUSH
NC **OBL** | MW **OBL** | GP **OBL**

Native perennial, spreading by rhizomes. **Stems** single or in small clusters, slender to stout, round in section, 1–8 dm long and 1–3 mm wide; sheaths red or purple at base. **Spikelets** long-ovate, 5–30 mm long and 2–4 mm wide, wider than stems; lowest scale sterile, encircling the stem; fertile scales lance-shaped to ovate, 2–5 mm long, brown or red-brown, with a green or pale midvein; sepals and petals reduced to usually 4, pale brown, barbed bristles; style 2-parted. **Achenes** lens-shaped, yellow to brown, 1–2 mm long; tubercle flattened-triangular, constricted at base. May–Aug. **Synonyms** A variable species known by a number of synonyms including *Eleocharis erythropoda, Eleocharis macrostachya, Eleocharis smallii*. **Habitat** Common species of shallow water of marshes, wet meadows, muddy shores, bogs, ditches, streambanks and swamps.

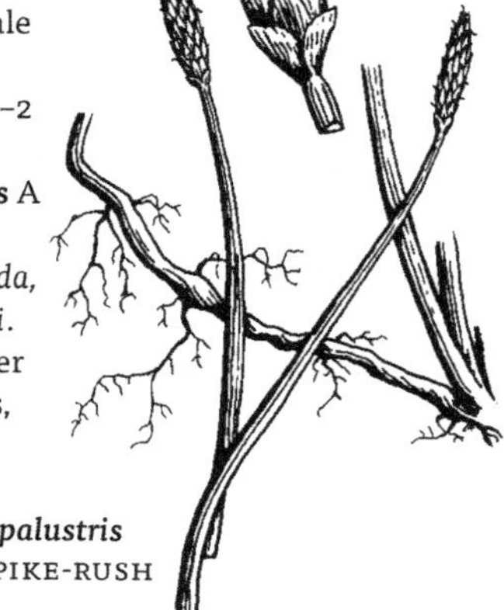

Eleocharis palustris
CREEPING SPIKE-RUSH

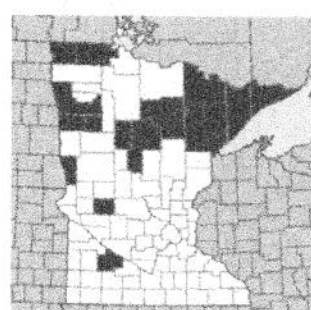

Eleocharis quinqueflora (Hartmann) Schwarz

FEW-FLOWER SPIKE-RUSH

NC **OBL** | MW **OBL** | GP **OBL** | *special concern*

Small, clumped native perennial, spreading by rhizomes. **Stems** threadlike, grooved, 1–3 dm long and less than 1 mm wide. **Spikelets** ovate, 4–8 mm long and 2–3 mm wide; scales ovate, brown, chaffy on margins, 2–5 mm long; sepals and petals reduced to bristles or absent; style 3-parted. **Achenes** gray-brown or brown, 3-angled, 1–3 mm long; tubercle slender, joined to the achene and beaklike. June–Aug. **Synonyms** *Eleocharis pauciflora*. **Habitat** Wet, sandy or gravelly shores and flats marshes and fens; often where calcium-rich. **Status** Special concern.

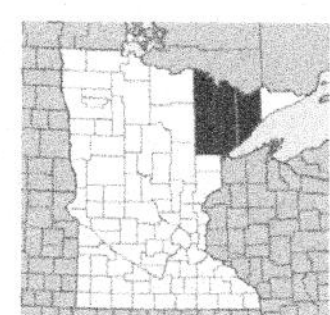

Eleocharis robbinsii Oakes

ROBBINS' SPIKE-RUSH

NC **OBL** | MW **OBL**

Clumped native perennial, spreading by rhizomes. Stems slender, 3-angled, 2–6 dm long and 1–2 mm wide; when underwater, plants often with numerous sterile stems from base; sheaths brown. Spikelets lance-shaped, 1–2 cm long and 2–3 mm wide, barely wider than stems; scales narrowly ovate, margins chaffy; sepals and petals reduced to 6 barbed bristles; style 3-parted. **Achenes** rounded on both sides, light brown, 2–3 mm long; tubercle flattened and cone-shaped, with a raised ring at base. July–Aug. **Habitat** Sandy or mucky lake and pond shores, exposed flats or in water to 1 m deep.

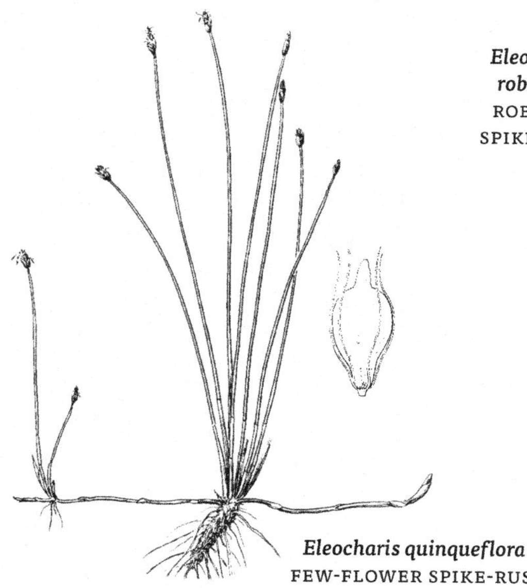

Eleocharis robbinsii
ROBBINS' SPIKE-RUSH

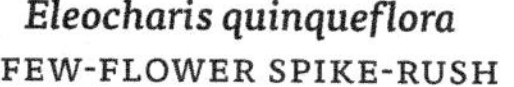

Eleocharis quinqueflora
FEW-FLOWER SPIKE-RUSH

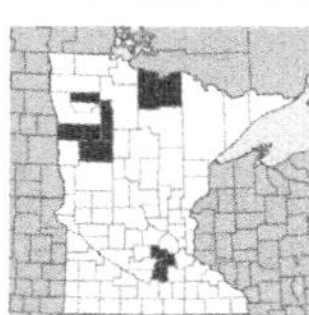

Eleocharis rostellata (Torr.) Torr.
BEAKED SPIKE-RUSH
NC **OBL** | MW **OBL** | GP **OBL** | *threatened*

Clumped native perennial, without creeping rhizomes. **Stems** flattened, wiry, 3–10 dm long and 1–2 mm wide; the fertile stems upright, the sterile stems often arching and rooting at tip; sheaths brown. **Spikelets** oblong, tapered at both ends, 5–15 mm long and 2–5 mm wide, wider than the stem; scales ovate, 3–5 mm long, green to brown with a darker midvein and translucent margins; sepals and petals reduced to 4–8 barbed bristles; style 3-parted. **Achenes** olive to brown, rounded 3-angled, 2–3 mm long; tubercle cone-shaped, joined with the achene body and beaklike. June–Aug. **Habitat** Shores, wet meadows, calcareous fens and mud flats; typically where calcium-rich and often associated with mineral springs. **Status** Threatened.

Eleocharis wolfii A. Gray
WOLF'S SPIKE-RUSH
NC **OBL** | MW **OBL** | GP **OBL** | *endangered*

Clumped native perennial, from slender rhizomes. **Stems** flattened, 2-edged, often twisted, 1–3 dm long and 1–2 mm wide; sheaths often purple at base, membranous at tip. **Spikelets** narrowly ovate, 4–10 mm long and 2–3 mm wide, wider than stem; scales green, tinged with purple, narrowly ovate, 2–3 mm long, with chaffy margins; bristles absent; style 3-parted. **Achenes** gray, ± round in section, about 1 mm long; tubercle cone-shaped, constricted at base where joins achene. June–July. **Habitat** Minnesota populations are mostly associated with bedrock pools at rock outcrops; other states report this species from a variety of open, wet sites.
Status Endangered.

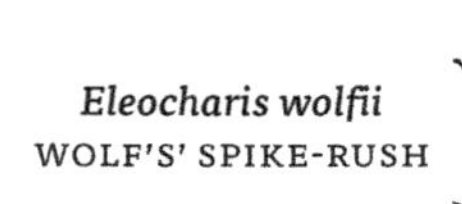

Eleocharis wolfii
WOLF'S' SPIKE-RUSH

ERIOPHORUM | *Cotton-grass*

Grasslike perennials. **Stems** clumped or single, round to rounded 3-angled in section. **Leaves** mostly at base of plant, the blades flat, folded or inrolled; upper leaves often reduced to bladeless sheaths. Flower heads at ends of stems, with 1 or several spikelets; scales many, spirally arranged, chaffy on margins; involucral bracts leaflike in species with several spikelets in the head, or reduced to scales in species with 1 spikelet at end of stems (*Eriophorum chamissonis, Eriophorum vaginatum*). **Flowers** perfect; sepals and petals numerous, reduced to long, cottony, persistent white to tawny brown bristles; stamens 3; styles 3-parted. **Achenes** brown, ± 3-angled, sometimes with a short beak formed by the persistent style.

ERIOPHORUM | COTTON-GRASS

1 Head a single spikelet at end of stem; leaflike bracts absent. **2**
1 Head of 2 or more spikelets; leaflike bracts present **3**

2 Plants forming colonies from rhizomes . . . . *Eriophorum chamissonis* . . . . RUSTY COTTON-GRASS
2 Plants densely clumped, rhizomes absent. . . . . *Eriophorum vaginatum* . . . . TUSSOCK COTTON-GRASS

3 Leaves 1–2 mm wide; leaflike bract 1, erect, the head appearing lateral from side of stem **4**
3 Leaves 3 mm or more wide; leaflike bracts 2 or 3, the head appearing terminal **5**

4 Blade of uppermost stem leaf much shorter than its sheath . . . . *Eriophorum gracile* | SLENDER COTTON-GRASS
4 Blade as long or longer than its sheath . . . . *Eriophorum tenellum* . . . . CONIFER COTTON-GRASS

5 Scales 3–7-nerved, copper-brown on sides . . . . *Eriophorum virginicum* . . . . TAWNY COTTON-GRASS
5 Scales with 1 nerve, sides olive-green to nearly black. **6**

6 Midvein of scale slender, fading before reaching tip of scale . . . . *Eriophorum angustifolium* | THIN-SCALE COTTON-GRASS
6 Midvein of scale widening toward tip of scale and reaching scale tip . . . . *Eriophorum viridicarinatum* | DARK-SCALE COTTON-GRASS

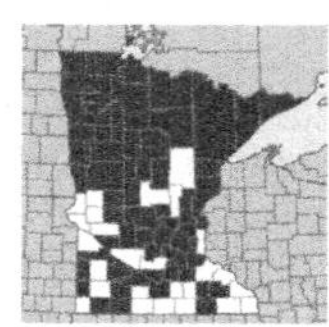

Eriophorum angustifolium Honck.
THIN-SCALE COTTON-GRASS
NC **OBL** | MW **OBL** | GP **OBL**
Grasslike native perennial, spreading by rhizomes and forming colonies. **Stems** mostly single, 2–8 dm long and 2–3 mm wide, ± round in section, becoming 3-angled below the head. **Leaves** few, flat or folded along midrib, 3–8 mm wide, often dying back from the tips; sheaths sometimes red, dark-banded at tip. **Spikelets** 3–10, clustered in heads 1–3 cm wide when mature, the heads drooping on weak stalks; involucral bracts leaflike, often black at base, the main bract upright and usually longer than the head; scales lance-shaped, brown or purple-green, 4–6 mm long, the midvein not extending to tip of scale; bristles bright white, 2–3 cm long. **Achenes** brown to nearly black, 2–3 mm long. May–July. **Synonyms** *Eriophorum polystachion*. **Habitat** Bogs, calcareous fens, wet meadows.

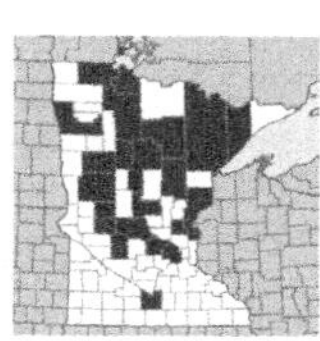

Eriophorum chamissonis C. A. Meyer
RUSTY COTTON-GRASS
NC **OBL** | MW **OBL** | GP **OBL**
Grasslike native perennial, spreading by rhizomes and forming colonies. **Stems** single or in small groups, stout, ± round in section, 2–6 dm long and 1–3 mm wide. **Leaves** few, mostly from base of plant and shorter than stems, the uppermost leaves from near middle of stem and often without blades, lower leaves round in section to 3-angled and channeled, 1–2 mm wide. **Spikelets** single, erect at end of stems, clustered in a ± round head 2–3 cm wide; involucral bracts not leaflike, reduced to black scales; flower scales narrowly ovate, black-green, with broad white margins and tips; bristles white to bright red-brown. **Achenes** dark brown, beaked, 2–3 mm long. June–July. **Habitat** Bogs.

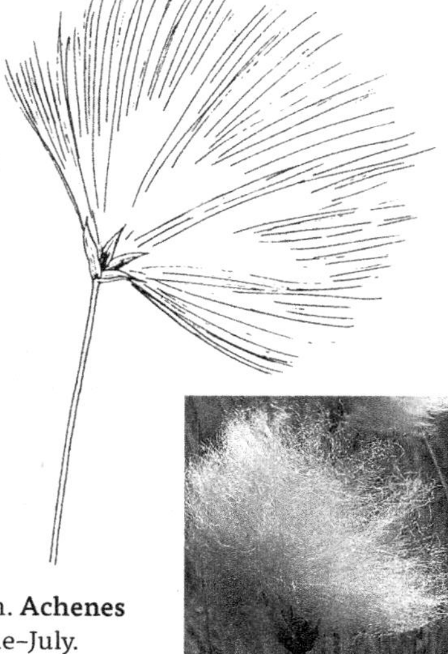

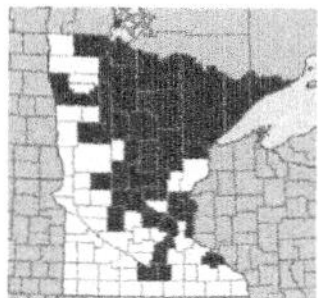

Eriophorum gracile Koch

SLENDER COTTON-GRASS

NC **OBL** | MW **OBL** | GP **OBL**

Grasslike native perennial, spreading from rhizomes. **Stems** single, spreading or reclining, slender, ± round in section, 2–6 dm long and 1–2 mm wide. **Leaves** few, channeled on upper side, 1–2 mm wide, the basal leaves often withered by flowering time, blades of uppermost leaves small. **Spikelets** in clusters of 2–5 at ends of stems, on spreading to nodding stalks 2–3 cm long; involucral bract leaflike and erect, shorter than spikelet cluster; scales ovate, pale to black-brown with a prominent midvein; bristles bright white. **Achenes** light brown, 3–4 mm long. May–July. **Habitat** Fens and bogs.

Eriophorum gracile
SLENDER COTTON-GRASS

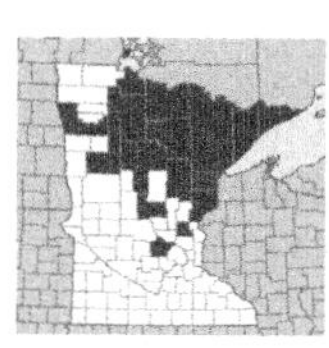

Eriophorum tenellum Nutt.

CONIFER COTTON-GRASS

NC **OBL** | MW **OBL** | GP **OBL**

Grasslike native perennial, with rhizomes and forming colonies. **Stems** single, slender, erect, 3–8 dm long, rounded 3-angled, rough-to-touch on upper angles. **Leaves** linear, 1–2 mm wide, channeled, not reduced and bladeless on upper stem. **Spikelets** 3–6, in short-stalked clusters at ends of stems, or with 1–2 rough, drooping stalks to 5 cm long; involucral bract leaflike, stiff and erect, usually shorter than the spikelet cluster; scales ovate, straw-colored to red-brown; bristles white. **Achenes** brown, 2–3 mm long. **Habitat** Bogs and conifer swamps.

Eriophorum tenellum
CONIFER COTTON-GRASS

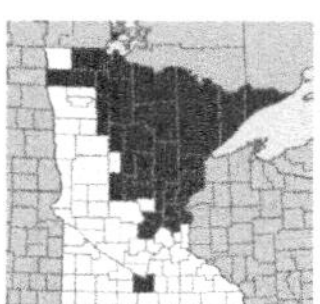

Eriophorum vaginatum L.

TUSSOCK COTTON-GRASS

NC **OBL** | MW **OBL** | GP **OBL**

Densely clumped, grasslike native perennial, forming large hummocks. **Stems** stiff, rounded 3-angled, 2–7 dm long. **Leaves** at base of stems, mostly shorter than stems, only 1 mm wide, with 1–3 inflated, bladeless sheaths on stem. **Spikelets** clustered in a single head at end of stems; involucral bracts absent; scales narrowly ovate, purple-brown to black, with white margins, spreading when mature; bristles usually white (rarely red-brown). **Achenes** obovate, 3–4 mm long. June. **Synonyms** *Eriophorum callithrix, Eriophorum spissum*. **Habitat** Sphagnum bogs and tamarack swamps.

Eriophorum vaginatum
TUSSOCK COTTON-GRASS

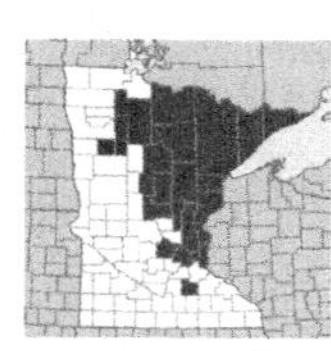

Eriophorum virginicum L.

TAWNY COTTON-GRASS

NC **OBL** | MW **OBL** | GP **OBL**

Large grasslike native perennial, with slender rhizomes. **Stems** single or in small groups, stiff, erect, to 1 m long, leafy, mostly smooth. **Leaves** flat, 2–4 mm wide, the uppermost often longer than the head. **Spikelets** in dense clusters of several to many at ends of stems, on short stalks of ± equal lengths, the clusters wider than long; involucral bracts 2–3, leaflike, spreading or bent downward, unequal, much longer than the head; scales ovate, thick, copper-brown with a green center; bristles tawny or copper-brown. **Achenes** light brown, 3–4 mm long. July–Aug. **Habitat** Sphagnum moss peatlands.

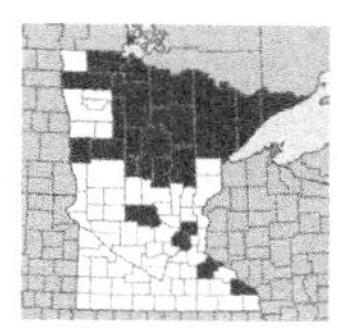

Eriophorum viridicarinatum (Engelm.) Fernald
DARK-SCALE COTTON-GRASS
NC **OBL** | MW **OBL** | GP **OBL**

Grasslike native perennial, forming colonies from spreading rhizomes. **Stems** mostly single, ± round in section, 3–7 dm long. **Leaves** flat except at tip, the uppermost leaves 10–15 cm long; sheaths green. **Spikelets** usually 20–30, clustered in heads at ends of stems, on short to long, finely hairy stalks; involucral bracts 2–4, not black at base, longer or equal to head; scales narrowly ovate, black-green, the midvein pale, extending to tip of scale; bristles white. **Achenes** brown, 3–4 mm long. May–July. **Habitat** Bogs and open conifer swamps.

Similar to **thin-scale cotton-grass** *(Eriophorum angustifolium), but usually with more spikelets, the scale midvein extending to the tip of scale, and the leaf sheaths not dark-banded at tip.*

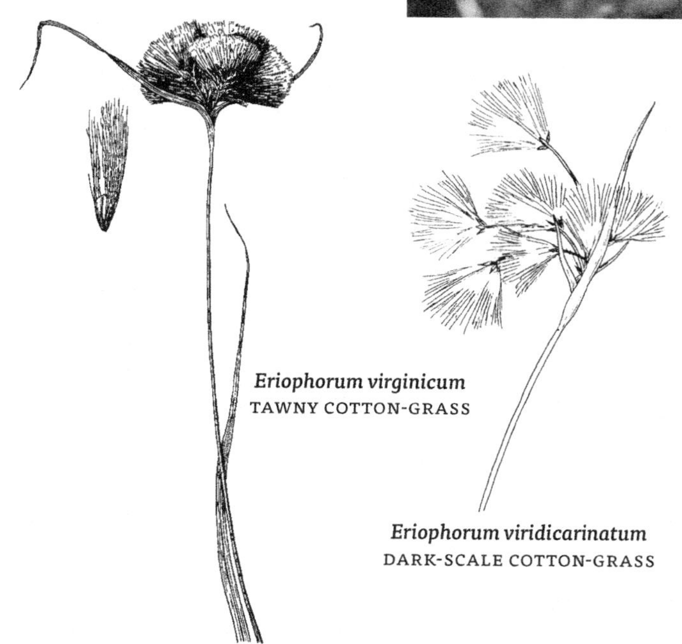

Eriophorum virginicum
TAWNY COTTON-GRASS

Eriophorum viridicarinatum
DARK-SCALE COTTON-GRASS

FIMBRISTYLIS | *Fimbry*

Annual or perennial grasslike plants. **Stems** slender, clumped or single. **Leaves** mostly at base of plants, narrowly linear, flat to inrolled. **Spikelets** many-flowered, in umbel-like clusters at ends of stems; involucral bracts 2–3, short and leaflike; scales spirally arranged and overlapping. **Flowers** perfect, sepals and petals absent, stamens 1–3; styles 2–3-parted, swollen at base, deciduous when mature. **Achenes** lens-shaped or 3-angled.

FIMBRISTYLIS | FIMBRY

1 Plants annual; achenes 3-angled; styles 3-parted .... *Fimbristylis autumnalis* | AUTUMN SEDGE

1 Plants perennial; achenes lens-shaped; styles 2-parted .... *Fimbristylis puberula* | CHESTNUT SEDGE

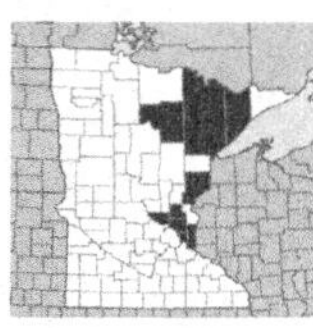

Fimbristylis autumnalis (L.) Roemer & Schultes

AUTUMN SEDGE

NC **FACW** | MW **OBL** | GP **OBL** | *special concern*

Clumped, grasslike native annual, with shallow fibrous roots. **Stems** flattened, slender, sharp-edged, 0.5–3 dm long. **Leaves** shorter than the stems, flat, 1–2 mm wide. **Spikelets** usually many in an open umbel-like cluster, the spikelets lance-shaped, 3–8 mm long, single or several at ends of threadlike, spreading stalks; involucral bracts 2–3, leaflike, usually shorter than the head; scales ovate, golden-brown with a prominent green midvein, 1–2 mm long; style 3-parted. **Achenes** ivory to tan, 3-angled and ribbed on the angles, to 0.5 mm long. July–Sept. **Habitat** Sandy or mucky shores (especially where seasonally flooded and then later drying), streambanks, wet meadows, ditches. **Status** Special concern.

Fimbristylis puberula (Michx.) Vahl

CHESTNUT SEDGE

NC **OBL** | MW **OBL** | GP **OBL** | *endangered*

Grasslike native perennial, with short rhizomes. **Stems** single or in small clumps, slender, stiff, 2–7 dm long, round to oval in section, sometimes swollen at base. **Leaves** shorter than the stems, usually inrolled, 1–3 mm wide, often hairy. **Spikelets** few to many in an umbel-like cluster; the spikelets ovate, 5–10 mm long, the central spikelet stalkless, the others on slender stalks; involucral bracts 2–3, leaflike, the longest equal or longer than the head; scales ovate, brown with a lighter midvein, 3–4 mm long, usually finely hairy, often tipped with a short awn; style 2-parted, the style branches finely hairy. **Achenes** light brown, lens-shaped, 1–2 mm long. June–Sept. **Habitat** Wet meadows, shores, and low prairie, often where sandy and calcium-rich; also in drier prairies. **Status** Endangered.

HEMICARPHA | *Dwarf-bulrush*

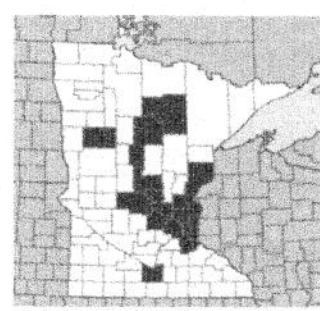

Hemicarpha micrantha (Vahl) Pax

DWARF-BULRUSH

NC **OBL** | MW **OBL** | GP **FACW**

Small, densely clumped grasslike native annual. **Stems** compressed, 3–15 cm long. slender, 2 per stem, to 10 cm long and 0.5 mm wide, mostly shorter than the stems. **Spikelets** in stalkless clusters of 1–3; the spikelets many-flowered, ovate, 2–5 mm long; involucral bracts 2–3, leaflike, the main bract upright and longer than the spikelets (the head appearing lateral); scales brown with a green midvein, 1–2 mm long, tipped with an awn; bristles absent; style 2-parted, not swollen at base. **Achenes** brown, oblong, to 1 mm long. Aug–Sept. **Synonyms** *Lipocarpa micrantha*. **Habitat** Sandy or muddy shores and streambanks, usually where seasonally flooded.

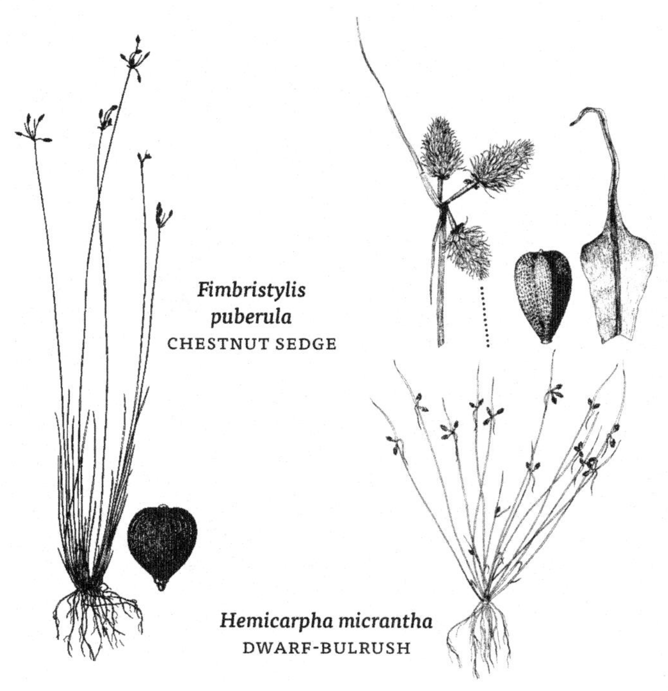

Fimbristylis puberula
CHESTNUT SEDGE

Hemicarpha micrantha
DWARF-BULRUSH

RHYNCHOSPORA | *Beak-rush*

Grasslike perennials, clumped or spreading by rhizomes. **Stems** erect, leafy, usually 3-angled or sometimes round. **Leaves** flat or rolled inward. **Spikelets** clustered in dense heads, the heads open to crowded; scales overlapping in a spiral. **Flowers** perfect, or sometimes upper flowers male only; sepals and petals reduced to usually 6 (1–20) bristles or sometimes absent; stamens usually 3; styles 2-parted, swollen at base and persistent on the achene as a tubercle. **Achenes** lens-shaped.

RHYNCHOSPORA | BEAK-RUSH

1 Spikelets white to tan; bristles 8 or more . . . . *Rhynchospora alba* . . . . WHITE BEAK-RUSH

1 Spikelets brown, dark olive-green or nearly black; bristles 5–6 . . . . **2**

2 Scales dark olive-green to black; bristles with upward-pointing barbs, at least some of the bristles longer than the tubercle . . . . *Rhynchospora fusca* | GRAY BEAK-RUSH

2 Scales brown; bristles with downward pointing barbs (rarely smooth), the bristles shorter to as long as the tubercle **3**

3 Stems narrow and threadlike; achene margins not translucent, achene body less than half as wide as long . . . . *Rhynchospora capillacea* . . . . NEEDLE BEAK-RUSH

3 Stems stout; achene with translucent margins, body more than half as wide as long . . . . *Rhynchospora capitellata* | BROWN BEAK-RUSH

Rhynchospora alba (L.) Vahl

WHITE BEAK-RUSH

NC **OBL** | MW **OBL** | GP **OBL**

Clumped, grasslike native perennial. **Stems** slender, erect, 1–6 dm long. **Leaves** bristlelike, 0.5–3 mm wide, shorter than the stems. **Spikelets** in 1–3 rounded heads, 5–20 mm wide, at or near ends of stems, the lateral heads usually long-stalked; the spikelets oblong, narrowed at each end, 4–5 mm long, white, becoming pale brown; bristles 8–15, downwardly barbed, about equaling the tubercle. **Achenes** lens-shaped, brown-green, 1–2 mm long; tubercle triangular, about half as long as achene. June–Sept. **Habitat** Bogs, fens, open conifer swamps of black spruce and tamarack.

Rhynchospora alba
WHITE BEAK-RUSH

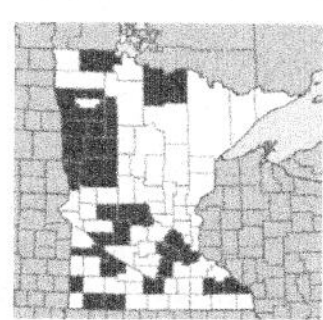

Rhynchospora capillacea Torr.
NEEDLE BEAK-RUSH
NC **OBL** | MW **OBL** | GP **OBL** | *threatened*

Small, clumped, grasslike native perennial. **Stems** slender, 0.5–4 dm long. **Leaves** threadlike, rolled inward, to only 0.5 mm wide, much shorter than the stem. **Spikelets** in 1–2 small, separated clusters, each cluster subtended by 1 to several short, bristlelike bracts; the spikelets ovate, 3–7 mm long; scales overlapping, ovate, brown with a paler, sharp-tipped midvein; bristles 6, downwardly barbed, longer than the achenes; style 2-parted. **Achenes** lens-shaped, satiny yellow-brown, 2 mm long; tubercle dull brown, narrowly triangular, about 1 mm long. June–Aug. **Habitat** Calcareous fens, interdunal flats, wet sandy or gravelly shores, seeps; usually where calcium-rich. **Status** Threatened.

Rhynchospora capitellata (Michx.) Vahl
BROWN BEAK-RUSH
[NC **OBL** | MW **OBL**]

Clumped, grasslike native perennial. **Stems** erect, 3-angled, 3–8 dm long. **Leaves** flat, 2–4 mm wide, rough on margins, shorter than stems. **Spikelets** 3–5 mm long, several to many in 2–7 rounded, ± loose clusters 1–1.5 cm wide, the lateral clusters often in pairs on slender stalks; bristles 6, 2–3 mm long, usually downwardly barbed, about equaling the achene; style 2-parted. **Achenes** lens-shaped, dark brown, 1–2 mm long; tubercle triangular, about as long as achene. June–Sept. **Synonyms** *Rhynchospora glomerata*. **Habitat** Wet sandy or mucky shores and flats, wet meadows, bogs, calcareous fens, ditches.

Presence in Minnesota not verified; known from Wisconsin.

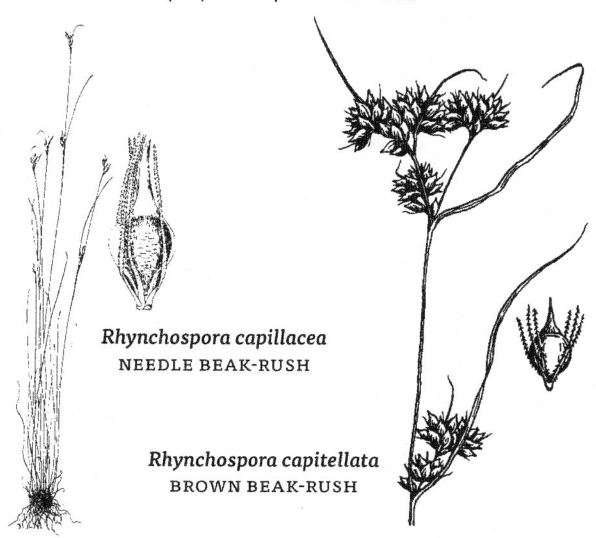

Rhynchospora capillacea
NEEDLE BEAK-RUSH

Rhynchospora capitellata
BROWN BEAK-RUSH

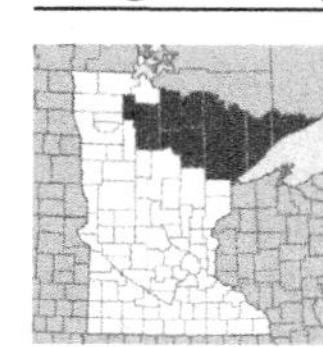

Rhynchospora fusca (L.) Aiton f.
GRAY BEAK-RUSH
NC **OBL** | MW **OBL** | GP **OBL** | *special concern*

Clumped, grasslike native perennial, spreading by short rhizomes and forming colonies. **Stems** slender, 3-angled, 1–3 dm long. **Leaves** very slender, rolled inward, mostly shorter than the stems. **Spikelets** spindle-shaped, dark brown, 4–6 mm long, in 1–4 loose clusters, the lower clusters on long stalks, each cluster subtended by an erect, leafy bract, the bract longer than the cluster; bristles 6, upwardly barbed; style 2-parted. **Achenes** light brown, 1–1.5 mm long; tubercle flattened-triangular, nearly as long as achene. **Habitat** Wet, sandy shores, interdunal wetlands, sedge meadows, bog mats; Red Lake peatland. **Status** Special concern.

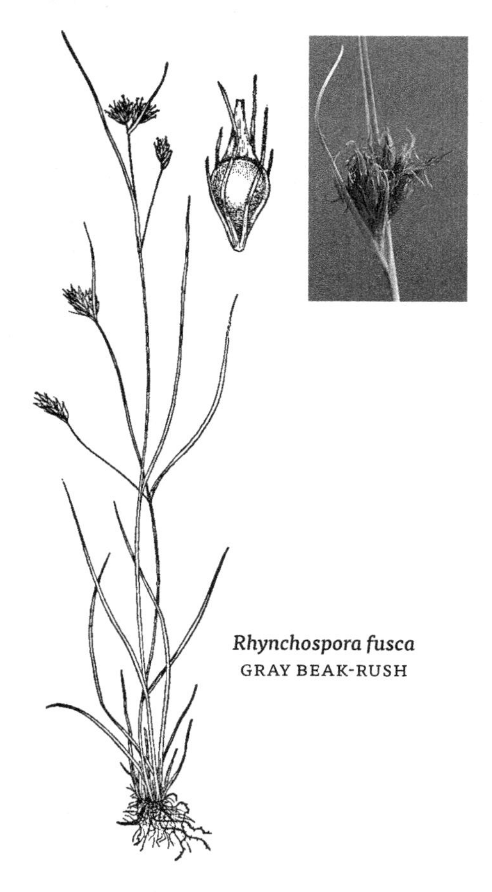

Rhynchospora fusca
GRAY BEAK-RUSH

SCIRPUS | *Bulrush*

Stout, rushlike perennials, mostly spreading by rhizomes. **Stems** unbranched, 3-angled or round in section, solid or pithy. **Leaves** broad and flat, to narrow and often folded near tip, or reduced to sheaths at base of stems; involucral bracts several and leaflike, or single and appearing like a continuation of the stem. **Spikelets** single, or in paniclelike or umbel-like clusters at ends of stems, or appearing lateral from the stem; the spikelets stalked or stalkless; scales overlapping in a spiral. **Flowers** perfect; sepals and petals reduced to 1–6 smooth or downwardly barbed bristles, or sometimes absent; stamens 2 or 3; styles 2–3-parted. **Achenes** lens-shaped, flat on 1 side and convex on other, or 3-angled, usually beak-tipped.

*The genus **Scirpus** has been reworked in the Flora of North America (Vol. 23). The traditional treatment of the genus is presented here, with newer names listed below the traditional name.*

SCIRPUS | BULRUSH

1 Spikelets single at end of slender stems; bracts below head absent or very short 2
1 Spikelets mostly more than 1; bracts conspicuous, leaflike or appearing to be a continuation of stem 4

2 Plants usually underwater, the leaves limp and floating; not clumped, spreading by rhizomes or stolons . . . . . *Scirpus subterminalis* . . . . . WATER-BULRUSH
2 Plants of peatlands, usually forming dense clumps 3

3 Stems ± round in section, smooth . . . . . *Scirpus cespitosus* . . . . . CLUMPED BULRUSH
3 Stems 3-angled, rough on angles . . . . . *Scirpus hudsonianus* . . . . . ALPINE COTTON-GRASS

4 Bract single, erect, similar to stem and appearing to be a continuation of it 5
4 Bracts 2 or more, flat and leaflike, usually spreading, appearing different than the stem 10

5 Stems sharply 3-angled 6
5 Stems round in section or rounded 3-angled 7

6 Leaves short, less than half length of stem; stigmas 2; achenes flat on 1 side, convex on other . . . . . *Scirpus pungens* | COMMON THREESQUARE
6 Leaves often longer than stem; stigmas 3; achenes 3-angled . . . . . *Scirpus torreyi* | TORREY'S THREESQUARE

7 Stems to 5 dm long; spikelets less than 10 . . . . *Scirpus smithii* . . . . BLUNTSCALE-BULRUSH
7 Stems more than 5 dm long; spikelets mostly more than 10 . . . . **8**

8 Scales pale brown, smooth; achenes 3-angled; stigmas 3; panicle branches flexible . . . . *Scirpus heterochaetus* | SLENDER BULRUSH
8 Scales red-brown, hairy; achenes flat on 1 side and convex on other; stigmas2; panicle branches stiff **9**

9 Stems firm, difficult to compress between fingers; spikelets oblong to cylindric, more than 10 mm long when mature . . . . *Scirpus acutus* . . . . HARDSTEM-BULRUSH
9 Stems spongy, easily compressed; spikelets ovate, mostly less than 10 mm long. . . . . *Scirpus validus* | SOFTSTEM-BULRUSH

10 Stems sharply 3-angled **11**
10 Stems round in section or upper stem rounded 3-angled **12**

11 Styles 3-parted; achenes 3-angled; leaf sheaths convex at tip . . . . *Scirpus fluviatilis* | RIVER-BULRUSH
11 Styles 2-parted; achenes lens-shaped; leaf sheaths straight or concave at tip . . . . *Scirpus maritimus* | ALKALI-BULRUSH

12 Lower sheaths red-tinged . . *Scirpus microcarpus* | SMALL-FRUIT BULRUSH
12 Sheaths green or brown **13**

13 Spikelets many in dense, ± round heads; bristles about as long as achene or shorter . . . . *Scirpus atrovirens* | BLACK BULRUSH
13 Spikelets few in open clusters; bristles much longer than achene. **14**

14 Mature bristles longer than scales, giving spikelets woolly appearance . . . . *Scirpus cyperinus* | WOOL-GRASS
14 Mature bristles equal or only slightly longer than scales, spikelets not woolly. . . . . *Scirpus pendulus* | DROOPING BULRUSH

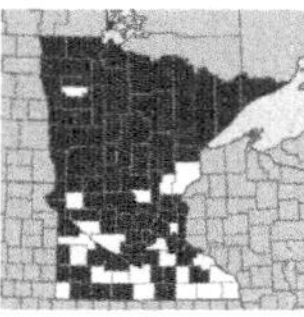

Scirpus acutus Muhl.
Schoenoplectus acutus (Muhl.) A. Löve & D. Löve
HARDSTEM-BULRUSH
NC **OBL** | MW **OBL** | GP **OBL**

Native perennial, from stout rhizomes and often forming large colonies. **Stems** round in section, 1–3 m long. reduced to 3–5 sheaths near base of stem, blades absent, or upper leaves with blades to 25 cm long; main bract erect, appearing as a continuation of stem, 2–10 cm long, eventually turning brown. **Spikelets** 5–15 mm long and 3–5 mm wide, in clusters of mostly 3–7, the clusters grouped into a branched head of up to 60 spikelets, the head appearing lateral from side of stem, the branches stiff and spreading; scales chaffy, mostly translucent, 3–4 mm long, often with red-brown spots, usually tipped with an awn to 1 mm long; bristles 6, unequal, usually shorter than achene; style 2-parted (rarely 3-parted). **Achenes** light green to dull brown, flat on 1 side and convex on other, 2–3 mm long, the style beak small, to 0.5 mm long. May–Aug. **Habitat** Common; plants usually emergent in shallow to deep water (1–2 m deep) of marshes, ditches, ponds and lakes; sometimes where brackish.

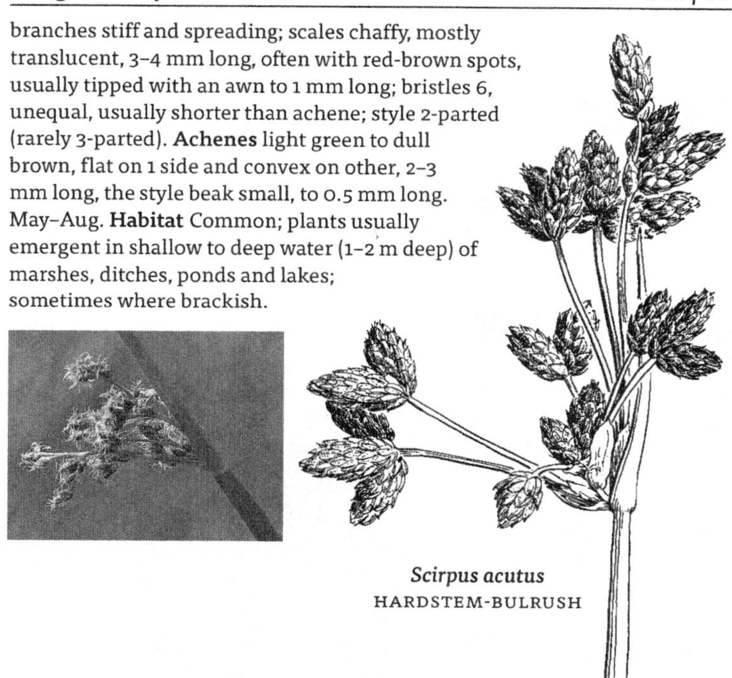

Scirpus acutus
HARDSTEM-BULRUSH

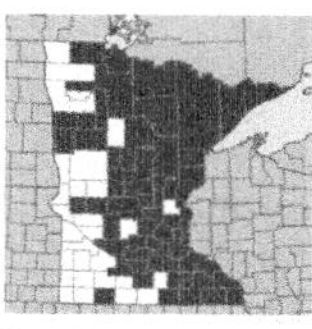

Scirpus atrovirens Willd.
BLACK BULRUSH
NC **OBL** | MW **OBL** | GP **OBL**

Loosely clumped native perennial, with short rhizomes. **Stems** 3-angled, leafy, 0.5–1.5 m long. **Leaves** mostly on lower half of stem, blades ascending, usually shorter than the head, 6–18 mm wide; bracts 3–4, leaflike, to 15 cm long, mostly longer than the head. **Spikelets** many, 2–8 mm long and 1–3 mm wide, crowded in rounded heads at end of stems, the heads on stalks to 12 cm long; scales brown-black, translucent except for the broad green midvein, 1–2 mm long, tipped by an awn to 0.5 mm long; bristles 6, white or tan, shorter or equal to the achene; style 3-parted. **Achenes** tan to nearly white, compressed 3-angled, about 1 mm long, with a short beak 0.2 mm long. June–Aug. **Synonyms** *Scirpus atrovirens* var.

pallidus, Scirpus hattorianus, Scirpus pallidus. **Habitat** Wet meadows, shores, ditches, streambanks, swamps, springs and other wet places.

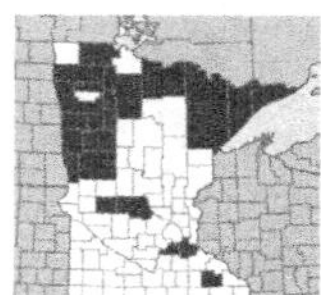

Scirpus cespitosus L.

Trichophorum cespitosum (L.) Schur
TUFTED BULRUSH
NC **OBL** | MW **OBL** | GP **OBL**

Densely clumped native perennial, rhizomes short. **Stems** slender, smooth, ± round in section, 1–4 dm long. **Leaves** light brown and scalelike at base of stems, and also usually 1 leaf upward on stem, the blade narrow, short, to 6 mm long. **Spikelets** 1 at end of stems, brown, 4–6 mm long, several-flowered; scales yellow-brown, deciduous, the lowest scale about as long as spikelet; bristles 6, usually slightly longer than achene; style 3-parted. **Achenes** brown, 3-angled, 1.5 mm long. **Habitat** Open bogs, cedar swamps, calcareous fens, wet swales between dunes; also Lake Superior rocky shores. A circumboreal species and an important plant of the arctic tundra.

Scirpus cespitosus
TUFTED BULRUSH

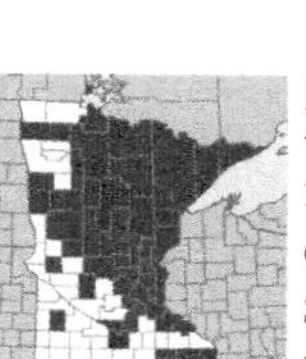

Scirpus cyperinus (L.) Kunth

WOOL-GRASS
NC **OBL** | MW **OBL** | GP **OBL**

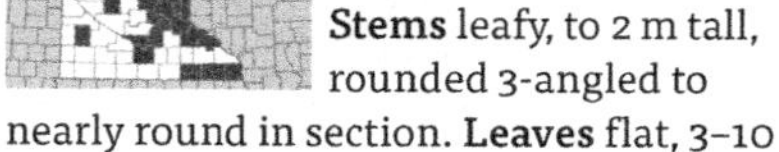

Coarse, densely clumped native perennial, rhizomes short. **Stems** leafy, to 2 m tall, rounded 3-angled to nearly round in section. **Leaves** flat, 3–10 mm wide, rough-to-touch on margins; sheaths brown; bracts 2–4, leaflike, spreading, usually drooping at tip, often red-brown at base. **Spikelets** numerous, ovate, 3–8 mm long and 2–3 mm wide, appearing woolly due to the long bristles, in clusters of 1 to several spikelets; the spikelet clusters grouped into large, spreading, branched heads at ends of stems; scales ovate, 1–2 mm long; bristles 6, smooth,

brown, much longer than achene and scale; styles 3-parted. **Achenes** white to tan, flattened 3-angled, 0.5–1 mm long, with a short beak. July–Sept. **Synonyms** *Scirpus atrocinctus, Scirpus pedicellatus*. **Habitat** Common species of wet meadows, marshes, swamps, ditches, bog margins, thickets; soils wet or in shallow standing water.

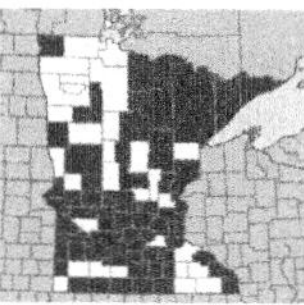

Scirpus fluviatilis (Torr.) A. Gray
Bolboschoenus fluviatilis (Torr.) Soják
RIVER-BULRUSH
NC **OBL** | MW **OBL** | GP **OBL**

Native perennial, spreading by rhizomes and often forming large colonies. **Stems** stout, erect, 6–15 dm long, sharply 3-angled, the sides ± flat. **Leaves** several on stem, smooth, 6–15 mm wide, upper leaves often longer than the head; bracts 3–5, leaflike, erect to spreading, to 3–4 dm long. **Spikelets** 1–3 cm long and 6–12 mm wide, clustered in an umbel with 10–20 spikelets at end of stem, several of the spikelets nearly stalkless in 1–2 clusters, others single or in groups of 2–5 at ends of spreading or drooping stalks to 8 cm long; scales gold-brown, short-hairy on back, 6–10 mm long, the midvein extended into a curved awn 1–3 mm long; bristles 6, unequal, white to copper-brown, downwardly barbed, persistent, about as long as body of achene; style yellow, 3-parted. **Achenes** 3-angled, dull, tan to gray-green, 3–5 mm long, with a beak to 0.5 mm long. June–Aug. **Synonyms** *Schoenoplectus fluviatilis*. **Habitat** Usually in shallow water of streams, ditches, marshes, lakes and ponds; sometimes where brackish.

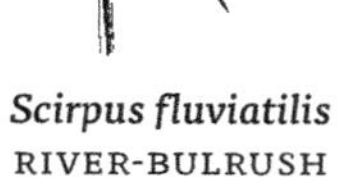

Scirpus fluviatilis
RIVER-BULRUSH

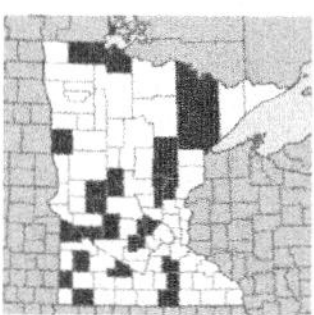

Scirpus heterochaetus Chase
Schoenoplectus heterochaetus (Chase) Soják
SLENDER BULRUSH
NC **OBL** | MW **OBL** | GP **OBL**

Native perennial, spreading by stout rhizomes. **Stems** slender, round in section, 1–2 m long. **Leaves** reduced to 3–4 sheaths at base of stem, upper sheaths with blades 6–8 cm long; main bract erect, 1–10 cm long, shorter than head. **Spikelets** mostly single at ends of stalks, the spikelets 5–15 mm long and 3–6 mm wide, in open, lax heads; scales chaffy, brown, 3–4 mm long, tipped with an awn to 2 mm long; bristles 2–4, unequal, about as long as achene; style 3-parted. **Achenes** 3-angled, light green to brown, 2–3 mm long, with a beak about

0.5 mm long. June–Aug. **Habitat** Emergent in shallow to deep water (1–2 m deep) of marshes, ponds and lakes, ditches.

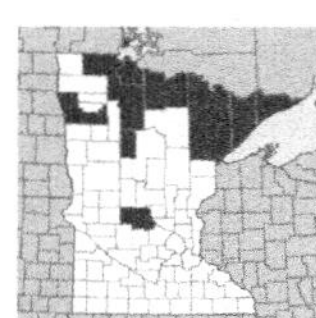

Scirpus hudsonianus (Michx.) Fern.

Trichophorum alpinum (L.) Pers.

ALPINE COTTON-GRASS

NC **OBL** | MW **OBL** | GP **OBL**

Native perennial, from short rhizomes. **Stems** single to clustered, slender, 1–4 dm long, sharply 3-angled, rough-to-touch on the angles. **Leaves** reduced to scales at base of stem, with 1–2 leaves upward on stem, these with short narrow blades 5–15 mm long. **Spikelets** single at ends of stems, brown, 5–7 mm long, with 10–20 flowers, involucral bract awl-shaped, shorter than spikelet, sometimes absent; scales ovate, blunt-tipped, yellow-brown; bristles 6, white, flattened, longer than the scales, when mature forming a white tuft 1–2 cm longer than the spikelet. **Achenes** 3-angled, dull brown, 1–4 mm long. **Synonyms** *Eriophorum alpinum*. **Habitat** Open bogs, conifer swamps, wet meadows, wet sandy shores; sometimes where calcium-rich.

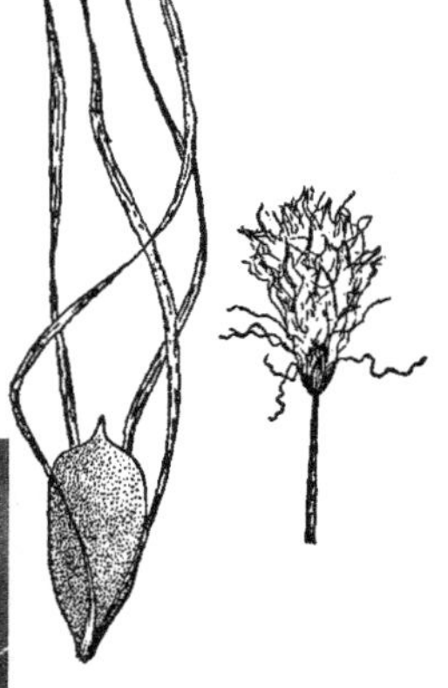

Scirpus hudsonianus
ALPINE COTTON-GRASS

Scirpus maritimus L.

Bolboschoenus maritimus (L.) Palla

ALKALI-BULRUSH

NC **OBL** | MW **OBL** | GP **OBL**

Native perennial, from tuber-bearing rhizomes. **Stems** single, sharply 3-angled, 5–15 dm long. several on stem, smooth, to 1 cm wide; bracts 3–5, the longest bract sometimes erect, to 30 cm long. **Spikelets** cylindric, 10–25 mm long and 6–9 mm wide, in clusters of 3–20 in a dense, stalkless head, or some spikelets single or in groups of 2–4 on stalks to 4 cm long; scales ovate, notched at tip, pale brown, 5–7 mm long, with an awn to 2 mm long; bristles 2–6, coppery, about half as long as achene; style 2-parted. **Achenes** lens-shaped, brown to black, 3–4 mm long, the beak small, to 0.3 mm long. June–Aug. **Synonyms** *Scirpus paludosus, Schoenoplectus maritimus*. **Habitat** Marshes, shores, ditches; especially where brackish.

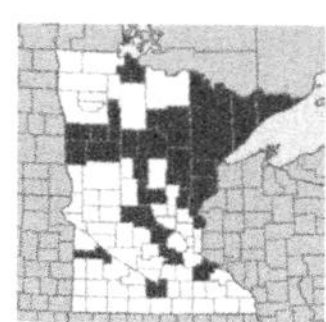

Scirpus microcarpus C. Presl
SMALL-FRUIT BULRUSH
NC **OBL** | MW **OBL** | GP **OBL**

Native perennial, from stout rhizomes. **Stems** single or few together, 5–15 dm long, weakly 3-angled. **Leaves** several along stem, flat, ascending, 7–15 mm wide, the upper leaves longer than the head, margins rough-to-touch; sheaths often red-tinged; bracts 3–4, leaflike, to 2–3 dm long. **Spikelets** numerous, 3–6 mm long and 1–2 mm wide; in a loose, spreading, umbel-like head, the head formed of clusters of 4–20 or more spikelets on stalks to 15 cm long; scales 1–2 mm long, brown and translucent except for green midvein; bristles 4–6, white to tan, downwardly barbed, longer than achene; style 2-parted. **Achenes** lens-shaped, pale tan to nearly white, about 1 mm long, the beak tiny. June–July. **Synonyms** *Scirpus rubrotinctus*. **Habitat** Streambanks, wet meadows, marshes, wet shores, thickets, swamps, springs; not in dense shade.

Scirpus maritimus
ALKALI-BULRUSH

Scirpus microcarpus
SMALL-FRUIT BULRUSH

Scirpus pendulus Muhl.
DROOPING BULRUSH
NC **OBL** | MW **OBL** | GP **OBL**

Loosely clumped native perennial, from short, thick rhizomes. **Stems** upright, rounded 3-angled, to 1.5 m long, lower stem covered by old leaf bases. **Leaves** several on stem, flat, 4–10 mm wide, shorter than head; bracts leaflike, 3 or more, shorter than the head, pale brown at base. **Spikelets** many, cylindric, 4–10 mm long and 2–4 mm wide; in an open, umbel-like head at end of stem, the spikelets drooping and clustered in groups of 1 stalkless and several stalked spikelets; scales about 2 mm long, red-brown with a green midvein; bristles 6, brown, smooth, longer than achene and about as long as scale; style 3-parted. **Achenes** compressed 3-angled, light brown, about 1 mm long, with a short, slender beak. June–Aug. **Synonyms** *Scirpus lineatus*. **Habitat** Marshes, wet meadows, streambanks, swamp openings and ditches.

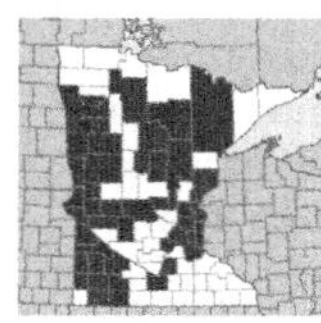

Scirpus pungens Vahl
Schoenoplectus pungens (Vahl) Palla
COMMON THREESQUARE
NC **OBL** | MW **OBL** | GP **OBL**

Native perennial, from slender rhizomes and forming colonies. **Stems** erect to somewhat curved, 2–12 dm long, 3-angled, the sides concave to slightly convex. **Leaves** mostly 1–3 near base of stem, usually folded, or channeled near tip, reaching to about middle of stem and 1–3 mm wide; main bract erect, sharp-tipped, resembling a continuation of the stem, 2–15 cm long. **Spikelets** 5–20 mm long and 3–5 mm wide, clustered in heads of 1–6 stalkless spikelets, the head appearing lateral; scales brown and translucent, 3–5 mm long, notched at tip, with a midvein extended into a short awn 1–2 mm long; bristles 4–6, unequal, shorter than achene; style 2–3-parted. **Achenes** light green or tan to dark brown, 3-angled or flat on 1 side and convex on other, 2–3 mm long, the beak to 0.5 mm long. May–Sept. **Synonyms** Includes *Scirpus americanus* and *Scirpus olneyi*. **Habitat** Shallow water, wet sandy, gravelly or mucky shores, streambanks, wet meadows, ditches, seeps and other wet places.

Scirpus pendulus
DROOPING BULRUSH

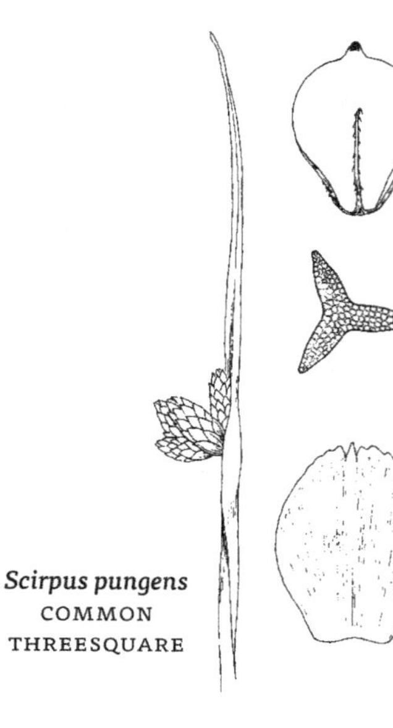

Scirpus pungens
COMMON THREESQUARE

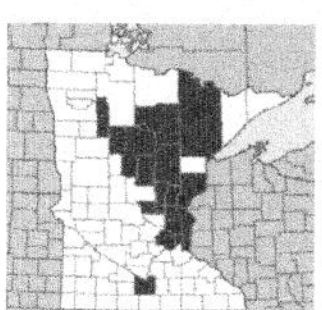

Scirpus smithii A. Gray

Schoenoplectus smithii (A. Gray) Soják

BLUNTSCALE-BULRUSH

NC **OBL** | MW **OBL** | GP **OBL**

Clumped native annual. **Stems** slender, smooth, round or rounded 3-angled, to 6 dm long. **Leaves** reduced to sheaths, or some with short blades; bract narrow, upright, 2–10 cm long, appearing to be a continuation of stem. **Spikelets** ovate, 5–10 mm long, in a single cluster of l–12 spikelets; scales yellow-brown with a green midvein; bristles 4–6, barbed or smooth, longer than achene, sometimes smaller or absent; style 2-parted. **Achenes** lens-shaped or flat on 1 side and convex on other, glossy brown to black, 1–2 mm long. July–Aug. **Synonyms** *Scirpus purshianus*. **Habitat** Sandy, gravelly or mucky shores, floating mats, bogs.

Scirpus smithii
BLUNTSCALE-BULRUSH

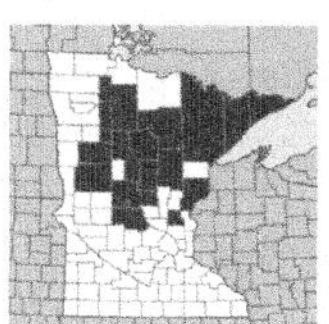

Scirpus subterminalis Torr.

Schoenoplectus subterminalis (Torr.) Soják

WATER-BULRUSH

NC **OBL** | MW **OBL** | GP **OBL**

Aquatic native perennial, spreading by rhizomes. **Stems** slender, weak, round in section, to 1 m or more long, floating or slightly emergent from water surface near tip. **Leaves** many, threadlike, channeled, from near base of stem and extending to just below water surface; bract 1–6 cm long, appearing to be a continuation of stem. **Spikelets** single at ends of stems, with several flowers, light brown, narrowly ovate, tapered at each end, 7–12 mm long; scales thin, 4–6 mm long, light brown with a green midvein; bristles shorter to about as long as achene, downwardly barbed; style 3-parted. **Achenes** 3-angled, brown, 2–4 mm long, tipped with a slender beak to 0.5 mm long. July–Aug. **Habitat** In water to about 1 m deep of lakes, ponds and bog margins.

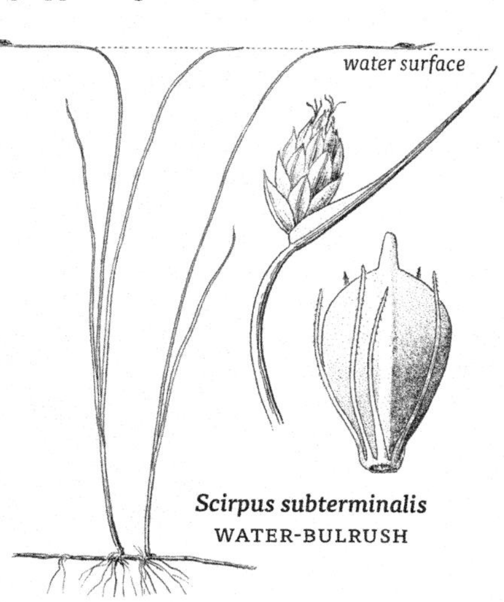

Scirpus subterminalis
WATER-BULRUSH

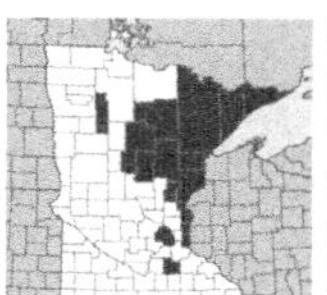

Scirpus torreyi Olney
Schoenoplectus torreyi (Olney) Palla
TORREY'S THREESQUARE
NC **OBL** | MW **OBL** | GP **OBL**

Native perennial, spreading by rhizomes and often forming colonies. **Stems** erect, sharply 3-angled, 5–10 dm long. **Leaves** several, narrow, often longer than the stem; bract erect, 5–15 cm long, appearing to be a continuation of stem. **Spikelets** ovate, light brown, 8–15 mm long, in a single head of 1–4 spikelets, the head appearing lateral from side of stem; scales ovate, shiny brown, with a greenish midvein sometimes extended as a short awn to 0.5 mm long; bristles about 6, downwardly barbed, longer than achene; style 3-parted. **Achenes** compressed 3-angled, shiny, light brown, 3–4 mm long, tipped by a slender beak to 0.5 mm long. June–Aug. **Habitat** Shallow water, wet sandy or mucky shores.

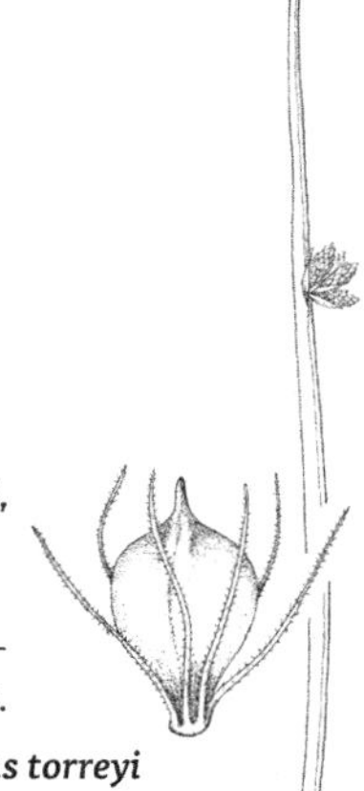

Scirpus torreyi
TORREY'S THREESQUARE

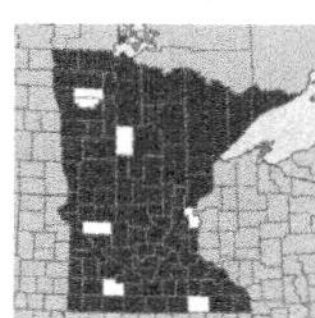

Scirpus validus Vahl
Schoenoplectus tabernaemontani (C.C. Gmel.) Palla
SOFTSTEM-BULRUSH
NC **OBL** | MW **OBL** | GP **OBL**

Native erennial, spreading by rhizomes and sometimes forming large colonies. **Stems** stout, smooth, erect, 1–3 m long, round in section. **Leaves** reduced to 4–5 sheaths at base of stem, or upper leaves with a blade to 7 cm long; main bract erect, 1–10 cm long, shorter than the head. **Spikelets** red-brown, 4–12 mm long and 3–4 mm wide, single or in clusters of 2–5 at ends of stalks, the stalks spreading or drooping, the clusters in paniclelike heads; scales ovate, light to dark brown, 2–3 mm long, the midvein usually extended into a short awn to 0.5 mm long; bristles 4–6, downwardly barbed, equal or longer than achene; style 2-parted. **Achenes** flat on 1 side and convex on other, brown to black, about 2 mm long, tapered to a very small beak to 0.2 mm long. June–Aug. **Habitat** Common, wide-ranging species of shallow water and shores of lakes, ponds, marshes, streams, ditches.

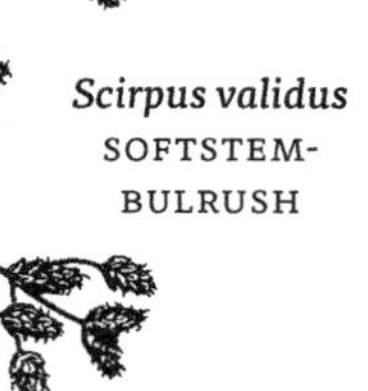

Scirpus validus
SOFTSTEM-BULRUSH

SCLERIA | *Nut-rush, Stone-rush*

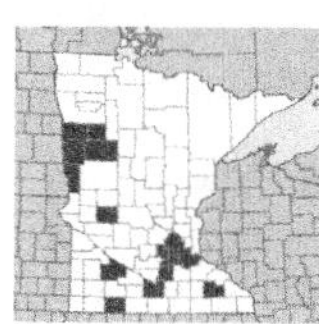

Scleria verticillata Muhl.
LOW NUT-RUSH
NC **OBL** | MW **OBL** | GP **OBL** | *threatened*

Clumped native annual, roots fibrous. **Stems** slender, smooth, 3-angled, 2–6 dm long. **Leaves** erect, linear, 1 mm wide, shorter than the stems; sheaths often hairy. **Flowers** either male or female, borne in separate spikelets on the same plant; male spikelets few-flowered; female spikelets with uppermost flower fertile, the lower scales empty; sepals and petals absent; **spikelets** in 2–8 separated heads, each head stalkless, 2–4 mm long, subtended by a small, bristlelike bract 4–6 mm long; scales lance-shaped. **Fruit** a hard, white achene, ± round,1 mm wide, covered with horizontal ridges, tipped with a short, sharp point. July–Sept. **Habitat** In Minnesota, restricted to calcareous fens in the prairie region; typically colonizing exposed marl and the margins of shallow pools. **Status** Threatened.

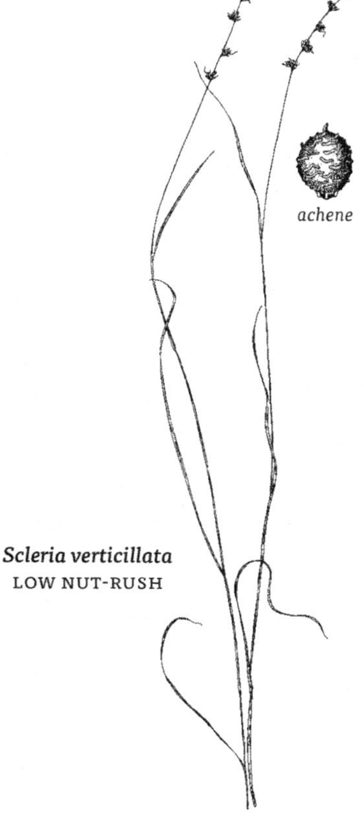

Scleria verticillata
LOW NUT-RUSH

Red Lake Peatland

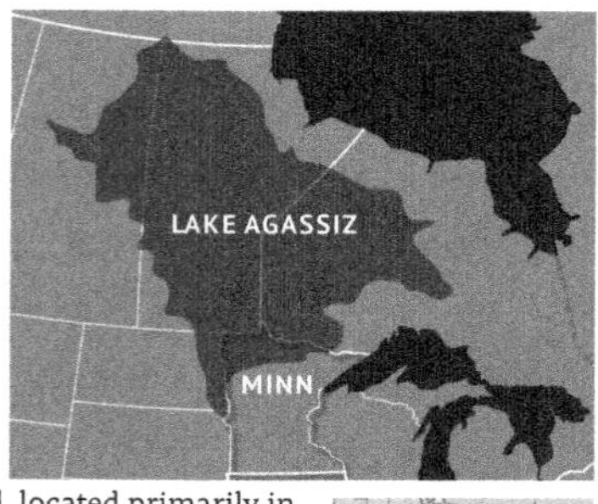

PEATLANDS occupy approx. 6 million acres (2.4 million hectares), or 10 percent of Minnesota's land area, more than any other state except Alaska. Red Lake Peatland is the largest intact ecosystem in Minnesota, and is considered the most diversely patterned peatland in the U.S.

Formation Lake Agassiz was an immense glacial lake, fed by glacial runoff at the end of the last glacial period, located primarily in what is now Manitoba, Ontario and northern Minnesota. Peatlands began to form in the Lake Agassiz basin following the onset of a cooler and wetter climate about 7,000 years ago. The peat build-up continues to the present day, but peat accumulates very slowly; in Minnesota, it has been measured at a rate of only 3–7 cm (1.5–3 inches) per 100 years.

Features Although the Peatland is relatively flat, there is enough slope to permit the slow movement of water across the landscape. The amount of water, its mineral content and pH, varies, and has produced a striking patterning of the vegetation. The resulting patterns can be classified into two general types: **Acid Peatlands** and **Rich Peatlands**.

Acid peatlands (or bogs) are convex landforms supporting a more or less continuous carpet of sphagnum moss. The surface rooting zone is above the influence of mineral-rich groundwater, and as a result, rainfall is the only source of nutrients for plants growing atop the sphagnum moss substrate; bogs are also very acidic. The most highly developed bogs form crests or domes that are sufficiently above the water table to allow better root growth and may be occupied by stunted trees of black spruce. The sloping, wetter sides of the bogs are dominated by Heath Family (Ericaceae) shrubs, and various sedges and herbs.

Rich peatlands are concave to level landforms fed by groundwater that has flowed through mineral soil and then discharged across the open or wooded peatlands. Wetter areas develop into treeless water tracks termed fens dominated by sedges and other grass-like plants. Slightly drier sites are dominated by swamps of tamarack and black spruce. Rich peatland landforms include water tracks, tear drop islands, and ribbed fens.

Rare flora and fauna Over two dozen rare plant species are known from Red Lake Peatland—including *Drosera linearis, Drosera anglica, Carex exilis, Juncus stygius, Rhynchospora fusca*, and *Xyris montana*. Rare animals include northern bog lemming, short-eared owl, yellow rail, and Wilson's phalarope.

ERIOCAULON | *Pipewort*

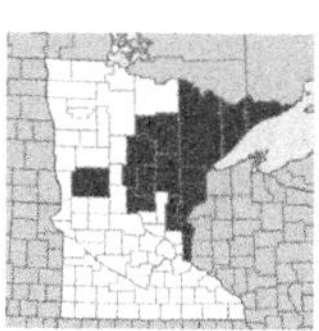

Eriocaulon aquaticum (Hill.) Druce
PIPEWORT
NC **OBL** | MW **OBL**

Native perennial, spongy at base, with fleshy roots. **Stems** usually single, leafless, slightly twisted, 5–7-ridged, 5–20 cm long (or reaching 2–3 m long when in deep water). **Leaves** grasslike, in a rosette at base of plant, thin and often translucent, 2–10 cm long and 2–5 mm wide, 3–9-nerved with conspicuous cross-veins. **Flowers** either male or female, grouped together in a single, ± round head at end of stem, the heads white-woolly, 4–6 mm wide. **Fruit** a 2–3-seeded capsule. July–Sept. **Synonyms** *Eriocaulon septangulare*. **Habitat** Shallow water, sandy or peaty shores.

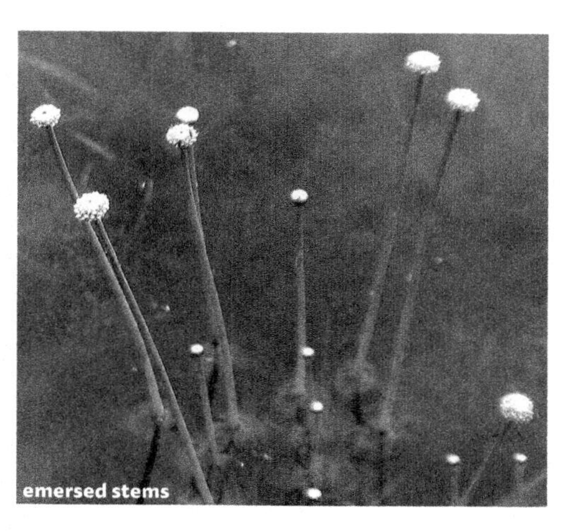
emersed stems

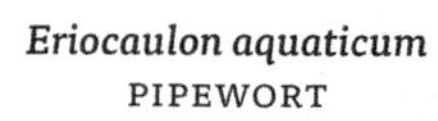
Eriocaulon aquaticum
PIPEWORT

AQUATIC, PERENNIAL HERBS. Stems leafy, the leaves usually whorled (*Elodea*), or plants stemless with clusters of long, linear, ribbonlike leaves (*Vallisneria*). **Flowers** usually either male or female and borne on separate plants, small and stalkless, or in a spathe at end of a stalk; sepals 3; petals 3 or absent; male flowers with 3 or more stamens; stigmas 3. **Fruit** several-seeded, maturing underwater.

HYDROCHARITACEAE | FROG'S-BIT FAMILY

1 Plants leafy-stemmed, the leaves short and whorled, to 3 cm long . . . . *Elodea* | WATER-WEED
1 Leaves all from base of plant, long and ribbonlike, to 1 m long . . . . *Vallisneria americana* | TAPE-GRASS; EEL-GRASS; WATER-CELERY

ELODEA | *Water-weed*

Aquatic perennial herbs, rooting from lower nodes or free-floating. **Stems** slender, leafy, branched. **Leaves** crowded near tip of stem, mostly in whorls of 3–4, or opposite, stalkless; margins finely sharp-toothed. **Flowers** either male or female and on separate plants, tiny, single from upper leaf axils, subtended by a 2-parted spathe, usually extended to the water surface by a long, threadlike hypanthium, or stalkless and breaking free to float to water surface in male flowers of *Elodea nuttallii*; sepals 3; petals 3 or absent, white or purple; male flowers with 9 stamens; female flowers with 3 stigmas, the stigmas entire or 2-parted. **Fruit** a capsule, ripening underwater. *Anacharis* or *Philotria* in older floras.

Two-leaf waterweed *(Elodea bifoliata) is reported from several locations in Minnesota, at the eastern edge of this species' main range of w USA; at least some of the leaves are in twos rather than mostly 3–4.*

ELODEA | WATER-WEED

1 Leaves mostly 2 mm or more wide; male flowers long-stalked in a spathe, the spathe more than 7 mm long, extended to water surface by a long, threadlike hypanthium . . . . . . *Elodea canadensis* | COMMON WATER-WEED
1 Leaves to 1.5 mm wide; male flowers stalkless in a spathe, the spathe 2–4 mm long, breaking free to float to water surface at flowering time . . . . *Elodea nuttallii* | FREE-FLOWERED WATER-WEED

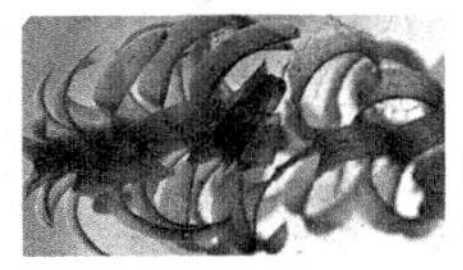

Elodea canadensis
COMMON WATER-WEED

Elodea canadensis Michx.
COMMON WATER-WEED
NC **OBL** | MW **OBL** | GP **OBL**

Submerged native perennial herb. **Stems** round in section, usually branched, 2–10 dm long. **Leaves** bright green, firm; lower leaves opposite, reduced in size, ovate or lance-shaped; upper leaves in whorls of 3, the uppermost crowded and overlapping, lance-shaped, 5–15 mm long and about 2 mm wide, rounded at tip. **Flowers** either male of female and on separate plants, at ends of threadlike stalks, 2–30 cm long; male flowers in spathes from upper leaf axils, the spathes about 10 mm long and to 4 mm wide; sepals green, 3–5 mm long; petals white, 5 mm long; stamens 9. Female flowers in spathes from upper leaf axils, the spathes 10–20 mm long, extended to water surface by a threadlike hypanthium; sepals 2–3 mm long; petals white, 2–3 mm long. **Fruit** a capsule, 5–6 mm long, tapered to a beak 4–5 mm long. June–Aug. **Synonyms** *Anacharis canadensis*. **Habitat** Common species of shallow to deep water of lakes (including Great Lakes), streams and ditches.

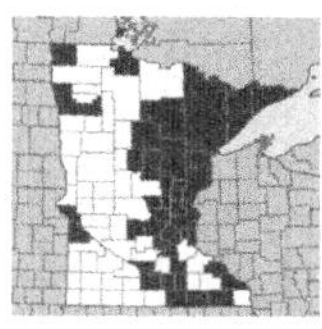

Elodea nuttallii (Planchon) St. John
FREE-FLOWERED WATER-WEED
NC **OBL** | MW **OBL** | GP **OBL**

Submerged native perennial herb. **Stems** slender, round in section, usually branched, 3–10 dm long. Lower leaves opposite, reduced in size, ovate to lance-shaped; upper leaves in whorls of 3 (or sometimes 4), not densely overlapping at tip, linear to lance-shaped, 6–13 mm long and 0.5–1.5 mm wide, tapered to a pointed tip. **Flowers** either male or female and on separate plants; **male flowers** in stalkless spathes from middle leaf axils, the spathes ovate, 2–3 mm long, the flowers single and stalkless in the spathe, breaking free and floating to water surface and then opening; sepals green or sometimes red, 2 mm long; petals absent or very short (to 0.5 mm long); stamens 9. **Female flowers** in cylindric spathes from upper leaf axils, the spathes 1–2.5 cm long, extended to

water surface by a threadlike stalk to 10 cm long; sepals green, about 1 mm long; petals white, longer than sepals. **Fruit** a capsule, 5–7 mm long. June–Aug. **Synonyms** *Anacharis nuttallii, Anacharis occidentalis, Philotria nuttallii.* **Habitat** Shallow to deep water of lakes, streams and ditches.

VALLISNERIA | *Tape-grass*

Vallisneria americana Michx.

TAPE-GRASS; EEL-GRASS; WATER-CELERY

NC **OBL** | MW **OBL** | GP **OBL**

Submerged native perennial herb, fibrous rooted, spreading by stolons and often forming large colonies. **Stems** absent. **Leaves** long and ribbonlike, in tufts from a small crown, to 1 m or more long and 3–10 mm wide, rounded at tip, margins smooth. **Flowers** either male or female and on separate plants; **male flowers** small, about 1 mm wide, in a many-flowered head, the head within a stalked spathe from base of plant, the stalk 3–15 cm long; sepals 3, petals 1, stamens 2; the male flowers released singly from the spathe and floating to water surface where they open. **Female flowers** single in a spathe, on long slender stalks that extend to water surface, the stalk contracting and coiling after flowering to draw the fruit underwater; sepals 3, petals small, 3; stigmas 3. **Fruit** a cylindric, curved capsule, 4–10 cm long. July–Sept. **Habitat** Shallow (sometimes deep) water of lakes and streams.

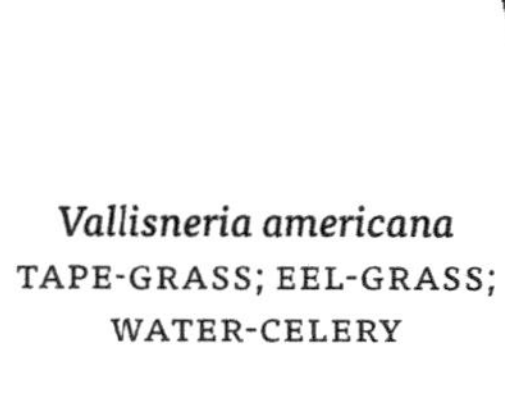

Vallisneria americana
TAPE-GRASS; EEL-GRASS; WATER-CELERY

PERENNIAL HERBS with rhizomes, bulbs, or fibrous roots. **Leaves** parallel-veined, narrow, 2-ranked, the margins joined to form an edge facing the stem (equitant). **Flowers** perfect, with 6 petal-like segments, single or in clusters at ends of stem, stamens 3, style 3-parted. **Fruit** a 3-chambered capsule.

IRIDACEAE | IRIS FAMILY

1 Flowers more than 2 cm wide; stems not winged; leaves more than 6 mm wide . . . . *Iris* | IRIS; FLAG
1 Flowers to 2 cm wide; stems winged; leaves to 6 mm wide . . . *Sisyrinchium* . . . . BLUE-EYED-GRASS

IRIS | *Iris, Flag*

Perennial herbs, spreading by thick rhizomes. **Stems** erect. **Leaves** swordlike, erect or upright, the margins joined to form an edge facing the stem. **Flowers** 1 or several at ends of **stems**; yellow or blue-violet; sepals 3, spreading or bent downward, longer and wider than the petals; petals 3, erect or arching; stamens 3; styles 3-parted, the divisions petal-like and arching over the stamens. **Fruit** an oblong capsule.

IRIS | IRIS, FLAG

1 Flowers yellow . . . . *Iris pseudacorus* | YELLOW FLAG
1 Flowers blue or violet 2

2 Stems as long or longer than leaves; base of plant often purple-tinged; sepal base unspotted, or with a with a hairless, green-yellow spot . . . . *Iris versicolor* | NORTHERN BLUE FLAG
2 Stems shorter than leaves; base of plant usually brown; sepal base with a hairy, bright yellow spot . . . . *Iris virginica* | SOUTHERN BLUE FLAG

Iris versicolor
NORTHERN BLUE FLAG

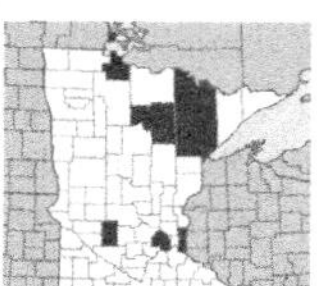

Iris pseudacorus L.
YELLOW FLAG
NC **OBL** | MW **OBL** | GP **OBL**

Introduced perennial herb, from thick rhizomes; may be invasive. **Stems** 0.5–1 m long, shorter or equal to the leaves. **Leaves** sword-shaped, stiff and erect, waxy, 1–2 cm wide. **Flowers** several at end of **stems**, yellow, 7–9 cm wide, sepals spreading, upper portion marked with brown; petals erect, narrowed in middle, 1–2.5 cm long. **Fruit** a 6-angled, oblong capsule, 5–9 cm long. May–June. **Habitat** Lakeshores, streambanks, marshes, ditches. Introduced from Europe, spreading throughout e USA and s Canada.

Iris pseudacorus (YELLOW FLAG)

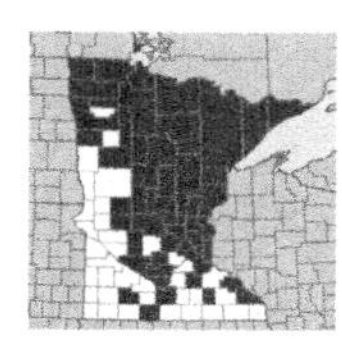

Iris versicolor L.
NORTHERN BLUE FLAG
NC **OBL** | MW **OBL** | GP **OBL**

Native perennial herb, from thick, fleshy rhizomes and forming colonies. **Stems** ± round in section, often branched above, 4–9 dm long. **Leaves** sword-shaped, erect or arching, somewhat waxy, 2–3 cm wide, usually shorter than stem. **Flowers** several on short stalks at ends of **stems**, blue-violet, 6–8 cm wide; sepals spreading, unspotted, or with a green-yellow spot near base, surrounded by white streaks and purple veins; petals erect, about half as long as sepals. **Fruit** an oblong capsule, 3–6 cm long. June–July. **Habitat** Marshes, shores, wet meadows, open bogs, swamps, thickets, forest depressions; often in shallow water.

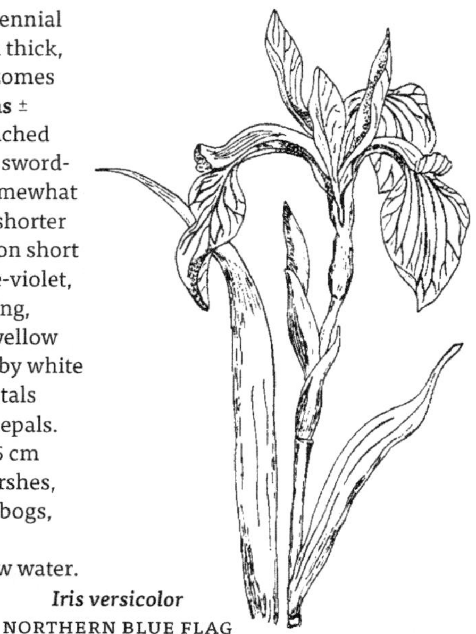

Iris versicolor
NORTHERN BLUE FLAG

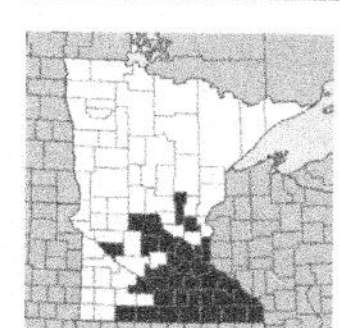

Iris virginica L.
SOUTHERN BLUE FLAG
NC **OBL** | MW **OBL** | GP **OBL**

Native perennial herb, from thick rhizomes, often forming large colonies. **Stems** ± round in section, to 1 m long. **Leaves** sword-shaped, erect or arching, 2–3 cm wide, usually longer than stems. **Flowers** several on short-stalks at ends of **stems**, blue-violet, often with darker veins, 6–8 cm wide; sepals spreading, curved backward at tip, with a hairy, bright yellow spot near base; petals shorter than sepals. **Fruit** an ovate to oval capsule, 4–7 cm long. May–July. **Synonyms** *Iris shrevei*. **Habitat** Swamps, thickets, shores, streambanks, marshes, ditches.

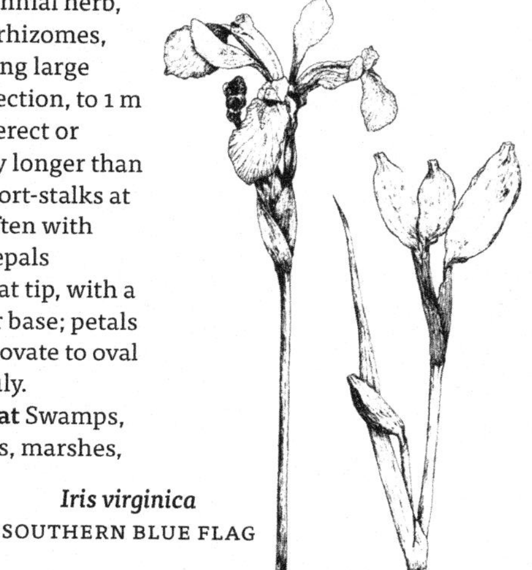

Iris virginica
SOUTHERN BLUE FLAG

SISYRINCHIUM | *Blue-eyed-grass*

Clumped perennial herbs, from fibrous roots. **Stems** slender, leafless, flattened or winged. **Leaves** narrow and linear, from base of plant, the margins joined and turned to form an edge facing the stem. **Flowers** in an umbel at end of stem, above a pair of erect green bracts (spathe), blue-violet (our species), with 6 spreading segments, the segments joined only at base, the tips rounded but with an small bristle. **Fruit** a rounded capsule; seeds round, black.

SISYRINCHIUM | BLUE-EYED-GRASS

1 Stems flattened and winged, 2–4 mm wide; larger leaves 2–3 mm wide . . . . *Sisyrinchium montanum* | MOUNTAIN BLUE-EYED-GRASS

1 Stem slender, barely winged, to 1 mm wide; larger leaves to 1.5 mm wide . . . . *Sisyrinchium mucronatum* | MICHAUX'S BLUE-EYED-GRASS

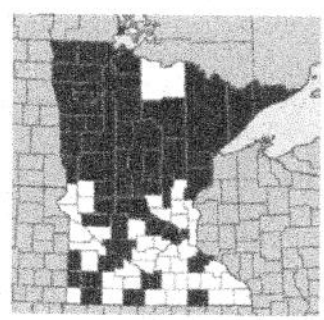

Sisyrinchium montanum Greene
MOUNTAIN BLUE-EYED-GRASS
NC **FAC** | MW **FAC** | GP **FAC**

Native clumped perennial, from fibrous roots; plants pale-green and waxy. **Stems** stiff and erect, leafless, flattened and winged, 1–5 dm tall and 2–4 mm wide. **Leaves** mostly from base of plant, narrow and grasslike, about half as long as stem, 1–3 mm wide. **Flowers** in head of 1 to several flowers at end of stem, subtended by a spathe, the spathe of 2 bracts, the outer bract 3–7 cm long, the inner bract about half as long; the flower segments (tepals) blue-violet with a yellow center, 5–15 mm long, with a short, slender tip. **Fruit** a ± round, pale brown capsule, 4–7 mm wide, on an erect stalk shorter than the inner bract. May–July. **Synonyms** *Sisyrinchium angustifolium*. **Habitat** Wet meadows, shores, thickets, ditches, swales; also in drier woods and fields.

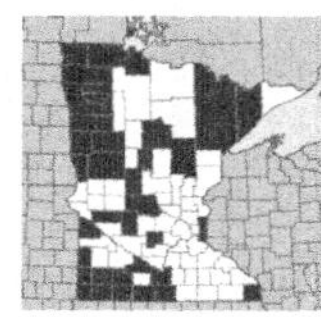

Sisyrinchium mucronatum Michx.
MICHAUX'S BLUE-EYED-GRASS
NC **FAC** | MW **FACW** | GP **FAC**

Native clumped perennial herb; plants dark green. **Stems** very slender, to 1 mm wide, leafless, margins not or barely winged. **Leaves** from near base of plant, narrow and linear, to 1.5 mm wide. **Flowers** in a single head at end of stem, subtended by a spathe, the spathe of 2 bracts, the bracts often purple-tinged, the outer bract 2–3 cm long, the inner bract shorter, 1–2 cm long; the segments (tepals) deep violet-blue, 8–10 mm long, tipped with a sharp point. **Fruit** a ± round, pale brown capsule, 2–4 mm long, on spreading stalks. May–June. **Habitat** Wet meadows, calcareous fens.

Sisyrinchium montanum
MOUNTAIN BLUE-EYED-GRASS

Sisyrinchium mucronatum
MICHAUX'S BLUE-EYED-GRASS

JUNCUS | *Rush*

Clumped or rhizomatous rushes, mostly perennial (annual in *Juncus bufonius*). **Stems** erect and unbranched. **Leaves** from base of plant or along stem, alternate, round in section, or flat to rolled inward, or reduced to sheaths at base of stem; leaves in some species with cross-partitions at intervals (septate). **Flowers** perfect, regular, in compact to open clusters of few to many flowers, subtended by 1 or several leaflike involucral bracts; sepals and petals of 6 chaffy, scalelike, green to brown tepals; stamens 6 or 3; stigmas 3, ovary superior, 1 or 3-chambered. **Fruit** a many-seeded capsule; seeds with a short slender tip or with a tail-like appendage at each end.

JUNCUS GROUPS (SUBGENERA) | RUSH

1 Flowers borne singly (each flower atop a stalk); bractlets present (except in *Juncus pelocarpus*) 2
1 Flowers borne in heads (flowers ± stalkless); bractlets absent 4

2 Inflorescences appearing lateral; inflorescence bract round in section, erect, appearing to be continuation of stem; basal leaves without blades, stem leaves absent . . . . **Group 1 (Subgenus Genuini)**
2 Inflorescences appearing terminal; inflorescence bract erect or upright, flat, involute or round in section; basal leaves (at least some) usually with blade, stem leaves present or absent 3

3 Leaves round in section, with cross-partitions at intervals (septate); capsules beaked . . . . **Group 2 (Subgenus Septati)**
3 Leaves flat, involute, or round in section, not septate; capsules not beaked . . . . **Group 3 (Subgenus Poiophylli)**

4 Leaves flat . . . . **Group 4 (Subgenus Graminifolii)**
4 Leaves round in section or compressed 5

5 Capsules large; seeds large, long tailed; leaves not noticeably septate . . . . **Group 5 (Subgenus Alpini)**
5 Capsules smaller; seeds not tailed or if tailed not long; leaves septate or not . . . . **Group 2 (Subgenus Septati)**

JUNCUS GROUP 1 (SUBGENUS GENUINI)

1 Stems densely clumped; stamens 3 . . . . *Juncus effusus* | SOFT RUSH p. 504
1 Stems single from rhizomes, the stems often in rows; stamens 6 2

2 Tepals dark brown; capsules red-brown; anthers equal to or longer than filaments............. *Juncus arcticus* | BALTIC RUSH; WIRE-RUSH p. 501
2 Tepals greenish; capsules green to light brown; anthers shorter than filaments .... *Juncus filiformis* | THREAD RUSH p. 505

JUNCUS GROUP 2 (SUBGENUS SEPTATI)

1 Seeds tailed, 0.7–2.6 mm long (including tails); seed body with a whitish translucent covering.. 2
1 Seeds not tailed, 0.3–0.7 mm long; seed body clear yellow-brown....... 3

2 Seeds 1.1–1.9 mm long; heads ovate, 5–50-flowered; inflorescence branches erect to ascending............. *Juncus canadensis* | CANADA RUSH p. 503
2 Seeds 0.7–1.2 mm; heads 2–8-flowered; inflorescence branches erect to spreading .... *Juncus brevicaudatus* | NARROW-PANICLE RUSH p. 502

3 Flowers 1–2(–4) at each node, not in heads; capsules fertile only below middle.. 4
3 Flowers in heads of 3–60; capsules fertile throughout or only below middle 5

4 Stems reclining and sprawling, to 10 cm long and forming mats; rare in ne Minnesota (Cook County) .... *Juncus subtilis* | SLENDER RUSH p. 508
4 Stems erect, to 70 cm tall, and not forming mats; n and e Minnesota .... *Juncus pelocarpus* | BROWN-FRUIT RUSH p. 507

5 Heads sphere-shaped or nearly so, 15–60-flowered 6
5 Heads obconic to hemispheric, 3–15-flowered 10

6 Stamens 3 7
6 Stamens 6.. 9

7 Plants clumped; tepals lance-shaped .... *Juncus acuminatus* .... TAPER-TIP RUSH p. 500
7 Plants rhizomatous; tepals narrowly lance-shaped 8

8 Capsules shorter than (and hidden by) tepals .... *Juncus brachycarpus* .... WHITE-ROOT RUSH p. 502
8 Capsule longer to slightly shorter than tepals .... *Juncus nodosus* .... KNOTTED RUSH p. 506

9 Outer tepals 2.4–4.1 mm long, equaling inner tepals; auricles 0.5–1.7 mm .... *Juncus nodosus* | KNOTTED RUSH p. 506
9 Outer tepals 4–6 mm, outer and inner tepals unequal in length; auricles 1–4 mm .... *Juncus torreyi* | TORREY'S RUSH p. 509

10 Stamens 3 11
10 Stamens 6 12

11 Heads 5–50; tepals 2.6–3.5 mm long, nearly equal; capsules 2.8–3.5 mm long .... *Juncus acuminatus* | TAPER-TIP RUSH p. 500
11 Heads 30–250; tepals 1.7–2.9 mm long, inner tepals shorter than outer tepals; capsules 1.9–2.9 mm ..... *Juncus nodosus* | KNOTTED RUSH p. 506

12 Stems sometimes creeping or floating; submersed leaves may be threadlike and formed before flowering; inflorescences with 1–9 heads .... *Juncus articulatus* | JOINTED RUSH p. 501
12 Stems erect; threadlike submersed leaves not formed except in *Juncus militaris*; heads 1–60 or more 13

13 Inner tepals blunt-tipped; inflorescence stiffly erect .... *Juncus alpinoarticulatus* | ALPINE RUSH p. 500
13 Inner tepals tapered to a sharp tip; inflorescence spreading .... *Juncus articulatus* | JOINTED RUSH p. 501

JUNCUS GROUP 3 (SUBGENUS POIOPHYLLI)

1 Plants annual, to 10 cm tall .... *Juncus bufonius* | TOAD-RUSH p. 503
1 Plants perennial, more than 10 cm tall 2

2 Auricles at summit of leaf sheath 3–6 mm long, membranous, transparent .... *Juncus tenuis* | PATH-RUSH p. 508
2 Auricles absent or very short membranous or hardened projections less than 2 mm long 3

3 Leaf blade flat, nw Minnesota... *Juncus compressus* | BLACK GRASS p. 504
3 Leaf blade round in cross-section or channeled and closed for ± entire length; more widespread in Minnesota .... *Juncus vaseyi* .... VASEY'S RUSH p. 509

JUNCUS GROUP 4 (SUBGENUS GRAMINIFOLII)

1 Stems single from long rhizomes; stamens 6, pale yellow; nw Minnesota only .... *Juncus longistylis* | LONG-STYLE RUSH p. 505
1 Stems single or in small clumps from short rhizomes; stamens 3, red-brown; se Minnesota ...... *Juncus marginatus* | GRASS-LEAF RUSH p. 505

JUNCUS GROUP 5 (SUBGENUS ALPINI)

One species in subgenus in Minnesota; leaf blades without cross-partitions at regular intervals .... *Juncus stygius* | MOOR RUSH p. 507

Juncus acuminatus Michx.
TAPER-TIP RUSH
NC **OBL** | MW **OBL** | GP **OBL**

Clumped perennial rush, native. **Stems** erect, slender, 2–8 dm tall, with 1–2 leaves. **Leaves** from stem and at base of plant, round to compressed in section, 5–40 cm long and 1–3 mm wide; auricles rounded, 1–2 mm long; bract erect, round, 1–4 cm long, shorter than the head. **Flowers** in an open, pyramid-shaped inflorescence, 5–12 cm long and less than half as wide, composed of 5–50 rounded heads 6–10 mm wide, each head with 5–30 flowers, the branches spreading, 1–10 cm long; tepals lance-shaped, green or straw-colored, 3–4 mm long; stamens 3, shorter than the tepals. **Capsules** oval, straw-colored to light brown, 3–4 mm long, about as long as the tepals, tipped with a short, blunt point. June–Aug. **Habitat** Wet sandy shores, streambanks and ditches; not in open bogs.

Juncus acuminatus
TAPER-TIP RUSH

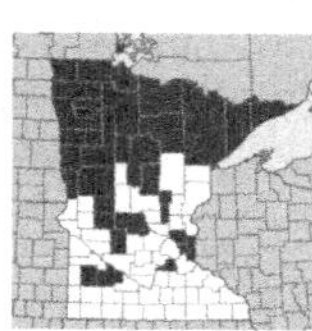

Juncus alpinoarticulatus Chaix.
ALPINE RUSH
NC **OBL** | MW **OBL** | GP **OBL**

Native perennial rush, spreading by rhizomes. **Stems** in small clumps, 1.5–4 dm long. **Leaves** mostly from base of plant and with 1–2 stem leaves, round in section, hollow, with small swollen joints, 2–12 cm long and 0.5–1 mm wide; sheaths green to red, auricles rounded, 0.5–1 mm long; bract round in section, 2–6 cm long and shorter than the head. **Flowers** in an open panicle of 5–25 heads, 2–15 cm long and 1–5 cm wide, the heads oblong pyramid-shaped, 2–6 mm wide, mostly 2–5-flowered, the branches upright, 1–7 cm long; tepals green to brown, 2–3 mm long, the inner tepals shorter, the margins chaffy; stamens 6. **Capsules** oblong, 3-angled, straw-colored to chestnut brown, satiny, 2–3 mm long, slightly longer than the tepals,

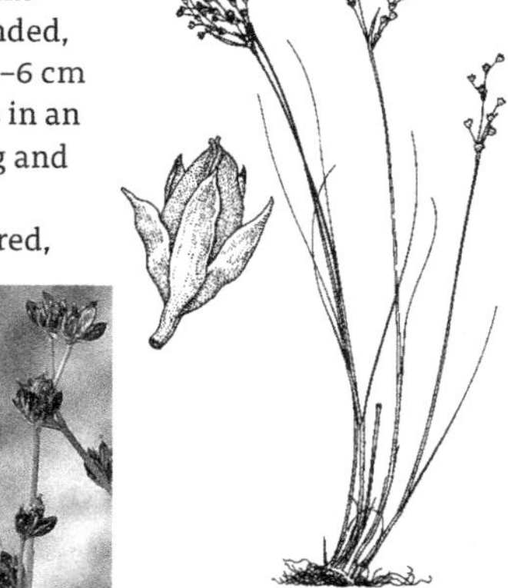

tapered to a rounded tip. June–Sept. **Synonyms** *Juncus alpinus*. **Habitat** Sandy or gravelly shores, streambanks, fens; often where calcium-rich.

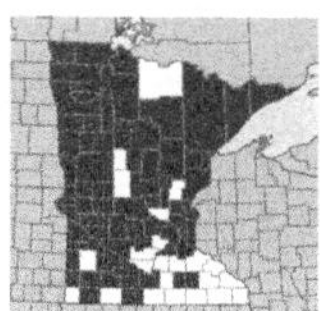

Juncus arcticus Willd.
BALTIC RUSH; WIRE-RUSH
NC **OBL** | MW **OBL** | GP **FACW**

Native perennial rush, spreading by stout, brown to black rhizomes. **Stems** slender and tough, dark green, 3–9 dm long, in rows from the rhizomes. **Leaves** reduced to red-brown sheaths at base of stem; bract erect, round in section, 1–2 dm long, longer than the head and resembling a continuation of stem. **Flowers** single on stalks, in dense to spreading heads, the heads appearing lateral, extending outward from stem 1–7 cm; tepals lance-shaped, dark brown, 3–5 mm long, margins chaffy; stamens 6. **Capsules** ovate, somewhat 3-angled, red-brown, 3–4 mm long, shorter to slightly longer than the tepals, tapered to a sharp point. May–Aug. **Synonyms** *Juncus balticus*. **Habitat** Common species of wet sandy or gravelly shores, interdunal wetlands near Great Lakes, meadows, ditches, marshes, seeps.

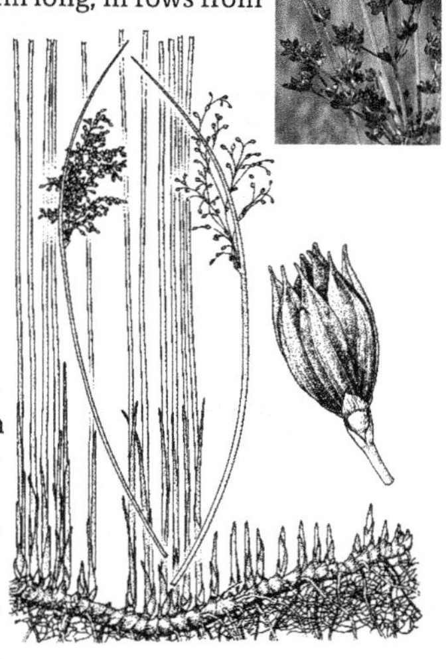

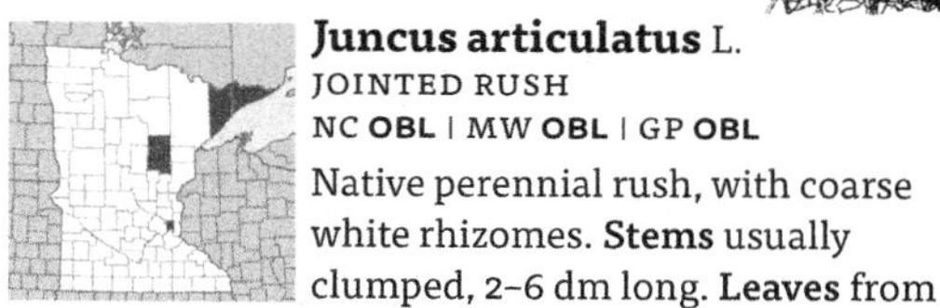

Juncus articulatus L.
JOINTED RUSH
NC **OBL** | MW **OBL** | GP **OBL**

Native perennial rush, with coarse white rhizomes. **Stems** usually clumped, 2–6 dm long. **Leaves** from stem and at base of plant, ± round in section, hollow, with small swollen joints, 4–12 cm long and 1–3 mm wide; sheaths green or sometimes red, auricles rounded, about 1 mm long; bract erect, round in section, 1–4 cm long, shorter than the head. **Flowers** in open panicles, 4–10 cm long and 3–6 cm wide, composed of 3–30 heads, the heads rounded, 6–8 mm wide, 3–10-flowered, panicle branches erect to widely spreading, 1–4 cm long; tepals green to dark brown, 2–3 mm long; stamens 6. **Capsules** oval, dark brown, shiny, 3–4 mm long, longer than the tepals, tapered to a tip. July–Sept. **Habitat** Sandy, gravelly or mucky shores, streambanks and springs.

Juncus brachycarpus Engelm.

WHITE-ROOT RUSH

NC **FACW** | MW **FACW** | GP **FACW**

Native perennial rush, spreading by stout white rhizomes. **Stems** erect, round in section, 3–8 dm long. **Leaves** from stem and base of plant, round in section, 3–50 cm long and 1–2 mm wide, cross-divided; auricles rounded, 0.5–2 mm long; bract erect, channeled, 1–3 cm long, shorter than the head. **Flowers** in an open or crowded raceme or panicle of 3–10 or more heads, 2–8 cm long and 1–3 cm wide, the heads round, 8–10 mm wide, with 30 or more flowers, the branches upright, 1–4 cm long; tepals narrowly lance-shaped, straw-colored, 2–4 mm long; stamens 3, shorter than the tepals. **Capsules** ovate, brown, 2–3 mm long, shorter than the tepals, abruptly tapered to a small point. June–Aug. **Habitat** Wetland margins, sandy swales and prairies. In Minnesota, known only from Blue Earth County; species more common in s USA.

Juncus brevicaudatus (Engelm.) Fernald

NARROW-PANICLE RUSH

NC **OBL** | MW **OBL** | GP **OBL**

Densely clumped native perennial rush. **Stems** erect, round in section, 1.5–5 dm long. **Leaves** from stem and base of plant, round in section, hollow, with small swollen joints, 3–20 cm long and 1–2 mm wide; sheaths green or sometimes red, auricles rounded, 1–2 mm long; bract erect, round in section, 2–7 cm long, shorter to longer than the head. **Flowers** in a raceme or panicle of 3–35 heads, 3–12 cm long and 1–4 cm wide, the heads oval, 2–6 mm wide, 2–7-flowered, branches upright, 0.5–3.5 cm long; tepals green to light brown, often red-tinged near tip, 3-nerved, 3–4 mm long, margins chaffy; stamens 3. **Capsules** oval, 3-angled, dark brown, 3–5 mm long, longer than the tepals, tapered to a sharp point. Aug–Sept. **Habitat** Wet meadows, marshes, fens, sandy lakeshores, streambanks, rocks along Lake Superior.

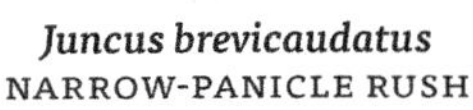

Juncus brevicaudatus
NARROW-PANICLE RUSH

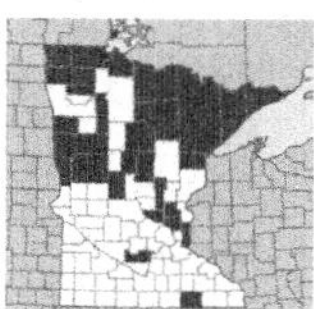

Juncus bufonius L.

TOAD-RUSH

NC **FACW** | MW **FACW** | GP **OBL**

Small annual rush, native. **Stems** clumped, erect to spreading, 5–20 cm long. **Leaves** from stem and at base of plant, flat or channeled, 1–7 cm long and to 1 mm wide, usually shorter than stem; sheaths green to red or brown, auricles absent; bract erect, 1–10 cm long, shorter than the head. **Flowers** single, mostly stalkless, with 1–7 flowers along each branch of the inflorescence, the inflorescence comprising half or more of the entire length of plant; tepals lance-shaped, green to straw-colored, 4–6 mm long, margins chaffy; stamens 6. **Capsules** ovate, brown or green, 3–4 mm long, rounded at tip, shorter than the tepals. June–Aug. **Habitat** Sandy or silty shores, mud flats, streambanks, wet compacted soil of trails and wheel ruts. Widespread species across most of North America; Eurasia.

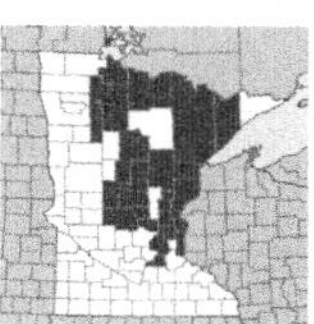

Juncus canadensis J. Gay

CANADA RUSH

NC **OBL** | MW **OBL** | GP **OBL**

Clumped perennial rush, native. **Stems** erect, rigid, round in section, 3–9 dm long. **Leaves** from stem and at base of plant, round in section, hollow, with small swollen joints, 3–20 cm long and 1–3 mm wide; sheaths green to red, auricles rounded, 1–2 mm long; bract erect, round in section, 3–7 cm long and shorter than the head. **Flowers** in an open or crowded raceme or panicle of few to many heads, 2–20 cm long and 1–10 cm wide, the heads ± round, 3–8 mm wide, with 5–40 or more flowers, the branches upright, 1–10 cm long; tepals narrowly lance-shaped, green to brown, 3–5 mm long; stamens 3. **Capsules** ovate, 3-angled, light to dark brown, 3–5 mm long, equal or longer than the tepals, rounded to a short tip. July–Sept. **Habitat** Sandy, muddy or mucky shores, marshes, streambanks, thickets, ditches.

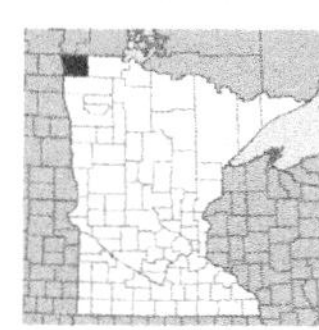

Juncus compressus Jacq.

BLACK GRASS

NC **FACW** | MW **OBL** | GP **FACW**

Clumped perennial rush, introduced. **Stems** erect, flattened, 2–7 dm long. **Leaves** from plant base and 1 or 2 along stem, flat or channeled, 5–20 cm long and to 1.5 mm wide; auricles rounded, to 1 mm long; bract erect, somewhat bent, flat or folded, 2–8 cm long, often longer than head. **Flowers** on short stalks 1–5 mm long, with 1–2 flowers along each branch of the inflorescence, the inflorescence 3–7 cm long and 1–3 cm wide, branches upright; tepals ovate, light to dark brown, 1–3 mm long, margins translucent; stamens 6. **Capsules** nearly round, light brown, 2–3 mm long, longer than tepals. June–Aug. **Habitat** Wet meadows, disturbed wet areas, ditches along highways where forming dark green colonies; often where salty. Introduced from Eurasia, naturalized in North America.

Juncus compressus
BLACK GRASS

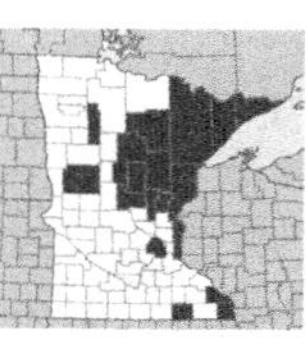

Juncus effusus L.

SOFT RUSH

NC **OBL** | MW **OBL** | GP **OBL**

Densely clumped native perennial. **Stems** erect, round in section, to about 1 m long. **Leaves** reduced to bladeless sheaths at base of stem, the sheaths to 2 dm long, mostly red-brown; bract round in section, 10–30 cm long, appearing like a continuation of stem, longer than the head. **Flowers** in a many-flowered inflorescence, with 2–4 flowers along each branch of the inflorescence, the inflorescence appearing lateral, the branches upright to spreading or bent downward, 2–6 cm long; tepals lance-shaped, green to straw-colored, 2–3 mm long; stamens 3. **Capsules** broadly ovate, olive-green to brown, 2–3 mm long, about as long as tepals, sometimes tipped with a short point. June–July. **Habitat** Marshes, shores, streambanks, bog margins, wet meadows. Common throughout much of e USA and se Canada.

Juncus effusus
SOFT RUSH

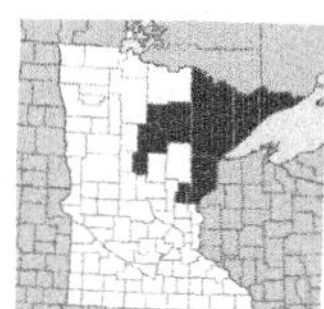

Juncus filiformis L.
THREAD RUSH
NC **FACW** | MW **FACW** | GP **OBL**

Native perennial rush, with short or long rhizomes. **Stems** clumped or in rows from the rhizomes, erect, round in section, 1–5 dm long. **Leaves** reduced to bladeless sheaths at base of stem, the sheaths pale brown, to 6 cm long; bract erect, round in section, 6–20 cm long, appearing to be a continuation of stem, longer than the head. **Flowers** in an branched inflorescence, 1–3 cm long, with 1–3 flowers along each branch of the inflorescence, the inflorescence appearing lateral, the branches erect to spreading, to 1 cm long; tepals lance-shaped, green to straw-colored, 2–3 mm long, margins chaffy; stamens 6. **Capsules** broadly ovate, light brown, 2–3 mm long, slightly longer than the tepals, tipped by a short beak. **Habitat** Sandy, mucky, or gravelly shores, streambanks, thickets.

Juncus filiformis
THREAD RUSH

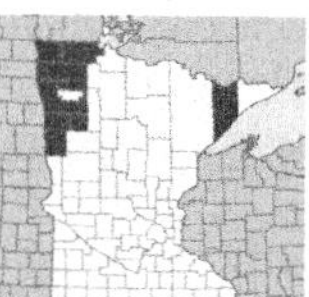

Juncus longistylis Torr.
LONG-STYLE RUSH
NC **FACW** | MW **FACW** | GP **FACW**

Native perennial rush, spreading by rhizomes. **Stems** 3–7 dm long. **Leaves** mostly at base of plant, flat and grasslike, 4–15 cm long and 1–3 mm wide, smaller upward; sheaths green, with broad membranous margins, auricles rounded, 1–2 mm long. **Flowers** in an inflorescence of 2–6 stalked, rounded heads, 5–7 cm long and 2–4 cm wide, the heads rounded, satiny chestnut brown, 1–2 cm wide, with 3–10 flowers, the branches erect, 1–4 cm long; tepals lance-shaped, green to brown, 4–6 mm long, margins translucent; stamens 6. **Capsules** oblong, brown, 2–3 mm long, slightly shorter than the tepals, rounded at tip, with a short beak. June–Aug. **Habitat** Shores, wet meadows, springs.

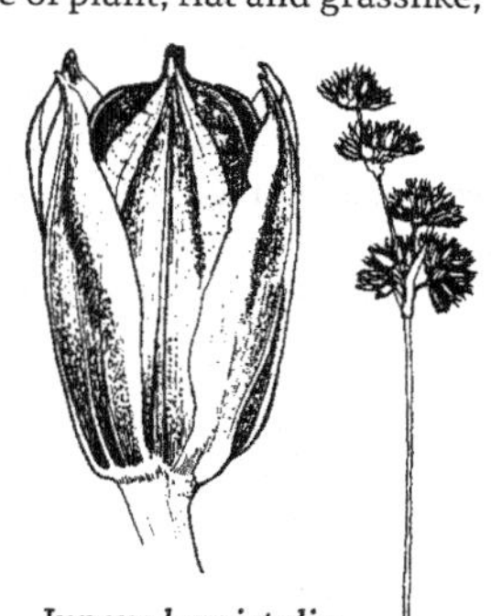

Juncus longistylis
LONG-STYLE RUSH

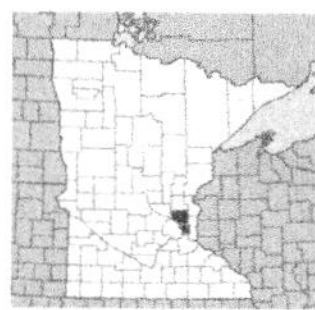

Juncus marginatus Rostk.
GRASS-LEAF RUSH
NC **FACW** | MW **FACW** | GP **FACW** | *special concern*

Native perennial rush, spreading by rhizomes. **Stems** single or in small clumps, erect, compressed, 2–5 dm long, bulblike at base. **Leaves** from base of plant and on stem, flat, grasslike, 2–30 cm long and 1–3 mm wide; sheaths green, membranous on margins, auricles rounded, to 0.5 mm long; bract erect to spreading, flat, 1–8 cm long, shorter to slightly longer than the head. **Flowers** in an open panicle, 2–8 cm long and 1–6 cm wide, composed of 5–15 heads, the heads rounded, 3–6 mm wide, 6–20-flowered, branches upright, 0.5–2.5 cm long; tepals lance-shaped, green with red spots, 2–3 mm long, margins chaffy; stamens 3. **Capsules** ± round, brown with red spots, 2–3 mm long, slightly longer than the tepals, rounded at tip. June–Aug. **Synonyms** *Juncus marginatus* var. *marginatus*; includes *Juncus biflorus*. **Habitat** Sandy shores and streambanks, wet meadows, marshes, low prairie, springs. **Status** Special concern.

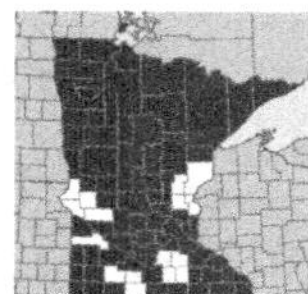

Juncus nodosus L.
KNOTTED RUSH
NC **OBL** | MW **OBL** | GP **OBL**

Native perennial rush, spreading by rhizomes. **Stems** erect, slender, round in section, 1.5–6 dm long. **Leaves** on stem and one at base of plant, round in section, hollow, with small swollen joints, 3–30 cm long and 1–2 mm wide, upper leaves usually longer than the head; sheaths green, their margins green, becoming yellow and membranous toward tip, auricles rounded, yellow, 0.5–1 mm long; bract erect to spreading, round in section, 2–12 cm long, usually much longer than head. **Flowers** in a raceme or panicle of several heads, 1–6 cm long and 1–3 cm wide, the heads ± round, 6–10 mm wide, 6–20-flowered, the branches erect to spreading, 0.5–3 cm long; tepals narrowly lance-shaped, green to light brown, 3–4 mm long; the margins narrowly translucent; stamens 6. **Capsules** awl-shaped, brown, 4–5 mm long, longer than the tepals, tapered to a sometimes curved beak. July–Sept. **Habitat** Sandy, gravelly or clayey shores and streambanks, wet meadows, fens, ditches, springs; often where calcium-rich.

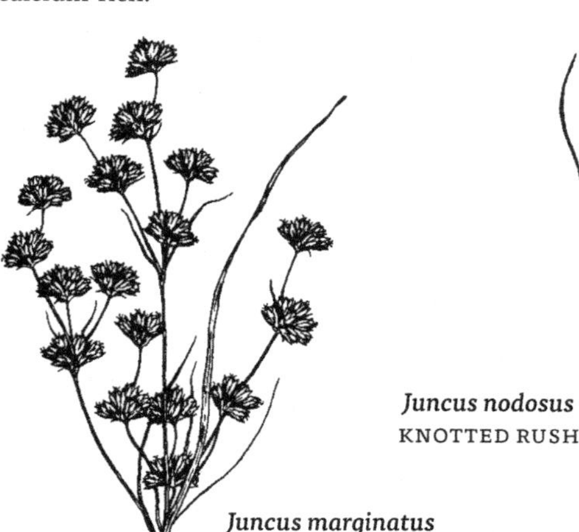

Juncus marginatus
GRASS-LEAF RUSH

Juncus nodosus
KNOTTED RUSH

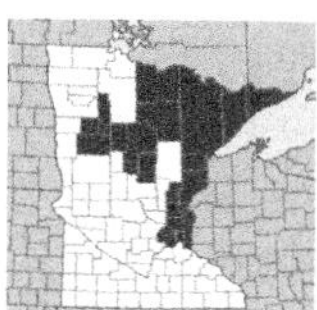

Juncus pelocarpus E. Meyer
BROWN-FRUIT RUSH
NC **OBL** | MW **OBL** | GP **OBL**

Native perennial rush, spreading by rhizomes and forming colonies. **Stems** erect, round in section, 1–4 dm long. **Leaves** from stem and at base of plant, round in section, very slender, 2–10 cm long and about 1 mm wide; auricles absent or short and straw-colored; bract erect, round in section, 2–4 cm long, shorter than the head. **Flowers** single or paired in a much-branched inflorescence, 5–15 cm long and 4–10 cm wide, the flowers on mostly 1 side of each branch, the branches upright to widely spreading, 1–4 cm long, with at least some of the flowers usually replaced by clusters of awl-shaped leaves; tepals ovate, dark brown, about 2 mm long, margins chaffy; stamens 6. **Capsules** narrowly ovate, dark brown, satiny, 2–3 mm long, equal or slightly longer than tepals, tapered to a slender beak. July–Aug. **Habitat** Shallow water, sandy or mucky shores, bog margins.

Juncus stygius L.
MOOR RUSH
NC **OBL** | GP **OBL** | *special concern*

Perennial rush, from slender rhizomes. **Stems** single or few together, erect, round in section, 1–4 dm long. **Leaves** 1–3 from near base of plant, with 1 leaf above middle of stem, round in section or somewhat flattened, 3–15 cm long and 0.5–2 mm wide; auricles short and rounded or absent; bract erect, round in section, 1–2 cm long, shorter than the head. **Flowers** in an inflorescence of 1–3 heads, the heads obovate, 5–10 mm wide, 1–4-flowered, branches erect, to 1 cm long; tepals lance-shaped, straw-colored to red-brown, 4–5 mm long, margins chaffy; stamens 6, nearly as long as the tepals. **Capsules** oval, 3-angled, green-brown, 6–8 mm long, longer than the tepals, tipped with a distinct point. **Habitat** Open bogs, marshes, and shallow water. **Status** Special concern.

Juncus subtilis E. Meyer

SLENDER RUSH

NC **OBL**

Native perennial rush, spreading by thin rhizomes and forming mats to 50 cm wide. **Stems** to 10 cm long, sprawling, floating or submersed, usually many- branched, to 1 mm in diameter. **Leaves** 1–5 from base of plant, 1–4 along stem, the blade round in section, to 3 cm long and less than 0.5 mm wide; auricles small, less tham 1 mm long, membranaceous. **Flowers** in group of 1–3, the flower branches spreading to erect; tepals reddish, oblong; outer tepals to 3 mm long; stamens 6. **Capsules** longer than tepals, chestnut brown, 2.5–5 mm long; seeds ovoid, to 0.5 mm long, not tailed. June–July. **Synonyms** *Juncus pelocarpus* var. *subtilis*. **Habitat** In Minnesota, the single known population in Cook County is along a sandy lakeshore in shallow water.

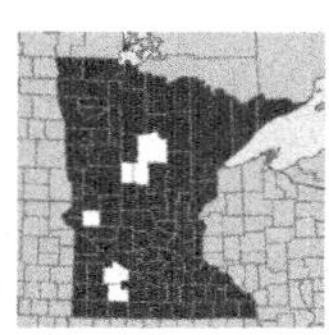

Juncus tenuis Willd.

PATH-RUSH

NC **FAC** | MW **FAC** | GP **FAC**

Clumped, perennial rush, native. **Stems** erect, round in section to slightly flattened, 1–6 dm long. **Leaves** near base of stem, flat to broadly channeled, 10–15 cm long and to 1 mm wide; sheaths green, the margins yellow and glossy, auricles triangular, 1–3 mm long; bracts 1–3 (usually 2), the lowest erect, flat, 6–10 cm long, longer than the head. **Flowers** stalkless or on short stalks to 3 mm long, on branches with 1–7 flowers, in a crowded to spreading head 2–5 cm long; tepals lance-shaped, green to straw-colored or light brown, 3–5 mm long, margins narrowly translucent; stamens 6. **Capsules** ovate, green to straw-colored, 2–5 mm long, shorter or equaling the tepals, rounded at tip. June–July. **Habitat** Common species of wet meadows, shores, streambanks, springs, common in disturbed places (often where soils compacted) such as trails, roadsides and ditches; also in drier woods and meadows.

Juncus tenuis
PATH-RUSH

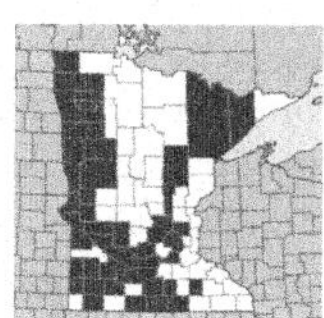

Juncus torreyi Coville
TORREY'S RUSH
NC **FACW** | MW **FACW** | GP **FACW**

Native perennial rush, from tuber-bearing rhizomes. **Stems** single, erect, round in section, 4–8 dm long. **Leaves** from stem and base of plant, round in section, hollow with small swollen joints, 15–30 cm long and 1–2 mm wide, the upper leaves often longer than the head; sheaths green, the margins white and translucent, auricles 1–3 mm long; bract erect or spreading, round in section, 4–12 cm long, longer than the head. **Flowers** in a crowded, rounded raceme or panicle of 3–23 heads, 2–5 cm long and as wide, the heads round, 10–15 mm wide, with 25 to many flowers, branches erect to spreading 1–4 cm long; tepals narrowly lance-shaped, green to brown, 3–5 mm long, margins narrowly translucent; stamens 6. **Capsules** awl-shaped, brown, 4–6 mm long, equal or longer than the tepals, tapered to a short beak. June–Sept. **Habitat** Sandy shores, streambanks, wet meadows, marsh borders, springs, ditches.

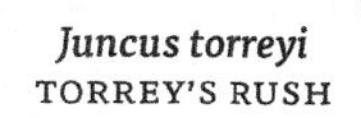

Juncus torreyi
TORREY'S RUSH

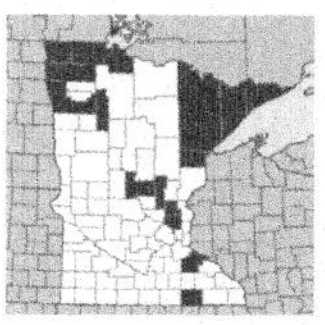

Juncus vaseyi Engelm.
VASEY'S RUSH
NC **FACW** | MW **FACW** | GP **FACW**

Clumped perennial rush, native. **Stems** erect, 2–6 dm long. **Leaves** all at base of plant, round in section, solid, narrowly channeled on upper surface, to 3 dm long and to 1 mm wide, usually shorter than stem; sheaths green or red, the margins membranous, auricles short or absent; bract upright, usually shorter than the head. **Flowers** single, stalkless or on short stalks, in a crowded inflorescence, 1–4 cm long; tepals lance-shaped, green to light brown, 4–6 mm long, margins narrowly translucent; stamens 6. **Capsules** cylindric, 1–2 mm long, equal or slightly longer than the tepals, blunt-tipped. July–Aug. **Habitat** Wet meadows, sandy shores.

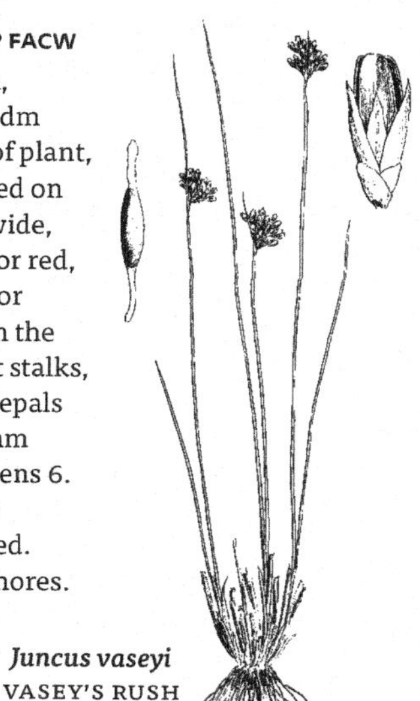

Juncus vaseyi
VASEY'S RUSH

TRIGLOCHIN | *Arrow-grass*

Grasslike perennial herbs, clumped from creeping rhizomes, often in brackish habitats. **Stems** slender, leafless. **Leaves** all from base of plant, slender, linear, round or somewhat flattened in section, sheathing at base. **Flowers** perfect, regular, on short stalks in a spikelike raceme at end of stem; flower segments (tepals) 6; stigmas 3 or 6, styles short or absent; stamens 6, anthers stalkless, nearly as large as tepals. **Fruit** of 3 or 6 carpels, these splitting when mature into 1-seeded segments.

TRIGLOCHIN | ARROW-GRASS

1 Plants generally small and slender; stigmas 3; fruits linear, clublike toward tip . . . . *Triglochin palustris* | MARSH ARROW-GRASS

1 Plants larger, usually 3 dm or more tall; stigmas 6; fruits short-cylindric. . . . . *Triglochin maritima* | COMMON OR SEASIDE ARROW-GRASS

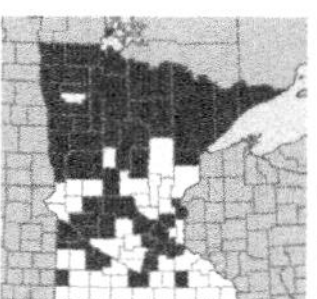

Triglochin maritima L.

COMMON OR SEASIDE ARROW-GRASS

NC **OBL** | MW **OBL** | GP **OBL**

Clumped perennial herb, from a thick crown and spreading by rhizomes, native. **Stems** 2–8 dm long. **Leaves** upright to spreading, somewhat flattened, to 5 dm long and 1–3 mm wide. **Flowers** 2–3 mm wide, in densely flowered, spikelike racemes 1–4 dm long; the flowers on upright stalks 4–6 mm long, the stalks extending downward on the stem as a wing; tepals 6, 1–2 mm long; stigmas 6; stamens 6. **Fruit** of 6 ovate carpels, 2–5 mm long and 1–3 mm wide, the carpel tips curved outward. June–Aug. **Habitat** Sandy, gravelly, or marly lakeshores and streambanks; marshes, brackish wetlands.

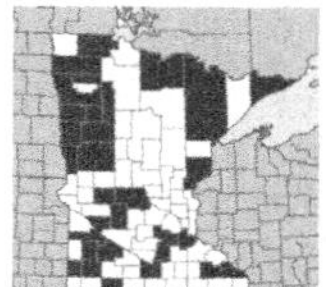

Triglochin palustris L.

MARSH ARROW-GRASS

NC **OBL** | MW **OBL** | GP **OBL**

Small, clumped perennial herb, native. **Stems** slender, 2–4 dm long. **Leaves** erect, round in section, to 3 dm long and 1–2 mm wide. **Flowers** small, 1–2 mm wide, in loosely flowered racemes, 10–25 cm long; the flowers on erect stalks, 2–5 mm long; tepals 6, 1–2 mm long; stigmas 3; stamens 6. **Fruit** of 3 narrow, clublike carpels, 5–8 mm long and 1 mm wide, splitting upward from base into 3 segments. June–Sept. **Habitat** Sandy, gravelly, or marly lakeshores and streambanks, calcareous fens, marshes, interdunal swales; often where calcium-rich.

Triglochin palustris
MARSH ARROW-GRASS

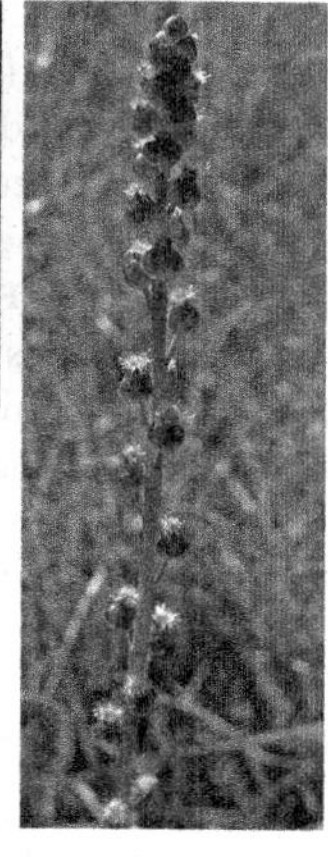

Muskgrass or Stonewort (*Chara* spp.) is a macro algae easily confused with vascular aquatic plants. Plants of *Chara* are gray-green, and have a gritty texture and strong musky odor when rubbed. When dried, *Chara* will turn whitish due to calcium deposits. Although not having true roots, *Chara* will loosely attach itself to the bottom of lakes and ponds via rhizoids (thread-like structures), and will sometimes form extensive colonies in shallow water, especially where calcium-rich. Another stonewort, ***Nitella***, is similar but plants have no skunky odor, and stems and branches are typically bright green and smooth to the touch.

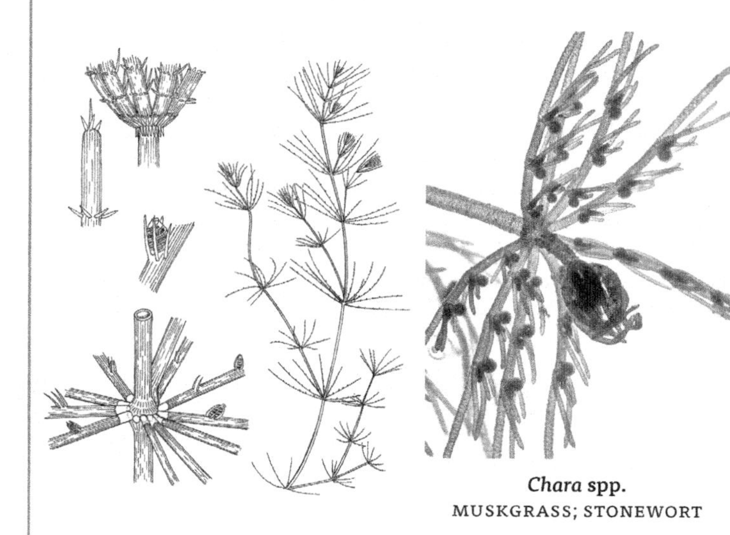

Chara spp.
MUSKGRASS; STONEWORT

SMALL PERENNIAL HERBS, floating at or near water surface, single or forming colonies. Plants thalluslike (not differentiated into stems and leaves), the thallus (or frond) flat or thickened; the roots, if present, unbranched, 1 or several from near center of leaf underside; reproducing vegetatively by buds from 1–2 pouches on the sides, the parent and budded plants often joined in small groups. **Flowers** rarely produced, either male or female, in tiny reproductive pouches on margins (*Lemna, Spirodela*) or upper surface (*Wolffia*) of the leaves, subtended by a small spathe within the pouch; sepals and petals absent; male flowers 1–2, consisting of 1 anther on a short filament; female flower 1 (a single ovary), in same pouch as male flowers. **Fruit** a utricle with 1 to several seeds.

LEMNACEAE | DUCKWEED FAMILY

1 Roots absent; leaves thickened, less than 1.5 mm long .... *Wolffia* .... WATER-MEAL
1 One to several roots present on leaf underside; leaves flat, mostly more than 1.5 mm long 2

2 Each leaf with 1 root; leaf underside green or purple-tinged .... *Lemna* .... DUCKWEED
2 Each leaf with 3 or more roots; leaf underside solid purple .... *Spirodela polyrhiza* | GREATER DUCKWEED

LEMNA | *Duckweed*

Small perennial floating herbs, with 1 root per frond (or roots sometimes absent on oldest and youngest leaves). **Fronds** single or 2 to several and joined in small colonies, floating on water surface or underwater (star-duckweed, *Lemna trisulca*), varying from round, ovate, to obovate or oblong, tapered to a long point (stipe) in star-duckweed; green or often red-tinged; upper surface flat to slightly convex or humped, underside flat or convex. Reproductive pouches 2, on margins of frond. **Flowers** uncommon, consisting of 2 stamens (male flowers) and a single pistil (female flower) in each pouch. **Fruit** an utricle with 1 to several seeds. Reproduction mostly by budding of new leaves from the reproductive pouches.

Lemna obscura *(Little duckweed) is reported from a single Minnesota location in Washington County. This species is more common in s USA.*

LEMNA | DUCKWEED

1 Frond tapered to a petiole-like base; lateral fronds usually attached to parent frond; plants typically forming tangled mats and floating below water surface .... *Lemna trisulca* | STAR DUCKWEED
1 Fronds sessile, ± round, floating on water surface 2

2 Fronds without reddish color or spots, mostly with 1 very distinct white spot (papilla) near tip on upper surface.; root sheath winged at base, root tip usually sharp pointed, roots to 3 cm long . . . . *Lemna perpusilla* . . . . MINUTE DUCKWEED

2 Fronds often reddish or with reddish spots, with or without distinct papilla near tip on upper surface; root sheath not winged, root tip mostly rounded, roots often longer than 3 cm **3**

3 Plants forming small, olive- to brown-colored, rootless turions (fleshy, overwintering structure that detaches, and then starts growth in the spring); frond underside strongly reddish . . . . *Lemna turionifera* . . . . TURION DUCKWEED

3 Turions absent; frond underside green or red **4**

4 Frond underside strongly reddish, usually with distinct white spots (papillae) on midline of upper surface of frond . . . . *Lemna turionifera* . . . . TURION DUCKWEED

4 Fronds not reddish on lower surface . . . *Lemna minor* | LESSER DUCKWEED

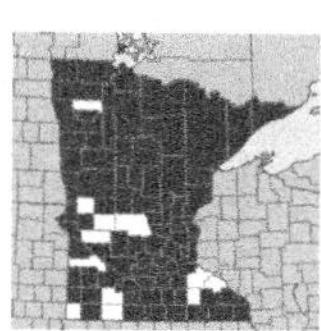

Lemna minor L.

LESSER DUCKWEED
NC **OBL** | MW **OBL** | GP **OBL**

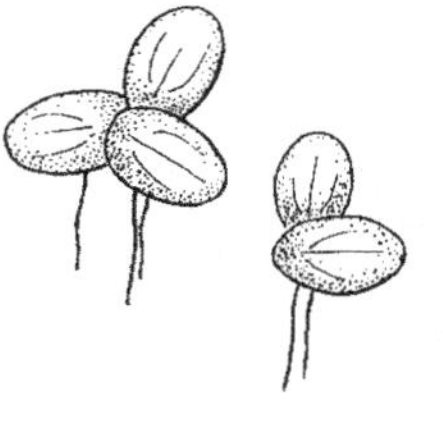

Native perennial floating herb, with 1 root from middle of underside of frond. **Fronds** ± round to oval, 2–4 mm long, often in groups of 2–5, underside often red-tinged; both sides flat or slightly convex. **Fruit** a 1-seeded utricle. July–Sept. **Habitat** Common species of quiet or stagnant water of ponds, oxbows, shores, slow-moving rivers, ditches.

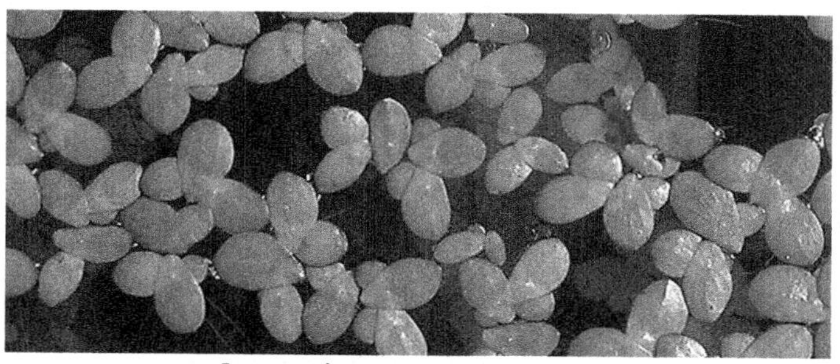

Lemna minor (LESSER DUCKWEED)

Lemna perpusilla Torr.

MINUTE DUCKWEED

NC **OBL** | MW **OBL** | GP **OBL**

Native perennial floating herb; roots single from underside of frond or absent. Fronds single or several together, obovate to elliptic, asymmetrical at base, 2–3 mm long, upper surface convex or keeled, with small bumps (papilla) near center and tip of frond. **Fruit** a 1-seeded utricle. **Habitat** Quiet water of ponds, ditches.

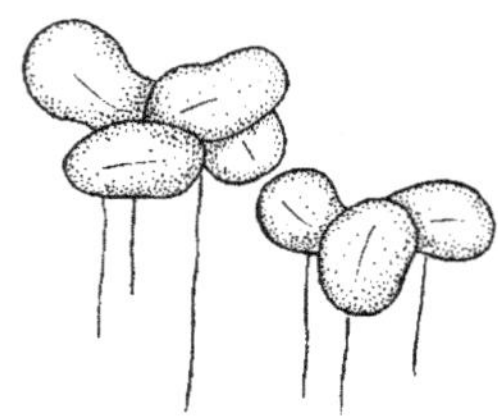

Lemna perpusilla
MINUTE DUCKWEED

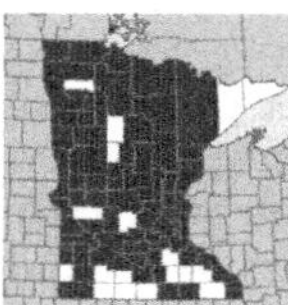

Lemna trisulca L.

STAR DUCKWEED

NC **OBL** | MW **OBL** | GP **OBL**

Native perennial floating herb, forming tangled colonies just below water surface, floating at surface only when flowering; roots single from underside of frond or absent. **Fronds** several to many, joined to form star-shaped colonies; fronds oblong lance-shaped, 5–20 mm long, tapered to a slender base (stipe), flat on both sides. **Fruit** a 1-seeded utricle. **Habitat** Ponds, streams, ditches.

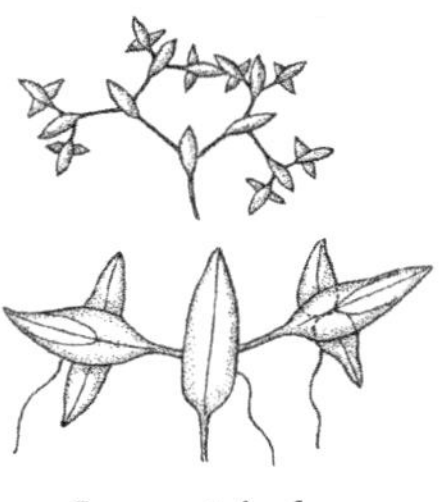

Lemna trisulca
STAR DUCKWEED

Lemna turionifera Landolt

TURION DUCKWEED

NC **OBL** | MW **OBL** | GP **OBL**

Small floating native perennial, single or in groups of several. **Fronds** obovate, usually flat and not humped, 1–4 mm long and 1–1.5 times longer than wide, veins 3, small white dots (papillae) present on midline of upper surface (visible with naked eye but cleared with 10x handlens); underside of frond usually red or purple and redder than upper side point of root, upper surface (especially near tip) sometimes red-spotted; **turions** sometimes present, dark-green to brown, 1–1.6 mm wide, without roots, sinking to bottom and forming new plants. **Habitat** Quiet water of lakes and ponds.

If turions and reddish color absent, plants of turion duckweed may not be separable from Lemna minor.

SPIRODELA | *Greater duckweed*

Spirodela polyrhiza (L.) Schleiden
GREATER DUCKWEED
NC **OBL** | MW **OBL** | GP **OBL**

Native perennial herb, floating on water surface; roots 5–12 per frond. **Fronds** usually in clusters of 2–5, flat, round to obovate, 3–6 mm long, upper surface green, underside red-purple. **Flowers** uncommon, comprised of 2–3 stamens (male flowers) and 1 pistil (female flower) in each pouch. **Fruit** a 1–2-seeded utricle. Reproduction mainly by budding of new leaves from reproductive pouches (1 pouch on each margin of frond). **Habitat** Cosmopolitan species of stagnant or slow-moving water of lakes, ponds, marshes and ditches, often with lesser duckweed (*Lemna minor*).

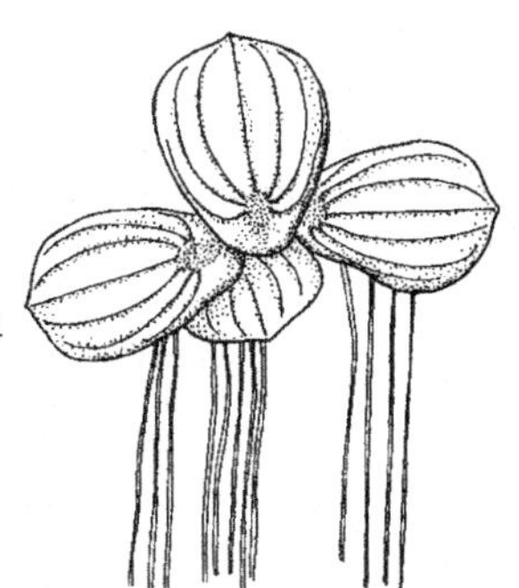

Spirodela polyrhiza
GREATER DUCKWEED

WOLFFIA | *Water-meal*

Tiny perennial herbs, without roots, floating at or just below water surface, sometimes abundant and forming a granular scum across surface, usually mixed with other members of Duckweed Family. **Fronds** single or often paired, globe-shaped or ovate, flat or rounded on upper surface. **Flowers** uncommon, consisting of 1 stamen (male flower) and 1 pistil (female flower) in the pouch. **Fruit** a round, 1-seeded utricle. Reproduction mainly by budding from the single pouch near base of frond.

Water-meals *are the world's smallest flowering plants. The fronds feel granular or mealy and tend to stick to skin. The three species in Minnesota often occur together, usually in stagnant or polluted water.*

WOLFFIA | WATER-MEAL

1 Leaves rounded on upper surface, not brown-dotted . . *Wolffia columbiana* . . . . COLUMBIA WOLFFIA

1 Leaves flattened on upper surface, brown-dotted (at 10x magnification) . . 2

2 Leaves rounded at tip, with a wartlike bump in center of upper surface . . . . *Wolffia brasiliensis* | POINTED DUCKWEED

2 Leaves with an upturned point at tip, wartlike bump absent . . . . *Wolffia borealis* | DOTTED WOLFFIA

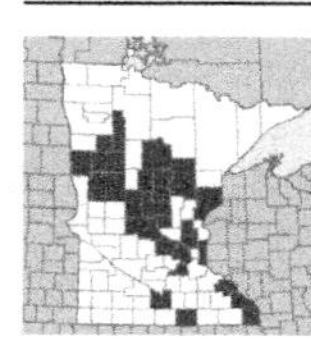

Wolffia borealis (Engelmann) Landolt
NORTHERN WATER-MEAL
NC **OBL** | MW **OBL** | GP **OBL**

Small native perennial floating herb. **Fronds** oval to oblong when viewed from above, 0.1–1 mm long and to 0.5 mm wide, with a raised pointed tip, usually brown-dotted; upper surface floating just above water surface, bright green, underside paler. **Synonyms** *Wolffia punctata*. **Habitat** Quiet water of ponds, marshes and ditches, often with other species of *Wolffia* and *Lemna*.

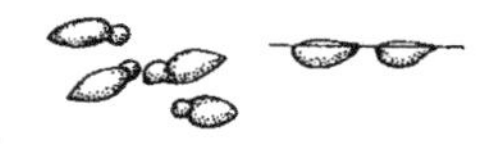

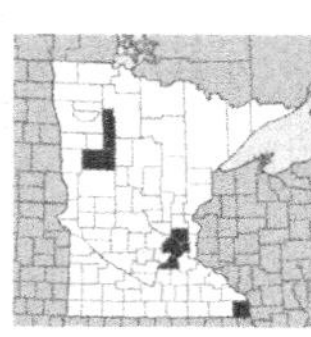

Wolffia brasiliensis Wedd.
BRAZILIAN WATER-MEAL
NC **OBL** | MW **OBL** | GP **OBL**

Small native perennial floating herb. **Fronds** broadly ovate, 0.5–1.5 mm long and 0.5–1 mm wide, rounded at tip, brown-dotted; upper surface floating just above water surface, with a wartlike bump near center. Aug–Sept. **Synonyms** *Wolffia papulifera*. **Habitat** Quiet water of ponds, often occurring with other species of *Wolffia*.

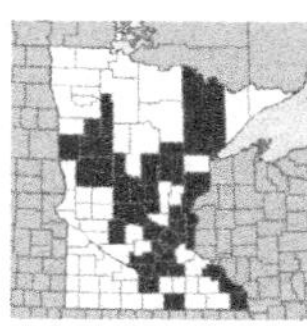

Wolffia columbiana Karsten
COLUMBIAN WATER-MEAL
NC **OBL** | MW **OBL** | GP **OBL**

Small native perennial floating herb. **Fronds** float low in water, only small upper surface exposed, round to broadly ovate and 1–1.5 mm long when viewed from above; nearly round when viewed from side, not raised and pointed at tip; green, not brown-dotted. **Habitat** Stagnant water of ponds and marshes.

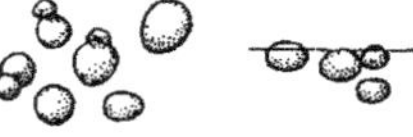

LEFT **Greater duckweed** *(Spirodela polyrhiza, larger plants) intermixed with* **water-meals** *(Wolffia spp.)*

PERENNIAL HERBS, from corms, bulbs or rhizomes. **Stems** leafy or leafless. **Leaves** linear to ovate, usually from base of plant, sometimes along stem, alternate to opposite or whorled. **Flowers** perfect (with both male and female parts), regular; sepals and petals of 6 petal-like tepals in 2 series of 3; stamens 6; ovary superior or inferior, 3-chambered. **Fruit** a capsule or round berry.

Recent treatments of the **Lily Family** *separate the various genera into new families; these are noted for each genus.*

LILIACEAE | LILY FAMILY

1 Leaves from stem 2
1 Leaves all from base of plant; flowers at ends of leafless stalks (scapes) . . 4

2 Plants large, 3–8 dm tall; flowers showy, orange. . . . *Lilium philadelphicum* . . . . WOOD-LILY
2 Plants smaller, mostly 0.5–4 dm tall; flowers small, white 3

3 Tepals 4; leaves usually 2 . . . . *Maianthemum canadense* . . . . CANADA MAYFLOWER; WILD LILY-OF-THE VALLEY
3 Tepals 6; leaves usually 3 . . . . . *Smilacina trifolia* | FALSE SOLOMON'S SEAL

4 Flowers yellow; ovary inferior. . . . . *Hypoxis hirsuta* | YELLOW STAR-GRASS
4 Flowers white or greenish; ovary superior 5

5 Stems sticky-hairy . . . . *Tofieldia glutinosa* | FALSE ASPHODEL
5 Stems smooth and waxy, not sticky-hairy . . . . *Zigadenus elegans* . . . . WHITE CAMAS

HYPOXIS | *Star-grass* (Hypoxidaceae)

Hypoxis hirsuta (L.) Cov.

YELLOW STAR-GRASS

NC **FAC** | MW **FAC** | GP **FACW**

Low native perennial herb, from a small, shallow corm. **Stems** leafless, lax, 1 to several, silky-hairy in upper part, shorter than leaves when flowering, to 4 dm long when mature. **Leaves** from base of plant, linear, hairy, to 6 dm long and 2–10 mm wide. **Flowers** 1–6 (usually 2), yellow, 1–2.5 cm wide, in racemes at ends of **stems**, tepals hairy on outside, 5–12 mm long, spreading in flower, closing and turning green after flowering, persistent. **Capsule** oval, 3–6 mm long; seeds black. May–July. **Habitat** Wet meadows, shores, moist prairie; often where calcium-rich.

LILIUM | *Lily* (Liliaceae)

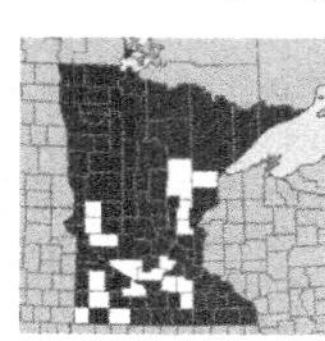

Lilium philadelphicum L.
WOOD-LILY
NC **FAC** | MW **FAC** | GP **FACU**

Native perennial herb from a scaly bulb. **Stems** erect, 3–8 dm long. **Leaves** all from stem, narrowly lance-shaped, 4–10 cm long and 3–9 mm wide, parallel-veined; lower leaves alternate, upper leaves opposite or whorled; petioles absent. **Flowers** 1–5, erect, large and showy, on stalks 1–8 cm long at ends of stem; tepals orange-red, yellow and dark-spotted toward base, lance-shaped, 4–8 cm long and 0.8–2.8 cm wide, stamens and pistil about as long as tepals; stigma 3-parted; ovary superior. **Capsule** oblong, 2.5–4 cm long; seeds flat. June–July. **Habitat** Wet meadows, low prairie, fens and open bogs, seeps, ditches; also in drier meadows, prairies and woods.

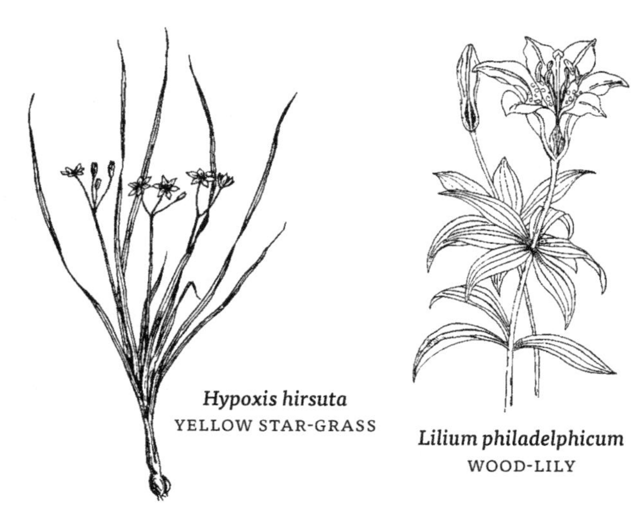

Hypoxis hirsuta
YELLOW STAR-GRASS

Lilium philadelphicum
WOOD-LILY

MAIANTHEMUM | *Wild lily-of-the-valley* (Asparagaceae)

Maianthemum canadense Desf.
CANADA MAYFLOWER; WILD LILY-OF-THE-VALLEY
NC **FACU** | MW **FAC** | GP **FACU**

Small native perennial herb, spreading by rhizomes. **Stems** erect, 5–20 cm long. **Leaves** usually 2 along stem, ovate, heart-shaped at base, 3–10 cm long; petioles short or absent. **Flowers** small, white, 4–6 mm wide, stalked, in a short raceme at end of stem, the raceme 3–6 cm long; tepals 4, spreading; stamens 4; style 2-lobed. **Fruit** a pale red berry, 3–4 mm wide; seeds 1–2. May–July. **Habitat** On hummocks in swamps, open bogs and thickets; also common in moist to dry woods.

Maianthemum canadense
CANADA MAYFLOWER;
WILD LILY-OF-THE-VALLEY

SMILACINA | *False Solomon's seal* (Asparagaceae)

Smilacina trifolia (L.) Desf.
FALSE SOLOMON'S SEAL
NC **OBL** | MW **OBL** | GP **OBL**

Native perennial herb, from long rhizomes. **Stems** erect, 1–5 dm long at flowering time. **Leaves** alternate, smooth, usually 3 (2–4), oval or oblong lance-shaped, 6–12 cm long and 1–4 cm wide; petioles absent. **Flowers** small, white, 8 mm wide, stalked, 3–8 in a raceme; tepals 6, spreading; stamens 6. **Fruit** a dark red berry, 3–5 mm wide; seeds 1–2. May–June. **Synonyms** *Maianthemum trifolium*. **Habitat** Open bogs, conifer swamps, thickets.

Smilacina trifolia
FALSE SOLOMON'S SEAL

TOFIELDIA | *False asphodel* (Tofieldiaceae)

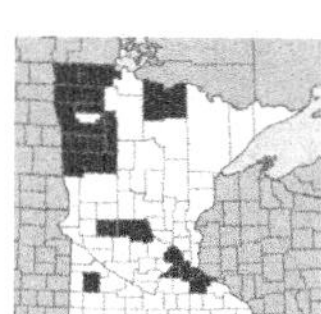

Tofieldia glutinosa (Michx.) Pers.
FALSE ASPHODEL
NC **OBL** | MW **OBL** | GP **OBL**

Native perennial herb, from a bulb. **Stems** erect, nearly leafless, 2–5 dm long, covered with sticky hairs. **Leaves** 2–4 from base of plant, linear, hairy, 8–20 cm long and to 8 mm wide, sometimes with 1 bractlike leaf near middle of stem. **Flowers** white, on sticky-hairy stalks 3–6 mm long, in a raceme 2–5 cm long when in flower, becoming longer when fruiting, 2–3 at each node of the raceme, upper flowers opening first; tepals 6, oblong lance-shaped, 4 mm long; stamens 6. **Fruit** an oblong capsule, 5–6 mm long; seeds about 1 mm long, with a slender tail at

each end. June–Aug. **Synonyms** *Triantha glutinosa*. **Habitat** Sandy or gravelly shores, calcareous fens.

ZIGADENUS | *Death camas* (Melanthiaceae)

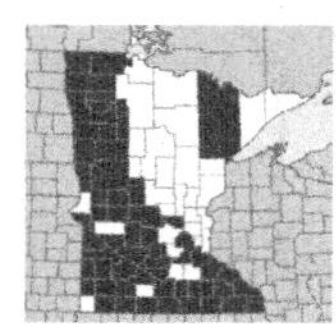

Zigadenus elegans Pursh

CAUTION - TOXIC

WHITE CAMAS

NC **FACW** | MW **FAC** | GP **FACW** | *Caution - Toxic*

Native perennial herb, from an ovate bulb; plants waxy, especially when young. **Stems** erect, 2–6 dm long. **Leaves** mostly from base of plant, linear, 2–4 dm long and 4–12 mm wide; stem leaves much smaller. **Flowers** green-yellow or white, in a raceme or panicle, 1–3 dm long, the branches upright, subtended by large, lance-shaped, green or purplish bracts; tepals 6, obovate, 7–12 mm long, usually purple-tinged near base; stamens 6. **Fruit** an ovate capsule, 10–15 mm long; seeds 3 mm long. July–Aug. **Synonyms** *Anticlea elegans*, *Zigadenus glaucus*. **Habitat** Sandy or rocky shores of Great Lakes, open bogs, calcareous fens.

All parts of this plant are toxic and should never be eaten.

Tofieldia glutinosa
FALSE ASPHODEL

Zigadenus elegans
WHITE CAMAS

NAJAS | *Naiad, Water-nymph*

Aquatic annual herbs, roots fibrous, rhizomes absent. **Stems** wavy, with slender branches. **Leaves** simple, opposite or in crowded whorls, stalkless, abruptly widened at base to sheath the stem; margins toothed to nearly entire, the teeth sometimes spine-tipped. **Flowers** either male or female, separate on same plant or on different plants, tiny, single and stalkless in leaf axils, enclosed by the sheathing leaf bases; male flowers a single anther within a membranous envelope (spathe), this surrounded by perianth scales, the scales sometimes joined into a tube; female flowers surrounded by 1–2 spathes, pistils 1, stigmas 2–4, style usually persistent. **Fruit** a 1-seeded achene.

Najas minor, *introduced and invasive in the e USA, is documented in Minnesota from a single Dakota County location.*

NAJAS | NAIAD, WATER-NYMPH

1 Leaves coarsely toothed and spine-tipped (spines visible without a lens), bright green; midvein of leaf underside and stems between nodes often prickly .... *Najas marina* | ALKALINE WATER-NYMPH
1 Leaves nearly entire or toothed (if spine-tipped, the spines not visible without a lens), often olive green; leaf surface and stems between leaves smooth **2**

2 Base of leaves lobed or clasping stem .... *Najas gracillima* .... SLENDER WATER-NYMPH
2 Leaves tapered to base, not lobed or clasping stem.................... **3**

3 Achenes smooth and glossy, widest above middle .... *Najas flexilis* .... NORTHERN WATER-NYMPH
3 Achenes rough and pitted, widest at middle and tapered to ends .... *Najas guadalupensis* | SOUTHERN WATER-NYMPH

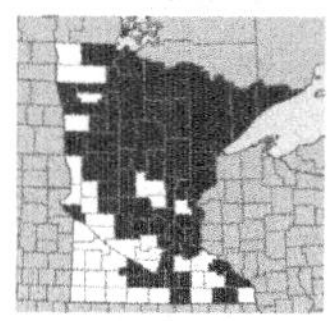

Najas flexilis (Willd.) Rostkov & Schmidt
NORTHERN WATER-NYMPH
NC **OBL** | MW **OBL** | GP **OBL**

Native annual aquatic herb. **Stems** branched, 5–40 cm long. **Leaves** densely clustered at tips of **stems**, linear, tapered to a long slender point, spreading or ascending, 1–4 cm long and to 0.5 mm wide; margins with tiny sharp teeth. **Flowers** either male and female, separate on same plant. **Achenes** oval, olive-green to red, the beak 1 mm or more long; seeds straw-colored, shiny, 2–4 mm long. July–Sept. **Habitat** Ponds, lakes and streams; common.

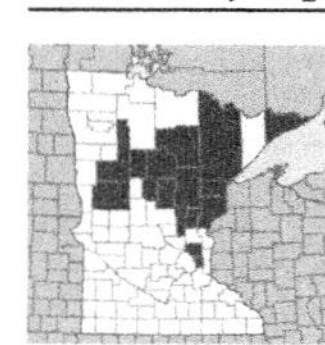

Najas gracillima (A. Braun) Magnus
SLENDER WATER-NYMPH
NC **OBL** | MW **OBL** | GP **OBL** | *special concern*

Native annual aquatic herb; plants light green. **Stems** very slender, branched, 0.5–5 dm long. **Leaves** opposite or in groups of 3 or more, bristlelike, 0.5–3 cm long and to 0.5 mm wide, spreading or ascending; margins with very small teeth. **Flowers** either male or female and on the same plant. **Achenes** cylindric, narrowed at ends; seeds light brown, 2–3 mm long. **Habitat** Shallow water of lakes, usually in muck; intolerant of polluted water. **Status** Special concern.

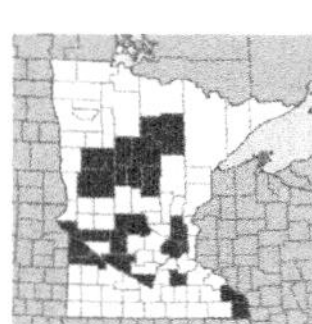

Najas guadalupensis (Sprengel) Magnus
SOUTHERN WATER-NYMPH
NC **OBL** | MW **OBL** | GP **OBL**

Native annual aquatic herb. **Stems** much branched, 1–6 dm long. **Leaves** numerous, linear, spreading and often curved downward at tip, 1–3 cm long and 0.5–2 mm wide; groups of smaller leaves also present in leaf axils; margins with very small teeth. **Flowers** either male or female, separate on same plant. **Achenes** cylindric, the beak to 0.5 mm long; seeds brown or purple, 1–3 mm long. July–Sept. **Habitat** Shallow to deep water of lakes, ponds and sometimes rivers; often with northern water-nymph (*Najas flexilis*) but less common.

Najas marina L.
ALKALINE WATER-NYMPH
S
NC **OBL** | MW **OBL** | GP **OBL**

Native annual aquatic herb. **Stems** stout, 1–5 dm long and 1–4 mm wide, compressed, branched, prickly. **Leaves** opposite or whorled, linear, 0.5–4 cm long and 1–4 mm wide, sometimes with spines on underside; margins coarsely toothed, the teeth 1–4 mm apart and spine-tipped. **Flowers** either male or female and on different plants. **Achenes** olive-green, the beak about 1 mm long; seeds dull, 2–5 mm long. July–Sept. **Habitat** Shallow water (to 1 m deep) of lakes and marshes. Nearly cosmopolitan, but local across North America. **Status** Special concern.

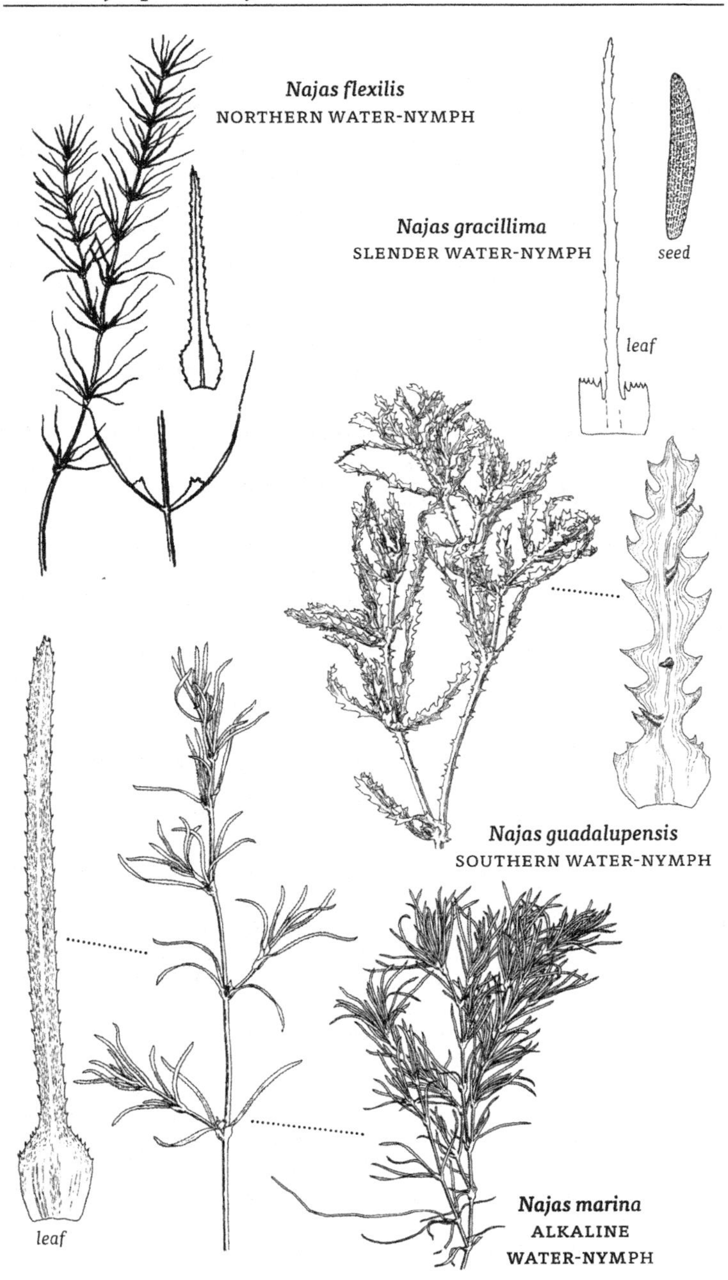
Najas flexilis
NORTHERN WATER-NYMPH
Najas gracillima
SLENDER WATER-NYMPH
seed
leaf
Najas guadalupensis
SOUTHERN WATER-NYMPH
leaf
Najas marina
ALKALINE
WATER-NYMPH

ORCHIDACEAE
Orchid Family

PERENNIAL HERBS, from fleshy or tuberous roots, corms or bulbs. **Leaves** simple, along the stem and alternate, or mostly at base of plant, stalkless and usually sheathing the stem, parallel-veined, often somewhat fleshy. **Flowers** perfect (with both male and female parts), irregular, showy in many species, in heads of 1 or 2 flowers at ends of stems, or with several to many flowers in a spike, raceme or panicle, each flower usually subtended by a bract; sepals 3, green or colored, sometimes resembling the lateral petals, the lateral sepals free, or joined to form an appendage below the lip (as in *Cypripedium*), or joined with the lateral petals to form a hood over the lip (*Spiranthes*); petals 3, white or colored, the 2 lateral petals alike, the lowest petal different and called the lip; stamens 1–2, attached to the style and forming a stout column; ovary inferior. **Fruit** a many-seeded capsule, opening by 3 or sometimes 6 longitudinal slits, but remaining closed at tip and base; seeds miniscule.

The **Orchid Family** *is believed to be the world's second largest flowering plant family (the Asteraceae is the largest), with over 900 genera and an estimated 27,000–28,000 species, most of which occur in the tropics.*

ORCHIDACEAE | ORCHID FAMILY

1 Plants without leaves; stems yellowish (chlorophyll absent) . . . . *Corallorhiza trifida* | YELLOW CORALROOT
1 Plants with 1 or more leaves; stems green 2

2 Lip of flower inflated and pouchlike 3
2 Lip not inflated or pouchlike 4

3 Leaf 1, lip hair-covered . . . . *Calypso bulbosa* | CALYPSO; FAIRY SLIPPER
3 Leaves 2 or more, lip hairless. . . . . *Cypripedium* | LADY'S-SLIPPER

4 Base of lip extended downward as a spur 5
4 Spur absent 6

5 Flowers showy, lip of flower white with purple spots . . . . *Amerorchis rotundifolia* | ROUND-LEAF ORCHID
5 Flowers white or yellow-green . . . . *Platanthera* | REIN-ORCHID

6 Lip above other flower parts 7
6 Lip below other flower parts 8

7 Flowers large and rose-purple; leaf 1 . . . *Calopogon tuberosus* | GRASS-PINK
7 Flowers small; yellow-green or cream-colored; leaves 1–5; very rare . . . . *Malaxis paludosa* | BOG ADDER'S MOUTH

8 Lip of flower bearded 9
8 Lip of flower not bearded 10

9 Leaf flat, present at flowering time. . . . . *Pogonia ophioglossoides* . . . . ROSE POGONIA; SNAKE-MOUTH

9 Leaf folded, absent at flowering (developing later) . . . . . . *Arethusa bulbosa* . . . . SWAMP-PINK; DRAGON'S MOUTH

10 Leaves in an opposite pair on stem. 11

10 Leaves not opposite 12

11 Leaves near middle of stem, lip of flower lobed at tip. . . . . *Listera* . . . . TWAYBLADE

11 Leaves basal, lip of flower not lobed at tip . . . . *Liparis loeselii* . . . . FEN-ORCHID; LOESEL'S TWAYBLADE

12 Upper stem leafless . . . . *Malaxis* | ADDER'S MOUTH

12 Upper stem with small, scalelike leaves . . . . *Spiranthes* | LADIES' TRESSES

AMERORCHIS | *Orchis*

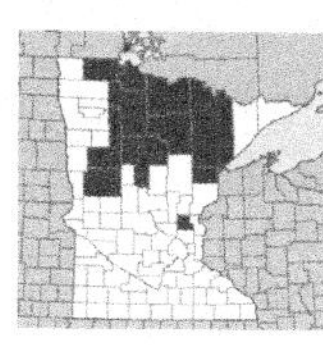

Amerorchis rotundifolia (Banks ex Pursh) Hultén
ROUND-LEAF ORCHID
NC **OBL** | MW **OBL**

Native perennial herb, roots few from a slender rhizome. **Stems** leafless, smooth, 15–30 cm long. **Leaves** single from near base of plant, oval, 4–15 cm long and 2–8 cm wide; usually with 1–2 bladeless sheaths below. **Flowers** 4 or more, in a raceme 3–8 cm long; sepals white to pale pink; petals white to pink or purple-tinged, the 2 lateral petals joined with the upper sepal to form somewhat of a hood over the column; lip white, with purple spots, 6–10 mm long and 4–7 mm wide, 3-lobed, the terminal lobe largest and notched at tip; spur about 5 mm long, shorter than lip. June–July. **Synonyms** *Orchis rotundifolia*. **Habitat** Conifer swamps (on moss under cedar, tamarack, or black spruce); southward in Minn in cold conifer swamps of balsam fir, black spruce and cedar; usually found over limestone and where sphagnum mosses not predominant.

ARETHUSA | *Dragon's mouth*

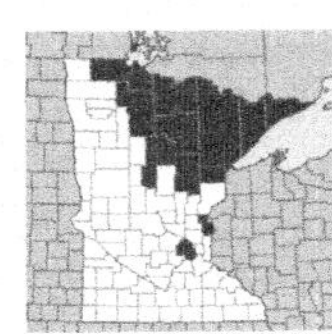

Arethusa bulbosa L.

SWAMP-PINK; DRAGON'S MOUTH

NC **OBL** | MW **OBL** | GP **OBL**

Native perennial herb; roots few, fibrous, from a corm. **Stems** leafless, smooth, 1–4 dm long. **Leaves** 1, linear, small and bractlike at flowering time, later expanding to 2 dm long and 3–8 mm wide; lower stem with 2–4 bladeless sheaths. **Flowers** single at ends of **stems**, sepals rose-purple, oblong, 2.5–5 cm long; petals joined and ± hoodlike over the column, lip pink, streaked with rose-purple, 2.5–4 cm long, curved downward near middle. June–July. **Habitat** Open bogs and conifer swamps (in sphagnum moss), floating mats around bog lakes, calcareous fens; often with grass-pink (*Calopogon tuberosus*) and rose pogonia (*Pogonia ophioglossoides*).

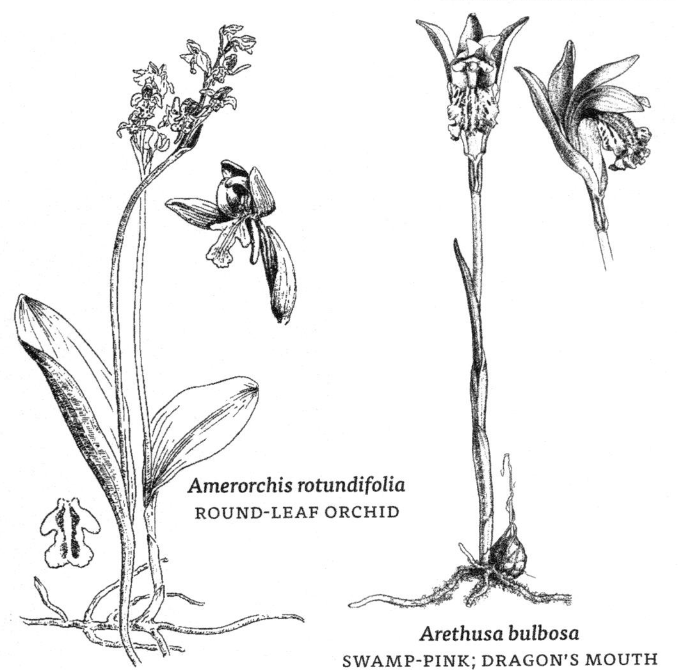

Amerorchis rotundifolia
ROUND-LEAF ORCHID

Arethusa bulbosa
SWAMP-PINK; DRAGON'S MOUTH

CALOPOGON | *Grass-pink*

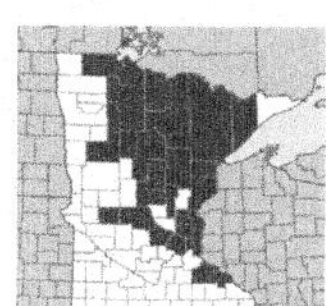

Calopogon tuberosus (L.) BSP.
GRASS-PINK
NC **OBL** | MW **OBL** | GP **OBL**

Native perennial herb, from a corm. **Stems** leafless, smooth, 2–7 dm long. **Leaves** 1 near base of plant, linear, 1–4 dm long and 2–15 mm wide. **Flowers** pink to purple, 2–15 in a loose raceme, 3–12 cm long; sepals ovate, 1–2.5 cm long; petals oblong, 1–2.5 cm long, the lip located above the lateral petals, 1–2 cm long, bearded on inside with yellow-tipped bristles. **Synonyms** *Calopogon pulchellus*. **Habitat** Open bogs and floating mats, openings in conifer swamps, calcareous fens.

Calopogon tuberosus
GRASS-PINK

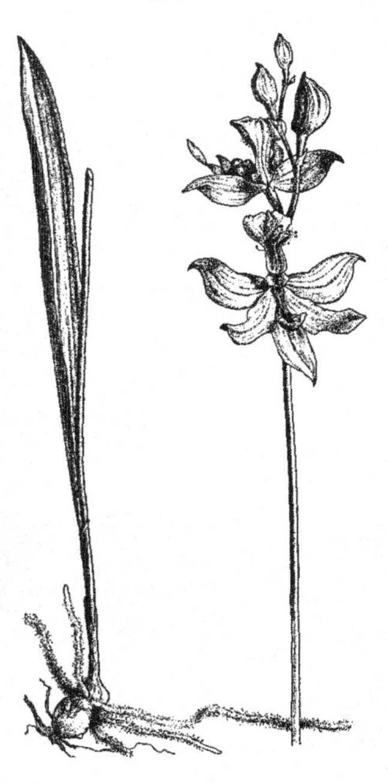

CALYPSO | *Calypso*

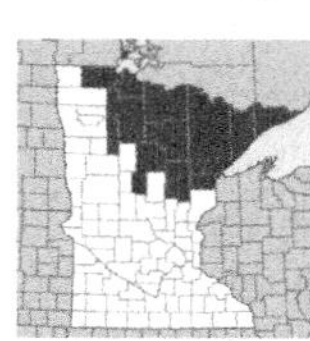

Calypso bulbosa (L.) Oakes
CALYPSO; FAIRY SLIPPER
NC **FACW** | MW **FACW** | GP **FACW**

Native perennial herb, from a corm. **Stems** 0.5–2 dm long, with 2–3 bladeless sheaths on lower portion. **Leaves** single from the corm, ovate, 3–5 cm long and 2–3 cm wide, petioles 1–5 cm long. **Flowers** 1, nodding at end of stem; sepals and lateral petals similar, pale purple to pink, lance-shaped, 1–2 cm long and 3–5 mm wide, lip white to pink, streaked with purple, 1.5–2 cm long and 5–10 mm wide, the lip extended to form a white "apron" with several rows of yellow bristles. May–June. **Habitat** Mature conifer forests or mixed forests of conifers and deciduous trees (such as balsam fir, hemlock and paper birch); soils rich in woody humus.

The single leaf of **calypso** *appears in late August or September, persists through the winter, and withers after flowering in spring. Between fruiting in June and July and the emergence of the new leaf in late summer of fall, no aboveground portions of the plant may be visible.*

Calypso bulbosa
CALYPSO; FAIRY SLIPPER

CORALLORHIZA | *Coralroot*

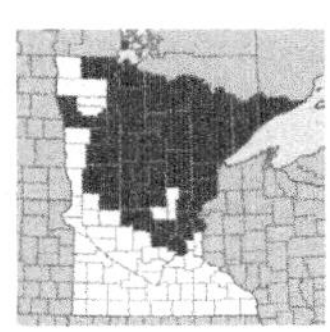

Corallorhiza trifida Chat.
YELLOW CORALROOT
NC **FACW** | MW **FACW** | GP **FAC**

Native perennial saprophytic herb, roots absent. **Stems** yellow-green, smooth, 1–3 dm long, single or in clusters from the coral-like rhizome. **Leaves** reduced to 2–3 overlapping sheaths on lower stem. **Flowers** yellow-green, 5–15 in a raceme 3–8 cm long; sepals and lateral petals yellow-green, linear, 3–5 mm long, lip white, sometimes with purple-spots, obovate, 3–5 mm long and 2–3 mm wide. **Capsules** drooping, 1–1.5 cm long and 3–7 mm wide. May–June. **Habitat** Moist to wet, mostly conifer woods, swamps (often under cedar) and thickets; usually where shaded.

Corallorhiza trifida
YELLOW CORALROOT

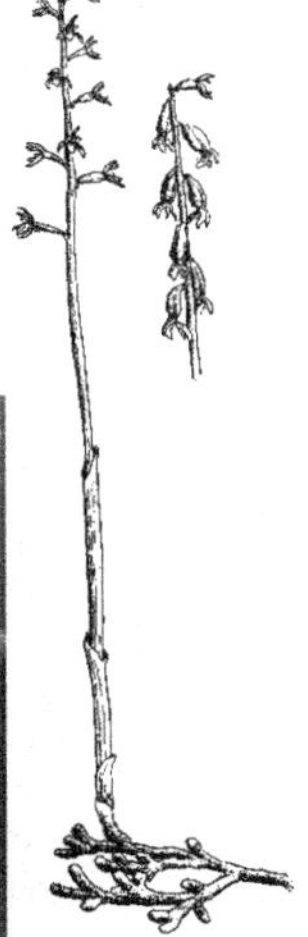

CYPRIPEDIUM | *Lady's-slipper*

Erect perennial herbs, from coarse, fibrous roots. **Stems** unbranched, often clumped, hairy. **Leaves** 2 or more at base of plant or along stem, broad. **Flowers** 1 or 2, large and mostly showy at ends of **stems**, white, pink or yellow; lateral sepals similar to lateral petals, the sepals joined to form a single appendage below the lip; lateral petals free and spreading, lip inflated and pouchlike, projecting forward; stamens 2, 1 on each side of column. **Fruit** a many-seeded capsule.

Lady's-slippers *should not be handled as they are covered with irritating hairs.*

CYPRIPEDIUM | LADY'S-SLIPPER

1 Lip pouch pink to purple; leaves 2 at base of stem .... *Cypripedium acaule*
.... MOCCASIN-FLOWER; PINK LADY'S-SLIPPER

1 Lip pouch yellow or whitish; leaves 3 or more on stem 2

2 Pouch yellow, sometimes brown- or purple-dotted 3

2 Pouch white to pink, or pink with white patches 4

3 Sepals and petals red-brown; lateral petals strongly twisted, brown-purple; pouch less than 4 cm long......... *Cypripedium parviflorum* var. *makasin*
.... SMALL YELLOW LADY'S-SLIPPER

3 Sepals and petals yellow to brown-green; lateral petals wavy, green with red-brown streaks; pouch more than 4 cm long .... *Cypripedium parviflorum* var. *pubescens* | LARGE YELLOW LADY'S-SLIPPER

4 Pouch projected downward into a cone-shaped spur .... *Cypripedium arietinum* | RAM'S-HEAD LADY'S-SLIPPER

4 Pouch not spurred... **5**

5 Sepals and lateral petals white; lip 3–5 cm long ...... *Cypripedium reginae* .... SHOWY LADY'S-SLIPPER

5 Sepals and lateral petals green; lip 1.5–2 cm long .... *Cypripedium candidum* | WHITE LADY'S-SLIPPER

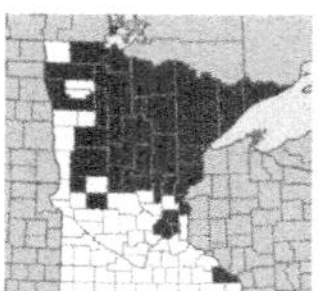

Cypripedium acaule Aiton
MOCCASIN-FLOWER; PINK LADY'S-SLIPPER
NC **FACW** | MW **FACW** | GP **FACW**

Native perennial herb, from coarse rhizomes; roots long and cordlike. **Stems** leafless, 2–4 dm long, glandular-hairy. **Leaves** 2 at base of plant, opposite, oval to obovate, 1–2 dm long and 3–10 cm wide, thinly hairy, stalkless. **Flowers** 1, nodding at end of stem; sepals and lateral petals yellow-green to green-brown, the 2 lower sepals joined to form a single sepal below the lip; lip drooping, pink with red veins, 3–5 cm long, cleft along the upper side and hiding the opening. May–June. **Habitat** Hummocks in conifer swamps; sites typically shaded, acidic and nutrient-poor, sometimes fairly dry; s in Minnesota also found on hummocks in open bogs.

Cypripedium acaule
MOCCASIN-FLOWER;
PINK LADY'S-SLIPPER

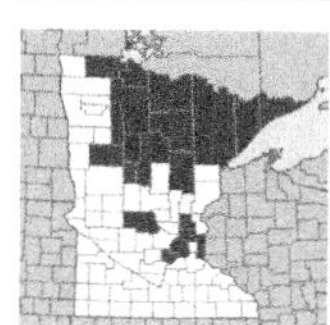

Cypripedium arietinum R. Br.

RAM'S-HEAD LADY'S-SLIPPER

NC **FACW** | MW **FACW** | GP **FACW** | *threatened*

Native perennial herb, from a coarse rhizome, roots long and cordlike. **Stems** slender, 1–4 dm long, thinly hairy. **Leaves** 3–5, above middle of stem, stalkless, oval, often folded, 5–10 cm long and 1.5–3 cm wide, finely hairy. **Flowers** 1 or sometimes 2 at ends of **stems**; sepals and lateral petals similar, green-brown; lip an inflated pouch, 1.5-2.5 cm long, white or pink-tinged, with prominent red-veins, extended downward to form a conical pouch. Late May–June. **Habitat** Conifer swamps, wet forest openings (often with northern white cedar); also in drier, sandy, conifer and mixed conifer-deciduous forests. **Status** Threatened.

Cypripedium arietinum
RAM'S-HEAD LADY'S-SLIPPER

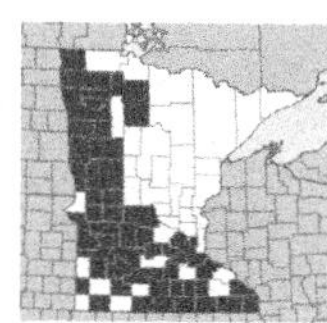

Cypripedium candidum Muhl.

WHITE LADY'S-SLIPPER

NC **OBL** | MW **OBL** | GP **OBL** | *special concern*

Native perennial herb, from a rhizome, roots long and cordlike. **Stems** 1.5–3 dm long, hairy. **Leaves** 2–4, upright, alternate along upper stem, oval, 5–15 cm long and 2–5 cm wide, sparsely glandular-hairy, stalkless; reduced to overlapping sheathing scales below. **Flowers** 1 at end of **stems**, the subtending bract leaflike, erect, 3–8 cm long; sepals and lateral petals green-yellow, often streaked with purple, the lateral sepals joined below lip, notched at tip; lateral petals linear lance-shaped, green-yellow, sometimes twisted, 2–4 cm long; lip a small inflated pouch, 1.5–2 cm long, white with faint purple veins. May–June. **Habitat** Calcium-rich wet meadows, low prairie, calcareous fens (often with shrubby cinquefoil, *Potentilla fruticosa*); usually where open and sunny. **Status** Special concern.

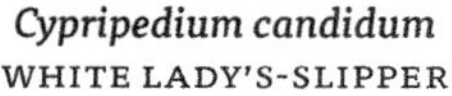

Cypripedium candidum
WHITE LADY'S-SLIPPER

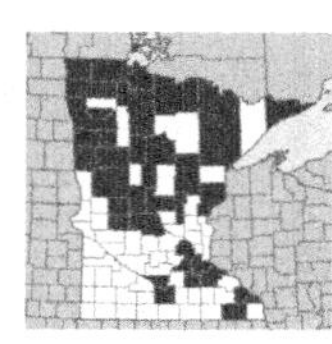

Cypripedium parviflorum var. makasin (Farw.) Sheviak
SMALL YELLOW LADY'S-SLIPPER
NC **FAC** | MW **FACW** | GP **FACW**

Native perennial herb, from rhizomes, roots long and numerous. **Stems** 1.5–6 dm long, glandular-hairy. **Leaves** 2–5, alternate along stem, ascending, oval, 5–18 cm long and 2–7 cm wide, sparsely hairy, stalkless. **Flowers** 1 (rarely 2) at ends of **stems**; sepals purple-brown, the lateral sepals joined below the lip, notched at tip; lateral petals linear, purple-brown, spirally twisted, 2–5 cm long; lip an inflated pouch, 1.5–3 cm long, yellow, often with purple veins and spots near opening. May–July. **Synonyms** *Cypripedium calceolus* var. *parviflorum*. **Habitat** Conifer swamps, wet meadows, fens, and moist forests (often under cedar); sphagnum mosses are usually sparse; sites are shaded or sunny, with organic or mineral, often calcium-rich soil; s in Minnesota also in open, calcium-rich swales.

Cypripedium parviflorum var. makasin
SMALL YELLOW LADY'S-SLIPPER

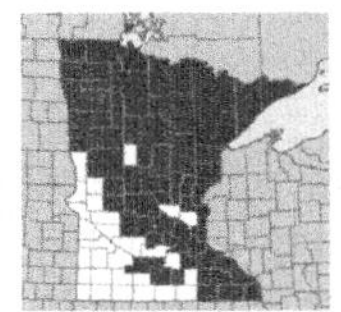

Cypripedium parviflorum var. pubescens (Willd.) O. W. Knight

LARGE YELLOW LADY'S-SLIPPER

NC **FAC** | MW **FACW** | GP **FACW**

Native perennial herb, from a rhizome, roots long and numerous. **Stems** 1.5–6 dm long, glandular-hairy. **Leaves** 3–6, alternate along stem, ascending, ovate to oval, 8–20 cm long and 3–8 cm wide, sparsely hairy. **Flowers** 1 (rarely 2) at ends of **stems**; sepals yellow-green, the lateral sepals joined below the lip, notched at tip; lateral petals linear, yellow-green, often streaked with red-brown, usually spirally twisted, 4–8 cm long; lip an inflated pouch, 3–6 cm long, yellow, often with purple veins near opening. May–July. **Synonyms** *Cypripedium calceolus* var. *pubescens*. **Habitat** Conifer swamps, bogs, fens, prairies and thickets, especially where soils derived from limestone; also in moist hardwood forests.

In **var. makasin**, *the lip is mostly 2–3 cm long, and the sepals and petals are dark red; in* **var. pubescens**, *the lip is mostly 3–6 cm long and the sepals and petals are yellow-green. However, hybrids between the two vars. occur.*

ABOVE *Cypripedium parviflorum var. pubescens*

LEFT *Cypripedium parviflorum var. makasin*

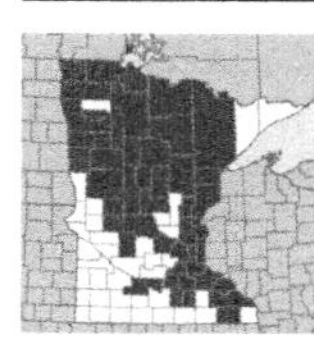

Cypripedium reginae Walter
SHOWY LADY'S-SLIPPER
NC **FACW** | MW **FACW** | GP **FACW**

Native perennial herb, from a coarse rhizome, roots many, long and cordlike. **Stems** 4–10 dm long, strongly glandular-hairy. **Leaves** 4–12, alternate along stem, spreading or ascending, broadly oval, 10–25 cm long, 4–12 cm wide, abruptly tapered to tip, nearly smooth to hairy, stalkless; reduced to sheaths at base. **Flowers** 1 or often 2 at ends of **stems**, the subtending bract leaflike, 6–12 cm long; sepals and lateral petals white, the lateral sepals joined to form an appendage under the lip, rounded at tip; lip an inflated pouch, 3–5 cm long, white, often infused with pink or purple. June–July. **Habitat** Conifer and hardwood swamps (especially balsam fir-cedar-tamarack swamps), bogs, calcareous fens, sedge meadows, floating mats, wet openings, wet clayey slopes, ditches; especially where open and sunny; most abundant in openings in wet forests and swamps not dominated by sphagnum mosses.

Cypripedium reginae
SHOWY LADY'S-SLIPPER

Showy lady's-slipper *is the official state flower of Minnesota.*

LIPARIS | *Twayblade*

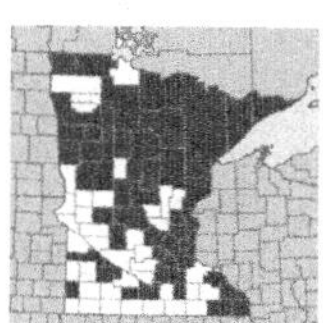

Liparis loeselii (L.) Rich.
FEN-ORCHID; LOESEL'S TWAYBLADE
NC **FACW** | MW **FACW** | GP **OBL**
Small, smooth, native perennial herb, from a bulblike base. **Stems** erect, 1–2.5 dm long, upper stem somewhat angled in section. **Leaves** 2 from base of plant, ascending, sheathing at base, shiny, lance-shaped to oval, 4–15 cm long and 1–4 cm wide. **Flowers** 2–15, yellow-green, small, upright, in an open raceme 2–10 cm long and 1–2 cm wide; sepals narrowly lance-shaped, 4–6 mm long and 1–2 mm wide; petals linear, 3–5 mm long, often twisted and bent forward under the lip; lip yellow-green, obovate, 4–5 mm long and 2–3 mm wide, tipped with a short point. **Capsules** persistent, short-cylindric, 8–12 mm long. June–Aug. **Habitat** Conifer swamps, fens, floating mats, streambanks, sandy shores, ditches; soils peaty to mineral, acid to calcium-rich.

Liparis loeselii
FEN-ORCHID;
LOESEL'S TWAYBLADE

LISTERA | *Twayblade*

Perennial herbs. **Stems** with a pair of opposite leaves near middle, **stems** smooth below leaves, hairy above. **Leaves** broad, stalkless. **Flowers** small, green to purple, in a raceme at end of stem, the lip 2-lobed or deeply parted.

LISTERA | TWAYBLADE

1 Lip 3–5 mm long, divided to about middle into 2 narrow segments .... *Listera cordata* | HEART-LEAVED TWAYBLADE
1 Lip 7–12 mm long, shallowly notched or divided 1|3 of length, the segments broad **2**

2 Lip wide at base, with a pair of auricles .... *Listera auriculata* .... AURICLED TWAYBLADE
2 Lip narrowed to base, auricles absent, reported from ne Minnesota .... *Listera convallarioides* | BROAD-LEAVED TWAYBLADE

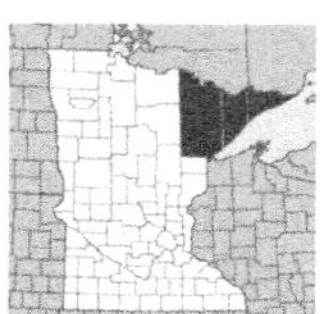

Listera auriculata Wieg.
AURICLED TWAYBLADE
NC **FACW** | *endangered*

Native perennial herb, roots fibrous. **Stems** 1–2 dm long, smooth below leaves, hairy above. **Leaves** 2 near middle of stem, opposite, ovate, 2–5 cm long and 2–4 cm wide. **Flowers** pale green, 8–15 in a raceme 4–8 cm long and 2–3 cm wide, on stalks 2–5 mm long; lip oblong, 6–10 mm long and 2–5 mm wide, the base with a pair of small clasping auricles, the tip cleft for about 1/4–1/3 of its length. June–Aug. **Habitat** Alluvial sand along rivers, often under alders; occasional in moist conifer or mixed conifer and deciduous forests; usually where shaded. **Status** Endangered.

Listera convallarioides (Swartz) Torr.
BROAD-LEAVED TWAYBLADE
NC **FACW** | GP **FACW** | *special concern*

Native perennial herb, roots fibrous. **Stems** 1–3 dm long, glandular-hairy above leaves, smooth below. **Leaves** 2, opposite near middle of stem, broadly ovate, 3–6 cm long and 2–5 cm wide, stalkless. **Flowers** yellow-green, 6–20 in a raceme 4–10 cm long and 2–3 cm wide; lip wedge-shaped, 9–11 mm long and to 6 mm wide at tip, usually with a small tooth on each side near the base, the tip shallowly 2-lobed. July–Aug. **Habitat** Seeps in forests, cedar swamps, wet, mixed conifer-deciduous woods, streambanks. **Status** Special concern; reported from Cook County in ne Minnesota but possibly now extinct from state.

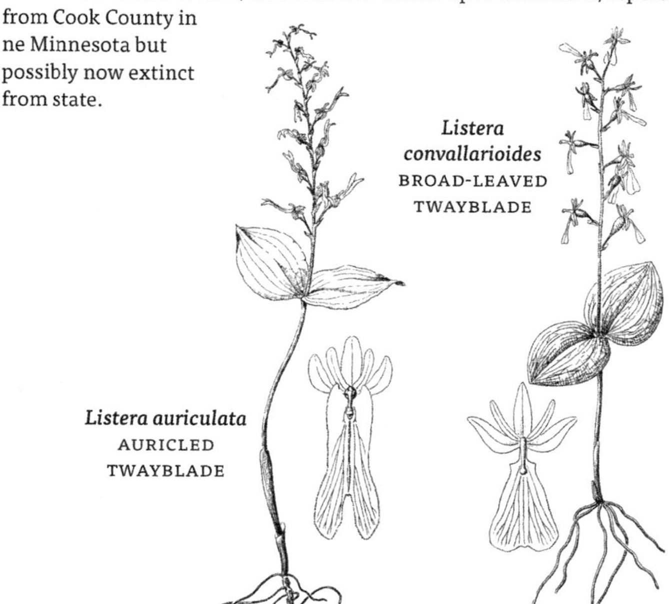

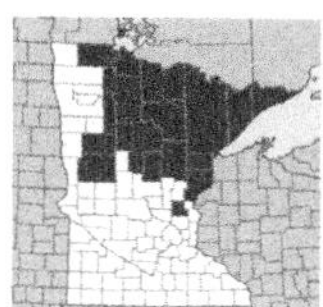

Listera cordata (L.) R. Br.
HEART-LEAVED TWAYBLADE
NC **FACW** | MW **FACW** | GP **FACU**
Native perennial herb, roots fibrous. **Stems** 1–3 dm long, glandular-hairy above leaves, smooth below. **Leaves** 2, opposite near middle of stem, 1–4 cm long and 1–3 cm wide, stalkless. **Flowers** green to red-purple, 6–20 in a raceme 3–12 cm long and 1–2 cm wide; lip slender, 3–5 mm long, with 2 teeth on side near base, the tip cleft halfway or more into spreading linear lobes. June–July. **Habitat** Open bogs and conifer swamps, where usually on sphagnum moss hummocks; hemlock groves.

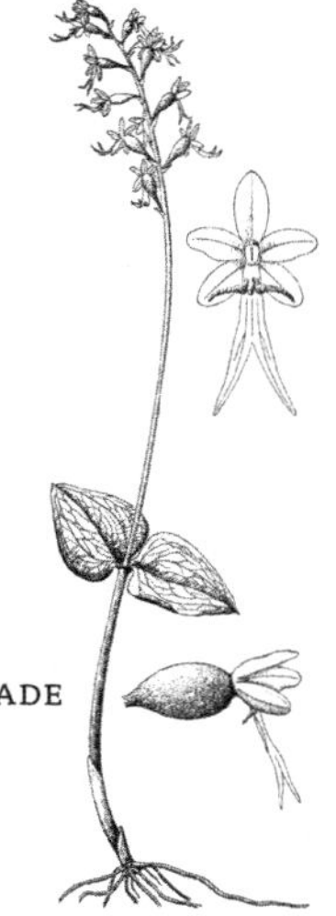

Listera cordata
HEART-LEAVED TWAYBLADE

MALAXIS | *Adder's mouth*

Small perennial herbs. **Leaves** 1–5 from base of plant or single along stem. **Flowers** green-white, spaced or crowded in slender racemes at ends of **stems**.

MALAXIS | ADDER'S MOUTH

1 Leaves 2 or more from base of plant, the leaves less than 2 cm long; rare in north-central Minnesota .... *Malaxis paludosa* | BOG ADDER'S MOUTH
1 Leaves 1 along stem, 2.5 cm or more long; more widespread 2

2 Flowers evenly spaced in a raceme 5–11 cm long .... *Malaxis monophyllos* | WHITE ADDER'S MOUTH
2 Flowers crowded near top of raceme, the raceme 2–5 cm long .... *Malaxis unifolia* | GREEN ADDER'S MOUTH

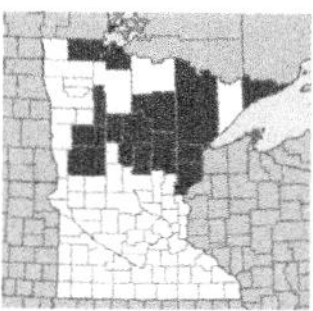

Malaxis monophyllos Swartz
WHITE ADDER'S MOUTH
NC **FACW** | MW **FACW** | GP **FACW** | *special concern*
Native perennial herb, from a bulblike base; roots few, fibrous. **Stems** smooth, 1–2 dm long. **Leaves** single, appearing to be attached well above base of stem, the leaf base clasping stem, ovate to oval, 3–7 cm long and 1.5–4 cm wide. **Flowers** small, green-white, 14–30 or more, in a long, slender, spikelike raceme 4–11 cm long and to 1 cm wide; on stalks 1–2 mm long, the flowers evenly spaced in the raceme; lip heart-shaped, bent downward, 2–3 mm long and 1–2 mm

wide, narrowed at middle to form a long, lance-shaped tip, with a pair of lobes at base. June–Aug. **Synonyms** Ours var. *brachypoda*. *Malaxis brachypoda*. **Habitat** Conifer swamps (cedar-balsam fir-spruce), especially in wet depressions and where soils are marly; sphagnum moss hummocks in conifer swamps, wet hardwood forests. **Status** Special concern.

Malaxis paludosa (L.) Swartz
BOG ADDER'S MOUTH
NC **OBL** | MW **OBL** | GP **OBL** | *endangered*

Small native perennial herb, from a bulblike base; roots few, fibrous. **Stems** leafless, smooth, 7–15 cm long. **Leaves** 2–5 from base of plant, obovate, 1–2 cm long and 0.5–1 cm wide, clasping stem at base. **Flowers** small, yellow-green, 10 or more in a slender, spikelike raceme 3–9 cm long and about 5 mm wide, the flowers evenly spaced in the raceme, twisted so that lip is uppermost in the flowers; lip very small, ovate, 1–1.5 mm long and 0.5 mm wide. July–Aug. **Synonyms** *Hammarbya paludosa*. **Habitat** Sphagnum moss hummocks in black spruce swamps, usually where somewhat open. **Status** Endangered, rare in nc Minnesota; a ± circumboreal species, south in USA to northern Minnesota.

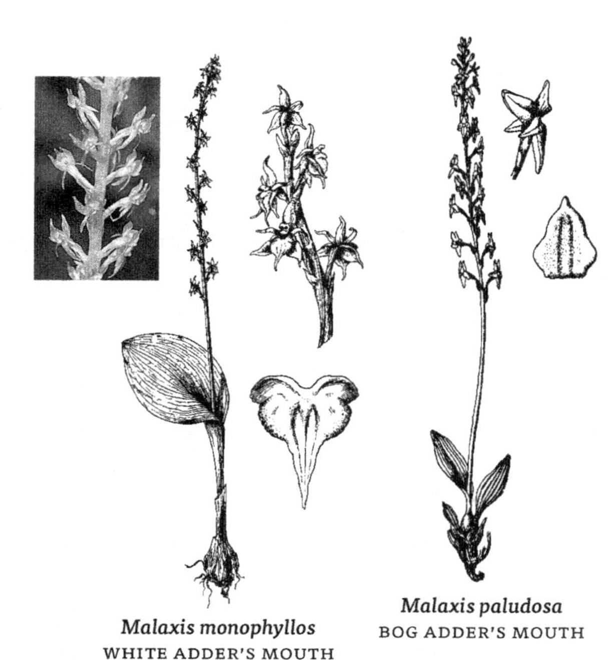

Malaxis monophyllos
WHITE ADDER'S MOUTH

Malaxis paludosa
BOG ADDER'S MOUTH

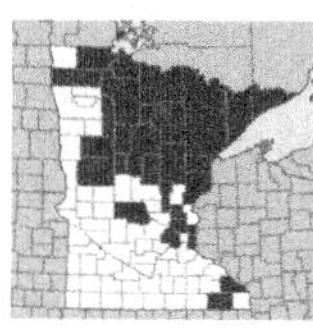

Malaxis unifolia Michx.
GREEN ADDER'S MOUTH
NC **FAC** | MW **FAC** | GP **FAC**

Small native perennial herb, from a bulblike base; roots few, fibrous. **Stems** smooth, 1–3 dm long. **Leaves** single, attached near middle of stem, ovate, 2–7 cm long and 1–4 cm wide. **Flowers** small, green, numerous in a cylindric raceme 1.5–6 cm long and 1–2 cm wide, the upper flowers crowded, the lower flowers more widely spaced; lowermost lip very small, 1–2 mm long, with 3 teeth at tip. June–Aug. **Habitat** Sphagnum moss hummocks in swamps, sedge meadows, thickets; also in moist to dry forests.

Malaxis unifolia
GREEN ADDER'S MOUTH

PLATANTHERA | *Rein-orchid*

Perennial herbs, from a cluster of fleshy roots. **Stems** erect, smooth. **Leaves** mostly along the stem, upright, reduced to sheaths at base and upward on stem; leaves basal in large round-leaf orchid (*Platanthera orbiculata*). **Flowers** white or green, several to many in a spike or raceme; upper sepal joined with petals to form a hood over the column; lateral sepals spreading; lip linear to ovate or 3-lobed, entire, toothed or fringed, extended backward into a spur, the spur commonly curved; stamens 1, the anther attached to the top of the short column. **Fruit** a many-seeded capsule.

PLATANTHERA | REIN-ORCHID

1 Margin of lip fringed; flowers often large and showy. 2
1 Margin of lip entire, not fringed; flowers small. 3

2 Flowers creamy or green-yellow, in narrow compact spikes to 3 cm wide; lip fringed nearly to base . . . . . *Platanthera lacera* | RAGGED FRINGED ORCHID
2 Flowers white, large, in spikes more than 3 cm wide; lip less deeply fringed . . . . *Platanthera praeclara* | WESTERN PRAIRIE FRINGED ORCHID

3 Stems leafless; leaves 1–2 at base of stem 4
3 Stems leafy; leaves 1 or more. 5

4 Leaves 1, ascending; stems to 3 dm long . . . . *Platanthera obtusata* . . . . BLUNT-LEAF ORCHID
4 Leaves 2, prostrate; stems 3–6 dm long . . . . *Platanthera orbiculata* . . . . LARGE ROUND-LEAF ORCHID

5 Stem leaf 1, sometimes with several small leaves on upper stem . . . . *Platanthera clavellata* | CLUB-SPUR ORCHID
5 Stem leaves 2 or more 6

6 Lip wide, 2-lobed and fringed at base . . . . *Platanthera flava* . . . . PALE GREEN ORCHID
6 Lip narrow, neither lobed nor fringed 7

7 Flowers white; lip widened at base . . . . *Platanthera dilatata* . . . . TALL WHITE BOG-ORCHID; BOG CANDLES
7 Flowers green-white; lip narrowly lance-shaped . . . *Platanthera huronensis* . . . . HURON GREEN ORCHID

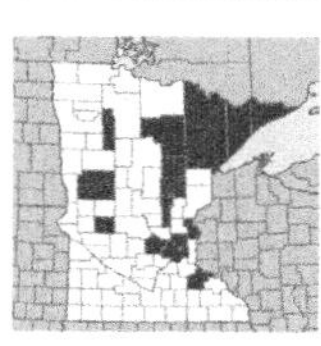

Platanthera clavellata (Michx.) Luer
CLUB-SPUR ORCHID
NC **FACW** | MW **OBL** | GP **OBL** | *special concern*

Native perennial herb, roots fleshy. **Stems** slender, 1–4 dm long. **Leaves** 1, near or just below middle of stem, oblong to lance-shaped, 5–15 cm long and 1–3 cm wide, usually with l–3 bractlike leaves above. **Flowers** 5–20, green-yellow, spreading, in a short raceme, 2–6 cm long and 1.5–3 cm wide; sepals and lateral petals broadly ovate, 3–5 mm long; lip oblong, 3–5 mm long, shallowly 3-lobed or toothed at tip; spur curved, widened at tip, 8–12 mm long. June–Aug. **Synonyms** *Habenaria clavellata*. **Habitat** In sphagnum moss of open bogs and floating mats, black spruce and tamarack swamps; also colonizing wet ditches. **Status** Special concern.

Platanthera clavellata
CLUB-SPUR ORCHID

Platanthera dilatata (Pursh) Lindl.
TALL WHITE BOG-ORCHID; BOG CANDLES
NC **FACW** | MW **FACW** | GP **FACW**

Native perennial herb, clove-scented, roots fleshy. **Stems** stout or slender, to 1 m long. **Leaves** 3–6, alternate along stem, upright, lance-shaped, to 10–20 cm long and 1–3 cm wide, with 1–2 small, bractlike leaves above and 1 bladeless sheath at base of stem. **Flowers** 10–60, bright white, upright, in a raceme 1–2.5 dm long; lateral sepals lance-shaped, 4–9 mm long and 1–3 mm wide; lateral petals similar but joined with upper sepal to form somewhat of a hood over the column; lip lance-shaped, widened at base, 6–8 mm long; spur slender, 4–8 mm long. June–July. **Synonyms** *Habenaria dilatata*. **Habitat** Wet, open bogs and floating mats, conifer swamps, streambanks, shores and seeps; often where sandy or calcium-rich (as in calcareous fens), not in deep sphagnum moss.

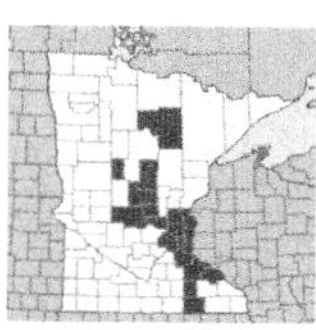

Platanthera flava (L.) Lindl.
PALE GREEN ORCHID
NC **FACW** | MW **FACW** | GP **FACW** | *endangered*
Native perennial herb, roots fleshy. **Stems** 3–7 dm long. **Leaves** 2–4, alternate along stem, lance-shaped or oval, to 5–15 cm long and 2–5 cm wide, with 1–3 bractlike leaves above. **Flowers** 15 or more, green-yellow or green, stalkless, in a raceme 5–15 cm long and 2–4 cm wide; sepals ovate, 2–3 mm long; lip bent downward, 3–6 mm long, the margin irregular, with a tooth near base on each side; spur 4–6 mm long. June–Aug. **Synonyms** Ours var. *herbiola*; *Habenaria flava*. **Habitat** Wet depressions in hardwood swamps, alder thickets, sedge meadows, moist sand prairies; often where calcium-rich, sometimes where disturbed. **Status** Endangered.

Platanthera dilatata
TALL WHITE BOG-ORCHID; BOG CANDLES

Platanthera flava
PALE GREEN ORCHID

Platanthera huronensis (Nutt.) Lindl.
HURON GREEN ORCHID
NC **FACW** | MW **FACW** | GP **OBL**

Native perennial herb, roots fleshy. **Stems** 2–8 dm long. **Leaves** 2–7, alternate on stem, linear to oblong, 5–30 cm long and 2–5 cm wide, with 1–3 smaller leaves above. **Flowers** small, green, erect, many in a raceme 4–25 cm long; lateral sepals ovate and spreading; lateral petals lance-shaped, curved upward and joined with upper sepal to form a loose hood over column; lip lance-shaped, 3–7 mm long, not abruptly widened at base; spur curved forward under the lip, about as long as lip, 3–7 mm long. June–Aug. **Synonyms** Formerly known as *Platanthera hyperborea*, but that species apparently restricted to Greenland and Iceland. *Habenaria hyperborea*. **Habitat** Moist to wet forests and swamps, thickets, streambanks, wet meadows, ditches.

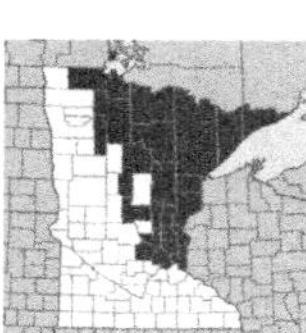

Platanthera lacera (Michx.) G. Don
RAGGED FRINGED ORCHID
NC **FAC** | MW **FAC** | GP **FAC**

Native perennial herb, roots fleshy. **Stems** 3–8 dm long. **Leaves** 3–7, alternate on stem, lance-shaped to oval, to 5–15 cm long and 1–4 cm wide; upper leaves much smaller. **Flowers** white or green-white, in a usually compact, many-flowered raceme, 5–20 cm long and 2–5 cm wide; sepals broadly oval, 4–7 mm long, the lateral ones deflexed behind the lip; lateral petals linear, entire; lip 10–16 mm long and 5–20 mm wide, deeply 3-lobed, each lobe fringed with a few long segments; spur curved, 1–2 cm long. June–Aug. **Synonyms** *Habenaria lacera*. **Habitat** Hummocks in open sphagnum bogs, conifer bogs, swamps, wet meadows, sandy prairie, thickets, ditches.

Platanthera lacera
RAGGED FRINGED ORCHID

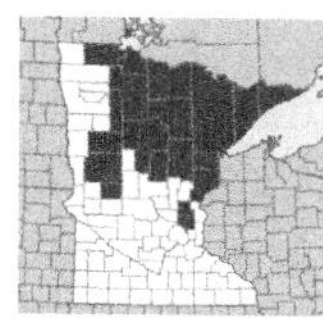

Platanthera obtusata (Banks ex Pursh) Lindl.
BLUNT-LEAF ORCHID
NC **FACW** | MW **FACW** | GP **FACW**

Native perennial herb, roots fleshy. **Stems** leafless, slender, 1–3 dm long. **Leaves** 1 at base of stem, ascending, persistent through flowering, obovate, 5–15 cm long and 1–4 cm wide, blunt-tipped, long-tapered to base. **Flowers** 4–20, green-white, in a raceme 3–12 cm long and 1–2 cm wide; lateral sepals ovate, spreading; petals ascending, widened below middle; lip lance-shaped, widened at base, 4–6 mm long; spur curved, tapered to a thin tip, 5–8 mm long. June–Aug. **Synonyms** *Habenaria obtusata*. **Habitat** Shaded hummocks in conifer swamps (especially under cedar, black spruce or balsam fir), wet mixed conifer-deciduous forests, alder thickets.

Platanthera obtusata
BLUNT-LEAF ORCHID

Platanthera orbiculata (Pursh) Lindl.
LARGE ROUND-LEAF ORCHID
NC **FAC** | MW **FAC** | GP **FAC**

Native perennial herb, roots fleshy. **Stems** 2–6 dm long, leafless apart from 1–6 small bracts. **Leaves** 2, opposite at base of plant, spreading or lying flat on ground, ± round, shiny, 6–15 cm long and 4–15 cm wide. **Flowers** green-white, several in a raceme 5–20 cm long and 3–6 cm wide; sepals ovate, to 1 cm long; petals ovate, 6–7 mm long; lip entire, rounded at tip, 10–15 mm long and 2 mm wide; spur 2–3 cm long, somewhat widened at tip. Late June–Aug. **Synonyms** *Habenaria orbiculata*. **Habitat** Shaded conifer swamps (white cedar, balsam fir, black spruce), especially where underlain by marl; also in drier conifer forests.

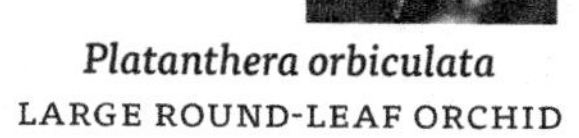

Platanthera orbiculata
LARGE ROUND-LEAF ORCHID

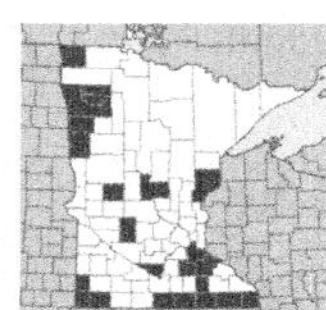

Platanthera praeclara Sheviak & Bowles

WESTERN PRAIRIE FRINGED ORCHID

NC **FACW** | MW **FACW** | GP **OBL** | *endangered*

Native perennial herb; roots thick, fleshy. **Stems** 4–8 dm long, smooth. **Leaves** 3–10 or more, upright, scattered and alternate on stem, smooth, lance-shaped to ovate lance-shaped, base of leaf clasping stem; lower leaves 10–15 cm long and 1.5–3.5 cm wide, the upper stem leaves much smaller. **Flowers** white to creamy white, large and showy, to 2.5 cm wide; up to 24 flowers in an open raceme; petals 3, the two lateral petals ragged at tip; the lip larger and deeply 3-lobed, the lobes fringed more than half way to base, 1.5–2.5 cm long and about as wide. June–Aug. **Synonyms** *Habenaria leucophaea* var. *praeclara*. **Habitat** Sedge meadows, low prairie, moist prairie swales; soils usually sandy loams and often calcium-rich. **Status** Endangered; federally listed as threatened across its entire range (mostly Great Plains region).

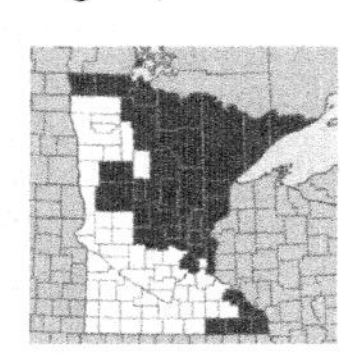

Platanthera psycodes (L.) Lindl.

PURPLE FRINGED ORCHID

NC **FACW** | MW **FACW** | GP **FACW**

Native perennial herb, roots thick and fleshy. **Stems** stout, 3–10 dm long. **Leaves** 4–12, alternate on stem, lance-shaped or oval, the upper much smaller and narrow. **Flowers** rose-purple, in a densely flowered, cylindric raceme 4–20 cm long and 3–5 cm wide; sepals oval to obovate, 4–6 mm long; petals spatula-shaped, finely toothed on margins; lip broad, 8–14 mm wide, deeply 3-lobed, the lobes fan-shaped, fringed to less than half way to base; spur curved, about 2 cm long. July–Aug. **Synonyms** *Habenaria psycodes*. **Habitat** Wetland margins, shores, wet forests, wet meadows, low prairie, roadside ditches; typically not on sphagnum moss.

POGONIA | *Pogonia*

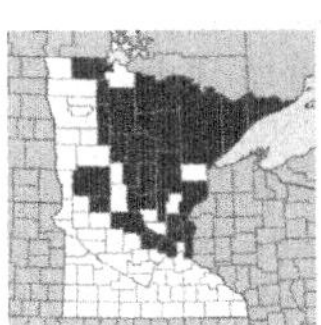

Pogonia ophioglossoides (L.) Ker Gawler
ROSE POGONIA; SNAKE-MOUTH
NC **OBL** | MW **OBL** | GP **OBL**

Native perennial herb, spreading by surface runners (stolons) which send up a stem every 10 cm or more apart. **Stems** slender, smooth, 1.5–4 dm long. **Leaves** single, attached about halfway up stem, narrowly oval, 3–10 cm long and 1–2.5 cm wide, stalkless. **Flowers** pink to purple, usually 1 at end of **stems**; sepals widely spreading, petals oval, angled over the column; lip pink with purple veins, 1.5–2 cm long and 5–10 mm wide, fringed at tip, bearded with yellow bristles. June–July. **Habitat** Conifer swamps and open bogs in sphagnum moss, floating sedge mats, sedge meadows, sandy interdunal wetlands.

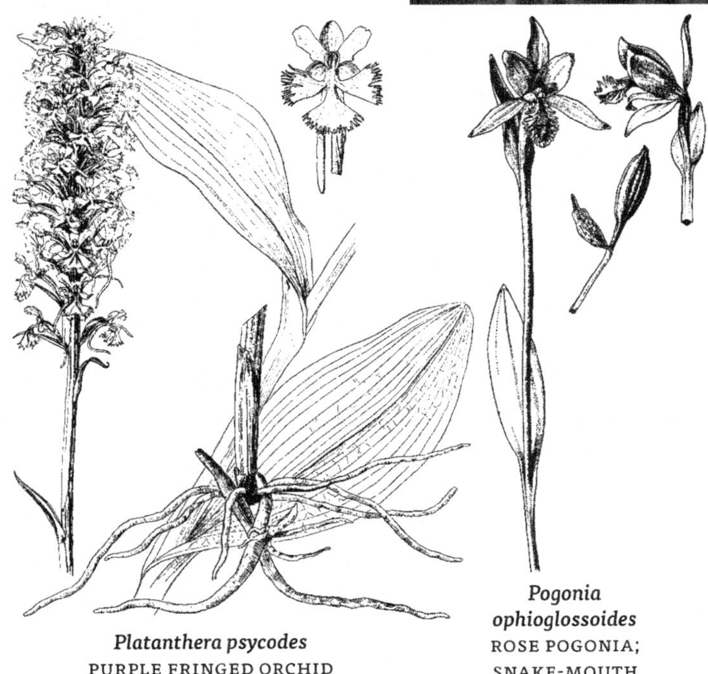

Platanthera psycodes
PURPLE FRINGED ORCHID

Pogonia ophioglossoides
ROSE POGONIA; SNAKE-MOUTH

SPIRANTHES | *Ladies' tresses*

Perennial herbs, from a cluster of tuberous roots. **Stems** slender, erect. **Leaves** largest at base of plant, becoming smaller upward on stem, the stem leaves erect and sheathing. **Flowers** small, white or creamy, spirally twisted in a densely flowered, spikelike raceme; sepals and lateral petals similar, the lateral petals joined with all 3 sepals or with only the upper sepal to form a hood over lip and column; lip folded upward near middle so that margins embrace the column, curved downward beyond the middle, with a pair of bumps or thickenings at base; anthers 1, from back of the short column.

SPIRANTHES | LADIES' TRESSES

1 Lip of flower violin-shaped (constricted near middle and widened near tip). .... *Spiranthes romanzoffiana* | HOODED LADIES' TRESSES
1 Lip not violin-shaped ..... *Spiranthes cernua* | NODDING LADIES' TRESSES

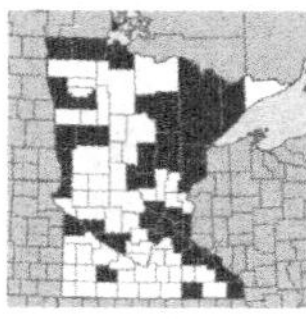

Spiranthes cernua (L.) L. C. Rich.

NODDING LADIES' TRESSES
NC **FACW** | MW **FACW** | GP **FACW**

Native perennial herb, roots fleshy. **Stems** 1–5 dm long, upper stem short-hairy, lower stem smooth. **Leaves** mostly at base of plant, usually present at flowering time, linear to oblong lance-shaped, 6–25 cm long and 5–15 mm wide; upper stem leaves 3–5, much smaller and bractlike. **Flowers** white, in a spikelike raceme 3–15 cm long, with 2–4 vertical rows of flowers, the rows spirally twisted; sepals and petals hairy on outside; lateral petals joined with upper sepal to form a hood; lip white, yellow-green at center, 6–10 mm long and 3–6 mm wide, slightly narrowed at middle, curved downward, the tip curved inward toward stem, the tip wavy-margined or with small rounded teeth, the base of lip with a pair of backward-pointing bumps. Aug–Oct. **Habitat** Open, usually sandy wetlands such as wet meadows, lakeshores, moist prairies, ditches and roadsides.

Spiranthes cernua
NODDING LADIES' TRESSES

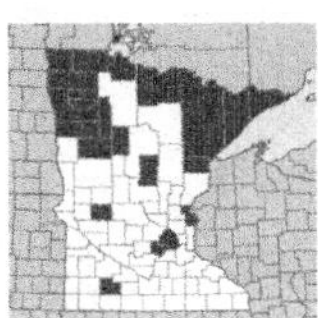

Spiranthes romanzoffiana Cham.
HOODED LADIES' TRESSES
NC **OBL** | MW **OBL** | GP **OBL**

Native perennial herb, roots thick and fleshy. **Stems** 1–4 dm long, upper stem finely hairy. **Leaves** mostly from base of plant, present at flowering time, upright, linear to narrowly lance-shaped, 5–20 cm long and 3–9 mm wide, the stem leaves becoming smaller and bractlike. **Flowers** white or cream-colored, in a spikelike raceme 3–10 cm long, with 1–3 vertical rows of flowers, the rows spirally twisted; sepals and lateral petals joined to form a hood over the lip; lip ovate, strongly constricted near middle (violin-shaped), curved downward, the tip ragged and bent inward toward stem, the bumps at base very small. July–Sept. **Habitat** Open wetlands including wet meadows, fens, lakeshores, open swamps, ditches, seeps; usually in neutral or calcium-rich habitats.

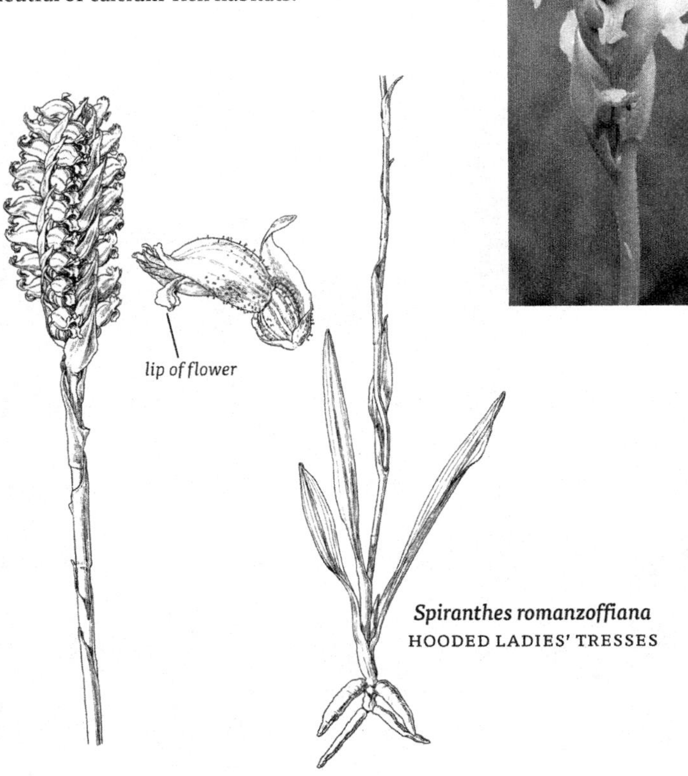

Spiranthes romanzoffiana
HOODED LADIES' TRESSES

POACEAE
Grass Family

PERENNIAL OR ANNUAL herbaceous plants, clumped or spreading by rhizomes. **Stems** (culms) usually hollow, with swollen, solid nodes. **Leaves** long, linear, parallel-veined, alternate in 2 ranks or rows, sheathing the stem, the sheaths usually split vertically, sometimes joined and tubular as in brome (*Bromus*) and mannagrass (*Glyceria*); with a membranous or hairy ring (**ligule**) at top of sheath between blade and stem, or the ligule sometimes absent; a pair of projecting lobes (**auricles**) sometimes present at base of blade.

Flowers (**florets**) small, usually perfect (with both male and female parts), or sometimes either male or female, the male and female flowers separate on the same or different plants. Florets grouped into **spikelets**, each spikelet with 1 to many florets, the florets stalkless and alternate along a small stem or axis (**rachilla**), with a pair of small bracts (**glumes**) at base of each spikelet (the glumes rarely absent); the glumes usually of different lengths, the lowermost (or first) glume usually smaller, the upper (or second) glume usually longer. Within the spikelet, each floret subtended by 2 bracts, the larger one (**lemma**) containing the flower, the smaller one (**palea**) covering the flower; the lemma and palea often enclosing the ripe fruit (grain or caryopsis); stamens usually 3 or sometimes 6, usually exserted when flowering; ovary superior, never enclosed in a sac (as in sedges); styles 2–3-parted, the stigmas often feathery.

Spikelets grouped in a variety of heads, most commonly in branching heads (**panicles**), or stalked along an unbranched stem (**rachis**) in a **raceme**, or the spikelets stalkless along an unbranched stem in a spike; spikelets breaking (disarticulating) either above or below the glumes when mature, the glumes remaining in the head if falling above the glumes, or the glumes falling with the florets if disarticulation is below the glumes.

RIGHT **Marsh-muhly** *(Muhlenbergia glomerata), a common grass of Minnesota wetlands, including calcareous fens. The spikelets are grouped into a narrow panicle.*

POACEAE
Grass Family

POACEAE GROUPS (GRASS FAMILY)

1 Stems often 2 m or more tall 2
1 Stems usually less than 2 m tall 3

2 Male and female flowers separate on same plant . . . . *Zizania palustris*
. . . . NORTHERN WILDRICE p. 586
2 Flowers perfect . . . . *Phragmites australis* | COMMON REED p. 578

3 Spikelets breaking below the glumes, the glumes falling with the florets .
. . . . **Group 1**
3 Spikelets breaking above the glumes, the glumes remaining in the head .
. . . . 4

4 Each spikelet with 1 fertile floret . . . . **Group 2**
4 Each spikelet with 2 or more fertile florets 5

5 Spikelets in spikes or spikelike racemes . . . . **Group 3**
5 Spikelets in open or dense panicles . . . . **Group 4**

POACEAE GROUP 1

Spikelets falling as entire unit (the glumes falling with the florets).

1 Spikelets falling with attached stalk and bristles . . . . *Hordeum jubatum*
. . . . FOXTAIL BARLEY p. 569
1 Spikelets falling separately, without attached stalk 2

2 Each spikelet with 2 florets . . . . *Sphenopholis obtusata*
. . . . WEDGE-GRASS p. 586
2 Each spikelet with 1 perfect floret (sterile or male florets sometimes present).
. . . . 3

3 Spikelets in a spike or raceme 4
3 Spikelets in a panicle 5

4 Spikelets ± round in outline; first and second glumes of equal lengths . . .
. . . . *Beckmannia syzigachne* | AMERICAN SLOUGHGRASS p. 555
4 Spikelets lance-shaped; first and second glumes of different lengths . . . . .
. . . . *Spartina* | CORDGRASS p. 584

5 Spikelets flattened on back sides 6
5 Spikelets flattened along margins 7

6 Spikelets awned; ligules absent . . . *Echinochloa* | BARNYARD-GRASS p. 562
6 Spikelets not awned; ligules present . . . . . . *Panicum* | PANIC-GRASS p. 576

7 Bracts below each spikelet 2; glumes absent . . . *Leersia* | CUT-GRASS p. 570
7 Bracts below each spikelet 3–4; glumes present 8

8 Panicle open . . . . *Cinna* | WOODREED p. 559
8 Panicle dense, cylinder-shaped and spikelike. *Alopecurus* | FOXTAIL p. 554

POACEAE GROUP 2

Spikelets with 1 fertile floret, disarticulating above glumes (the glumes remaining attached to the head).

1 Florets stiff and shiny . . . . *Phalaris arundinacea* . . . . REED CANARY-GRASS p. 577
1 Florets soft and papery, not shiny 2

2 Glumes much smaller than floret 3
2 At least 1 glume about same length as spikelet 4

3 Lemma 5-veined . . . . *Leersia* | CUT-GRASS p. 570
3 Lemma 3-veined . . . . *Muhlenbergia* | MUHLY p. 572

4 Florets 3, the 2 lower florets male or reduced to scales 5
4 Florets 1, with both male and female parts 6

5 Panicle dense and spikelike . . . . *Phalaris arundinacea* . . . . REED CANARY-GRASS p. 577
5 Panicle open . . . . *Hierochloe odorata* . . . . SWEET GRASS; HOLY GRASS; VANILLA GRASS p. 569

6 Stalk within spikelet (rachilla) elongate and bristlelike behind the palea . . . . *Calamagrostis* | REEDGRASS p. 558
6 Rachilla not elongated 7

7 Lemmas with 3 pronounced veins and tipped with an awn . . . . *Muhlenbergia* | MUHLY p. 572
7 Lemmas with 5 faint veins, awn (if present) from back of lemma . . . . *Agrostis* | BENTGRASS p. 553

POACEAE GROUP 3

Head a spike.

1 Spikes several to many at ends of stems; 1-sided, the spikelets all in 2 rows on lower side of rachis . . . . *Leptochloa fusca* | SPRANGLETOP p. 572
1 Spikes single; spikelets on opposite sides of rachis 2
2 Spikelets in groups of 3 at nodes of spike, the 2 side spikelets reduced and

on short stalks about 1 mm long . . . . *Hordeum jubatum* . . . . FOXTAIL BARLEY p. 569

2 Spikelets mostly 2 at each node, the spikelets all alike . . . . *Elymus* . . . . WILD RYE p. 564

POACEAE GROUP 4

Head a panicle; fertile florets more than 1; spikelets disarticulating above the glumes (the glumes remaining attached to head).

1 Glumes about as long as spikelet 2
1 Glumes much shorter than spikelet. 3

2 Florets 2. . . . . *Deschampsia cespitosa* | TUFTED HAIRGRASS p. 561
2 Florets 3 or more *Scolochloa festucacea* | SPRANGLETOP; WHITETOP p. 584

3 Lemmas with 3 prominent veins . . . . *Eragrostis* | LOVEGRASS p. 564
3 Lemmas with 5 or more prominent veins. 4

4 Edges of leaf sheaths joined for at least half of their length 5
4 Edges of leaf sheaths joined only at base 6

5 Lemmas awned . . . . *Bromus* | Brome p. 556
5 Lemmas awnless . . . . *Glyceria* | MANNAGRASS p. 565

6 Male and female flowers on separate plants; brackish habitats . . . . *Distichlis spicata* | INLAND SALTGRASS p. 561
6 Flowers perfect, with both male and female parts. 7

7 Lemmas with parallel veins extending to a blunt tip . . . . *Puccinellia* . . . . ALKALI-GRASS p. 582
7 Lemmas tapered to a point, the veins converging at tip . . . . *Poa* . . . . BLUEGRASS p. 579

RIGHT **Big bluestem** *(Andropogon gerardii), a dominant grass of tallgrass prairie, sometimes occurs in wet meadows and occasionally in calcareous fens. Its spikelets are grouped into a head of three or more spike-like clusters, leading to one of its common names–turkeyfoot.*

AGROSTIS | *Bentgrass*

Perennial grasses, clumped or spreading by rhizomes or sometimes by stolons. **Leaves** soft, flat. **Head** an open panicle. **Spikelets** small, 1-flowered, breaking above glumes; glumes ± equal length, 1-veined; floret shorter than glumes; lemma awnless or with a short straight awn; palea small or absent; stamens usually 3.

AGROSTIS | BENTGRASS

1 Plants clumped; palea ± absent. . . . . *Agrostis hyemalis* | TICKLEGRASS
1 Plants with rhizomes and|or stolons; palea present, about half as long as lemma . . . . *Agrostis stolonifera* | REDTOP; SPREADING BENTGRASS

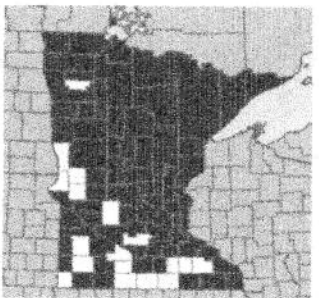

Agrostis hyemalis (Walter) Britton, Sterns & Poggenb

ROUGH BENTGRASS; TICKLEGRASS
NC **FAC** | MW **FAC** | GP **FACW**

Clumped native perennial grass. **Stems** slender, erect to reclining, 2–6 dm long. **Leaves** mostly at or near base of plant, upright to spreading, flat to inrolled, 1–2 mm wide, smooth or somewhat rough-to-touch; sheaths smooth, the ligule translucent, 1–2 mm long, rounded and usually ragged at tip. **Head** an open panicle, 1–3 dm long, the branches threadlike and spreading, the branches themselves branched and with spikelets only above their middle. **Spikelets** 1-flowered, often purple, 1–3 mm long; glumes lance-shaped, 1–3 mm long; lemma 1–2 mm long, unawned or with a short straight awn; palea absent. June–Aug. **Synonyms** *Agrostis scabra*. **Habitat** Wet meadows, bogs, ditches, streambanks, shores; more commonly in dry, sandy places.

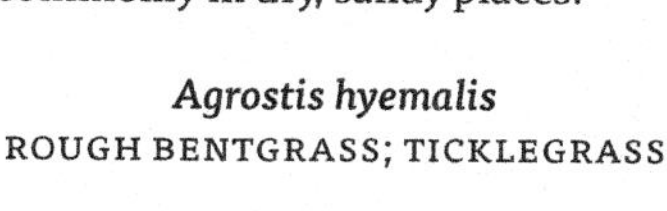

Agrostis hyemalis
ROUGH BENTGRASS; TICKLEGRASS

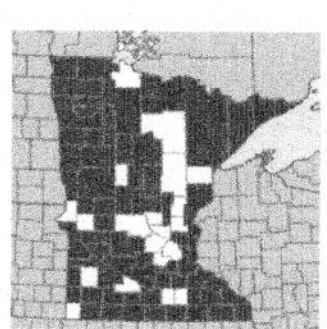

Agrostis stolonifera L.

REDTOP; SPREADING BENTGRASS
NC **FACW** | MW **FACW** | GP **FACW**

Introduced perennial grass, spreading by rhizomes and also sometimes by stolons. **Stems** erect or ± horizontal at base, 3–10 dm or more long. **Leaves** ascending, 2–8 mm wide, rough-to-touch; sheaths smooth, the ligule translucent, usually splitting at tip, 2–5 mm long. **Head** an open panicle, 3–20 cm long, the branches spreading, branched and with spikelets along their entire length. **Spikelets** 1-flowered, usually purple, 2–4 mm long; glumes lance-shaped, 1.5–2.5 mm long; lemma 2/3 length of glumes, 1–2 mm long; palea present, about half as long as lemma. July–Sept. **Synonyms** *Agrostis alba, Agrostis gigantea, Agrostis palustris*. **Habitat** Wet meadows, ditches, streambanks and shores; disturbed areas. Common; introduced from Europe as a pasture grass, now naturalized throughout most of USA and s Canada.

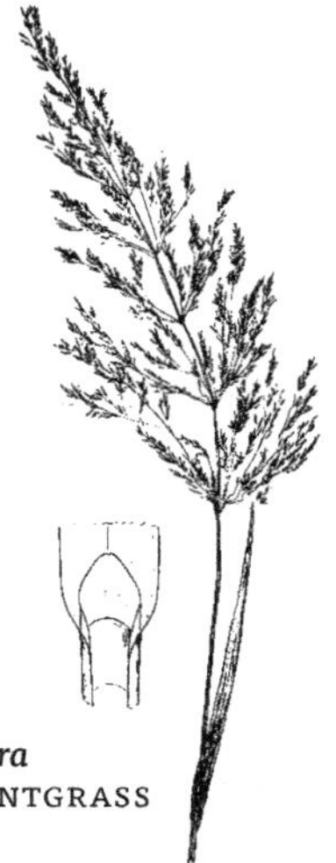

Agrostis stolonifera
REDTOP; SPREADING BENTGRASS

ALOPECURUS | *Foxtail*

Annual or perennial grasses. **Leaves** soft, flat. **Stems** erect or ± horizontal at base. Heads densely flowered, cylindric, spikelike panicles. **Spikelets** 1-flowered, flattened, breaking below the glumes; glumes equal length, 3-nerved, often silky hairy on back, awnless; lemma about as long as glumes or shorter, awned from the back, the awn shorter to longer than the glume tips; palea absent.

ALOPECURUS | FOXTAIL

1 Plants single or in small clumps; spikes not bristly in outline; awns straight, shorter or to 1 mm longer than tip of glumes . . . . *Alopecurus aequalis* | SHORT-AWN FOXTAIL

1 Plants densely clumped; spikes bristly in outline; awns bent, 2–4 mm longer than glume tips . . . . *Alopecurus carolinianus* | CLUMPED FOXTAIL

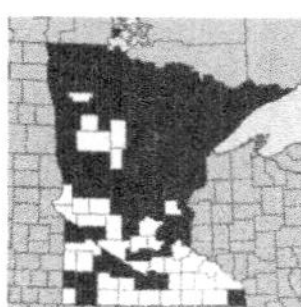

Alopecurus aequalis Sobol.

SHORT-AWN FOXTAIL
NC **OBL** | MW **OBL** | GP **OBL**

Native annual or short-lived perennial grass. **Stems** single or in small clumps, slender, erect to ± horizontal, 2–6 dm long, often rooting at the nodes. **Leaves** 1–5 mm wide, finely rough-to-touch above; ligule membranous, rounded to elongate, 2–7 mm long. **Head** an erect, spikelike panicle, 2–7 cm long and 3–5 mm wide. **Spikelets** 1-flowered; glumes 2–3 mm long, blunt-tipped, hairy on the keel and veins; lemma about equaling the glumes, awned from back, the awn straight, to 1.5 mm longer than glume tips. June–Aug. **Habitat** Shallow water or mud of wet meadows, marshes, ditches, springs, open bogs, fens, shores and streambanks; sometimes where calcium-rich.

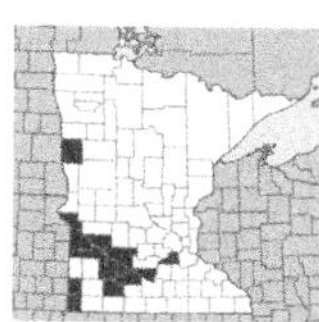

Alopecurus carolinianus Walter
CLUMPED FOXTAIL
NC **FACW** | MW **FACW** | GP **FACW**

Densely clumped native annual grass. **Stems** erect to upright, 1–4 dm long. **Leaves** 1–3 mm wide, finely rough-to-touch above; ligule membranous, rounded to elongate, 1–5 mm long. **Head** a cylindric, spikelike panicle, 1–5 cm long and 3–5 mm wide. **Spikelets** 1-flowered; glumes 2–3 mm long, blunt- tipped, hairy on keel; lemma about as long as glumes, awned from back, the awn bent near middle and 2–3 mm longer than glume tips. May–July. **Habitat** Mud flats, temporary ponds, wet meadows, marshes, low prairie, fallow fields.

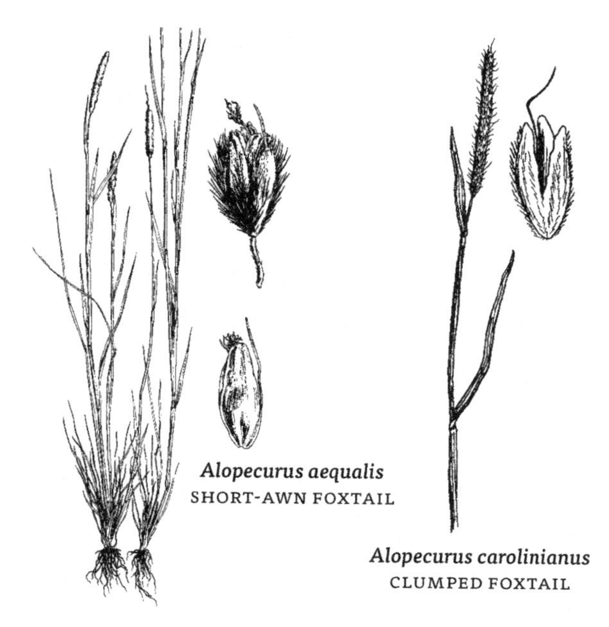

Alopecurus aequalis
SHORT-AWN FOXTAIL

Alopecurus carolinianus
CLUMPED FOXTAIL

BECKMANNIA | *Sloughgrass*

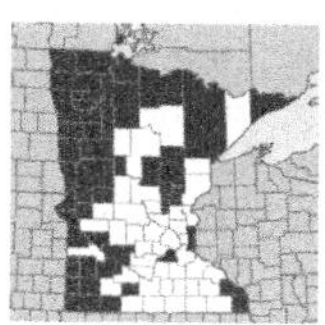

Beckmannia syzigachne (Steud.) Fernald
AMERICAN SLOUGHGRASS
NC **OBL** | MW **OBL** | GP **OBL**

Stout native annual grass. **Stems** single or in small clumps, 4–10 dm long. **Leaves** flat, 3–10 mm wide, rough-to-touch; sheaths overlapping, smooth, the upper sheath often loosely enclosing lower part of panicle; ligule membranous, rounded to acute, 3–6 mm long. **Head** of many 1-sided spikes in a narrow panicle 10–30 cm long, the panicle branches erect, overlapping, 1–5 cm long; each spike 1–2 cm long, with several to many spikelets in 2 rows on the rachis. **Spikelets** 1–2-flowered, overlapping, nearly round, 2–4 mm long, straw-colored when mature, breaking below the glumes; glumes equal, broad, inflated along midvein, with a short, slender tip; lemma about as long as glumes but narrower; palea nearly as long as lemma. June–Sept. **Habitat** Wet meadows, marshes, ditches, shores and streambanks.

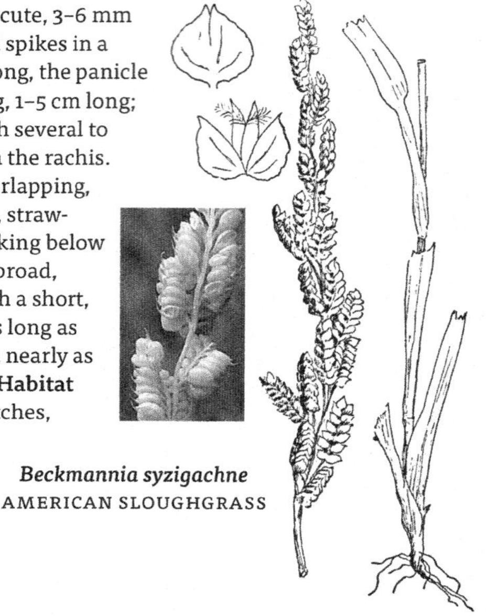

Beckmannia syzigachne
AMERICAN SLOUGHGRASS

BROMUS | *Brome*

Perennial grasses. **Leaves** flat; sheaths closed to near top. **Head** a panicle of lax spikelets. **Spikelets** with several to many flowers, breaking above the glumes; glumes shorter than lemmas; lemmas awned (in species included here); stamens usually 3.

BROMUS | BROME

1 Lemmas hairy across back ..... *Bromus altissimus* | EAR-LEAVED BROME
1 Lemmas smooth on back, lemma margins fringed with hairs .... *Bromus ciliatus* | FRINGED BROME

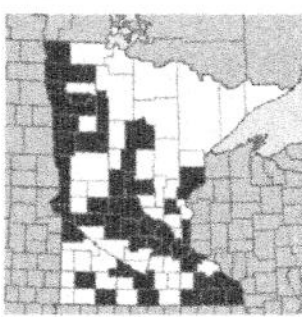

Bromus altissimus Pursh.

EAR-LEAVED BROME
NC **FACW** | MW **FACW** | GP **FACW**

Native perennial grass. **Stems** single or in small clumps, ± smooth, 6–15 dm long. **Leaves** flat, 8–20 along stem; 10–15 mm wide, with a pair of auricles at base; sheaths with a dense ring of hairs at top. **Head** a panicle 1–2 dm long, the branches spreading or drooping. **Spikelets** several-flowered, 2–3 cm long, first glume 5–8 mm long, awl-shaped, second glume wider, 6–10 mm long; lemmas 10–12 mm long, hairy, awned, the awn 2–7 mm long. **Synonyms** *Bromus latiglumis*. **Habitat** Floodplain forests, thickets and streambanks, sometimes in rocky woods and slopes.

Similar to **Canada brome** *(Bromus pubescens), a species of mostly drier woods, but top of sheaths of ear-leaved brome have a ring of dense hairs, and leaf blades have well-developed auricles.*

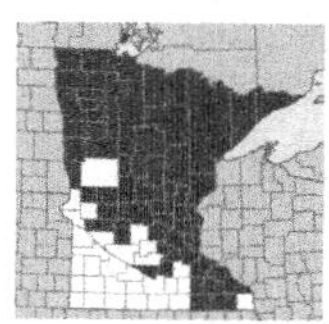

Bromus ciliatus L.
FRINGED BROME
NC **FACW** | MW **FACW** | GP **FAC**

Native perennial grass, rhizomes absent. **Stems** single or few together, smooth or hairy at nodes, 5–12 dm long. **Leaves** flat, 4–10 mm wide, usually with long, soft hairs mainly on upper surface; sheaths usually with long hairs; ligule membranous, short, to 2 mm long, ragged across tip. **Head** a loose, open panicle 1–3 dm long, the branches usually drooping. **Spikelets** large, 4–10-flowered, 1.5–3 cm long and 5–10 mm wide; glumes usually ± smooth, lance-shaped, the first glume 4–9 mm long, the second glume 6–10 mm long, often tipped with a short awn; lemma 10–15 mm long, ± smooth on back, usually long-hairy along lower margins, tipped with an awn 2–6 mm long; palea about as long as body of lemma. July–Sept. **Habitat** Streambanks, shores, thickets, sedge meadows, fens, marshes; also in moist woods.

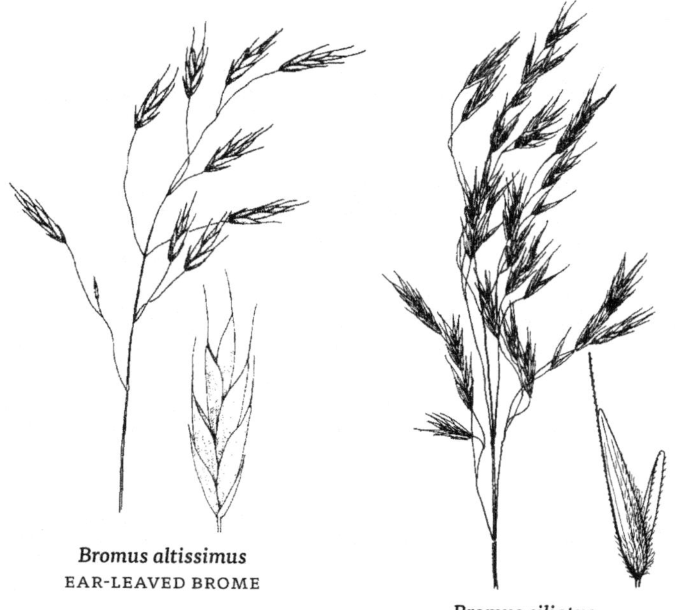

Bromus altissimus
EAR-LEAVED BROME

Bromus ciliatus
FRINGED BROME

CALAMAGROSTIS | *Reedgrass*

Perennial grasses, spreading by rhizomes. **Stems** single or in clumps. **Leaves** flat or inrolled, green or waxy blue-green, smooth or rough-to-touch; sheaths smooth; ligule large, membranous, usually with an irregular, ragged margin. **Head** a loose and open, or dense and contracted panicle. **Spikelets** 1-flowered, breaking above glumes; glumes nearly equal, lance-shaped; lemma shorter than glumes, lance-shaped, awned from back, the awn about as long as lemma, the base of lemma (callus) bearded with a tuft of hairs, these shorter to as long as lemma; palea shorter than lemma; stamens 3.

CALAMAGROSTIS | REEDGRASS

1 Panicle ± loose and open, the branches ascending to spreading; leaves ± lax, flat, 2–6 mm wide. . . . . *Calamagrostis canadensis* | BLUEJOINT

1 Panicle contracted, the branches short, ascending to appressed; leaves stiff, often inrolled, 1–4 mm wide when flattened . . . . *Calamagrostis stricta* . . . . NARROW-SPIKE REEDGRASS

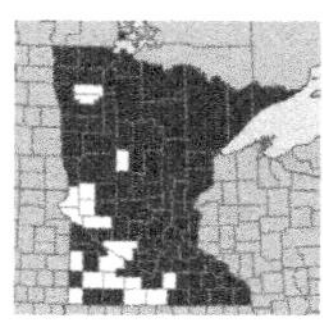

Calamagrostis canadensis (Michx.) P. Beauv.

BLUEJOINT

NC **OBL** | MW **OBL** | GP **FACW**

Native perennial grass, from creeping rhizomes. **Stems** erect, in small clumps, 6–15 dm long, often rooting from lower nodes when partly underwater. **Leaves** flat, green to waxy blue-green, 3–8 mm wide, rough-to-touch on both sides; sheaths smooth; ligules 3–7 mm long. **Head** a ± open panicle, 8–20 cm long, the branches upright or spreading. **Spikelets** 1-flowered, 2–6 mm long; glumes ± equal, 2–4 mm long, smooth or finely rough-hairy on back; lemma ± smooth, awned from middle of back, the awn straight, base with dense callus hairs about as long as lemma. June–Aug. **Habitat** Common to abundant species of wet meadows, shallow marshes, calcareous fens, streambanks, thickets.

Calamagrostis canadensis
BLUEJOINT

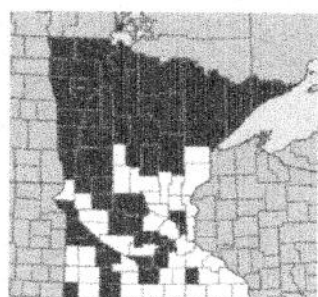

Calamagrostis stricta (Timm) Koeler
NARROW-SPIKE REEDGRASS
NC **FACW** | MW **FACW** | GP **FACW**

Native perennial grass, spreading by rhizomes; plants waxy blue-green. **Stems** erect, 3–12 dm long. **Leaves** stiff, often inrolled, 1–4 mm wide when flattened. **Head** a narrow panicle, 5–15 cm long, the branches short, upright to erect. **Spikelets** 1-flowered; glumes 3–6 mm long, smooth or rough-hairy on back; lemma rough-hairy, 2–4 mm long, awned, the awn straight, from near middle of back, base with many callus hairs, half to as long as lemma. June–Sept. **Synonyms** *Calamagrostis inexpansa, C. lacustris, C. neglecta*. **Habitat** Wet meadows, shallow marshes, shores, streambanks; rocky shore of Lake Superior.

Similar to **bluejoint** *(Calamagrostis canadensis), but the head narrow and crowded and the leaves often inrolled.*

Calamagrostis stricta
NARROW-SPIKE REEDGRASS

CINNA | *Woodreed*

Tall perennial grasses, rhizomes weak or absent. **Leaves** wide, flat and lax; ligule brown, membranous, with an irregular, jagged margin. **Head** a large, closed to open panicle, the branches upright to spreading or drooping. **Spikelets** small, 1-flowered, laterally compressed, breaking below the glumes; glumes nearly equal, lance-shaped, keeled; lemma similar to glumes, with a short awn from just below the tip; palea shorter than lemma; stamens 1.

CINNA | WOODREED

1 Panicle ± crowded and narrow, the branches upright; second glume 4–6 mm long .... *Cinna arundinacea* | COMMON WOODREED
1 Panicle open, the branches spreading to drooping; second glume 2–4 mm long .... *Cinna latifolia* | DROOPING REEDGRASS

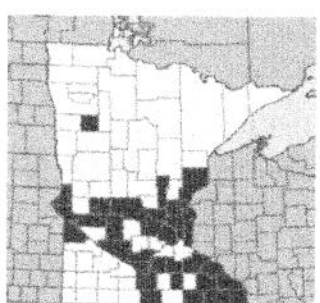

Cinna arundinacea L.
COMMON WOODREED
NC **FACW** | MW **FACW** | GP **FACW**

Native perennial grass, rhizomes weak or absent. **Stems** 1 or few together, erect, 6–15 dm long, often swollen at base. **Leaves** 4–12 mm wide, margins rough-to-touch; sheaths smooth; ligule red-brown, 3–10 mm long. **Head** a narrow panicle, dull gray-green, 1–3 dm long, the branches upright. **Spikelets** 1-flowered; glumes narrowly lance-shaped, 3–5 mm long, the first glume 1-veined, the second glume 3-veined, usually rough-hairy; lemma 3–5 mm long, rough-hairy on back, usually with an awn to 0.5 mm long, attached just below tip and mostly shorter than lemma tip. Aug–Sept. **Habitat** Swamps, floodplain forests, streambanks, pond margins, moist woods.

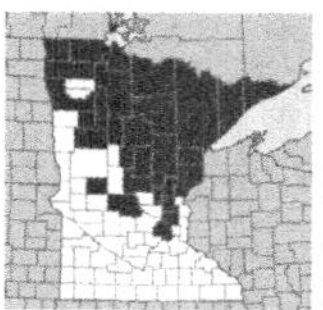

Cinna latifolia (Trevir.) Griseb.
DROOPING REEDGRASS
NC **FACW** | MW **FACW** | GP **OBL**

Native perennial grass, with weak rhizomes. **Stems** single or in small groups, erect, 5–13 dm long, not swollen at base. **Leaves** 5–15 mm wide, usually rough-to-touch; sheaths smooth to finely roughened; ligule pale, 2–7 mm long. **Head** a loose, open panicle, pale green, satiny, 1–3.5 dm long, the branches spreading to drooping. **Spikelets** 1-flowered; glumes narrowly lance-shaped, 1-veined, 2–4 mm long; lemma 2–4 mm long, finely rough-hairy on back, usually with an awn to 1.5 mm long from just below the tip, the awn usually longer than the tip. July–Aug. **Habitat** Wet woods, swamps, springs.

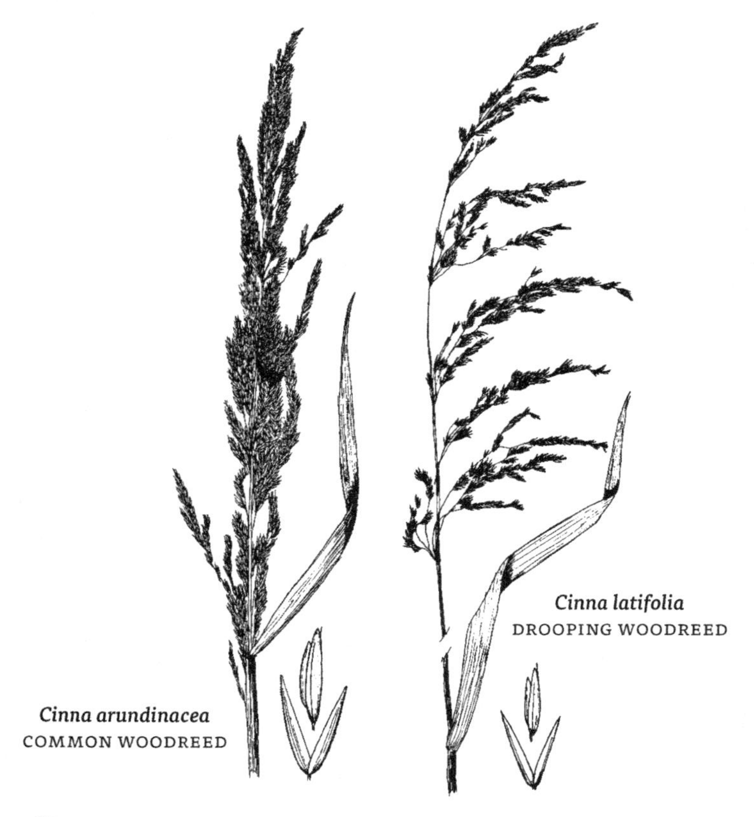
Cinna arundinacea
COMMON WOODREED

Cinna latifolia
DROOPING WOODREED

DESCHAMPSIA | *Hairgrass*

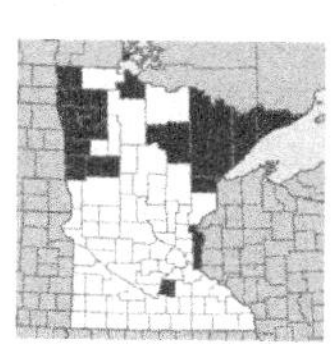

Deschampsia cespitosa (L.) P. Beauv.
TUFTED HAIRGRASS
NC **FACW** | MW **FACW** | GP **FACW**

Densely clumped native perennial grass. **Stems** stiff, erect, 3–10 dm long. **Leaves** mostly from base of plant, usually shorter than head, flat or inrolled, 2–4 mm wide; sheaths smooth; ligule white, translucent, 3–10 mm long. **Head** a narrow to open panicle, 1–4 dm long, the panicle branches threadlike, upright to spreading, the lower branches in groups of 2–5, flowers mostly near branch tips. **Spikelets** 2-flowered, purple-tinged, fading to silver with age, 2–5 mm long, breaking above the glumes; glumes shiny, 2–5 mm long, the first glume slightly shorter than second glume; lemma smooth, 2–4-toothed across the flat tip, awned from near base on back, the awn shorter to about as long as lemma. June–July. **Habitat** Wet meadows, streambanks, shores, calcium-rich seeps and springs, rocky shores of Great Lakes.

Deschampsia cespitosa
TUFTED HAIRGRASS

DISTICHLIS | *Saltgrass*

Distichlis spicata (L.) Greene
INLAND SALTGRASS
NC **FACW** | MW **FACW** | GP **FACW**

Short, native perennial grass, spreading by scaly rhizomes and forming patches; the male and female flowers on separate plants. **Stems** stiff, erect, 1–3 dm long. **Leaves** upright, the upper often longer than the head, mostly inrolled, 5–10 cm long and 0.5–3 mm wide, smooth or with sparse hairs; sheaths overlapping, smooth or sparsely hairy, usually long-hairy at collar; ligule small. **Head** an unbranched, narrow, spikelike panicle, 3–7 cm long. **Spikelets** several to many, upright, 8–20 mm long; male spikelets straw-colored, female spikelets green-gray, breaking above the glumes; glumes unequal, 1–5 mm long; lemmas ovate, 3–6 mm long. June–Sept. **Synonyms** *Distichlis stricta*. **Habitat** Seasonally wet, brackish flats, shores and disturbed areas.

ECHINOCHLOA | *Barnyard-grass*

Large, weedy, annual grasses. **Stems** single or several together, erect to ± horizontal, to 1 m or more long. **Leaves** flat, wide and smooth; sheaths smooth or hairy; ligules absent. **Head** a dense panicle, the branches crowded with spikelets forming racemes or spikes. **Spikelets** with 1 terminal fertile floret and 1 sterile floret, breaking below the glumes, nearly stalkless; glumes unequal, the first glume 3-veined, to half the length of second glume, the second glume 5-veined; sterile lemma similar to second glume, awned or awnless; fertile lemma smooth and shiny.

ECHINOCHLOA | BARNYARD-GRASS

1 Lower leaf sheaths rough-hairy; spikelets each with 2 awns . . . . *Echinochloa walteri* | SALTMARSH COCKSPUR GRASS
1 Leaf sheaths smooth; spikelets with usually 1 awn (from sterile lemma) . **2**

2 Fertile lemma rounded or broadly tapered to a thin, membranous, withered beak . . . . *Echinochloa crusgalli* | BARNYARD-GRASS
2 Fertile lemma tapered to a stiff, persistent beak . . . . . *Echinochloa muricata* . . . . BARNYARD GRASS

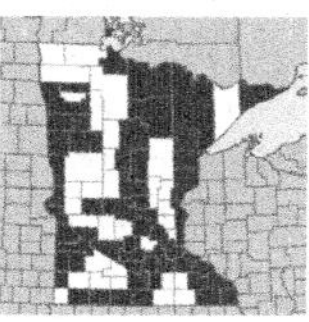

Echinochloa crusgalli (L.) P. Beauv.

BARNYARD-GRASS
NC **FAC** | MW **FACW** | GP **FAC**

Introduced, weedy annual grass. **Stems** 1 m or more long. **Leaves** 7–30 mm wide; sheaths smooth; ligule absent. **Head** an erect, green to purple panicle, 1–2.5 dm long; panicle branches spreading to erect, long-hairy, some of the hairs as long or longer than spikelets (excluding spikelet awns). **Spikelets** 3–5 mm long (excluding awns); glumes awnless; sterile lemma awnless or with an awn to 4 cm or more long; tip of fertile lemma firm, shiny, rounded or broadly tapered to a point, the beak usually green and withered, the lemma body and beak separated by a line of tiny hairs. July–Sept. **Habitat** Shores, wet meadows, ditches, streambanks, mud flats, moist disturbed areas. Introduced from Europe, naturalized across much of s Canada and most of USA.

Echinochloa crusgalli
BARNYARD-GRASS

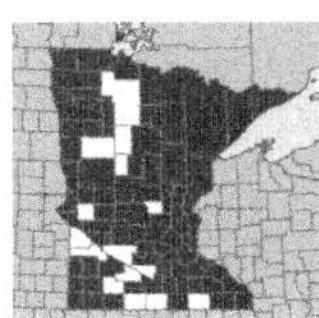

Echinochloa muricata (P. Beauv.) Fernald
BARNYARD-GRASS
NC **OBL** | MW **OBL** | GP **FACW**

Native, often weedy, annual grass. **Stems** 1 m or more long. **Leaves** 5–30 mm wide; sheaths smooth; ligule absent. **Head** a green to purple panicle, sometimes strongly purple, 1–3 dm long, panicle branches spreading, hairs on branches absent or to 3 mm long and shorter than spikelets. **Spikelets** 2–4 mm long (excluding awns); glumes awnless; sterile lemma awnless or with an awn 5–10 mm long; tip of fertile lemma firm, shiny, gradually tapered to the stiff beak, the lemma body and beak not separated by a line of tiny hairs (the beak itself often short-hairy). July–Sept. **Synonyms** *Echinochloa pungans*. **Habitat** Shores, streambanks and ditches, where sometimes in shallow water.

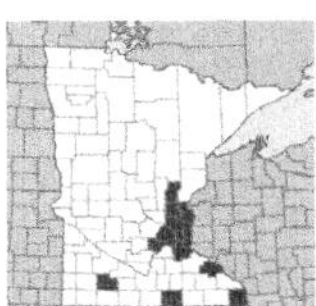

Echinochloa walteri (Pursh) Heller
SALTMARSH COCKSPUR GRASS
NC **OBL** | MW **OBL** | GP **OBL**

Tall native annual grass. **Stems** usually erect, 1–2 m long. **Leaves** 10–25 mm wide; lower sheaths usually rough-hairy; ligule absent. **Head** a dense panicle, often nodding, 1–3 dm long. **Spikelets** ± hidden by awns, the awns 1–3 cm long from sterile lemmas and 2–10 mm long from second glume; fertile lemma oval, with a small, withering tip, but not separated by a line of hairs as in *Echinochloa crusgalli*. Aug–Sept. **Habitat** Streambanks, lakeshores, ditches.

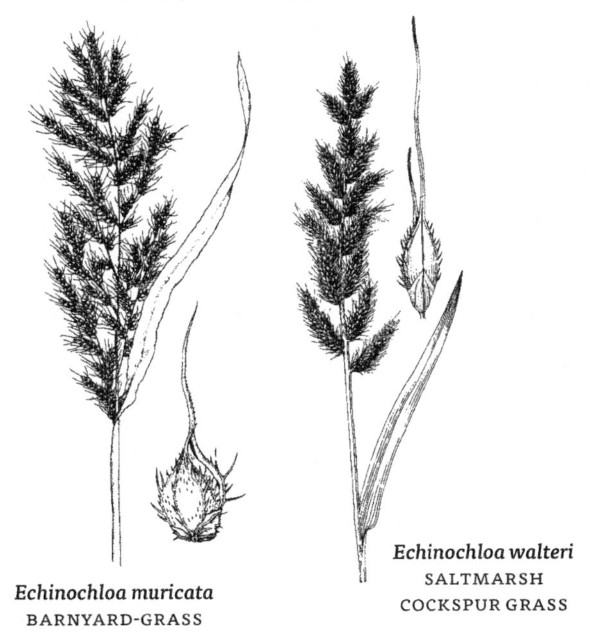

Echinochloa muricata
BARNYARD-GRASS

Echinochloa walteri
SALTMARSH COCKSPUR GRASS

ELYMUS | *Wildrye*

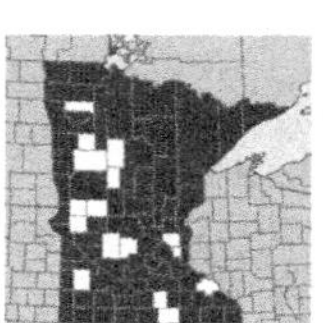

Elymus virginicus L.
VIRGINIA WILDRYE
NC **FACW** | MW **FACW** | GP **FAC**

Clumped native perennial grass. **Stems** 6–12 dm long. **Leaves** flat, 5–15 mm wide, rough-to-touch on both sides; sheaths smooth; ligules short. **Head** an erect, densely flowered spike, 5–15 cm long, the base of spike often covered by top of upper sheath. **Spikelets** usually 2 at each node, 2–4-flowered, breaking below glumes; glumes firm, 1–2 mm wide, yellowish, bowed-out at base, tapered to a straight awn about 1 cm long; lemmas 6–9 mm long, smooth to hairy, usually with a straight awn to 3 cm long; stamens 3. July–Aug. **Habitat** Floodplain forests, thickets, streambanks.

Elymus virginicus
VIRGINIA WILDRYE

ERAGROSTIS | *Lovegrass*

Annual grasses (those included here), perfect-flowered or with male and female flowers on different plants. **Stems** clumped, or spreading and rooting at lower nodes and with creeping stolons. **Leaves** with short, flat to folded blades; sheaths hairy near top; ligule a ring of short hairs. Heads usually many, in an open or narrow panicle. **Spikelets** few- to many-flowered, breaking above glumes, laterally compressed, the florets overlapping; glumes unequal; lemmas 3-veined; palea shorter than lemma, 2-veined.

ERAGROSTIS | LOVEGRASS

1 Plants mat-forming; base of stems lying along ground, rooting at lower nodes, the nodes bearded with hairs; spikelets 3–10 mm long . . . . *Eragrostis hypnoides* | CREEPING LOVEGRASS

1 Plants not mat-forming; stems erect, not rooting at lower nodes, the nodes smooth; spikelets to 3 mm long . . . . *Eragrostis frankii* . . . . SANDBAR LOVEGRASS

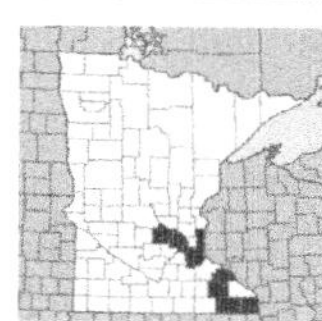

Eragrostis frankii C. A. Meyer
SANDBAR LOVEGRASS
NC **FACW** | MW **FACW** | GP **FACW**

Densely clumped native annual grass. **Stems** branched, 1–5 dm long. **Leaves** smooth, 1–4 mm wide; sheaths smooth but long-hairy at top; ligule short-hairy. **Head** an open panicle, 5–20 cm long, the branches mostly ascending. **Spikelets** 3–6-flowered, 2–3 mm long and 1–2 mm wide. Aug–Sept. **Habitat** Wet, muddy areas, streambanks, sandbars, roadside ditches, cultivated fields.

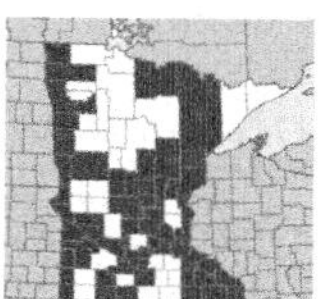

Eragrostis hypnoides (Lam.) BSP.
CREEPING LOVEGRASS
NC **OBL** | MW **OBL** | GP **OBL**

Mat-forming native annual grass. **Stems** mostly spreading and rooting at lower nodes, 5–15 cm long, smooth but short-hairy at nodes. **Leaves** flat to folded, 1–5 cm long and 1–3 mm wide, upper surface hairy; sheaths smooth except for hairs at top and sometimes along margins; ligule of short hairs about 0.5 mm long. **Head** a loose panicle, 2–6 cm long. **Spikelets** 10–35-flowered, linear, 3–10 mm long; glumes 1-veined, 0.5–1.5 mm long; lemma smooth and shiny, 1–2 mm long. July–Sept. **Habitat** Wet, sandy or muddy shores and streambanks, sand bars, mud flats.

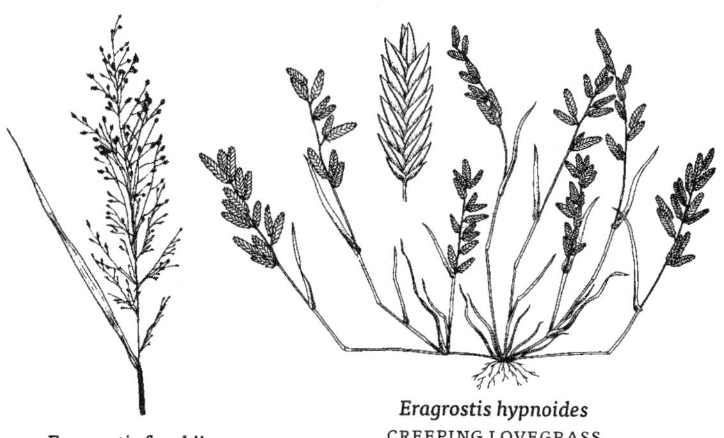

Eragrostis frankii
SANDBAR LOVEGRASS

Eragrostis hypnoides
CREEPING LOVEGRASS

GLYCERIA | *Mannagrass*

Perennial grasses, loosely clumped or spreading by rhizomes. **Stems** upright, or reclining at base and often rooting at lower nodes. **Leaves** flat; sheaths tubular, the margins mostly closed. **Head** an open panicle. **Spikelets** 3-flowered, ovate to linear, ± round in section or somewhat flattened, breaking above the glumes; glumes unequal, shorter than lemmas, 1-veined; lemmas unawned, usually 7-veined; palea about as long as lemma; stamens 3 or 2.

GLYCERIA | MANNAGRASS

1 Spikelets linear-cylindric, 10 mm long or longer 2
1 Spikelets ovate, 2 to 7 mm long. 3

2 Leaves less than 5 mm wide; lemmas ± smooth. . . . . *Glyceria borealis* . . . . NORTHERN MANNAGRASS
2 Leaves 5 mm or more wide; lemmas finely hairy. . . . . *Glyceria septentrionalis* | EASTERN MANNAGRASS

3 Spikelets 3–4 mm wide; veins of lemma not raised . . . . *Glyceria canadensis* . . . . RATTLESNAKE-MANNAGRASS
3 Spikelets 2–2.5 mm wide; veins of lemma raised 4

4 Spikelets 4–7 mm long . . . . . . *Glyceria grandis* | AMERICAN MANNAGRASS
4 Spikelets 2–4 mm long. . . . . *Glyceria striata* | FOWL-MANNAGRASS

Glyceria borealis (Nash) Batchelder
NORTHERN MANNAGRASS
NC **OBL** | MW **OBL** | GP **OBL**

Native perennial grass. **Stems** erect or reclining at base, often rooting from lower nodes, 6–12 dm long. **Leaves** flat or folded, 2–5 mm wide, smooth; sheaths smooth; ligule 3–10 mm long. **Head** a panicle, 2–4 dm long, with stiff, erect to ascending, branches to 8–12 cm long, each with several spikelets. **Spikelets** linear, mostly 6–12-flowered, 1–1.5 cm long; glumes rounded at tip, 2–3 mm long; lemmas 3–4 mm long, 7-veined. June–Aug. **Habitat** Marshes, ponds, stream, ditches, often in shallow water or mud.

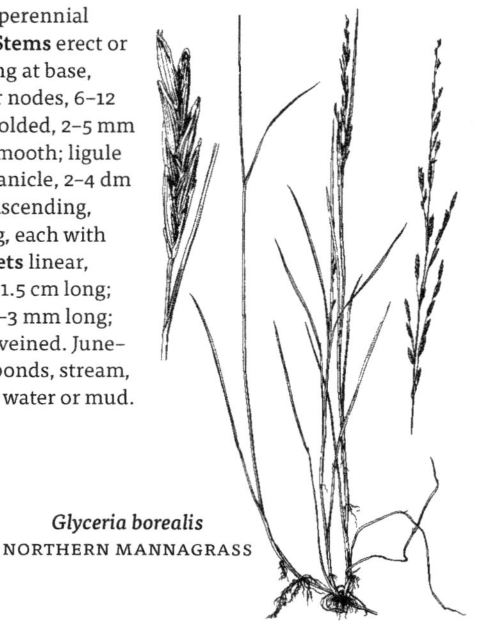

Glyceria borealis
NORTHERN MANNAGRASS

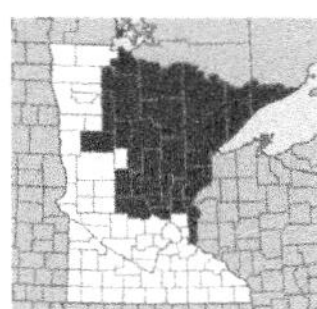

Glyceria canadensis (Michx.) Trin.
RATTLESNAKE-MANNAGRASS
NC **OBL** | MW **OBL** | GP **OBL**

Native perennial grass. **Stems** single or few together, erect, 6–15 dm long. **Leaves** 3–7 mm wide, upper surface rough; ligule 2–5 mm long. **Head** an open panicle, 1–3 dm long, the branches drooping, with spikelets mostly near tips. **Spikelets** ovate, 5–10-flowered, 5–7 mm long, the florets spreading; glumes 2–3 mm long, the first glume lance-shaped, the second glume ovate; lemma veins not raised. **Habitat** Marshes, swamps, thickets, open bogs.

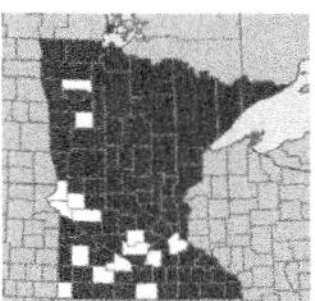

Glyceria grandis S. Wats.
AMERICAN MANNAGRASS
NC **OBL** | MW **OBL** | GP **OBL**

Loosely clumped native perennial grass. **Stems** erect, stout, 1–1.5 m long and 4–6 mm wide. **Leaves** flat, smooth, 6–12 mm wide; sheaths smooth; the ligule translucent, 3–6 mm long. **Head** a large, open, much-branched panicle, 2–4 dm long, usually nodding at tip, branches lax and drooping when mature. **Spikelets** ovate, purple, slightly flattened, 5–9-flowered, 4–7 mm long; glumes pale or white, 1–3 mm long; lemmas purple, 2–3 mm long. June–Sept. **Habitat** Marshes, ditches, streams, lakes and ponds, open bogs, fens; usually in shallow water or mud.

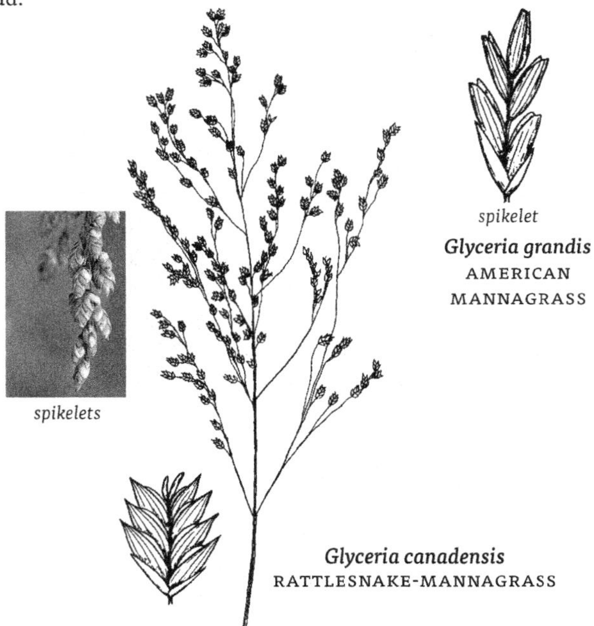

Glyceria grandis
AMERICAN MANNAGRASS

Glyceria canadensis
RATTLESNAKE-MANNAGRASS

Glyceria septentrionalis A. Hitchc.
EASTERN MANNAGRASS
[NC **OBL** | MW **OBL**]

Native perennial grass. **Stems** somewhat fleshy, often ± horizontal at base and rooting from lower nodes, 1–1.5 m long. **Leaves** 6–10 mm wide; sheaths smooth; ligule large. **Head** a narrow panicle, 2–4 dm long, the branches to 10 cm long, each with several spikelets. **Spikelets** 1–2 cm long, 8–14-flowered; glumes 2–4 mm long; lemmas green or pale, 4–5 mm long, spreading when mature; palea often longer than lemma. June–Aug. **Habitat** Swamps, thickets, shallow water of pond margins, wet depressions in forests.

Eastern mannagrass *may occur in se Minnesota, but presence in state not verified.*

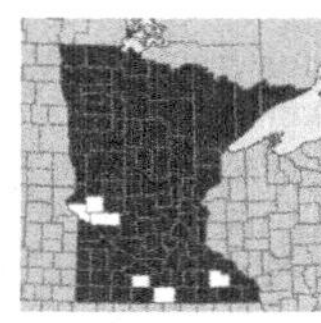

Glyceria striata (Lam.) A. Hitchc.
FOWL-MANNAGRASS
NC **OBL** | MW **OBL** | GP **OBL**

Loosely clumped native perennial grass; plants pale green. **Stems** erect, slender, 3–10 dm long. **Leaves** flat or folded, smooth, 2–6 mm wide; sheaths smooth; ligule 1–3 mm long. **Head** an open, loose panicle, 1–2 dm long, the branches lax, drooping. **Spikelets** ovate, often purple, 3–7-flowered, 3–4 mm long; glumes 0.5–1.5 mm long; lemma 2 mm long, strongly 7-veined. June–Aug. **Habitat** Common species of swamps, thickets, low areas in forests, wet meadows, springs, streambanks.

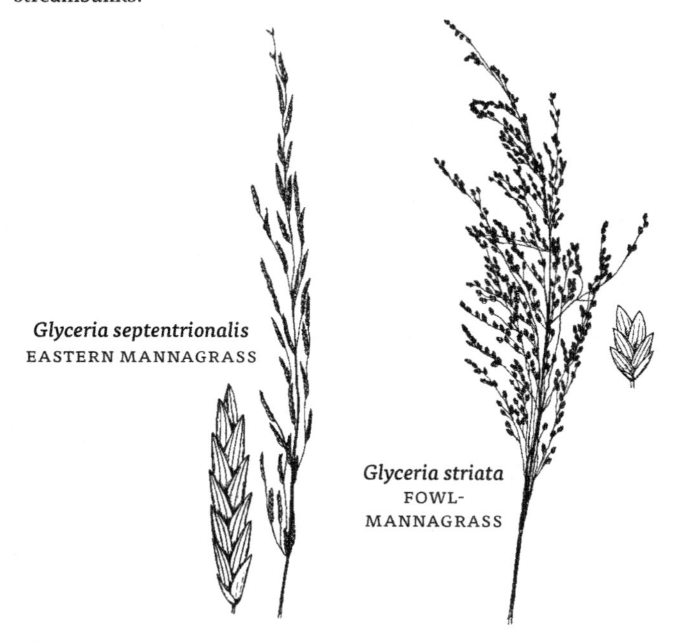

Glyceria septentrionalis
EASTERN MANNAGRASS

Glyceria striata
FOWL-MANNAGRASS

HIEROCHLOE | *Sweetgrass*

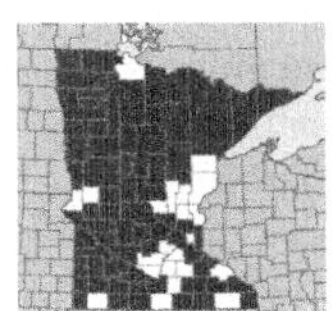

Hierochloe hirta (Schrank) Borbás

SWEETGRASS; HOLY GRASS; VANILLA GRASS

NC **FACW** | MW **FACW** | GP **FACW**

Native perennial grass, from creeping rhizomes; plants sweet-scented, especially when dried. **Stems** erect, 2–6 dm tall, smooth. **Leaves** flat, 2–6 mm wide, smooth or short-hairy; stem leaves short, 1–4 cm long, leaves on sterile shoots much longer; sheaths smooth or short-hairy at top; ligule membranous, 1–4 mm long. **Head** a pyramid-shaped panicle, 5–10 cm long, the branches spreading to drooping. **Spikelets** 3-flowered, the lower 2 florets male, the terminal spikelet perfect, golden brown, or green or purple at base and golden near tips, 5 mm long, breaking above the glumes; glumes ovate, shiny, 4–6 mm long; lemmas 3–4 mm long, the male lemma hairy. May–July. **Synonyms** *Hierochloe odorata*. **Habitat** Wet meadows, shores, low prairie; often where sandy.

Hierochloe hirta
SWEETGRASS; HOLY GRASS; VANILLA GRASS

HORDEUM | *Barley*

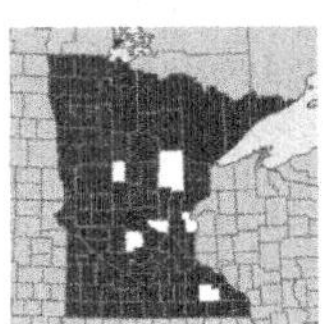

Hordeum jubatum L.

FOXTAIL BARLEY

NC **FAC** | MW **FAC** | GP **FACW**

Clumped native perennial grass; plants smooth to densely hairy. **Stems** erect or reclining at base, 2–7 dm long. **Leaves** usually flat, 2–5 mm wide; ligule less than 1 mm long. **Head** a terminal spike, erect to nodding, 3–10 cm long, appearing bristly due to the long, spreading awns from glumes and lemmas. **Spikelets** 1-flowered, 3 at each node, the center spikelet fertile, stalkless, the 2 lateral spikelets sterile, short-stalked, reduced to 1–3 spreading awns; the 3 spikelets at each node falling as a unit; glumes of fertile spikelet awnlike; lemma lance-shaped, tipped by a long awn; the glume and lemma awns 2–7 cm long. June–Sept. **Habitat** Wet meadows, ditches, shores, shallow marshes, disturbed areas; often where brackish.

LEERSIA | *Cut-grass*

Perennial grasses, spreading by long rhizomes. **Stems** slender, somewhat weak. **Leaves** flat, smooth to hairy or rough-to-touch; ligules membranous, short. **Head** an open panicle. **Spikelets** 1-flowered, laterally compressed, falling as a unit from the stalk; glumes absent; lemmas smooth to bristly hairy, 5-veined; palea narrow, about as long as lemma; stamens 2–3 (our species).

LEERSIA | CUT-GRASS

1 Spikelets ovate, 3–4 mm wide...... *Leersia lenticularis* | CATCHFLY GRASS
1 Spikelets linear, 1–2 mm wide.. 2

2 Stems round in section; leaves very rough-to-touch; spikelets 4–6 mm long .... *Leersia oryzoides* | RICE CUT-GRASS
2 Stems flattened in section; leaves smooth or finely roughened; spikelets to 3.5 mm long .... *Leersia virginica* | WHITE GRASS

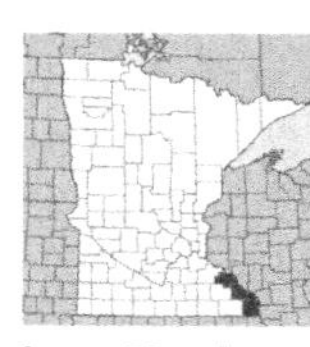

Leersia lenticularis Michx.

CATCHFLY GRASS

NC **OBL** | MW **OBL** | GP **OBL** |*special concern*

Native perennial grass, from creeping rhizomes. **Stems** 1–1.5 m long. **Leaves** lax, smooth to soft-hairy, 1–2 cm wide; sheaths smooth, or hairy at top; ligule flat-topped, 1 mm long. **Head** a panicle, 1–2 dm long, often drooping, the branches spreading, each branch with 1–4 spikelike racemes 1–2 cm long. **Spikelets** 1-flowered, pale, flat, nearly round in section, 4–5 mm long, short-stalked, closely overlapping one another; glumes absent; lemma 4–6 mm long, the veins and keel fringed with bristly hairs. **Habitat** River floodplains in se Minnesota. **Status** Special concern.

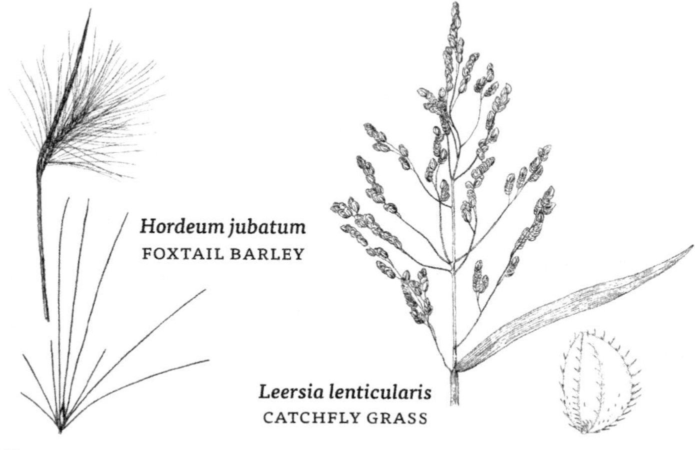

Hordeum jubatum
FOXTAIL BARLEY

Leersia lenticularis
CATCHFLY GRASS

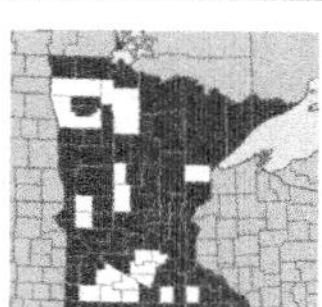

Leersia oryzoides (L.) Swartz
RICE CUT-GRASS
NC **OBL** | MW **OBL** | GP **OBL**

Loosely clumped native perennial grass, from creeping rhizomes. **Stems** weak and sprawling, rooting at nodes, 1–1.5 m long. **Leaves** flat, 2–3 dm long and 5–10 mm wide, rough-to-touch, the margins fringed with short spines; sheaths rough-hairy; ligule flat-topped, 1 mm long. **Head** an open panicle at end of stem and from leaf axils (these often partly enclosed by leaf sheaths), 1–2 dm long, the branches ascending to spreading. **Spikelets** 1-flowered, oval, 5 mm long and 1–2 mm wide, compressed, pale green, turning brown with age; glumes absent; lemma covered with bristly hairs. July–Sept. **Habitat** Muddy or sandy streambanks, shores, swales and marshes; sometimes forming large patches.

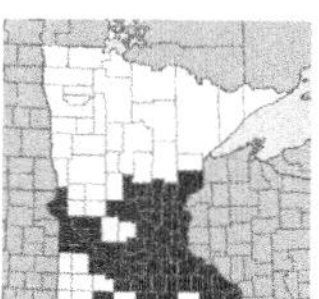

Leersia virginica Willd.
WHITE GRASS
NC **FACW** | MW **FACW** | GP **FACW**

Native perennial grass, spreading by rhizomes. **Stems** slender and weak, often ± horizontal at base and rooting at nodes, 5–12 dm long. **Leaves** rough-hairy, especially along margins, 5–20 cm long and 5–15 mm wide; sheaths smooth or finely hairy; ligule short, flat-topped. **Head** an open panicle, 1–2 dm long, the branches separated along the rachis, stiffly spreading, the spikelets from middle to tip of branches. **Spikelets** oblong, barely overlapping one another, 3 mm long and 1 mm wide, sparsely hairy; glumes absent; lemma 3–4 mm long, the keel and margins sparsely hairy. July–Sept. **Habitat** Swamps, floodplain forests, shaded forest depressions, streambanks.

Leersia oryzoides
RICE CUT-GRASS

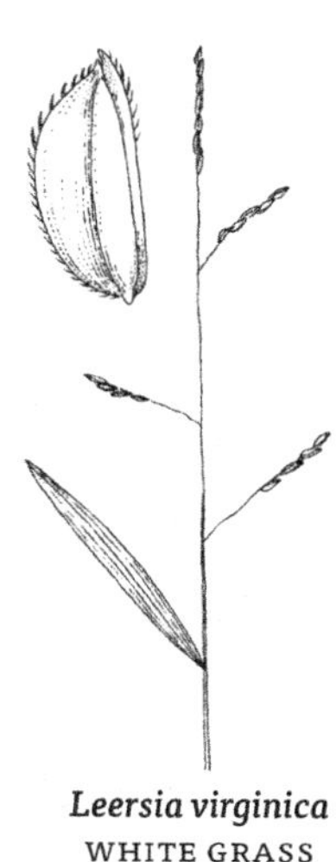

Leersia virginica
WHITE GRASS

LEPTOCHLOA | *Sprangletop*

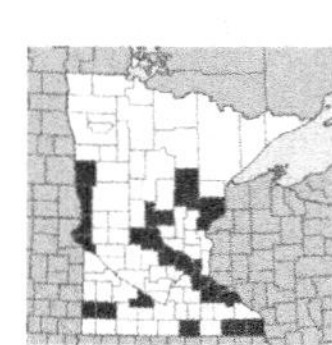

Leptochloa fusca (L.) Kunth
SPRANGLETOP
NC **OBL** | MW **OBL** | GP **FACW**

Clumped native annual grass. **Stems** erect to spreading, branched from base, 2–10 dm long, somewhat fleshy. **Leaves** flat to loosely inrolled, 1–3 mm wide, finely rough-to-touch; sheaths ± smooth, often purple, the upper sheath often partly sheathing the head; ligule 3–5 mm long. **Head** a ± cylindric panicle, 5–20 cm long and 2–5 cm wide, composed of several to many branches, the branches upright and bearing spikelets in racemes. **Spikelets** 6–12-flowered, 5–10 mm long, breaking above the glumes; glumes unequal, lance-shaped, 1-veined, the first glume 2–4 mm long, the second glume 4–5 mm long; lemma 4–5 mm long, 3-veined, tipped with an awn 4–5 mm long; palea about as long as lemma. July–Sept. **Synonyms** *Diplachne acuminata, Leptochloa fascicularis*. **Habitat** Shores, streambanks, muddy or sandy flats, usually where flooded part of year, often where brackish.

Leptochloa fusca
SPRANGLETOP

MUHLENBERGIA | *Muhly*

Perennial grasses, clumped or with creeping rhizomes. **Stems** erect or reclining at base, often branching from base. **Leaves** smooth to hairy, ligules membranous. **Head** a panicle, usually narrow and spikelike, sometimes open and spreading, at ends of stems and sometimes also from leaf axils. **Spikelets** 1-flowered, breaking above glumes; glumes usually nearly equal in length, 1-veined, the tip often awned; lemma lance-shaped, 3-veined, sometimes awned, some species with long, soft hairs at lemma base; palea about as long as lemma.

MUHLENBERGIA | MUHLY

1 Panicle open and loose, 4 cm wide or more **2**
1 Panicle slender and densely flowered, less than 2.5 cm wide **3**

2 Plants spreading by rhizomes . . *Muhlenbergia asperifolia* | ALKALI MUHLY
2 Plants clumped, rhizomes absent . . . . *Muhlenbergia uniflora*
. . . . BOG MUHLY

3 Leaf blades usually inrolled, to 1 mm wide; panicles few-flowered, the heads not round in outline .... *Muhlenbergia richardsonis* | MAT-MUHLY
3 Leaf blades flat, 2–7 mm wide; panicles usually densely flowered, the heads ± round in outline 4

4 Stems smooth and shiny ..... *Muhlenbergia frondosa* | WIRESTEM MUHLY
4 Stems dull, finely hairy (at least below the nodes)..................... 5

5 Panicle stiffly erect, 5–10 mm wide; glumes longer than lemma .... *Muhlenbergia glomerata* | MARSH-MUHLY
5 Panicle bent or nodding, less than 5 mm wide; glumes shorter or equal to lemma.. 6

6 Panicle silvery green; lemmas with awn 5 mm or more long .... *Muhlenbergia sylvatica* | FOREST MUHLY
6 Panicle pale green or purple-tinged; lemmas unawned or with short awn less than 5 mm long .... *Muhlenbergia mexicana* | Wirestem-muhly

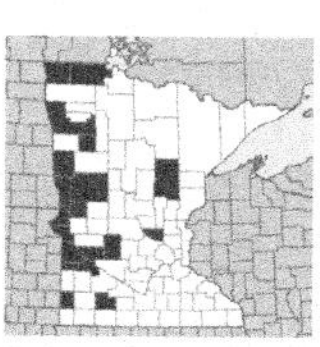

Muhlenbergia asperifolia (Nees & Meyen) L. Parodi
ALKALI MUHLY
NC **FACW** | MW **FACW** | GP **FACW**

Native perennial grass, from slender, scaly rhizomes. **Stems** 1–5 dm long, becoming ± horizontal near base, rooting and branching from lower nodes, the branches spreading, waxy. **Leaves** upright, flat, 2–6 cm long and 1–3 mm wide, rough-to-touch; sheaths smooth; ligule ragged, to 0.5 mm long. **Head** an open panicle, 5–15 cm long, the branches threadlike, widely spreading. **Spikelets** 1-flowered (sometimes 2-flowered), single on the branches, purple or dark gray, 1–2 mm long; glumes nearly equal, half to nearly as long as spikelet; lemma unawned, 1–2 mm long. July–Sept. **Habitat** Wet meadows, seeps, shores and mudflats, often where brackish.

Muhlenbergia asperifolia
ALKALI MUHLY

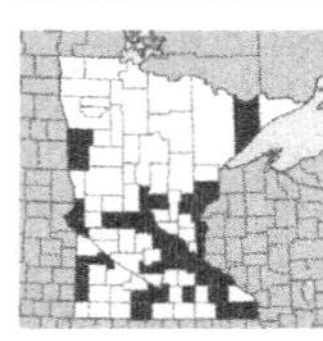

Muhlenbergia frondosa (Poiret) Fern.

WIRESTEM MUHLY

NC **FACW** | MW **FACW** | GP **FACW**

Native perennial grass, from stout, scaly rhizomes. **Stems** 4–10 dm long, unbranched and erect when young, becoming branched and sprawling with age, smooth and shiny between nodes. **Leaves** lax, smooth, 3–10 cm long and 2–6 mm wide; ligule fringed, 1–2 mm long. **Head** a narrow panicle, to 10 cm long, from ends of stems and leaf axils (where partly enclosed by sheaths), the branches erect to spreading, with spikelets from near base to tip. **Spikelets** 1-flowered; glumes 2–3 mm long, tipped with a short awn; lemma 3–4 mm long, usually with an awn to 1 cm long, short-hairy at base. Aug–Sept. **Habitat** Floodplain forests, streambanks, thickets, shores; also somewhat weedy in disturbed areas such as along railroads.

Muhlenbergia frondosa
WIRESTEM MUHLY

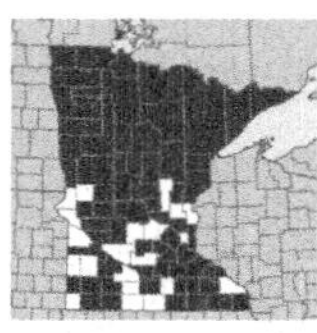

Muhlenbergia glomerata (Willd.) Trin.

MARSH-MUHLY

NC **OBL** | MW **FACW** | GP **FACW**

Native perennial grass, spreading from rhizomes. **Stems** upright, 3–9 dm long, sometimes with a few branches from base, dull and finely hairy between nodes. **Leaves** flat, lax, 5–15 cm long and 2–6 mm wide; sheaths smooth; ligule fringed, to 0.5 mm long. **Head** a narrow, crowded, cylindric panicle, 2–10 cm long and 5–10 mm wide, the lower clusters of spikelets often separate from one another. **Spikelets** 1-flowered, often purple-tinged, 5–6 mm long; glumes nearly equal, longer than the floret, tipped with an awn 1–5 mm long; lemma lance-shaped, 2–3 mm long, with long, soft hairs at base. Aug–Sept. **Habitat** Swamps, wet meadows, marshes, springs, open bogs, fens, calcareous shores.

Muhlenbergia glomerata
MARSH-MUHLY

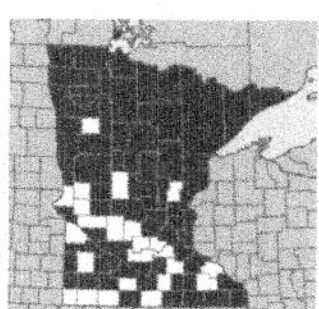

Muhlenbergia mexicana (L.) Trin.

WIRESTEM MUHLY

NC **FACW** | MW **FACW** | GP **FACW**

Native perennial grass, from scaly rhizomes. **Stems** upright, 2–8 dm long, sometimes branched from base; dull and finely hairy between nodes. **Leaves** flat, lax, 5–20 cm long and 2–5 mm wide; sheaths smooth; ligule entire to fringed, to 1 mm long. **Head** a narrow, densely flowered panicle, 5–15 cm long and 2–10 mm wide, from ends of stems and leafy branches. **Spikelets** 1-flowered, green or purple, 2–3 mm long; glumes nearly equal, lance-shaped, 3–4 mm long, about as long as floret, tipped with a short awn about 1 mm long; lemma lance-shaped, 2–3 mm long, unawned or with an awn to 7 mm long. Aug–Sept. **Synonyms** *Muhlenbergia foliosa*. **Habitat** Swamps, floodplain forests, thickets, wet meadows, marshes, springs, fens and streambanks.

Muhlenbergia mexicana
WIRESTEM MUHLY

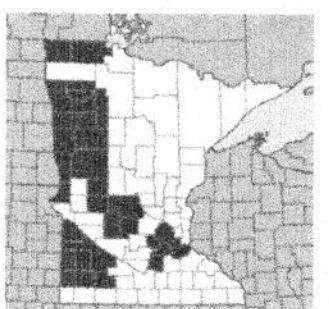

Muhlenbergia richardsonis (Trin.) Rydb.

MAT-MUHLY

NC **FACW** | MW **FAC** | GP **FAC**

Loosely clumped, native perennial grass, rooting from lower nodes and forming mats. **Stems** very slender, erect or ± horizontal at base, 2–6 dm long. **Leaves** upright, usually inrolled, 1–5 cm long and 1–2 mm wide; sheaths smooth; ligule 2–3 mm long. **Head** a narrow panicle, 2–8 cm long. **Spikelets** 1-flowered, uncrowded, green or gray-green, 2–3 mm long; glumes nearly equal, ovate, to half as long as floret; lemma lance-shaped, smooth, 2–3 mm long tipped with a short point. July–Sept. **Habitat** Low prairie, wet meadows, marshes and seeps; often where brackish.

Muhlenbergia richardsonis
MAT-MUHLY

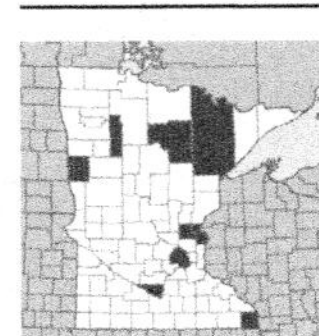

Muhlenbergia sylvatica Torr.
FOREST MUHLY
NC **FACW** | MW **FACW** | GP **FACW**

Native perennial grass, spreading by rhizomes. **Stems** erect, or sprawling when old, 4–10 dm long, coarse-hairy between nodes. **Leaves** flat, lax, upright to spreading, 5–15 cm long and 2–6 mm wide; sheaths smooth; ligule fringed, 1–3 mm long. **Head** a slender panicle, often nodding, 5–20 cm long and 2–7 mm wide. **Spikelets** 1-flowered, 2–4 mm long, at ends of stalks about 3 mm long; glumes nearly equal, sharp-tipped, shorter than lemma; lemma 2–4 mm long, short hairy at base, tipped with an awn 5–15 mm long. Aug–Sept. **Habitat** Streambanks, shaded wet areas.

Muhlenbergia uniflora (Muhl.) Fernald
BOG MUHLY
NC **OBL** | MW **OBL** | *special concern*

Clumped native perennial grass. **Stems** very slender, 2–4 dm long, often ± horizontal and rooting at base. **Leaves** flat, crowded near base of plant, 5–10 cm long and to 1 mm wide; sheaths ± smooth, compressed; ligule ragged, about 1 mm long. **Head** a loose, open panicle, 7–20 cm long and 2–4 cm wide, the branches threadlike. **Spikelets** 1-flowered (rarely 2-flowered), oval, purple-tinged, 1–2 mm long; glumes about equal, ovate, to half the length of spikelet; lemma 1–2 mm long, unawned. **Habitat** Wetland margins, exposed sandy shores. **Status** Special concern.

Muhlenbergia uniflora
BOG MUHLY

PANICUM | *Panic-grass*

Annual or perennial grasses. Heads narrow to open panicles (ours). **Spikelets** small, with 1 fertile flower; glumes usually unequal, the first glume membranous, usually very small, second glume green, about as long as spikelet; sterile lemma similar to second glume, enclosing the palea and sometimes a male flower, fertile lemma whitish, smooth.

PANICUM | PANIC-GRASS

1 Spikelets smooth . . . . *Panicum flexile* | WIRY WITCH-GRASS
1 Spikelets at least sparsely hairy . . . . *Panicum boreale* . . . . NORTHERN PANIC-GRASS

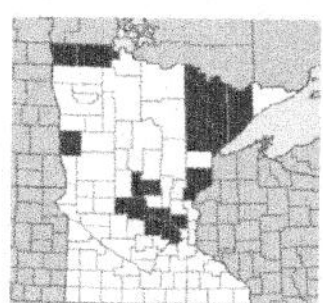

Panicum boreale Nash
NORTHERN PANIC-GRASS
NC **FAC** | MW **FAC** | GP **FACU**

Native perennial grass, in small clumps. **Stems** upright, 2–6 dm long. **Leaves** upright to spreading, 5–20 cm long and 1–2 cm wide, smooth or sometimes hairy on underside, base of leaf often fringed with hairs; sheaths hairy; ligule absent or a fringe of short hairs. **Head** an open panicle, 5–12 cm long, the branches spreading or upright. **Spikelets** oval in outline, finely hairy, about 2 mm long, on long stalks with 1 fertile flower; first glume to half as long as second glume; second glume and lemma purple-tinged, about equal, and as long as fruit. June–Aug. **Synonyms** *Dichanthelium boreale*. **Habitat** Local in wet prairies and tamarack bogs.

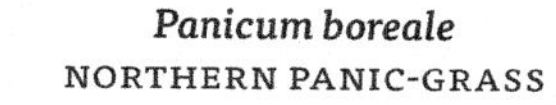

Panicum boreale
NORTHERN PANIC-GRASS

Panicum flexile (Gattinger) Scribn.
WIRY WITCH-GRASS
[NC **FACW** | MW **FACW**]

Slender native annual grass. **Stems** erect, 2–7 dm long, branched from base, hairy at nodes. **Leaves** erect, smooth or sparsely hairy, 10–30 cm long and 2–6 mm wide. **Head** a narrow panicle, 10–20 cm long and about a third as wide, the branches threadlike, upright to spreading. **Spikelets** lance-shaped, 3–4 mm long, with 1 fertile flower; first glume about half as long as second glume and sterile lemma. Aug–Sept. **Habitat** Sandy and gravelly shores, marshes; often where calcium-rich.

***Panicum* flexile** *may occur in Minnesota, but presence not verified; known from adjacent states.*

Panicum flexile
WIRY WITCH-GRASS

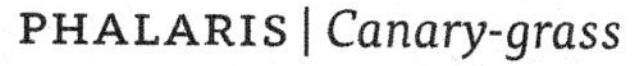

PHALARIS | *Canary-grass*

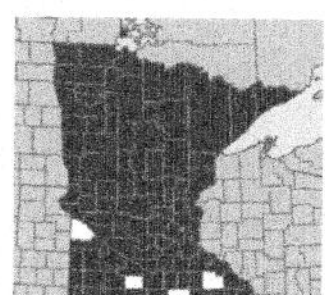

Phalaris arundinacea L.
REED CANARY-GRASS
NC **FACW** | MW **FACW** | GP **FACW**

Tall perennial grass, spreading by rhizomes and typically forming large, dense colonies; introduced and often invasive. **Stems** stout, smooth, 5–15 dm long. **Leaves** flat, smooth, 1–2 dm long and 1–2 cm wide; sheaths smooth; ligule membranous, 3–8 mm long. **Head** a narrow, densely flowered panicle, 5–25 cm long, often purple-tinged, becoming straw-colored with age, the branches short and upright to ascending. **Spikelets** 4–6 mm long, breaking above glumes, with 1 fertile flower and 2 small sterile lemmas below; glumes nearly equal, longer than fertile floret, lance-shaped, tapered to tip or short-awned, 3-veined; fertile lemma ovate, 3 mm long, shiny; palea as long as lemma. June–July. **Habitat** Common to abundant species of wet meadows, shallow marshes, ditches, shores and streambanks.

Reed canary-grass *is an aggressive, competitive wetland species, to the detriment of other plants. Most populations are probably non-native strains, originally introduced from Eurasia as a pasture plant and now widely naturalized.*

Phalaris arundinacea
REED CANARY-GRASS

PHRAGMITES | *Common reed*

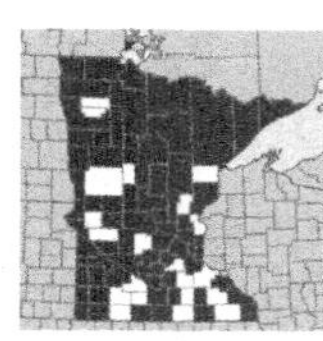

Phragmites australis (Cav.) Trin.
COMMON REED
NC **FACW** | MW **FACW** | GP **FACW**

Tall, stout perennial reed, from deep, scaly rhizomes, or the rhizomes sometimes exposed and creeping over the soil; often forming large colonies. **Stems** erect, hollow, 2–4 m long and 5–15 mm wide near base, the internodes often purple. **Leaves** flat, long, 1–3 cm wide; sheaths open; ligule white, 1 mm long. **Head** a large, plumelike panicle, purple when young, turning yellow-brown with age, 15–40 cm long, much-branched, the branches angled or curved upward. **Spikelets** 3–7-flowered, linear, 10–15 mm long, breaking above the glumes; the stem within the spikelet (rachilla) covered with long silky hairs, these longer than the florets and becoming exposed as the lemmas spread after flowering; glumes unequal, the first glume half the length of second glume. Grain (seed) seldom produced.

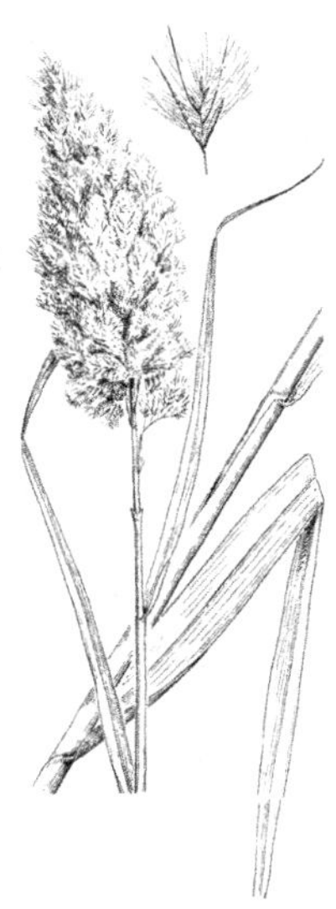

Aug–Sept. **Synonyms** Native and introduced subspecies occur and may hybridize: subsp. *americanus* and subsp. *australis*. *Phragmites communis*. **Habitat** Fresh to brackish marshes, shores, streams, ditches, occasional in tamarack swamps; sometimes in shallow water; found nearly worldwide.

Common reed (*Phragmites australis*)

POA | *Bluegrass*

Perennial, loosely clumped or rhizomatous grasses. **Leaves** mostly near base, flat to folded, the tip keeled similar to the bow of a boat; sheaths partly closed, ligules membranous. **Head** an open panicle. **Spikelets** small, with 2 to several flowers breaking above the glumes; glumes nearly equal, the first glume usually 1-veined, the second glume 3-veined; lemmas often with a tuft of distinctive cobwebby hairs at base; palea nearly as long as lemma.

POA | BLUEGRASS

1 Keel of lemma silky-hairy; lemma nerves without hairs ...... *Poa alsodes* .... GROVE BLUEGRASS
1 Keel of lemma and some or all nerves hairy 2

2 Plants with rhizomes .... *Poa pratensis* | KENTUCKY BLUEGRASS
2 Plants without rhizomes 3

3 Panicle branches single or in groups of 2; sheaths rough-to-touch; ligules less than 2 mm long; rare .... *Poa paludigena* | MARSH BLUEGRASS
3 Panicle branches in groups of 3–5; sheaths smooth; ligules 3–5 mm long; widespread species .... *Poa palustris* | FOWL BLUEGRASS

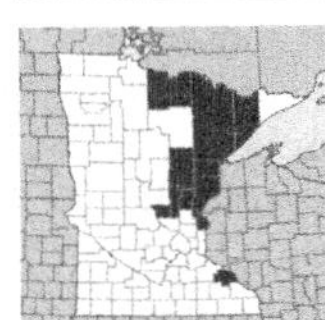

Poa alsodes A. Gray
GROVE BLUEGRASS
NC **FAC** | MW **FACW**

Loosely clumped, native perennial grass, rhizomes absent. **Stems** slender, 3–8 dm long. **Leaves** lax, 5–20 cm long and 2–5 mm wide; sheaths smooth, ligule 1–3 mm long. **Head** a lax, open panicle, 10–20 cm long, the branches becoming widely spreading, mostly in groups of 4–5, with 1 to few spikelets near tip of branch; base of panicle sometimes remaining enclosed by sheath. **Spikelets** ovate, 2–3-flowered, 3–5 mm long; glumes nearly equal, 2–4 mm long; lemmas 2–4 mm long, with cobwebby hairs at base. May–July. **Habitat** Alder thickets, swamp hummocks, most common in moist deciduous or mixed conifer-deciduous forests. In Minnesota, most common along Lake Superior.

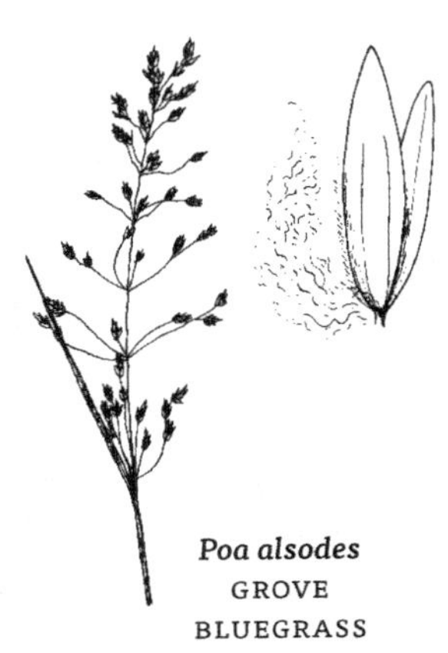

Poa alsodes
GROVE BLUEGRASS

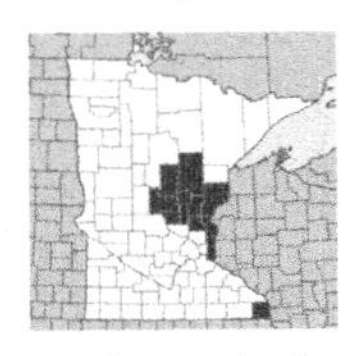

Poa paludigena Fernald & Wieg.
MARSH BLUEGRASS
NC **OBL** | MW **OBL** | *threatened*

Native perennial grass, without rhizomes. **Stems** single or in small clumps, slender and weak, 2–6 dm long. **Leaves** upright, to 10 cm long and 1–2 mm wide; sheaths finely rough-hairy; ligule flat-topped, about 1 mm long. **Head** a loose, open panicle, 5–12 cm long, the lower branches in groups of 2, with a few spikelets above middle. **Spikelets** 2–5-flowered, 4–5 mm long, glumes lance-shaped, the first glume to 2 mm long, the second glume 2–3 mm long; lemma 3–4 mm long, with cobwebby hairs at base. June–July. **Habitat** Swamps, alder thickets, sedge meadows, open bogs, cold springs; usually in sphagnum moss and often under black ash (*Fraxinus nigra*). **Status** Threatened.

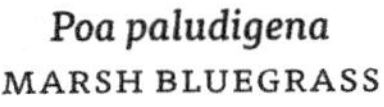

Poa paludigena
MARSH BLUEGRASS

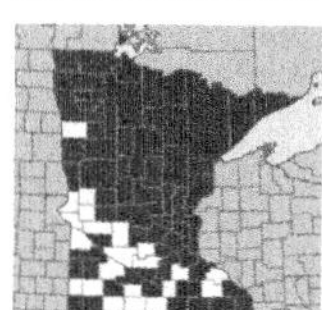

Poa palustris L.
FOWL BLUEGRASS
NC **FACW** | MW **FACW** | GP **FACW**

Loosely clumped, native perennial grass. **Stems** smooth, 4–12 dm long, reclining at base and rooting from lower nodes, lower portion often purple-tinged. **Leaves** flat, upright to spreading, 1–4 mm wide, rough-to-touch; sheaths smooth; ligule 2–5 mm long. **Head** a loosely spreading panicle (narrow when emerging from sheath), 1–3 dm long, the branches in mostly widely separated groups along panicle stem (rachis). **Spikelets** 2–4-flowered, 2–5 mm long and 1–2 mm wide; glumes nearly equal, lance-shaped, 2–3 mm long, often purple; lemma 2–3 mm long, often purple on sides, with cobwebby hairs at base. June–Sept. **Habitat** Wet meadows, marshes, shores, streambanks, ditches and low prairie; also moist woods.

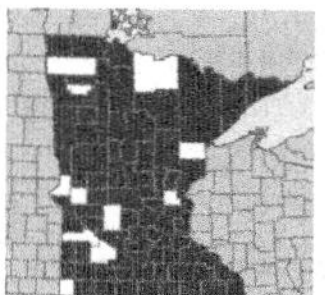

Poa pratensis L.
KENTUCKY BLUEGRASS
NC **FACU** | MW **FAC** | GP **FACU**

Introduced perennial grass, spreading by rhizomes and forming a sod. **Stems** erect, 3–10 dm long. **Leaves** flat or folded, 1–4 mm wide, margins sometimes somewhat rough-to-touch; sheaths smooth; ligule 0.5–2 mm long. **Head** an open, pyramid-shaped panicle, 5–15 cm long, the branches spreading to ascending, the lowest branches in groups of 4–5. **Spikelets** 2–5-flowered, green or purple-tinged, compressed, 3–5 mm long and 2–3 mm wide; glumes unequal, lance-shaped, 2–4 mm long, roughened on keels; lemma 2–4 mm long, with an obvious tuft of cobwebby hairs at base, often purple-tinged on sides. May–Aug. **Habitat** All types of moist to dry places; not usually in very wet situations.

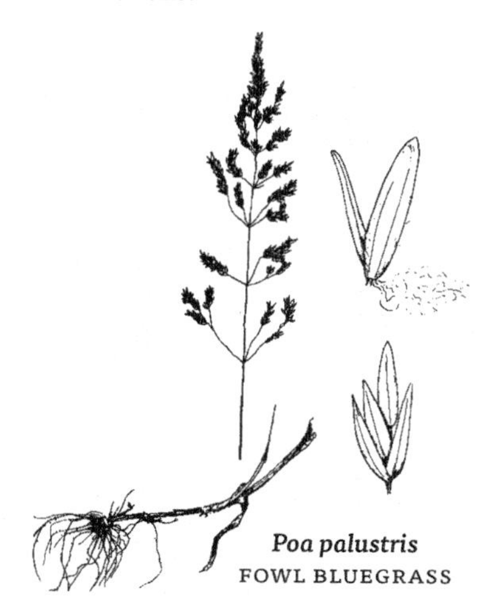

Poa palustris
FOWL BLUEGRASS

Poa pratensis
KENTUCKY BLUEGRASS

Kentucky bluegrass, *introduced from Europe as a lawn and pasture grass, is now widely established in North America.*

PUCCINELLIA | *Alkali-grass*

Clumped, smooth perennial grasses, usually in brackish habitats (except *Puccinellia pallida*). **Leaves** mostly from base of plants, flat to inrolled. **Head** an open panicle, the branches upright to spreading. **Spikelets** several-flowered, oval to linear, nearly round in section, breaking above the glumes; glumes unequal, the first glume 1-veined, the second glume 3-veined; lemmas rounded on back, often short-hairy at base; palea shorter to about as long as lemma.

PUCCINELLIA | ALKALI-GRASS

1 Plants not in brackish wetlands; lemma with 5 prominent veins . . . . *Puccinellia pallida* | PALE ALKALI-GRASS

1 Plants of brackish wetlands and along salted highways; lemma only faintly veined 2

2 Lower panicle branches horizontal or angled downward when mature; lemma broad, not tapered to the blunt or rounded tip . . . . *Puccinellia distans* | EUROPEAN ALKALI-GRASS

2 Lower panicle branches usually angled upward; lemma narrow, tapered to a rounded tip . . . . *Puccinellia nuttalliana* | NUTTALL'S ALKALI-GRASS

Puccinellia distans (Jacq.) Parl.
EUROPEAN ALKALI-GRASS
NC **FACW** | MW **OBL** | GP **FACW**

Clumped, smooth perennial grass, introduced. **Stems** erect or reclining at base, 1–5 dm long. **Leaves** flat to slightly inrolled, 1–3 mm wide; ligule about 1 mm long. **Head** a loose, pyramid-shaped panicle, 5–15 cm long, the branches in groups, the lower branches angled downward. **Spikelets** 3–7-flowered, 4–6 mm long; glumes ovate, 1–2 mm long; lemmas about 2 mm long, smooth or short-hairy at base. May–Aug. **Habitat** Occasional in brackish waste areas and ditches along salted highways; introduced from Eurasia.

Puccinellia distans
EUROPEAN ALKALI-GRASS

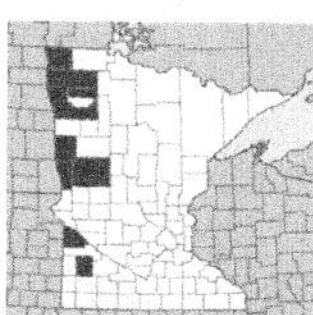

Puccinellia nuttalliana (Schultes) A. Hitchc.
NUTTALL'S ALKALI-GRASS
NC **OBL** | MW **OBL** | GP **OBL**

Clumped native perennial grass. **Stems** slender, erect, 2–8 dm long. **Leaves** flat or often inrolled, 1–3 mm wide; ligule 1–3 mm long, **Head** an open panicle, 5–25 cm long, the branches ascending to spreading, rough-to-touch, to 10 cm long, the spikelets mostly above middle of branch. **Spikelets** 3–9-flowered, slender, 4–7 mm long, glumes lance-shaped, 1–3 mm long; lemmas oblong, 2–3 mm long, with tiny hairs at base. June–July. **Synonyms** *Puccinellia airoides*. **Habitat** Moist flats, sometimes in shallow water, often where salty or disturbed.

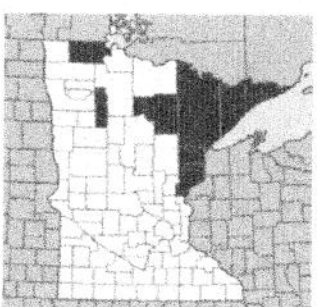

Puccinellia pallida (Torr.) R. T. Clausen
PALE ALKALI-GRASS
NC **OBL** | MW **OBL** | GP **OBL**

Native perennial grass. **Stems** slender, weak, usually reclining at base, 3–10 dm long. **Leaves** flat, soft, 3–8 mm wide; sheaths open; ligule 3–9 mm long. **Head** a pale green, open panicle, 5–15 cm long, the branches upright, becoming spreading. **Spikelets** 4–7-flowered, oval in outline, 5–7 mm long; glumes rounded at tip, 1–3 mm long; lemmas 2–3 mm long, 5-veined, finely hairy, the tip rounded and ragged. June–Aug. **Synonyms** *Glyceria pallida, Puccinellia fernaldii, Torreyochloa pallida*. **Habitat** Marshes, pond margins, alder thickets, forest depressions; often in shallow water.

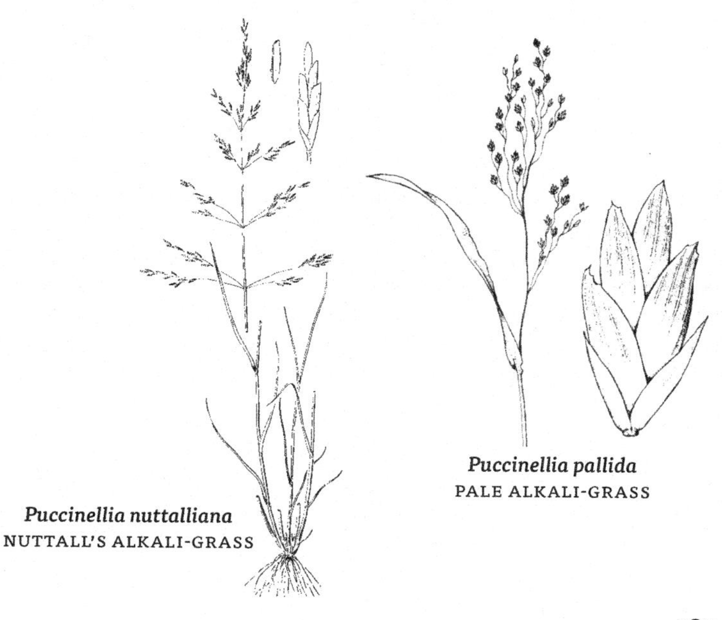

Puccinellia nuttalliana
NUTTALL'S ALKALI-GRASS

Puccinellia pallida
PALE ALKALI-GRASS

SCOLOCHLOA | *Sprangletop*

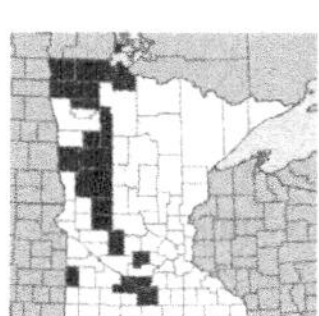

Scolochloa festucacea (Willd.) Link
SPRANGLETOP; WHITETOP
NC **OBL** | MW **OBL** | GP **OBL**

Tall native perennial grass, spreading by thick rhizomes and forming colonies. **Stems** erect, hollow, 1–2 m long and 3–5 mm wide near base, usually with a few suckers and roots from lower nodes. **Leaves** flat or slightly inrolled, 3–10 mm wide, tapered to a sharp tip, upper surface rough-to-touch; sheaths smooth; ligule white, ragged at tip, 4–7 mm long. **Head** a loose, open panicle, 15–20 cm long, the branches ascending, the lowest branches much longer than upper. **Spikelets** 3–4-flowered, purple or green, becoming straw-colored, 7–10 mm long, breaking above glumes; glumes unequal, lance-shaped, the first glume 3-veined, 4–7 mm long, the second glume 5-veined, 6–9 mm long; lemmas lance-shaped, about 6 mm long; palea as long as lemma. June–July. **Synonyms** *Fluminea festucacea*. **Habitat** Shallow water, marshes; sometimes mowed for hay during dry periods.

Scolochloa festucacea
SPRANGLETOP; WHITETOP

SPARTINA | *Cordgrass*

Coarse perennial grasses, spreading by long scaly rhizomes. **Stems** stout and erect. **Leaves** flat to inrolled, tough, rough-to-touch; sheaths smooth; ligule a fringe of hairs. **Head** of several to many 1-sided spikes in racemes at ends of stem, the spikes upright to appressed. **Spikelets** 1-flowered, flattened, overlapping in 2 rows on 1 side of the rachis, breaking below the glumes; glumes unequal, 1–2-veined, with rough hairs on the keel; lemma with pronounced midvein and 2 faint lateral veins; palea about as long as lemma.

SPARTINA | CORDGRASS

1 Plants 1–2 m tall; leaf blades flat (at least near base), more than 5 mm wide . . . . *Spartina pectinata* | PRAIRIE CORDGRASS

1 Plants to 1 m tall; leaf blades inrolled or flat, 2–5 mm wide . . . . *Spartina gracilis* | ALKALI CORDGRASS

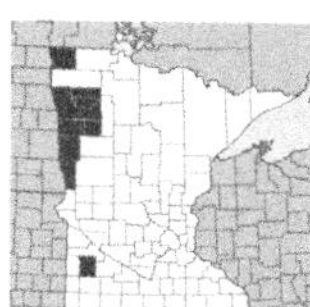

Spartina gracilis Trin.

ALKALI CORDGRASS

NC **FACW** | MW **FACW** | GP **FACW**

Native perennial grass, from rhizomes. **Stems** 4–8 dm long. **Leaves** usually inrolled, 10–20 cm long and 2–4 mm wide. **Head** a spikelike raceme of 4–8, 1-sided spikes, the spikes 2–5 cm long, appressed to the raceme stem (rachis). **Spikelets** 1-flowered, 6–9 mm long; glumes and lemma fringed with hairs on keel, the first glume half as long as second; lemma nearly as long as second glume. July–Sept. **Habitat** Wet meadows, shores, flats and seeps; often where brackish.

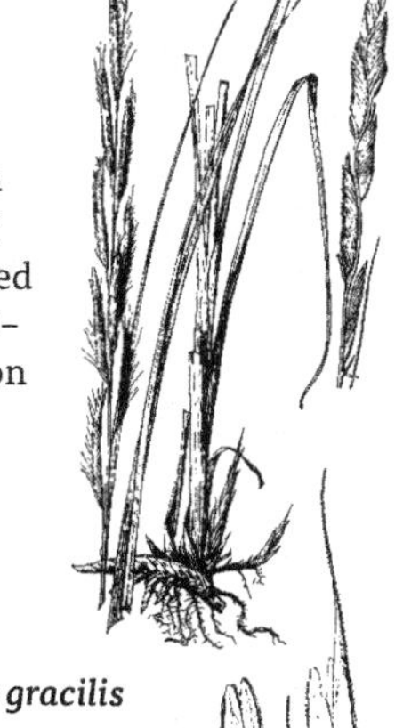

Spartina gracilis
ALKALI CORDGRASS

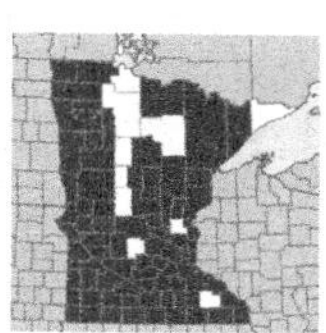

Spartina pectinata Link

PRAIRIE CORDGRASS

NC **FACW** | MW **FACW** | GP **FACW**

Native perennial grass, with scaly rhizomes. **Stems** tough, 1–2 m long. **Leaves** flat to inrolled, 3–10 mm wide, margins very rough. **Head** a spikelike raceme of mostly 10–30, 1-sided spikes, the spikes upright to sometimes appressed, 3–10 cm long. **Spikelets** 1-flowered, 8–11 mm long, fringed with hairs on keels; first glume nearly as long as floret, tapered to tip or with an awn 1–5 mm long, second glume longer than floret, tipped with an awn 2–8 mm long; lemma 7–9 mm long, shorter than second glume. July–Sept. **Habitat** Shallow marshes, wet meadows, sandy shores, ditches, low prairie.

Spartina pectinata
PRAIRIE CORDGRASS

SPHENOPHOLIS | *Wedge-grass*

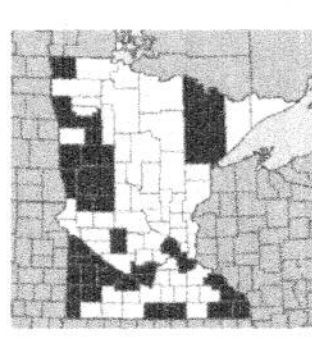

Sphenopholis obtusata (Michx.) Scribn.
WEDGE-GRASS
NC **FAC** | MW **FAC** | GP **FAC**

Clumped native perennial (sometimes annual) grass; plants smooth to rough-hairy. **Stems** slender, 2–10 dm long. **Leaves** upright to spreading, flat, rough-to-touch, 2–7 mm wide; ligule membranous, ragged at tip, 1–4 mm long. **Head** a dense, shiny, spikelike panicle, 5–20 cm long, the spikes often (in part) separate from one another. **Spikelets** 2-flowered, 3–4 mm long, unawned, breaking below the glumes; glumes 2–3 mm long, the first glume linear, 1-veined, the second glume broader, 3–5-veined; lemma 2–3 mm long, 1-veined; palea linear, about as long as lemma. June–Aug. **Synonyms** *Sphenopholis intermedia*. **Habitat** Low prairie, wet meadows, gravelly shores, streambanks, wetland margins; also in moist woods.

Sphenopholis obtusata
WEDGE-GRASS

ZIZANIA | *Wildrice*

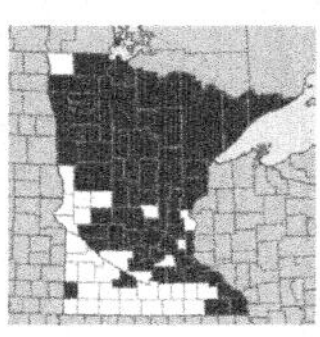

Zizania palustris L.
NORTHERN WILDRICE
NC **OBL** | MW **OBL** | GP **OBL**

Large native annual emergent grass, with fleshy yellow roots. **Stems** single or few together, 1–3 m long. **Leaves** flat, 1–4 cm wide, smooth or finely hairy, usually floating on water surface early in season, becoming upright; sheaths short-hairy at top, smooth below; ligule membranous, entire or with a jagged margin, 10–15 mm long. **Head** a panicle, 3–6 dm long, the branches 10–20 cm long; male and female flowers separate on same plant, the male flowers on lower panicle branches, female flowers on upper branches, the male portion becoming spreading, branches of female portion remaining upright. **Spikelets** 1-flowered, round in section, breaking as a unit from the stalk; glumes absent; male spikelets straw-colored to purple, 6–12 mm long, hanging downward from branches, lemma linear, tapered to tip or tipped with an awn to 3 mm long, early deciduous; female spikelets linear, purple or light green, lemma awl-shaped, 1–2 cm long, tapered to a slender awn 3–6 cm long. Grain cylindric, dark brown to black, 1–2 cm long. July–Sept. **Habitat** Shallow water (up to 1 m deep) or mud of streams, rivers, lakes, ponds; where water is slightly flowing and not stagnant; soils vary from muck to silt, sand, or gravel, with best establishment of plants on a layer of soft silt or muck several cm thick.

Zizania palustris *is the source of commercial wild rice. and large areas of lakes and shallow marshes may be dominated by this plant. The grain is also an excellent food for waterfowl. Many populations are introductions to intentionally spread the species.*

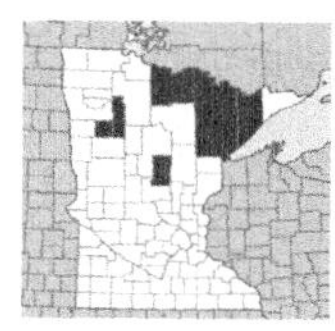

Annual wildrice *(Zizania aquatica var. aquatica) is sometime considered a separate species and is reported to occur in ne and nc Minnesota (map, left). Plants are taller (2–3 m), with broader leaves (1–4.5 cm wide) than Zizannia palustris. However, the distinctions are not always clear.*

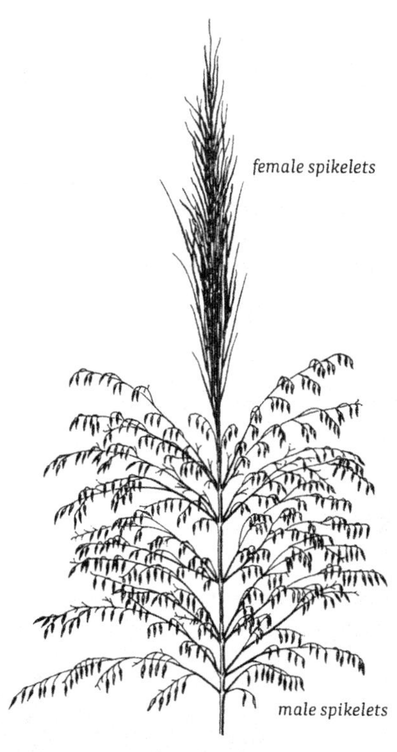

Zizania palustris
NORTHERN WILDRICE

Inflorescence, with spreading male-flowered spikelets (lower) and upright female-flowered spikelets above.

MOSTLY PERENNIAL, aquatic or emergent herbs. **Leaves** alternate, stalkless and straplike, or with a petiole and broad blade. **Flowers** perfect (with both male and female parts), regular or irregular, single from leaf axils or in spikes or panicles, subtended by leaflike bracts (spathes), light yellow, white or blue-purple, perianth of 6 petal-like lobes, usually joined near base to form a tube; stamens 3–6, the filaments attached to throat of perianth tube; ovary superior, 3-chambered, style 1. **Fruit** a many-seeded capsule inside the spathe, or a 1-seeded, achene-like utricle.

PONTEDERIACEAE | WATER-HYACINTH FAMILY

1 Flowers 2-lipped, each lip 3-lobed, the 3 lower lobes spreading; stamens 6, 3 longer than petals, 3 shorter; fruit 1-seeded. . . . . *Pontederia cordata* . . . . PICKEREL-WEED
1 Flowers regular, the lobes ± equal; stamens 3, all longer than petals; fruit a many-seeded capsule . . . . *Heteranthera* | MUD-PLANTAIN

HETERANTHERA | *Mud-plantain*

Annual or perennial aquatic herbs, rooting in mud. **Stems** underwater and growing to surface, or emersed. **Leaves** linear or ovate. **Flowers** single, elongating over several days, subtended by leaflike bracts (spathes), yellow to blue-purple to white; stamens 3, unequal; style 3-lobed. **Fruit** a capsule.

HETERANTHERA | MUD-PLANTAIN

1 Flowers blue-purple or white; leaves with a petiole and blade, emersed or floating . . . . *Heteranthera limosa* | SMALLER MUD-PLANTAIN
1 Flowers light yellow; leaves linear and straplike, not differentiated into petiole and blade, usually underwater. . . . . *Heteranthera dubia* . . . . WATER STAR-GRASS

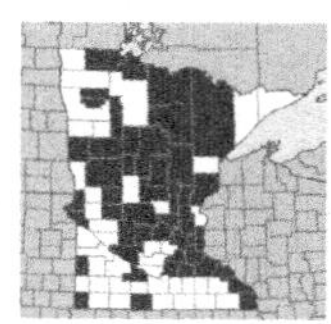

Heteranthera dubia (Jacq.) MacMill.

WATER STAR-GRASS
NC **OBL** | MW **OBL** | GP **OBL**

Native aquatic perennial herb, with lax stems and leaves, or plants sometimes exposed and forming small, leafy rosettes. **Stems** slender, forked, often rooting at lower nodes, to 1 m long. **Leaves** alternate, linear, flat, translucent, rounded at tip or tapered to a point, 2–12 cm long and 2–6 mm wide, the midrib and veins inconspicuous; petioles absent. **Flowers** 1, opening on water surface, light yellow, enclosed in a spathe from upper leaf axils, the spathe membranous, 2–5 cm long, surrounding

flower

much of the slender perianth tube; perianth tube often curved, 2–8 cm long, the 6 perianth segments linear, 4–6 mm long; stamens 3, all alike. **Fruit** a many-seeded capsule about 1 cm long. July–Sept. **Synonyms** *Zosterella dubia*. **Habitat** Shallow water, muddy shores of ponds, lakes, streams and marshes.

Water star-grass *is distinguished from the pondweeds (Potamogeton) by lack of a leaf midrib. Plants are sometimes used in aquariums.*

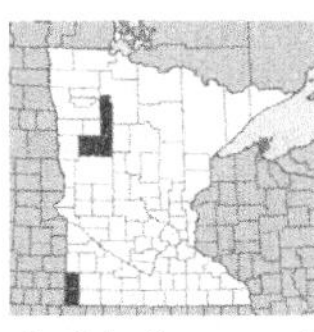

Heteranthera limosa (Swartz) Willd.
SMALLER MUD-PLANTAIN
NC **OBL** | MW **OBL** | GP **OBL** | *threatened*

Small native annual herb. **Stems** much-branched from base, 1–3 dm long, short when exposed, longer and sprawling when in water. **Leaves** with blade and petiole, the blades usually emersed, ovate to oval, 2–6 cm long and 1–3 cm wide, tapered to a rounded tip, base rounded or flat across; petioles 5–15 cm long, with a membranous sheath at base. **Flowers** 1, enclosed by a spathe; spathe folded, 2–4 cm long, abruptly narrowed at tip, enclosing the tubular portion of the perianth, the flower and spathe at end of a stout stalk arising from stem; perianth segments usually blue-purple, sometimes white, lance-shaped, 5–10 mm long, the perianth tube 1–4 cm long, the lobes ± equal, 5–15 mm long, the 3 upper lobes with a yellow spot at base; stamens 3, the 2 lateral stamens short, yellow, the center stamen longer and blue or yellow. June–July. **Habitat** In Minnesota, known from seepage or rainwater pools on granitic or quartzite rock outcrops. **Status** Threatened.

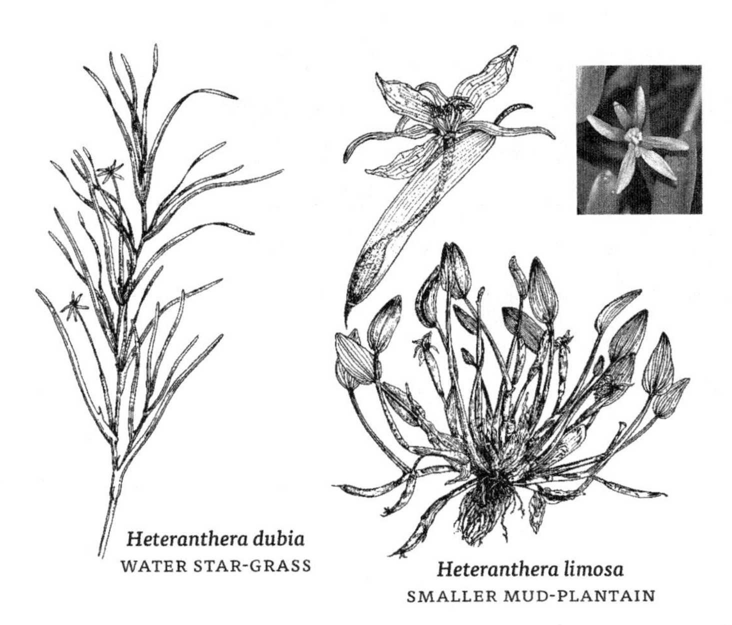

Heteranthera dubia
WATER STAR-GRASS

Heteranthera limosa
SMALLER MUD-PLANTAIN

PONTEDERIA | *Pickerel-weed*

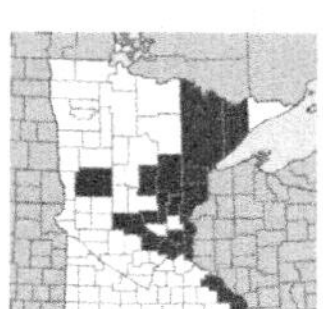

Pontederia cordata L.
PICKEREL-WEED
NC **OBL** | MW **OBL** | GP **OBL**

Native perennial emergent herb, spreading from rhizomes and forming colonies. **Stems** stout, upright, to 12 dm long, with 1 leaf. **Leaves** lance-shaped to ovate, 5–20 cm long and 2–15 cm wide, heart-shaped at base; petioles 3–7 cm long, sheathing on stem. **Flowers** blue-purple (rarely white), many in a spike 5–15 cm long, subtended by a bractlike spathe 3–6 cm long; perianth funnel-like, the tube 6 mm long, 2-lipped above, upper lip with 3 ovate lobes, lower lip with 3 slender, spreading lobes, the lobes 7–10 mm long. **Fruit** a 1-seeded utricle, 5–10 mm long. June–Sept. **Synonyms** *Pontederia lanceolata*. **Habitat** Shallow water (to 1 m deep) of lakes, ponds, rivers and swamps.

Pontederia cordata
PICKEREL-WEED

POTAMOGETON | *Pondweed*

Aquatic perennial herbs; **leaves** all submerged or with leaves both submerged and floating, from rhizomes or tubers, sometimes reproducing and over-wintering by free-floating winter buds (turions). **Stems** long, wavy, anchored to bottom by roots and rhizomes. **Leaves** alternate, or becoming opposite upward in some species, simple, with an open or closed sheath at their base. **Underwater leaves** usually linear and threadlike, sometimes broader, margins often wavy, usually stalkless. **Floating leaves**, if present, oval or ovate, stalked, with a waxy upper surface. **Flowers** perfect, regular, green to red, in stalked spikes at ends of stems or from leaf axils, usually raised above water surface, the spikes with few to many small flowers; perianth of 4 sepal-like bracts; stamens 4. **Fruit** a group of 4 beaked achenes.

Some species of Potamogeton have been separated into a new genus **Stuckenia***; these new names are listed below the traditional name.*

POTAMOGETON | PONDWEED

1 Plants with underwater leaves only, these all alike . . . . **Group 1**
1 Plants with 2 kinds of leaves: broad floating leaves and broad or narrow underwater leaves . . . . **Group 2**

POTAMOGETON GROUP 1

Plants with underwater leaves only, these all alike.

1 Leaves broad, lance-shaped to oval or ovate, never linear 2
1 Leaves linear 7

2 Leaf margins wavy-crisped, finely toothed . . . . *Potamogeton crispus* . . . . CURLY PONDWEED p. 597
2 Leaf margins flat or sometimes wavy, entire (or rarely finely toothed at tip) 3

3 Base of leaf blade tapered, not clasping stem. 4
3 Base of leaf blade clasping stem 5

4 Plants green, upper leaves stalked, leaf margins finely toothed near tip . . . . *Potamogeton illinoensis* | ILLINOIS PONDWEED p. 601
4 Plants red-tinged, upper leaves ± stalkless, leaf margins entire . . . . *Potamogeton alpinus* | RED PONDWEED p. 595

5 Stems whitish; leaves 10–30 cm long; fruit 4–5 mm long. . . . . *Potamogeton praelongus* | WHITESTEM-PONDWEED p. 604
5 Stems green; leaves 1–12 cm long; fruit 2–4 mm long 6

6 Leaves ovate, mostly 1–5 cm long, margins flat; stipules small or absent; plants drying olive-green .... *Potamogeton perfoliatus* .... REDHEAD-GRASS; CLASPING-LEAVED PONDWEED p. 604

6 Leaves lance-shaped, mostly more than 5 cm long; margins wavy-crisped; stipules conspicuous, persisting as shreds; plants drying light green..... .... *Potamogeton richardsonii* | CLASPING-LEAVED PONDWEED p. 606

7 Stipules joined with lower part of leaf to form a sheath at least 1 cm long 8

7 Stipules free from leaf, or rarely joined to leaf base for only 1–2 mm ... 11

8 Leaves 4–8 mm wide, auricled at base, margins finely toothed .... *Potamogeton robbinsii* | FERN-PONDWEED p. 606

8 Leaves threadlike, rarely to 3 mm wide, not auricled, margins entire.... 9

9 Leaves gradually tapered to tip; rhizomes tuber-bearing; stigmas raised on a tiny style............ *Potamogeton pectinatus* | SAGO-PONDWEED p. 603

9 Leaves rounded, blunt-tipped or tipped with a short, sharp point, stigmas inconspicuous, broad and not raised 10

10 Plants short, to 0.5 m long; sheaths tight around stem; spikes with 2–5 whorls of flowers .... *Potamogeton filiformis* .... THREADLEAF-PONDWEED p. 599

10 Plants large and coarse, 2–5 m long; sheaths enlarged to 2–5 times diameter of stem; spikes with 5–12 whorls of flowers. ...... *Potamogeton vaginatus* .... BIGSHEATH-PONDWEED p. 608

11 Plants with slender creeping rhizomes................................ 12

11 Plants with short rhizomes or rhizomes absent (plants often rooting at lower nodes of stem) 13

12 Flower clusters on stalks at ends of stems, the stalks mostly 5–25 cm long; leaves threadlike, narrower than stems; rare .... *Potamogeton confervoides* .... ALGA PONDWEED p. 596

12 Flower clusters on stalks from leaf axils, the stalks less than 3 cm long; leaves linear, wider than stems.................... *Potamogeton foliosus* .... LEAFY PONDWEED p. 599

13 Leaves 9- to many-veined (with 1–2 main veins and many finer ones) .... *Potamogeton zosteriformis* | FLATSTEM-PONDWEED p. 609

13 Leaves 1–7-veined 14

14 Leaves without glands at base .... *Potamogeton foliosus* .... LEAFY PONDWEED p. 599

14 At least some of leaves with pair of glands at base 15

15 Leaves with 5–7 veins ..... *Potamogeton friesii* | FRIES' PONDWEED p. 600
15 Leaves with 3 (rarely 1 or 5) veins 16

16 Leaves gradually tapered to a bristlelike tip...... *Potamogeton strictifolius* .... STRAIGHT-LEAVED PONDWEED p. 607
16 Leaves rounded at tip or tapered to a point, not bristle-tipped 17

17 Leaves 1–4 mm wide, rounded at tip; body of achene 2.5–4 mm long .... *Potamogeton obtusifolius* | BLUNTLEAF-PONDWEED p. 603
17 Leaves to 2.5 mm wide, usually tapered to a sharp tip; body of achene to 2 mm long... *Potamogeton pusillus* | SLENDER OR SMALL PONDWEED p. 605

POTAMOGETON GROUP 2

Plants with 2 kinds of leaves: broad floating leaves and broad or narrow underwater leaves.

1 Underwater leaves broad, never narrowly linear...................... 2
1 Underwater leaves linear or threadlike 7

2 Floating leaves with 30–55 veins; underwater leaves with 30–40 veins ... . *Potamogeton amplifolius* | BIGLEAF PONDWEED p. 595
2 Floating leaves with fewer than 30 veins; underwater leaves with less than 30 veins 3

3 Underwater leaves with more than 7 veins, all leaves stalked........... 4
3 Underwater leaves mostly with 7 veins, at least the lower leaves stalkless 5

4 Base of floating leaves ± heart-shaped.............. *Potamogeton pulcher* .... SPOTTED PONDWEED p. 605
4 Base of floating leaves tapered or rounded, not heart-shaped .... *Potamogeton nodosus* | LONGLEAF PONDWEED p. 602

5 Margins of underwater leaves finely toothed near tip .... *Potamogeton illinoensis* | ILLINOIS PONDWEED p. 601
5 Margins of underwater leaves entire.............................. 6

6 Plants red-tinged; underwater leaves 5–20 cm long and at least as wide as floating leaves, mostly on main stem............... *Potamogeton alpinus* .... RED PONDWEED p. 595
6 Plants green; underwater leaves 3–8 cm long and narrower than floating leaves, often numerous on short branches from leaf axils............... .... *Potamogeton gramineus* | VARIABLE PONDWEED p. 600

7 Spikes of 1 kind only; fruits not (or only slightly) compressed; stipules not joined with leaf base **8**

7 Spikes of 2 kinds: those in axils of lower underwater leaves on short stalks; those in axils of upper or floating leaves often emersed on long stalks; fruit flattened; stipules of leaves (or at least some of lower leaves) joined with leaf base **12**

8 Floating leaves less than 1 cm wide and less than 2 cm long . . . . *Potamogeton vaseyi* | VASEY'S PONDWEED p. 608

8 Floating leaves more than 1 cm wide and more than 2 cm long **9**

9 Underwater leaves flat and tapelike, 2–10 mm wide . . . . *Potamogeton epihydrus* | RIBBONLEAF-PONDWEED p. 598

9 Underwater leaves round in cross-section, often reduced to a petiole, mostly less than 1.5 mm wide **10**

10 Blade of floating leaves oval, tapered to base; fruit 3-keeled . . . . *Potamogeton nodosus* | LONGLEAF PONDWEED p. 602

10 Blade of floating leaves ovate to nearly heart-shaped at base; fruit barely keeled **11**

11 Floating leaves mostly 3–10 cm long; spikes 3–6 cm long . . . . *Potamogeton natans* | FLOATING PONDWEED p. 601

11 Floating leaves 2–5 cm long; spikes 1–3 cm long . . . . *Potamogeton oakesianus* | OAKES' PONDWEED p. 602

12 Underwater leaves hair-like, to only about 0.3 mm wide, acute to long-tapering at tip; tips of floating leaves acute . . . . . *Potamogeton bicupulatus* . . . . SNAILSEED PONDWEED p. 596

12 Underwater leaves hair-like but slightly wider (more than about 0.5 mm), leaf tips obtuse to acute; floating leaf tips rounded. **13**

13 Underwater leaves blunt-tipped, floating leaves with a small notch at tip; achene beak absent . . . . *Potamogeton spirillus* . . . . NORTHERN SNAILSEED PONDWEED p. 607

13 Underwater leaves tapered to a pointed tip; floating leaves not notched at tip; achene beak tiny; rare. . . . . *Potamogeton diversifolius* . . . . COMMON SNAILSEED PONDWEED p. 598

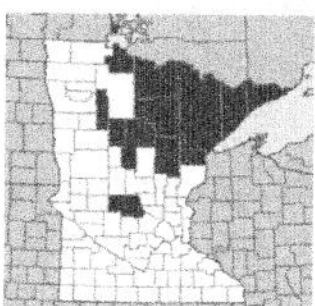

Potamogeton alpinus Balbis.
RED PONDWEED
NC **OBL** | MW **OBL** | GP **OBL**

Native aquatic perennial herb; plants red-tinged. **Stems** round in section, unbranched or sometimes branched above, to 1 m long and 1–2 mm wide. **Underwater leaves** linear lance-shaped, 4–20 cm long and 5–15 mm wide, 7–9-veined, usually rounded at tip, narrowed to a stalkless base. **Floating leaves** often absent, if present, thin, obovate, 4–6 cm long and 1–2 cm wide, 7- to many-veined, rounded at tip, tapered to a narrow base; stipules not joined to leaf base, membranous, 1–3 cm long and to 1.5 cm wide. **Flowers** in cylindric spikes, 1–3 cm long, with 5–9 whorls of flowers, on stalks 6–15 cm long and about as thick as stem. **Achene** yellow-brown to olive, flattened, 3 mm long, the beak short. July–Sept. **Habitat** Shallow to deep (and usually cold) water of lakes and streams.

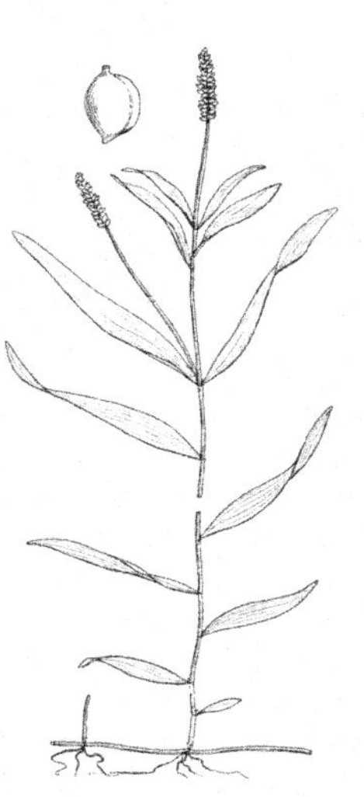

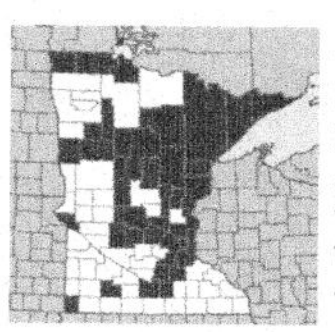

Potamogeton amplifolius Tuckerman
BIGLEAF PONDWEED
NC **OBL** | MW **OBL** | GP **OBL**

Native aquatic perennial herb. **Stems** round in section, usually unbranched, to 1 m or more long and 2–4 mm wide. **Upper underwater leaves** ovate, folded and sickle-shaped, 8–20 cm long and 2–7 cm wide, many-veined; **lower underwater leaves** lance-shaped, to 2 cm wide, often not folded, usually decayed by fruiting time, many-veined; petioles 1–5 cm long. **Floating leaves** usually present at flowering time, ovate 5–10 cm long and 3–6 cm wide, many-veined, rounded at tip or abruptly tapered to a sharp tip, rounded at base; petioles 5–15 cm long; stipules open and free of the petioles, 5–12 cm long, long-tapered to a sharp tip. **Flowers** in dense cylindric spikes, 3–6 cm long in fruit, on stalks 6–20 cm long, widening near tip. **Achene** green-brown to brown, 4–5 mm long, beak to 1 mm long. July–Aug. **Habitat** Shallow to deep water of lakes and rivers.

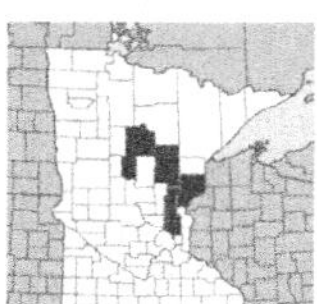

Potamogeton bicupulatus Fern.

SNAILSEED PONDWEED

NC **OBL** | MW **OBL** | *endangered*

Native aquatic perennial herb; plants delicate, spreading by rhizomes. **Stems** compressed. **Leaves** with both underwater and floating leaves, or the floating leaves sometimes absent; ± spirally arranged. **Underwater leaves** light green to rarely brown, hair-like, linear, 1.5–11 cm long and less than 0.5 mm wide, base slightly tapered, basal lobes absent, not clasping stem, stalkless, margins entire, not crisped; veins 1; stipules joined to blade for less than 1/2 stipule length. **Floating leaves** variable, stalked, oval, to 2 cm long and 1 cm wide, upper surface light green, base tapered or rounded, apex acute to long tapered; veins 3–7. **Flowers** in unbranched heads. **Achenes** greenish brown, somewhat keeled, beak absent; very small but have a noticeably bumpy surface (visible to the naked eye) due to 3 rows of sculpted ridges around the rim of the tiny, disk-shaped seed, similar to the closely related *Potamogeton diversifolius*. Flowering early summer–fall. **Synonyms** *Potamogeton diversifolius* var. *trichophyllus*. **Habitat** Soft water lakes (water low in dissolved minerals). **Status** Endangered.

Potamogeton bicupulatus *is similar to* ***Potamogeton diversifolius*** *(also state endangered). Both species have extremely fine, hair-like underwater leaves, but in Potamogeton diversifolius, these leaves very slightly wider and the floating leaves somewhat rounded at tip.*

Potamogeton confervoides Reichenb.

ALGA PONDWEED

[NC **OBL**]

Native aquatic perennial herb, from a long rhizome. **Stems** slender, to 8 dm long, branched, the branches forking. **Leaves** many, all underwater, flat, bright green, 2–5 cm long and about 0.3 mm wide, tapered to a hairlike tip, l-veined; stipules short-lived, 1–5 cm long. **Leaves** so delicate that they resemble greenish colored hair in the water. **Flowers** in a short spike 5–10 mm long, at end of an erect stalk 5–20 cm long. **Achene** 2–3 mm long, with a sharp keel. June–Aug. **Habitat** In Minnesota, known from n St Louis County in shallow water of a small, acidic, bog lake.

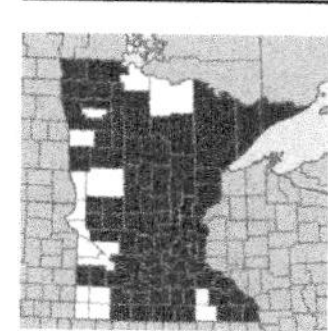

Potamogeton crispus L.

CURLY PONDWEED

NC **OBL** | MW **OBL** | GP **OBL**

Introduced aquatic perennial herb, sometimes invasive. **Stems** compressed, with few branches, to 8 dm long and 1–2 mm wide. **Leaves** all underwater, oblong, 3–9 cm long and 5–10 mm wide, rounded at tip, slightly clasping at base, stalkless, 3–5-veined, margins wavy-crisped, finely toothed; stipules 4–10 mm long, slightly joined at base, early shredding. **Flowers** in dense cylindric spikes, 1–2 cm long, appearing bristly in fruit from long achene beaks; on stalks 2–6 cm long. **Achene** brown, 2–3 mm long, with a beak 2–3 mm long. April–June. **Habitat** Shallow to deep water of lakes (including Great Lakes) and rivers; pollution-tolerant. Native of Europe; established across most of North America.

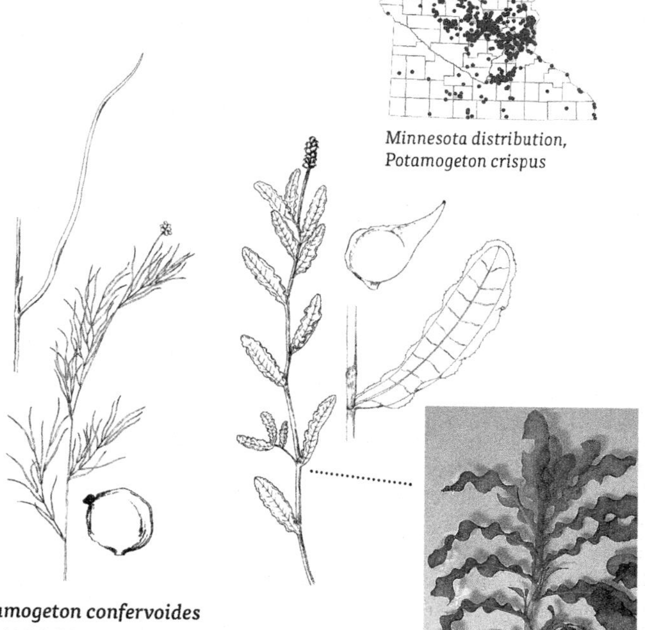

Minnesota distribution, Potamogeton crispus

Potamogeton confervoides
ALGA PONDWEED

Potamogeton crispus
CURLY PONDWEED

Potamogeton diversifolius Raf.

COMMON SNAILSEED PONDWEED

NC **OBL** | MW **OBL** | GP **OBL** | *endangered*

Native aquatic perennial herb. **Stems** slender, flattened or round in section, branched, to 15 dm long and 1 mm wide. **Underwater leaves** linear, flat, 1–10 cm long and mostly about 0.5 mm wide, 1-veined (sometimes 3-veined), stalkless; stipules 2–18 mm long, joined to leaf blade for less than half their length. **Floating leaves** variable, sometimes absent, oval, leathery, 5–40 mm long and 4–20 mm wide, acute to rounded at tip, rounded at base, 3- to many-veined, the veins sunken on leaf underside; petioles 1–3 cm long; stipules not joined to leaf base, 5–20 mm long. **Flowers** in spikes of 2 types, the underwater spikes 3–6 mm long, on stalks 2–10 mm long; emersed spikes cylindric, 1–3 cm long, on stalks 5–30 mm long. **Achene** olive to yellow, round and flattened, spiraled on surface, winged, the beak tiny. June–Sept. **Synonyms** Includes *Potamogeton capillaceus*. **Habitat** Soft water lakes (water low in minerals). **Status** Endangered.

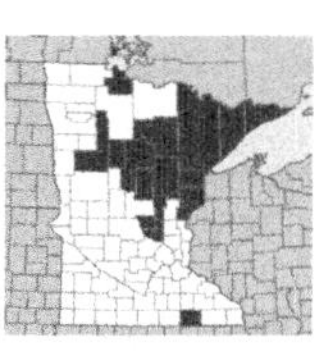

Potamogeton epihydrus Raf.

RIBBONLEAF-PONDWEED

NC **OBL** | MW **OBL** | GP **OBL**

Native aquatic perennial herb. **Stems** slender, compressed, sparingly branched, to 2 m long and 1–2 mm wide. **Underwater leaves** linear, ribbonlike, 10–20 cm long and 3–8 mm wide, with a translucent strip on each side of midvein forming a band 1–3 mm wide, 5–13-veined, stalkless; stipules 1–3 cm long, not joined to leaf. **Floating leaves** usually present and numerous, opposite, oval to obovate, 3–8 cm long and 1–2 cm wide, mostly obtuse to bluntly abruptly short-awned at the tip, 11–25-veined, tapered to flattened petioles; stipules free, 1–3 cm long. **Flowers** in dense, cylindric spikes 2–3 cm long, on stalks 2–6 cm long and about as thick as stem. **Achene**

olive to brown, 2–3 mm long; beak tiny. July–Sept. **Habitat** Water to 2 m deep in lakes, ponds and rivers.

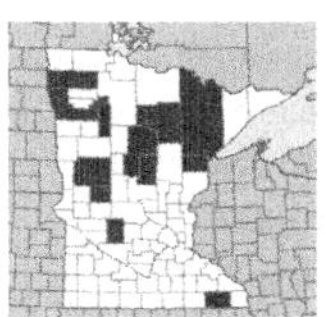

Potamogeton filiformis Pers.

Stuckenia filiformis (Pers.) Börner
THREADLEAF-PONDWEED
NC **OBL** | MW **OBL** | GP **OBL**

Native aquatic perennial herb, from a long, tuber-bearing rhizome. **Stems** ± round in section, branched from base, mostly unbranched above, 1–5 dm or more long and 1 mm wide. **Leaves** all underwater, narrowly linear, 5–10 cm long and 0.2–2 mm wide, 1-veined; stipules 1–3 cm long, joined to base of leaf blade, forming a tight sheath around stem. **Flowers** in underwater spikes, 1–5 cm long, with 2–5 separated whorls of flowers, on slender stalks 2–12 cm long. **Achene** olive-green, 2–3 mm long, the beak flat, tiny. July–Aug. **Habitat** Mostly shallow water (to 1 m) in lakes and rivers.

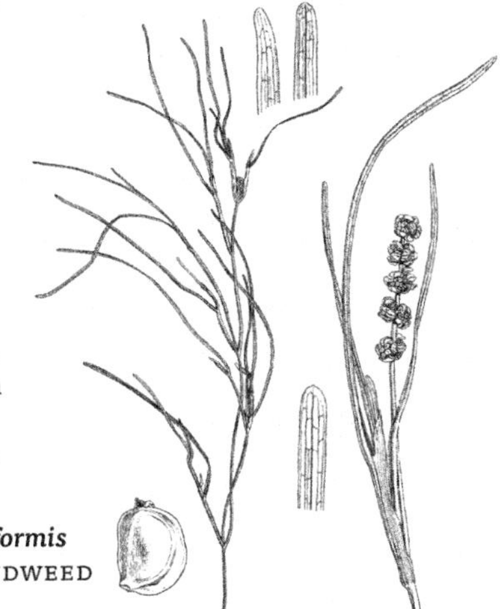

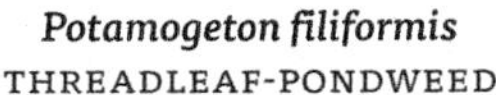

Potamogeton filiformis
THREADLEAF-PONDWEED

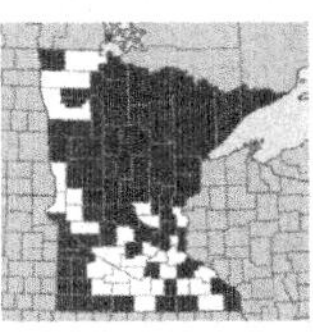

Potamogeton foliosus Raf.

LEAFY PONDWEED
NC **OBL** | MW **OBL** | GP **OBL**

Native aquatic perennial herb. **Stems** compressed, much-branched, to 8 dm long and 1 mm wide. **Leaves** all underwater, linear, 1–8 cm long and 1–2 mm wide, 1–3-veined, stalkless; stipules free, 0.5–2 cm long, glands usually absent at base of stipules. **Flowers** in rounded to short-cylindric spikes, 2–7 mm long, with 1–2 whorls of flowers, on stalks 5–12 mm long, widened at tip. **Achene** green-brown, 1.5–3 mm long, winged, the beak to 0.5 mm long. June–Aug. **Habitat** Shallow to deep water of lakes, ponds, rivers and streams.

Potamogeton foliosus
LEAFY PONDWEED

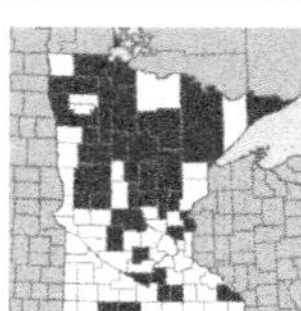

Potamogeton friesii Rupr.
FRIES' PONDWEED
NC **OBL** | MW **OBL** | GP **OBL**

Native aquatic perennial herb. **Stems** compressed, branched, 1–1.5 m long and to 1 mm wide. **Leaves** all underwater, linear, 3–7 cm long and 1.5–3 mm wide, tip rounded with a short slender point, tapered to the base, 5–7-veined, stalkless, margins flat or becoming rolled under; stipules free, 5–20 mm long, fibrous, often shredding above, 2 glands present at base of stipule. **Flowers** in cylindric spikes, 8–16 mm long, with 2–5 whorls of flowers, on stalks 1.5–6 cm long. **Achene** olive-green to brown, 2–3 mm long, beak flat, short. June–Aug. **Habitat** Shallow to deep water of lakes, ponds, rivers and streams.

Potamogeton gramineus L.
VARIABLE PONDWEED
NC **OBL** | MW **OBL** | GP **OBL**

Native aquatic perennial herb. **Stems** slender, slightly compressed, much-branched, to 8 dm long and 1 mm wide. **Underwater leaves** variable, linear to lance-shaped or oblong lance-shaped, 3–9 cm long and 3–12 mm wide, 3–7-veined, tapered to a stalkless base. **Floating leaves** usually present, oval, 2–6 cm long and 1–3 cm wide, 11–19-veined, rounded at base; petioles 2–10 cm long, shorter to longer than blade; stipules free, persistent, 1–4 cm long. **Flowers** in dense, cylindric spikes, 1.5–4 cm long, the stalks thicker than stem, 2–10 cm long. **Achene** dull green, 2–3 mm long. June–Aug. **Habitat** Shallow to deep water of streams, ponds and lakes.

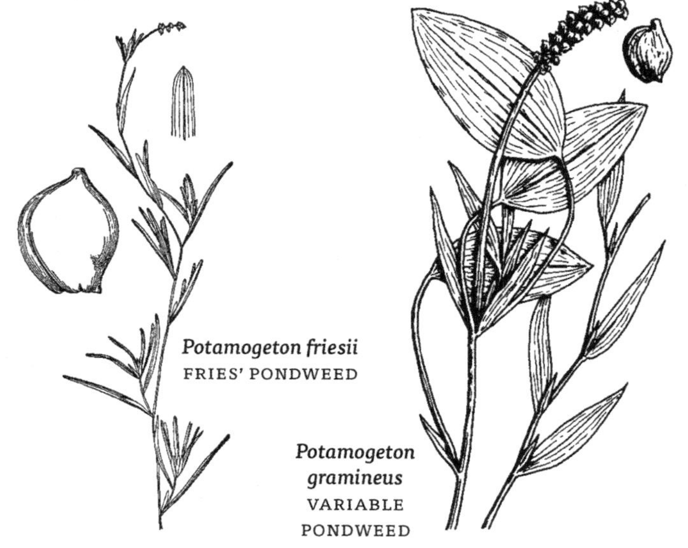

Potamogeton friesii
FRIES' PONDWEED

Potamogeton gramineus
VARIABLE PONDWEED

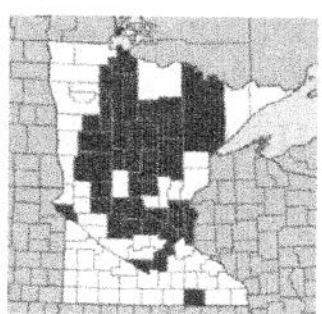

Potamogeton illinoensis Morong

ILLINOIS PONDWEED

NC **OBL** | MW **OBL** | GP **OBL**

Native aquatic perennial herb. **Stems** nearly round in section, usually branched, to 2 m long and 2–5 mm wide. **Underwater leaves** lance-shaped to obovate, 6–20 cm long, 2–4 cm wide, 9–17-veined, tapered to a broad, flat petiole, 2–4 cm long; stipules free, persistent, 3–8 cm long. **Floating leaves** sometimes absent, opposite, lance-shaped to oval, 5–14 cm long and 2–6 cm wide, 13- to many-veined, often short-awned from the rounded tip, rounded to wedge-shaped at base; petioles 3–10 cm long, shorter than blades. **Flowers** in dense cylindric spikes, 2–6 cm long, on stalks 4–20 cm long, usually wider than stem. **Achene** olive-green, 3–4 mm long, the beak short, blunt. July–Sept. **Synonyms** *Potamogeton angustifolius*. **Habitat** Shallow to deep water of lakes and rivers.

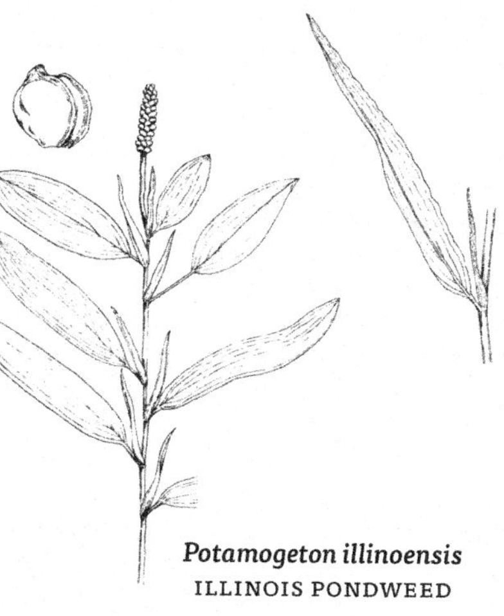

Potamogeton illinoensis
ILLINOIS PONDWEED

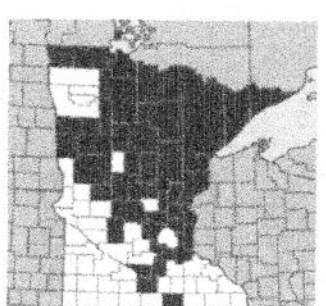

Potamogeton natans L.

FLOATING PONDWEED

NC **OBL** | MW **OBL** | GP **OBL**

Native aquatic perennial herb. **Stems** slightly compressed, usually unbranched, 0.5–2 m long and 1–2 mm wide. **Underwater leaves** reduced to linear, bladeless, expanded petioles (phyllodes), these often absent by flowering time, 10–30 cm long and 1–2 mm wide. **Floating leaves** ovate to oval, 4–10 cm long and 2–5 cm wide, usually tipped with a short point, rounded to heart-shaped at base, many-veined; petioles usually much longer than blades, the blade often angled at juncture with petiole; stipules free, 4–10 cm long, persistent

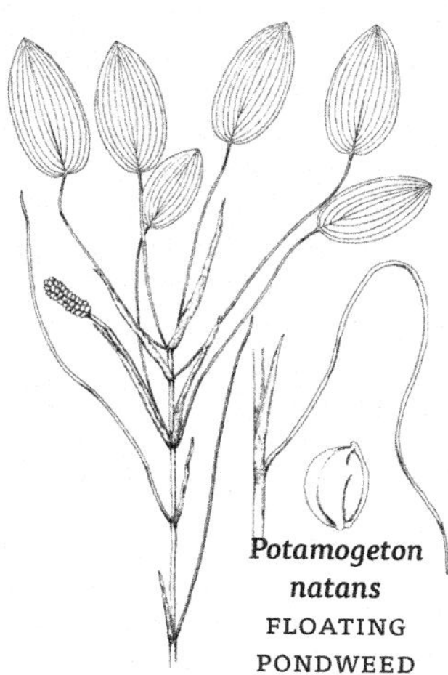

Potamogeton natans
FLOATING PONDWEED

or shredding with age. **Flowers** in dense cylindric spikes, 2–5 cm long, stalks thicker than the stem, 6–14 cm long. **Achene** green-brown to brown, 3–5 mm long, with a loose, shiny covering, the beak short. June–Aug. **Habitat** Usually shallow water (to 2 m deep) of ponds, lakes, rivers and peatlands.

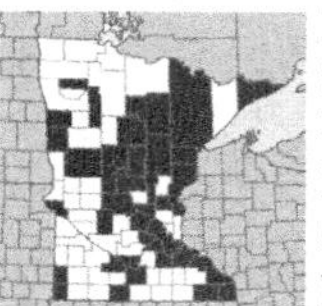

Potamogeton nodosus Poiret
LONGLEAF PONDWEED
NC **OBL** | MW **OBL** | GP **OBL**

Native aquatic perennial herb. **Stems** round in section, branched, to 2 m long and 1–2 mm wide. **Underwater leaves** commonly decayed by fruiting time, lance-shaped to linear, translucent, 10–30 cm long and 1–3 cm wide, 7–15-veined, gradually tapered to a petiole 4–10 cm long. **Floating leaves** oval, thin, 5–12 cm long and 1–5 cm wide, tapered at both ends, many-veined; petioles somewhat winged, 5–20 cm long and 2–3 mm wide, usually longer than blades; stipules free, those of underwater leaves often absent by flowering time, those of floating leaves persistent, 3–10 cm long. **Flowers** in dense cylindric spikes, 2–6 cm long, on stalks 3–15 cm long and thicker than stem. **Achene** red-brown to brown, 3–4 mm long, the beak short. July–Aug. **Habitat** Shallow water to 2 m deep, mostly in rivers; lakes.

Potamogeton oakesianus J. W. Robbins
OAKES' PONDWEED
NC **OBL** | MW **OBL**

Native aquatic perennial herb. **Stems** slender, often much-branched, to 1 m long. **Underwater leaves** bladeless, petiolelike, 0.5–1 mm wide, often persistent. **Floating leaves** oval, 3–6 cm long and 1–2 cm wide, rounded at base, 12- to many-veined; petioles 5–15 cm long; stipules free, 2.5–4 cm long. **Flowers** in cylindric spikes, 1.5–3 cm long, on stalks 3–8 cm long and wider than stem. **Achene** 2–4 mm long, with a tight, dull covering, the beak flat.
Habitat Ponds and streams, peatland pools.

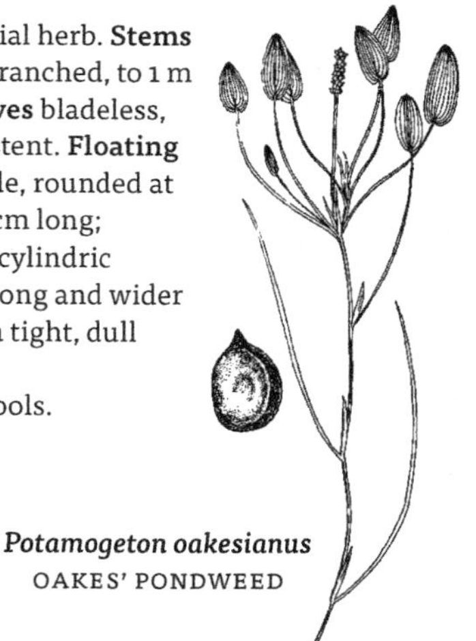

Potamogeton oakesianus
OAKES' PONDWEED

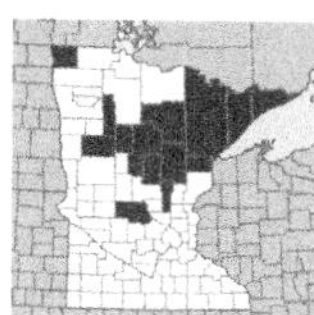

Potamogeton obtusifolius Mert. & Koch

BLUNTLEAF-PONDWEED

NC **OBL** | MW **OBL** | GP **OBL**

Native aquatic perennial herb, rhizomes ± absent. **Stems** very slender, compressed, much-branched, to 1 m long. **Leaves** all underwater, linear, stalkless, often red-tinged, 3–10 cm long and 1–4 mm wide, rounded at tip, the midvein broad, base usually with pair of translucent glands; stipules free, white, 1–2 cm long. **Flowers** in thick cylindric spikes, 8–14 mm long, on slender, upright stalks 1–3 cm long. **Achene** 2–3 mm long, the beak rounded, 0.5 mm long **Habitat** Lakes, ponds and streams, pools in bogs and fens.

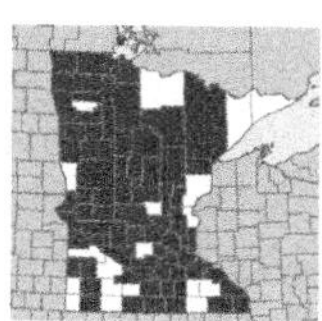

Potamogeton pectinatus L.

Stuckenia pectinata (L.) Börner

SAGO-PONDWEED

NC **OBL** | MW **OBL** | GP **OBL**

Native aquatic perennial herb, the rhizomes tipped with a white tuber important in the diet of waterfowl. **Stems** slender, round in section, 3–10 dm long and 1–2 mm wide much-branched and forking above, fewer branched near base. **Leaves** all underwater, threadlike to narrowly linear, 3–12 cm long and 0.5–1.5 mm wide, stalkless; stipules joined to base of blade for 1–3 cm, forming a sheath around stem. **Flowers** on underwater, cylindric spikes 1–5 cm long, with 2–5 whorls of flowers, on lax, threadlike stalks to 15 cm long. **Achene** yellow-brown, 3–4 mm long, the beak to 0.5 mm long. June–Sept. **Habitat** Shallow to deep water of lakes, ponds and streams; tolerant of brackish water. Common, found nearly worldwide.

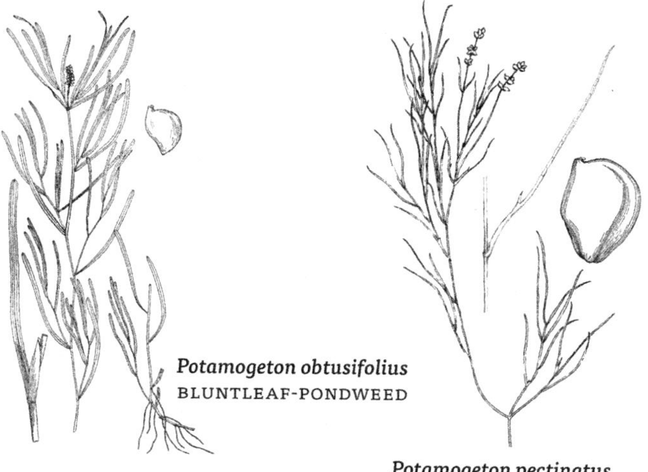

Potamogeton obtusifolius
BLUNTLEAF-PONDWEED

Potamogeton pectinatus
SAGO-PONDWEED

Potamogeton perfoliatus L.
REDHEAD-GRASS;
[NC **OBL** | MW **OBL** | GP **OBL**]
CLASPING-LEAVED PONDWEED

Native aquatic perennial herb. **Stems** slender, to 2.5 m long, often much-branched. **Leaves** all underwater, ovate to nearly round or sometimes lance-shaped, 1–7 cm long and 5–30 mm wide, tip often very finely toothed, base heart-shaped and clasping stem, stalkless; stipules free, soon decaying. **Flowers** on underwater cylindric spikes, 1–5 cm long, on upright stalks 1–7 cm long and about as wide as stem. **Achene** 2–3 mm long, the beak short, curved. **Habitat** Lakes and streams.

Reported for Minnesota but distribution unknown; more common in ne USA.

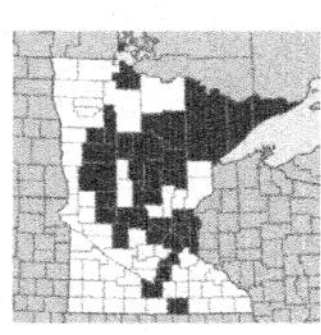

Potamogeton praelongus Wulfen
WHITESTEM-PONDWEED
NC **OBL** | MW **OBL** | GP **OBL**

Native aquatic perennial herb. **Stems** white-tinged, compressed, branched, to 2–3 m long and 2–4 mm wide, the shorter internodes often zigzagged. **Leaves** all underwater, lance-shaped, 10–30 cm long and 1–4 cm wide, with 3–5 main veins, rounded and hoodlike at tip, base ± heart-shaped and clasping stem, stalkless, margins entire and gently wavy; stipules free, white, 1–3 cm long, fibrous at tip. **Flowers** in dense, cylindric spikes 2–5 cm long; stalks erect, 1–4 dm long, as wide as stems. June–Aug. **Habitat** Shallow to deep water of lakes, streams.

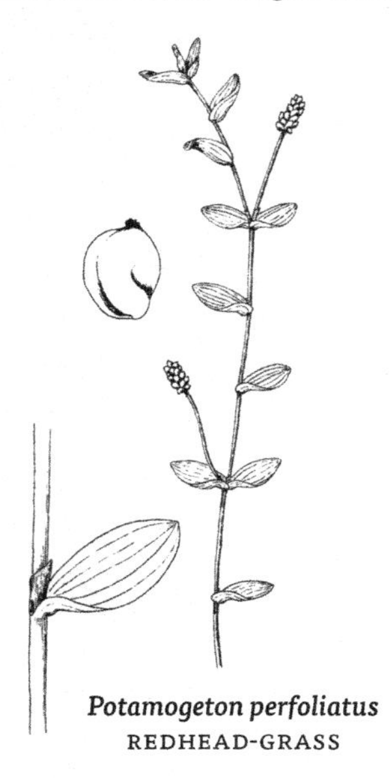

Potamogeton perfoliatus
REDHEAD-GRASS

Potamogeton praelongus
WHITESTEM-PONDWEED

Potamogeton pulcher Tuckerman

SPOTTED PONDWEED

NC **OBL** | MW **OBL** | GP **OBL**

Native aquatic perennial herb. **Stems** round in section, unbranched, black-spotted, usually less than 5 dm long. **Underwater leaves** thin, narrowly lance-shaped, 8–15 cm long and 1–3 cm wide, base tapered to a short petiole, margins wavy; the lowest leaves often thick and spatula-shaped. **Floating leaves** alternate, clustered at top of stem on short branches, ovate, 4–8 cm long and 2–5 cm wide, many-veined, the base somewhat heart-shaped; petioles black-spotted, 2–8 cm long; stipules free, to 6 cm long. **Flowers** in dense cylindric spikes, 2–4 cm long, on stalks 5–10 cm long and slightly wider than stem. **Achene** 4–5 mm long, the beak broad and blunt. **Habitat** Muddy shores and shallow water of lakes.

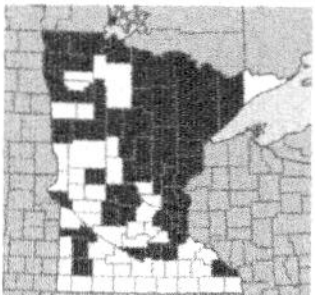

Potamogeton pusillus L.

SLENDER OR SMALL PONDWEED

NC **OBL** | MW **OBL** | GP **OBL**

Native aquatic perennial herb, rhizomes ± absent. **Stems** very slender, round in section, usually freely branched, 2–10 dm long and about 0.5 mm wide. **Leaves** all underwater, linear, 1–7 cm long and 0.5–2 mm wide, tapered to a stalkless base, the midvein broad; stipules free, boat-shaped, brown-green, 4–10 mm long and 2x width of leaf base, soon decaying, glands sometimes present at stipule base. **Flowers** in short-cylindric spikes 2–10 mm long, the flowers in 1–3 whorls, on slender, upright stalks 1–5 cm long. **Achene** green to brown, 1–2 mm long, the beak flat. June–Aug. **Habitat** Shallow water (to 2 m deep) of lakes and ponds, occasionally in streams.

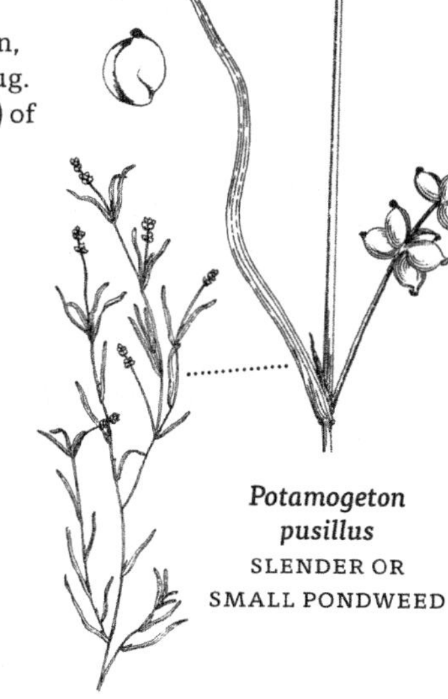

Potamogeton pulcher
SPOTTED PONDWEED

Potamogeton pusillus
SLENDER OR SMALL PONDWEED

Potamogeton richardsonii (Ar. Benn.) Rydb.
CLASPING-LEAVED PONDWEED
NC **OBL** | MW **OBL** | GP **OBL**

Aquatic perennial herb. **Stems** brown to yellow-green, round in section, sparingly to freely branched, mostly 3–10 dm long and 1–2.5 mm wide, the shorter internodes rarely zigzagged. **Leaves** all underwater, lance-shaped, 5–12 cm long and 1–2.5 cm wide, with 13 or more prominent veins, base heart-shaped and clasping stem, stalkless, margins entire and gently wavy; stipules free, 1–2 cm long, soon shredding into white fibers. **Flowers** in dense cylindric spikes 1.5–4 cm long, on stalks 2–20 cm long, the stalks strongly curved when in fruit. **Achene** green to brown, 2–4 mm long, the beak short. July–Aug. **Habitat** Shallow to deep water of lakes and streams.

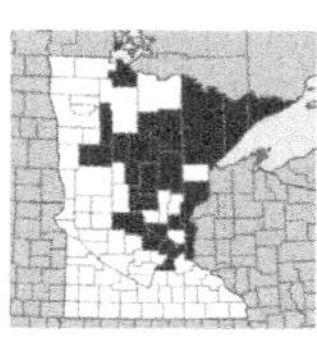

Potamogeton robbinsii Oakes
FERN-PONDWEED
NC **OBL** | MW **OBL** | GP **OBL**

Native aquatic perennial herb, rhizomes not tuberous. **Stems** few-branched below, much-branched above, to 1 m long. **Leaves** all underwater, crowded in 2 ranks, linear, 4–10 cm long and 3–7 mm wide, tapered to a pointed tip, abruptly narrowed at base, with rounded auricles where joined with stipule, midvein pronounced, margins pale; stipules joined to leaf for 5–15 mm, soon decaying into fibers. **Flowers** on underwater, cylindric spikes 1–2 cm long, with 3–5 separated whorls of flowers, the inflorescence often branched into 5–20 stalks, 2–5 cm long, at ends of stems. **Achenes** rarely produced, 3–5 mm long, the beak thick, somewhat curved; reproduction most commonly by stem fragments which root from the nodes. July–Aug. **Habitat** Shallow to deep water of lakes, ponds and streams.

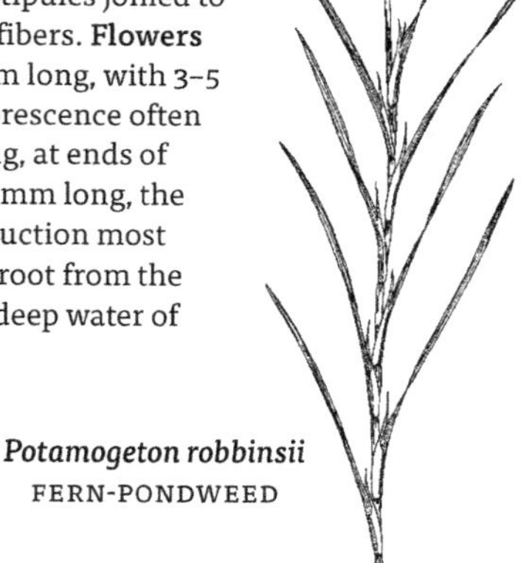

Potamogeton robbinsii
FERN-PONDWEED

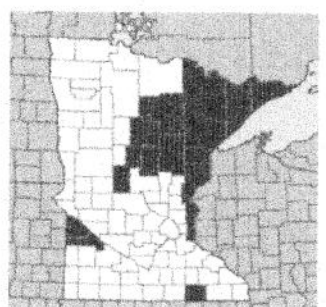

Potamogeton spirillus Tuckerman
NORTHERN SNAILSEED PONDWEED
NC **OBL** | MW **OBL** | GP **OBL**

Native aquatic perennial herb. **Stems** compressed, to 1 m long, branched, the branches short and often curved. **Underwater leaves** 1–8 cm long and 0.5–2 mm wide, rounded at tip, stalkless; stipules joined for most of length. **Floating leaves**, if present, 1–4 cm long and 5–12 mm wide, 5–13-veined, the veins sunken on underside of blade, petioles 2–4 cm long; stipules free. **Flowers** in 2 types of spikes, the underwater spikes round, with 1–8 fruits, ± stalkless in the leaf axils; emersed spikes longer, cylindric, to 8–12 mm long, on stalks from leaf axils. **Achene** 1–3 mm long, flattened, winged, spiraled on surface, the beak absent. **Habitat** Shallow water of lakes and ponds.

Potamogeton spirillus
NORTHERN SNAILSEED PONDWEED

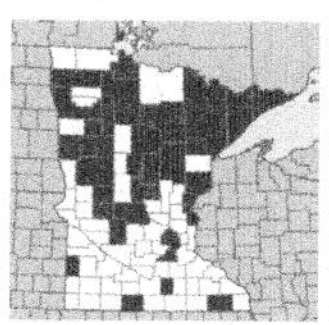

Potamogeton strictifolius Ar. Bennett
STRAIGHT-LEAVED PONDWEED
NC **OBL** | MW **OBL** | GP **OBL**

Native aquatic perennial herb. **Stems** slender, slightly compressed, unbranched or branched above, to 1 m long and 0.5 mm wide. **Leaves** all underwater, linear, upright, 1–6 cm long and 0.5–2 mm wide, 3–5-veined, the veins prominent on underside, tapered to stalkless base, margins often rolled under; stipules free, white, shredding at tip, 5–20 mm long; 2 glands present at base of stipules. **Flowers** in cylindric spikes 6–15 mm long, with 3–5 whorls of flowers, on stalks 1–5 cm long. **Fruit** green-brown, 2 mm long, the beak broad, rounded. June–Aug. **Habitat** Shallow to deep water of lakes and rivers.

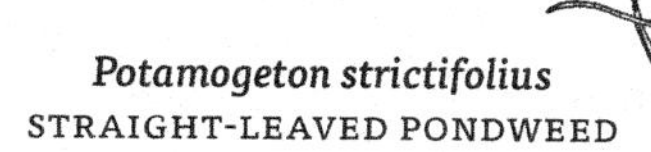

Potamogeton strictifolius
STRAIGHT-LEAVED PONDWEED

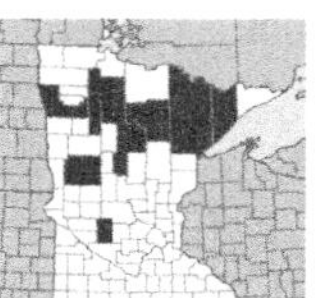

Potamogeton vaginatus Turcz.
Stuckenia vaginata (Turcz.) Holub
BIGSHEATH-PONDWEED
NC **OBL** | MW **OBL** | GP **OBL** | *special concern*

Native aquatic perennial herb, rhizomes tipped by a tuber 3–5 cm long. **Stems** round in section, much-branched above, to 1.5 m long and 1–2 mm wide. **Leaves** all underwater, crowded in 2 ranks, threadlike to narrowly linear, 2–20 cm long and 0.5–2 mm wide, with 1 main vein; stipules joined to base of leaf for 1–5 cm and sheathing stem, the sheaths on main stem inflated 2–4x wider than the stem. **Flowers** in spikes 3–6 cm long, with 5–12 spaced whorls of flowers, on lax, slender stalks to 10 cm long, the stalks often much shorter than upper leaves. **Fruit** dark green, 3 mm long, the beak short or nearly absent. July–Aug. **Synonyms** *Potamogeton interruptus*. **Habitat** Cold-water streams and lakes. **Status** Special concern.

Potamogeton vaginatus
BIGSHEATH-PONDWEED

Potamogeton vaseyi Robbins
VASEY'S PONDWEED
NC **OBL** | MW **OBL** | GP **OBL** | *special concern*

Native aquatic perennial herb. **Stems** threadlike, 2–10 dm long, much-branched, the upper branches short. **Underwater leaves** transparent, linear, 2–6 cm long and to 1 mm wide, tapered to a sharp tip, 1-veined or rarely with 2 weak lateral nerves, stalkless; stipules free, linear, white, 1–2 cm long, sometimes with 2 glands at base. **Floating leaves** on flowering plants only, opposite, obovate, leathery, 8–15 mm long and 4–7 mm wide, 5–9-veined, the veins sunken on underside, petiole about as long as blade. **Flowers** in cylindric spikes 3–8 mm long, with 1–4 whorls of flowers, on stems 1–3 cm long. **Fruit** 2–3 mm long, the beak short. **Synonyms** *Potamogeton lateralis*. **Habitat** Shallow to deep water of ponds. **Status** Special concern.

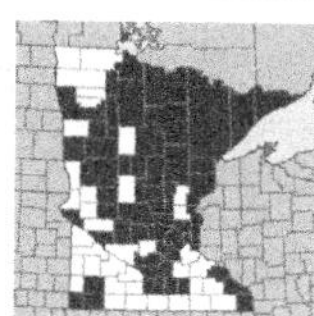

Potamogeton zosteriformis Fernald
FLATSTEM-PONDWEED
NC **OBL** | MW **OBL** | GP **OBL**

Native aquatic perennial herb, rhizomes ± absent. **Stems** strongly flattened, sometimes winged, freely branched, to 1 m long and 1–3 mm wide. **Leaves** all underwater, linear, 5–20 cm long and 3–5 mm wide, 15- to many-veined, tapered to a tip, or sometimes with a short, sharp point, slightly narrowed to the stalkless base; stipules free, white, shredding with age, 1–4 cm long. **Flowers** in cylindric spikes, 1–2.5 cm long, with 7–11 whorls of flowers, on curved stalks 2–6 cm long. **Fruit** dark green to brown, 4–5 mm long, the beak short and blunt. July–Aug. **Habitat** Shallow to deep water of lakes and streams.

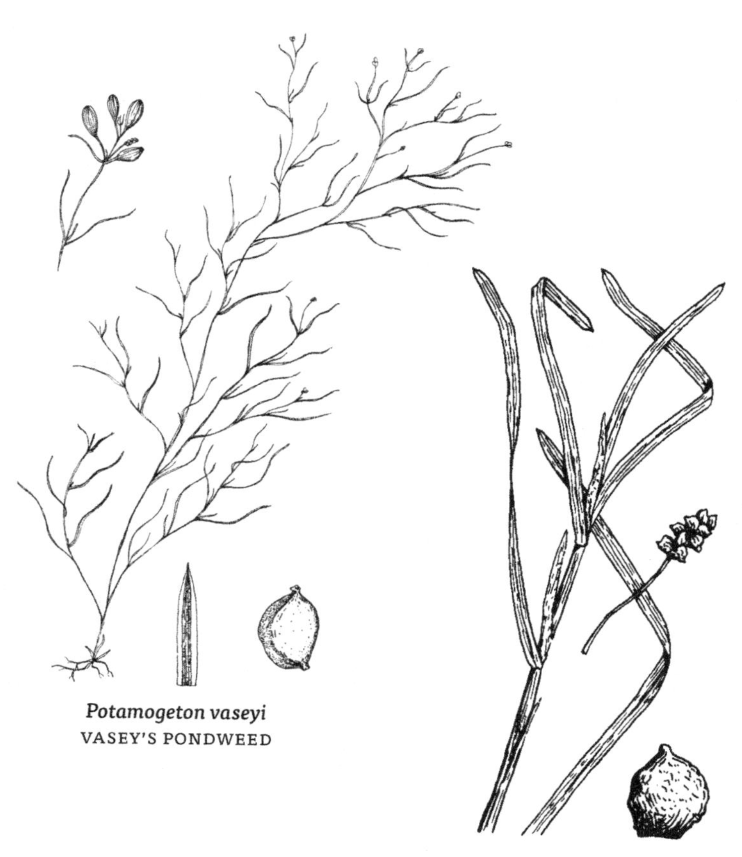

Potamogeton vaseyi
VASEY'S PONDWEED

Potamogeton zosteriformis
FLATSTEM-PONDWEED

RUPPIA | *Ditch-grass, Widgeon-grass*

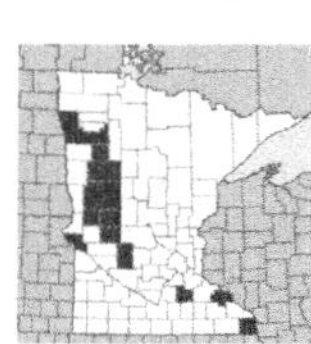

Ruppia cirrhosa (Petagna) Grande
DITCH-GRASS; WIDGEON-GRASS
NC **OBL** | MW **OBL** | GP **OBL** | *special concern*

Native aquatic perennial herb. **Stems** slender, round in section, white-tinged, wavy, to 6 dm long, branching at base and with short branches above, the internodes often zigzagged. **Leaves** simple, alternate or opposite, stalkless, thread- like, mostly 5–25 cm long and 0.5 mm wide, 1-veined, with a sheathing stipule at base. **Flowers** very small, perfect, in small, 2-flowered spikes from leaf axils, the spikes enclosed by the leaf sheath at flowering time, the flower stalks elongating and usually coiling as fruits mature; sepals and petals absent; stamens 2; pistils typically 4 (varying from 2–8), raised on a slender stalk in fruit and becoming umbel-like. **Fruit** an olive-green to black, ovate drupelet, 2–3 mm long. July–Aug.
Habitat Lakes and ponds, often with high mineral content. **Status** Special concern.

*Most inland populations of this plant best classified as **Ruppia cirrhosa**; plants of ocean coasts separated as **Ruppia maritima**.*

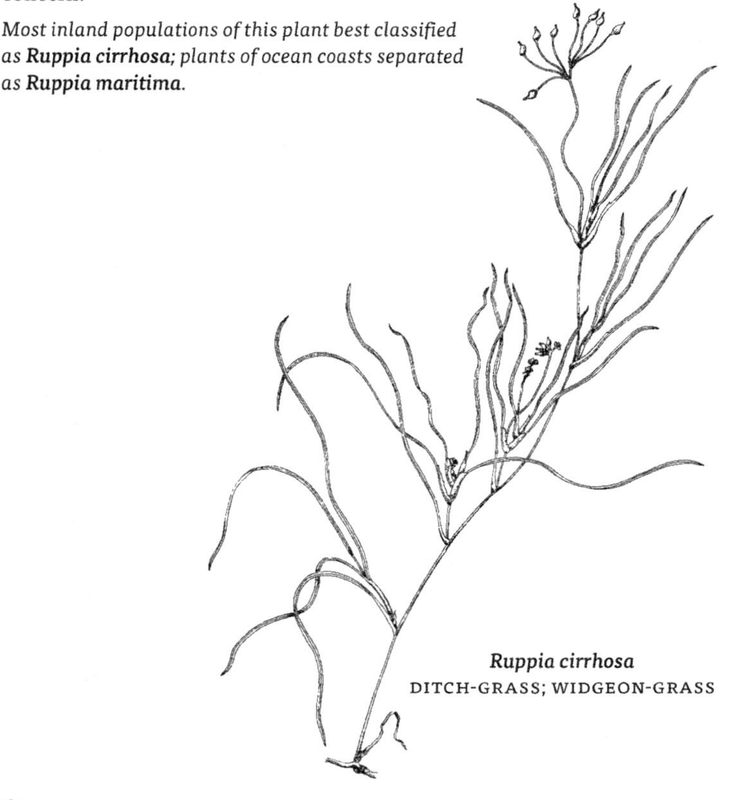

Ruppia cirrhosa
DITCH-GRASS; WIDGEON-GRASS

SCHEUCHZERIA | *Pod-grass*

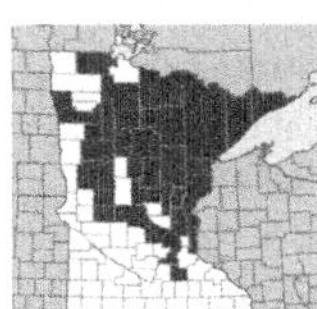

Scheuchzeria palustris L.

POD-GRASS

NC **OBL** | MW **OBL** | GP **OBL**

Native perennial rushlike herb, from creeping rhizomes. **Stems** 1 to several, 1–4 dm long, remains of old leaves often persistent at base of plant. **Leaves** alternate, several from base and 1–3 along stem, 1–3 dm long and 1–3 mm wide, the stem leaves smaller; lower part of blade half-round in section, with an expanded sheath at base, upper portion of blade flat, with a small pore at leaf tip. **Flowers** perfect, regular, green-white, in a several-flowered raceme 3–10 cm long, the flowers on stalks 1–2.5 cm long; tepals 6, in 2 series, ovate, 2–3 mm long; stamens 6. **Fruit** a group of 3 (rarely to 6) spreading follicles, 5–10 mm long, each with 1–2 seeds; seeds brown-black, 4–5 mm long. May–June. **Habitat** Wet, sphagnum moss peatlands.

Scheuchzeria palustris
POD-GRASS

SPARGANIUM | *Bur-reed*

Perennial sedgelike herbs, floating or emergent in shallow water, from rhizomes and forming colonies. **Stems** stout, usually erect, unbranched, round in section. **Leaves** long, broadly linear, sheathing stem at base. **Flowers** crowded in round heads, the heads with either male or female flowers; **male heads** few to many, borne above female heads in a unbranched or sparsely branched inflorescence; the **female heads** 1 to several, from leaf axils or borne above axils on upper stem; sepals and petals reduced to chaffy, spatula-shaped scales, these appressed to the achenes in the mature female heads; male flowers with mostly 3–5 stamens; female flowers with a 1–2-chambered pistil, stigmas 1 or 2. **Fruit** a beaked, nutlet-like achene, stalkless or short-stalked.

SPARGANIUM | BUR-REED

1 Plants large, about 1 m tall; leaves usually erect; stigmas 2; achenes broadly oblong pyramid-shaped .... *Sparganium eurycarpum* .... COMMON OR GIANT BUR-REED
1 Plants smaller, leaves erect or floating and limp; stigmas 1; achenes slender 2

2 Leaves and inflorescence emergent, erect, stiff; leaves keeled entire length or at least near base 3
2 Leaves and inflorescence floating, limp; leaves ± flat , not keeled or keeled only near base 6

3 Inflorescence rachis (flowering stem) branched 4
3 Inflorescence rachis not branched 5

4 Rachis branches (all or at least some), with 1–3 female heads; fruiting heads 1.5–2.5 cm wide; fruit dull .... *Sparganium americanum* .... AMERICAN BUR-REED
4 Rachis branches without female heads; fruiting heads 2.5–3.5 cm wide; fruit ± shiny .... *Sparganium androcladum* | BRANCHED BUR-REED

5 Male heads 1(rarely 2); fruiting heads to 1.6 cm wide; fruit beak 1.5–2 mm long .... *Sparganium glomeratum* | CLUSTERED BUR-REED
5 Male heads mostly 3–7; fruiting heads 1.6–3.5 cm wide; fruit beak 2–4.5 mm long .... *Sparganium emersum* | DWARF BUR-REED

6 Male heads 1(rarely 2); inflorescence rachis unbranched .... *Sparganium natans* | SMALL BUR-REED
6 Male heads 2 or more; inflorescence rachis branched or unbranched ... 7

7 Rachis usually branched; tepals at middle of fruit stipe (stalk) .... *Sparganium fluctuans* | FLOATING BUR-REED

7 Rachis unbranched; tepals at or near base of fruit stipe **8**

8 Male heads (at least in part) not contiguous; fruit beak 2–4.5 mm long; leaves keeled near base . . . . *Sparganium emersum* | DWARF BUR-REED

8 Male heads contiguous, appearing as single elongated head; fruit beak 1.5–2 mm long; leaves not keeled . . . . *Sparganium angustifolium* . . . . NARROW-LEAVED BUR-REED

Sparganium americanum Nutt.

AMERICAN BUR-REED

NC **OBL** | MW **OBL** | GP **OBL**

Native perennial herb. **Stems** stout, erect, mostly unbranched, 3–10 dm long. **Leaves** linear, flat to somewhat keeled, to 1 m long and 4–12 mm wide; leaflike bracts on upper stem shorter than leaves, widened at base. **Inflorescence** usually unbranched, or with a few, straight branches; female heads stalkless, 2–4 on main stem, sometimes with 1–3 on branches, 2 cm wide when mature; scales widest at tip; male heads 3–10 on main stem, sometimes with 1–5 on branches. **Achenes** widest at middle, tapered to both ends, dull brown, 3–5 mm long, the beak straight, 2–4 mm long. July–Aug. **Habitat** Marshes, shallow water, streambanks.

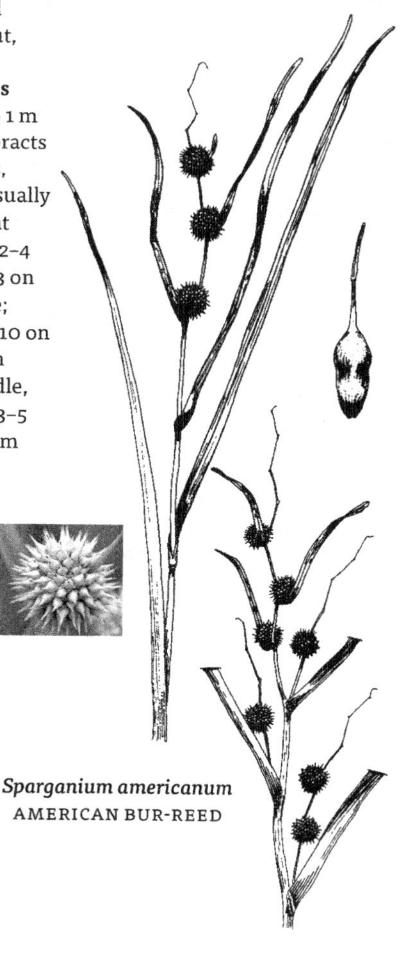

Sparganium americanum
AMERICAN BUR-REED

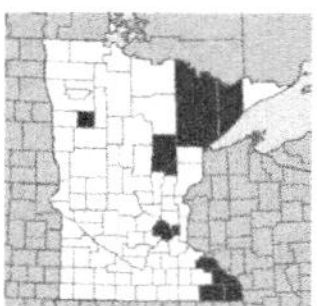

Sparganium androcladum (Engelm.) Morong

BRANCHED BUR-REED
NC **OBL** | MW **OBL** | GP **OBL**

Native perennial herb. **Stems** stout, erect, branched, 4–10 dm long. **Leaves** linear, keeled, triangular in section near base, 4–8 dm long and 5–12 mm wide; bracts leaflike, upright, shorter than leaves, slightly widened at base. **Inflorescence** often branched, the branches zigzagged; female heads stalkless, 2–4 on main stem, absent or occasionally 1 near base of branches, 3 cm wide when mature; scales spatula-shaped, widest at tip; male heads 5–8 on main stem, 3 or more on branches. **Achenes** oval, shiny light brown, 5–7 mm long, often slightly narrowed at middle, the beak straight, 4–6 mm long. July–Aug. **Synonyms** *Sparganium lucidum*. **Habitat** Marshes, lakeshores, fens.

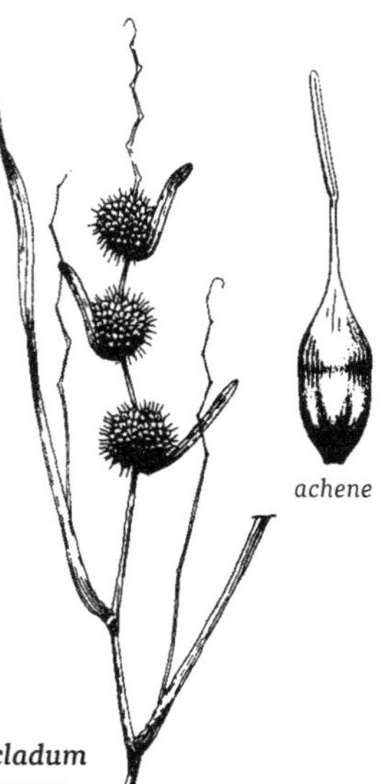

Sparganium androcladum
BRANCHED BUR-REED

Sparganium angustifolium Michx.

NARROW-LEAVED BUR-REED
NC **OBL** | MW **OBL** | GP **OBL**

Native perennial herb. **Stems** long and usually floating. **Leaves** floating, mostly 2–3 mm wide, often wider at base. Inflorescence unbranched; female heads 1–3, shiny, about 2 cm wide, the lowest stalked, the upper female heads stalkless; scales spatula-shaped, ragged at tip; male heads 2–6, close together above female heads. **Achenes** spindle- shaped, 5–7 mm long, dull brown except at red-brown base, abruptly contracted to a beak 1–3 mm long. July–Aug. **Habitat** Lakes, ponds and shores.

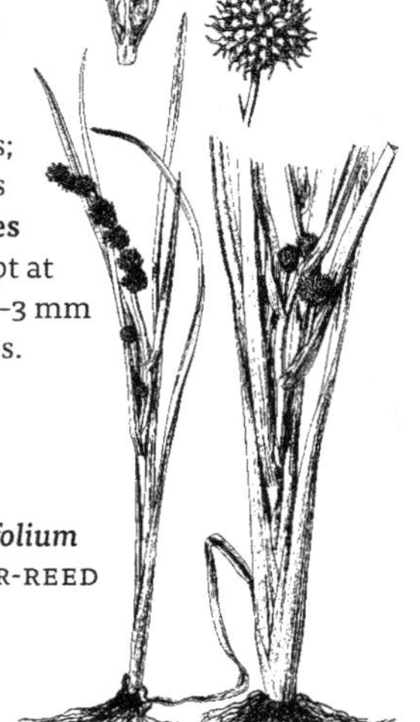

Sparganium angustifolium
NARROW-LEAVED BUR-REED

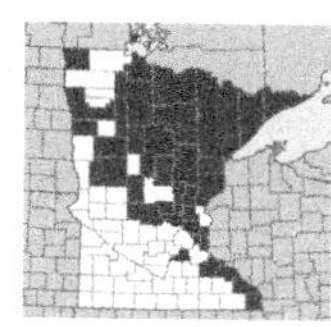

Sparganium emersum Rehmann
DWARF BUR-REED
NC **OBL** | MW **OBL** | GP **OBL**

Native perennial herb. **Stems** usually erect, sometimes lax and trailing in water, 2–6 dm long. **Leaves** linear, yellow-green, flat to keeled, 3–7 dm long and 3–6 mm wide, usually longer than stems; bracts leaflike, erect, barely widened at base. **Inflorescence** unbranched, 1–2 dm long; female heads 1–4, stalkless or lowest head often stalked, at least 1 head on stem above leaf axils, 1.5–2.5 cm wide when mature; scales spatula-shaped, widest at tip; male heads usually 2–5, 1.5–2 cm wide at flowering time. **Achenes** widest at middle, tapered to both ends, 4–5 mm long, shiny olive-green, the beak 3–5 mm long. June–Aug. **Synonyms** *Sparganium chlorocarpum*. **Habitat** Shallow water or mud of marshes, streams, ditches, open bogs, ponds.

Sparganium emersum
DWARF BUR-REED

Sparganium eurycarpum Engelm.
COMMON OR GIANT BUR-REED
NC **OBL** | MW **OBL** | GP **OBL**

Native perennial herb. **Stems** stout, branched, 4–10 dm long. **Leaves** linear, bright green, keeled, 8–10 dm long and 5–12 mm wide; bracts leaflike, slightly widened at base. Inflorescence 1–3 dm long, branched from the bract axils; lower branches with 1 female head and several male heads, main stem and upper branches with 6–10 male heads; female heads 2–6, 1.5–2.5 cm wide in fruit, scales spatula-shaped; male heads numerous, 1–2 cm wide. **Achenes** oblong pyramid-shaped, 6–8 mm long, the top flattened, 4–7 mm wide, brown to golden-brown, the beak 2–4 mm long. June–Aug. **Habitat** Usually in shallow water of marshes, streams, ditches, ponds and lakes, often with cat-tails (*Typha*).

Sparganium eurycarpum
COMMON OR GIANT BUR-REED

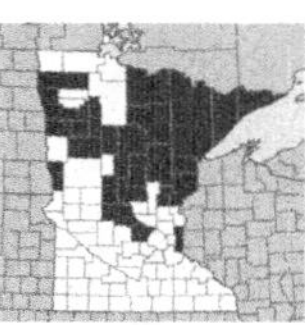

Sparganium fluctuans (Morong) Robinson
FLOATING BUR-REED
NC **OBL** | MW **OBL** | GP **OBL**

Native perennial herb. **Stems** slender, floating, to 15 dm long. **Leaves** floating, linear, flat, translucent, 3–10 mm wide, underside with netlike veins; bracts leaflike, short, widened at base. **Inflorescence** usually branched, the main stem with 2–4 male heads, the branches with 1 female head near base and 2–3 male heads above; female heads 2–4, 1.5–2 cm wide when mature, scales oblong; male heads to 1 cm wide. **Achenes** obovate, 3–4 mm long, sometimes narrowed near middle, brown, the beak curved, 2–3 mm long. **Habitat** In shallow water of ponds and lakes.

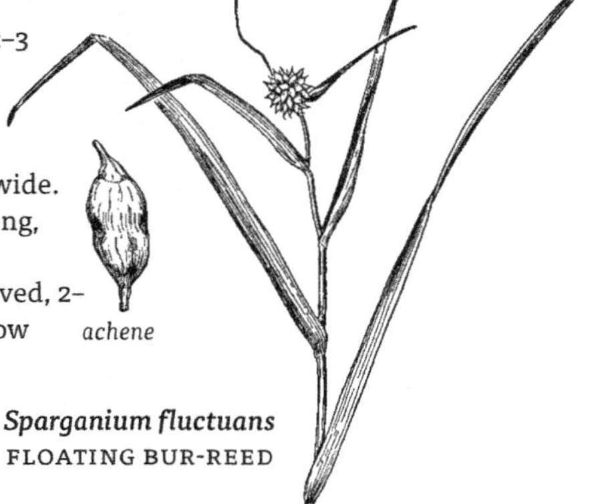

Sparganium fluctuans
FLOATING BUR-REED

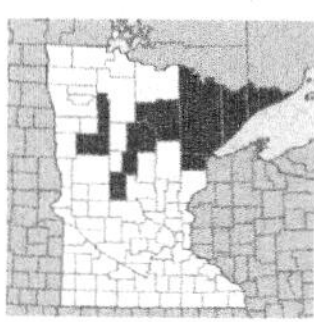

Sparganium glomeratum Laest.
CLUSTERED BUR-REED
NC **OBL** | MW **OBL** | GP **OBL** | *special concern*

Native perennial herb. **Stems** stout, floating or erect, 2–4 dm long. **Leaves** linear, ± flat, 3–8 mm wide; bracts leaflike, widened at base. **Inflorescence** usually unbranched; female heads several, clustered on the stem, stalkless, 1.5–2 cm wide when mature, scales narrowly oblong; male heads 1–2 above the female heads and continuous with them on stem. **Achenes** widest at middle, tapered to both ends, 3–8 mm long, slightly narrowed below the middle, shiny brown, the beak ± straight, 1–2 mm long. **Habitat** Shallow water of marshes and bogs. **Status** Special concern.

achene

Sparganium natans L.
SMALL BUR-REED
NC **OBL** | MW **OBL** | GP **OBL**

Native perennial herb. **Stems** usually long and floating, sometimes shorter and upright, 1–3 dm or more long. **Leaves** linear, dark green, thin, flat, 2–6 mm wide; bracts leaflike, short, somewhat widened at base. **Inflorescence** unbranched; female heads 2–3, from bract axils, stalkless or the lowest sometimes short-stalked, 1 cm wide when mature; scales spatula-shaped, widest at tip; male heads usually 1 (rarely 2). **Achenes** broadly oval, 3–4 mm long, dull green-brown, the beak 1–2 mm long. **Synonyms** *Sparganium minimum*. **Habitat** Shallow water, pond margins.

Sparganium natans
SMALL BUR-REED

TYPHA | *Cat-tail*

Large reedlike perennials, from fleshy rhizomes and forming colonies. **Stems** erect, unbranched, round in section, sheathed for most of length by overlapping leaf sheaths. **Leaves** mostly near base of plant, alternate in 2 ranks, erect, linear, spongy. **Flowers** tiny, either male or female, separate on same plant; petals and sepals reduced to bristles. **Male flowers** usually of 3–5 stamens, bristles absent or 1–3 or more. **Female flowers** intermixed with some sterile flowers; pistil 1, raised on a short stalk (gynophore), with numerous bristles near base, the bristles longer than pistil; small bracts (bractlets) also sometimes present, these intermixed with the bristles, slender but with a widened brown tip. **Heads** in a single, dense, cylindric spike, with male flowers above female, the male and female portions of the spike unalike, contiguous in common cat-tail (*Typha latifolia*) or separated in narrow-leaved cat-tail (*Typha angustifolia*); the mature spike brown and fuzzy in appearance due to the crowded stigmas and gynophore bristles. **Fruit** a yellow-brown achene, 1–2 mm long, the style persistent, long and slender with an expanded stigma.

TYPHA | CAT-TAIL

1 Male and female portions of spike usually separated; leaves to 1 cm wide; stigmas long and slender, pale brown .... *Typha angustifolia* .... NARROW-LEAVED CAT-TAIL

1 Male and female portions of spike usually contiguous, not separated; leaves mostly 1–2 cm wide; stigmas broad and flattened, dark brown.......... .... *Typha latifolia* | COMMON CAT-TAIL

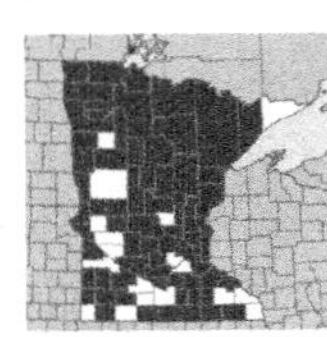

Typha angustifolia L.

NARROW-LEAVED CAT-TAIL
NC **OBL** | MW **OBL** | GP **OBL**

Introduced perennial emergent herb. **Stems** erect, 1–2 m long. **Leaves** upright, flat, 4–10 mm wide. **Flowers** either male or female, on separate portions of the spike, separated by an interval of 2–10 cm; male portion 7–20 cm long and 7–15 mm wide, male bractlets brown; female portion of spike dark brown, 10–20 cm long and 1–2 cm wide; each flower with 1 bristlelike bractlet, these flat and brown at the widened tip, gynophore hairs brown-tinged at tips; stigmas pale brown, linear, 1 mm long. **Fruit** 5–7 mm long, subtended by many fine hairs, the hairs slightly widened and brown at tip. June. **Habitat** Marshes, lakeshores, streambanks, roadside ditches, pond margins, usually in shallow water; more tolerant of brackish conditions than common cat-tail (*Typha latifolia*); nearly cosmopolitan.

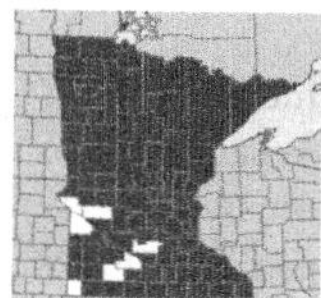

Typha latifolia L.

COMMON CAT-TAIL

NC **OBL** | MW **OBL** | GP **OBL**

Native perennial emergent herb. **Stems** erect, 1–2.5 m long. **Leaves** upright, mostly 1–2 cm wide. **Flowers** either male or female, the male and female portions of spike normally contiguous, rarely separated by 3–4 mm; male portion 5–15 cm long and 1.5–2 cm wide at flowering time, male bractlets white; female portion of spike dark brown, 10–15 cm long and 2–3 cm wide when mature, female bractlets absent, gynophore hairs white; stigma lance-shaped, becoming dark brown, less than 1 mm long. **Fruit** 1 cm long, with many white, linear hairs from base. June. **Habitat** Marshes, lakeshores, streambanks, ditches, pond margins, usually in shallow water; less tolerant of brackish conditions than narrow-leaved cat-tail (*Typha angustifolia*).

A hybrid between Typha angustifolia and Typha latifolia is termed **Typha x glauca** *Godr. Usually larger than either parent, male and female portions of hybrid plants are usually separated by a space to 4 cm long. The male portion of the spike is light brown, 0.5–2 dm long and about 1 cm wide at flowering time; the female portion is dark brown, 10–20 cm long and 1–2 cm wide. Since Typha x glauca is sterile, reproduction is vegetative from rhizomes. The hybrid may be found wherever populations of Typha angustifolia and Typha latifolia overlap and is common (and invasive) throughout Minnesota.*

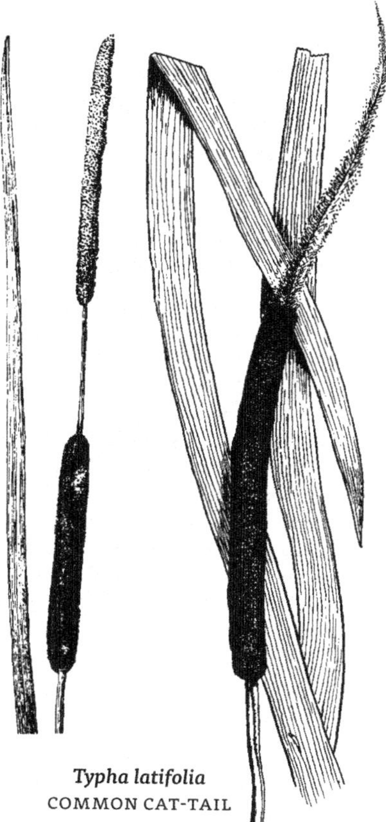

Typha angustifolia
NARROW-LEAVED CAT-TAIL

Typha latifolia
COMMON CAT-TAIL

XYRIS | *Yellow-eyed grass*

Perennial rushlike herbs. **Stems** erect, leafless, straight or sometimes ridged. **Leaves** all from base of plant, upright to spreading, linear, often twisted, usually dark green. **Flowers** small, perfect, yellow, from base of tightly overlapping bracts or scales, in rounded or cylindric heads at ends of stems; sepals 3, petals 3; stamens 3; style 3-parted. **Fruit** an oblong, 3-chambered capsule.

XYRIS | YELLOW-EYED GRASS

1 Plants swollen and hard at base . . . . *Xyris torta*
. . . . TWISTED YELLOW-EYED GRASS

1 Plants flattened and soft at base . . . . *Xyris montana*
. . . . NORTHERN YELLOW-EYED GRASS

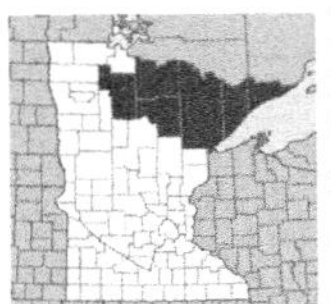

Xyris montana H. Ries
NORTHERN YELLOW-EYED GRASS
NC **OBL** | MW **OBL** | GP **OBL** | *special concern*

Densely clumped, native perennial herb. **Stems** leafless, 0.5–3 dm long, round in section, straight or lower part of stem slightly twisted. **Leaves** narrowly linear, flat or only slightly twisted, 5–20 cm long and 1–2 mm wide, rough, dark green, red-purple at base. **Flowers** yellow, in ovate spikes less than 1 cm long; scales obovate, finely fringed at tip; lateral sepals about as long as scales, linear, margins entire or finely hairy near tip. **Seeds** 1 mm long. **Habitat** Wet sandy shores, pools in sphagnum peatlands. **Status** Special concern.

Xyris torta J. E. Smith
TWISTED YELLOW-EYED GRASS
NC **OBL** | MW **OBL** | GP **OBL** | *endangered*

Native perennial herb. **Stems** leafless, 1.5–8 dm long, spirally twisted, ridged. **Leaves** linear, twisted, 2–5 dm long and 2–5 mm wide; outer leaves shorter, tinged purple-brown and swollen and bulblike at base. **Flowers** yellow, in cylindric spikes 1–2.5 cm long; scales oblong; lateral sepals linear, about as long as scales, tips of scales and lateral sepals with tuft of short, red-brown hairs; petals obovate, 4 mm long. **Seeds** 0.5 mm long. June–Aug. **Habitat** Wet sandy shores. **Status** Endangered.

flower

Xyris montana
NORTHERN YELLOW-EYED GRASS

Xyris torta
TWISTED YELLOW-EYED GRASS

ZANNICHELLIA | *Horned pondweed*

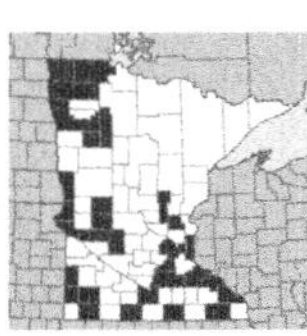

Zannichellia palustris L.

HORNED PONDWEED
NC **OBL** | MW **OBL** | GP **OBL**

Native perennial aquatic herb, with creeping rhizomes and often forming extensive underwater mats. **Stems** slender and delicate, wavy, 0.5–5 dm long, branched from base. **Leaves** simple, opposite (or upper leaves appearing whorled), threadlike, 2–8 cm long and 0.5 mm wide, stalkless; stipules membranous and soon deciduous. **Flowers** small, produced underwater, either male or female, separate on plant but from same leaf axil, with 1 male flower and usually 4 (varying from 1–5) female flowers at each node, surrounded by a membranous, spathelike bract; petals and sepals absent; male flower a single anther. **Fruit** a brown to red-brown, crescent-shaped nutlet, gently wavy on margins, 2–3 mm long, tipped by a beak 1–2 mm long; the fruits mostly 2–6 per node. June–Aug. **Habitat** Submerged in fresh or brackish water of streams, reservoirs, muddy lake and pond bottoms, marshes and ditches.

Similar to **Potamogeton** *and sometimes placed in the Pondweed Family (Potamogetonaceae).*

Zannichellia palustris
HORNED PONDWEED

Wetland Indicator Status

- **OBL (Obligate Wetland)**—Plants that almost always occur in wetlands (i.e. almost always in standing water or seasonally saturated soils.
- **FACW (Facultative Wetland)** Plants that usually occur in wetlands, but may occur in non-wetlands.
- **FAC (Facultative)** Plants that occur in wetlands and non-wetland habitats.
- **FACU (Facultative Upland)** Plants that usually occur in non-wetlands but may occur in wetlands.
- **UPL (Obligate Upland)** Plants that almost never occur in wetlands (or in standing water or saturated soils).

achene A one-seeded, dry, indehiscent fruit with the seed coat not attached to the mature wall of the ovary.

acid Having more hydrogen ions than hydroxyl (OH) ions; a pH less than 7.

acute Gradually tapered to a tip.

adventive Not native to and not fully established in a new habitat.

alkaline Having more hydroxyl ions than hydrogen ions; a pH > 7.

alluvial Deposits of rivers and streams.

alternate Borne singly at each node, as in leaves on a stem.

ament Spikelike inflorescence of same-sexed flowers (either male or female); same as catkin.

angiosperm A plant producing flowers and bearing seeds in an ovary.

annual A plant that completes its life cycle in one growing season, then dies.

anther Pollen-bearing part of stamen.

appressed Lying flat to or parallel to a surface.

aquatic Living in water.

areole In leaves, the spaces between small veins.

aromatic Strongly scented.

ascending Angled upward.

asymmetrical Not symmetrical.

auricle An ear-shaped appendage to a leaf or stipule.

awl-shaped Tapering gradually from a broad base to a sharp point.

awn A bristle-like organ.

axil Angle between a stem and the attached leaf.

barb Downward pointing projections.

basal From base of plant.

basic A pH greater than 7.

beard Covering of long or stiff hairs.

berry Fruit with the seeds surrounded by fleshy material.

biennial A plant that completes its life cycle in two growing season, typically flowering and fruiting in the second year, then dying.

blade Expanded, usually flat part of a leaf or petiole.

bog A wet, acidic, nutrient-poor peatland characterized by sphagnum and other mosses, shrubs and sedges. Technically, a type of peatland raised above its surroundings by peat accumulation and receiving nutrients only from precipitation.

boreal Far northern latitudes.

brackish Salty.

bract An accessory structure at the base of some flowers, usually appearing leaflike.

bractlet A secondary bract (*Typha*).

branchlets A small branch.

bristle A stiff hair.

bulblet Small bulb borne above ground, as in a leaf axil.

calcareous fen An uncommon wetland type associated with seepage areas, and which receive groundwater enriched with primarily calcium and magnesium bicarbonates.

calcium-rich Refers to wetlands underlain by limestone or receiving water enriched by calcium compounds.

calyx All the sepals of a flower.

capsule A dry, dehiscent fruit splitting into 3 or more parts.

carpel Fertile leaf of an angiosperm, bearing the ovules. A pistil is made up of one or more carpels.

caryopsis The dry, indehiscent seed of grasses.

catkin Spikelike inflorescence of same-sexed flowers (either male or female); same as ament.

chaff Thin, dry scales; in the Asteraceae, sometimes found as chaffy bracts on the receptacle.

circumboreal Refers to a species distribution pattern which circles the earth's boreal regions.

clasping Leaves that partially encircle the stem at the base.

cleistogamous Type of flower that remains closed and is self-pollinated.

clumped Having the stems grouped closely together; tufted.

colony-forming A group of plants of the same species, produced either vegetatively or by seed.

column The joined style and filaments in the Orchidaceae.

coma A tuft of fine hairs, especially at the tip of a seed.

composite An inflorescence that is made up of many tiny florets crowded together on a receptacle; members of the Composite Family (Asteraceae).

compound A leaf with two or more leaflets.

concave Curved inward.

cone The dry fruit of conifers composed of overlapping scales.

conifer Cone-bearing woody plants.

convex Curved outward.

corm An enlarged, rounded, underground stem, usually covered with papery scales or modified leaves.

corolla Collectively, all the petals of a flower.

corymb A flat-topped or convex inflorescence.

crisped An irregularly crinkled or curled leaf margin.

crown Persistent base of a plant, especially a grasses.

culm The stem of a grass or grasslike plant, especially a stem with the inflorescence.

cyme A type of inflorescence in which the central flowers open first.

deciduous Not persistent.

dehiscent Splitting open at maturity.

dicots One of two main divisions of the Angiosperms (the other being the Monocots); plants having 2 seed leaves (cotyledons), net-venation, and flower parts in 4s or 5s (or multiples of these numbers).

dioecious Bearing only male or female flowers on a single plant.

disarticulation Spikelets breaking either above or below the glumes when mature, the glumes remaining in the head if disarticulation above the glumes, or the glumes falling with the florets if disarticulation is below the glumes.

discoid In composite flowers (Asteraceae), a head with only disk (tubular) flowers, the ray flowers absent.

disjunct A population of plants widely separated from its main range.

disk In the Asteraceae, the central part of the head, composed of tubular flowers.

dissected Leaves divided into many smaller segments.

disturbed Natural communities altered by human influences.

divided Leaves which are lobed nearly to the midrib.

dolomite A type of limestone consisting of calcium magnesium carbonate.

driftless area Portions of sw Wisconsin, ne Iowa, and se Minnesota that are not covered by glacial drift.

drupe A fleshy fruit with a single large seed such as a cherry.

elliptic Broadest at the middle, gradually tapering to both ends.

emergent Growing out of and above the water surface.

emersed leaf Growing above the water surface or out of water.

endangered A species in danger of extinction throughout all or most of its range if current trends continue.

endemic A species restricted to a particular region.

entire With a smooth margin.

erect Stiffly upright.

escape A cultivated plant which establishes itself outside of cultivation.

evergreen Plant retaining its leaves throughout the year.

exserted Extending beyond the mouth of a structure such as stamens extending out from the mouth of the corolla.

fen An open wetland usually dominated by herbaceous plants, and fed by in-flowing, often calcium- and/or magnesium-rich water; soils vary from peat to clays and silts.

fern Perennial plants with spore-bearing leaves similar to the vegetative leaves and bearing sporangia on their underside, or the spore-bearing leaves much modified (Pteridophyta order).

fibrous A cluster of slender roots, all with the same diameter.

filament The stalk of a stamen which supports the anther.

floating mat A feature of some ponds where plant roots form a carpet over some or all of the water surface.

floodplain That part of a river valley that is occasionally covered by flood waters.

floret A small flower in a dense cluster of flowers; in grasses the flower with its attached lemma and palea.

follicle A dry, dehiscent fruit that splits along one side when mature.

genus The first part of the scientific name for a plant or animal (plural genera).

gland An appendage or depression which produces a sticky or greasy substance.

glaucous Having a bluish appearance.

glumes A pair of small bracts at base of each spikelet the lowermost (or first) glume usually smaller the upper (or second) glume usually longer.

grain The fruit of a grass; the swollen seedlike protuberance on the fruit of some Rumex.

gymnosperm Plants in which the seeds are not produced in an ovary, but usually in a cone.

gynophore The central stalk of some flowers, especially in cat-tails (*Typha*).

hardwoods Loosely used to contrast most deciduous trees from conifers.

herb A herbaceous, non-woody plant.

herbaceous Like an herb; also, leaflike in appearance.

hummock A small, raised mound formed by certain species of sphagnum moss.

humus Dark, well-decayed organic matter in soil.

hybrid A cross-breed between two species.

hypanthium A ring, cup, or tube around the ovary; the sepals, petals and stamens are attached to the rim of the hypanthium.

indehiscent Not splitting open at maturity.

indusium In ferns, a membranous covering over the sorus (plural indusia).

inferior The position of the ovary when it is below the point of attachment of the sepals and petals.

inflorescence A cluster of flowers.

insectivorous Refers to the insect trapping and digestion habit of some plants as a nutrition supplement.

interdunal swale Low-lying areas between sand dune ridges.

internode Portion of a stem between two nodes.

introduced A non-native species.

invasive Non-native species causing significant ecological or economic problems.

involucral bract A single member of the involucre; sometimes called phyllary in composite flowers (Asteraceae)

involucre A whorl of bracts, subtending a flower or inflorescence.

irregular flower Not radially symmetric; with similar parts unequal.

joint A node or section of a stem where the branch and leaf meet.

keel A central rib like the keel of a boat.

lance-shaped Broadest near the base, gradually tapering to a narrower tip.

lateral Borne on the sides of a stem or branch.

lax Loose or drooping.

leaf axil The point of the angle between a stem and a leaf.

leaflet One of the leaflike segments of a compound leaf.

lemma In grasses, the lower bract enclosing the flower (the upper, smaller bract is the palea).

lens-shaped Biconvex in shape (like a lentil).

lenticel Blisterlike openings in the epidermis of woody stems, admitting gases to and from the plant, and often appearing as small oval dots on bark.

ligulate Having a ligule; in the Asteraceae, the strap-shaped corolla of a ray floret.

ligule In grasses and grasslike plants, the membranous or hairy ring at top of sheath between the blade and stem.

linear Narrow and flat with parallel sides.

lip Upper or lower part of a 2-lipped corolla; also the lower petal in most orchid flowers.

lobed With lobes; in leaves divisions usually not over halfway to the midrib.

local Occurring sporadically in an area.

low prairie Wet and moist herbaceous plant communities, typically dominated by grasses.

margin The outer edge of a leaf.

marl A calcium-rich clay.

marsh Wetland dominated by herbaceous plants, with standing water for part or all the growing season, then often drying at the surface.

megaspore Large, female spores.

microspore Small, male spores.

midrib The prominent vein along the

main axis of a leaf.

mixed forest A type of forest composed of both deciduous and conifer trees.

moat The open water area ringing the outer edge of a peatland or floating mat.

monecious Having male and female reproductive parts in separate flowers on the same plant.

monocots One of two main divisions of the Angiosperms (the other being the Dicots); plants with a single seed leaf (cotyledon); typically having narrow leaves with parallel veins, and flower parts in 3s or multiples of 3.

muck An organic soil where the plant remains are decomposed to the point where the type of plants forming the soil cannot be determined.

mucro A sharp point at termination of an organ or other structure.

naked Without a covering; a stalk or stem without leaves.

native An indigenous species.

naturalized An introduced species that is established and persistent in an ecosystem.

needle A slender leaf, as in the Pinaceae.

nerve A leaf vein.

neutral A pH of 7.

node The spot on a stem or branch where leaves originate.

nutlet A small dry fruit that does not split open along a seam.

oblanceolate Reverse lance-shaped; broadest at the apex, gradually tapering to the narrower base.

oblong Broadest at the middle, and tapering to both ends, but broader than elliptic.

obovate Broadly rounded at the apex, becoming narrowed below.

ocrea A tube-shaped stipule or pair of stipules around the stem; characteristic of the Smartweed Family (Polygonaceae).

opposite Leaves or branches which are paired opposite one another on the stem.

organic Soils composed of decaying plant remains.

oval Elliptical.

ovary The lower part of the pistil that produces the seeds.

ovate Broadly rounded at the base, becoming narrowed above; broader than lanceolate.

palea The uppermost of the two inner bracts subtending a grass flower (the lower bract is the lemma).

palmate Divided in a radial fashion, like the fingers of a hand.

panicle An arrangement of flowers consisting of several racemes.

pappus The modified sepals of a composite flower which persist atop the ovary as bristles, scales or awns.

parallel-veined With several veins running from base of leaf to leaf tip, characteristic of most monocots.

peat An organic soil formed of partially decomposed plant remains.

peatland A wetland whose soil is composed primarily of organic matter (mosses, sedges, etc.); a general term for bogs and fens.

pepo A fleshy, many-seeded fruit with a tough rind, as a melon.

perennial Living for 3 or more years.

perfect A flower having both male (stamens) and female (pistils) parts.

perianth Collectively, all the sepals and petals of a flower.

perigynium A sac-like structure enclosing the pistil in *Carex* (plural perigynia).

petal An individual part of the corolla, often white or colored.

petiole The stalk of a leaf.

phyllary An involucral bract subtending the flower head in composite flowers (Asteraceae).

phyllode An expanded petiole.

pinna The primary or first division in a fern frond or leaf (plural pinnae).

pinnate Divided once along an elongated axis into distinct segments.

pinnule The pinnate segment of a pinna.

pistil The seed-producing part of the flower, consisting of an ovary and one or more styles and stigmas.

pith A spongy central part of stems and branches.

pollen The male spores in an anther.

prairie An open plant community dominated by herbaceous species, especially grasses.

prostrate Lying flat on the ground.

raceme A grouping of flowers along an elongated axis where each flower has its own stalk.

rachilla A small stem or axis.

rachis The central axis or stem of a leaf or inflorescence.

radiate heads In composite flowers, heads with both ray and disk flowers (Asteraceae).

ray flower A ligulate or strap-shaped flower in the Asteraceae, where often the outermost series of flowers in the head.

receptacle In the Asteraceae, the enlarged summit of the flower stalk to which the sepals, petals, stamens, and pistils are usually attached.

recurved Curved backward.

regular Flowers with all the similar parts of the same form; radially symmetric.

rhizome An underground, horizontal stem.

rib A pronounced vein or nerve.

rootstock Similar to rhizome but referring to any underground part that spreads the plant.

rosette A crowded, circular clump of leaves.

samara A dry, indehiscent fruit with a well-developed wing.

saprophyte A plant that lives off of dead organic matter.

scale A tiny, leaflike structure; the structure that subtends each flower in a sedge (Cyperaceae).

scape A naked stem (without leaves) bearing the flowers.

section Cross-section.

secund Flowers mostly on 1 side of a stalk or branch.

sedge meadow A community dominated by sedges (Cyperaceae) and occurring on wet, saturated soils.

seep A spot where water oozes from the ground.

sepal A segment of the calyx; usually green in color.

sheath Tube-shaped membrane around a stem, especially for part of the leaf in grasses and sedges.

shrub A woody plant with multiple stems.

silicle Short fruit of the Mustard Family (Brassicaceae), normally less than 2x longer as wide.

silique Dry, dehiscent, 2-chambered fruit of the Mustard Family (Brassicaceae), longer than a silicle.

simple An undivided leaf.

sinus The depression between two lobes.

smooth Without teeth or hairs.

sorus Clusters of spore containers (plural sori).

spadix A fleshy axis in which flowers are embedded.

spathe A large bract subtending or enclosing a cluster of flowers.

spatula-shaped Broadest at tip and tapering to the base.

sphagnum moss A type of moss common in peatlands and sometimes forming a continuous carpet across the surface; sometimes forming layers several meters thick; also loosely called peat moss.

spike A group of unstalked flowers along an unbranched stalk.

spikelet A small spike; the flower cluster (inflorescence) of grasses (Poaceae) and sedges (Cyperaceae).

sporangium The spore-producing structure (plural sporangia).

sporophyll A modified, spore-bearing leaf.

spreading Widely angled outward.

spring A place where water flows naturally from the ground.

spur A hollow, pointed projection of a flower.

stamen The male or pollen-producing organ of a flower.

staminode An infertile stamen.

stem The main axis of a plant.

stigma The terminal part of a pistil which receives pollen.

stipe A stalk.

stipule A leaflike outgrowth at the base of a leaf stalk.

stolon A horizontal stem lying on the surface of the soil.

style The stalklike part of the pistil between the ovary and the stigma.

subspecies A subdivision of the species forming a group with shared traits which differ from other members of the species (subsp.).

subtend Attached below and extending upward.

succulent Thick, fleshy and juicy.

superior Referring to the position of the ovary when it is above the point of attachment of sepals, petals, stamens, and pistils.

swale A slight depression.

swamp Wooded wetlands dominated by trees or shrubs; soils are typically wet for much of year or sometimes inundated.

talus Fallen rock at the base of a slope or cliff.

taproot A main, downward-pointing root.

tendril A threadlike appendage from a stem or leaf that coils around other objects for support (as in *Vitis*).

tepal Sepals or petals not differentiated from one another.

terminal Located at the end of a stem or stalk.

thallus A small, flattened plant structure, without distinct stem or leaves.

thicket A dense growth of woody plants.

threatened A species likely to become endangered throughout all or most of its range if current trends continue.

translucent Nearly transparent.

tree A large, single-stemmed woody plant.

tuber An enlarged portion of a root or rhizome.

tubercle Base of style persistent as a swelling atop the achene different in color and texture from achene body

tundra Treeless plain in arctic regions, having permanently frozen subsoil.

turion A specialized type of shoot or bud that overwinters and resumes growth the following year.

umbel A cluster of flowers in which the flower stalks arise from the same level.

umbelet A small, secondary umbel in an umbel, as in the Apiaceae.

upright Erect or nearly so.

utricle A small, one-seeded fruit with a dry, papery outer covering.

valve A segment of a dehiscent fruit; the wing of the fruit in *Rumex*.

variety Taxon below subspecies and differing from other varieties within the same subspecies (var.).

vein A vascular bundle, as in a leaf.

velum The membranous flap that partially covers the sporangium in *Isoetes*.

vine A trailing or climbing plant, dependent on other objects for support.

whorl A group of 3 or more parts from one point on a stem.

wing A thin tissue bordering or surrounding an organ.

woody Xylem tissue (the vascular tissue which conducts water and nutrients).

Composite flower, Asteraceae, with disk and ray flowers

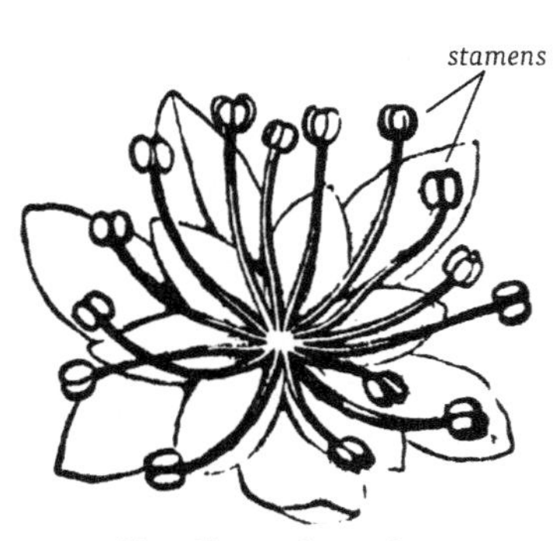

Dicot flower, imperfect, with only male parts

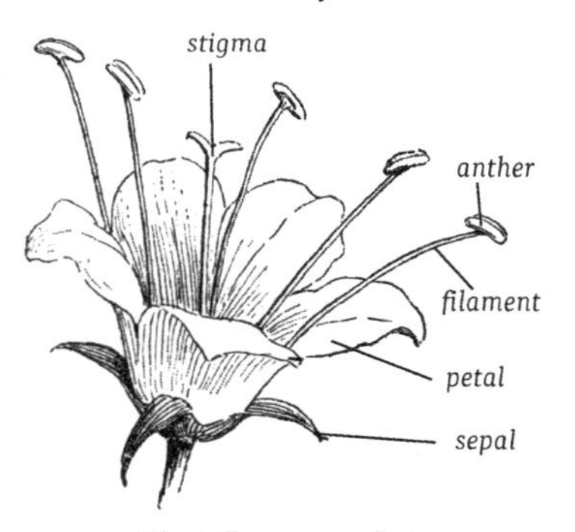

Dicot flower, perfect, with male and female parts

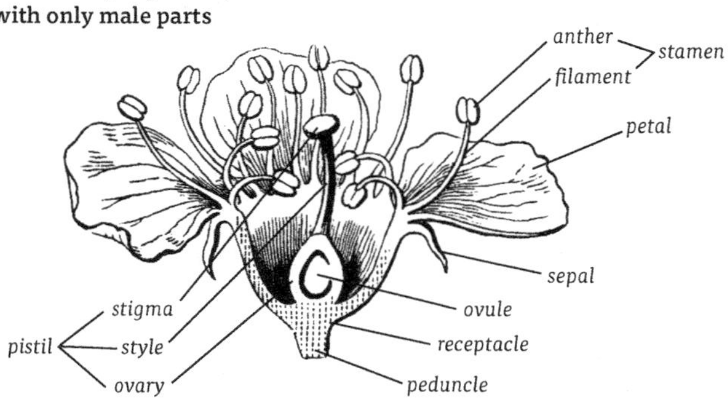

Dicot flower, perfect, with male and female parts

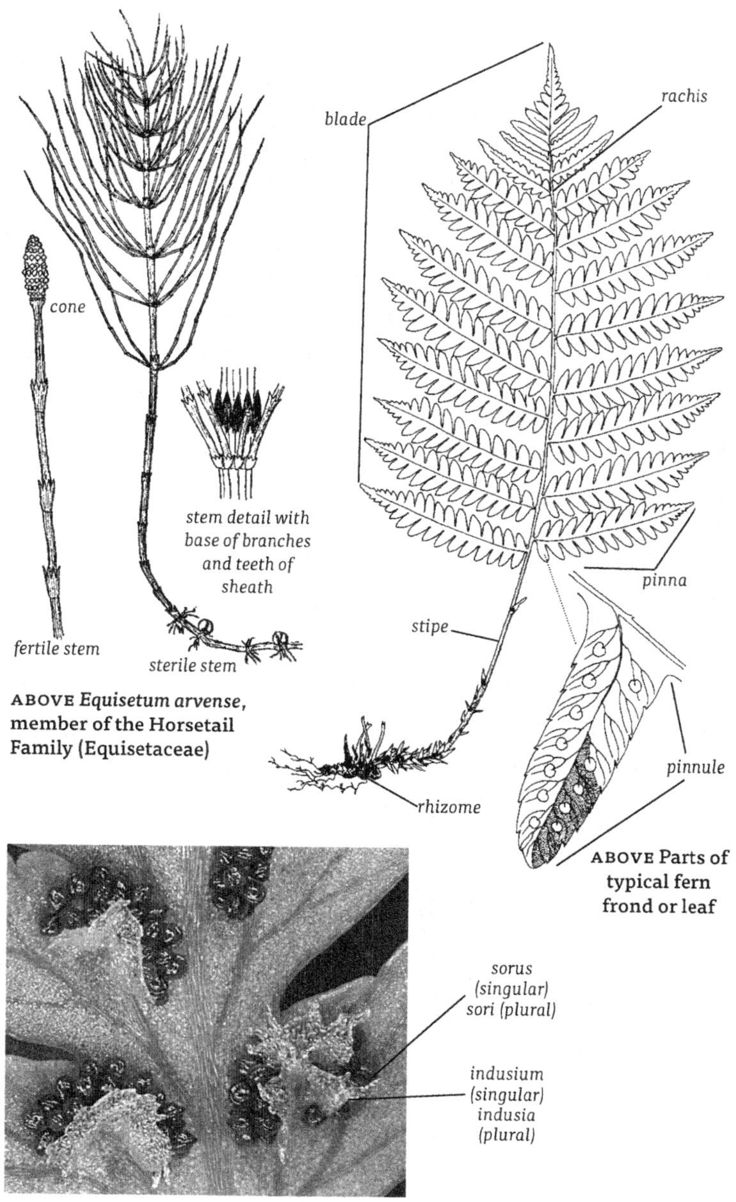

ABOVE ***Equisetum arvense*****, member of the Horsetail Family (Equisetaceae)**

ABOVE Parts of typical fern frond or leaf

Underside of lady fern (***Athyrium filix-femina*****) pinnule showing sori and indusia.**

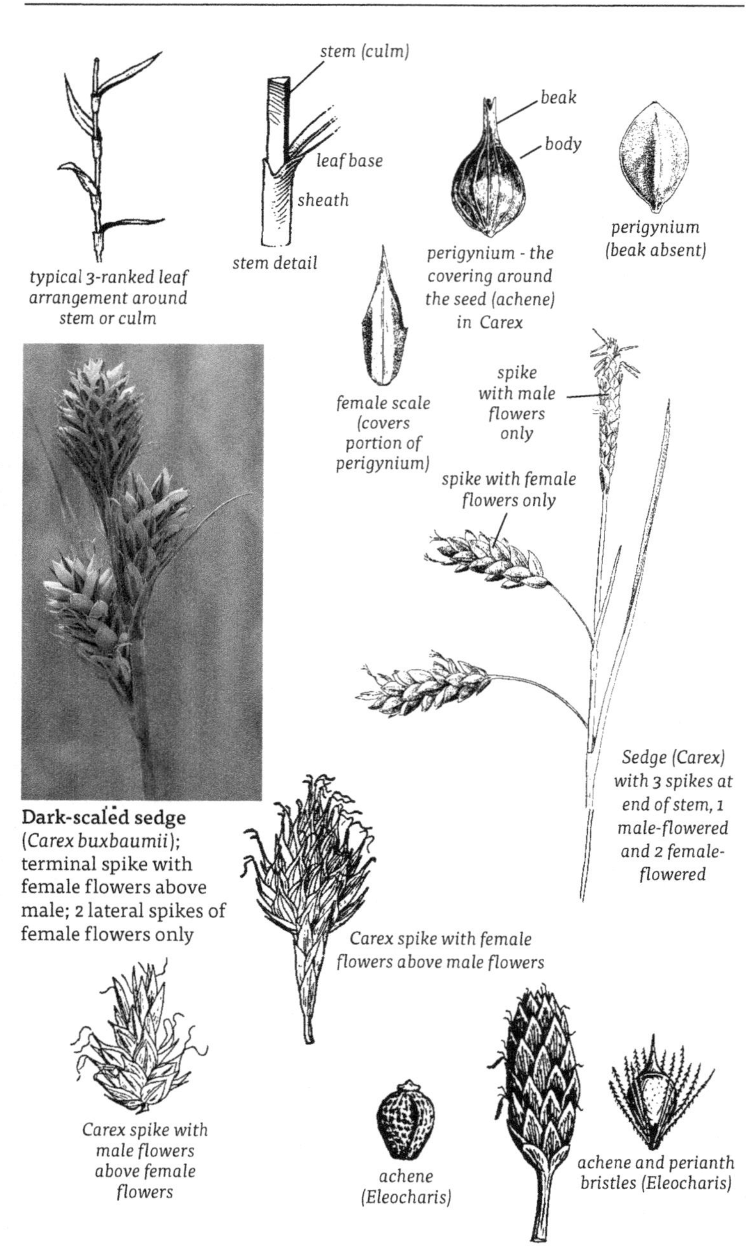

Dark-scaled sedge (*Carex buxbaumii*); terminal spike with female flowers above male; 2 lateral spikes of female flowers only

Typical spikelet in spike-rush (*Eleocharis*)

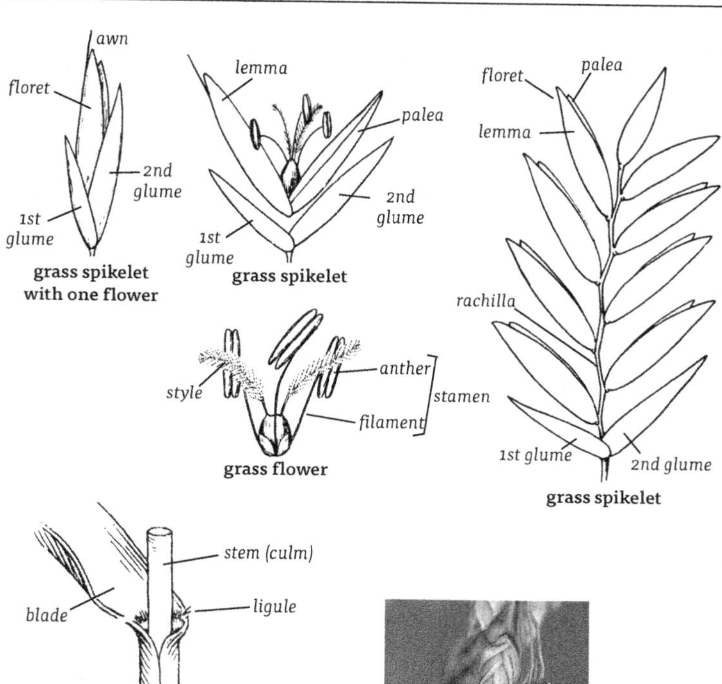

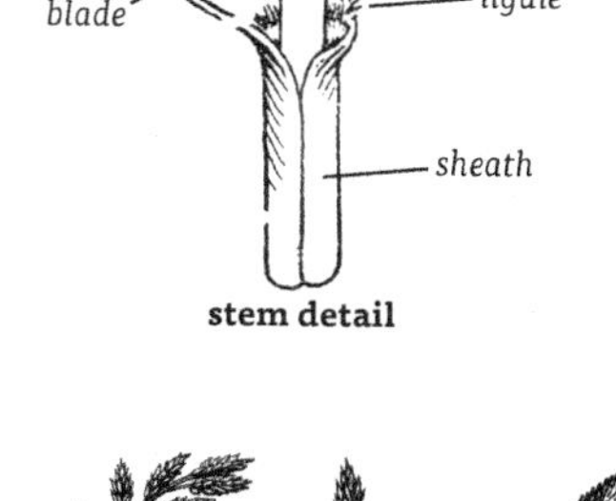

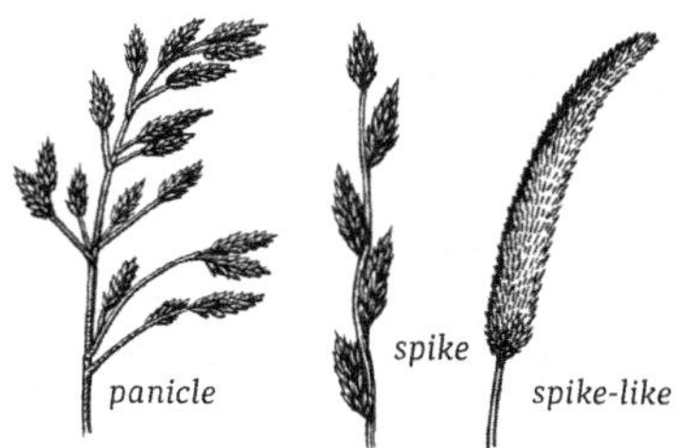

Primary types of grass inflorescences

ABOVE Inflorescence of **rattlesnake-mannagrass** (*Glyceria canadensis*); spikelets grouped into a panicle (often drooping); each spikelet with 5-10 flowers subtended by pair of glumes.

ABBREVIATIONS

Botanical
subsp. subspecies
var. variety

Measurements
mm millimeter
cm centimeter
dm decimeter
m meter
x times
± more or less

Geographical
c central
e east
n north
s south
w west

States and Provinces
Ala Alabama
Appal Appalachia
Ariz Arizona
Ark Arkansas
BC British Columbia
c Amer central America
Calif California
Can Canada
Colo Colorado
Conn Connecticut
Daks North and South Dakota
DC District of Columbia
Del Delaware
Fla Florida
Ga Georgia
Ill Illinois
Ind Indiana
Kans Kansas
Ky Kentucky
La Louisiana
Lab Labrador
LP Lower Peninsula of Michigan
Man Manitoba
Mass Massachusetts
Md Maryland
Mex Mexico
Mich Michigan
Minn Minnesota
Miss Mississippi
Mo Missouri
Mont Montana
N Amer North America
NB New Brunswick
NC North Carolina
ND North Dakota
Neb Nebraska
New Eng New England
Nev Nevada
Nfld Newfoundland
NH New Hampshire
NJ New Jersey
NM New Mexico
NS Nova Scotia
NWT Northwest Territories
NY New York
Okla Oklahoma
Ont Ontario
Ore Oregon
Pa Pennsylvania
PEI Prince Edward Island
Que Quebec
RI Rhode Island
S Amer South America
Sask Saskatchewan
SC South Carolina
SD South Dakota
Tenn Tennessee
Tex Texas
UP Upper Peninsula of Michigan
USA United States
Va Virginia
Vt Vermont
Wash Washington
Wisc Wisconsin
WVa West Virginia

Minnesota
Land area: 79,610 sq mi. (206,190 sq km)
KITTSON
ROSEAU
LAKE OF THE WOODS
MARSHALL
KOOCHICHING
PENNINGTON
RED LAKE
BELTRAMI
COOK
POLK
CLEARWATER
ITASCA
ST. LOUIS
LAKE
NORMAN
MAHN-OMEN
HUBBARD
CLAY
BECKER
CASS
WILKIN
WADENA
CROW WING
AITKIN
CARLTON
OTTER TAIL
TODD
MORRISON
MILLE LACS
KANABEC
PINE
TRAVERSE
GRANT
DOUGLAS
BENTON
BIG STONE
STEVENS
POPE
STEARNS
SHER-BURNE
ISANTI
CHISAGO
SWIFT
KANDIYOHI
ANOKA
RAMSEY
WASHINGTON
LAC QUI PARLE
CHIPPEWA
MEEKER
WRIGHT
HENNE-PIN
Minnea-polis
St. Paul
YELLOW MEDICINE
RENVILLE
McLEOD
CARVER
SCOTT
DAKOTA
SIBLEY
LINCOLN
LYON
REDWOOD
NICOLLET
LE SUEUR
RICE
GOODHUE
WABASHA
BROWN
PIPE-STONE
MURRAY
COTTON-WOOD
WATON-WAN
BLUE EARTH
WASECA
STEELE
DODGE
OLMSTED
WINONA
ROCK
NOBLES
JACKSON
MARTIN
FARIBAULT
FREEBORN
MOWER
FILLMORE
HOUSTON

REFERENCES

Barnes, B., and W. Wagner. 1981. *Michigan Trees*. The University of Michigan Press. Ann Arbor, MI. 383 p.

Black, M., and E. Judziewicz. 2009. *Wildflowers of Wisconsin and the Great Lakes Region: A Comprehensive Field Guide*. The University of Wisconsin Press. Madison, WI. 320 p.

Case, F., Jr. 1987. *Orchids of the Western Great Lakes Region*. Cranbrook Institute of Science Bulletin No. 48. Bloomfield Hills, MI. 240 p.

Chadde, S. 2012. *A Great Lakes Wetland Flora*. A complete, Illustrated Guide to the Aquatic and Wetland Plants of the Upper Midwest (4th ed.). 683 p.

Cholewa, Anita F. 2010. *Comprehensively Annotated Checklist of the Flora of Minnesota, version 2010.3, October 2010*. The University of Minnesota Herbarium

Cody, W., and D. Britton. 1989. *Ferns and Fern Allies of Canada*. Publication 1829/E. Research Branch, Agriculture Canada. Ottawa, Canada. 430 p.

Coffin, B., and L. Pfannmuller, ed. 1989. *Minnesota's Endangered Flora and Fauna*. MN Dept. of Natural Resources.

Cowardin, L., V. Carter, F. Golet, and E. LaRoe. 1979. *Classification of Wetlands and Deepwater Habitats of the United States*. U.S. Department of the Interior, Fish and Wildlife Service. Washington, DC. 103 p.

Crow, G., and C. Hellquist. 2000. *Aquatic and Wetland Plants of Northeastern North America* (2 vols.). University of Wisconsin Press. Madison, WI.

Crum, H. 1976. *Mosses of the Great Lakes Forest*. University Herbarium, University of Michigan, Ann Arbor, MI. 404 p.

Crum, H. 1988. *A Focus on Peatlands and Peat Mosses*. The University of Michigan Press. Ann Arbor, MI. 306 p.

Crum, H., and L. Anderson. 1981. *Mosses of Eastern North America* (2 vols). Columbia Univ. Press. New York, NY.

Curtis, J. 1971. *The Vegetation of Wisconsin*. The University of Wisconsin Press. Madison, WI. 657 p.

Eastman, J. 1995. *The Book of Swamp and Bog: Trees, Shrubs, and Wildflowers of Eastern Freshwater Wetlands*. Stackpole Books. Mechanicsburg, PA. 237 p.

Eggers, S., and D. Reed. 1997. *Wetland Plants and Plant Communities of Minnesota and Wisconsin* (2nd ed.). U.S. Army Corps of Engineers, St. Paul District. 264 p.

Fassett, N. 1957. *A Manual of Aquatic Plants*. The University of Wisconsin Press. Madison, WI. 405 p.

Flora of North America Editorial Committee. 1993. *Flora of North America North of Mexico*. Set, partially published. Oxford University Press. New York, NY.

Gleason, H., and A. Cronquist. 1991. *Manual of Vascular Plants of Northeastern United States and Adjacent Canada* (2nd Ed.). The New York Botanical Garden. Bronx, NY. 910 p.

Hipp, A. 2008. *Field Guide to Wisconsin Sedges: An Introduction to the Genus Carex (Cyperaceae)*. The University of Wisconsin Press. Madison, WI. 280 p.

Holmgren, N. (editor). 1998. *Illustrated Companion to Gleason and Cronquist's Manual*. New York Botanical Garden. Bronx, NY. 937 p.

Kartesz, J.T. 2010. *Floristic Synthesis of North America, Version 1.0*. Biota of North America Program (BONAP). Chapel Hill, NC.

Lichvar, R.W, and J. T. Kartesz. 2012. North American Digital Flora: National Wetland Plant List, v.2.4.0 (wetland_plants.usace.army.mil). U.S. Army Corps of Engineers, Engineer Research and Development Center, Cold Regions Research and Engineering Laboratory, Hanover, NH, and BONAP, Chapel Hill, NC.

Ownbey, G., and T. Morley. 1991. *Vascular Plants of Minnesota: A Checklist and Atlas*. The University of Minnesota Press. Minneapolis, MN. 306 p.

Reed, P. 1988. *National List of Plant Species that Occur in Wetlands: North Central (Region 3)*. Biological Report 88(26.3). U.S. Department of the Interior, Fish and Wildlife Service. Washington, DC. 99 p.

Smith, W. 2012. *Native Orchids of Minnesota*. The University of Minnesota Press. Minneapolis, MN. 254p.

Smith, W. 2008. *Trees and Shrubs of Minnesota*. The University of Minnesota Press. Minneapolis, MN.

Soper, J., and M. Heimburger. 1982. *Shrubs of Ontario*. The Royal Ontario Museum. Toronto, Ontario. 495 p.

Swink, F., and G. Wilhelm. 1994. *Plants of the Chicago Region* (4th Ed.). Indiana Academy of Science. Indianapolis, IN. 921 p.

Tryon, R. 1980. *Ferns of Minnesota* (2nd Ed.). The University of Minnesota Press. Minneapolis, MN. 165 p.

Tryon, R., N. Fassett, D. Dunlop, and M. Diemer. 1953. *The Ferns and Fern Allies of Wisconsin*. The University of Wisconsin Press. Madison, WI. 158 p.

Voss, E.G. 1972. *Michigan Flora, Part I Gymnosperms and Monocots*. Cranbrook Institute of Science Bulletin 55 and University of Michigan Herbarium. 488 p.

Voss, E.G. 1985. *Michigan Flora, Part II Dicots (Saururaceae–Cornaceae)*. Cranbrook Institute of Science Bulletin 59 and University of Michigan Herbarium. 724 p.

Voss, E.G. 1996. *Michigan Flora, Part III Dicots (Pyrolaceae–Compositae)*. Cranbrook Institute of Science Bulletin 61 and University of Michigan Herbarium. 622 p.

Wetter, A.W, T.S. Cochrane, M.R. Black, H.H. Iltis, P.E. Berry. 2001. *Checklist of the Vascular Plants of Wisconsin*. Tech. Bulletin No. 192. Dept. Natural Resources, Madison, WI. 258 p.

Wright, H., B. Coffin, and N. Aaseng (editors). 1992. *The Patterned Peatlands of Minnesota*. University of Minnesota Press. Minneapolis, MN. 327 p.

REFERENCES

Maps were generated with permission of and from data of the Biota of North America Program (BONAP): *www.bonap.org*

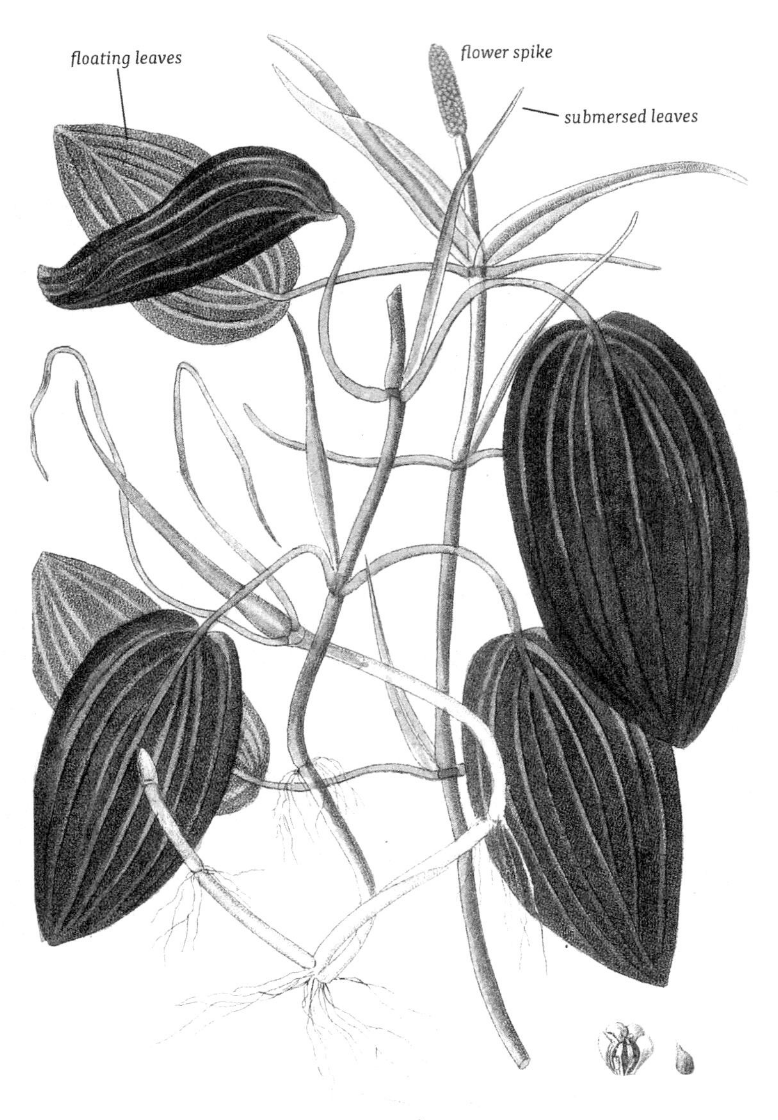

Floating pondweed (*Potamogeton natans*), p. 601
Pondweed Family, Potamogetonaceae

A

B

C

F

H

M

Q

R

T

CPSIA information can be obtained
at www.ICGtesting.com
Printed in the USA
LVOW13s2226090717
540781LV00007B/386/P